Electricity

Electricity

Efficient End-Use and New Generation Technologies, and Their Planning Implications

Thomas B. Johansson,
Birgit Bodlund, and
Robert H. Williams, Editors

Lund University Press

Lund University Press
Box 141 S-221 00 Lund
Sweden
Art nr 20090
ISBN 91-7966-065-7

Chartwell-Bratt
ISBN 0-86238-208-4

Printed in Sweden
Studentlitteratur
Lund 1989

Contents

List of Contributors 8

Foreword 15

Editors' Introduction 17

End-Use of Electricity

Energy-Efficient Electric Motor Drive Systems 21
Samuel F. Baldwin

Energy-Efficient Lighting 59
Terry McGowan

Energy-Efficient Windows: Options with Present and Forthcoming Technology 89
Claes Göran Granqvist

Low Electricity Appliances – Options for the Future 125
Jørgen S. Nørgård

Heat Pumps 173
Thore Berntsson

Use of Electricity in Commercial Buildings 217
Enno Abel

Building Design and Electricity Use in Single-family Houses 235
Arne Elmroth

Electrotechnologies and Steelmaking 261
Sven Eketorp

New Electrothermal Processes 297
Helmut Schaefer and Manfred Rudolph

Keeping Score in Electricity Conservation Programs 353
Gautam S. Dutt and Margaret F. Fels

Superconductivity and the Efficient Use of Electricity 389
Nils-Johan Bergsjö and Lars Gertmar

Electronic Office Equipment: The Impact of Market Trends and Technology on End-Use Demand for Electricity 427
Les Norford, Ari Rabl, Jeffrey P. Harris and Jacques Roturier

The Role of Sectoral Shift in Trends in Electricity Use in United States and Swedish Manufacturing and in Comparing Forecasts 461
Gale A. Boyd and Marc H. Ross

Technological Change and Industrial Electricity Use 489
Towards Better Generalizations
Adam Kahane

New Generation Technologies

Expanding Roles for Gas Turbines in Power Generation 503
Robert H. Williams and Eric D. Larson

Pressurised Fluidised Bed Combustion 555
Krishna K. Pillai

Low-Cost Power From Thin-Film Photovoltaics 595
David E. Carlson

Fuel Cells 627
A review of fuel cell technology and its applications
Leo J. M. J Blomen

Coal Gasification Technologies for Combined Cycle Power Generation 665
Werner Peters

Biomass Gasification for Gas Turbine Power Generation 697
Eric D. Larson, Per Svenningsson, Ingemar Bjerle

Implications for Planning

Implementing Electricity Conservation Programs: Progress Towards Least-Cost Energy Services Among U.S. Utilities 741
Howard S. Geller

The Coming Reformation of the Electric Utility Industry 765
Måns Lönnroth

Planning Implications of Electricity Conservation: The Case of Denmark 787
Niels I. Meyer and Jørgen S. Nørgård

Energy Service Companies: The French Experience 811
Olivier de la Morinière

Innovative Approaches to Marketing Electric Efficiency 831
Robert H. Williams

The Importance of Government-Supported Research and Development in Advancing Energy Efficiency in the United States Buildings Sector 863
Howard S. Geller, Jeffrey Harris, Marc R. Ledbetter, Mark D. Levine, Evan Mills, Robert Mowris, Arthur H. Rosenfeld

The Challenge of Choices: Technology Options for the Swedish Electricity Sector 883
Birgit Bodlund, Evans Mills, Tomas Karlsson and Thomas B. Johansson

Index 949

List of Contributors

Enno Abel, Ph.D.
Professor, Department of Building Services Engineering
Member of the Royal Academy of Engineering Sciences
Chalmers University of Technology
S-412 96 Gothenburg
Sweden

Samuel F. Baldwin, Ph.D.
Office of Technology Assessment
U.S. Congress
Washington, D.C. 20510
USA

Nils-Johan Bergsjö, M.Sc.
Director of the Laboratory for Electrical Engineering
ABB Corporate Research
S-721 78 Västersås
Sweden

Thore Berntsson, Ph.D.
Professor, Dept. of Heat and Power Technology
Chalmers University of Technology
S-412 96 Gothenburg
Sweden

Ingemar Bjerle, Ph.D.
Professor, Department of Chemical Engineering II
University of Lund
Box 124
S-221 00 Lund
Sweden

L.J.M.J. Blomen, Dr. Ir.
Group Vice President R&D
Kinetics Technology International Group B.V.
26, Bredewater, P.O. Box 86,
2700 AB Zoetermeer
The Netherlands

Birgit Bodlund, Ph.D.
Head of the Energy Systems Section of the
Development and Environment Division at Vattenfall
S-162 87 Vällingby
Sweden

Gale A. Boyd, Ph.D.
Economist in the Energy and Environmental Systems Division
Argonne National Laboratory
Argonne, IL 60439
USA

David E. Carlson, Ph.D
Vice President & General Manager
826 Newtown-Yardley Road
Newtown, PA 18940
USA

Gautam S. Dutt, Ph.D.
Depto. de Fisica, 3er. piso
Facultad de Ciencias
Universidad Nacional Autonoma de Mexico
04510 Mexico D.F.
Mexico

Sven Eketorp, Ph.D.
Professor emeritus, Royal Institute of Technology
S-100 44 Stockholm
Sweden

Arne Elmroth, Ph.D.
Professor, Department of Building Technology
Lund University
Box 118
S-221 00 Lund
Sweden

Margaret F. Fels, Ph.D.
Senior Research Scientist
Center for Energy and Environment Studies
Princeton University
Princeton, NJ 08544
USA

Howard S. Geller, M.Sc.
Associate Director
American Council for an Energy-Efficient Economy
1001 Connecticut Avenue, N.W.
Washington, D.C. 20036
USA

Lars Gertmar, Ph.D.
D.Sc., is Manager of Electrical Drives and
Measurements
Laboratory for Electrical Engineering
ABB Corporate Reseach
S-721 78 Västerås
Sweden

Claes G. Granqvist, Ph.D.
Researcher in Energy–related Applied Physics
Swedish Natural Science Research Council
Physics Department
Chalmers University of Technology
S-412 96 Gothenburg
Sweden

Jeffrey P. Harris
Staff Scientist
Applied Science Division
Bldg. 90 H
Lawrence Berkeley Laboratory
University of California
Berkely, CA 94720
USA

Thomas B. Johansson, Ph.D.
Professor, Environmental and Energy Systems Studies
University of Lund
Gerdagatan 13
S-223 62 Lund
Sweden

Adam Kahane
Group Planning
Shell International Petroleum Company
PL/12 Shell Centre
London SE1 7NA
United Kingdom

Tomas Karlsson, M.Sc.
Analyst, Planning Division
Vattenfall
S-162 87 Vällingby
Sweden

Eric D. Larson, Ph.D.
Visiting Research Engineer
Department of Environmental and Energy Systems Studies
University of Lund
Gerdagatan 13
S-223 62 Lund
Sweden

on leave from the Center for Energy and Environmental Studies
Princeton University Princeton, NJ USA

Marc R. Ledbetter, Ph.D.
Senior Associate
American Council for an Energy-Efficient Economy
1001 Connecticut Avenue, N.W.
Washington, DC 20036
USA

Mark Levine, Ph.D.
Program Leader
Energy Analysis Program
Bldg. 90-3125
Lawrence, Berkeley Laboratory
Berkeley, CA 94720
USA

Måns Lönnroth, Ph.D.
Advisor, Prime Minister's Office
Rosenbad 4
S-103 33 Stockholm
Sweden

Terry K. McGowan, FIES
Manager, Lighting Technical Support
GE Lighting
Nela Park – 4162
Cleveland, Ohio 44112
USA

Niels I. Meyer, Ph.D.
Professor, Physics Laboratory III
The Technical University of Denmark
DK 2800 Lyngby
Denmark

Evan Mills, M.Sc.
Department of Environment and Energy Systems Studies
University of Lund
Gerdagatan 13
S-223 62 Lund
Sweden

Robert Mowris
Senior Research Associate
Center for Building Science
Bldg. 90-3058
Lawrence Berkeley Laboratory
Berkeley, CA 94720
USA

Olivier de la Morinière, Ph.D.
Operating Manager of the MC International Group
Compagnie Generale de Chauffe
Les Bureaux du Palais
61/69 Rue de Bercy
75012 Paris
France

Leslie Norford, Ph.D.
Professor, Department of Architecture
MIT
Rm 7-403
77 Massachusetts Ave.
Cambridge, MA 02139
USA

Jørgen S. Nørgård, Ph.D.
Lecturer of physics
Physics Laboratory III
The Technical University of Denmark
DK 2800 Lyngby
Denmark

Werner Peters, Professor Dr.
Member of the Board of Directors
Bergbau Forschung GmbH
Research Center of the Coal Mining Industry of West Germany, Essen
apl. Professor University of Essen
FRG

Krishna K. Pillai, Ph.D.
Project Development Manager
ABB Carbon AB
S-612 20 Finspång
Sweden

Ari Rabl, Ph.D.
Centre d'Energétique
Ecole des Mines
60, BD. St Michel
75272 Paris CEDEX 06
France

Arthur H. Rosenfeld, Ph.D.
Professor, Director, Center for Building Science
Bldg. 90-3058
Lawrence Berkeley Laboratory
Berkeley, CA 94720
USA

Marc H. Ross, Ph.D.
Professor, Senior Scientist at Argonne National Laboratory
and at University of Michigan
Ann Arbor, MI 48109
USA

Jacques Roturicr, Ph.D.
Professor, Institut Universitaire de Technologie
Université de Bordeaux I
33400 Talence
CEDEX
France

Manfred Rudolph, Dr. -Ing.
Lehrstuhl für Energiewirtschaft und
Kraftwerkstechnik
Technische Universität München
Arcisstraße 21
8000 München 2
FRG

Helmut Schaefer, Prof. Dr. -Ing.
Lehrstuhl für Energiewirtschaft und
Kraftwerkstechnik
Technische Universität München
Arcisstraße 21
8000 München 2
FRG

Per Svenningsson, M.Sc.
Research Associate
Department of Environmental and Energy Systems Studies
University of Lund
Gerdagatan 13
S-223 62 Lund
Sweden

Robert H. Williams, Ph.D.
Senior Research Scientist
Center for Energy and Environmental Studies
Princeton University
Princeton, New Jersey 08544
USA

Foreword

Power producers all over the world are currently undergoing a process of transformation into energy service companies. This is particularly true of Vattenfall, the Swedish State Power Board. Intensive efforts are being made to develop new technology with the aim of creating techniques that are efficient and compatible with the environment, in both the end-use and production of electricity.

In Sweden, the demands made by society have more impact on the electricity sector than in many other countries. We have decided to decommission nuclear power by the year 2010, and nuclear power accounts for approximately 50% of the electricity currently produced. We have strict requirements regarding environmental protection, including an undertaking not to increase emissions of carbon dioxide. There are four rivers that are protected by law against development for hydro power production.

These external demands make it necessary for Vattenfall to extend its horizons and acquire know-how relating to long-term technical developments at a more intensive rate than has previously been the case. New plants to replace current electricity production will be more expensive than those we use today. As a consequence of this, Vattenfall is directing its efforts towards end-use and rational utilization of electricity. New ways of producing electricity are also of considerable interest, as are the possibilities of known techniques being improved economically and technically.

Following disussions between T.B. Johansson, R.H. Williams, and Vattenfall, we have agreed to organize a Congress – "Electricity" – with the aim to widen our perspectives and search for new options, and to intensify contact with the research sector. This book has been produced as a basis for the Congress. Since it is not possible to make a total review of all development that is in progress, it is necessary to make a choice. Recognized researchers in areas where interesting developments are in progress were commissioned to contribute chapters for this book. Their manuscripts have received peer review and have been revised before being accepted for publication. The authors are responsible for the content and conclusions of their chapters.

At the Congress to be held in Gothenburg in Spring 1989, the various chapters will be discussed and debated, as well as the significance of the findings in helping to select the design of energy systems in the future.

The work on the book and the preparations for the Congress have been a source of inspiration for all those involved, and it is our hope that the Congress will prove to be a rewarding event for all participants.

Stockholm, December, 1988

Ingvar Wivstad
Vice President, Research and Development
Vattenfall.

Editors' Introduction

The purpose of this book is: (1) to provide an up-to-date overview of important new energy-efficient end-use and generating technologies for electricity and the ongoing structural changes in industrial economies which affect electricity demand, and (2) to gain insights into the policy and planning implications of these new technologies for the electricity sector.

Electricity accounted for about 10% of the global final energy demand in 1980. Most projections of future total energy demand–including low energy futures that emphasize energy efficiency[1]–are characterized by an increasing electricity fraction. This reflects the high value of electricity as an energy carrier.

Continuing electricity production growth could be problematic, however. Traditional electricity generating technologies pose a series of economic, environmental, and security challenges. In many parts of the world, new power plant are much more costly than existing units, in part due to changes in power plant design which were made to mitigate environmental and safety concerns. Additionally, dependence on oil gives rise to security concerns because the world's remaining oil resources are concentrated in the politically volatile Middle East. Dependence on fossil fuels generally, but especially coal, aggravates the global greenhouse warming. And widespread dependence on nuclear power runs the risk of nuclear weapons proliferation

Fortunately, more electricity-efficient end-use technologies and new generating technologies are emerging from research and development, and they could improve the outlook for electricity. Moreover, in industrialized countries structural changes away from materials-intensive activities and towards more knowledge-intensive ones are leading to declining relative demand for both energy and electricity per unit of economic output. Thus, while the ability of electricity to deliver heating, lighting, electrolytic,

[1] José Goldemberg, Thomas B. Johansson, Amulya K.N. Reddy and Robert H. Williams, 1988. *Energy for a Sustainable World*, New Delhi: Wiley Eastern.

mechanical, and information services in a wide variety of ways makes the availability of electricity crucial to the growth of these new industries, this growth will tend to be extensive rather than intensive. (The situation in developing countries is quite different, because the need for materials creates demand for intensive electricity use.)

This book represents an attempt to better understand the new opportunities for meeting electric service needs from a societal perspective. To this end, the economic evaluations herein were carried out using a real discount rate of 6% – and all direct taxes and subsidies were purposely ignored. In many instances, presently available technologies that are socially desirable under such economic calculations are *not* attractive to many decisionmakers in current market and institutional conditions. What changes are needed to improve these conditions and how such changes might be implemented are questions this book also explores.

Not all opportunities for improving end-use efficiency or new generation are treated in this volume. The energy-efficient technologies discussed here are limited to those that are commercially available (though in most instances are not widely used under current market and institutional conditions) or could be commercialized within a decade. The fossil fuel technologies described here are those that would reduce carbon dioxide emission rates by using fossil fuels more efficiently and by emphasising fossil fuels with low carbon dioxide emission rates. Biomass and photovoltaic sources are particularly promising renewable energy supply options for the electricity sector which are discussed here. Nuclear technologies are not considered, however. Nuclear fission is not included in part because this exercise is intended to explore the alternative electricity options open to Sweden as it phases out nuclear power, fulfilling the mandate of the Swedish electorate. But from a global perspective nuclear fission is not included because the risk of nuclear weapons proliferation would become severe in a world in which fission played a major role, as would be the case if fission were pursued as a major strategy for coping with the greenhouse warming. Nuclear fusion is not considered here because fusion will not be ready for commercial application for several decades.

Will exploiting the technological opportunities identified here solve the serious global problems posed by energy? Do they provide a sufficient basis for a truly sustainable long-term energy strategy? Comprehensive answers to these questions are beyond the scope of this effort, but some insights are provided here in the detailed case study for Sweden. This study, which makes use of the findings from many chapters in this book, suggests that even under the especially challenging conditions facing the Swedish electricity sector, it might be possible to find paths for the development of electricity that are compatible with economic, security, and environmental goals.

We hope that utility, industry, and government officials, as well as students and researchers, from around the world, will find this book useful, particularly in terms of understanding the prospects for least-cost electricity futures consistent with environmental, security and economic concerns. And we also hope it will encourage policymakers to find ways to turn these opportunities into reality.

Lund, Stockholm, and Princeton, January 1989

Thomas B. Johansson Birgit Bodlund Robert H. Williams

Acknowledgements

The editors thank Ingvar Wivstad and other Vattenfall officials for their support and encouragement of the *Electricity* project, the authors for the creativity and enthusiasm they brought to their chapters, the many reviewers for helping assure that the finalized manuscripts meet the highest technical and expository standards, and David Sheridan for his diligence and care in editing the draft chapters.

Energy-Efficient Electric Motor Drive Systems[1]

Samuel F. Baldwin

Abstract

Electric motors typically account for 2/3 of all electricity use. Although individual components in electric motor drive systems perform relatively well, passing through the many devices of a system can reduce the delivered energy to as little as 5–10% of the input fossil fuel. Numerous technologies are being developed to reduce these losses, including energy efficient motors, improved magnetic materials, and electronic adjustable speed drives. These technologies offer significant improvements in overall system performance as well as numerous operational advantages.

1 Introduction

Electric motors are the workhorses of modern industrial society. They run home refrigerators; drive office heating and ventilation systems; power industries' pumps, fans, and compressors; and keep cities' water supplies flowing. In 1986, the fraction of U.S. primary energy (fossil heat equivalent basis) devoted to the electric sector was 36%, and of the electricity generated, nearly two-thirds (Table 1) was used in electric motors. The total annual cost of electricity to U.S. consumers is currently about $150 billion, and for electric motors alone therefore approaches $100 billion[2] (1,2).

[1] Portions of this paper have been reproduced, with permission, from the Annual Review of Energy, Vol. 13 – Copyright 1988 by Annual Reviews Inc.

[2] The $-value for electricity is less than 2/3 of the total because of the higher than average percentage of motor drive in industry and the corresponding lower average cost of electricity.

Table 1. Estimated motor drive as a fraction of total U.S. electricity use.

	LITTLE, 1972	DOE, 1977	SRI, 1980
Residential	7.1%	2.8%	19.6%
Refrigeration*	–	–	8.2
Space Heating*	–	–	4.1
Air Conditioning*	–	–	4.5
Other	–	–	2.8
Commercial	11.9	14.5	14.7
Air Conditioning*	6.5	–	8.0
Refrigeration*	4.2	–	5.2
Other	1.2	–	1.5
Industrial	27.9	26.1	30.0
Pumps*	8.5	–	9.6
Compressors*	4.9	–	5.3
Blowers and Fans*	4.3	–	4.9
Machine Tools	2.4	–	2.5
Other Integral Hp	3.1	–	3.5
DC Drives	2.1	–	3.1
Fractional Hp	1.2	–	1.1
HVAC	0.7	–	–
Public and Miscellaneous	17.3	14.4	3.5
Municipal Water Works*	6.9	0.7	–
Electric Utilities*	8.7	6.5	–
Mining and Construction*	1.7	2.0	–
Other	–	5.2	–
TOTAL	64.2	57.8	67.8
Source	Ref. (5)	Ref. (7)	Ref. (33)

* Primarily consists of pumps, fans, and/or compressors.

Despite their important role in the economy, however, electric motors have been technologically quiescent for several decades[3], in sharp contrast to the revolutionary changes that have occurred in communication and information technologies. Now these very technologies are being combined with new developments in power electronics, magnetic materials, and system design to

3 This is not to say that there have been no advances at all. Major advances include vacuum pressure impregnation with synthetic resins, more rugged stator and rotor construction, and the extensive use of computer design techniques. See, for example, references (3,4).

dramatically transform electric motor drive. These changes promise to improve significantly the manner and the efficiency with which we use electricity.

1.1 Electric Motor Fundamentals

Electric motors have traditionally been classified as direct current (dc), synchronous, or induction (Figure 1). DC motors have been used where the value of their easy and precise control outweighed their expensive maintenance[4]. Synchronous motors have been reserved mainly for large installations where their higher efficiencies compensated for their higher installed costs. In contrast, induction motors have been used where the need for reliability and low cost was most important, and have consequently been the predominant choice in residential (one-phase), commercial and industrial (three-phase) applications for decades.

As seen in Figure 2, standard three-phase ac induction motors have efficiencies ranging from about 73% for 1 hp (0.75 kW[5]) motors to 93% at 150 hp (112 kW). The losses that occur can be categorized as conduction, magnetic core, friction and windage, and stray losses; they are depicted in Figure 3. Conduction losses are due to resistive (I2R) heating in the stator and rotor. Magnetic core losses are due to a combination of eddy current and hysteresis losses induced by the 60-Hz field in the stator and rotor iron. Friction losses occur in the motor bearings, and the windage losses are caused by the ventilation fan and other rotating parts. The remaining losses are usually lumped together in the catch-all term 'stray load' losses[6] (5,7,9–11).

With the increasing cost of electricity in recent years, many manufacturers have begun to produce lines of energy-efficient motors with typical improvements in efficiency as depicted in Figure 2. Techniques used to achieve these savings include: using higher conductivity and larger cross section conductors in the stator and rotor to reduce resistive heating losses; both increasing the core length and using thinner laminations of better magnetic materials to reduce eddy current and hysteresis losses in the core; and reducing

4 This includes the difficulty of developing the high level of expertise needed to work with brushes and slip rings.

5 Both american and metric units are listed here for convenience. Horsepower (hp) is a measure of motor shaft output power and is converted to mks units using 0.7457 kW/hp. To determine motor input power, the shaft output must be divided by the motor efficiency. All motor powers listed will be shaft output unless otherwise specified.

6 Stray load losses include leakage fluxes, space harmonics, and saturation effects in the rotor and stator iron, among others.

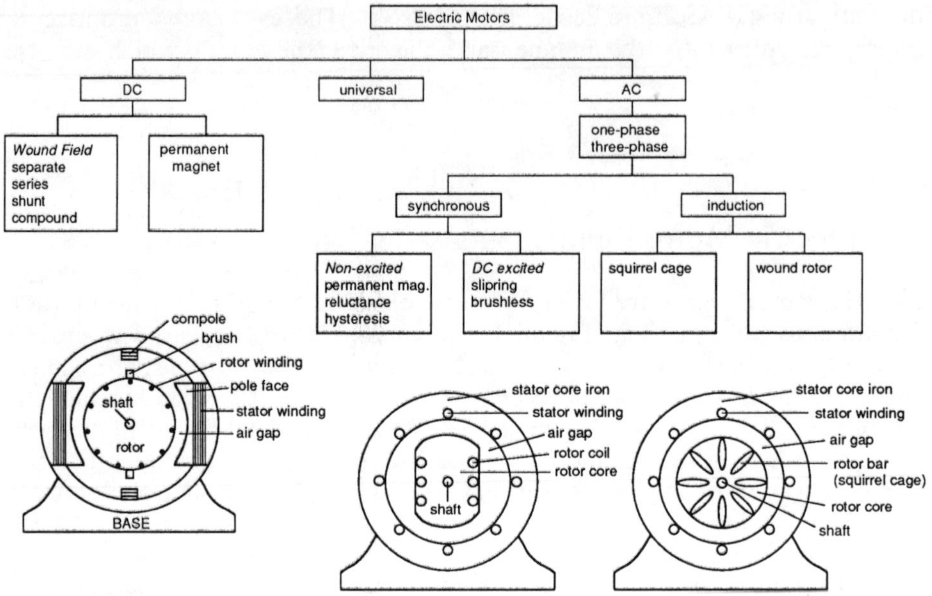

Figure 1. Classification and characteristics of electric motors. Nearly all motors establish stator and rotor magnetic fields at an angle to each other to generate torque (8). In dc motors, constant magnetic fields are generated in the stator by dc or by permanent magnets. Carbon brushes are used to switch mechanically between coils as the rotor turns so as to keep its dc generated magnetic field stationary in space and with respect to the stator; compensating poles, or compoles, are often used to minimize the transient switching voltages. The stator and rotor coils can be connected to separate power supplies; or in series, in parallel (shunt), or in various combinations of series and parallel (compound) to achieve the desired performance characteristics. The speed and torque of dc motors are easily controlled (6) by varying the stator (field) and/or rotor (armature) current and/or voltage.

AC motors use: (A) three-phase alternating current offset proportionately in time (phase) and space (on the stator winding) to produce, in effect, a single magnetic flux wave rotating around the stator at the line synchronous speed; or (B) single-phase ac (which cannot produce an effectively rotating magnetic flux) with a motor having auxiliary starting windings (split phase, capacitor start) or windings which produce an asymmetric torque (shaded pole). Once a single phase motor rotor begins turning, however, it is subject to a net positive rotating stator flux in its reference frame.

Synchronous motors typically use three-phase ac to generate a constant magnetic field rotating in the stator as described above and have a constant magnetic field fixed and rotating with the rotor. This rotor field is generated by: (A) dc using slip rings to carry current to the rotor; (B) dc using a brushless exciter (current induced in auxiliary coils and rectified); or (C) permanent magnets. Alternatively, (D) the rotor reluctance (the magnetic circuit dual of ohmic resistance) and an asymmetric (salient pole) rotor; or (E) hysteresis (the lag of magnetization {of the rotor} behind the magnetizing field {of the stator}) can be used to provide a preferential rotor orientation. The rotor of a synchronous motor turns in lock-step with the rotating flux field of the stator.

Induction motors are ac motors in which the rotor turns at a slightly slower speed (slip when expressed in percent difference) than the effectively rotating stator flux. By Faraday's law of induction, this relative motion between the stator field and rotor induces a current, and thus a magnetic field, in the rotor as well. Squirrel-cage motors use bars of copper in the shape of a "squirrel-cage" in the rotor. Varying the shape (to take advantage of the ac skin effect) and the resistivity of these bars adjusts the torque-speed characteristics as desired. Wound rotors have a winding similar to that of the stator, which is connected via slip rings to the outside – allowing the rotor characteristics to be changed.

Universal motors have brushes and can be operated on either single-phase ac or on dc.

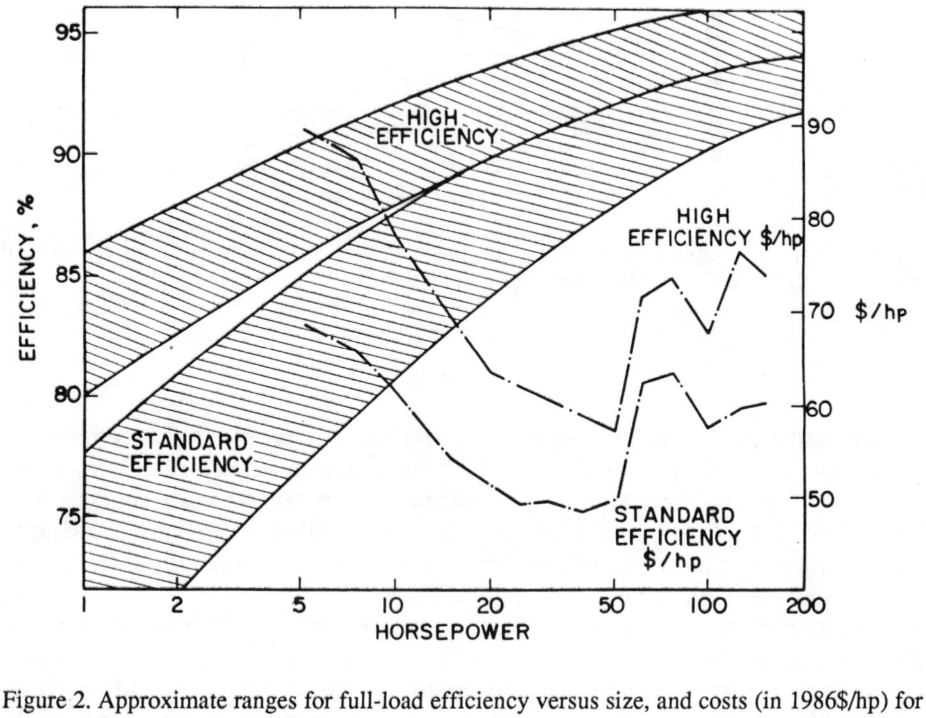

Figure 2. Approximate ranges for full-load efficiency versus size, and costs (in 1986$/hp) for one manufacturer versus size for National Electrical Manufacturers Association (NEMA) design B standard and energy-efficient 1800-rpm, three-phase induction motors. Adapted from (10) and motor catalogue data.

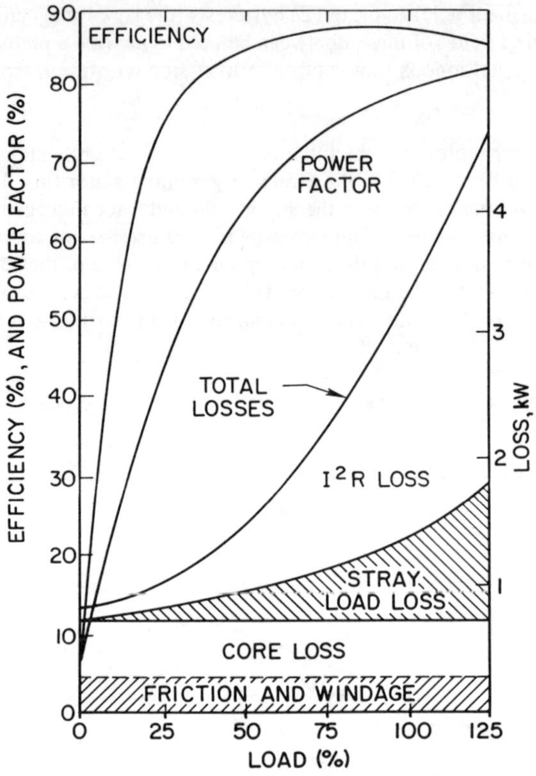

Figure 3. Typical components of loss, approximate efficiency, and power factor versus load for a three-phase, design B, 1800-rpm induction motor (5).

the air gap between rotor and stator to reduce the required magnetomotive force and its generating currents (10,12). Optimization studies examining more than 100 induction motor design parameters have detailed the lifetime savings possible under various conditions. In the case of single-phase motors, simply optimizing motor dimensions could raise efficiencies an average of 4.5% over present-day standard motors at no additional cost (13–15).

Although there is a price premium for energy-efficient motors (Figure 2), they often pay for themselves quickly by lowering the high cost of operation. A standard 50-hp (37 kW) motor, for example, might cost \$2500 to purchase and \$2.05 per hour to operate at full load (91% efficient @ \$0.05/kWhr). In contrast, a comparable energy-efficient motor might cost \$2875 to purchase and \$1.96 per hour to operate (95% efficient). Thus, in a single 6000 hour (full-load) operating year, the difference in operating costs is $540, for a net saving of $165 more than the premium price. Alternatively, the price premium can be described as equivalent to $317 per kW-capacity saved[7], including the duty cycle. Even scrapping a working motor and replacing it with an efficient model would have a simple payback of just over 5 years. Of course, shorter operating hours would stretch out these payback times.

It is important to note, however, that improving motor efficiency can have side-effects that require special attention. Decreasing resistance in the rotor circuit in order to reduce resistive heating losses in energy-efficient motors, for example, will also reduce the starting torque and increase the rotor speed at its load point (Figure 4). And high-efficiency motors can have higher transient starting currents–increasing the load on the power supply (16). Low starting torques and thermal capacities may require choosing oversized motors to start high-inertia loads such as fans (18,19). This can be particulary impor-

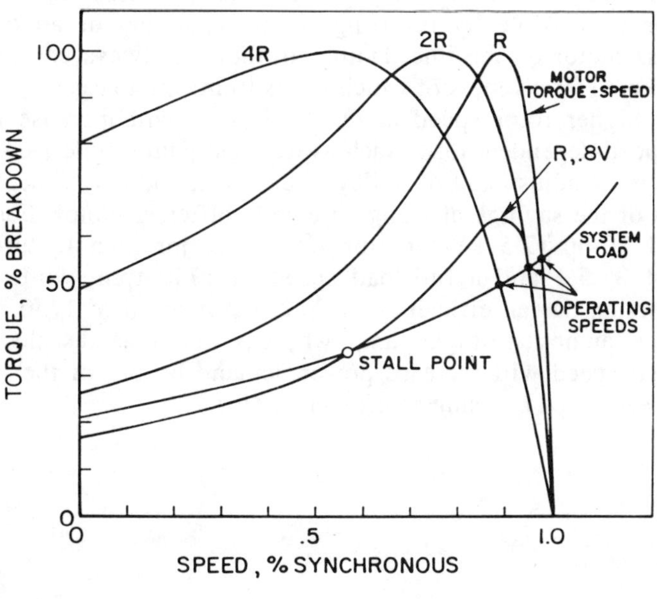

Figure 4. Motor speed versus torque curves for various values of rotor resistance and applied voltages. Also shown is a system load curve and the corresponding operating points for each motor torque-speed curve.

[7] This is per kW capacity at the motor input, not per kW of shaft output. Savings in generation capacity will be slightly larger as compounded transmission and distribution losses, etc., are avoided.

tant where there is significant variation in the line voltage and the combination of low voltage and large load inertia can cause the motor to stall (Figure 4) and even burn out. Motor oversizing itself may have little or no direct efficiency penalty. For instance, a standard 5-hp (3.7 kW) motor at full load may have an efficiency of 87.5%, compared to 87% for a 10-hp (7.5 kW) motor at half load. Similarly, a 75-hp (56 kW) motor may have an efficiency of 94% compared to 94.5% for a 125-hp (93 kW) motor at 60% load (Figure 2). At very low load, of course, the efficiency drops rapidly (Figure 3), but the amount of power that is wasted is relatively small. An over-sized motor, however, does lead to a lower power factor[8] (Figure 3).

A load with a low power factor requires larger generation, transmission, and distribution equipment to handle the increase in kVA (kilovoltamperes), increases resistive heating losses in this equipment, and can lower system voltage regulation (7). Up to one-fifth of the total losses in power distribution in the United States, or as much as 1.5% of total US electrical generation (2), can probably be attributed to a lagging power factor (5), and utilities often penalize users who have excessively low ones (17). Correction for a lagging power factor is possible by inserting shunt capacitors or an overexcited synchronous[9] motor on the line. Doing so, though, always entails a capital expense, the cost-effectiveness of which varies from case to case.

Finally, a higher rotor speed at the load point will increase the power demand of speed-dependent loads such as fans and pumps unless compensated for by appropriate adjustment of pulleys, gears, etc, and may cancel a significant fraction of the savings of using an energy-efficient motor. For example, the standard 10-hp (7.5 kW) motor from a major manufacturer has an efficiency of 87.5% and a full-load speed of 1740 rpm, while the high-efficiency version has an efficiency of 90.2% at a speed of 1750 rpm. For a load, such as a fan or centrifugal pump, whose power demand scales as $(rpm)^3$, this additional speed will increase power demand by 2/3 of the power that might have been saved by using an efficient motor.

[8] The power factor is the cosine of the phase angle between the voltage and current.

[9] A typical synchronous motor application for power factor correction would be to drive a constant load, low-speed air compressor which provides process compressed air to a plant, and raising the level of voltage/current to the rotor (overexcite) until it operates with a leading power factor (10).

1.2 Conventional System Design

Electric motors are just one element of very large and complex systems, each component of which must be taken into account to optimize overall energy-efficiency. Table 2 lists estimated national average efficiencies and net available power at different points in pumping systems in the U.S. As seen, each kWhr saved by reducing piping system losses saves roughly 9 kWhr of power plant fuel. These average values, however, include many systems which are not throttled. As shown below, the losses in specific cases can be much higher. Careful design and operation of end-use equipment can thus have enormous leverage on energy savings upstream (7,20). In addition, reducing the load can save considerable capital investment at each point of the system upstream.

Pumps, fans, and compressors are the most common motor driven equipment (Table 1), and each specific design has a unique performance map (Figures 5,6) determined by the detailed fluid mechanics of its operation (21). The system operating point is then the intersection of this performance map with the system load curve (Figures 5,6). The system load, in turn, is the power embodied in the flow of the moving fluid (liquid or gas) and is given in the most simple form by the sum of kinetic energy, static pressure drop, and gravitational potential[10]. In principle, equipment is chosen so that the optimal

Table 2. U.S. average pumping system losses*.

Location	Efficiency, output as % of input at device	Remainder, % of starting total
Coal	–	100.
Electric Power Plant	30	30.
Power Transmission Lines	92	27.6
Power Transformers	96	26.5
Electric Motors	88	23.3
Shaft Couplings	98	22.9
Pump	72	16.5
Pump Throttling Valve	86	14.1
Piping System	80	11.3

* Includes unthrottled systems.
Source: Reference 7.

[10] The system load in this simple form is then given by:

$$P = 1/2\rho AV^3 + pAV + \rho AVgZ = \tau\omega = \rho QH \qquad (1)$$

where ρ is the fluid density, A is the duct or pipe cross-sectional area, V is the fluid velocity, p is the static plus dynamic (pipe friction) pressure the flow is working against, g is the gravitational acceleration, and Z is the change in vertical elevation. The torque-speed, $\tau\omega$, and flow-head, QH=AVH, are useful comparative forms for describing the system.

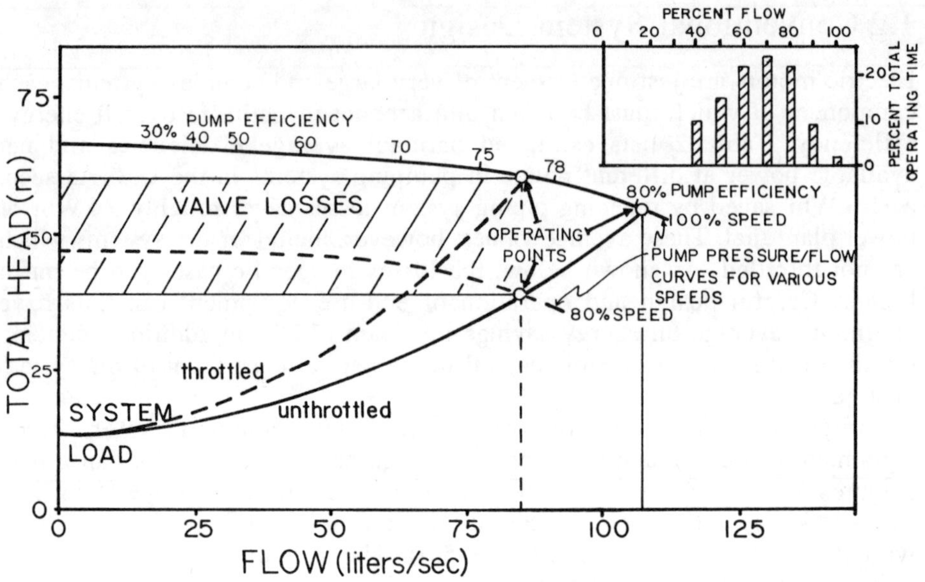

Figure 5. Throttling losses in a variable-flow, variable-pressure pumping system at 80% flow as compared to reduced speed operation. The energy losses due to throttling are roughly proportional to the shaded area. Inset is a typical pump duty cycle. Note that throttling also causes the pump to operate at a lower efficiency. From (27).

efficiency is matched to the desired operating point. In practice, there are numerous complications when systems are designed.

1.2.1 System Sizing

Systems are usually deliberately designed to be oversized, that is, to have excess capacity. Pipes, ducts, pumps, fans, and other driven equipment and components are oversized, partly because it is extremely difficult to predict system flow rates and friction factors accurately in advance, and to allow for the effect of the buildup of deposits on duct and pipe walls over time (7,22,23). Motors are oversized to handle starting electrical, mechanical, and thermal stresses, particuarly with high-inertia loads such as fans (18); to provide a safety margin for the worst-case load over their lifetimes; and occasionally to handle plant expansion or perhaps to be widely interchangeable within the plant (5). In developing countries, oversizing motors is important for preventing motor stall and possible burnout when the line voltage drops (24). More generally, systems are oversized because the increased energy and capital

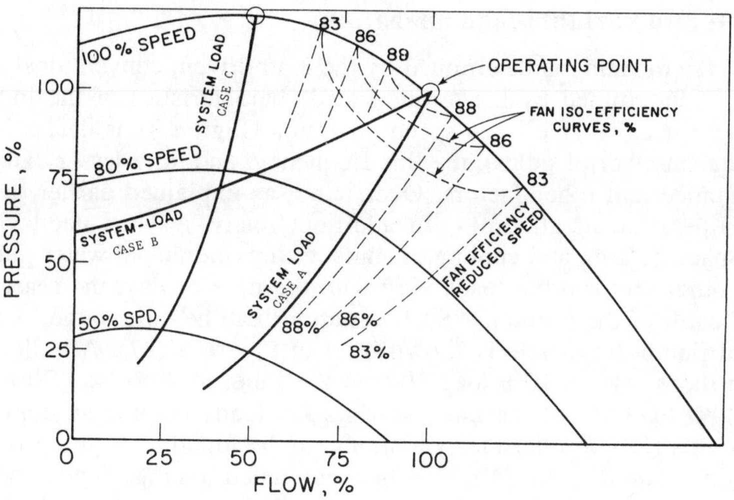

Figure 6. Characteristic pressure and flow curves for a fan at different speeds; fan iso-efficiency curves; and system loads for: Case A – unthrottled, low-static pressure drop, high dynamic pressure drop; Case B – unthrottled, high-static pressure drop, low dynamic pressure drop; and Case C – throttled, low-static pressure drop, high dynamic pressure drop as in Case A. Note that for a high static pressure drop, changing fan speeds does not allow the highest fan efficiency point to track the load curve. Adapted from (23).

costs are usually less than the risk of equipment failure. In manufacturing, for example, average electricity costs in the US are equivalent to just 2.8% of the value-added (25); the cost of motor failure and unplanned shutdown of the entire process line can be a much more severe penalty (5).

The intent of this design process is understandable and reasonable. Manufacturers, design engineers, and users all want to ensure that the system can meet the demands placed on it at every stage of the process. Although oversizing can provide direct safety margins for equipment, it can have unintended negative side-effects. For example, when each successive element of a drive system is sized to handle the load presented by the previous component plus a safety margin, oversizing can quickly become excessive and require throttling valves or vanes to limit flow. Throttling not only wastes power itself, but may also move the fan or pump off its optimal operating efficiency (design) point (Figures 5,6), making its operation inefficient as well. Standard engineering design rules, together with manufacturing safety margins, sometimes automatically lead to significant inefficiencies of this type which the user may be unaware of (26). Oversizing motors also leads to a low power factor (Figure 3). Re-evaluating traditional safety margins may therefore offer significant savings of both capital and operating costs.

1.2.2 Part- and variable-load operation

Although system sizing is determined by the load driven, conventional system operation is determined by the torque-speed characteristics of the induction motor. These usually give fixed-speed operation (Figure 4) as determined by the winding (number of poles), the line frequency, and, to a lesser extent, the rotor resistance and other factors. Oversizing, as explained earlier, leads to part-load operation at all times. In addition, many systems need variable outputs. Space heating and cooling, manufacturing, municipal water pumping, and most other motor drive loads vary with the time of day, the season, and even the health of the economy. Such variations can be quite large. A typical centrifugal pump duty cycle is shown inset in Figure 5 (27). A boiler might vary even more – from 10% load 10% of the time, to 40% load 20% of the time, to 80% load 5% of the time, with its fan loads varying an appropriate fraction of this (28). Similar load variations can be found for compressors (29) and machine tools (29,30). Table 3 lists estimated average loads and other characteristics of motors in the United States. Driving such part- or variable-load systems with fixed speed motors has often led to *extremely inefficient control strategies.*

In some systems, output is controlled by turning the motor on and off as needed. In an air-handling system with turbulent flow, for example, the power loss due to pressure drop in the ducts varies nearly as the cube of the flow rate. A system that has a duty cycle of 50% could instead provide the same total flow by running continuously at half the flow rate. This would reduce the power loss due to duct friction by 87.5%, and would reduce the corresponding total energy consumption by 75%.

Other systems run the motor continuously at part-load. Both motor efficiency and power factor, however, decrease for part-loads as shown in Figure 3 (5). This is particularly important for loads less than 50%.

Table 3. Characteristics of U.S. motor population, 1977.

	Electric Motor Size; Horsepower					
	<1	1–5	5.1–20	21–50	51–125	>126
Total Number, millions	660.	55.	10.5	3.3	1.7	1.0
Average Life, years	12.9	17.1	19.4	21.8	28.5	29.3
Weighted Average Size, hp	0.26	2.1	11.9	32.5	86.7	212.
Average Efficiency, %	65.	77.	82.5	87.5	91.	94.
Average Load, % full load	70.	50.	60.	70.	85.	90.
Average Capacity Factor, %	3.5	7.	17.4	27.7	37.5	43.2
Total Annual Use, 10^9 kWhr	30.5	33.9	103.4	155.2	337.7	573.3

Source: Reference 7.

Finally, in many systems control is exercised by letting the motor run at its normal speed and then throttling the input or output with a control valve or vane. Throttling valves (on pumps) or outlet dampers (on fans) reduce system flow by raising the static pressure the pump or fan is working against (Figures 5,6). Inlet vanes (fans) impart a spin to air entering a fan and thereby effectively change the pitch of the fan vane and thus change its characteristic curves. These strategies allow flow reductions typically of 50–60% and sometimes more (31).

Industrial and commercial pumps, fans, and compressors have estimated average losses of 20–25% or more when throttled or otherwise inefficiently controlled (20,22,32–34). Figure 5 illustrates such losses. In reviews of entire plants, losses in motors and drives have been estimated to be as large as 40–80% (35). To reduce these enormous energy losses, a method of varying induction motor speed without sacrificing the reliability of brushless operation is needed.

The control of motor speed has been a perennial problem for engineers. In November, 1896, Harry Ward Leonard said to the American Institute of Electrical Engineers, "The control of the speed of an electric motor from a state of rest to a state of full speed is a problem of rapidly growing importance to the electrical engineer" (36).

In the 100 years since Tesla first developed the principles of induction motors, a variety of techniques have been used to control induction motor speed; some of these are listed in Figure 7 (10,36). Unfortunately, until recently no drive or coupling technique proved sufficiently reliable, efficient, flexible, and widely applicable or low enough in cost to be generally accepted; recent advances in power electronics are now changing this situation.

2 The Materials Revolution

Materials technologies have improved rapidly in recent years as understanding of their physics has advanced. Improvement has been most dramatic in solid state electronics devices – mainly computer and communications technologies. But similar advances are being made in technologies of use in motor-driven systems. Among these technologies are power electronics that allow precise control and matching of motor speed and torque to the load – thereby eliminating, for example, throttling losses; soft magnetic materials to reduce core losses in transformers and motors; improved permanent magnets to eliminate rotor resistive heating losses; and many others. The recent advances in high-temperature superconductors (see Bergsjö and Gertmar in this book) may one day lead to the elimination of resistive heating losses throughout electric power systems as well as providing numerous other advantages.

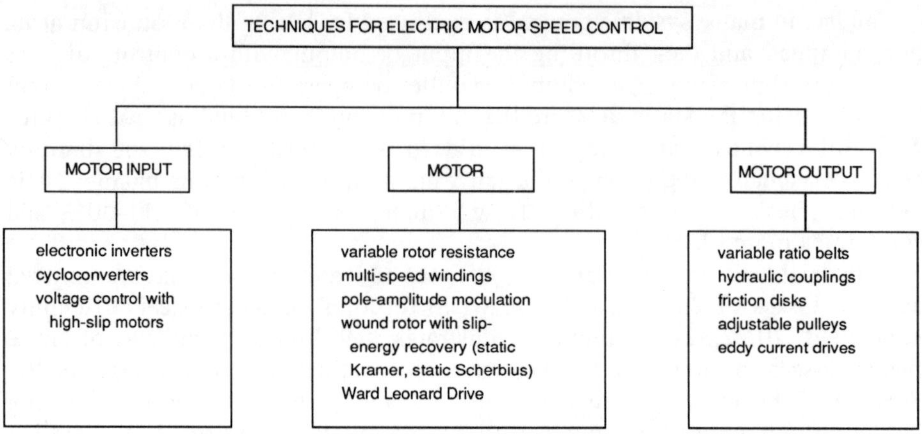

Figure 7. Classification of various techniques for electric motor speed control. Motor input technologies such as ASDs and cycloconverters are used to adjust the effective input frequency and voltage to ac motors. Voltage control with a high-slip motor simply adjusts where the motor torque-speed curve (Figure 4) intersects the load curve. Motor technologies such as multi-speed windings and pole-amplitude modulation effectively change the number of poles, and hence the synchronous speed, of the stator winding. Variable rotor resistance and slip-energy recovery systems change the characteristics of the rotor circuit (Figure 4). The Ward Leonard system consists of an ac motor driving a dc generator, which in turn drives a dc motor which produces the output. The speed of the dc motor is controlled by varying the armature (rotor) voltage. Finally, speed changing systems on the motor output include direct mechanical transmission systems such as variable ratio belts and gear boxes, and adjustable pulleys; mechanical systems with adjustable slip such as friction disks and hydraulic couplings; and electromechanical systems with adjustable slip such as eddy current drives. These are but a few of the many systems developed.

2.1 Power Electronics, Sensors, and Controllers

As discussed above, a crucial factor in achieving higher efficiencies in motor-drive systems is the ability to precisely control motor speed. One technique is to resynthesize the power going into an induction motor at an adjustable frequency. Adjustable-speed electronic drives (ASD) driven by mercury-arc rectifiers such as thyratrons and ignitrons were designed and tested in the 1930s (36,37), but found little acceptance because of their high cost and inconvenience. The development of solid- state electronic devices[11] with large power-handling capabilities, reductions in their costs, and rising electricity

[11] By "device" is meant the transistors, thyristors, MOSFETs, and similar active components (as opposed to passive components such as resistors, capacitors, inductors, transformers, etc.) used in electronics.

prices have made solid state electronic motor drives both possible and increasingly desirable. By 1986, over 200,000 electronic ASDs ranging in size from approximately 0.25 to 67,000 hp (0.20–50,000 kW) had been sold in the United States (34,38).

During the 40 years since the invention of the point-contact transistor in 1947 and the 30 years since the invention of the thyristor in 1957, the capabilities of modern power electronic devices have improved dramatically[12] (Figure 8) (39,40). Today, thyristors 10 cm in diameter, weighing 2.3 kg packaged, and capable of handling 3000 A at 4000 V have been commercialized, and 3000 A, 8000 V devices are under development (41,42).

Recently, bipolar junction transistors and Gate Turn-Off thyristors have similarly undergone significant improvements in their high-power capability. These devices greatly simplify the circuitry of ASDs, and correspondingly, can also reduce the cost and increase the reliability of ASDs. In the near future, new power devices such as the insulated gate bipolar transistor and the MOS-controlled thyristor may provide further improvements in ASD performance (126).

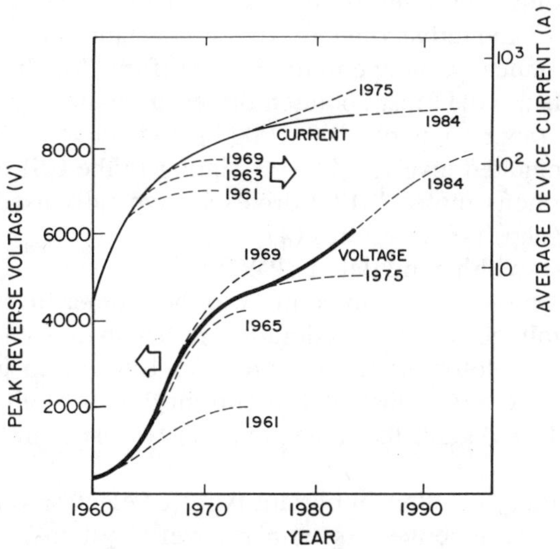

Figure 8. Maximum voltage and average current ratings of thyristors, and performance predictions by year. Adapted from (40).

12 The development of these devices has demanded numerous technical tradeoffs. For example, the series resistance of the power bipolar transistor increases as the square of the blocking voltage capability, correspondingly lowering its current limit (43). Similarly, to increase the breakdown voltage of a thyristor, it must be made thicker, which slows its turn-off time (44). Each type of device has a different set of characteristics that determine the best choice in any particular ASD.

2.1.1 Adjustable speed electronic drives

Numerous types of motor speed control systems have been developed (Figures 7,9). Electronic systems can be divided primarily into those that change the frequency of the power entering the motor and those which effectively change the rotor voltage and current (Figure 9). Of particular interest here are inverters, a block diagram of which is shown in Figure 9. A variety of other classifications have also been used for inverters, including: by the type of semiconductor device, circuit, or motor used; by the control strategy; or by the voltage or power level (32,36,45–48,130–132).

In the voltage-source inverter (VSI) drive, line power is converted to dc, smoothed by a capacitive filter, and then inverted to variable-frequency and variable-voltage ac power. The frequency is adjusted as desired to control the motor speed; the voltage is adjusted to keep the ratio of voltage to frequency constant (except at low speed, when the voltage is raised to compensate for the rotor resistance) and thus maintain constant motor torque over the desired speed range. The VSI produces a stepped-voltage waveform, an example of which is shown in Figure 9 (10,32,36,45–47).

The current-source inverter (CSI) also converts line power to dc, then smooths it with an inductive filter to give a constant current to the inverter, which in turn produces a stepped-current waveform for the motor[13] (Figure 9). VSIs and CSIs are the most common drives at medium (approximately 50–200 hp) and high (approximately 200–1000 hp and above) powers (38,45).

The load-commutated inverter (LCI) is similar to the CSI but is used with a modified synchronous motor[14]. LCI drives are typically used in systems with very high (>1000 hp) power ratings (47).

Although pulse-width modulated (PWM) drives similarly adjust the frequency to control motor speed, in contrast to these other inverters the signal is a chopped full voltage of varying duration rather than a continuous reduced voltage (Figure 9). Motor inductance provides filtering to leave the average voltage-to-frequency ratio constant.The current drawn by the motor is nearly sinusoidal. PWM drives are the most common type now in use at low power (20,38,47,48).

There are a variety of other important types of electronic motor controllers in use, including cycloconverters, slip recovery systems, and power factor

[13] The output voltage of a CSI is determined by the properties of the motor. Because of the high motor inductance, switching produces large voltage spikes, which essentially limits the choice of driving device to thyristors. The cost and complexity of the circuitry needed to control these thyristors then limits the application of CSI drives to larger motors. Finally, a CSI must be operated in a closed loop with a particular motor to avoid instabilities due to the strong dependence of inverter voltage on motor load (32,45–47).

[14] Modifications include a special field exciter and oversized damper windings, as well as a shaft encoder for slow-speed operation. The LCI uses the electromotive forces generated back by the motor to switch off the electronic devices in the inverter.

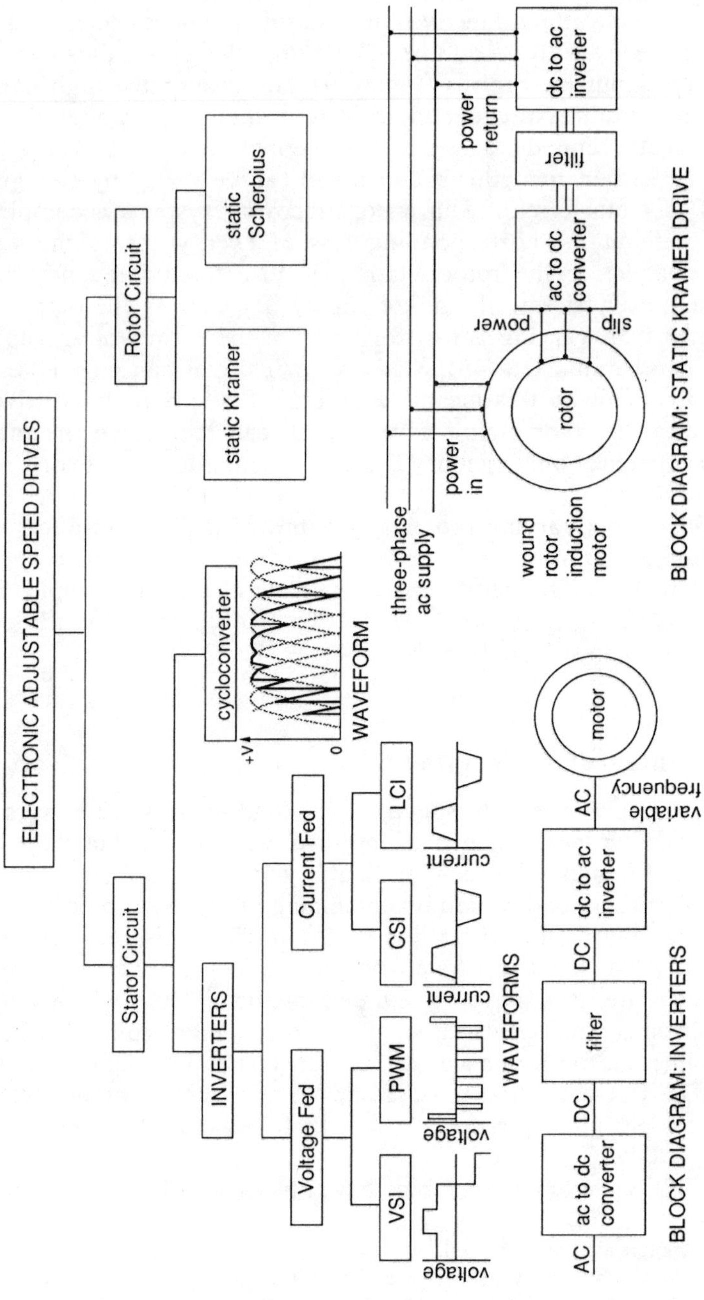

Figure 9. Classification of electronic adjustable speed drives, and block diagrams and waveforms for selected cases. Descriptions are in the text.

controllers. Cycloconverters directly switch ac line frequency to a synthesized variable ac frequency with no dc link. The output waveform thus consists of segments of the input waveforms (Figure 9). Because of the high harmonic content of the output voltage and input currents, the maximum output frequency is usually limited to less than half the input frequency (46).

As discussed earlier, induction motor speed can be varied by changing the resistance of the rotor circuit. Slip energy recovery systems accomplish the same thing, without the corresponding loss of energy across the variable resistor, by connecting the rotor windings[15] to a fixed frequency inverter which returns power back to the motor supply. By varying the output of the inverter, rather than varying a resistor's value, the rotor voltage and thus motor speed can be adjusted (46). Static Kramer drives can only take power from the rotor circuit by this means (Figure 9). Static Scherbius drives can feed power into the rotor circuit as well, and can thus drive the motor at higher than synchronous speed. These systems do, however, require connections to the rotor.

Finally, rather than varying motor speed, power factor controllers reduce the voltage applied to a motor (as a function of its power factor) when it is operating at part load. This results in energy savings of, for example, 25% at 40% load for single-phase motors and 10% at 40% load for three-phase motors. Savings increase rapidly for small loads (10,32,33).

2.1.2 Power integrated circuits

By precisely matching motor speed to its load, ASDs offer significant improvements in the performance of motor-drive systems, but their cost is often too high for general use – particularly in low- and medium-power applications. Costs can be lowered by combining all the electronic circuitry of the ASD into a few integrated circuits (and sometimes discrete devices) and mounting them on a single substrate ('power hybrid') or by other techniques. More significant are new 'power integrated circuits' (PICs) which put all of the ASD electronic components into a single integrated circuit (127,133). The overall principle and performance of these devices is identical to an ASD – they are ASDs. But compared to ASDs made of discrete components, their size[16] and cost are greatly reduced while their reliability is increased (49–51,127).

These new PIC technologies combine power electronic devices[17] operating at

[15] Wound rotor motors are typically used.

[16] A PIC could be easily placed inside, for example, the motor housing itself. In contrast, conventional ASDs require small to large (according to power rating) cabinet-sized installations which are then located conveniently and cables run to the motor.

[17] Most common in PICs are MOSFETs or bipolar transistors.

hundreds of volts and up to ten or more amperes on the same chip as their microprocessor controller operating at low voltages and currents. In addition to inputs from external sensors, pressure, temperature, acceleration, magnetic field, or other sensors can be placed directly on the chip, enhancing system control and reliability (52,127). For example, some PICs monitor their own temperatures and regulate them by reducing operating currents or even shutting down if temperatures get too high (51). Adding such intelligent functions can be done at little or no additional cost (127).

PIC technology is progressing rapidly. Devices capable of handling 1.5 kVA are available; by 1995 devices handling 100 kVA are expected. By 1992, this is expected to be a $1.2 billion-per-year industry (53). As volume increases and prices drop, PICs can be expected to have a significant impact on low- to medium-power motor-drive systems.

2.1.3 Alternative motor designs

The use of electronic ASD technologies allows several new motor designs to be considered. Most significant of these is the switched reluctance motor (126,134–136). This motor uses an ASD which sequentially switches between stator coils using just one polarity of current – greatly simplifying the circuitry compared to the above ASDs. This motor is also noted for its simple, reliable, and robust construction (there are no rotor coils), and for its relatively high efficiency over a wide operating range – comparable to induction motors. It suffers drawbacks, however, in its generation of high levels of acoustic noise and excessive torque pulsation.

2.2 Soft Magnetic Materials

Significant efficiency improvements in motors and transformers are possible through the use of improved soft magnetic materials to reduce hysteresis and eddy current losses in their cores. Core losses in generators, transformers (distribution + other), and motors have been estimated to be about 3, 48 (31+17), and 30 billion kWhr respectively during 1977. Combined, these losses were 3.8% of total net US electricity generation (54–56). Core losses in lamp ballasts cost a further 16 billion kWhr annually or 0.75% of net US electricity generation (54).

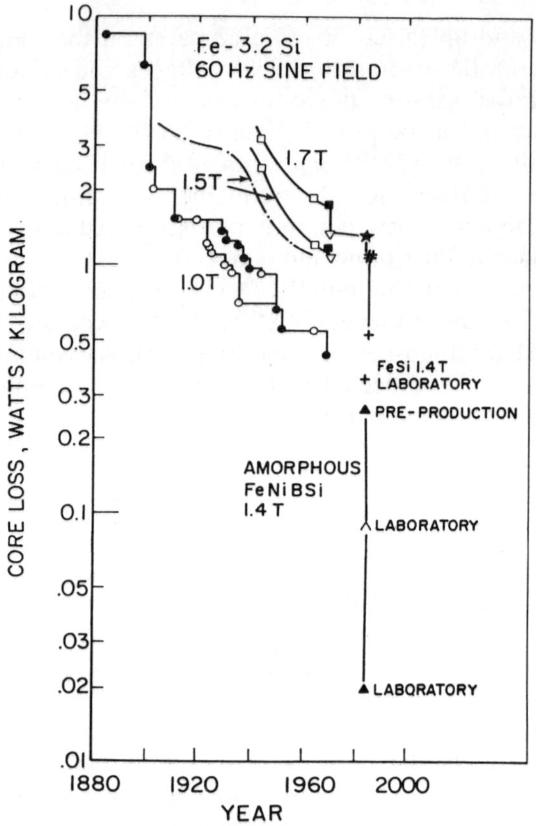

Figure 10. Power loss for magnetic cores versus year. Values marked * are from (60), # (61), + (62), ^ (66). Other values are from (56,59,60).

Soft magnetic materials have been steadily and dramatically improved[18] during the past 100 years as indicated in Figure 8. Today, non-oriented electrical steels, the predominant type used in standard motors and transformers, have losses of 2–4 W/kg at 60 Hz and 1.5 Tesla maximum flux density. Grain-oriented electrical steels have typical losses of 1.3 W/kg at 1.7 T (60). Recent innovations[19] in Japan have reduced losses in production units to

[18] Among the innovations that have made this improvement possible are the addition of silicon to iron and the control of other impurities to reduce eddy current and hysteresis losses, increase permeabilities, and reduce aging effects; precise orientation of the crystals within the steel sheet to maximize permeability and minimize hysteresis; stress coatings that interact favorably with the magnetostriction; and others (56–59).

[19] These include use of a laser-scribing technique to nucleate domains, among others, in production units (60, 61); and in the laboratory, precise control of the annealing process and secondary recrystalization to produce almost perfectly oriented crystals (62).

1.1 W/kg at 1.7 T (60,61) and in the laboratory to as low as 0.57 W/kg at 1.7 T and approximately 0.34 W/kg at 1.4 T (62).

The recent development of amorphous alloys of FeNiBSi, such as "Metglas," offers further performance gains[20]. Amorphous alloys of FeNiBSi have now achieved losses as low as 0.25 W/kg (1.4 T) in preproduction units of 25 kVA distribution transformers and 0.02–0.09 W/kg (1.4 T) for the lowest laboratory measurements (60,66). Extensive work has been done on the use of amorphous materials to reduce core losses in transformers (67–69) and motors (70). And with the reduction in core losses there is a reduction in operating temperature, leading to longer insulation life and further reductions in the losses of the copper windings (65). Advanced amorphous alloys could all but eliminate the annual $7 billion in magnetic core losses in the United States and could potentially dramatically restructure the $1 billion-per-year US soft magnetic materials industry (60).

2.3 Permanent Magnets

Permanent magnets have seen similarly dramatic improvements in recent years (Figure 11). In the 1930s, Japanese discovered alnicos (57,60,71). In the 1940s ferrites were discovered and developed in the Netherlands (57). Rare-earth permanent magnets (REPMs) were first demonstrated by Strnat (US) in the 1960s, and the Japanese introduced Sm_2Co_{17} – the second generation of these REPMs – in the 1970s (60). In 1983, both General Motors (US) and Sumitomo Metals (Japan) announced $Nd_2Fe_{14}B$ permanent magnets. These $Nd_2Fe_{14}B$ permanent magnets, according to the National Materials Advisory Board of the National Research Council, have an energy product "so large that it will revolutionize the design of motors" (60).

Electronically commutated, or brushless, permanent magnet (PM) motors differ from conventional motors in two fundamental ways. First, their rotors are made of permanent magnet material. This provides a significant increase in

[20] The lack of structure in amorphous materials minimizes magnetocrystalline anisotropy and thus hysteresis losses. Further, these alloys have a moderately high electrical resistivity, which minimizes eddy current losses. Finally, the very rapid cooling required to freeze in an amorphous state, 10^5–10^6 °C per minute, requires that they be produced in very thin laminations, leading, in principle, to inherently simple, fast, and less capital-intensive production systems: one simply squirts the material out to freeze on a rotating drum with few special precautions and all in one step. This is in marked contrast to today's oriented silicon steels, which require as many as 6–10 steps of careful melting, ingot casting, and precise schedules of hot and cold rolling and high-temperature annealing (1,56,59,63–65).

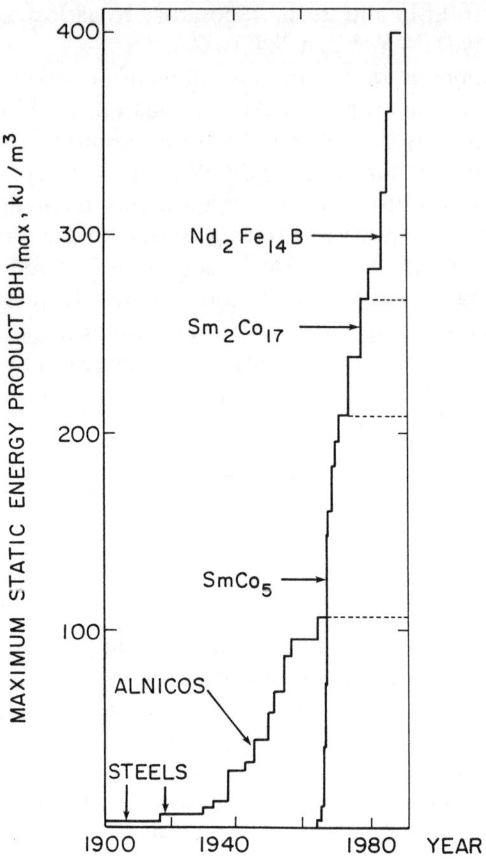

Figure 11. Energy products for selected permanent magnets since 1900. Note that the increase in magnetic energy density allows decreases (though not directly proportional) in motor size; and increases the ratio of motor torque to inertia. From (81, 129).

motor efficiency because the resistive heating losses of conventional ac or dc motor rotors are eliminated, as is the brush friction of dc motors. PM motors have as much as 18–35% lower losses than comparable high-efficiency induction motors (72,73). Lower rotor losses also mean cooler operation. Brushless operation and solid PM rotors may make such motors as reliable as their induction counterparts and far more reliable than dc motors, with their brush maintenance problems (72,74–77).

Second, sequential switching of the stator coils, based on the rotor position as determined by optical encoders on the rotor shaft, is used to drive the rotor. This technique provides better speed and torque control than ASD/induction motor systems, yet with somewhat less complicated circuitry (72,74–77).

These two fundamental differences lead to other advantages as well. In comparison to induction motors, brushless PM motors can often provide higher starting torque (which is also less sensitive to voltage fluctuations), greater torque to inertia, and higher power factor, yet are as much as 30% smaller in size and weight (72,74–77). The increase in PM energy density (Figure 11) is a measure (though not directly proportional) of these improvements. Unlike dc motors, PM motors cannot overspeed, yet they can operate at very high speeds. PM motors that can operate at more than 20,000 rpm are now available (78). In some cases this high speed will allow gears, belt drives, and other such devices to be eliminated. The speed range of PM motor and controller systems is also large, limited by cogging at the low end and centrifugal strength at the high end (75).

Finally, because PM materials have an intrinsically low permeability, the airgap in PM motors is less critical than in ac or dc machines. Consistent motor performance is then less costly to maintain in production. Parts counts can also be reduced. Combined with their other advantages, including all those of ASDs as discussed below, these have made brushless PM motors very attractive (75).

PM motors, however, do have drawbacks. Like ASDs used with induction motors, their controllers generate harmonics (see below). When started directly off the line and while accelerating to synchronous speed, PM motors suffer torque pulsation over a larger speed range than their induction motor counterparts (72). Excessive overloads, even transient ones, can cause a loss in performance by demagnetizing the PM rotor (79). And due to the constant field of the PM itself, driving the motor at high speeds can induce large back voltages in the stator (4). Finally, the magnetic materials are temperature-sensitive; they can suffer long-term irreversible demagnetization if operated at moderately high temperatures, above 150–200 °C (77). This is particularly true for the new $Nd_2Fe_{14}B$ magnets, although the addition of small amounts of cobalt may raise the operating temperatures considerably (80, 81).

Even before the development of PM materials with the potential for relatively low cost and high performance (such as $Nd_2Fe_{14}B$ in 1983), manufacturers of appliances, computer peripherals and office equipment, machine tools, and robotics had responded avidly to the potential presented by PM motors (74–76). The world market in 1985 for PMs was $1 billion of which Japan produced 40%, the United States about 32%, and Europe 20% (60,82). Sales of PM motors and controllers are expected to grow 20% annually (75). Due to their many attractive features, including plentiful raw materials and easy machinability, $Nd_2Fe_{14}B$ magnets are expected to rapidly become a major force in the PM market (83–85).

3 Modern System Design

ASD technology allows a significant change in system design and operation. In contrast to the conventional practice of oversizing systems, using constant speed motors to drive them, and throttling excess flow with vanes or valves, ASDs and their associated sensors and controllers allow precise matching of motor speed to load. This can significantly improve overall system efficiencies and process control, as well as provide many other advantages as discussed below. Furthermore, this flexibility may allow design rules that lead to extreme over-sizing to be relaxed, potentially reducing capital investments.

Under part-load or variable flow operation, ASDs provide significant energy savings by taking advantage of the affinity laws for fans, (centrifugal) pumps, and certain other equipment. That is, changing the rotational speed of the fan or pump changes the flow rate proportionally to the speed, the pressure drop or head by the square of the speed, and the power by the cube of the speed. Thus, to meet the demands of a system[21] at half its maximum flow rate, an ASD could be used to reduce the motor speed to half its line-synchronous speed, with a corresponding drop in power demand to $(0.5)^3 = 0.125$ of its maximum (Figure 6). In contrast, reducing flow with a throttling valve generates large back pressures and similarly large energy losses. An example of this is shown schematically in Figure 5. In addition to these direct energy losses across the throttling valve, however, the lower pump efficiency (Figure 5, see also Figure 6 for fans) at the throttled operating point and the lower motor efficiency at part load (Figure 3) must be taken into account. For systems with highly variable loads and/or duty cycles; steep system and/or equipment pressure-flow curves; and large powers and annual operating hours, the resulting savings with ASDs can be quite large (20,27,86,87).

Under constant flow applications, ASDs may also be able to provide savings. The past 100 years have been spent optimizing pumps, fans, and compressors to run at the maximum synchronous speed of the induction motor, 3000/3600 rpm. ASDs greatly expand this operational speed range; and the use of these "electric" gears[22] may allow equipment operation at more efficient speeds[23]. For example, pumping efficiency tends to rise with speed over a broad range[24]. Thus, for a system with a flow rate of 23 m^3/hr and a head of 100 m,

21 Assuming no static pressure drop or gravitational energy terms (see Note 10 above).

22 Personal communication, Gunnar Mellgren, ABB Corporate Research, Västerås, Sweden.

23 Mechanical gears also offer this option, but may be more awkward to use, particularly in retrofits.

24 The pump efficiency increases asymptotically with the pump specific speed which is proportional to $NQ^{1/2}/H^{3/4}$, where N is the pump shaft speed, Q is the volumetric flow rate, and H is the pump total head. Thus, low flow, high head systems have a low specific speed and are far from the asymptotic efficiency limit (128).

increasing the shaft speed from 3600 to 7200 rpm raises the pumping efficiency (for good design practice) from about 53% to 66%. A further increase to 10800 rpm would raise the efficiency to 70% (128). Of course, additional friction, pump shaft strength, and many other factors may limit the practicality of such high speed operation. Low flow, high head applications are an obvious opportunity, but there may be many others.

In practice, the application of ASDs requires attention to a variety of factors. The system load may demand constant pressure and variable flow; constant flow and variable pressure; or both variable flow and variable pressure. Different analyses are required for each case (27). For systems with large static pressure drops and/or gravitational energy to overcome (see Note 10 above), the system load will not reduce to zero at zero flow and speed – in sharp contrast to the affinity laws for pumps and fans. The result can be a large overestimate of the potential energy savings of an ASD (87). This characteristic also makes it very difficult to match the maximum efficiency point of the fan or pump to the system load across the entire range of operation (Figure 6). Such difficulties can arise even in heating, ventilation, and air-conditioning (HVAC) systems, for example, where a static pressure must be maintained at the terminal boxes to allow proper flow control to each space (31,88).

System efficiency calculations must now include losses within the ASDs themselves as well as additional losses in the motors (32,114,140). Induction motor losses, for example, can be expected to increase by 20% (or 2% of total energy consumption for a 90% efficient motor) when the motor is used with a six-step VSI because of harmonics (89). Most manufacturers recommend that standard ac motors be torque derated 10–15% when used with VSI or CSI (36). When operating near full-load, bypass switches can sometimes help avoid the excess losses in both ASDs and motors.

Finally, adjustable-speed drives can also be used with intelligent controls, and offer the potential to greatly change the way motive power is used. For example, an intelligent air-handling system could measure in real time the motor and fan efficiencies as well as the pressure drops in the duct as a function of volume blown, then choose how fast to run to meet ventilation needs and minimize energy use. At the expense of lower furnace or heat-exchanger efficiency, air of a larger temperature difference might allow slower flow rates using less energy overall. Even the quality of the energy could be taken into account, sacrificing more lowgrade thermal heat for smaller gains in high-grade electrical energy.

4 Case Studies

Adjustable-speed drives are being successfully used in numerous applications. In industry they are used in boiler fans in chemical plants (28), utility plants (90), packaging equipment (91), glass-blowing machines (92), cement factories (93), and many others. In commerce they are used in commercial refrigeration systems (94), HVAC systems (95), and other applications. In residences, they are used in air-conditioning and heat-pump systems (96–98,137,138,141).

For systems with variable loads, these electronic ASDs have much better efficiency and performance than eddy current couplings, variable rotor-resistance systems, hydraulic couplings, and other drives (32), and have shown remarkable savings in practice. Numerous case studies in the literature document typical energy savings of 30–50% in fan and boiler feedpumps, 20–25% in compressors, 30–35% in blowers and fans, 20–25% in pumps, 25–35% in central refrigeration systems, and 20% in air-conditioning and heat pumps (7,28,33,34,47,99–101). A residential heat pump system with ASD/permanent magnet motor driven compressor and fans, due on the US market in 1989, is expected to offer heating and cooling efficiencies 30–40% better than today's average model (141). Of course, ASDs will not provide savings in constant speed full-load applications, so national energy savings will be less than might be calculated by assuming universal adoption at the above efficiencies.

Payback times of 2–3 years or less are common for ASDs. For example, a 1000 hp (750 kW) ASD might have capital and installation costs of \$100/hp (\$133/kW) each, or \$200,000 total (38). Retrofitted to a 1000 hp (750 kW) motor (assuming a motor efficiency of 95%) running 80% of the time; assuming 30% total system energy savings by use of an ASD to avoid throttling; and at an electricity cost of \$0.05/ kWhr, then this ASD would save \$83,000 per year – a simple payback of just 2.5 years. This corresponds to an installed capacity-saved cost of \$1060/kW, including the duty cycle.

At the other end of the scale, the cost of mass produced ASDs used with residential heat pumps and other appliances is down to \$25/hp (\$33/kW) (132). For a standard induction motor with an efficiency of 75% (Figure 1) running on a 30% annual duty cycle; an electricity cost of \$0.05/kWhr; and arbitrarily assuming 20% savings, such an ASD would have a payback time of just one year. This corresponds to an installed cost of \$420/kW capacity saved, including the duty cycle. Of course, the actual savings may well be higher due to the ASD alone; moveover, other measures such as use of permanent magnet motors and improved compressors and other equipment might incease savings even further. Also, the cost of electricity is usually closer to \$0.10/kWhr.

In addition to higher efficiency and frequently lower lifetime costs, electronic ASDs offer a number of advantages over operating induction motors directly off the line. Intelligent controls can be incorporated to improve process quality. By permitting slow, controlled starts, ASDs reduce electrical stresses on motors, transformers, and switchgear, and reduce mechanical stresses on motors, gears, and associated equipment (34,47,48,99). In contrast, starting an induction motor directly off the line typically leads to current inrushes six times the normal full-load operating current, and for larger motors overheating can limit line starts to twice per hour. Operation at reduced speeds can further extend motor and equipment lifetimes directly, but supplemental cooling may be necessary at low speeds, since direct cooling by the motor's fan is then less effective (20). Indirectly, ASDs increase lifetimes of motor-driven equipment by avoiding the back-pressures generated by conventional throttle valves or dampers used to limit the output of pumps or fans, and by permitting constant bearing lubrication – in contrast to equipment operated in an on-off mode. Users who have converted to ASDs have often saved 10–20% on maintenance (99).

Other ASD advantages include isolation of the motor from the power line, which can reduce problems caused by varying or unbalanced line voltage. In some cases, power failure for a few cycles allows fixed-speed motors to slow to a point at which they cannot be reaccelerated without decreasing their loads. In contrast, ASDs can often compensate for such interruptions and thus provide a "ride-through" capability (20). ASDs can also allow operation at higher speeds than the 60-Hz line frequency allows. Further, ASDs are both relatively compact and can be located conveniently, wherever the power cables to the motor can be run; thus they are easy to retrofit. In contrast, mechanical speed controls must be located in-line with the motor shaft, often requiring extensive repositioning of the motor and driven equipment. Although ASDs have demonstrated outstanding reliability (125), there nevertheless remain critical applications where it may be desirable to provide bypass switches around the ASD and related controls to ensure uninterrupted operation in case of failure (28).

5 Utility Operation

ASDs can improve power plant operation and may someday assist load management efforts. ASDs, however, may also increase uncertainty about future load growth and may cause problems by their generation of power line harmonics.

ASDs have been used with a variety of power plant motors – such as forced-draft and induced-draft fans, boiler-feed and condensate pumps, and many others – to provide significant energy savings as well as other operational advantages discussed above. Use of a 2000-hp ASD on the #2 boiler-feed pump at the Ft. Churchill, Nevada, power plant (a 110 MW gas fired unit used for spinning reserve), for example, allowed the minimum load to be reduced from 16 to 12 MW and the auxiliary power demands to be reduced by 0.9 MW – saving $1.6 million per year at an installed cost of just $375,000 (90,125,142,143).

ASDs will be used increasingly in household and commercial refrigeration and air conditioning systems due to their efficiency and other advantages. By appropriately modifying the design of these ASDs and by establishing standard protocols, direct implementation of load management techniques via power-line carriers or other means (144) may then be possible at very low marginal costs. For example, circuitry to detect a power-line carrier signal and then to turn down or turn off a refrigerator or air conditioner could be incorporated directly into a power integrated circuit driving the motor. Time-of-day electricity cost controls could be similarly implemented (124). Simple on-off load management techniques are already being used in large scale systems in the U.S. and Europe (145,146). Southern California Edison, for example, currently has 100,000 air conditioners in a load management program networked via a VHF-FM radio system. Despite the high cost of establishing the system – approximately $120 per participating household just for the control devices, their installation, and marketing – the cost benefits to the utility in terms of reduced peak-load demand, etc., have been almost 4 times greater. Further, the program has been very well received; just 1.8% of the participants have withdrawn voluntarily.

The total electricity savings possible by the use of ASDs is not yet well understood and will increase uncertainty in load growth forecasts in the near-term. Total savings will depend on a variety of factors including the rate of cost reduction of ASDs and corresponding market penetration, and the types of part- or variable-loads driven. Further, the individual savings achieved may depend on the development of new engineering design rules that fully exploit the opportunities presented by ASDs. As noted above, in reviewing entire plants, the losses in motors and drives have been found to be as much as 40–80% (35).

The harmonics generated by electronic devices can also cause problems for utilities; problems that have long been with us. One of the first applications of rectifiers was for electrochemical processing at a copper refinery in Utah during the 1920s. When energized, the rectifier's harmonics induced voltages in telephone lines paralleling the power lines sufficient to disrupt communications (102,103).

Harmonics are in fact generated by many sources. The nonlinear characteristics of magnetic cores, for example, introduce harmonics. In recent times, competitive pressures have forced manufacturers to design equipment more and more critically, pushing the operating points further into the nonlinear characteristics of the magnetic materials and thereby greatly increasing harmonic generation (104,105).

Similarly, the nonlinear characteristics of the electronic devices in ASDs generate harmonics. Converters that use thyristors cause line notching – a brief short circuit that sends the line voltage abruptly to zero at that point. Furthermore, the current drawn is closer to a square wave than a sine wave. Converters that use diode bridges draw current in a periodic pulse and also cause distortion of the line voltage. A six-step inverter will generate characteristic harmonics at $nK \pm 1$ = 5,7,11,13,17,... times the line frequency, each with maximum theoretical amplitudes equal to (1/harmonic) of the fundamental. Twelve-step inverters can, in principle, significantly reduce the generation of harmonics, but in practice system reactance and converter phase shifts somewhat reduce their advantage (32,34,47,48,106–109).

The harmonics generated can cause a variety of problems (32,34,47,48, 107,108,110–117). Within the motor, they can increase total motor losses by several percent; increase voltage stress on windings and their insulation; sometimes cause torque pulsation at low speeds, localized heating, and acoustic noise; excite resonant torsional vibrations (shortening equipment shaft lifetimes); and cause other problems. On the line, they can cause localized heating within the distribution transformer, and increase losses throughout the generation and distribution system. Harmonics can also excite series or (more often) parallel resonance in the power circuits, particularly in the presence of distributed power factor correction capacitors. Even the induction watthour meters used to measure household electricity consumption are affected, consistently recording values as much as 1% too high when harmonics are present (111). Finally, harmonics on the line can interfere with computers, communications, and other equipment. The harmonics generated by ASDs may be particularly dangerous because the amounts of power being handled are so large.

Numerous technologies are being developed to reduce these harmonics (118–120). For example, harmonics can be minimized by appropriately adjusting switching patterns according to precomputed values in look-up tables stored in read-only memories (120), or even by doing real-time sampling and appropriate adjustment of the synthesized waveform (121,122). Of course, standard harmonic filters can also be used (104).

Many countries have established legislation limiting the line harmonics allowable from converters (104,107,108,123). Total harmonic distortion is typically limited to 5% with sublimits on individual harmonics. In the United States, FCC rules regulate only those devices that generate or use timing signals greater than 10 kHz. Devices that run at a slower clock rate are exempt even if they generate unacceptable line interference. In contrast, IEEE standard 519 specifically addresses line noise generated by converters on the basis of line notch size and voltage distortion, and has been used with considerable success (108). A standard specifically for ASDs, P958, will soon be proposed by an IEEE working group now being formed (113).

6 Conclusion

These various opportunities for reducing electricity use can now be compared as shown in the hypothetical estimates of Table 4, which in turn are based on the pumping system of Figure 5.

As seen in Table 4 for 80% flow, just 5% of the input coal is converted to useful energy (to overcome static pressure drop) at the end of the piping, and losses are spread throughout the system. ASDs achieve the greatest individual savings, and do so by eliminating shaft coupling and throttle valve losses, and by raising pump efficiencies by operating closer to the design point. Surprisingly, simply increasing the pipe diameter by 25% also offers large savings due to the excessive piping losses of this particular system (Figure 5) and due to the great leverage such reductions have upstream in the system[25].

Repeating the calculations for 65% flow (see the inset load profile of Figure 5) gives similar savings: the increased throttling losses and reduced pump efficiency (72% due to being further from the design point) are roughly counterbalanced by decreased piping losses. For this particular system, combining all of the various improvements listed leads to fuel savings of some 60% (Table 4). Achieving such large savings generally, however, is highly unlikely: many systems operate at, or near full-load most of the time; many will have appropriately sized piping – balancing the capital and energy costs. But by thinking of these as systems rather than as collections of individual components, numerous opportunities can be found for reducing electricity use. This is an area worthy of much more research.

The technologies described here are already available commercially, with sales growing rapidly, or in field testing. In Japan, for example, sales of ASD driven heat pumps are reportedly greater than those of constant speed systems (137). Energy savings are, however, but one factor driving the dissemination of these technologies, ASDs in particular. As important are their numerous other advantages noted above, including better control of process, often reduced maintenance, and relatively easy retrofits. Intelligent controls will be increasingly prominent and important in the future.

[25] This does not, however, take into account changes in pump or motor efficiency, etc. It also does not provide for the complex tradeoffs in capital costs for larger pipe (less the smaller motor and pump) versus the energy savings.

Table 4. Hypothetical savings for various modifications to the pumping system of figure 5.

LOCATION	BASE-LINE eff	BASE-LINE rmn	EN EF[f] MOTOR eff	AMOR[g] CORE eff	POWER[h] FACTOR eff	PM[i] MOTOR eff	PUMP[j] SPEED eff	PIPE[k] SIZE eff	ASD[l] eff	ALL[m] eff
COAL	–	100	–	–	–	–	–	–	–	–
GENERATION	33[a]	33	–	*	33.8	–	–	–	–	33.8
TRANSMISSION &DISTRIBUTION	90[b]	30	–	91.5	*	–	–	–	–	91.5
ASD	–	–	–	–	–	–	–	–	95	95
ELECTRIC MOTOR	91[c]	27	94	91.5	*	95.5	–	–	90	94.5
SHAFT COUPLING	98[d]	26	–	–	–	–	–	–	100	100
PUMP	77[e]	20	–	–	–	–	80	–	79	83
THROTTLE VALVE	66[e]	13	–	–	–	–	–	–	100	100
PIPING SYSTEM	35[e]	5	–	–	–	–	–	56	35	56
COAL REQD.		100	96.8	97.8	97.6	95.3	96.2	63	67	37
SAVINGS		0	3.2	2.2	2.4	4.7	3.8	37	33	63

"eff" is the device efficiency in %. "rmn" is the remaining energy as a % of the starting total. COAL REQD is the input coal energy required and "SAVINGS" are the percentage point reductions in coal use made possible for the different component and system improvements while still providing the same useable energy at the end of the piping system.

REFERENCES: BASELINE: a–(139); b–(54); c–(27) and Figure 2; d–(7); e–(27) for the case of 80% flow under variable pressure and flow, see Figure 5. The piping system efficiency is given here by the ratio of system static pressure at zero flow to total head less valve losses. f–ENERGY EFFICIENT MOTOR (27) and Figure 2. g–AMORPHOUS CORE transformers and motor (54) and Figures (3, 10), *generator savings are additional but are not detailed here. h–POWER FACTOR IMPROVEMENTS all *power factor losses (5) have been lumped together and assumed cut in half. i–PERMANENT MAGNET MOTOR losses assumed to be 25% less than an energy efficient motor (72,73) and Figure 2. This is for a PM motor run directly off the line so that there are no controller or harmonic losses. With a controller, losses will be slightly less than the ASD case. j–PUMP SPEED twice the baseline specific speed (128). k–PIPE SIZE 25% greater than baseline reduces pipe friction by 59%, assuming laminar flow (128). l--ADJUSTABLE SPEED DRIVE (27), note that the expected decrease in electric motor efficiency due to ASD harmonics (32,114,140) is in part compensated for by operating the motor at a higher effective load (Figure 3) and that the pump efficiency is increased by running closer to the design point than is possible by throttling (Figure 5), but is still less than the design point efficiency of 80% due to the pump affinity laws not tracking the load curve with static pressure drop; shaft couplings and throttling valves are assumed to be eliminated. m–ALL, combining the PM motor (with efficiency reduced slightly for controller harmonics), amorphous core transformers, ASD, 25% larger pipe, operating at the higher power factor, and running the pump at twice the specific speed. Note that other potential synergistic savings or losses (such as lower efficiency due to reduced size) between components are not considered.

Actual savings may differ significantly from these estimates.

Modern solid-state electronics, magnetic materials, and other technologies are revolutionizing motor-drive system design and operation worldwide. Although these technologies are already proving irresistible in many applications, much work remains to be done: to further improve equipment performance and reduce costs; to develop standards for the use of ASDs (113); and to develop new rules for system design and operation so that the full potential of these technologies can be realized.

Acknowledgements

The author wishes to acknowledge the Hewlett Foundation, the seven New Jersey gas and electric utilities, the New Jersey Department of Energy, and especially the Center for Energy and Environmental Studies, Princeton University for their support while much of this work was being done. The views expressed are those of the author alone and do not reflect any endorsement by the Office of Technology Assessment.

References

1. Baldwin, S.F. 1988. *Energy-Efficient Electric Motor Drive Systems: I. High Performance Materials, Adjustable Speed Drives, and System Design.* Center for Energy and Environmental Studies, Princeton University.
2. USDOE Energy Info. Admin. 1987. *Annual Energy Review 1986.*
3. Staub, A.O., Owen, E.L. 1986. *Solid State Motor Controllers.* IEEE Trans. Ind. Applic. IA-22, 6:1113–20.
4. Owen, E.L. 1984. Power Electronics and Rotating Machines – Past, Present, and Future, 15th Annual Power Electronics Specialist Conference, pp. 3–11.
5. A.D. Little, Inc., 1976. *Energy Efficiency and Electric Motors.* U.S. Dept. of Commerce, NTIS PB-259 129.
6. Nasar, S.A. 1970. *Electromagnetic Energy Conversion Devices and Systems.* Prentice-Hall, Inc.
7. USDOE, A.D. Little, Inc. 1980. *Classification and Evaluation of Electric Motors and Pumps,* DOE/CS-0147.
8. Fitzgerald, A.E., Kingsley, C. Jr., Umans, S.D. 1983. *Electric Machinery.* New York: McGraw Hill.
9. McDonald, W.J., Hickok. H.N. 1985. *Energy Losses in Electrical Power Systems.* IEEE Trans. Ind. Applic. IA-21:124–36.
10. Andreas, J.C. 1982. *Energy-Efficient Electric Motors.* Marcel Dekker, Inc. New York.
11. Jimoh, A.A., Findlay, R.D., Poloujadoff, M. 1985. *Stray Losses in Induction Machines: Part I. Definition, Origin and Measurement; Part II. Calculation and Reduction.* IEEE Trans. Power App. Sys., PAS-104:1500-12.
12. Reason, J. 1986. *Powerplant Motors. Power,* pp. S1-S16

13. Appelbaum, J., Fuchs, E.F., White, J.C. 1987. *Optimisation of Three – Phase Induction Motor Design. Part I: Formulation of the Optimization Technique* and with I.A. Khan, *Part II: The Efficiency and Cost of an Optimal Design.* IEEE Trans. En. Conv. EC-2:407–22.
14. Fuchs, E.F., Appelbaum, J., Khan, I.A., Hall, J., Frank, U.V. 1987. *Optimisation of Induction Motor Efficiency*, two volumes, EPRI EL-4152, July 1985, May 1987, Palo Alto, Ca.
15. Fetih, N.H., El-Shewy, H.M. 1986. *Induction Motor Optimum Design, Including Active Power Loss Effect*, IEEE Trans. En. Conv. EC-1:155–60.
16. Scheda, F.A. 1986. *Transient Inrush Currents in High-Efficiency and Standard Motors*, IEEE Trans. Ind. Applic. IA-22:145–47.
17. Energy User News, 1985. *Computing Power Factor Penalties*, p. 3.
18. Mullen, D.M. 1985. Specify Motors By Starting Duty, Not By Running Duty. *Power*, pp. 55–57.
19. Griffith, J.W., Mckay, R.M. 1986. *Improved Motors For Utility Applications: V. 6: Squirrel-Cage Rotor Analysis*, EPRI EL-4286.
20. Hickok, H.N. 1985. *Adjustable Speed – A Tool for Saving Energy Losses in Pumps, Fans, Blowers, and Compressors.* IEEE Trans. Ind. Applic., IA-21:124–36.
21. Stepanoff, A.J. 1957. *Centrifugal and Axial Flow Pumps*, 2nd Edition. New York: John Wiley and Sons.
22. Settles, G.S., Hamrick, J.T., Barr, W.J., Summerfield, M., Gunn, M., 1977. Energy-Efficient Pump Utilization. J. *Energy*, 1:65–72.
23. Reason, J. 1983. Fans, *Power*, pp. S1-S24.
24. Geller, H.S. 1986. *End-Use Electricity Conservation: Options for Developing Countries.* World Bank En. Dept. Paper 32, Washington, D.C.
25. Ross, M. 1986. *Trends in the Use of Electricity in Manufacturing.* IEEE Technology and Society, pp. 18–22.
26. Baldwin, S.F., Finlay, E. 1987. *Analysis of Electric Motor Driven Systems in the Sugar Industry: Electricity Conservation in Boiler Fans and Irrigation Pumps.* Presented at Meeting of the Jamaican Assoc. of Sugar Technologists, Ocho Rios, Jamaica, Nov. 5–6.
27. Armintor, J.K., Connors, D.P. 1987. *Pumping Applications in the Petroleum and Chemical Industries.* IEEE Trans. Ind. Applic. IA-23:37–48.
28. Rozzner, J.D., Myers, P.W., Robb, C.J. 1985. *The Application of Adjustable Frequency Controllers to Forced Draft Fans for Improved Reliability and Energy Savings*, IEEE Trans. Ind. Applic., IA-21:1482–89.
29. Price, A.C., Ross, M.H. 1988. *Reducing Electricity Costs in Automotive Assembly and Stamping Plants.* University of Michigan, unpublished.
30. Ladomatos, N., Lucas, N.J.D., Murgatroyd, W., Wilkins, B.C. 1979. Industrial Energy Use – III. The Prospects For Providing Motive Power In A Machine Tool Shop From A Centralized Hydraulic System, *Energy Research* 3:19–28.
31. Brothers, P.W., Warren, M.L. 1986. *Fan Energy Use In Variable Air Volume Systems.* ASHRAE Transactions, V.92, pt.2.
32. Mohan, N. 1981. *Techniques for Energy Conservation in AC Motor- Driven Systems*, EPRI EM-2037, Palo Alto, Ca.
33. Masher, D.P., Smith, J.S. 1981. *Impact of Advanced Power Semiconductor Systems on Utilities and Industry.* EPRI, EM-2112, Palo Alto, Ca.
34. *EPRI Journal*, 1984. Pacing Plant Motors For Energy Savings, p. 22-28.
35. Murgatroyd, W., Wilkins, B.C. 1976. The Efficiency of Electric Motive Power In Industry, *Energy* 1:337–45.

36. Jones B.L., Brown, J.E. 1984. *Electrical Variable-Speed Drives*. IEE Proceedings, 131A:516–58.
37. Owen, E.L., Morack, M.M., Herskind, C.C., Grimes, A.S. 1984. *AC Adjustable-Speed Drives With Electronic Power Converters – The Early Days*. IEEE Trans. Ind. Applic: 20:298–307.
38. EPRI, 1987. *Adjustable Speed Drives: Directory, Manufacturers and Applications*, Second Edition, Palo Alto, Ca.
39. Chen, D.Y. 1987. *Power Semiconductors: Fast, Tough, and Compact*. IEEE Spectrum, pp. 30–35.
40. Adler, M.S., Owyang, K.W., Baliga, B.J., Kokosa, R. 1984. *The Evolution of Power Device Technology*. IEEE Trans. Elec. Dev. ED-31:1570–91.
41. Ikeda, Y., Yatsuo, T. 1987. Recent Progress in Power Electronic Devices, *Hitachi Review* 36:5–12.
42. Douglas, J. 1986. Sealed in Silicon: The Power Electronics Revolution. EPRI Journal, pp. 5–15.
43. Baliga, B.J., Chen, D.Y. 1984. *Power Transistors: Device Design and Applications*. N.Y: IEEE Press.
44. Kaplan, G. 1982. Thyristors: *Future Workhorses in Power Transmission*. IEEE Spectrum, pp. 40–52.
45. Jarc, D.A., Novotny, D.W. 1987. *A Graphical Approach to AC Drive Classification*. IEEE Trans. Ind. Applic. IA-23:1029–35.
46. Dewan, S.B., Slemon, G.R., Straughen, A. 1984. *Power Semiconductor Drives*. John N.Y: Wiley and Sons.
47. Reason, J. 1985. Large Multi-Speed Motor Drives – The Alternatives, *Power*, pp. 68–71.
48. Cleaveland, P. 1984. *Motor Controls: A Technology Update*, Chiltons I&CS, pp. 50–55.
49. Brody, H. 1985. *Smart Power*, *High Technology*, pp. 18–24.
50. Robb S.P., Sutor, J.L. 1985. Comparing Power ICs. *Machine Design*, pp. 119–22.
51. Rumennik, V. 1985. *Power Devices Are In The Chips*. IEEE Spectrum, pp. 42–48.
52. Allan, R. 1984. Sensors in Silicon, *High Techn.*, pp. 43–50.
53. Marshall, B. 1986. *The Future of Smart PICS*, in "Recent Advances In Discrete MOS-Gated Power Structures", Electro/86, Boston, May 13–15.
54. Werner, F.E. 1980. *Electrical Steels: 1970–1990*, in "Energy Efficient Electrical Steels" Eds. A.R. Marder and E.T. Stephenson, Proc. AIME Metal. Soc. Pittsburgh, Oct. 5–9.
55. USDOE En. Info. Admin. 1987. *Monthly Energy Review*.
56. Luborsky, F.E., Frischmann, P.G., Johnson, L.A. 1978. The Role of Amorphous Materials In The Magnetics Industry, *J. Magn. Magn. Mater*. 8:318–29.
57. Bickford, L.R. 1985. *Magnetism During the IEEE's First One Hundred Years (1884–1984)*. IEEE Trans. Mag. MAG-21:2–9
58. Littman, M.F. 1971. "Iron and Silicon-Iron Alloys", IEEE Trans. Mag. MAG-7:48–60
59. Jacobs, I.S. 1979. Magnetic Materials and Applications – A Quarter Century Overview, *J. Appl. Phys*. 50:7294–306.
60. National Materials Advisory Board 1985. Magnetic Materials, NMAB-426, Washington, D.C: National Academy Press.
61. Fukuda, T., Hoshi, M. 1984. *New Technology for Large Power Transformers*, Hitachi Review, 33:137–40.

62. Takahashi, N., Ushigami, Y., Yabumoto, M., Suga, Y., Kobayashi, H., et. al. 1986. *Production of Very Low Core Loss Grain-Oriented Silicon Steel.* IEEE Trans. Mag. MAG-22:490–95.
63. Gilman, J.J. 1980. Magnetic Glasses, *Science*, 208:856–61.
64. Chen, H.S. 1980. Glassy Metals, *Rep. Prog. Phys.* 43:353–432.
65. Moorjani, K., Coey, J.M.D. 1984. *Magnetic Glasses*, Elsevier.
66. Tan, K.S. 1986. *Amorphous Ribbon Domain Refinement by Scribing.* IEEE Trans. Mag. Mag-22:188–91.
67. Takach, D.S., Boggavarapu, R.L. 1985. *Distribution Transformer No-Load Losses.* IEEE Trans. Power App. Sys. PAS-104:181–93.
68. Whitley, D.W. 1987. *Type Testing of Amorphous Metal Distribution Transformers.* IEEE Trans. Power Deliv. PWRD-2:827–30.
69. Saied, M.M. 1987. *Feasibility of Using Magnetic Amorphous Metals As Core Materials in Power Transformers*, Elec. Mach. Power Sys. 12:325–41.
70. Johnson, L.A., Cornell, E.P., Bailey, D.J., Hegyi, S.M. 1982. *Application of Low Loss Amorphous Metals in Motors and Transformers.* IEEE Trans. Power App. Sys. PAS-101:2109–14.
71. Chikazumi, S. 1982. Evolution of Research In Magnetism in Japan. *J. Appl. Phys.* 53:7631–36.
72. Richter, E., Miller, T.J.E., Neumann, T., Hudson, T.L. 1985. *The Ferrite Permanent Magnet AC Motor – A Technical and Economical Assessment.* IEEE Trans. Ind. Appl. 21:644–50.
73. Cornell, E.P. 1983. *Permanent Magnet AC Motors.* Proc. Conf. Drives/Motors/Controls 83, eds. P. Lawrenson, A. Hughes.
74. Klein, F.N., Kenyon, M.E. 1984. Permanent Magnet DC Motors Design Criteria and Operation Advantages. IEEE Trans. Ind. Appl. IA-20:1525-31.
75. *Appliance*, 1984. Special Issue, 41(4): 2.
76. Barber, J. 1984. New DC Brushless Entries Touted For Low-Speed Uses. *Energy User News.*
77. Miller, T.J. 1983. Rare Earth Magnets Contribute to Small Size, High Torque of New Motor Designs. *Control Engineering*, pp. 90–91.
78. Strnat, K. 1985. F. M. Tait Professor of Electrical Engineering, University of Dayton, Dayton, Ohio. Personal Communication, November 19.
79. Demerdash, N.A., Nyamusa, T.A., Nehl, T.W. 1985. *Comparison of Effects of Overload On Parameters and Performance of Samarium-Cobalt and Strontium-Ferrite Radially Oriented Permanent Magnet Brushless DC Motors.* IEEE Trans. Power App. Systems, PAS-104:2223–31.
80. Robinson, A.L. 1984. Powerful New Magnetic Material Found. *Science* 223:920–22.
81. Strnat, K.J. 1986. Rare-Earth Permanent Magnets: *Two Decades of New Magnet Materials.* Symp. Soft and Hard Mag. Mater. with Applic., Amer. Soc. for Metals, Orlando, Fla. 4–9 Oct.
82. Graff, G. 1986. Remaking the Magnet, *New York Times*, Sunday, March 30, p. F23.
83. Strnat, K. 1983. *Study and Review of Permanent Magnets for Electric Vehicle Propulsion Motors*, DOE/NASA/0189–83/2.
84. Rahman, M.A., Slemon, G.R. 1985. *Promising Applications of Neodymium Boron Iron Magnets in Electrical Machines.* IEEE Trans. Mag. MAG-21:1712–1716.
85. *New York Times*, 1986. Supermagnets: More Pull, Less Weight, March 30, p. F23.
86. Pottebaum, J.R. 1984. *Optimal Characteristics of a Variable-Frequency Centrifugal Pump Motor Drive.* IEEE Trans. Ind. Applic., IA-20:23–31.

87. Wood, W.R. 1987. Beward of Pitfalls when Applying Variable-Frequency Drives. *Power*, pp. 47–49.
88. Norford, L., Englander, S., personal communication, CEES Princeton.
89. Scholey, D. 1982. *Induction Motors for Variable Frequency Power Supplies*. IEEE Trans. Ind. Applic., IA-18:368–72.
90. Mohan, N., Brooks, K.F., Montgomery, W. 1985. *Energy Conservation In AC Motor Driven Systems By Means of Solid State Adjustable Frequency Controllers*. Dept. Elec. Engn., U. of Minn., Minneapolis.
91. Wasilewski, J.E. 1985. Solid-State AC Drives: New Reliability for Packaging Equipment. IEEE Trans. Ind. Applic. IA-21:107–11.
92. Cuffe, S.S., Hammond, P.W. 1985. *A Variable Frequency AC Blower Drive Installation for Efficient and Accurate Control of Glass Tempering*. IEEE Trans. Ind. Applic. IA-21:1047–52.
93. Nilsson, T.R., Sinner, B., Volden, O.V. 1986. *Optimized Production and Energy Conservation*. IEEE Trans. Ind. Applic. IA-22:442–45.
94. Ponczak, G. 1987. Energy Management in Supermarkets: Variable Speed Compressor System to Save $3000/year. *Energy User News*, May 4.
95. Raffaele, P. 1987. New Controls; HVAC Retrofit to Save Firm $93K per Year. *Energy User News*, 12, April 27.
96. Geller, H.S. 1986. *Energy-Efficient Residential Appliances: Performance Issues and Policy Options*. IEEE Tech. and Soc. 5:4–10.
97. Raymond, G. 1987. *EPRI/Carrier Advanced Heat Pump*. ACEEE First National Conference on Appliance Efficiency, Oct. 27, Arlington, Va.
98. Mohan, N., Ramsey, J.W. 1986. *Comparative Study of Adjustable-Speed Drives for Heat Pumps*, EPRI EM-4704.
99. Watson, F. 1984. AC Speed Drives Catching On In Industrial Process Use. *Energy User News*, 9, Aug. 20.
100. Harrer, B.J., Kellogg, M.J., Lykes, A.J., Imhoff, K.L., Fisher, Z.J. 1986. *Electric Energy Savings From New Technologies*, PNL 5665-rev 1.
101. Oliver, J., Ferraro, R. 1986. Reduce Plant Auxiliary Load, Save Fuel with ASDs, *Electrical World*, pp. 42–44.
102. Stratford, R.P. 1981. *Analysis and Control of Harmonic Current in Systems with Static Power converters*. IEEE Trans. Ind. Applic. IA-17, N.1.
103. Douglas, J. 1985. Quality of Power in the Electronics Age. *EPRI Journal*, pp. 6–13.
104. Arrillaga, J., Bradley, P.A., Bodger, P.S. 1985. Power System Harmonics. New York: John Wiley and Sons.
105. Mahmoud, A.A. 1984. *Power System Harmonics: An Assessment*, in "Power System Harmonics", Eds. A.A. Mahmoud, W.M. Grady, M.F. McGranaghan. Piscataway, N.J: IEEE Press.
106. Jarc, D.A., Connors, D.P. 1985. Variable Frequency Drives and Power Factor. IEEE Trans. Ind. Appl. IA-21:771–77.
107. Bradley, D.A., Morfee, P.J., Wilson, L.A. 1985. *The New Zealand Harmonic Legislation*. IEE Proceedings 132A:177–84.
108. Jarc, D.A., Schieman, R.G. 1985. *Power Line Considerations for Variable Frequency Drives*. IEEE Trans. Ind. Appl. IA-21:1099–1105.
109. Schieman, R.G. 1985. *Power System Requirements for Thyristor Drives*. IEEE Trans. Ind. Applic. IA-21:849–52.
110. Gonzalez, D.A., McCall, J.C. 1987. *Design of Filters to Reduce Harmonic Distribution In Industrial Power Systems*. IEEE Trans. Ind. Applic. IA-23:504–11.

111. Fuchs, E.F., Roesler, D.J., Alashhab, F.S. 1987. *Sensitivity of Electrical Appliances to Harmonics and Fractional Harmonics of the Power System's Voltage. Part I: Transformers and Induction Machines; Part II: Television Sets, Induction Watthour Meters and Universal Machines.* IEEE Trans. Power Deliv. PWRD-2:437–53.
112. Appiarius, J.C., Houghtaling, D.W., Lackey, D.F., McCoy, R.M., Miller, N.W. 1986. *Improved Motors for Utility Applications. V.4: Impact of Harmonics.* EPRI EL-4286.
113. Hassan, I.D. 1987. *Specifying Adjustable-Speed AC Drive Systems and Currently Available Industry Standards.* IEEE Trans. Ind. Applic. IA-23:581–85.
114. Rice, D.E. 1986. *Adjustable Speed Drive and Power Rectifier Harmonics – Their Effect on Power System Components.* IEEE Trans. Ind. Applic., IA-22:161–77.
115. Nash, H.O., Wells, F.M. 1985. *Power Systems Disturbances and Considerations for Power Conditioning.* IEEE Trans. Ind. Applic. IA-21:1472–81.
116. Philibert, C.L. 1986. Watch Out For Harmonics When Specifying SCR Motor Drives. *Power*, pp. 49–50.
117. Wolff, F.H., Molnar, A.J. 1985. Variable-Frequency Drives Multiply Torsional Vibration Problems. *Power*, pp. 83–86.
118. Hombu, M., Ueda, S. Ueda, A., Matsuda, Y. 1985. *A New Current Source GTO Inverter with Sinusoidal Output Voltage and Current.* IEEE Trans. Ind. Applic. IA-21:1192–98.
119. Divan, D.M. 1985. *Optimum PWM Waveform Synthesis – A Filtering Approach.* IEEE Trans. Ind. Applic. IA-21:1199–1205.
120. Boys, J.T., Walton, S.J. 1985. *A Loss Minimized Sinusoidal PWM Inverter.* IEE Proc. 132B:260–68.
121. Bowes, S.R., Midoun, A. 1985. *Sub-Optimal Switching Strategies for Micro-processor Controlled PWM Inverter Drives.* IEE Proc. 132B:133–45.
122. Hamman, J., DuToit, L.P. 1986. *A New Microcomputer Controlled Modulator for PWM Inverters.* IEEE Trans. Ind. Applic. IA-22: pp. 281–85.
123. Pileggi, D.J., Emanuel, A.E. 1984. *An Examination of Existing Harmonic Recommended Limits, Guides and Standards*, in Power System Harmonics, Eds. A.A. Mahmoud, W.M. Grady, M.F. McGranaghan, Piscataway, N.J: IEEE Press.
124. Rosenfeld, Arthur H., Bulleit, Douglas A., Peddie, Robert A. Smart Meters and Spot Pricing: Experiments and Potential. *IEEE Technology and Society Magazine*, 1986, pp. 23–28.
125. Oliver, J.A., Banerjee, B. Ben. 1988. *Power Measurement and Harmonic Analysis of Large Adjustable Speed Drives.* IEEE Trans. Energy Conversion, EC-3:384–90.
126. Lipo, Thomas A. 1988. *Recent Progress in the Development of Solid-State AC Motor Drives.* IEEE Trans. Power Electronics, PE-3:105–17.
127. Fury, Art. 1988. Smart Power: Changing the Face of Power Control. Electronics, pp. 135–46.
128. Karassik, I.J., Krutzsch, W.C., Fraser, W.H., Messina, J.P. 1986. *Pump Handbook.* New York: McGraw-Hill.
129. Sagawa, M., Hirosawa, S., Yamamoto, H., Fujimura, S., Matsuura, Y. 1987. Nd-Fe-B Permanent Magnet Materials. *Jap. Jour. Appl. Phys.* 26:785–800.
130. Staub, A.O., Owen, E.L. 1986. *Solid-State Motor Controllers.* IEEE Trans. Ind. Applications, IA-22:1113–120.
131. Bose, Bimal K. Ed. 1981. Adjustable Speed AC Drive Systems. IEEE Press, pp. 1–21.
132. Greenberg, Steve, Harris, Jeffrey P., Akbari, Hashem., Almeida, Anibal de. 1988. *Technology Assessment: Adjustable-Speed Motors and Motor Drives (Residential and Commercial Sectors).* Lawrence Berkeley Laboratory.

133. Wolley, E. Duane. 1986. *Power Semiconductors: a Perspective of the Power Semiconductor Committee.* IEEE Trans. Ind. Applic. IA-22:970–71.
134. Dawson, Graham E., Eastham, Anthony R., Mizia, Jerzy. 1987. *Switched-Reluctance Motor Torque Characteristics: Finite-Element Analysis and Test Results.* IEEE Trans. Ind. Applic. IA-23:532–37.
135. Ray, W.F., Lawrenson, P.J., Davis, R.M., Stephenson, J. Michael, F., Norman N., Blake, R.J. 1986. *High-Performance Switched Reluctance Brushless Drives.* IEEE Trans. Ind. Applic. IA-22:722–29.
136. Rahman, M. Azizur, Osheiba, Ali M. 1985. *Steady-State Performance Analysis of Polyphase Hysteresis-Reluctance Motors.* IEEE Trans. Ind. Applic. IA-21: 659–663.
137. Jeter, S.M., Wepfer, W.J., Fadel, G.M., Cowden, N.E., Dymek, A.A. 1987. Variable Speed Drive Heat Pump Performance. *Energy,* 12:1289–98.
138. Eto, Joseph H., Almeida, Anibal De. 1988. *Saving Electricity in Commercial Buildings With Adjustable-Speed Drives.* IEEE Trans. Ind. Applic. IA-24:439–43.
139. Williams, Robert H., Larson, Eric D. 1986. *Steam-Injected Gas Turbines and Electric Utility Planning.* IEEE Technology and Society, pp. 29–38.
140. Erickson, C. James. 1988. *Motor Design Features for Adjustable-Frequency Drives.* IEEE Trans. Ind. Applic. IA-24:192–98.
141. The Advanced Heat Pump: All the Comforts of Home ...and Then Some. *EPRI Journal.* See also ref. 97.
142. Oliver, Jim, Ferraro, Ralph. 1986. *Reduce Plant Auxiliary Load, Save Fuel With ASDs.* Electrical World, pp.42–45.
143. Montgomery, Wilbur, Tatar, George, Oliver, James A., Ferraro, Ralph J. 1985. *Ft. Churchill Station Boiler Feed Pump Conversion to Adjustable Speed Operation.* PCEA Engineering and Operating Conference, Los Angeles.
144. Gaushell, Dennis J. 1985. *Automating the power grid.* IEEE Spectrum, pp.39–45.
145. Strickler, G.F., Noell, S.K. 1988. *Residential Air Conditioner Cycling: A Case Study.* IEEE Trans. Power Systems PWRS-3:207–12.
146. Bohlin, P., Edvinsson, M., Lindbergh, G., Lundqvist, C.G. *Successful Implementation of A Nation-Wide Load Management System.* IEEE Trans. Power Systems PWRS-1(4): 90–95.

Energy-Efficient Lighting

Terry McGowan

Abstract

Lighting remains a significant load for electric utilities representing from 8–17 percent of total electric consumption in industrialized countries according to recent estimates. However, new lighting technology using a systems approach involving the lamp, ballast, lighting fixture, energy management controls and operational parameters has the potential to substantially improve lighting efficiency and reduce total lighting energy use. In some systems, a 3-fold increase in efficiency per socket can be achieved. Implementing effective strategies to utilize this potential and putting it into the hands of lighting users will be the key to any significant lighting energy reductions.

1 Introduction

The efficient end-use of electricity for lighting has been a traditional goal of lighting planners. While lighting typically has relatively low first costs, it has relatively high operating costs of which electric energy can be 70–90 per cent of the total[1]. This strong energy component has tended to focus attention on lighting system efficiency because any efficiency improvements can be almost directly equated to cost savings. Efficiency improvements in lighting systems – especially in the light sources and auxiliaries – are therefore normal and even expected. Efforts to reduce lighting energy use during the 1970's directed new attention to the lighting system. As a result, lighting fixtures, lamps, lighting controls, lighting design, and system operating and maintenance procedures have been subjected to careful scrutiny. A large body of new information on lighting efficiency is available and, more importantly, it is finding its way into common lighting design practice. (See, for example, the Illuminating

[1] A comprehensive listing of the performance and economic characteristics of general lighting systems may be found in reference 1.

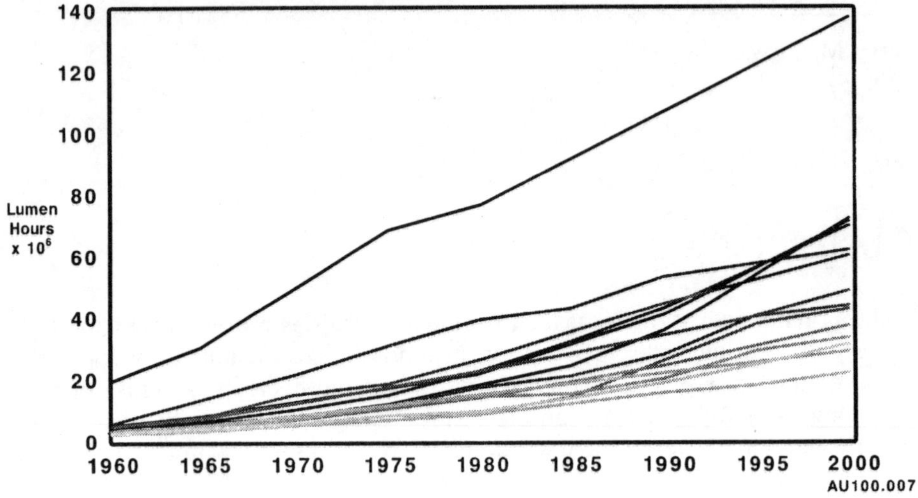

Figure 1. Light production per capita. Lumen-hours for 16 industrialized countries as reported by the International Commission on Illumination (CIE) 1987. Note: the practice of the CIE is not to specify the data by country; however, the countries contributing the data are listed in the text. (also applies to Figs. 2, and 3).

Engineering Society of North America (IESNA) Lighting Energy Management (LEM) Series of publications (2).

Prescriptive and non-prescriptive ways to reduce lighting energy have been developed and some are now official policy or legal requirements in countries around the world. The recommendations can be complex and highly detailed, but the objectives may be generally outlined as follows:

1 Provide illumination levels suited to the person and the visual work being done.
2 Utilize the most efficient light sources and systems.
3 Control where and when the light is to be provided.
4 Integrate electric and daylighting systems.
5 Maintain and operate the systems properly to minimize losses due to dirt and depreciation.

Lighting was the first major load of centralized electric generating facilities and it continues to utilize a substantial portion of the generating capacity of

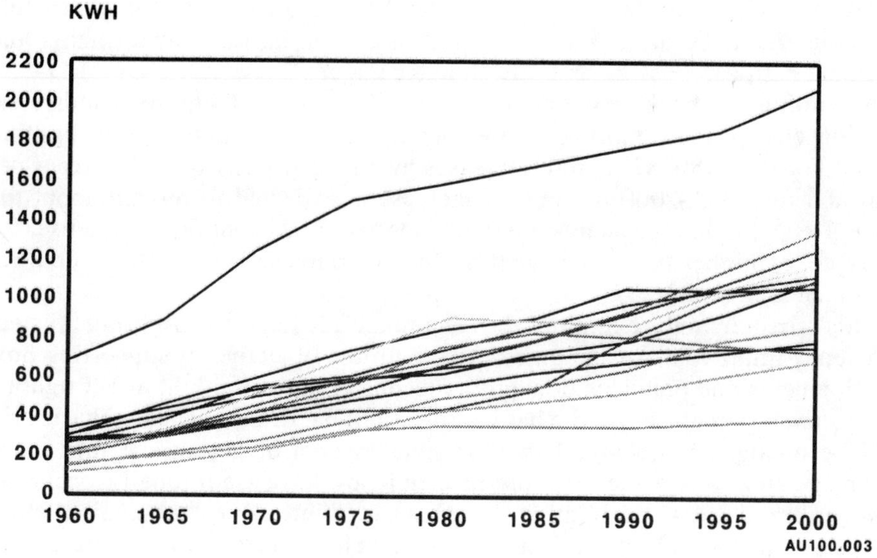

Figure 2. Lighting energy consumption per capita.

industrialized countries. International Commission on Illumination (CIE) data[2] for 16 member countries indicate that lighting energy as a percentage of *total electric energy consumption* ranges from 8 to 17 percent (1985).

The *use of lighting,* which really means the use of luminous energy as measured in lumen-hours, is not declining in those countries reporting, however. On a per capita basis, the lumen-hours have grown at an average rate of somewhat more than 5 percent per year (1980–87), but the rate is expected to be slightly below 5 percent per year for the 1980–2000 period (Figure 1). As near as can be determined from the rather sparse historical data available, no society's desire for light has ever been satisfied, but certainly there is a saturation point. Nevertheless, during the period of electric lighting – the last 100 years – it seems clear that if energy and equipment resources are available and if lighting can be provided at lower cost or with greater quality, people will want and use more light.

[2] Countries reporting include: Austria, Czechoslovakia, Denmark, Federal Republic of Germany, Finland, France, German Democratic Republic, Great Britain, Hungary, Italy, The Netherlands, Poland, USA, USSR and Yugoslavia (3).

A central question, therefore, is: can the demand for light be satisfied while reducing the consumption of electric energy for lighting by utilizing the optimum hardware, design and operating techniques.

According to the CIE data, the answer is "yes". As Figure 2 indicates, lighting energy consumption is not increasing at the same rate as lighting usage. For the 1980–87 period, increases averaged slightly over 2 percent per year and for 1987–2000 the rate of increase is expected to remain about the same. Hence, lighting systems are becoming more efficient due to the use, for example, of higher lumen per watt discharge sources. The CIE data confirm such increases (Figure 3).

This chapter will examine the technical reasons for lighting system efficiency improvements in some detail, with the objective of suggesting how developments and practices now on the technical horizon might affect lighting energy use – perhaps beyond what is shown in the CIE data. The emphasis here will be on light sources and their associated control gear since this is the technical area where major gains in efficiency have traditionally occurred. Sources are also, as one lighting designer has pointed out, "the engine which drives the whole lighting industry". The development of a new light source typically results in the development of new luminaires, applications, and designs.

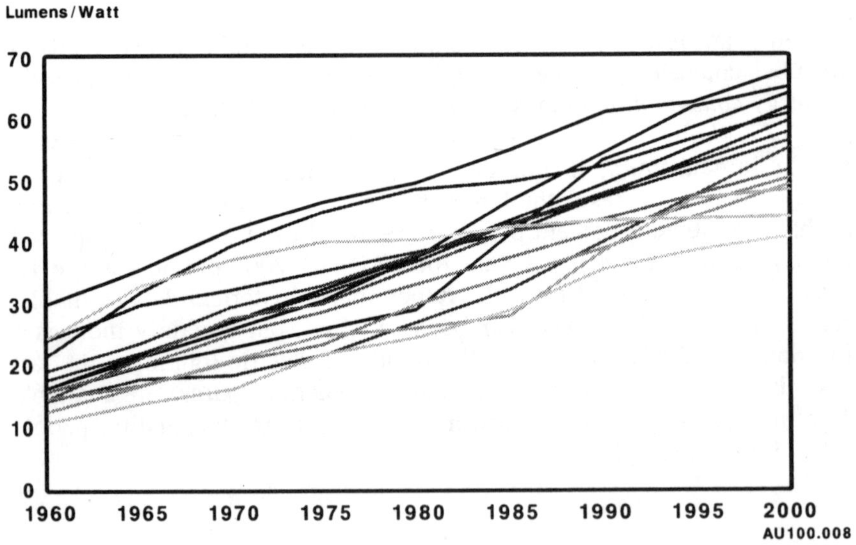

Figure 3. Total lighting system luminous efficacy (lumens/watt) change over time.

2 The Lighting System

The process by which electricity is turned into light and finally becomes a signal which reaches the eye is fairly complex and not fully understood. There has been considerable debate, for instance, about how much light is needed to perform visual work. In several countries, standards based upon a CIE visibility model are used[3] Other countries, however, temper the CIE model with a consensus approach which takes into account national expertise, economics and practice.

It has been shown that the eye can see effectively at fairly low levels of illuminance; however, visual performance continues to improve as illumination is increased so higher levels may be cost effective, depending upon the importance of the visual task. Also, many people seem to prefer substantially higher levels – as much as 2 to 5 times more illumination than might appear to be required for a given visual task. Whether this preference is learned, or is physiologically based is not clear. More importantly, what effect depriving people of their preferred illumination has upon them has not been determined. The point is that illumination levels will continue to be a matter of speculation and mass changes in practice cannot be relied upon as a way to substantially reduce lighting energy use.

Figure 4 is a schematic of a lighting system based upon the idea that visual work is to be done. This approach assumes that light, defined as luminous energy and measured in lumen-hours, is the commodity of interest.

Luminous flux (measured in lumens) distributed over the visual task and its background produces contrast which is the major component of visibility. Contrast can be calculated and measured fairly easily. So can lumens per unit area which is defined as illuminance. Illuminance is usually expressed as lux or lumens/square meter in metric units, and common practice has been to use illuminance as an indicator of visibility.

The link can be tenuous, however, because of other effects such as glare, color, and reflections which affect contrast. These factors are important, sometimes even critical, in certain seeing situations and a good deal of research has been devoted to quantifying their influence. But they have not been shown in this simplified model (Figure 4) since contrast still appears to be the most important quantity, considering the range of visual tasks and seeing conditions normally encountered. Contrast, and therefore illumination, is the purpose of the electric lighting system which in turn determines the use of electric energy for lighting.

[3] There is an abundance of material on this subject. A good place to start is CIE Publication #19 (4)

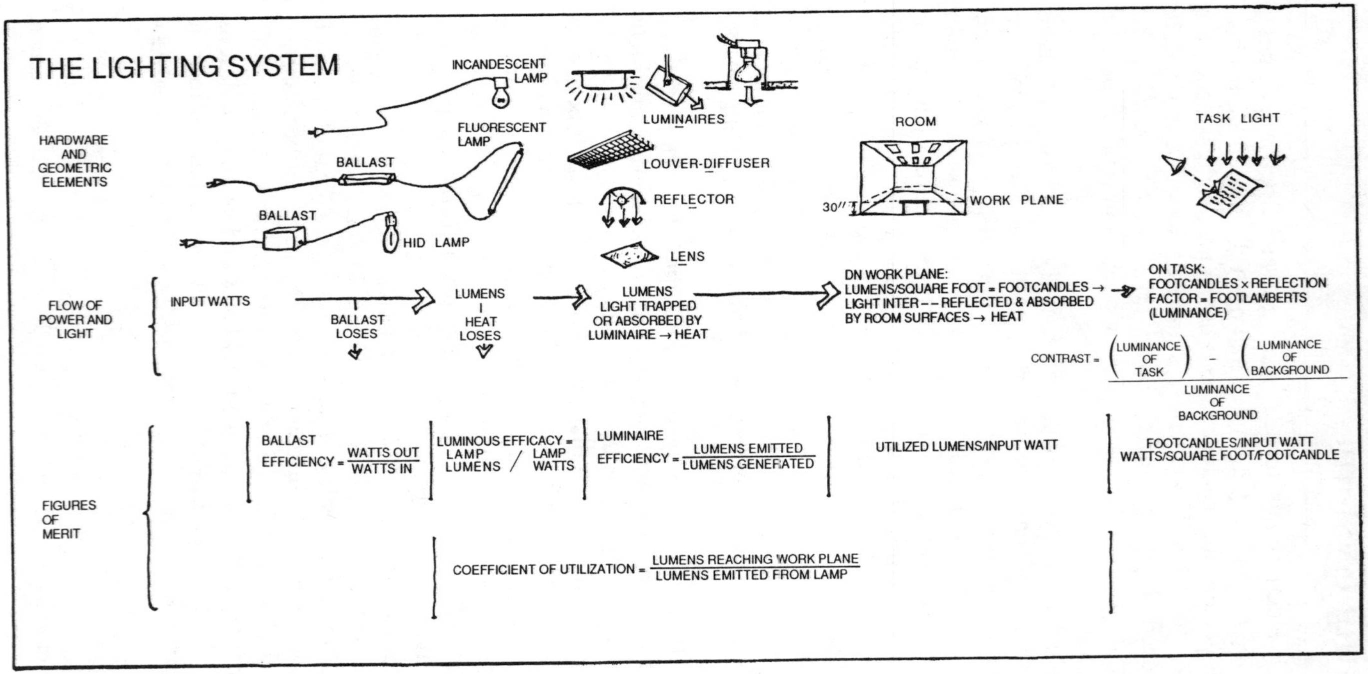

Figure 4. Elements of the lighting system based upon the idea that visibility is a function of contrast. Note: 1 lux = 10.76 footcandles.

Contrast involves the visual task. Usually the task's characteristics – size, reflectance, color, texture, shape – are considered to be fixed, but that is not necessarily the case. Altering the visual characteristics of a task can make a substantial difference in the amount of light needed to see it. This is an active subject and there are examples such as the efforts to improve visual display unit (VDU) visibility which have yielded good results. Unfortunately, though, there are no generally agreed-upon estimates of how much lighting energy might be saved by improving the broad range of visual tasks.

The rest of the lighting system schematic (Figure 4) can be understood in classical illuminating engineering or electrical engineering terms. It is here that most of the hardware improvements have occurred and here where there are still numerous opportunities for practical efficiency gains that will probably have the greatest impact in the near term on the end-use of electrical energy for lighting.

3 The Light Source

The light source is the point at which electrical power is transformed into radiant power. The figure of merit for the process is efficacy. It is measured in lumens per watt. A simple measure of radiant output versus electrical input (watts out/watts in) will not do because the sensitivity of the human eye must be taken into account and the eye response varies across the visible spectrum. All radiant watts are therefore unequal with regard to their ability to provide visibility.

The light source is the critical element of the system because it strongly affects operating and maintenance costs. A lamp is typically short lived compared to the rest of the hardware in the system. Moreover, a lamp's performance characteristics are dynamic. For example, lamp efficacy changes over time – sometimes rapidly, environmental temperatures may affect output, color characteristics can shift as the source operates, dirt causes light output loss, variations in electrical input may cause losses in life, light output or efficacy.

Practical light sources exhibit a wide range of luminous efficacies. Discharge sources (see below) usually have higher efficacies than incandescent sources because less power is turned into heat, but the best incandescent sources today have higher efficacies than the poorest discharge sources.

Table 1. Typical power distribution for general lighting electric lamps.[a]
This table indicates the general power distribution characteristics for the three main types of light sources: incandescent, fluorescent, and high intensity discharge (HID). The two major HID lamps are the metal-halide (MH) and high pressure sodium (HPS).

	Incandescent 100W		Fluorescent		HID	
	120V	230V[b]	40W	SL–18	MH(400W)	HPS(400W)
Visible light (%)	10	5	20	17.2	21	30
Ultraviolet (%)	–	–	–	–	3	–
Infrared (IR) (%)	72	83	33	28.9	32	35
Non-radiated IR (%)	18	12	30	41.7	31	20
Ballast losses (%)	–	–	17	12.2	13	15
Luminous efficacy (initial lamp lumens/watt)	17.5	13.8	78.7	61.1	100	125

a Adapted from IEEE Std. 739–1984 (Reference 5), Table 25. The SL-18 is an electronically-ballasted self-contained screw-base fluorescent lamp designed to replace a general service incandescent lamp of 60 or 75 watts. Data are from Dorleijn, J.W.E. and Jack, A.G., *Power Balances for Some Fluorescent Lamps.* Journal of the IES, 15, Fall, 1985, pp. 75–84.

b 115–125 volts is the standard socket voltage in North America while 220–240 volts is normal in Europe. Within regions, however, considerable variance may be found and in some cases incandescent lamps may be specially designed and distributed for local power distribution systems.

Clearly, heat management, i.e., the control and reduction of both radiated and non-radiated infrared, is the factor which will have the most significant impact on lamp efficacy.

From a practical standpoint, the most efficient light source may not always be suitable for a given lighting application because it does not fit the lighting requirements. Higher wattage lamps generally have higher efficacies than lower wattage lamps, but the use of, say, a 1000 watt metal-halide lamp in a commercial office situation is problematical because it is difficult and expensive to distribute the light from such a powerful source uniformly and comfortably.

Figure 5 suggests a way of taking applicability into account. It shows system efficacy – lamp or lamps plus ballasts (if necessary) versus lamp lumen output. Changes in efficacy as a function of lamp output can easily be determined and, knowing the "lumen package" required for a certain lighting application, the most efficient system can be selected.

For example, the shaded area indicates a "range of applicability" for practical applications in commercial lighting such as offices, health care facilities and public buildings. Similar ranges could be defined for industrial or outdoor lighting. The figure indicates that a "Fluorescent Electronic" system –

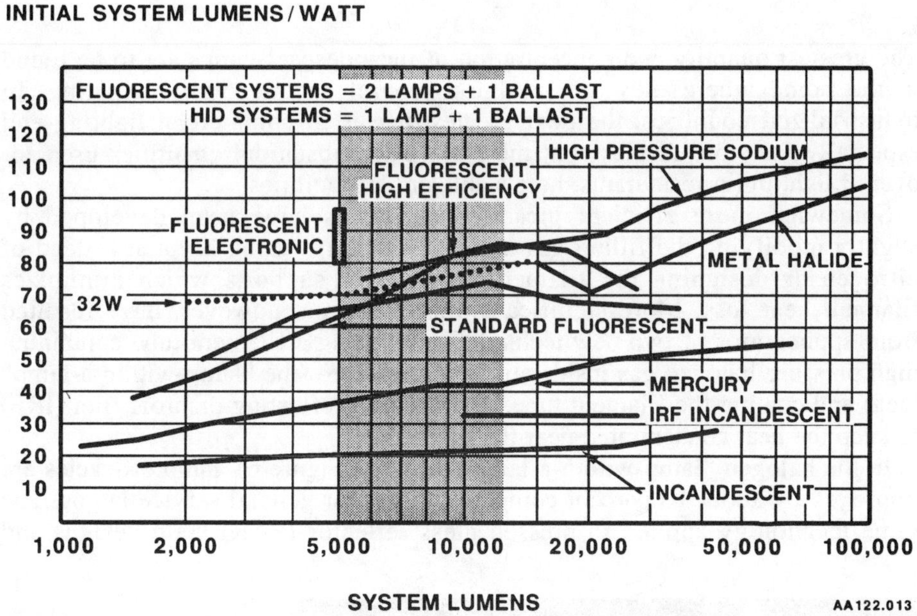

Figure 5. System efficacy as a function of system lumens. For general interior commercial lighting, the normal range of application is approximately 5000–10000 lumens/luminaire (shaded area).

one which utilizes an electronic high-frequency ballast – would be worth considering, but that the poorest of those systems would have the same efficacy and deliver less light than would a "Fluorescent High Efficiency" system – one which utilizes low-loss electromagnetic ballasts.

The figure emphasizes the widening gap between older conventional fluorescent systems and newer high efficacy systems. In addition, it shows the large differences in efficacy between mercury and other discharge lamp systems. New low wattage metal-halide lamps are now becoming available and these have more than double the efficacy of low wattage mercury lamps. A new incandescent lamp development (IRF), which will be described below, stands out as potentially important for systems with relatively high lumen requirements.

4 Incandescent Lamps

The greatest quantity and concentration of incandescent lamps are to be found in the home where they are used for general and decorative lighting. In industrialized countries, the next greatest use is in commercial lighting and especially for store display, but there are also substantial quantities used for overall lighting in restaurants, hotels and similar facilities.

Somewhat more efficient incandescent lamps have been developed by slightly modifying the filling gas mixture – perhaps using krypton instead of nitrogen or designing the filament with fewer supports which minimizes filament heat loss. More significant improvements, however, have resulted from applications of two new technologies: putting a filament tube containing high pressure halogen gas inside an outer enclosure (the "lamp-within-a-lamp" idea) and coating the filament tube with infrared reflecting dichroic film (IRF) to keep the heat concentrated near the filament.

In the halogen "lamp-within-a-lamp" designs (Figure 6), lamp efficacies are improved by about 10 percent compared to similar general service lamps. The same technology applied to pressed glass reflector (PAR) lamp designs and

Figure 6. Halogen general lighting service filament tube (front) and lamp – 90 watts.

coupled with the use of more accurate reflectors is even more potent, i.e. 90 watt lamps can now be used for applications where 150 watt lamps have been used previously without reductions in the lighted area or the amount of light.

The price of halogen general service lamps has come down, but unless replacement labor costs are considered, they may not be economical since only 10 watts are saved versus standard designs. However, since the performance characteristics of incandescent lamps are functionally interrelated, the lamp designer can choose to take the improved performance of halogen lamps as longer life, higher light output or lower watts, or any combination of these to optimize the economic parameters. In Table 2, for example, conventional and reduced-wattage lamps are compared with a halogen lamp designed for reduced wattage, increased life and the same light output as the standard version.

4.1 IRF Incandescent Lamps

Incandescent lamps with infrared reflecting film (IRF) coatings are presently the most efficient type of incandescent source. This development takes halogen technology a step further by combining a quartz halogen filament tube with a new heat reflecting film coating which is visibly transparent. The coating is applied in molecularly-thin layers to the outside surface of the lamp to make an interference film which passes visible light (Figure 7), but reflects infrared

Table 2. Incandescent lamp comparisons (120 volt general service designs).[a]

		Conventional (GLS)	Reduced Wattage	Halogen
Watts		100	90	90
Initial lumens		1750	1620	1750
Rated life (hours)		750	750	2000
Initial lumens/watt		17.5	18.0	19.4
User price	$	0.77	0.78	3.94
Cost-of-light[b]	$	9.59	9.89	7.41

a For these data "Conventional" refers to standard GLS lamps; "Reduced Wattage" means GLS lamps which incorporate minor modifications such as lighter filament supports and/or filling gases with lower heat transfer characteristics; "Halogen" is the lamp-within-a-lamp technology as discussed in the text.

b Cost-of-Light is expressed as $/million lumen-hours and calculated according to the definition published by the Illuminating Engineering Society of North America. See Reference 6, 1981 Application Volume, Chapter 3. For these studies a commercial situation is assumed. The cost of energy is taken as $.08/kWh, systems are operated for 3,000 hours/year, the labor cost to replace one lamp is $5.00 and the luminaires are assumed to be existing.

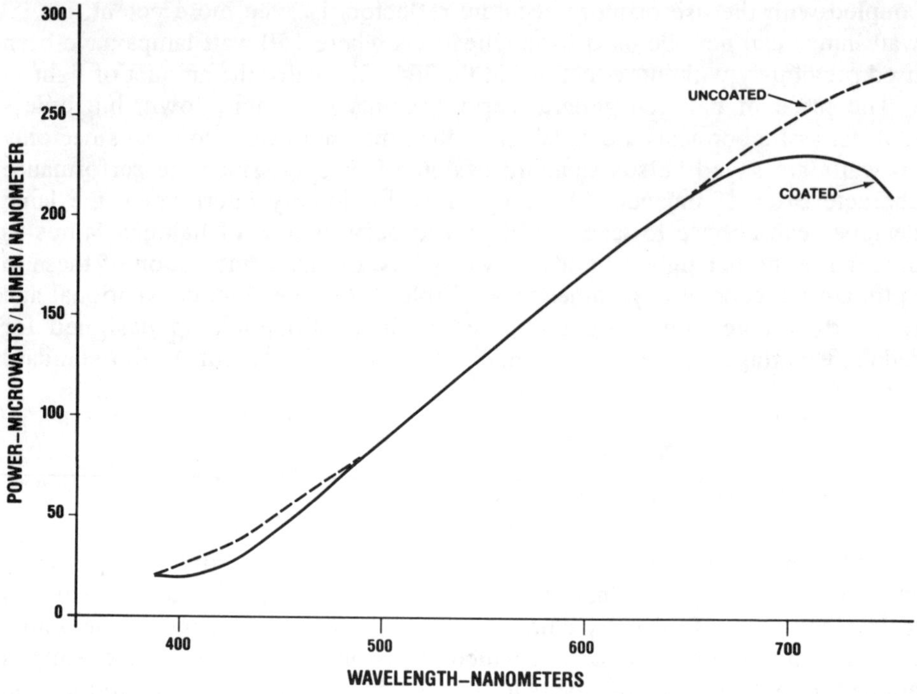

Figure 7. Spectral power distribution of an infrared reflecting film (IRF) coated tungsten-halogen lamp.

energy back onto the filament surface. The result is that much less electrical power is required to keep the filament hot. Lamps utilizing this technology are similar in appearance to and interchangeable with conventional linear tungsten-halogen lamps (Figure 8) which are in use world-wide. Thus, there are significant cost-saving and energy-saving retrofit opportunities as shown in Table 3. Substituting a 900 watt IRF lamp for a conventional 1500 watt tungsten-halogen lamp saves $96.00 over the life of the lamp (at $.08/kWh) compared to only about $26.00 added cost for the IRF lamp.

Because of certain technical problems, IRF coatings are not well suited to low wattage lamps and they must be applied to cylindrical filament tube geometry at present, but those limitations may change. Lamp applications include flood and area lighting, sports facilities, stage and studio lighting (where lower wattage requirements result in lower heat loads) and some interior commercial general lighting (7).

The IRF development, which represents one of the largest one-time improvements in the history of incandescent lighting, has moved incandescent

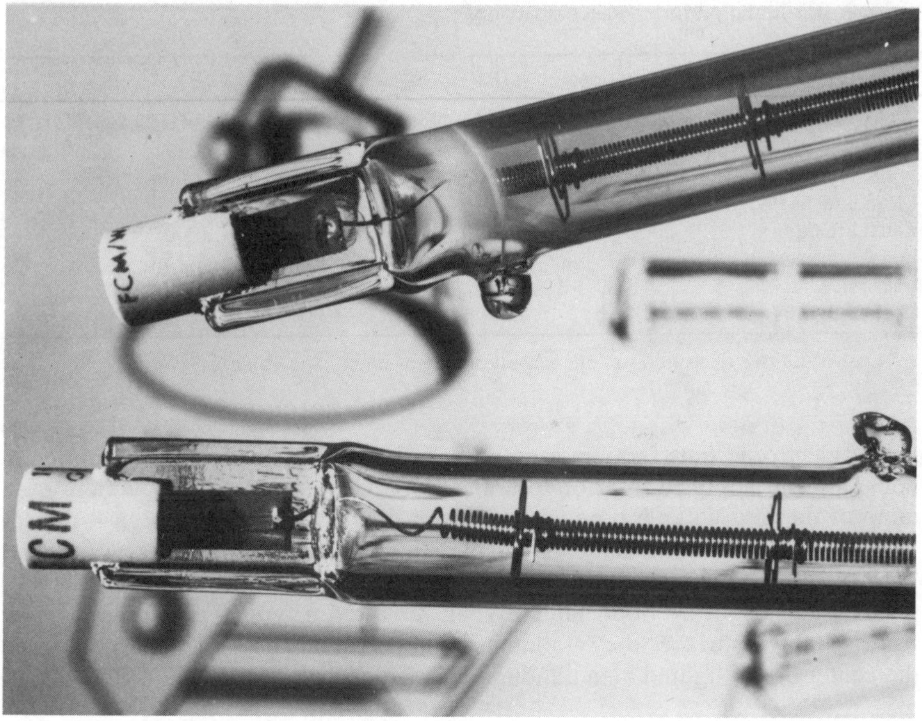

Figure 8. High wattage tungsten-halogen lamps. IRF lamp (top) compared to a conventional lamp.

lamp efficacy into the discharge lamp range. The challenge now is to apply the technology to general lighting service lamps at a cost low enough to be utilized in the billions of existing incandescent sockets.

4.2 Low Voltage Incandescent Lamps

Another technical development which is helping to reduce the energy used by incandescent sources is the substitution of low voltage (12 volt) compact filament lamps for standard voltage types. A low voltage filament design, which can also incorporate the halogen technology, is physically smaller for the same wattage and acts more like a point source from the optical standpoint than does a standard voltage lamp. Low voltage lamps have been found particularly useful on 220–240 volt systems because electromagnetic and the

Table 3. IRF Lamp performance comparisons.

	Conventional	IRF	Conventional	IRF
Watts	500	350	1500	900
Volts	120	120	240,277	240,277
Initial lumens	11100	10000	35800	32000
Mean lumens (%)	97	95	97	95
Rated life (hours)	2000	2000	2000	2000
Initial lumens/watt	22.2	28.6	23.9	35.6
User price	$ 8.89	22.14	12.59	38.51
Cost-of-light[a]	$ 4.36	4.37	3.70	3.08

a Cost-of-light is measured in dollars/million lumen- hours. See Table 2.

newer electronic transformers are relatively easy to incorporate into luminaire housings and apply to interior spot and floodlighting projects. A widely-used lamp of the low voltage type is the 50 mm. integral reflector design – the so-called "MR16". For accent and display lighting in retail stores, a 75 watt "MR16" 12 volt lamp is typically used to replace a standard voltage 150 watt pressed glass spot or flood lamp. When mounted a typical 3–5 meters away from a lighted surface, the two lamps produce about the same illumination in the center of the lighted area although the size of the area will be different.

5 Fluorescent Systems

Fluorescent lamps have generated most of the light in commercial and industrial applications for some years, so system efficiency improvements – especially those which result in lower input watts and can be retrofited into existing installations – have a large built-in market base. Depending upon the supply voltage and ballast type, improvements involving increased lamp efficacy, optimized lamp temperature characteristics, reduced ballast losses, and greater luminaire optical efficiency may be possible. The large installed luminaire base also means that improvements must be simple to implement and they must not involve significant time or labor; otherwise the costs will quickly outweigh the savings. Users are concerned about system reliability and potentially higher maintenance costs. They are used to having fluorescent systems in which lamps typically last 3–5 years in commercial service and ballasts which last 10–15 years or longer. Energy-saving modifications which are untried or compromise reliability are difficult to sell even though substantially lower energy costs are promised.

Fluorescent lamps are temperature sensitive. Both light output and lamp power are a function of the tube wall temperature which controls the amount of mercury present in vapor form inside the lamp. This makes lamp light output and luminaire power input difficult to measure and sensitive to the thermal design of the luminaire as well as the ambient temperature surrounding the equipment. In existing installations, as luminaire watts are reduced, perhaps by removing lamps or modifying the lamp or ballast circuitry, luminaire temperatures fall. Hence, there is concern with some lamp types that bulb wall temperatures may be below optimum (usually about 38 degrees C. at the coldest point on the tube surface). If fluorescent lamps are too cool, light output drops, but more rapidly than power input, so lamp efficacy suffers.

Depending upon the main supply voltage, line frequency, and particular ballast circuit, there are a number of options available for modifying an existing fluorescent lighting system to take advantage of the various energy-saving lamp and control gear developments. Possibilities include:

I Substitute lower wattage "krypton" lamps (limited applicability).

II Change out the standard electromagnetic (EM) ballast to a more efficient (low-loss) electromagnetic unit.

III Substitute a lamp with built-in automatic cathode switches (rapid start circuits only).

IV Retain standard lamps, but modify the ballast circuit with an additional impedance device.

V Install a high frequency electronic ballast in place of the EM unit.

VI Improve the luminaire efficiency by installing a high- reflectance aluminum or silver reflector. Alternatively, new luminaires with such surfaces may be used. Estimated luminaire efficiency improvement is 25% compared to luminaires with depreciated white-painted metal reflecting surfaces.

VII Install a controllable electronic ballast instead of a fixed-output unit to take advantage of manual adjustment possibilities or external sensor inputs for daylighting or lumen maintenance integration. It is estimated that this option will reduce energy use (kWh) by 35%.

Table 4 shows performance characteristics of fluorescent systems modified by the above options. A number of additional options are discussed in Reference 8.

5.1 Electronic Ballasts

Fluorescent lamps become more efficient when supplied with high frequency power. Presently available commercial equipment usually operates in the range from 25–40 kHz to be above the human hearing range but below the range where the ballast and wiring act efficiently as a source of radio emissions. In this region, 38 mm fluorescent lamps have an intrinsic efficacy gain of about 10%. Coupled with lower ballast losses that are possible with electronic units, system efficacy rises to about 30% more than with a standard lamp and electromagnetic ballast.

Still more significant are the control capabilities which can be designed into electronic ballasts. Solid state electronic control circuitry adds little to the complexity of an electronic ballast. Hence, adjustable output and a full range of programed functions are practical even on a luminaire-by-luminaire basis. Control strategies that might be employed include: compensating for normal light output depreciation due to dirt and loss of lamp efficacy (minimizing overdesign); reacting to daylight; adjusting the output of individual luminaires for the task; personnel sensing; lighting scheduling and architectural dimming. As noted, option VII in Table 4 estimates the electric energy savings potential at 35% (without sacrificing lighting quality or quantity), but this estimate may be rather conservative. Reported studies (11) have measured lighting energy savings from 32–52 percent with the implementation of just two (scheduling and daylight integration) of the several strategies listed above.

Table 4. Energy-saving options for fluorescent lighting systems – lamp and ballast combinations.[a]

Option:	Base	I	II	III	IV	V	VI	VII
Initial Lamp Lumens	6300	5500	5500	5400	6300	5500	5500	5500
Input Watts	90.5	79.5	70.0	63.5	64.0	57.7	57.7	57.7
Lumens/watt	65.4	64.2	72.3	75.5	66.0	87.3	87.3	87.3
Relative Light (%)	100	86	85	81	71	85	106	106
Annual kWh	271	239	210	191	192	173	173	113

a The performance data in this table have been taken from a number of sources including manufacturers' specification sheets, laboratory test reports and published papers. See especially References 9 and 10. The data are believed to indicate the typical performance of commercially available equipment, but may not apply to any particular combination of lamp-ballast-luminaire. In some countries the options may be limited by the types of lamps and ballasts normally used; however, Options V. and VI. (electronic high-frequency ballasts and high-reflectance luminaire reflecting surfaces) are widely available. Initial lamp lumens refer to the lamp manufacturer's ratings for 2–40 watt 38 mm lamps. Luminaire input watts and Relative light are measured with enclosed luminaires at 25°C. ambient temperature; Annual energy use is based upon 3000 operating hours per year.

5.2 High Efficiency Phosphors

One of the factors presently responsible for increasing fluorescent lighting system efficacy is phosphor chemistry. The new phosphors described as tristimulus or "tri-phosphors" emit in fairly narrow regions in the red, green and blue portions of the spectrum. Older halophosphate phosphors emitted rather broadly throughout the spectrum. Lamps can be manufactured with layers of both types. These double-layer coatings provide improved color rendering and efficacy, but at substantially lower cost per lumen. Table 5 lists some representative values. Overall, double coat phosphors add about 1% to operating costs (the total of lamps+replacement labor+electric energy)

What is not apparent from the table is the fact that the tri-phosphors can withstand higher power densities (watts per square cenitmeter) than halophosphate phosphors with less degradation over time. This makes tri-phosphors superior for smaller-diameter higher-current lamps which also tend to have higher efficacies. Double-coat and full tri-phosphor coatings are virtually the only phosphor types found on the newer compact fluorescent lamps.

Table 5. Comparison of fluorescent lamp tri-phosphor and standard phosphor coatings.[a]

	Standard	Double Coat	Tri-phosphor
Initial lumens	3150	3250	3275
Mean lumens (%)	87.9	88	87.9
Initial lumens/watt	78.7	81.3	81.9
Life (hours)	– – 20,000 @ 3 hours or more/start – –		
Color rendering index[b]	62	70	82
User price	$2.27	4.05	8.28

a The table values apply to a 40 watt rapid start 38 mm fluorescent lamp. The "Double Coat" phosphor is a two-layer coating consisting of a layer each of the standard and tri- phosphor materials. Chromaticity (apparent color temperature) which is a measure of the visual warmth or coolness of a particular lamp color is rated as the same for the three designs.

b The color rendering index measure is the CIE Color Rendering Index (R_a) which is expressed according to the definition of the Test Color Method. CIE Committee TC-3.2, *Method of Measuring and Specifying Color Rendering Properties of Light Sources,* 2nd Edition, CIE Publication No. 13.2, Paris, 1972.

5.3 Compact Fluorescent Lamps

Another area of fluorescent system efficacy improvement is the development of smaller diameter 26 mm (T8) and 16 mm (T5) lamps. The 26 mm designs are manufactured as straight or curved (U-shaped) lamps in common modular lengths. They are inherently about 3% more efficient than the 38 mm designs, and because of the smaller tube diameter and resulting concentration of the light, luminaires are able to utilize the light more efficiently and in a smaller physical enclosure. The overall advantage of 26 mm lamps is not yet clear; however, comparison tests involving U-shaped 38 mm and 26 mm lamps in prismatic and parabolic luminaires (12) showed that 26 mm lamp systems (lamp + ballast) are between 5 and 9 percent more efficient than conventional fluorescent systems when used with electromagnetic ballasts and 16 to 19 percent more efficient when high frequency electronic ballasts are used.

The smaller 16 mm lamps (T5) offer even greater efficiency advantages. They can be formed with tight bends into U-shapes and even double bent into a relatively small volume. This group of lamps is known by several names: compact fluorescent, PL, or biaxial. The double-bend lamps are being referred to as cluster lamps, double twin tube or quadtubes. Lamp efficacies for the line of biaxial lamps shown in Figure 9 range from 57 to 81 lumens/watt.

Which system should be chosen? The answer is more complex than ever before and depends more upon the application and upon the size and shape of luminaire desired than previously. Both small- diameter 26 mm and 16 mm

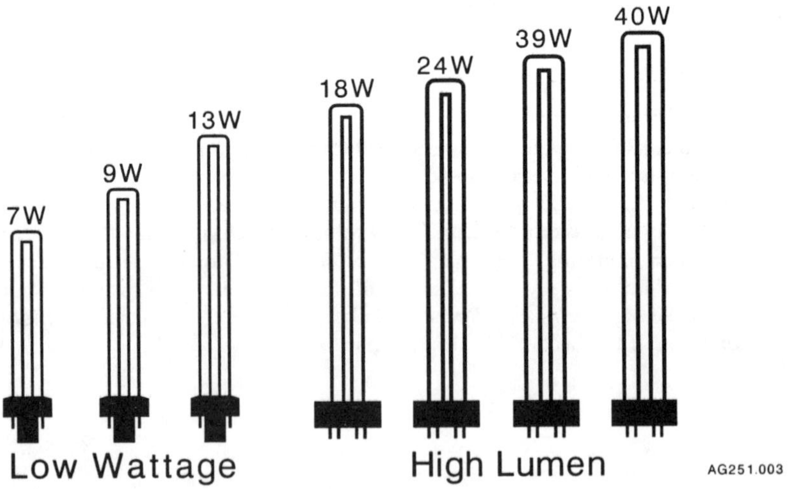

Figure 9. The range of compact biaxial fluorescent lamps. Initial light output ratings go from 400 lumens (7 w) to 3150 lumens (40 w).

biaxial systems are relatively efficient, although the biaxial system appears to have an advantage when operated on high-frequency electronic ballasts (Table 6).

5.4 Integral Compact Fluorescent Lamps

With their common goal of converting the millions of existing incandescent sockets, lamp manufacturers are in strong competition to find satisfactory high-efficiency substitutes for the common range of general service (GLS) lamps which are typically 25, 40, 60, 75, 100, and 150 watts, with the most popular types varying somewhat by country and region. A good substitute must have low initial cost, good color, similar lumen output, a size comparable to the lamp it replaces, and long life (recognizing that there is usually some relationship between lamp life, value and product price in the user's mind).

Compact fluorescent lamps (Figure 10) meet many of the requirements listed above, but the number of sockets available to these lamps has been limited by the larger size of the lamp and the necessity to packaging a ballast, screw base, wiring and diffusing glass or plastic enclosure into one assembly. Relatively high initial prices (about $25/unit to the user) when the first models were introduced early in the 1980's slowed sales in the residential market. Hence, manufacturers have been concentrating on the commercial and

Table 6. Comparison of efficacies: 38 mm versus 26 mm and 16mm fluorescent lamp-ballast combinations in two luminaire types.[a]

Luminaire type	Ballast type	Lamp	Lumens/watt
2-Lamp parabolic	Electromagnetic	38mm	32.2
	(60 Hz.)	26mm	38.9
		16mm[b]	42.8
	Electronic	38mm	38.5
	(28 kHz.)	26mm	41.9
		16mm	51.5
3-Lamp prismatic	Electromagnetic	38mm	30.6
		26mm	36.8
		16mm	41.3
	Electronic	38mm	40.8
		26mm	48.9
		16mm	52.9

a Table data from References 9 and 12.

b The 38 and 26 mm lamps are conventional straight tubular designs. The 16mm lamp is a compact biaxial type.

Figure 10. "Compax" Integral Compact Fluorescent Lamp. An electromagnetically-ballasted self-contained unit rated at 15 watts, 700 initial lumens and 9000 hours life. It is designed to replace a 60 watt general lighting service (GLS) incandescent lamp.

industrial markets in which users are more used to evaluating equipment on overall economics so as to reduce costs. Meanwhile, additional manufacturers have developed products and, as a consequence, there are several dozen lamp types now available ranging from 7–20 watts in power and from 700 to over 1200 lumens in output.

The data (Table 7) suggest that it makes economic sense to use the compact integral units if light has value and the user is concerned about the energy and operating costs. For the user or householder where low to zero replacement labor costs apply, the economics are less compelling, but there is still a clear advantage for the fluorescent units. The stumbling block in the way of greater acceptance of the integral units for residential lighting is apparently the relatively high initial cost compared to what householders are used to paying for standard incandescent lamps. While mass production, higher volumes and less expensive electronic ballasts are likely to bring user prices into the $5–10 range, so energy cost savings will pay for the higher initial cost, it is still difficult for residential users to see the connection between the cost of the lamp and savings on their monthly electric energy bills. Energy officials and electric utilities who have promoted compact fluorescent units have found that subsidizing the price of the lamps or even providing them at no charge are the only reliable ways to significantly increase acceptance and usage. One electric utility which has provided no- cost compact fluorescent units as replacements for standard incandescent household lamps is the Southern California Edison Co. in Rosemead, California. The program's objective was load reduction and was based upon the idea that the cost of providing the lamps at no cost to residential customers would be less than the cost of building new generating capacity (see Geller in this volume).

5.5 New Fluorescent Technology

There is considerable developmental potential remaining in fluorescent lamps and systems. Laboratory work is presently concentrating on two promising areas: isotopic enrichment of mercury for higher arc efficiency and phosphor physics which may lead to "two photon" phosphors in the future (present phosphors emit one photon for every ultra-violet photon generated in the lamp arc). In addition, researchers are working on electrodeless lamps, which have already been demonstrated and which operate at much higher frequencies (mHz rather than kHz). If any of these technical pathways lead to a commercial

Table 7. Economic and performance comparisons of standard incandescent versus integral compact fluorescent lamps[a].

	60 Watt Inc.	COMPAX Fluor.	75 Watt Inc.	SL-18 Fluor.
Watts	60	15	75	18
Initial lumens	870	700	1190	1100
Rated life (hours)	1000	9000	750	7500
Initial lumens/watt	14.5	46.7	15.9	61.1
User Price $	0.77	12.75	0.77	18.20
Cost-of-light[b] $	13.06	6.04	12.37	5.49
Cost-of-light[c] $	6.88	4.98	6.35	4.69

a Commercially available lamps – 120 volt designs. The "Compax" unit has an electromagnetic ballast; the "SL-18" an electronic ballast.

b Cost-of-light definitions and assumptions are shown in Table 2. Values on this line include $5.00/lamp for replacement labor when the lamp fails.

c No replacement labor cost included for values on this line.

product, practical lamp efficacy will rise to over 100 lumens/watt in useful sizes for general lighting. If systems are able to incorporate several of these potential developments, system efficacy could push well into the 100+ lumen/watt range.

6 High Intensity Discharge Lamp Systems

For general lighting applications, the HID group of lamps includes the mercury, metal-halide (M-H) and high pressure sodium (HPS) lamp types. Figure 5 illustrates the lumen output range and associated system efficacies. Mercury is presently of consequence only as a replacement lamp and its use is declining since, for new installations, M-H and HPS can do the lighting task more efficiently and at least as well as mercury in most other performance categories.

A number of countries still utilize mercury extensively for outdoor lighting. For such systems, a conversion from mercury to HPS always reduces electric energy use if the same lumens are generated before and after the conversion, and the effort is almost always cost effective if energy cost savings can be used to liquidate the conversion costs. Both mercury and HPS lamps are rated for 24000+ hours of operation. They share similar optical characteristics so maintenance costs are only minimally affected and capital-intensive changes involving wiring, trenching, pole spacings and mounting heights are usually not required. In some instances, just ballast and lamp changes are made; in others, luminaires are exchanged. Table 8 lists some typical mercury to HPS system conversion possibilities which result in about the same before-and-after illumination levels. These apply primarily to the outdoor lighting of roadways, parking areas and safety and security installations.

As an alternative to fluorescent lighting in interiors, HID systems are generally not more efficient at nominal ceiling heights (2.5–4.5 meters), but they may have lower operating costs because of the longer rated life for HPS lamps and because fewer lamps and luminaires are generally required for equal illumination compared to fluorescent. At higher ceiling heights in both commercial and industrial applications as well as outdoors for roadway and area illumination, M-H and HPS sources are usually superior to mercury-based systems from both the maintenance and energy cost standpoints.

For general lighting, HID systems using either M-H or HPS sources are always more efficient compared to the use of incandescent. However, for many sockets compact fluorescent lamps offer another choice particularly if the application involves converting incandescent downlighting sockets. These

Table 8. Potential power savings – mercury to high pressure sodium lamp conversions.

Mercury Lamp Watts	Input Watts	High pressure sodium Lamp Watts	Input Watts	Savings/lamp Watts
100	131[a]	50	60	71[b]
175[c]	216	70	85	131
–	–	100	117	99
250	301	150	176	125
400	474	200	234	240
1000	1135	400	451	684

a Input watts are the sum of the lamp watts and ballast losses.
b The power savings depend upon the particular ballast type used. The values shown are typical for 60 Hz electromagnetic types.
c Both the 70 and 100 watt high pressure sodium (HPS) lamps are used for 175 watt mercury to HPS conversions depending upon the illumination required.

sockets represent the major use of incandescent lamps in the commercial segment. HID sources offer superior optical control versus fluorescent, but their color and initial cost attributes are typically inferior. Compact fluorescent lamps are physically large, but more types are available, especially in the lower lumen sizes. Currently, a choice between using HID or fluorescent is quite application specific and may ultimately be decided on the basis of the lighting designer's aesthetic preference rather than clear-cut economic or energy calculations.

Lighting designers will have more and more choices as the number of low wattage M-H lamps continues to expand. A 70 watt and a 100 watt lamp are available and a 32 watt design, rated at 2500 initial lumens, reached the marketplace in 1986 (Figure 11). These sources, which provide white light similar in color rendering and chromaticity values to fluorescent lamps, now offer a high- efficiency alternative to incandescent sources rated at 100 watts and above – a lumen region not covered by the quadtube or cluster type of compact fluorescent lamp. Low wattage M-H lamps are particularly suitable for lighting systems where small physical size and good optical control are necessary. This includes downlighting, show window, display, and low level indirect lighting applications.

In the same time frame as the development of lower wattage M-H lamps, "deluxe color" and "white" HPS lamps, also in low wattages, have become available. There are at least three competing technologies involved in this development, but it is now clear that HPS lamps are able to offer another good-color option in the 2000–5000 lumen application region.

The greatest opportunities for energy savings with HID light sources lie in the areas of outdoor and indoor lighting where high-wattage lamps are applied

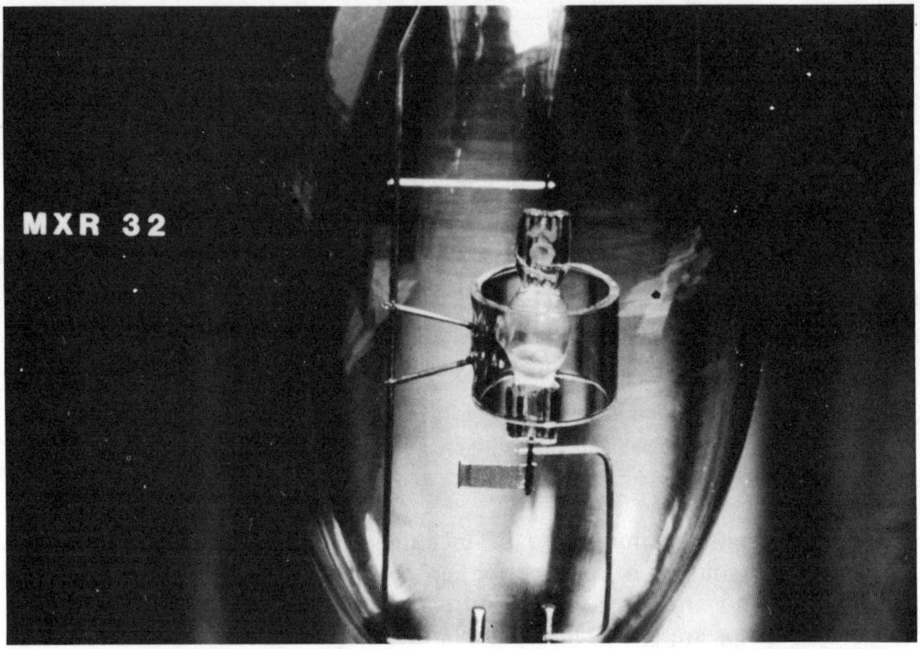

Figure 11. Near view of the arc tube assembly inside a 32 watt metal-halide lamp. Rated at 2500 initial lumens and 7500 hours life, this lamp is designed to replace 150 watt GLS incandescent lamps in commercial downlighting applications. The overall length of the lamp is 134 mm.

– roughly in the range of 2000 lumens and higher – and where incandescent and mercury lamps have been used in the past. There are also some savings possibilities in high bay (4.5 meters and above) industrial lighting installations where mercury and fluorescent lamps are presently used. Typically, a mercury to HPS or mercury to M-H conversion will reduce energy costs without reducing quality or quantity of illumination – providing the color of HPS is acceptable. But the benefits of converting from fluorescent systems to HID are mainly reduced maintenance costs.

HID lamp systems can be made more efficient using technologies similar to those applied to fluorescent systems such as improved phosphors (M-H only) and the application of electronic ballasts. The system efficacy improvements with electronics are not as great as with fluorescent systems, however, and

Table 9. Performance Comparisons of representative general lighting metal-halide and high pressure sodium lamps.

Metal-halide			High pressure sodium		
Lamp watts	Lumens	Lumens/watt	Lamp watts	Lumens	Lumens/watt
32	2500[a]	78.1[b]	35	2250	64.3
–	–	–	50	4000	80.0
70	5000	71.4	70	6400	91.4
100	8500	85.0	100	9500	95.0
150	11250	75.0	150	16000	106.7
175	14000	80.0	200	22000	110.0
250	20500	82.0	250	27500	110.0
–	–	–	310	37000	119.3
400	36000	90.0	400	50000	125.0
1000	110000	110.0	1000	140000	140.0

a Initial rating.
b Lamp only.

accrue mainly from reduced internal ballast losses. Long term, higher lamp efficacies are expected – perhaps reaching the 150–200 lumen/watt range – but the development of practical lamps is dependent upon improved arc tube materials, cathode construction and techniques for arc tube heat management. It is not clear at this point whether the high pressure sodium or the metal-halide technology will eventually lead to the superior product when efficacy as well as overall performance are taken into consideration.

7 Lighting System Operational Considerations

Reducing the electric energy requirements of lighting systems requires a systems approach. It is not enough to use a more efficient source, say, in a retrofit application, if that source reduces luminaire efficiency or performs less well on the existing ballast. This chapter has examined ways to improve lighting system efficiency primarily by utilizing improved components: lamps, ballasts and luminaires. The second step which has been found effective in practice is to apply lighting controls to ensure that light is being provided only when and where it is required. (An excellent summary of control hardware and application techniques may be found in Reference 13.) Controls should take into account (at least) scheduling, available daylighting and occupancy sensing.

8 Daylighting

Daylighting is receiving new attention as ways are sought to heat, light and improve the appearance and function of interior environments. While it is not possible here to review all of the ramifications that daylighting could potentially have on lighting energy use, note should be taken of recent work which has helped to concentrate efforts on daylighting technology (14 and 15).

Daylighting cannot have a massive impact on the near-term use of electric energy for lighting because the buildings which will determine the base lighting electrical load for the next 10–30 years are already built. These buildings can be modified somewhat, however, and architectural systems including shields, blinds, reflectors and the fast-developing technology of "light pipes" (16) are being examined.

Programs to modify existing buildings have been tried and there are incentives for new buildings in some regions to take advantage of daylighting, but the programs are quite limited and how widespread the effect of these efforts will be is not clear. Lovins and Sardinsky (10) offer the tantalizing possibility of reducing electric lighting energy by more than 50% in new commercial buildings even if only the perimeter spaces are daylit.

9 Maintenance of Lighting Systems

Lighting maintenance is a neglected part of efforts to reduce lighting energy perhaps because the task is so diffuse. Frequently, the lighting designer has little or no information about maintenance with which to work in the design phase, even to the point that the selection of room surfaces, colors and furnishings may even be delayed until after the lighting is installed. The designer, therefore, may have to assume a worst case to avoid having the illumination levels drop below recommended minimum values.

Worst case design usually assumes spot relamping, cleaning at relamping (or no cleaning at all), minimal room reflectances and an assumption of one category worse for dirt conditions than expected. Fortunately, this rather pessimistic scenario seems to be changing as lighting designers and their clients insist on tighter design criteria in an effort to better control costs.

What is closer attention to maintenance worth in terms of lighting energy use? The range illustrated by the commonly-used dirt depreciation curves is over 70 percent (6) considering just the loss in luminaire output from accumulated dirt on the lamp and luminaire surfaces, cleaning cycles, and the assumption that the lighting system must be planned to compensate for those losses.

A more realistic view can be obtained by examining the various losses that are related to maintenance and assigning some typical values. The IESNA procedure recommends the calculation of a light loss factor (LLF) which is the product of the loss factors for the individual parts of the system. The procedure divides the LLF into two parts: those factors which cannot be recovered through maintenance such as low voltage to the luminaire and those factors which can be returned to their new or original values through cleaning and maintenance. The recoverable factors are: room surface dirt depreciation (RSDD), lamp lumen depreciation (LLD), lamp burnout factor (LBO) and luminaire dirt depreciation (LDD). These are all assumed to be time dependent.

If the unrecoverable factors have been taken into account, then the light loss factor is:

$$LLF = RSDD \times LLD \times LBO \times LDD$$

In commercial and industrial situations where there is fluorescent lighting with good shielding, a moderate cleaning program, prompt spot relamping, dirt conditions commensurate with air conditioned spaces and reasonable attention to room surfaces, an estimated range for the factors would be: RSDD = 0.94–0.96; LLD = 0.85–0.88; LBO = 1.0; and LDD = 0.75–0.88. The total LLF would then range from 0.62 to 0.72. That translates into a compensating factor of about 15% which applies to both initial and operating conditions. Stated another way: the "penalty" for *not* considering and committing to a maintenance program in the design stage is up to 15% of the initial cost of the system as well as a continuing 15% (or more) of the operating costs including energy.

10 Reducing the Use of Energy for Lighting

A lighting system is a unique mixture of art and science. People are their own lighting experts and tend to have definite feelings about the lighting in their home or workplace. They seem to like light whether it emanates from natural or electric sources and lumen-hours per capita continue to increase even in industrialized countries where abundant electric lighting has been the norm and facilities are built according to modern lighting practice. Thus, there is a continuing demand for the commodity and the saturation point has not been found.

Parts of the lighting system can be made more efficient. In particular for lamps, ballasts and luminaires, improvements which are already in existence or expected near term can raise the efficacy of lamps from 10–20 percent for incandescent, about 15–30 percent for HID systems and perhaps 30–60 percent for fluorescent systems. Better optical design and the use of high reflectance surfaces can increase luminaire efficacy by 25–40 percent. These efficiency improvements translate directly into lighting energy reductions in new systems where they can be incorporated into the design process. In existing systems, with greater physical, electrical and (usually) cost restrictions, perhaps half of the potential energy savings can be realized.

Of greater importance, however, is how to implement these improvements. The large inventory of incandescent sockets used in commerce, industry and the home will eventually be equipped, not with another type of incandescent lamp, but, most likely, with a much higher-efficacy fluorescent device. Therefore, the potential efficacy improvement per socket could be almost 300 percent. Applying controls to commercial lighting systems can reduce kWh in the range of 25–50 percent depending upon the strategies which are employed. Controls, however, even the simple presence sensing and scheduling types which have a relatively high return on investment, are still not widely used. Energy-saving incandescent and fluorescent lamps represent less than 50 percent of the volume of lamps sold (USA data) some 10–15 years after market introduction and in spite of rather vigorous promotional efforts on the part of manufacturers, government, electric utilities and others.

The purpose of presenting this information is to aid energy planners. The above data are intended to help define and measure the parameters involved in the efficient use of electricity for lighting and show how these parameters are likely to change in the future. Understood is the fact that energy supplies are costly both in terms of economic and environmental costs and it is necessary to determine how the use of energy for lighting can be reduced while still providing the lighting required for the functioning of a modern industrialized society. For any country or utility authority working toward this goal, a good first step is to determine where the lighting is, obtain some measure of quantity either in numbers of sockets or by power density; then inventory what kind of sources and systems are being used to generate it. The cooperation of equipment manufacturers and distributors who have hardware shipment information is crucial to this process. From the resulting data, lumen-hours by light source type can be determined along with overall luminous efficacy. Segmented data (stores, industries, offices, homes) are helpful in determining where to implement and emphasize action programs.

Lighting energy reduction efforts appear to be successful when there is a desire to reduce costs or where education and economic incentives are coupled. Mandated programs also appear to work, but they are always controversial and tend to limit the freedom of choice and practice. Legislating the use of high efficiency equipment may reduce energy, for example, but cause unforeseen maintenance or operational problems.

As an illustration, suppose that the use of high pressure sodium lamps were to be required for industrial storage areas. Normally, the use of such equipment would be considered an energy-efficient lighting solution. However, suppose further that the lighting is not needed all of the time because of limited occupancy. Turning the lamps on only when needed if occupancy periods are short, dramatically shortens lamp life which would increase maintenance costs and, because of the 3–5 minute warm-up time required for the lamps to reach full output, sufficient lighting may not be available when the users need it. Depending upon the burning period, a less efficient, but more practical solution might be to use incandescent lamps. Keeping mandated requirements up-to-date and administering enforcement policies and procedures are also difficult.

The data in this report suggest that *substantially less energy can be used for lighting than at present and that future technology developments offer continuing significant energy-saving opportunities*. How to take full advantage of the potential has become a key component of present efforts to implement energy- saving hardware and techniques. Some organizational control is required particularly to determine priorities; otherwise, those solutions which offer the greatest savings will tend to be done first and that practice might make impossible lower level or more expensive solutions. A systems approach to control planning and coordination may be the most important element of all.

References

1. *General Lighting Cost Analysis – Industrial and Commercial, 1985/86*. Publication 205–51201, GE Lighting, Nela Park, Cleveland, Ohio 44112 USA.
2. IES Lighting Energy Management Series (LEM 1,2,3,4 & 6) Illuminating Engineering Society of North America, 345 East 47th Street, New York, NY 10017 USA.
3. International Commission on Illumination (CIE) Committee TC 7- 03 *Development of Lighting, Trends of lighting Development Parameters – Issue 7*, Committee Report, May, 1987. International Commission on Illumination, A-1030 WIEN, Kegelgasse 27, Austria.
4. International Commission on Illumination (CIE), *An Analytic Model for Describing the Influence of Lighting Parameters Upon Visual Performance*, 2 Volumes. CIE publication 19.2, 2nd Edition, 1981. See Reference 3.
5. The Institute of Electrical and Electronics Engineers, *IEEE Recommended Practice for Energy Conservation and Cost- Effective Planning in Industrial Facilities*, IEEE Std. 739 1984. IEEE, Inc. 345 East 47th St., New York, NY 10017 USA.
6. Kaufman, J.E. (Editor). *IES Handbook* Illuminating Engineering Society of North America (IESNA), 345 East 47th Street, New York, NY 10017 USA.
7. Hoegler, L.E., McGowan, T.K. 1984. *Practical high- efficiency tungsten-halogen lamps using IR reflecting films*. Journal of the Illuminating Engineering Society (USA) 14:165–74.

8. Chapman, D.F., Slater, A.I., Isaacs, A.L. 1987. *Energy limiters: Their place in lighting control.* Lighting Research and Technology. (UK) 19:53–60
9. Dobras, Q.D., Hart, A.L. 1986. *Comparative performance of four-foot fluorescent systems*, Lighting Design + Application, (USA) 16:43–53.
10. Lovins, A.B., Sardinsky, R. 1988. *The State of the Art: Lighting* Rocky Mountain Institute, 1739 Snowmass Creek Road, Snowmass, Colorado 81654 USA.
11. Rubenstein, F., Karayel, M. 1982. *The Measured Energy Savings from Two Lighting Control Strategies* LBL Report 14651, Lawrence Berkeley Laboratory, Berkeley, CA 94720 USA.
12. Hammer, E.E. 1987. *Commparison of Bent Tube Fluorescent Lamps in a 2'x 2' Luminaire.* To be published. A paper presented at the 1987 Annual Conference of the IESNA Scottsdale, Arizona.
13. Rea, M.S., ed. 1984. *Proceedings of Lighting Control Symposium,* National Research Council, Ottawa, Ontario K1AOR6 Canada.
14. Vonier, Thomas, AIA, (editor) 1983. *General Proceedings - 1983 International Daylighting Conference* Phoenix, Arizona, USA.
15. Zdepski, M.S., McCluney, Ross, (Editors) 1986. *Proceedings – 1986 International Daylighting Conference* Long Beach, California, USA.
16. Gilmore, V.E. 1988. *Piping Light* Popular Science, 6:77- 79, 117. Popular Science Pub. Co., 380 Madison Ave., New York, NY 10017.

Energy-Efficient Windows: Options with Present and Forthcoming Technology

Claes Göran Granqvist

Abstract

Current research and development offers important opportunities for improving the energy efficiency of architectural windows. Thin surface coatings can, in principle, cut the solar energy input in half, and likewise cut the thermal loss of double-glazed windows by half. Such coatings exist on the market, and superior products are likely to appear within a short time. In the future, the throughput of radiant energy may be regulated in accordance with dynamic needs of heating and lighting by "smart" windows incorporating electrochromic materials. Vacuum windows with aerogel spacers – currently being researched – may yield k-values down to ~ 0.5 $Wm^{-2}K^{-1}$. These developments will lead to lower energy demands for buildings, but the extent of the energy savings to be achived has not yet been quantified.

1 Introduction

Space conditioning of residential and commercial buildings in Sweden account for about 40% of the annual energy consumption. Approximately, one-sixth of this is required to offset heat losses through windows, i.e., roughly 7% of Sweden's annual energy consumption is tied to the fenestration performance. Hence, it is obvious that the window properties have a significant effect not only on the thermal and visual comfort but also on the peak and average electrical demand at the regional and national level. The purpose of this paper is to review current research and development on materials for energy efficient windows and to point at options with present and forthcoming technology. We will see how superior thermal insulation can be combined with good visual transmittance, how overheating can be avoided by reflecting off the infrared part of the solar spectrum, and how today's research on materials for optical modulation may lead to tomorrow's "smart" windows which are able to regulate the throughput for radiant energy in accordance with dynamic needs. Earlier reviews on materials for energy-efficient windows, with a narrower scope than the present one, have been given recently by Granqvist (1,2) and Lampert (3).

The function of the window, in general terms, is to transmit a controlled amount of luminous radiation (for vision) and solar radiation (for space heating) at a specific – usually minimized – heat transfer. The heat transfer comprises additive contributions from thermal radiation, conduction in solids and gases, and gas convection. The ubiquitous radiation is of particular significance for at least four reasons: the most fundamental object of a window is to transmit light; radiative transfer often accounts for as much as half of the heat loss; radiative properties can be conveniently modified by thin coatings and surface treatments on the glass; and dynamic control of the radiative properties can be achieved in smart windows. A key concept for energy efficiency is spectral selectivity, implying that the radiative properties should be qualitatively different for different wavelength ranges. It is possible to control conduction and convection through a material interposed between the panes.

There are a great many ways to improve the energy efficiency of windows. This paper does not purport to cover all of them in detail. Instead, I focus on research and development which, in my view, holds particular promise for the future. It should be noted that the present paper has at least two major limitations. First, economic estimates related to new fenestration technology are given only at a few places. The ongoing research on windows, and on the materials required for their energy efficiency, is very vigorous and carried out worldwide, and I feel that it would be premature to try to set down costs of products, such as the previously mentioned smart window, which are in early stages of development and for which no generally agreed upon technology has been established. Second, the implications of the ongoing research on the energy balance of actual buildings, though of obvious practical interest, is only touched upon. The reason for this unfortunate omission is simple: the pertinent research has not yet been done. In fact, one of my goals in this paper is to bring researchers with the appropriate background in the energy balance of buildings abreast of some of the work in physics and materials science, so that the results of the basic studies can be rapidly integrated in more applications-oriented work.

2 Energy Efficiency in Different Climates

2.1 Ambient Radiation

The key to energy efficiency of windows lies in a clear understanding of the radiative properties of our natural ambience. These are illustrated in Figure 1 with a common logarithmic wavelength scale on the abscissa. The wavelength scale embraces visible light ($0.4 < \lambda < 0.7$ µm, where λ signifies wavelength), ultraviolet radiation ($\lambda < 0.4$ µm) and infrared radiation ($\lambda > 0.7$ µm). The most fundamental property is that matter emits radiation. Its character is conveniently discussed by starting with the ideal blackbody, whose emitted spectrum – known as the Planck spectrum – is uniquely defined if the absolute temperature is known. Part (a) of Figure 1 depicts Planck spectra for four temperatures bracketing those of practical significance for windows. The vertical scale denotes power per unit area and wavelength increment (hence the queer-looking unit GW/m^3). The spectra are bell-shaped and confined to the $2 < \lambda < 100$ µm range. The peaks in the spectra are displaced towards shorter wavelength as the temperature goes up; at room temperature the peak lies at about 10 µm. Thermal radiation from a material is obtained by multiplying the Planck spectrum by a dimensionless function – the emittance.

Figure 1 (b) reproduces a solar spectrum for radiation outside the earth's atmosphere (4). The curve has a bell shape defining the sun's surface temperature (~ 6000°C). The integrated area under the curve gives the solar constant (1353 ± 21 Wm^{-2}); this is the largest possible power density on a surface oriented perpendicular to the sun in the absence of atmospheric damping. Solar radiation is seen to be confined to the $0.25 < \lambda < 3$ µm range.

Since windows are close to ground level, it is of obvious interest to consider to what extent atmospheric absorption influences solar irradiation and thermal emission. Figure 1 (c) illustrates a typical absorption spectrum vertically across the full atmospheric envelope at clear weather conditions (5). The spectrum is found to be quite complicated with bands of large absorption – caused mainly by water vapour, carbon dioxide, and ozone – and intervening bands of high transparency. The actual solar irradiation at the ground is obtained by multiplying, at each wavelength, curves (b) and (c). It is evident that the majority of the solar radiation can be transmitted down to ground level, and only parts of the ultraviolet and infrared tails are strongly damped. The maximum power density perpendicular to the sun is limited to about 1000 Wm^{-2}. Thermal radiation from a surface exposed to the clear sky is seen to be strongly absorbed, except in the $8 < \lambda < 13$ µm range where the transmittance can be large provided that the humidity is moderately low. Thus, part of the thermal radiation can go almost unperturbed from an object at ground level into space, which gives a source of passive cooling with a power density of about 100 Wm^{-2} at ambient temperature (6).

Figure 1(d) illustrates the relative spectral sensitivity of the human eye in its light-adapted (photopic) state. The bell-shaped curve extends across the $0.4 < \lambda < 0.7$ μm interval with its peak at 0.555 μm (7). In its darkness-adapted (scotopic) state, the eye's sensitivity is displaced about 0.05 μm toward shorter wavelengths.

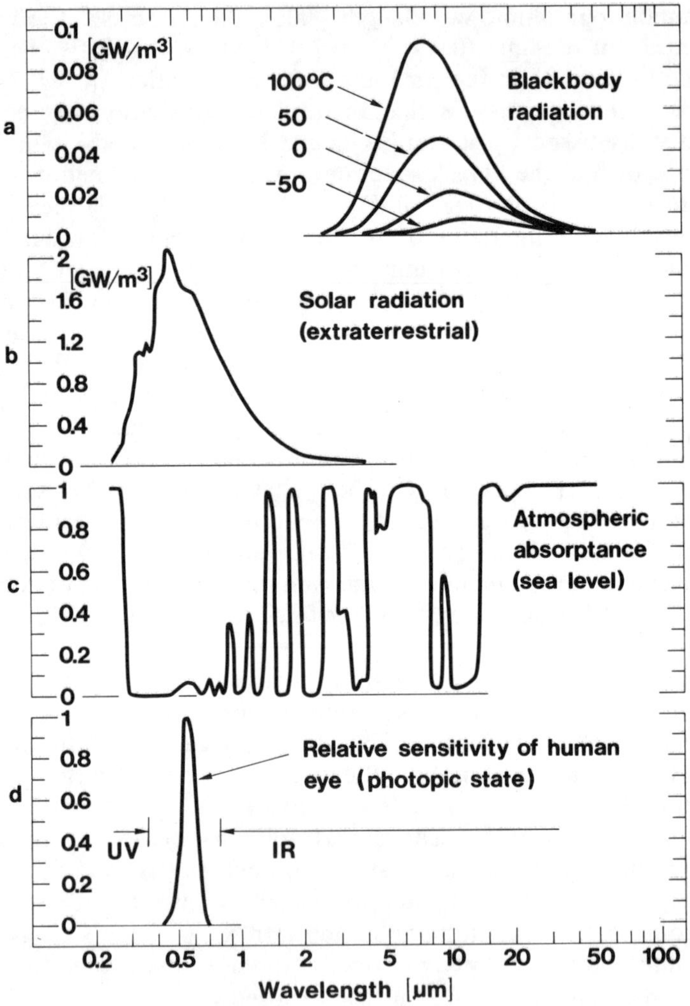

Figure 1. Spectra for (a) blackbody radiation at four temperatures, (b) solar irradiation outside the earth's atmosphere, (c) typical absorptance across the full atmospheric envelope, and (d) relative sensitivity of the human eye. Arrows denote the spectral range for ultraviolet (UV) and infrared (IR) radiation. The intervening spectral interval pertains to luminous (i.e., visible) light.

2.2 What is Energy Efficiency?

An energy-efficient window is a device capable of providing good lighting during the day and good thermal comfort both during day and night at mimimum demand of paid energy. Thermal comfort implies that overheating as well as excessive cooling should be avoided, that draught should be small, etc. Window energy efficiency, therefore, involves control of radiative inflow and of heat losses due to radiation, conduction and convection. The radiative component to energy efficiency is conveniently discussed with reference to the just introduced ambient radiation. It is suitable to make a separation into the requirements imposed by a warm, a cold, and a temperate climate.

In a *warm* climate, it is frequently the case that the solar energy which enters through the windows and is absorbed in the room causes overheating. Space conditioning then requires electrically-driven air cooling equipment. Clearly it is energy effective to have "solar control" windows which block the infrared part of the spectrum ($0.7 < \lambda < 3$ μm), without excessive lowering of the luminous transmittance ($0.4 < \lambda < 0.7$ μm). From the spectral distribution of solar energy, it is inferred that, in principle, it is possible to *exclude about half of the solar energy at no decrease in luminous transmittance.*

In a *cold* climate, a window frequently causes an undesired loss of energy, and hence space conditioning involves heating provided by electricity, fossil fuel, firewood, etc. It is obvious that energy efficiency hinges on a decrease of the heat losses. Convection can be diminished by use of multiply-glazed windows incorporating one or more slabs of, essentially, still gas. The heat transfer can be further lowered by diminishing the thermal radiation. In order to give some indication of the magnitude of improvement that can be accomplished, the results of a computer modelling (8) concerning windows with one, two, or three panes separated by air gap(s) of 12.7 mm are shown in Figure 2. The glass surfaces are designated by consecutive numbers, with the outside surface labeled 1. One of the surfaces, whose number is given at the pertinent curve, is assigned an emittance in the 0–85% range, where the upper limit refers to normal glass, and the thermal conductance, i.e., the k-value, is computed. The data refer to winter conditions with an outside temperature of −18°C and a wind speed of 6.7 ms^{-1}. The upper two curves show that if the emittance of one of the surfaces in a single-pane window is lowered, there is a drop in the k-value from its magnitude ~ 6 $Wm^{-2}\,K^{-1}$ for normal glass. A coating on surface 2 theoretically is able to diminish the k-value to ~4.4 $Wm^{-2}K^{-1}$. The difference between curves 1 and 2 is due to the wind. In the double-glazed unit, the k-value can drop from ~ 2.8 $Wm^{-2}\,K^{-1}$ for normal glass to ~ 1.4 $Wm^{-2}K^{-1}$ when the emittance of either of the surfaces facing the air gap (2 or 3) is brought to zero. For triple glazing, the corresponding improvement is from 1.8 to 1.2 $Wm^{-2}\,K^{-1}$. Note that all of these improvements deal with radiation at $\lambda > 3$ μm, i.e., outside the solar range. It is possible, in principle, to *improve the thermal insulation of a double-glazed window by about a factor two at no decrease in solar transmittance.*

In a *temperate* climate, there is sometimes a need for preventing excessive solar heating, whereas at other times one wishes to let in as much as possible to provide free heating. Hence one wants to have a window which enables a dynamic throughput of radiant energy. With established technology, one can regulate solar heating mechanically by having movable shades, blinds or

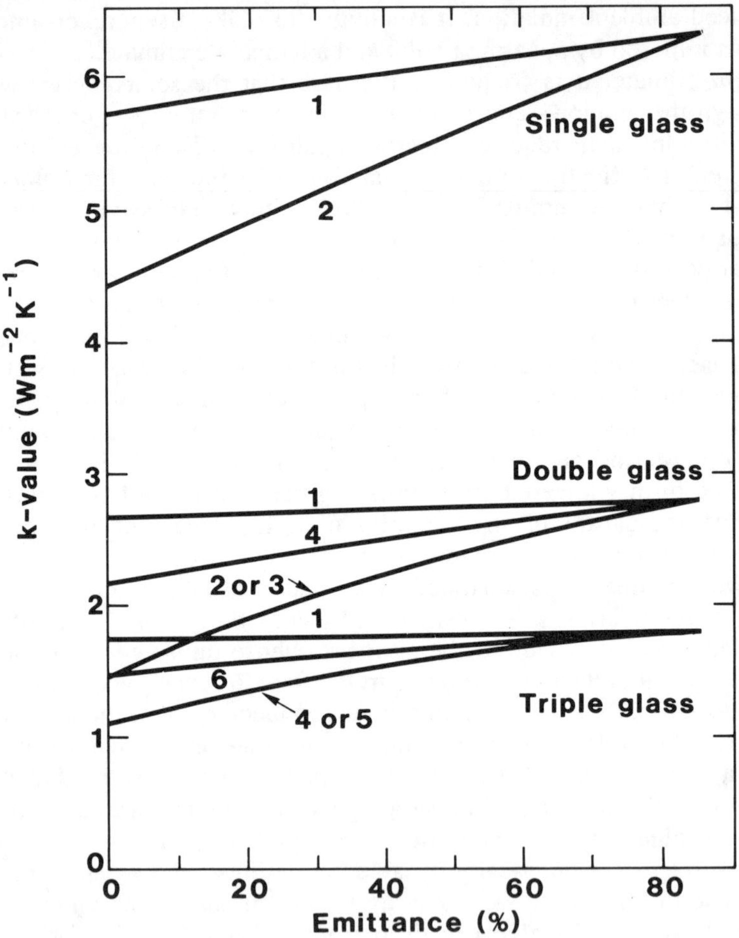

Figure 2. Computed k-values of various windows designs, as reported by Rubin et al. (8). The curves refer to single, double, and triple glazing; the glass surfaces are designated by consecutive numbers, with the outside surface labeled 1. One of the glass surfaces, whose designation is given at the pertinent curve, has an emittance between 0 and 85%.

(roller) curtains, etc. However, a superior solution is to invoke chromogenic materials, which offer the possibility of changing the inflow of luminous and/or solar radiation in accordance with demands which can vary over the day or season. Devices of this type are called smart windows[1];. The concept, introduced only a few years ago (9), seems to have gained acceptance as a generic term. Smart windows can be user operated or automatic. Among the many conceivable possibilities is a system in which a thermometer senses the indoor air and activates a low-voltage electric pulse to the window which sets its solar transmittance to a desired level. Chromogenic fenestration is readily combined with good thermal insulation through multiple glazing and low emittance. *By incorporating materials which enable the throughput of radiant energy to be controlled between widely separated limits, smart windows achieve energy efficiency.*

Good thermal insulation improves energy efficiency also in a more indirect way than through an improvement of the k-value. Note, for instance, that when the heat flow through the window is low, the surface temperature of the inner glass will remain close to that of the indoor air, even if the outdoor air is very cold. This significantly decreases the downward stream of cold air at the window, which improves the comfort noticeably. If this source for draught is eliminated, not only can the comfort temperature be decreased by a few °C, but space utilization can be enhanced because the area at the windows needs no longer be avoided; and heating installations can be simplified.

2.3 Quantitative Performance Parameters for Energy-Efficient Windows.

The idealized properties of energy-efficient windows, stated above, cannot be obtained with practical materials, and hence there is a need to specify suitable performance parameters which tell how far off a window is from certain design goals. The radiative properties comprise the absorptance A, reflectance R, transmittance T and emittance E, each signifying a certain fraction of incident, or maximum emitted, radiative power. Energy conservation dictates that at each wavelength and for directionally uniform radiation one has

$$A(\lambda) + R(\lambda) + T(\lambda) = 1$$

$$E(\lambda) = A(\lambda).$$

1 "Switchable windows" is another term used in current literature. In my opinion, this gives the impression that the window can attain either of two – or perhaps more – specified transmittance levels. However, a characteristic feature of the "smart windows" discussed below is that they can adjust their transmittance to any level between widely separated extrema.

The latter equation applies to a body in thermodynamical equilibrium.

Pertinent wavelength-averaged radiative properties can be defined by integrating over the eye's sensitivity curve, which gives the luminous (lum) performance, a typical solar (sol) spectrum, and a blackbody spectrum, which gives the thermal (therm) radiative performance. Thus, we denote luminous transmittance by T_{lum}, solar transmittance by T_{sol}, and thermal transmittance by T_{therm}, and analogous symbols are used for the other radiative properties. For the luminous properties, one integrates over the efficiency function in Figure 1(d). For the solar properties, one can integrate by use of tabulated spectra appropriate to a certain air mass (AM) (4). In the case of vertical windows, it is often appropriate to use the AM2 spectrum, corresponding to the sun being 30° above the horizon. For the thermal properties, one integrates over a Planck spectrum – cf. Figure 1(a) – defined by the temperature τ. In most cases, it is suitable to specify properties with regard to radiation having perpendicular incidence. However, the radiation of interest for thermal insulation takes place for all directions within a hemisphere (10), so that E_{therm} should be obtained from an integration also over angles.

A distinctive feature for the radiative properties of ambience is their spectral selectivity, i.e., their confinement to well-defined and sometimes non-overlapping wavelength intervals. Thus, it is possible to have a window with a T_{sol}/T_{lum} ratio significantly less than unity, as desired for a warm climate; a large T_{sol} and a small E_{therm}, as desired for a cold climate; and a variable T_{lum} or T_{sol} together with a small E_{therm}, as desired for a temperate climate.

In a non-evacuated double-glazed window, the heat transfer can be diminished by roughly 50% through a reduction of E_{therm}. The remaining heat transfer is associated with conduction and convection in the space between the panes. These latter aspects of heat transfer are not elaborated here, and I refer to the literature (11, 12) for detailed treatises. Also this issue will be discussed further in Section 4.1.

3 Coated Window Glass

3.1 Window glass: A Baseline for Improvements

The purpose of this introductory section is to present selected optical data on window glass, mainly of standard types, in order to provide a baseline for subsequent discussions of means to improve the energy efficiency by use of surface coatings with thicknesses in the range 0.01 to 1 μm.

Normal windows are made by the float process in which the glass is solidified on a bath of molten tin. The uniformity and flatness of this glass are excellent. Figure 3 illustrates spectral transmittance in the solar range for three types of float glass. It is seen that T_{lum} is large. The transmittance in the infrared as well as in the ultraviolet is significant and dependent on the glass type. In the thermal infrared – not shown in Figure 3 – glass is virtually opaque. The major difference among the glass types in Figure 3 is their metal oxide content. With regard to energy efficiency, the most salient influence of the metal oxides is to produce a broad absorption band centered at $\lambda \approx 1$ µm. If a maximum value of T_{sol} is desired, a low content of Fe^{2+} ions is preferrable; the upper curve in Figure 3 refers to such a glass which has $T_{lum} \approx T_{sol} \approx 91\%$. The middle curve in Figure 3 is valid for a normal type of float glass, which has $T_{lum} \approx 87\%$ and $T_{sol} \approx 78\%$. The bottom curve shows the transmittance through a glass with a large content of Fe^{2+} ions. The absorption band extends somewhat into the visible and gives a greenish tint. This glass may be suitable for avoiding excessive solar heating. It is characterized by $T_{lum} \approx 72\%$ and $T_{sol} \approx 47\%$. Thus, the T_{sol}/T_{lum} ratio is 0.65, which seems to be about the smallest value one can reach in glass coloured by metal-type additives. Some of the absorbed energy will lead to a heating of the glass and a concomitant reemission of thermal energy, so that the total energy transmission is larger than T_{sol}, particularly for a tinted glass. For the normal float glass, according to the middle curve in Figure 3, the total energy transmission (calculated with certain assumptions) is 83%.

Let me now consider the reflectance of the glass. In the spectral range where the absorption is weak and at normal incidence, each air/glass interface

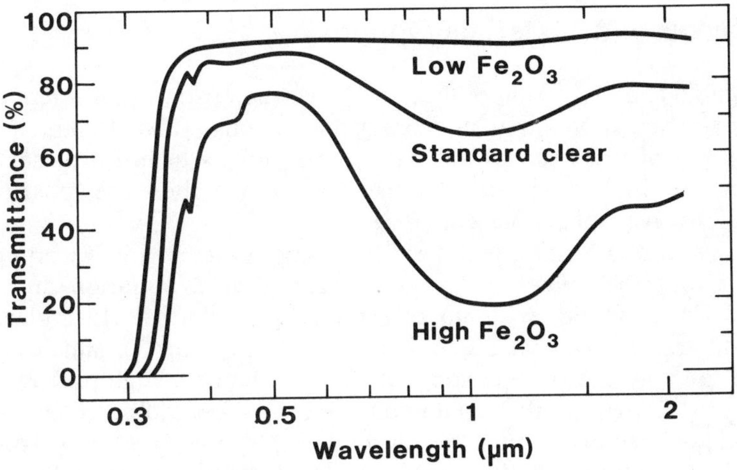

Figure 3. Spectral normal transmittance for float glass of three different qualitites. Specifically the curves refer to 6 mm Flachglas Optiwhite (upper), 6 mm Pilkington float glass (middle), and 4 mm Flachglas Flag Plus 4 (lower).

produces ~ 4% reflectance. It is inferred that $T_{lum} < 92\%$ is valid for a single pane. The overall transmittance is further diminished at multiple glazing. In the thermal infrared, the reflectance is low, which leads to a high emittance – in practice $E_{therm} \approx 85\%$.

Photochromic glass (13) is well known, and widely used in sun-glasses. It deserves some comment here, even if it is not presently (1988) produced for architectural applications. Essentially, ultraviolet irradiation onto these special glasses leads to a reversible formation of anisotropic silver specs, which are strongly absorbing particularly at short wavelengths. The optical properties of photochromic glass are governed by the darkened and cleared transmittance and by the darkening and clearing rates. These are dependent on the glass composition, and a cleared transmittance up to 90%, or a darkened transmittance down to 5%, are possible. The photochromism is somewhat temperature dependent, and a temperature rise yields enhanced transmittance and dynamics. Generally, darkening to ~ 80% of the full range takes place in 1 minute, but a further minor darkening is measurable for times exceeding 1 hour. Clearing in the abscence of irradiation progresses slower than darkening and is incomplete even after 1 hour. The photochromic modulation of T_{sol} is significantly less than for T_{lum}. The cost of photochromic glass is higher than for the standard glass owing to the manufacturing process and the special heat treatment required. The float process has not been demonstrated, as far as I know.

3.2 Objects and Materials Options

As a peamble to a discussion of the specific characteristics of different coatings for energy-efficient windows in subsequent sections, Table 1 summarizes the general goals of the coatings, the principle solutions to fulfill these, and the pertinent coating materials. It also states whether the coatings are commercially available at present (1988).

Diminished solar heating is desired for energy efficiency in a warm climate. About half of the solar energy comes as infrared radiation and can be excluded, in principle, with no effect on T_{lum}. Normal float glass has a significant transmission at $0.7 < \lambda < 3$ µm (cf. Figure 3), and it is clearly effective to apply a surface coating whose reflectance is high preferentially at $\lambda > 0.7$ µm. Extremely thin continuous layers of the "noble" metals copper, silver and gold can be used for this purpose. T_{lum} can be boosted by embedding the metal layer between high-refractive-index dielectric layers. A less efficient alternative is to start with a tinted glass – such as the one represented by the lower curve in Figure 3 – and prevent thermal radiative inflow by a coating with low E_{therm}.

Table 1. General properties of coatings for energy-efficient windows

Goal	Principle solution	Coating materials*	Status**
Diminished solar heating	Reflect at $0.7<\lambda<3$ µm	M or D/M/D	C
Thermal insulation	Reflect at $3<\lambda<50$ µm	D/M/D, SnO_2:F, In_2O_3:Sn ZnO:Al, ...	C
Dynamic radiation control	Absorb or reflect in electrochromic material	Li_xWO_3, NiO_xH_y,... in multilayer design with transparent ion conductor	R
	Reflect at $0.7<\lambda<3$ µm in thermochromic material	VO_2-based	R
Higher transmittance	Antireflect at $\lambda \approx 0.55$ µm	AlO_xF_y, ...	R

* M is Ag, Cu, Au(or Al); D is Bi_2O_3, SnO_2, TiO_2, ZnO or ZnS.

** C denotes that coated glass is commercially available, even if products with superior properties are likely to appear in the future. R denotes that only research-type samples are available.

Good thermal insulation is imperative in a cold climate. Ordinary glass is characterized by a high thermal emittance, and it is effective to apply a surface coating with high reflectance at $3 < \lambda < 50$ µm. Two different types of coatings can be used to provide a low E_{therm}: extremely thin metal films embedded between high-refractive-index dielectric layers with thicknesses chosen so as to maximize T_{sol}, and certain oxide semiconductor layers. Among the latter are SnO_2 doped by F, In_2O_3 doped by Sn, and ZnO doped by Al.

A dynamic throughput of radiant energy is highly desirable in a temperate climate, which calls for coatings of chromogenic materials. Two important approaches exploit materials exhibiting electrochromism (9,14) and thermochromism (15). An electrochromic material is characterized by its optical properties being able to show a reversible and persistent change under the action of an electric field. The pertinent materials are transition metal (tungsten, vanadium, nickel, etc.) oxide layers, whose optical properties are altered by varying their content of small mobile ions (H^+, Li^+, ...). Additional layers serving as transparent ion storage, ion conductor, and electric conductors are required too. A thermochromic material has optical properties which depend reversibly on the temperature. VO_2-based coatings can have a transmittance which goes down once a certain temperature is exceeded and are hence of interest for automatic irradiation control.

Each air/glass interface yields a ~ 4% reduction in T_{lum} and T_{sol}. Still larger reductions can be caused by surface coatings, particularly if they consist of materials with high refractive indices. A decrease of T_{sol} is undesirable in cold and, perhaps, temperate climates. A large R_{lum} may lead to visually disturbing effects of various kinds. The remedy to these problems is to apply an antireflection coating. Certain novel high-rate-sputtered metal oxyfluoride coatings seem to have a great potential for antireflecting coated and uncoated window glass (16).

It should also be noted that coatings can be used to achieve glare and color, thus producing interesting architectural effects. As an example, silicon coatings can yield a durable reflecting glass. This kind of coating, however, has almost no effect on E_{therm} and may, in fact, under certain climatic conditions be detrimental to energy efficiency.

3.3 Noble-Metal Based Coatings

Very thin noble-metal coatings are, at least in principle, the simplest solution for reaching a significant short-wavelength transmittance combined with a long-wavelength reflectance. The best optical properties are obtainable with copper, silver, and gold; alternative materials may be titanium nitride and aluminium. Thin silver coatings stand out as the superior material on account of its low absorption of luminous and solar radiation. The other metals are characterized by absorption, which lies at $\lambda < 0.5$ μm for copper, gold and titanium nitride (hence their colour), and at $\lambda \approx 0.8$ μm for aluminium. Chemical and mechanical ruggedness is a critical issue for all noble-metal based window coatings, and results of laboratory experiments and of field tests dictate that the coatings be used only on glass surfaces facing the gas enclosure(s) in hermetically sealed multiply-glazed units. Windows of this type are commercially available and, in fact, widely used in new buildings and for retrofitting old ones. Reviews of metal based window coating are given in references 17–19.

Calculations show that extremely thin uniform noble metal coatings would have excellent properties for energy-efficient windows. The crux, though, is that such coatings tend to form non-uniform (agglomerate) structures at small thicknesses. This gives a significant limitation in their performance as window coatings. Requiring that the infrared reflectance be high, one is confined to $T_{lum} \approx T_{sol} < 60\%$. If a still higher transmittance is desired, one must use a multilayer coating.

The limited transmittance through reasonable uniform noble-metal layers largely is caused by reflection at the surfaces of the coating, and hence it is possible to improve the transmittance by additional layers which act so as to antireflect the metal. One is then led to consider dielectric/metal and dielectric/metal/dielectric multilayers. Dielectrics with high refractive indices – such as Bi_2O_3, In_2O_3, SnO_2, TiO_2, ZnO and ZnS – give the largest enhancement of the transmittance. By selecting the thicknesses of the three-layer configuration properly, one can optimize for a warm climate (maximum reflectance for $0.7 < \lambda < 3$ µm) or a cold climate (maximum T_{sol}; minimum E_{therm}).

Figure 4 depicts measured transmittance spectra, compiled from different sources, for coatings of the titanium dioxide/silver/titanium dioxide (TiO_2/Ag/TiO_2) type on glass. These data are characteristic for what one can achieve by exploiting the three-layer design. It is inferred that $T_{lum} > 80\%$ and $E_{therm} << 20\%$ are valid for all coatings. The dashed curve pertains to a coating with maximum infrared reflectance; it yields $T_{sol} \approx 50\%$ and $R_{sol} \approx 42\%$. The dotted curve corresponds to $T_{sol} \approx 67\%$ and $R_{sol} \approx 26\%$. Clearly, it is feasible

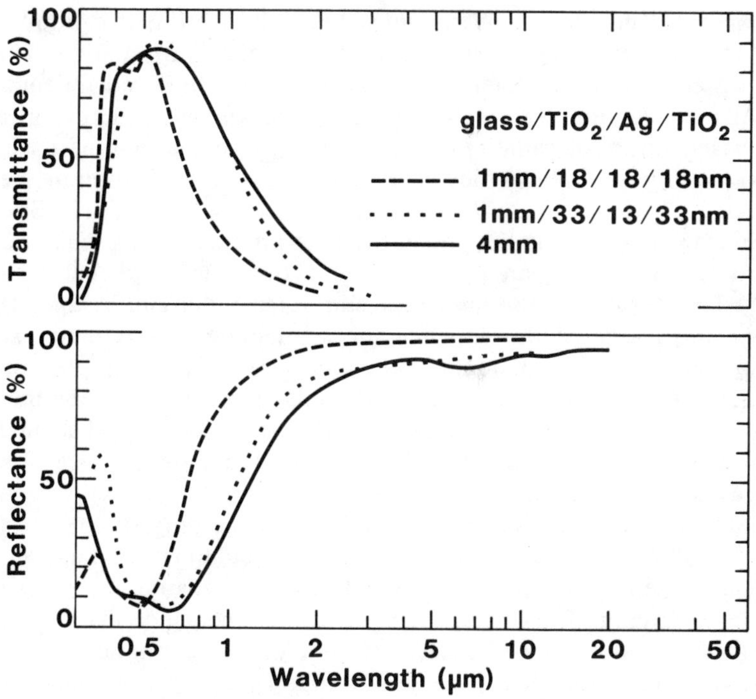

Figure 4. Spectral normal transmittance and near-normal reflectance measured for TiO_2/Ag/TiO_2 based coatings on glass. Pertinent thicknesses for the glass and for the constituent layers of the coatings are shown. Solid curve stems from a commercially produced sample (Glaverbel Superlight), dotted curve from Jarvinen (20), and dashed curve from Fan and Bachner (21).

to construct coatings for different climates simply by altering the thicknesses. The solid curve was measured for an optimized commercial coating on float glass of the type represented by the middle curve in Figure 3. This latter coating yields $T_{lum} \approx 87\%$ and $T_{sol} \approx 72\%$. Noble-metal based three layer coatings can be produced, using large-scale automatic equipment, at a cost of about \$10 US/m^2 in the case of $TiO_2/Ag/TiO_2$, provided that materials utilization is optimized. The cost is related to the processing – rather than to the raw materials – which is generally the case even for coatings which incorporate expensive materials such as gold.

3.4 Doped Oxide Semiconductor Coatings

Doped oxide semiconductor coatings offer an alternative to the noble-metal based coatings. The two classes of coatings have specific advantages and disadvantages. The semiconductor must be of a material that allows good transmission in the luminous and solar ranges. Further it must allow doping to a level which makes the material metallic and hence infrared reflecting and electrically conducting. The materials which are known to be useful are all oxides based on zinc, cadmium, indium and tin and alloys of these. The required doping is often achieved by the addition of a foreign element; particularly good properties have been obtained with SnO_2:F, SnO_2:Sb, In_2O_3:Sn and ZnO:Al. Another possibility is to provide doping via a moderate oxygen deficiency. If prepared properly, the above mentioned coatings can be virtually non-absorbing for luminous and solar radiation. A specific and important advantage of the doped oxide semiconductors is their excellent chemical and mechanical durability, which enables their use on glass surfaces exposed to the air. Window glass with SnO_2:F coating is commercially available. Reviews on doped oxide semiconductor coatings and their uses as transparent infrared reflectors and transparent conductors are found in references 17–19 and 23–29.

Recent work (29) has led to a detailed understanding of the physics of heavily doped oxide semiconductors, and it is feasible to perform a quantitative modelling of their optical properties as a function of coating thickness (t) and doping concentration. From such analyses, one can conclude that doped oxide semiconductor coatings are useful for giving a low E_{therm} but not for reflecting infrared solar radiation. It can be shown that $T_{lum} \approx 80\%$ is obtainable for coated glass; T_{sol} decreases significantly with enhanced doping concentration and/or thickness; and a low E_{therm} can be reached only at $t > 0.2$ µm. As an optimum performance, one may put $E_{therm} \approx 20\%$ and $T_{sol} \approx 78\%$ at $t \approx 0.2$ µm.

The required coating thickness equals a fraction of the wavelength for visible light, and it follows that optical interference effects must be given serious consideration. In particular, minor variations in t may cause significant iridescence, i.e., a rainbow exhibition of colours at different parts of the coated window. Such effects are not important for noble-metal based layers since they have t << 0.2 μm. Iridescent windows generally are experienced as aesthetically unpleasant, and it is concluded that extreme thickness control must be exercised during practical manufacturing of the coated glass. One possibility to avoid iridescence is to work with t ~1 μm. Such thick coatings yield numerous interference-induced peaks and dips of the transmittance and reflectance across the $0.4 < \lambda < 0.7$ μm interval. A thickness-dependent displacement of these is not recognized by the eye, which hence senses a uniform coloration. However, the use of a much larger thickness than the one demanded for a low E_{therm} is clearly inefficient in terms of materials utilization, coating time, and cost. Further, thick oxide semiconductor coatings can display some light scattering, sometimes referred to as haze.

Figure 5 illustrates transmittance and reflectance spectra as measured for three different doped oxide semiconductor coatings. The curves show rather high T_{lum}, moderately high T_{sol}, and indicate low E_{therm}. The upper part refers to a In_2O_3:Sn coating prepared for research. This coating matches the theoretical predictions for an optimized material very well. The middle part concerns a 0.30 μm thick ZnO:Al coating again prepared for research. The bottom part was recorded for a SnO_2:F coating produced commercially on 4 mm standard float glass. The latter coating is considerably thicker than the other two, as evidenced by the densely spaced peaks in the spectra. It has $T_{lum} \approx 74\%$, which is lower than for the In_2O_3:Sn and ZnO:Al coatings, and $T_{sol} \approx 77\%$. Some data on doped SnO_2 coatings have been summarized recently (22).

The cost for large-scale doped oxide semiconductor coatings is dependent on production technology, as well as material, and is difficult to estimate. With regard to technology, it should be pointed out that glass manufacturers have had success in their development of SnO_2-based coatings prepared by so called spray pyrolysis directly onto the hot glass on a float line. This technology holds promise for very inexpensive coatings, having costs of only a few US dollars/m^2. With regard to material, one should note that indium is much more scarcer and more expensive than zinc and tin, although detailed information on its anbundancy is hard to obtain. Hence, In_2O_3:Sn coatings should be used only when the ultimate in electrical conductivity or infrared reflectance are required.

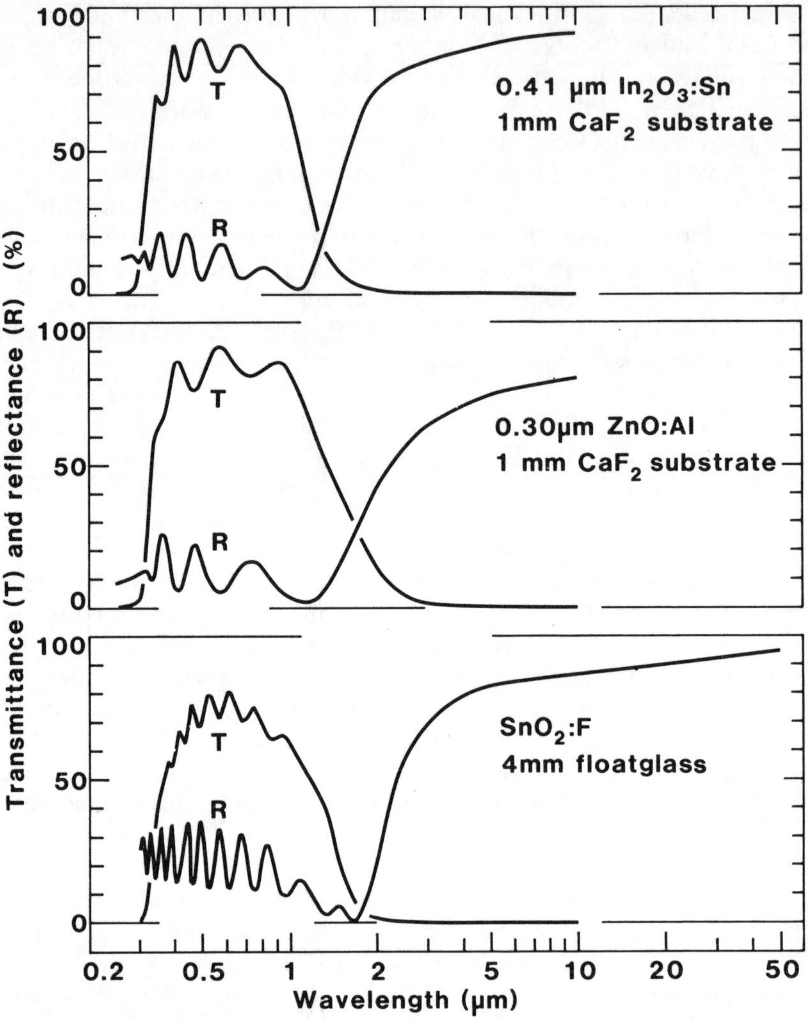

Figure 5. Spectral normal transmittance and near-normal reflectance measured for In_2O_3:Sn on 1 mm CaF_2, ZnO:Al on 1 mm CaF_2, and SnO_2:F on 4 mm float glass (Glaverbel Comfort Glass). For our present purposes, the substrate material and the substrate thickness are is of little significance. Coating and substrate thickness are as given. Partly compiled from Hamberg and Granqvist (29) and Jin et al. (30).

3.5 Electrochromic Coatings

Electrochromism is a multi-faceted phenomenon which is well known in metal oxides based on tungsten, vanadium, nickel, molybdenum, titanium, iridium, etc., and in numerous organic substances. The change in the optical properties is caused by the injection or extraction of mobile ions. A material colouring under injection (extraction) of ions is referred to as cathodic (anodic). Absorptance modulation as well as reflectance modulation are feasible. Electrochromism has been studied since the 1960's for high-contrast non-emissive display applications (31–36). Since the mid-1980's, research and development on transparent electrochromic materials and devices for smart window applications have burgeoned both in academia and in industry (9,14).

An electrochromic coating must encompass several layered materials. Figure 6 sketches a basic five-layer configuration which is convenient for discussing the operating principles. It comprises two outer transparent conductors (required for setting up a distributed electric field) an electrochromic layer, an ion conductor, and an ion storage. Colouration and bleaching are accomplished when ions are moved from the ion storage, via the ion conductor, into the electrochromic layer, or when the process is run in reverse. The ion storage can be another electrochromic layer, preferrably anodic if the base electrochromic layer is cathodic, or vice versa. One may also combine the conductor and storage media into one layer of electrolyte. Further, it is possible to exclude the ion storage and instead rely on a replenishment of H^+ ions (protons), originating from the dissociation of water molecules diffusing in from an ambience with controlled humidity. Use of a purely ionic conductor means that the electric field has to be applied only when the optical properties are to be altered. The design in Figure 6 has a low value of E_{therm} because of the external transparent conductor. Alternatively, the electrochromic coating can be combined with a lamination material which joins two glass panes; in this case an additional transparent and infrared-reflecting coating is required to have a low E_{therm}.

The state of the art in electrochromic smart windows is believed to be represented in Figure 7. The investigated device incorporates two glasses with low Fe_2O_3 content, each having a transparent conducting layer and an electrochromic layer, laminated together by a solid transparent polymeric ion conductor (37). The electrochromic layers are of Li_xWO_3 and $Li_yV_2O_5$, and the polymer adhesive is a newly developed methyl methacrylate network incorporating poly (propylene glycol) complexed with lithium perchlorate. The device is mechanically rugged. A low E_{therm} can be obtained by having a layer of SnO_2:F, for example, on either of the exterior glass surfaces.

The time for going from fully bleached to fully coloured state, here called the response time r, is of obvious significance. The required time scale is set

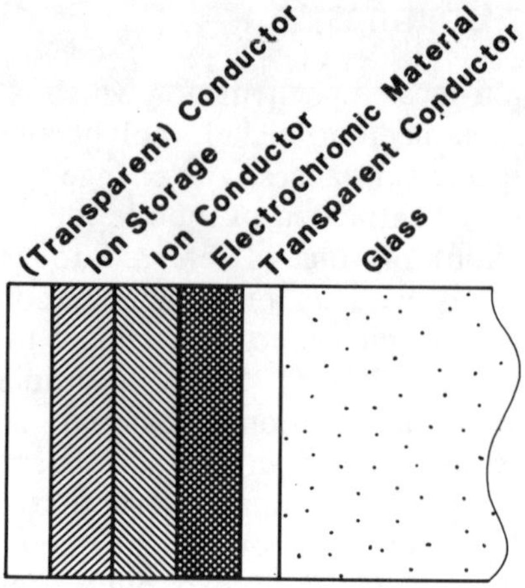

Figure 6. Basic design of an electrochromic coating for smart windows. From Svensson and Granqvist (9).

by the eye's ability to accomodate, which is on the order of minutes. A response time of this magnitude can be reached in WO_3-based electrochromic layers up to ~ 1 m^2 in size (38). In fact the magnitude of r is critically dependent on the electrical resistance in the transparent conductor(s). It is known (39) that r rises at low temperatures, but the effect is of little practical significance for an electrochromic coating placed on the inner glass of a double-glazed window. Full colour uniformity over an extended surface cannot be expected during changes of T_{lum} and T_{sol}, but the unevenness can be made invisible at the expense of a long response time. Iridescence may be a concern for electrochromic multilayer coatings placed on a single glass surface, just as it is for doped oxide semiconductor coatings. Electrochromic coatings operated in a laminate configuration, however, are expected to be less prone to show iridescence, owing to the matching of refractive indices for adjacent layers.

It is not possible to give even a rough estimate of the cost of a practical smart window. However, it should be noted that inherently cheap techniques are presently being developed for efficient deposition of electrochromic layers. The electrical power needed for operating an electrochromic smart window can be assessed from relations between charge density and colouration. A simple analysis indicates that the electrical energy requirement is as low as << 1 kWh per year and square meter window area.

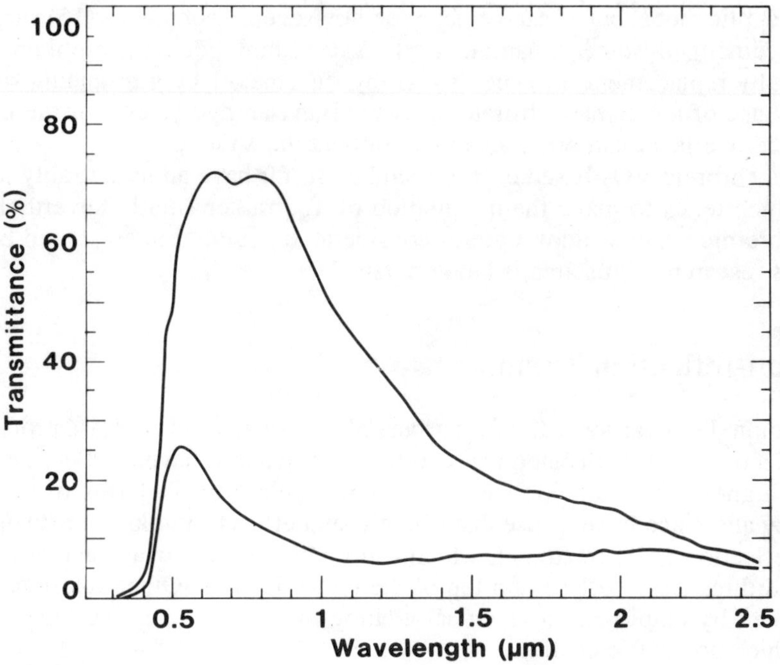

Figure 7. Spectral normal transmittance measured for an all-solid-state research-type electrochromic smart window in fully bleached and fully coloured states. From Andersson et al. (37).

3.6 Thermochromic Coatings

Thermochromism is well known for inorganic materials in the liquid and solid state (40) as well as in many organic materials. Among the many thermochromic transition metal oxides (41), vanadium dioxide stands out as particularly interesting for applications on smart windows. VO_2 undergoes a structural transformation at a "critical" temperature τ_c, below which the material is semiconducting and relatively non-absorbing in the infrared, and above which the material is metallic and infrared reflecting. A window with a VO_2 coating, therefore, has a T_{sol} which drops at $\tau > \tau_c$, i.e., the window is capable of an automatic control of the throughput of radiant energy. This control may be achieved by use of a single layer, i.e., with a coating design which is simpler than the one in an electrochromic smart window. However, the control is built into the material and is not easily modified by an operator. Crystals of VO_2 are characterized by $\tau_c \approx 68°C$, which is too high for normal window applications, but τ_c can be depressed by several techniques (15) such as by replacement of some vanadium by tungsten, molybdenum, niobium and rhenium, by replacement of some oxygen by fluorine, or by introducing stress either by use of a suitable substrate or by applying an overlayer. For the latter option, the overlayer can serve also to antireflect the VO_2 coating.

Thermochromic VO_2-based coatings studied so far have an undesirably large A_{lum}, which tends to make the modulation of T_{sol} rather small. Nevertheless, thermochromic smart windows are of considerable potential interest, and basic materials research in this area is being pursued.

3.7 Anti-reflection Treatments

It was pointed out earlier that each surface of a normal window glass produces ~4% reflection, and that coated glass surfaces, in general, have a reflection that is even higher. This reflection is often undesirable both in terms of energy efficiency and since it can cause disturbing visual effects. A solution is to apply an antireflection (AR) treatment, which can invoke a layer of refractive index ~ 1.4 and thickness ~ 0.1 μm on top of the (coated) glass. Other solutions can be obtained by employing a chemical etching of the glass surface (42) or by having thick polymeric coatings.

Dielectric AR layers of MgF_2, etc., have been used on lenses and other optical components for many years. These layers are hard and chemically inert but, unfortunately, do not lend themselves to efficient large-scale deposition. In fact, until recently there were no durable low-refractive-index coatings which could be applied by practically useful techniques. Currently, the situation is changing, and a range of newly developed metal oxyfluoride coatings have given very promising results (16, 43). Aluminium oxyfluoride layers have shown good durability in a humid atmosphere, under ultraviolet irradiation, and during heating to high temperatures in air. Figure 8 shows spectral transmittance and reflectance for a 1 mm thick glass with and without an AR layer on one side. It is seen that T_{lum} is increased by a few percent and that R_{lum} is strongly decreased by the aluminium oxyfluoride. A similar AR layer, put on 4 mm float glass having a commercial $Bi_2O_3/Ag/Bi_2O_3$ coating, enhanced T_{lum} by 7% and T_{sol} by 6%, and decreased R_{lum} by 7% (16).

Dielectric AR layers tend to enhance E_{therm} for noble-metal based as well as for doped oxide semiconductor coatings, but the effect is on the ~ 1% level, which is of little practical significance. In addition, the AR layers can be used to avoid iridescence – particularly for specified viewing conditions – and to make light scattering (haze) less apparent.

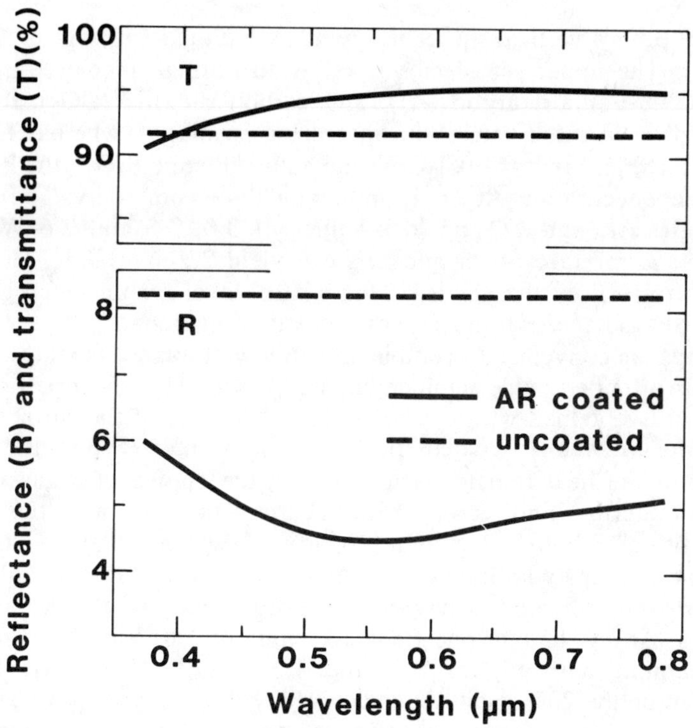

Figure 8. Spectral normal transmittance (T) and near-normal reflectance (R) measured for uncoated (dashed curves) and aluminium-oxyfluoride-coated (solid curves) glass. From Harding (44).

4 Materials Interposed Between the Glass Panes

4.1 Gases

Hermetically sealed, doubly or multiply glazed windows make it possible to incorporate a gas other than air between the glass panes. Such windows have been available on the market for many years. By selecting a suitable gas, one can alter the heat transfer by thermal radiation, conduction and convection – thereby enhancing the energy efficiency.

The possible improvements are illustrated in Figure 9, which gives k-value measured as a function of the distance between two glasses for air and five different, chemically specified, gases (45). The data were recorded in the center of a ~ 0.6 m^2 vertical square test window with gas spacings of 6,9,12,16 and 20 mm. The upper set of curves refers to normal uncoated glass with $E_{therm} \approx 85\%$. For air, argon (Ar), and carbon dioxide (CO_2), there is a monotonic drop in the k-value with increasing spacing s, whereas for sulfur hexafluoride (SF_6), Freon 12 (Cl_2CF_2) and sulfur dioxide (SO_2), the k-value is much less dependent on s. At s = 12 mm, which is a common value for sealed windows, air, Ar, and CO_2 yield k-values of 3.0, 2.8 and 2.6 $Wm^{-2}K^{-1}$, respectively. A mixture of Ar and SF_6 can yield 2.7 $Wm^{-2}K^{-1}$. Hence, the improvement given by the gas is on the ~ 10% level for uncoated glass. An addition of SF_6 is beneficial in terms of acoustic insulation.

Gas fillings are conveniently combined with low-emittance coatings based on noble metals or doped oxide semiconductors (45,46). The lower part of Figure 9 shows results obtained with a window for which one of the surfaces facing the gas enclosure was Au coated and had $E_{therm} \approx 6.5\%$. When the radiative component to the heat transfer is minimized, the choice of a gas with low thermal conduction and convection becomes critical, and the relative improvement possible by selecting a proper gas becomes enlarged. It is found that the k-value drops with increasing s for air and Ar, and that the other gases give a more complicated behaviour or even a rise of the k-value with increasing s. At s = 12 mm one finds a k-value of 1.8 $Wm^{-2}K^{-1}$ for air – in good agreement with Figure 2 – and 1.5 $Wm^{-2}K^{-1}$ for Ar. Thus, an improvement on the 20% level can be achived with gas between panes and low emittance coated glass. For krypton (Kr) gas, the k-value can be as low as 0.98 $Wm^{-2}K^{-1}$ at s = 12 mm. Xenon (Xe) is expected to be even better.

In units of US dollars per m^2 window area, the cost of gases ranges from about 0.1 for Ar, 1 for SF_6, 10 for Kr to 100 for Xe, assuming an annual consumption of < 1000 m^3 at the filling site. For Kr and Xe, global scarcity preclude large-scale uses. The filling cost for a standard window is ~ \$0.25 US on the production line and ~\$1 US off the production line. The sealing quality is obviously important if the the window is to hold the gas for many years. The experience from practical tests is somewhat mixed – severe leaking as well as adequate gas retention have been reported. It appears that gas leakage is not a problem for properly designed and manufactured windows.

4.2 Solid Transparent Insulation Materials

There is a wide range of solid materials which can be interposed between glass panes in order to improve the thermal insulation of windows. Figure 10 subdivides the materials into four groups: (1) thin flexible polymer foils; (2) polymer honeycomb materials; (3) bubbles, foams and fibres, and; (4) inorganic microporous materials, especially silica aerogels. If the honeycomb

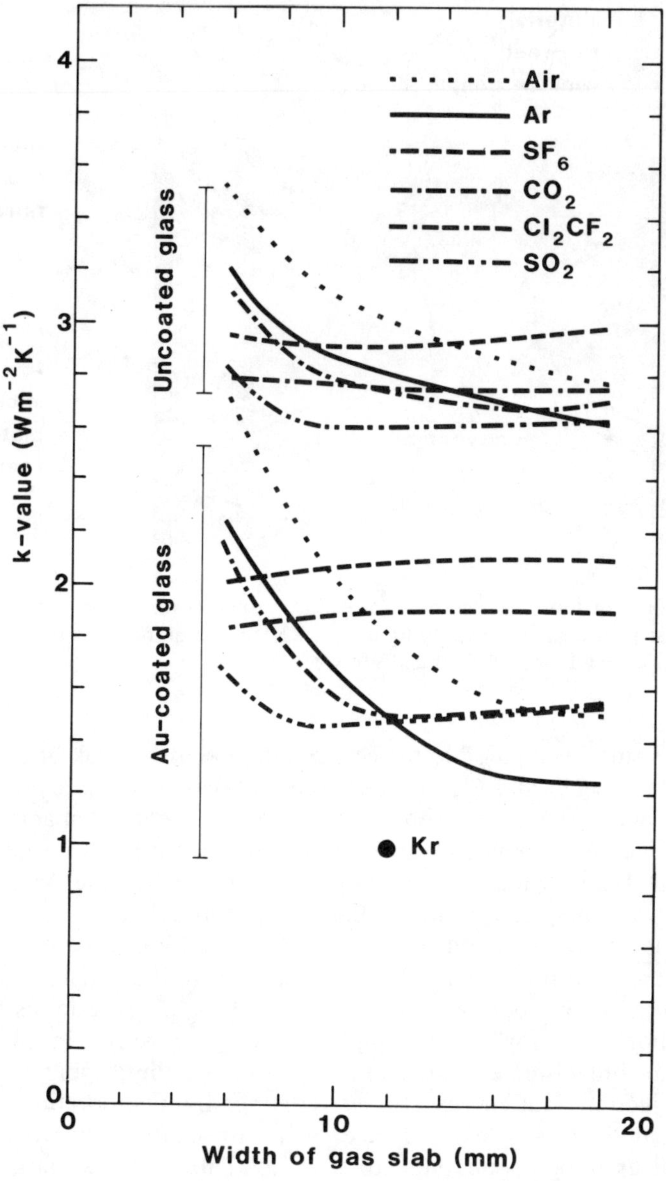

Figure 9. k-value vs. glass spacing measured for double-glazed test windows. Variously dashed and dotted curves refer to the shown gases used as fillings between the glass panes. Upper set of curves were recorded with uncoated glass. Lower set of curves corresponds to one of the glass surfaces having an Au coating. A single data point is shown for Kr gas. From Gläser (45).

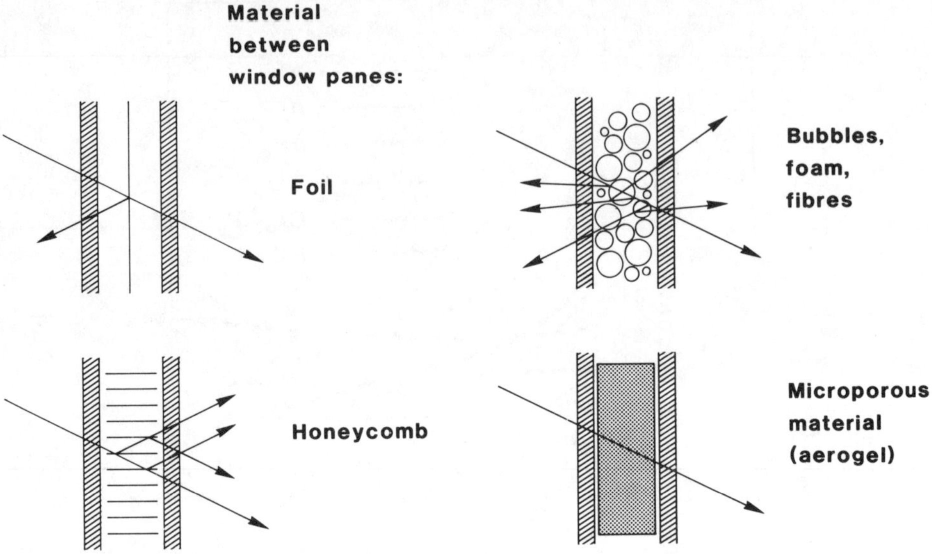

Figure 10. Principles of four different types of solid transparent insulation materials placed between two glass panes. Arrows signify light rays. Reflections at the glass surfaces are not shown. The classification follows Platzer and Wittwer (47).

cross-section is small compared to the cell length, one may speak of a capillary structure. Materials (1) and (4) can be almost invisible to the eye, whereas materials (2) and (3) cause strong scattering. Hence the "macroporous" honeycombs, bubbles, foams and fibres do not provide a good visual indoor-outdoor contact, but they may be used to provide diffuse lighting. Materials (2) and (4) allow, in principle, very high solar energy throughput.

Flexible foils can be mounted between glass panes by use of frames which maintain the material in permanent tension. Foils have their main application as a carrier for a coating devised for giving a high T_{lum}/T_{sol} ratio, as required for a warm climate; or for giving a high T_{sol} together with a low E_{therm}, as required for a cold climate. Thus, one can create a light-weight multiple glazing in conjuction with energy efficient coatings by incorporating only two panes of uncoated glass. Clear polyester foil with coatings based on noble metals as well as doped semiconductors are available on the market. Their optical properties are not quite as good as those for glass with corresponding coatings (cf. Figures 4 and 5). The cost for coating plastic foil is comparable to the one for coating glass panes, at least for noble-metal based layers.

Transmittance for diffuse solar irradiance[2], denoted T_{sol}^{d}, and k-value have been reported for numerous configurations (48,49). Honeycombs of different types – made of polystyrol, polyamide, polyvinylchloride and polycarbonate – typically gave T_{sol}^{d} between 71 and 45% and k-value between 1.3 and 0.9 $Wm^{-2}K^{-1}$, when the thickness was ~ 10 cm. The thickness dependence of T_{sol}^{d} is an interesting property. It is very different for different types of macroporous materials. For example, a material consisting of 1.7 mm polycarbonate capillaries had $T_{sol}^{d} \approx 73\%$ at small thickness and $T_{sol}^{d} \approx 60\%$ at 10 cm thickness. Its k-value at 6 and 10 cm thickness was 1.45 and 0.75 $Wm^{-2}K^{-1}$, respectively. A material comprising poly-methyl-methacrylate foam had a much stronger decrease of T_{sol}^{d} with increasing thickness. At 1.6 cm thickness, different foam types showed T_{sol}^{d} between 60 and 48% and k-value between 3.1 and 3.6 $Wm^{-2}K^{-1}$.

By going from a macroporous to a microporous structure – with inhomogeneities much less than the wavelengths of visible light – one can achieve a transparent solid material with superior thermal insulation (50). A particularly interesting material can be obtained by supercritical drying of colloidal silica gel (51). The ensuing substance, called a silica aerogel, consists of silica particles of size ~ 1 nm interconnected so that a loosely packed structure with pore sizes of 1 to 100 nm is formed. The aerogel density is 70 to 270 kg m^{-3}, compared with 2200 kg m^{-3} for non-porous silica glass, implying a porosity up to 97%. Many properties of silica aerogels are compiled in reference 52. Silica aerogel can be prepared both as transparent tiles and as a translucent granular material. Large-scale production is possible (53,54). A granular filling is expected to have a production cost of ~$20 US per m^2 window area.

A high T_{lum} and T_{sol} together with a low k-value makes silica aerogel of great interest for energy-efficient windows, with and without low-emittance coating. Further, the mechanical rigidity permits its use as a spacer in windows with a low gas pressure, or vacuum, between the panes. Figure 11 illustrates spectral transmittance for a 4 mm thick tile (55). The transmittance exceeds 80% at $0.6 < \lambda < 2.2$ μm but drops at $\lambda < 0.6$ due to scattering from density variations. Transmittance spectra for granular aerogels have also been reported (56): at 1 cm one finds $T_{sol} \approx 69\%$ and $T_{sol}^{d} \approx 55\%$. Smaller thicknesses may be used in practical windows. The k-value of this kind of material has been measured as a function of the gas pressure in the aerogel and for two magnitudes of the emittance of the surrounding surfaces (57). The measurements were done at ~40°C. With nitrogen gas at atmospheric pressure, the k-value was ~1.8 $Wm^{-2}K^{-1}$ for a 1.5 cm thick layer of granules.

2 Diffuse solar irradiation normally prevails over direct impingement during overcast weather. Under a clear sky, the diffuse component is only a few percent.

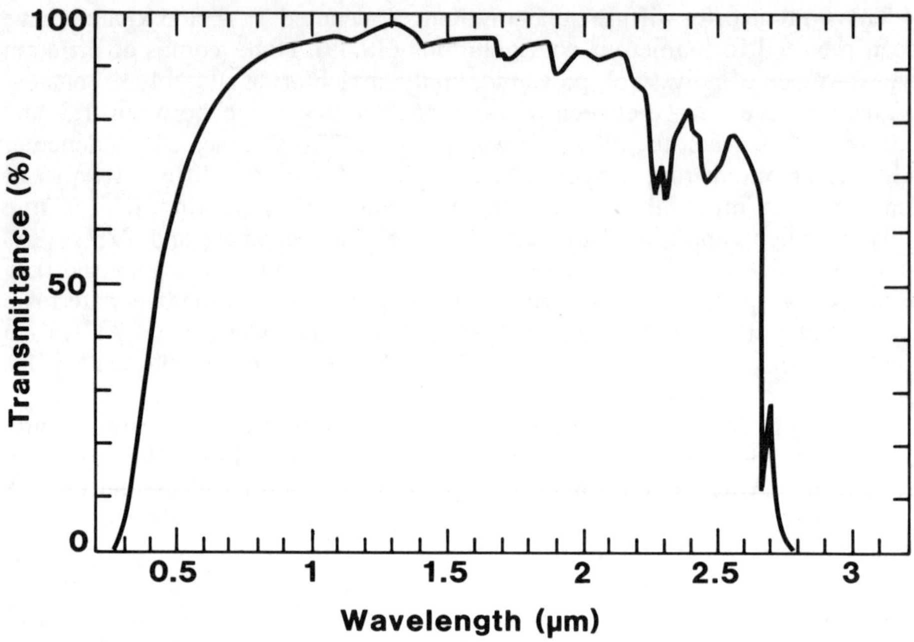

Figure 11. Spectral transmittance measured for a 4 mm thick silica aerogel tile. From Tewari et al. (55).

Decreasing the gas pressure made the k-value drop. For pressures below 1 mbar, the radiative heat exchange became significant – particularly at $3 < \lambda < 5$ µm – and hence an enclosure with low emittance yielded a depressed k-value. At 10^{-3} mbar, the k-value was ~ 0.9 $Wm^{-2}K^{-1}$ for an emittance of 0.9 (representative of uncoated glass) and ~0.6 $Wm^{-2}K^{-1}$ for an emittance of 0.05 (representative of low-emittance-coated glass). At 0°C, one can extrapolate that the corresponding k-values would be 0.7 and 0.5, $Wm^{-2}K^{-1}$ respectively.

Instead of having aerogel spacers in evacuated windows, it is possible to use small spherical glass supports. Such a window, with low emittance coating and evacuated to < 10^{-3} mbar, can yield a k-value of 0.6 $Wm^{-2}K^{-1}$ in a compact (< 1 cm thick) and light-weight (~ 14 kg m^{-2}) design (58). Laser edge sealing of the evacuated window is a promising technique (59). Vacuum windows are unproven, though, and it remains to be demonstrated that a sufficiently low pressure can be maintained for the required number of years.[3]

[3]The requirements on good sealing and low outgassing are much more servere for a vacuum window than for a window containing an inert gas.

4.3 Chromogenic Materials

Chromogenic coatings on glass were discussed previously with a focus on electrochromics-based and thermochromics-based smart windows. In this section, I discuss some recent work on "thick" materials which allow dynamic throughput of radiant energy when interposed between window panes. Specifically, I cover photochromic, thermochromic, and electrically controlled liquid-crystal-based materials.

Plastic sheet or foil with *photochromic* additives offer interesting possibilities (60,61), and detailed results have been given for spirooxazine in a host of cellulose acetate butyrate. The material can be produced by injection molding. Darkening and clearing dynamics of T_{lum} have been reported for a 1.5 mm thick sheet at 20°C. Solar exposure makes T_{lum} drop from 82 to 23% in about one minute owing to the formation of an absorption band at $0.5 < \lambda < 0.7$ μm. Clearing is slower, and it takes about 20 minutes to regain full transmittance. The dynamics is much more rapid, however, than for photochromic glass. The spirooxazine-induced photochromism is temperature dependent to an undesired extent.

Thermochromic control of radiation throughput can be accomplished with different kinds of materials. Polymeric "cloud gels" are well known in this context (62,63). Clouding – i.e., transition to a diffusely scattering state – can set in above a certain temperature due to a reversible thermochemical dissolution of water and a thermally induced modification in the length of the polymer molecules. The clouding point can be regulated to within 1.5°C in the 9 to 90°C range. Both T_{lum} and T_{sol} can drop by ~50% when the clouding point is exceeded in a 1 mm thick layer interposed between two glass panes. The material has to be used in a sealed window. Its cost in a window can be less than \$1.80 US/m^2. On the negative side, if thermal gradients exist over a cloud-gel-containing window, they may manifest themselves as areas with clear and cloudy appearances.

Electrically controlled *liquid-crystal-based materials* can switch between a transparent state and a highly scattering state in a way similar to the one for cloud gels (64,65). One such material, recently commercialized, consists of a polymer layer with μm-sized cavities containing a nematic liquid crystal. This layer is laminated between two In_2O_3:Sn-coated polyester foils serving as extended electrodes. Applying a voltage over the electrodes, one can align the liquid crystal molecules and thereby obtain refractive-index-matching between the inclusions and the surrounding polymer matrix across the solar range. This represents the transparent state of the material. In the absence of an electric field, the liquid crystal molecules become more randomly oriented, which creates refractive-index-mismatch and hence strong scattering. A switching between 82 and 62% transmission has been demonstrated. An electric power of < 20 Wm^{-2} is needed for keeping the window in its transparent state.

5 Conclusion

5.1 Summary of Key Results, and Some Remarks

This review has embraced a large number of possibilities for achieving energy efficiency of architectural windows. This multiplicity of options is easily understood since:

- energy efficiency means different things in different climates,
- there are several principle approaches to energy efficiency which can be exploated, either separately or in conjunction, and
- many different materials may be useful depending on the climate and the energy efficiency approach being pursued.

An energy efficient window must include at least two *glass* panes. Float glass is a rather standardized product, but it should be kept in mind that a low content of Fe^{2+} ions in otherwise normal float glass can yield a significant improvement in solar energy transmission.

Thin *coatings* can modify the radiative properties of the glass surfaces in many different ways. Thus, one can use noble-metal based coatings for blocking the transmission of infrared solar radiation, thereby cutting down the solar heating in a warm climate, or one can use noble-metal based or doped oxide semiconductor coatings for combining high transmission of solar radiation with low emission of thermal radiation in a cold climate. These coatings are produced commercially on a large scale. Noble-metal based coatings must be used in hermetically sealed environments, whereas doped oxide semiconductors are inert and rugged. Iridescence and haze may be problems for doped oxide semiconductor coatings, albeit not unsurmountable ones.

Coatings with dynamic properties – to be used in smart windows – are subject to vigorous research efforts and, in my opinion, are rapidly approaching a stage where commercialization is feasible. All-solid-state electrochromics-based coatings are of particular interest. Laboratory studies indicate that their transmittance of visible light and solar energy can be altered gradually and reversibly between ~10 and ~80% by electric pulses; the energy requirement is very small. Thermochromic coatings is another, more remote, possibility. Novel antireflection coatings seem to be able to increase the transmittance and decrease the reflectance of window glass with and without various types of coatings. There is no ground for assuming that a coated glass necessarily should have a lower transmittance of visible or solar radiation than an uncoated one; in fact, the converse may well come true in the future.

Materials interposed between the window panes can diminish the heat transfer due to conduction and convection. Gases, particularly Ar and SF_6, as well as low-emittance-coated polymer foils are well known in this context. Low-density silica aerogel tiles is a novel material capable of giving an extremely low heat transfer.

The k-value is one of the parameters which govern energy efficiency. It is illustrative to follow its decrease through the measures listed in Table 2. A double-pane window with low-emittance coating and gas filling can give a k-value of ~ 1.5 $Wm^{-2}K^{-1}$. Even lower values, down to ~ 1.2 $Wm^{-2}K^{-1}$, can be reached with triple glazing – however at the expense of a lower transmission and a larger weight. The limitation in transmittance and the weight penalty become even more pronounced in windows with four or more panes, and such windows are of little practical interest. In future evacuated double-pane windows one may reach an ultimate k-value of ~ 0.5 $Wm^{-2}K^{-1}$. It should be stressed that the k-value does not include the input of radiant energy through the window, and hence it is appropriate only to north-facing windows or night conditions. When a solar inflow is considered, a well insulated window often provides a net input of energy (sometimes phrased in terms of a negative k-value) particularly for south-facing windows. One concludes that a comparison of the k-value for a window with the k-value for an opaque material, such as a wall, is difficult and has to be done with caution. It should also be remarked that the stated k-values refer to the centers of windows[4].

Table 2. k-values for some different types of windows

Window type	k-value ($Wm^{-2}K^{-1}$)	Status*
Single-pane	6	C
Double-pane	3	C
+ low emittance coating	1.8	C
+ gas filling	1.5	C
Evacuated double-pane (silica aerogel spacer)	~ 0.5	R

* C is commercial; R is research-type.

[4]The influence of the window frame on the overall k-value was treated in reference. 66. The significance of proper framing was stressed, and it was shown that if aluminium frame spacers were replaced by fiberglass, butyl rubber or stainless steel, the total heat loss through the window was reduced by 3 to 6%. The area of the glass closest to the frame must be described by a k-value higher than the one pertinent to the center of the window. A multiplicative factor of 1.2 for uncoated glass and 1.5 for low-emittance-coated glass has been used (67) to account for the thermal performance of a ~ 6 cm wide band at the frame.

5.2 Effects on the Energy Economy of Buildings: A Plea for Research

This article has discussed many different types of windows, with a focus on promising designs which are one or more steps away from commercialization. It seems self-evident that these novel windows, with superior properties, will have a significant impact on the energy economy of the buildings in which they are used. A realistic assessment of their impact will have to include consideration of seasonal and daily variations of (average) climatic conditions, building design and orientation, influences of inhabitants, physical properties of the actual windows, etc. Research into energy implications of windows has been conducted for many years. However, this research has focused primarly on windows available on the market. Very little is known about the role of the superior windows discussed here. What are, for example, the benefits of smart windows? How important is it to use large-scale antireflection coatings to boost the solar transmittance? These and many other relevant questions can be asked, but few answers are known.

A theoretical study of the effect of some different innovative fenestrations on the energy use in buildings was carried out a few years ago (68). It specifically regarded an office building in a southern US climate. The results of the study certainly are not applicable to other climates and building types in their details, but I expect that the qualitative conclusions are quite relevant. Figure 12 shows the total monthly energy consumption with standard windows and with electrochromic smart windows that are actively controlling the radiative throughput. The improvement caused by the smart windows is striking. Fenestrations with photochromic glass (i.e., passive radiative control), shading in conjunction with standard glass, and low-transmission glass, for most months provided some improvement over standard windows, but they were considerably less effective than the smart windows.

Analogous modelling efforts for other building types, climates, etc., are reasonably straight-forward using available computer codes, and it is my hope that such work will be conducted in the near future. Only then can the potential benefits of smart windows and other innovative fenestrations be quantified.

5.3 Whither Windows?

Current research and development on energy efficient windows is likely to lead to significant changes in fenestration. An assessment of these changes must be a subjective one – based on my personal experiences (and prejudices) – but it is worth attempting.

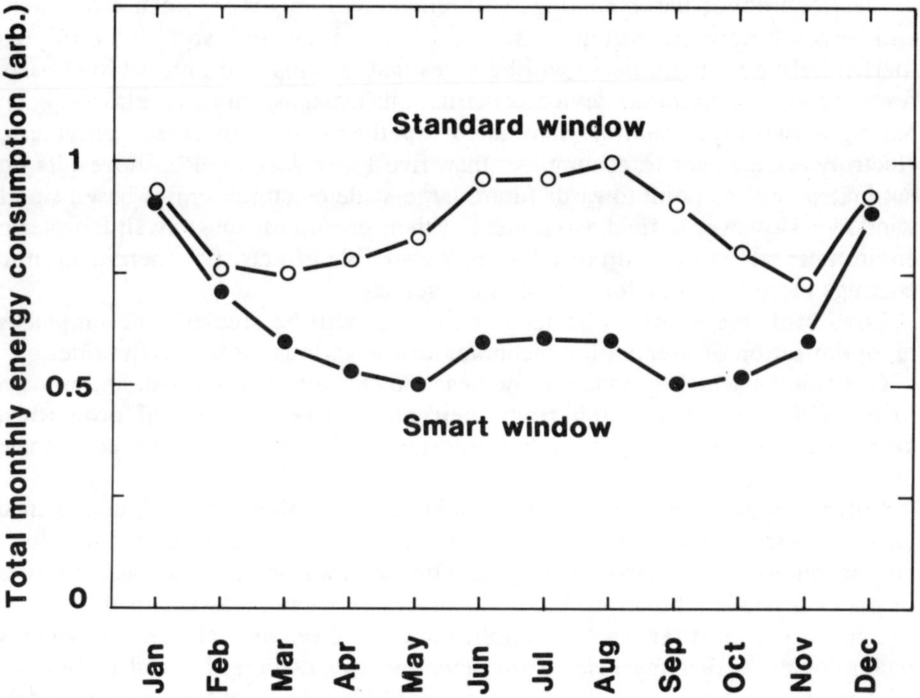

Figure 12. Total monthly energy consumption for all end uses modeled for an office in a southern US climate. Open and filled circles denote results for standard clear-glass windows and for electrochromic smart windows. The energy consumption (shown in arbitrary units) is peaked during winter and summer owing to heating and cooling requirements, respectively. The large difference between the sets of data signifies the potential gain by using new fenestration. It should be noted that the shown data refer to a special building type and a special climate, and should not be applied indiscriminately to other conditions. Replotted from Johnson et al. (68).

Noble-metal based coatings are presently produced with properties approaching the theoretical optimum. Research on technologies for making thinner continuous metal layers than those now available may lead to a marginal improvement in transmittance and drop in cost.

Doped oxide semiconductor coatings, produced by large-area techniques, do not yet have optical properties which quite match the theoretical limits. Refinements in deposition technology may lead to low-cost coatings with < 1% luminous absorptance, no discernible haze, low thermal emittance, and electrical resistivity down to ~ 10^{-4} Ω cm. Novel ZnO-based coatings are of interest since they can block some of the ultraviolet solar radiation which may otherwise cause degradation of plastics and textiles.

Electrochromics-based coatings will remain as subjects for intense research, and several new transition metal oxide coatings and solid electrolytes (particularly polymeric ones) will be investigated. This will give a broad basis for selecting an optimum device construction. Designs with two glasses, each having a two-layer coating, laminated together by a transparent polymeric electrolyte are easier to accomplish than five-layer designs. Recent results on laminated devices point towards future large-scale electrochromics-based smart windows. However, a final assessment of their usefulness must await long-term environmental testing with a focus on irreversible effects. For thermochromic coatings there is a need for more basic research.

Low-refractive-index antireflection coatings will be studied with emphasis on optimization of preparation technologies for various metal oxyfluorides.

Gas fillings which produce a low heat transfer are well known, and remain to be widely used. Judging from their physical and chemical properties, krypton and xenon are the best, but unfortunately scarcity preludes large-scale uses.

Silica aerogel is a very promising thermal insulation material, and is also useful as a spacer, when interposed between window panes. Research into low-cost technologies for large-scale production of transparent silica aerogel tiles or plates is needed, however, before its practicality can be ascertained.

Finally, the most far-reaching implication of the ongoing changes in window technologies is that the *role of the window may change*. Instead of being a passive building component, which commonly lets in or out too much energy, it can become an active component which regulates the inflow of radiant energy in response to dynamic needs while remaining thermally well insulated. Such a development is in keeping with today's tendencies towards "intelligent buildings" equipped with advanced measuring and control technologies. In addition, the energy requirement for smart windows is small enough to allow self-contained designs suitable for retrofit windows.

Acknowledgement

During the writing of this review I have benefitted from correspondence and discussions with many colleagues. Among those I would like to mention C.M. Lampert and S. Selkowitz at Lawrence Berkeley Laboratory, USA, V. Wittwer at Fraunhofer-Institut für Solare Energiesystme, Freiburg, Germany, B. Karlsson, Linköping University, Sweden, J. Karlsson, Window Barrier AB, Gothenburg, Sweden, and C.G. Ribbing, Uppsala University, Sweden.

References

1. Granqvist, C.G. 1987. in *Physics of Non-Conventional Energy Sources and Materials Science for Energy*, edited by G. Furlan, N.A. Mancini, A.A.M. Sayigh and B.O. Seraphin. Singapore: World Scientific, p. 217.
2. Granqvist, C.G. 1987. in *Physics and Technology of Solar Energy*, edited by H.P. Garg et al. (Riedel, Dordrecht, The Netherlands), Vol. 2, p. 191.
3. Lampert, C.M. 1987. in *Physics of Non-Conventional Energy Sources and Material Science for Energy*, edited by G. Furlan, N.A. Mancini, A.A.M. Sayigh and B.O. Seraphin. Singapore: World Scientific, p. 143.
4. Thekaekara, M.P. 1977. in *Solar Energy Engineering*, edited by A.A.M. Sayigh, New York: Academic, p. 37.
5. Kondratyev, J. Ya. 1969. *Radiation in the Atmosphere*, New York: Academic.
6. Eriksson, T.S., Granqvist, C.G. 1982. *Appl. Opt.* 21: 4381.
7. Wyszecki, G., Stiles, W.S. 1982. *Color Science*, 2nd edition. New York: Wiley, p. 256.
8. Rubin, M., Creswick, R., Selkowitz S. 1980. *Proc. Fifth National Passive Solar Conf.* New York: Pergamon, p. 990.
9. Svensson, J.S.E.M., Granqvist, C.G. 1985. *Solar Energy Mater.* 12: 391.
10. Siegel, R., Howell, J.R. 1981. *Thermal Radiative Heat Transfer*, 2nd edition. New York: McGraw-Hill.
11. Kreith, F., Black, W.Z. 1980. *Basic Heat Transfer*. New York: Harper & Row.
12. Rohsenow, W.M., Hartnett, J.P., Ganic', E.N. 1985. *Handbook of Heat Transfer Fundamentals*, 2nd edition. New York: McGraw-Hill.
13. Araujo, R.J. 1980. *Contemp. Phys.* 21: 77.
14. Lampert, C.M., 1984. *Solar Energy Mater.* 11: 1.
15. Babulanam, S.M., Eriksson, T.S., Niklasson, G.A., Granqvist, C.G. 1987. *Solar Energy Mater.* 16: 347.
16. Harding, G.L., Hamberg, I., Granqvist, C.G. 1985. *Solar Energy Mater.* 12: 187.
17. Valkonen, E., Karlsson, B., Ribbing, C.-G. 1984. *Solar Energy* 32: 211.
18. Köstlin, H. 1982. *Festkörperprobleme* 22: 229.
19. Berning, P.M. 1983. *Appl. Opt.* 22: 4127.
20. Jarvinen, P.O. 1978. *J. Energy* 2: 95.
21. Fan, J.C.C., Bachner, F.J. 1976. *Appl. Opt.* 15: 1012.
22. Karlsson, T., Ribbing, C.-G., Roos, A., Valkonen, E. 1988. *Int. J. Energy Res.* 12:23.
23. Jarzebski, Z.M., Marton, J.P. 1976. *J. Electrochem.* Soc. 123: 199c, 299c, 333c.
24. Haacke, G. 1977. *Ann. Rev. Mater. Sci.* 7: 73.
25. Lampert, C.M. 1981. *Solar Energy Mater.* 6: 1.
26. Jarzebiski, Z.M. 1982. *Phys. Stat. Sol.* 71: 13.
27. Manifacier, J.C. 1982. *Thin Solid Films* 90: 297.
28. Chopra, K.L., Major, S., Pandya, K. 1983. *Thin Solid Films* 102: 1.
29. Hamberg, I., Granqvist, C.G., 1986. *J. Appl. Phys.* 60: R123.
30. Jin, Z.-C., Hamberg, I., Granqvist, C.G. 1988. *J. Appl. Phys* 64:5117.
31. Faughnan, B.W., Crandall, R.S. 1980. in *Display Devices,* edited by J.I. Pankove, Topics in Applied Physics, Vol. 40. Berlin: Springer, p. 181.
32. Beni, G., Shay, J.L. 1982. *Adv. Image Pickup and Display* 5: 83.
33. Dautremont-Smith, W.C. 1982. *Displays,* January 3, April 67.
34. Agnihotry, S.A., Saini, K.K., Chandra, S. 1986. *Indian J. Pure Appl. Phys.* 24: 19, 24.
35. Deb, S.K. 1986. *Proc. Soc. Photo-Opt. Instrum. Engr.* 692: 19.

36. Oi, T. 1986. *Ann. Rev. Mater. Sci.* 16: 185.
37. Andersson, A.M., Granqvist, C.G., Stevens, J.R. 1988. *Proc. Soc. Photo-Opt. Instrum. Engr.,* to be published.
38. Viennet, R., Randin, J.-P., Raistrick, I.D. 1982. *J. Electrochem. Soc.* 129: 2451.
39. Randin, J.P. 1981. in *Proc. First European Display Research Conf.,* Eurodisplay '81, Berlin: VDE-Verlag, p. 94.
40. Sone, K., Fukuda, Y. 1987. *Inorganic Thermochromism,* Springer Series on Inorganic Chemistry Concepts, Vol 10. Berlin: Springer.
41. Goodenough, J.B. 1971. *Progr. Solid State Chem.* 5: 145.
42. Cathro, K.J., Constable, D.C., Solaga, T., 1981. *Solar Energy* 27: 491.
43. Jiang, S.-J. , Jin, Z.-C., Granqvist, C.G., 1988. *Appl. Opt.* 27, 2847.
44. Harding, G.L. 1985. *Solar Energy Mater.* 12: 169.
45. Gläser, H.J. 1977. *Glastechn. Ber.* 50: 248.
46. Arasteh, D., Selkowitz, S., Hartmann, J. 1986. Proc. *ASHRAE/DOE/BTECC Conference on Thermal Performance of the Exterior Envelopes of Buildings III,* Clearwater Beach, Florida, USA, 2-5 December 1985; also Lawrence Berkeley Laboratory Report LBL-20348.
47. Platzer, W.J., Wittwer, V. 1987. *Proc. Workshop on Optical Measurement Techniques,* Ispra, Italy, 27–29 October.
48. Pflüger, A., Platzer, W.J., Wittwer, V. 1988. *Proc. ISES Solar World Congress 1987,* to be published.
49. Platzer, W.J., Wittwer, V. 1988. *Transparente Wärmedämmematerialen für den Einsatz im Solarenergiebereich,* to be published.
50. Pflüger, A. 1987. *Solar Energy Mater.* 16: 185.
51. Fricke, J. 1988. J. Non-Cryst. Solids 100: 169; *Sci. Am.* 258(5): 92.
52. Fricke J. editor. 1986. *Aerogels*, Springer Proc. in Physics, Vol. 6. Berlin: Springer.
53. Henning, S., Svensson, L. 1981. *Phys. Scripta* 23: 697.
54. Henning, S. 1986. in *Aerogels,* edited by J. Fricke. Berlin: Springer, p. 38.
55. Tewari, P.H., Hunt, A.J., Lieber, J.G., Lofftus, K. 1986. in *Aerogels,* edited by J. Fricke, Berlin: Springer. p. 142.
56. Platzer, W.J., Wittwer, V. and Mielke, M., 1986. in *Aerogels*, edited by J. Fricke. Berlin: Springer, p. 127.
57. Fricke, J., Caps, R., Büttner, D., Heinemann, U., Hümmer, E., Kadur, A. 1987. *Solar Energy Mater.* 16: 267.
58. Benson, D.K., Tracy, C.E., Jorgensen, G.J. 1984. *Proc. Soc. Photo-Opt. Instrum. Engr.* 502: 146.
59. Benson, D.K., Tracy, C.E., Potter, T., Christensen, C., Soule, D.E. 1988. *Solar Energy Research Institute Report* SERI/SP-255–3318.
60. Chu, N.Y.C. 1986. *Solar Energy Mater.* 14: 215.
61. Novinson, T. 1987. *Proc. Soc. Photo-Opt. Instrum. Engr.* 823: 138.
62. Germer, J. 1984. *Solar Age,* October, p.20.
63. Boy, E., Meinhardt, S., 1988. *in Proc. Second International Workshop on Transparent Insulation Materials in Solar Energy Conversion for Buildings and Other Applications.* Freiburg, Germany, 24–25 March.
64. Anjaneyulu, Y., Yoon, D.W. 1986. *Solar Energy Mater.* 14: 223.
65. van Konynenburg, P., Marsland, S., McCoy, J. 1987. *Proc. Soc. Photo-Opt. Instrum. Engr.* 823: 143.

66. Enermodal Engineering Ltd., Waterloo, Canada. *The Effect of Frame Design on Window Heat Loss*. Phase I. Report prepared for the Renewable Energy Branch, Energy, Mines and Resources Canada, Ottawa, Canada, September 1987.
67. Peterson, Jr., C.O. 1987. *Glass Digest,* p. 70.
68. Johnson, R., Connell, D., Selkowitz, S., Arasteh, D. 1985. *Lawrence Berkeley Laboratory Report LBL-20080*.

Low Electricity Appliances – Options for the Future

Jørgen S. Nørgård

Abstract

A substantial portion of the electricity in industrialized countries goes into running appliances, both in the home and in the service sector. In Denmark it is about 45% of the electricity, if ventilation is included. Here the potentials for saving electricity by improving the efficiency of appliances are analyzed, exemplified by the major domestic appliances. The results of this analysis show that the present level of electricity services – such as cooling, cleaning, cooking and provision of clean air can be maintained while using only 26% of present electricity consumption, if efforts are directed towards development and implementation of efficient technology. The estimated added cost for the advanced efficient technology corresponds to an average price of approximately 2.5 US cents per kWh electricity saved. At a crude oil price of 20 dollars per barrel, the fuel cost alone at an oil-fired power plant amounts to 3.5 cents per produced kWh electricity.

1 Introduction

This chapter deals with electricity consuming domestic appliances as well as similar small appliances used in the service sector. However, electricity used to produce space heat, hot water, light, and space cooling are not considered here. These uses are analysed in other chapters. Typically, the appliances analysed here account for 30 to 50% of an industrialized nation's total electricity consumption. For Denmark, it is found that approximately 45% of electricity is used to run appliances, i.e. – 2500 kWh/year per capita. Out of this around half is used in domestic appliances, when lighting and hot water are left out (1,2,3,4).

Approximately 55% of Denmark's electricity – or 3200 kWh/year per capita – is used by appliances if lighting is included, see Table 1.

Table 1. Denmark's electricity consumption in 1988 for various appliances plus lighting. First column shows how large a percentage of Danish household possesses each appliance. The consumption is expressed in kWh per capita, not per household. Besides the domestic consumption is also shown the consumption for similar uses in other sectors as well as the total. Specific consumption of various types of appliances appears in the figures in this chapter. There are 5.1 million people in Denmark and the average household consists of 2.3 persons. The two first columns are based on 1986 figures plus trends from Danish Electricity Utilities (3,4). The third column is estimates, derived from (1,2,5).

Type of End-use technology	Domestic saturation level %	Domestic sector's consumption kWh/capita	Other sectors' consumption kWh/capita	All sectors' consumption kWh/capita
Refrigerator	74	100	50	150
Freezers	67	165	70	235
Combined Refr./Fr.	35	90	0	90
Clothes Washer	66	125	50	175
Dish Washer	28	50	40	90
Clothes Dryer	19	35	30	65
Cooking	8	260	80	340
Heat Distribution	35	70	200	270
Ventilation	~0	0	500	500
Mischellaneous	...	125	300	425
Lighting	...	260	600	860
All appliance + light	...	1280	1920	3200

2 Methodology and Concepts

This chapter takes an *active scenario* approach to future development in electricity consumption and conservation. This is in contrast to the more common *passive forecasting* which involves estimating the types of appliances that are *likely* to be in use in the future, how energy efficient they are *likely* to be, etc. Many studies are addressing these questions, basically considering electricity demand as an exogenous parameter in energy planning. The active approach assumes the future demand for electricity will be an integrated part of energy policy and energy planning, and it will, to a certain extent, be shaped by them. For appliances electricity demand is basically determined by the three factors:

- future *stock* of appliances,
- future *behaviour* in using appliances, and
- future energy *efficiency* of the appliances.

Obviously, all three factors are decided by humans, and therefore depend on political as well as individual choices. Hence, the future demand is a flexible element in energy planning. Actually, as we shall see, it is a very flexible element.

If society decide that a reduction in electricity demand is desirable for one reason or another, it is essential to know to which degree a reduction is technically and economically feasible. Taxes, tariff structures, standards, information, and other measures (as discussed in other chapters) can *always* achieve the savings. Concrete understanding of what is feasible is a prerequisite, however, for setting quantitative and explicit goals.

All three factors mentioned above – the stock, the behaviour, and the efficiency – are important. Over the long haul, i.e., beyond 25 years, the stock and behaviour will probably be the most decisive factors determing electricity demand. But the technical efficiency of the appliances is important both in the short and long term. Most of this chapter will be devoted to what *can* be achieved through improved technical efficiency, which of course is not necessarily what is going to happen. For this reason, the focus here is *not* on what manufacturers *presently* plan to develop or on present *trends* in the market. Rather it is on what consumers and politicians can reasonably ask for in the years to come.

As indicated above, savings which can be achieved through basic changes in the daily use of the appliances are in general not included in this analyses. In some cases, however, the behaviour and the technology are so closely interconnected, as in the case of cooking, that it is difficult to separate them. But no changes in behaviour which could be construed as a decline in the material standard of living, are considered. In many instances, the proposed technical changes actually leads to an improved comfort and living standards, such as making cooking faster.

2.1 Electricity Services and Efficiencies

In an industrialized society today, electricity is used to provide a wide range of various *electricity services*. Typical examples from the private household are: cooled food storage space, cleaned clothes, entertainment, and warm meals. The level of energy services can usually be derived from the two first factors described above – the *stock* of appliances and the *behaviour* in using them. The satisfaction of human needs through the delivery of a service is often difficult to define quantitatively. Electricity service is no exception. Consequently, the efficiency of the electricity end-use technology, the appliances, is also hard to define and quantify in a precise way.

Despite these definition problems, the end-use efficiency should not be neglected, since it is essential for any future electricity planning. In this chapter, *electricity efficiency* for the appliances will be defined as a ratio between the output in the form of desired electricity services and the associated input of electricity. Units for efficiency can be expressed in terms of "kg of cleaned laundry per kWh", "litre of storage space at +5°C per kWh/year" or "hours of TV-watching per kWh". The reciprocal of electricity efficiency is here termed *electricity intensity*. Throughout the chapter, the saving potentials by annual electricity consumption for a typical domestic service level, for instance for a 200 litre refrigerator, will be presented. In the graphs, dimensionless scale of an *intensity index*, based on average technology in use in 1988 as 1.0 will be provided. An intensity index of, say, 0.75 for refrigerators implies that the type described *as well as* other types and sizes of refrigerators could provide the 1988-level of service with 25% less electricity. (In that case, *efficiency* has improved by 33%).

The technical electricity efficiency as defined above might not be the most appropriate measure of technological progress, when aiming at saving electricity. For instance, one can increase refrigerator efficiency, measured as litres per kWh/year, by choosing a *larger* refrigerator, which actually consumes more electricity per year and which might provide more storage space than a household reasonably needs. This calls for considering how efficient *society* and *lifestyle* are in providing true satisfaction of human needs. Such considerations, however, are not generally taken in this chapter, but they are extremely important for future electricity consumption.

2.2 Electricity Versus Heat

In analysing the potentials for electricity conservation, we are repeatedly faced with questions of *low temperature heat* versus electricity. Even though space heating is dealt with in other chapters, the question comes up here in the context of:

- the loss of free heat gains from electric appliances, when these are made more efficient,
- the options of substituting low temperature heat for electricity in some appliances.

In most countries, the price per unit of heat based on oil, gas or coal, or based on cogeneration of heat and electricity, is lower than that of electricity. In fuel consumption (and hence in many environmental costs), electricity is 2–3 times more costly than heat. It takes three units of fuel energy to provide one unit of electricity, and one unit of electricity can, by means of a heat pump, provide around three units of heat. In the cases coming up in this chapter, *marginal* costs are considered, since it is assumed the heating systems is already there or is installed and run for other reasons. The losses in a heating system are dominated by permanent losses, that is losses independent of the consumption of heat. Marginal efficiency is therefore high. In the case of district heat from a cogenerating power plant, a unit of heat can actually be provided marginally by only half a unit of fuel, or at one-sixth of fuel cost of heat from resistance heating by electricity generated in a conventional thermal powerplant.

Another benefit of using heat instead of electricity is that since heat can more easily be stored than electricity, such a shift can contribute to a better load management.

The above illustrates why the economic and environmental costs of low temperature heat are less than for electricity, a fact that should be taken into account in long-term energy planning. In this chapter, it is conservatively assumed that the marginal production cost of heat is half that of electricity in Denmark – 2 US cents per kWh of heat.

Almost all the electricity used to run appliances ends up as heat sooner or later. This heat can often contribute to heating rooms during cold seasons in a cold climate, such as in parts of Europe, North America and Soviet Union. How much of the electricity that ends up as useful heat also depends on the insulation standards of the buildings, the heating system used, etc.

For Denmark, it has been estimated that on the average around 40% of the electricity used domestically actually saves heat. The remaining 60% either: dissipates during the one third of the year when no heating is required; causes excessive room temperatures as in a kitchen during cooking; dissipates outdoor or in rooms where no heat is wanted; or leaves the house as waste water or air from cleaning appliances. With better insulation standards in the future, the percentage contributing as free heat gain will be lower than the 40%, since the season during which heating is required, will be shorter and excessive indoor temperatures more frequent.

Based on the 40%, the calculation for saving a kWh of electricity is as follows, when marginal cost for heat is assumed half of that for electricity, p:

1.0 kWh electricity saved:	+ 1.0·p
0.4 kWh heat at 0.5*p lost:	– 0.2·p
Actual saving per kWh electricity:	+ 0.8·p

In this case, the economic saving is 20% less than what the electricity price directly indicates. For houses with electric resistance heating, the economic saving would be 40% less when free heat is included in the calculations and for district heating from cogeneration only 5–10% less.

The free heat, however, is not always a blessing. Certainly not in warmer climates where it can add to the cost of electricity for cooling whether this is provided by a simple ventilation system or an air conditioning system. Even in a climate like the Danish, many buildings require cooling by ventilation for several months of the year, and in these cases electricity savings in appliances also results in ventilation savings.

2.3 End-Use Technology

In this analysis of saving potentials, the whole chain of technical components involved in the end-use technology, i.e., the technology which converts electricity into electricity services, are considered. It is essential to look at both the single components and the system as a whole.

The base year for comparison is 1988 and hence the *average* efficiency of *technology in use in 1988* is the reference point. Often, a Danish average which seems to be reasonably representative for industrialized countries will be used.

The *average* efficiency of the *technology sold in 1988* can give some clues about future electricity demand, assuming no change in electricity conservation policy.

The most electricity efficient technology on the market is referred to as *best available technology* in 1988.

Finally, for each type of appliance, the options for further improving efficiency, are extensively discussed, and are referred to as *efficiency advanced technology*. They illustrate the potential for reduced electricity demand if an active R&D policy is pursued.

The efficiency advanced technology should not be confused with a technical or theoretical limit to what can be achieved. The advanced technology presented in this chapter is based on known technological elements and components. No technological break-throughs are assumed in this analysis. There is no doubt that potentials for higher efficiency than those presented here will occur, especially if research into more electricity-efficient appliances is encouraged.

2.4 Cost-Effectiveness

The shift towards more efficient appliances which has been observed since 1973 has generally not indicated a higher retail price for more efficient models, despite the fact that often there are some extra cost associated with the production (3,7,60). This can be explained by general increase in productivity at the factories, which might otherwise have resulted in declining prices. Only in a few cases, when for instance a freezer has been marketed specificly as a low energy model, has a higher price been charged. In such cases, the extra price is usually set to be reasonably balanced with the savings in the consumer's electricity bill for, say, a chest freezer (8). Hence, it makes little sense to base future cost-effectiveness of the savings on such prices. After some years the selling price will more likely reflect a possible *extra production cost*, assuming free competition. This extra cost should be considered the cost the parameter of importance in long-term electricity planning.

Every change in production of appliances will involve a certain initial cost, often quite large. However, changes are made also for other reasons than electricity savings, and not all initial costs should be ascribed to that. More important is the extra cost associated with the running production of each unit related to extra material as well as to extra labor. Despite difficulties and uncertainties, we have estimated such extra cost of producing the more efficient appliances, instead of the average sold in 1988. This extra cost is based on introducing the efficient units in the *natural course of replacement*, not the more expensive accelerated discarding of well functioning appliances. This means a 10–20 years time horizon in electricity conservation strategy and planning. An *accelerated replacement* to save electricity might in some cases be justifiable on economic, environmental or energy grounds, but it is not included here.

Efficiency advanced technologies suggested in this chapter are all estimated to be cost-effective from a socioeconomic point of view, especially if external benefits like environmental protection are taken into account.

The cost-effectiveness calculation made for each type of appliance reflects only the direct dollars and cents cost of saving the electricity. A depreciation of 7% per year and a rate of interest of 6% per year is used.

In the following the potentials for efficiency improvements of the major appliances are described. No complete and detailed technical description is included, but some characteristic examples plus concluding results are presented.

3 Refrigerators and Freezers

Historically, mechanical refrigeration has replaced: food storage in cellars, ice boxes, and more frequent shopping for fresh food. Today, two different storage temperatures dominates. A cabinet with a temperature around +5°C will in this chapter be referred to as a *refrigerator*. A unit with about –18°C will be referred to as a *freezer*. Many combinations of the two exist, and are here termed refrigerator/freezer.

3.1 Efficiency

Standard procedures for testing electricity consumption of refrigerators exist in most countries. The most commonly used European test implies an ambient temperature of +25°C and no door openings. Kitchen temperature will usually be lower, but the 25°C is chosen partly to compensate for no door openings during the test. Electricity consumption figures in the following refer to these test conditions. The efficiency of a refrigerator or a freezer can be expressed in terms of intensity, that is, electricity consumed per year per litre of storage volume.

Compared to other appliances like washing machines and cooking ranges, the electricity consumption of refrigerators and freezers is relatively independent of the daily behaviour of the users. Today, typically less than 20% of electricity consumption by a refrigerator is caused by door openings and insertion of food. However, the choices of temperature in the room, where the unit is placed, is very important to the electricity consumption. For a typical refrigerator, lowering the kitchen temperature by 5°C will result in an electricity savings of approximately 30%. Consequently, keeping a low temperature in the kitchen, for instance during the night, not only saves space heat but also electricity for refrigerators and freezers.

Refrigeration is a typical example of a chain of technology, which through several links converts the electricity into electricity service, in this case cooled storage space.

3.2 Thermal Insulation of Cabinets

The main task of a refrigerator or freezer is to maintain a temperature lower than the ambient temperature. Thermal insulation is therefore an essential element of the system.

Today's dominant material for thermal insulation in these appliances is polyurethane foam expanded with the CFC-gas, CFC11. Heat conductivity of this foam is around 35% lower than that of the mineral wool it replaced. In Europe, the switch to foam occurred in 1960s and was used by manufacturers as an opportunity to reduce wall thicknesses in refrigerators and freezers to as low as 25 and 40 mm respectively. In USA, however, the foam insulation came into widespread use only after 1973 and was used as a means of reducing electricity consumption rather than wall thickness. Consequently, American refrigerators today typically have a higher insulation standard than the European models. In this chapter, it is assumed that changes in insulation thickness do not affect the inside volume of the refrigerators and freezers and the volume remains unchanged. On this basis, an increased thickness will always result in a reduced electricity consumption per litre of storage volume (9). The changes in outside volume, and hence in extra space taken up in the house, is accounted for in the cost estimates.

The CFC11 used for foaming is, however, one of the chlorofluoro-carbons to be gradually banned in the future because of its environmental impact on the ozone layer and the "greenhouse effect". Other insulation materials will have to be used, either old or new types. Plastic foam with its excellent structural advantages, can be expanded by other gases. Some, like CO_2., will result in higher thermal conductivity, which can then be compensated by thicker walls. New artificial gases, for instance CFC123, are under development for foam insulation, but they are still in the testing stage, and there is always the risk that they will contain unforseen long-term environmental and health hazards. Heat conductivity for foam with CFC123, and hence electricity consumption for a refrigerator, is estimated to be 10% higher than with CFC11, if thickness is not changed (10).

Vacuum insulation is another option. It can be either "soft" vacuum with a powder to reduce heat radiation and to act as spacers to keep the walls apart, or it can be "hard" vacuum with small solid spacers between the two walls (11). Layers of reflecting curtains between the walls can in hard vacuum panels reduce the radiation of heat. One technical solution could consist in embedding some standard vacuum panels in plastic foam in the walls of refrigerators, thereby maintaining the structural advantages of the foam. Refrigerators with soft vacuum panels applied in this way are being developed in Japan (12). Despite the heat bridges at edges and through spacers, it seems technologically possible to reduce the heat transmission in refrigerators an freezers substantially. The economy and durability of these options in practice, however, have yet to be seen.

No matter how well insulated the walls are, some heat is still admitted into the cabinet by door openings and insertion of food. The test procedures with no door openings therefore do not represent very well the actual use of superinsulated cabinets. Here a procedure with door openings like the Japanese might be more appropriate.

To maintain constant temperature inside the cabinet, the various heat loads have to be counteracted by a refrigeration system, which removes the heat.

3.3 Compressors and Refrigerants

Absorption systems have been in widespread use and are still marketed to run on gas or kerosene, but total energy efficiency is low and seem to remain low compared to compressor systems. Absorption systems might, however, be a future option with solar *heat* or other renewable *heat* as the energy source.

Conversion of *electricity* directly into temperature differences can be achieved by the *Peltier-element*. This seems like an attractive solution from a durability point of view, but efficiency is still too low for general use. The focus here will be on the *compressor* systems which dominate the market today and which are superior in efficiency.

Hermetically-sealed compressor units are used in essentially all domestic refrigerators and freezers, and in a large part of the units used in the commercial sector as well. *Reciprocating compressor* units have been improved in efficiency over recent years by introducing run capacitors, semi-direct intake, etc. Both these compressors and the motors could still be designed to be more efficient. *Rotary compressors* are another option now being marketed, claiming a 15% higher efficiency than the best reciprocal type (13).

The choice of refrigerant also affects the system efficiency. Presently the CFC-gasses, CFC12 and CFC22, dominates the domestic units. But substitutes will have to be found due to environmental problems. CFC12 in particular posesses ozone depleting properties similar to those of CFC11 used for foam.

Which refrigerant will be used in the future is uncertain. Efficiency could benefit or suffer from new refrigerants, but for this analysis no change in efficiency due to new refrigerants is anticipated.

3.4 Heat Exchangers

Besides the compressor, the refrigeration system also includes two heat exchangers, i.e., the *evaporator* which cools the cabinet and the *condenser* which dissipates the heat to the ambient air. The efficiency of the whole refrigeration system is inversely proportional to the temperature difference between these two heat exchangers. The COP, coefficient of performance, is defined as the ratio between kWh of heat removed from the cabinet and the kWh of electricity used. For typical domestic refrigerators, the COP is between 1.0 and 1.5.

In order to reduce the temperature differences between the heat exchangers, and hence increase COP, the following steps can be taken: 1) improve ventilation of the heat exchangers, 2) enlarge and redesign surface of them to facilitate the rate of heat transfer, and 3) increase their thermal capacity to extend the time available for heat transfer (14). The latter seems to be an attractive and simple alternative to variable speed compressors, because it allows a relatively large compressor to run efficiently at low heat load while still maintaining the ability to handle peak heat loads. The mechanical ventilation of heat exchangers often consumes a relatively large amount of electricity. But even without such forced ventilation, a COP around 2.0 has been achieved in some systems – including those with very low refrigeration capacity, which normally have a low COP (15).

3.5 Separate Refrigerator and Freezer

In Northen Europe, it is rather common to have refrigerator (+5°C) and freezer (–18°C) as two separate units. This makes good sense from an electricity conservation point of view. It is thermodynamically inefficient to create the +5°C temperature through a cycle reaching minus 20°C or even lower.

There are other reasons why a combined refrigerator/freezer system is problematic energywise. If one evaporator serves both freezer and refrigerator compartment, as in typical American models, the evaporator will be subjected to moisture from both. Since the refrigerator door is opened more frequently than the door in the freezer, the frost built-up will typically be more than ten times higher than for the freezer alone (16,17). A European chest freezer requires defrosting (manually) only every one or two years, whereas a combined American model typically is set to defrost automatically more than once a day! Defrosting of a refrigerator evaporator requires no energy, since the ice melts off between compressor runs.

Another advantage of separating refrigerator and freezer is a better temperature control in both, which also results in electricity savings. Finally, the units can be separately switched off, repaired and replaced.

The disadvantage of separate refrigerator and freezer units is mainly the extra cost for another compressor. A solution could therefore be to run two refrigeration circuits alternately with one compressor.

3.6 Refrigerators

The various options for electricity consumption in refrigerators are shown in Figure 1, represented by a 200 litre unit with no freezer compartment. The *intensity indices* found for this unit are estimated to be fairly representative of most brands and sizes available.

A prototype low energy refrigerator (LER) developed in 1984 (14) reached the market late 1988 as LER200 (18). This 200 litre refrigerator includes many of the electricity savings features described earlier, including increased thermal capacities of the two heat exchangers. They are invisible and integrated into the inner and outer case. The cabinet appears with a smooth back wall outside as well as inside. This unit has an electricity consumption of about 90 kWh/year and is shown in Figure 1 as the best available.

The advanced technology version with 50 kWh/year is based on the LER200 model, but it is equipped with a more efficient motor and rotary compressor and without integrating the condenser. An *alternative* way to reach a similar

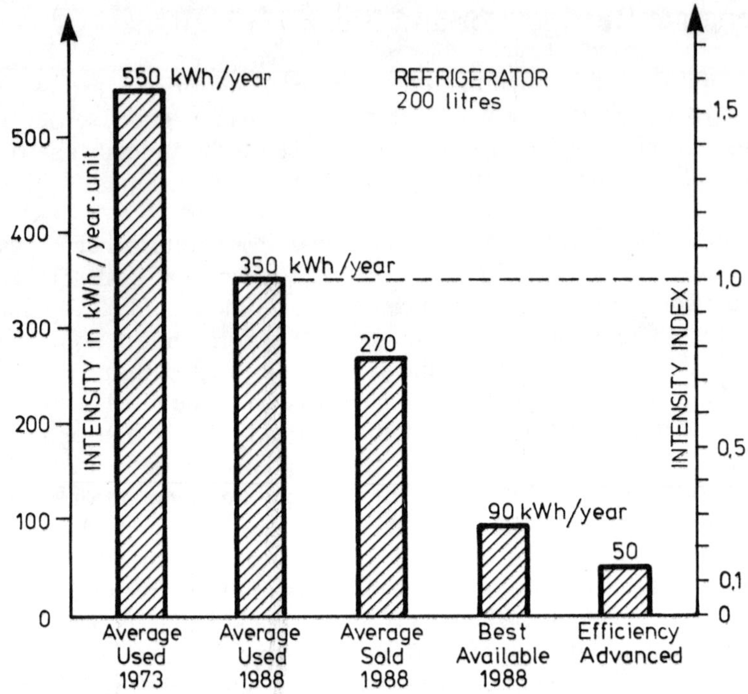

Figure 1. Electricity consumption at European standard test condition for various versions of a 200 litre refrigerator with no freezer compartment. The intensity index on the scale to the right illustrates the relative improvements, and it is approximately transferable to other refrigerators as well as to units with a small freezer compartment and to commercial units in shops, etc.

low level is by having the condenser, and maybe part of the cabinet, exposed to outdoor temperatures, which in regions such as Scandinavia is below 5°C most of the year. Combining the two measures could bring consumption well below the 50 kWh/year shown in Figure 1. Vacuum insulation, however, is not assumed in the advanced technology example shown here.

3.6.1 Cost-effectiveness for refrigerators

Generally, a higher efficiency refrigerator has not meant a higher sales price. The extra cost of producing the LER200 compared to a similar average model sold is estimated to be about $25 per unit, plus another 5$ for marginal cost of the extra 0.1m^3. space it takes up in the house. Hence, the LER200's additional cost totals $30. Compared to the average appliances sold, the electricity saving is 180 kWh/year. Based on the assumption in the section on methodology and concepts, i.e. a 7% depreciation per year and a 6% interest per year, the price of the saved electricity is 2.2 cents/kWh.

The efficiency advanced level of 50 kWh/year can be reached in two ways, as described above. A better motor and rotary compressor will cost an estimated $ 15 extra. The solution with outdoor cooling can best be introduced in connection with construction or renovation of houses. It requires one or two small vent holes and costs about $ 10. The price for the 40 kWh electricity saved per year will in the two cases be 4.8 and 3.3 cents/kWh, respectively. The price of the saved electricity for the efficiency advanced level is assumed to be 4.5 cents/kWh.

3.7 Freezers

Manufacturers were for long very reluctant to increase insulation thickness in refrigerators. For freezers, especially chest freezers, this option for saving electricity was introduced in the early 1980s (21). Today, the most efficient freezer with a typical size on the market is a 250 litre chest freezer with rotary compressor which apparently consumes 180 kWh/year at standard test, (20,21) as shown in Figure 2.

A 50% further increase in insulation thickness is assumed acceptable for the efficiency advanced model shown. It is a chest freezer with 250 litre volume, equipped with improved heat exchangers as well as rotary compressors. This analysis is based on experiences from the development of LER200 with high thermal masses in the heat exchangers and increased insulation thickness.

As mentioned already, defrosting is required very rarely in chest freezers, and manual defrost is acceptable. If, for upright freezers, automatic defrosting is wanted, experiments with a prototype has shown that it needs as little as 2–10 kWh extra per year if the defrost cycle is properly controlled according to needs (16). This is much less than the 200–300 kWh/year difference between models with automatic and manual defrosting on the market today (19).

3.7.1 Cost-effectiveness for freezers

The added costs associated with producing the best on the market, plus the extra space it takes up in the house, is estimated to be about $ 30, mainly for the rotary compressor, extra foam and steel. The advanced model mentioned above will add another estimated $20 to that for heavier heat exchangers. The price for the saved electricity is 1.8 cents/kWh for the best on the market and around 6 cents/kWh for saving the additional 80 kWh/year with the advanced technology.

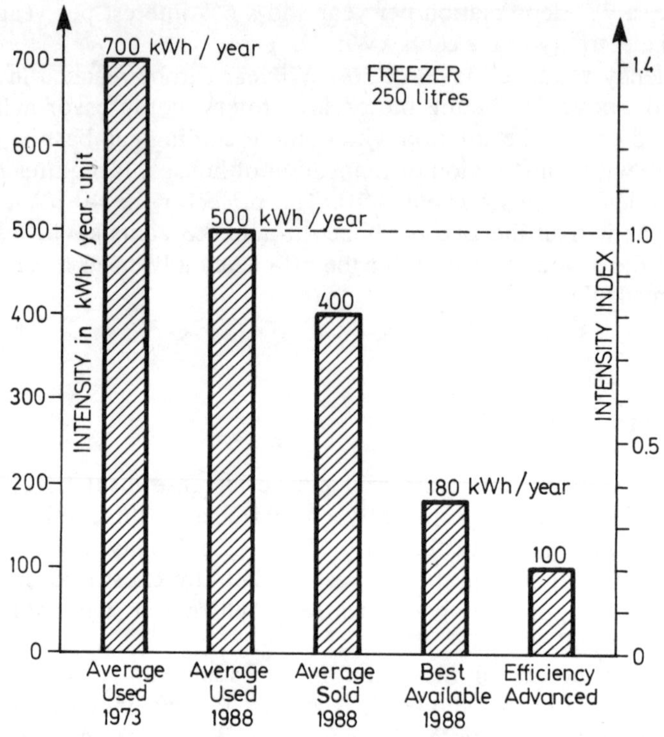

Figure 2. Electricity consumption at European standard test for various versions of a chest freezer around 250 litre on the Danish market. The intensity indices from this are estimated to reflect the relative potentials for improvements of various sizes and brands as well as the various types used in the commercial and service sector.

3.8 Combined Refrigerator/Freezers

Units with both a large refrigerator compartment and a large freezer compartment are common in the USA. As described earlier, there are some disadvantages of running such combined units on one refrigerator system as is usually done. A typical American size could be 500 litre with a 350 litre refrigerator compartment and 150 litre as an automaticly defrosted freezer. *Average stock* of such units in use today would, with the European standard test, consume about 1000 kWh/year, see Figure 3. A prototype which consumes about 400 kWh/year at European test condition was built in 1986 (16,17), but further improvements can be implemented (12). An efficiency advanced model could bring consumption down to *200 kWh*, for instance by simply adding an advanced refrigerator and freezer on top of each other.

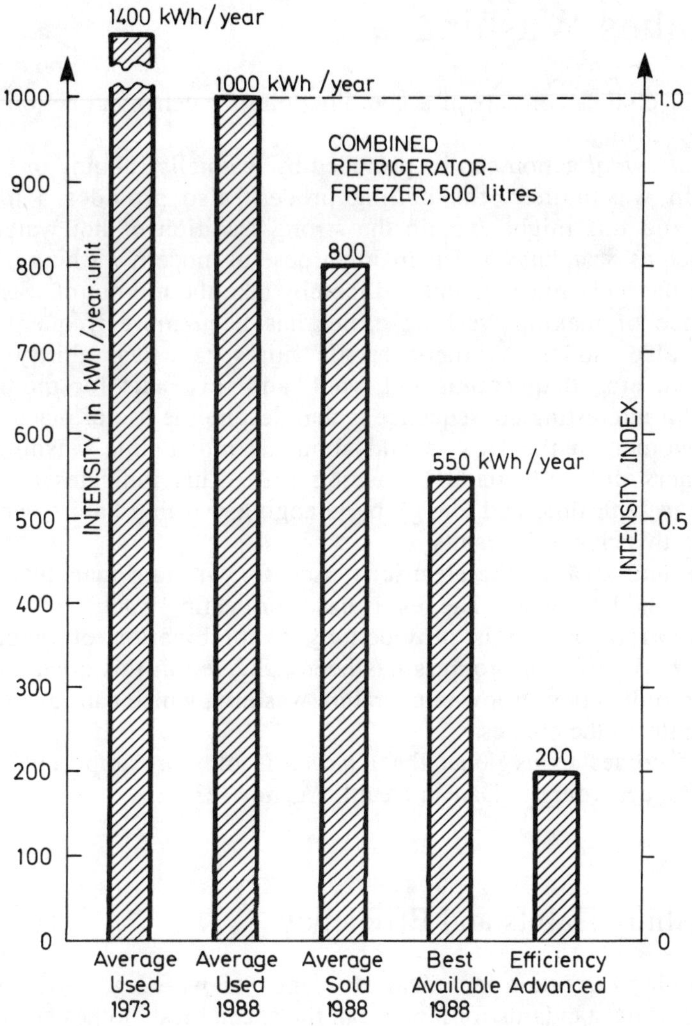

Figure 3. Electricity consumption for various versions of a 500 litre combined refrigerator/freezer, in which the freezer takes up about one-third of the volume and has automatic defrosting. Numbers are mainly based on US data (12,19) but transferred to European stan-dard test conditions.

Cost-effectiveness of the best available and the advanced combined refrigerator/freezer are 2.0 cents/kWh and 5.3 cents/kWh respectively, based on the estimates for separate units.

4 Clothes Washing

Cleaning of clothes consists of a combination of mechanical action, chemical action, and rinsing.

The *mechanical* action can be provided by manually rubbing or beating the laundry. In washboilers, the boiling process also provides a mechanical agitation, and this might explain the strong tradition of hot water wash in regions such as Scandinavia. The main purpose of modern washing machines is to provide the mechanical action and thereby take the toil out of washing. One consequence of making washing easier has been more frequent washing. However, also modern synthetic textile fibres have less ability to absorb perspiration, etc., than cotton and wool, and this calls for more frequent washing. An interesting consequence is that despite the use of machines, home working women in the USA spend about as much time washing as their grandmothers did with manual washing (22). This indicates a substantial flexibility in both time and energy by changing washing habits, which is not included in this chapter's results.

Only around 20% of the electricity used for washing machines goes into substituting the hard work. The rest is used for heating water.

The *chemical* action is provided by a combination of detergent and temperature. Significant progress has been achieved in this area, and more is on its way in direction of low temperature washing, which can save energy and also be gentle on the clothes.

When it comes to hygiene, the *rinsing* is of more importance than the temperature, according to Danish Health Agency (23).

4.1 Washing Habits and Efficiency

It is difficult to define a technical efficiency of washing, partly because of problems setting standards for how clean the special test clothes should be. It is assumed here that the output wanted is a laundry with today's typical cleanness, especially from a hygienic point of view. Efficiency improvements are discussed as reduction in energy and particularly electricity consumption to provide this output. A typical washing pattern for a Danish household is used to illustrate annual consumptions and savings, as shown in Figure 4.

Despite strong traditions, washing habits have changed towards lower temperatures over the last 15 years, partly to save energy and partly to protect the new textiles. Also time is saved for heating the water. In Denmark, the laundries are today roughly equally distributed between the three major washing temperatures: 90°C (hot wash), 60°C (warm wash) and 30°C (fine

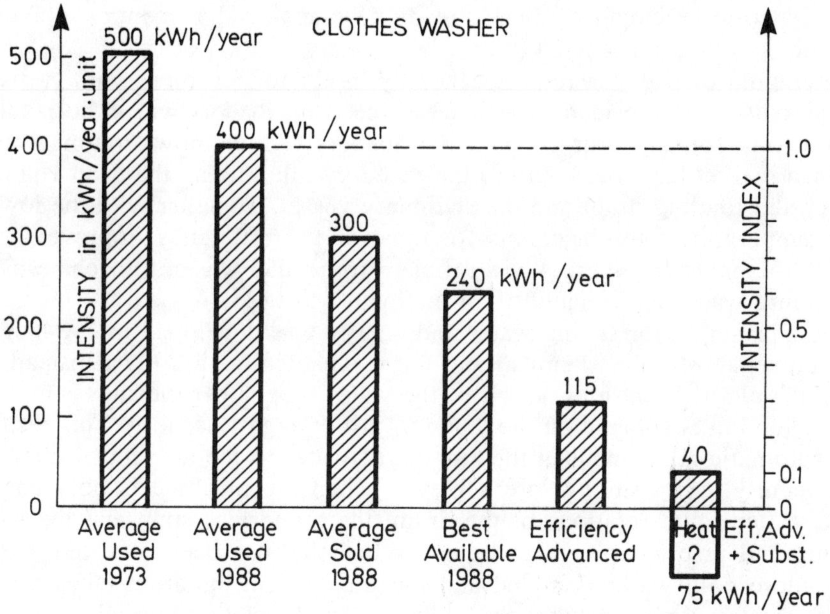

Figure 4. Annual electricity consumption for various washing principles in 4 kg washing machines for the 200 washings per year typical in Denmark. The differences are due to both direct technical improvements and changes in washing pattern promoted by new detergents and new fabrics.

wash) (24). For Germany, slightly more wash is done at 60°C (7). Ambient water temperatures (8–15°C) is hardly used for washing in Scandinavia, but they are quite widespread in Japan and not uncommon in USA and Canada, often for all laundry in a household, but usually combined with stronger and bleaching detergents.

The use of hot water wash is hardly necessary for any domestic laundry today because of: better detergents, less dirty work, and more frequent washing.

4.2 Recent Improvements

The 80% or so of the electricity consumption which is used for heating water can in general be reduced by:

- lowering the temperature for washing,
- using less water in the main wash, and
- heating water by sources other than electricity.

Some reduction in temperature has already been achieved as mentioned above and more options are discussed below.

The amount of water needed is generally larger for machines with vertical rotation axes – the top loaders – than for the front loaders with a horizontal rotation axis for the drum. The front loaders, which now dominate the European market have been further improved by minimizing the dead volume between the rotating drum and the stationary vessel. Together with the lower water temperatures, this accounts for most of the efficiency improvements since 1973 shown in Figure 4. An efficient front load washing machine uses 1.5–2.5 litre water per kg laundry for the main wash (25,26).

A new principle for saving water and energy was developed in 1984. It is based on separating the chemical and the mechanical action (27). Instead of having the clothes splashing in water, the water – or rather the suds – for the chemical action is sprayed on the laundry, just enough to soak it. This results in an automatic adjustment in the amount of water to the amount of clothes. Consequently, extra savings are achieved when the machine is only partly loaded. After some rotations of the drum, the laundry is spun and the suds reheated and resprayed on the clothes. When this has been repeated a few times, more cold water is added and the mechanical agitation takes place. Finally, the laundry is rinsed thoroughly with cold water as normal. This new system saves water, energy and detergent without lowering the temperature. A model based on it was marketed recently (28), but in Danish standard tests this version did not seem to provide any energy savings compared to improved versions of conventional European machines (26). However, these standard tests do not reflect the saving achieved during washing with part load only.

Also other models, including some top loaders, can adjust the level of water to the amount of laundry and thus save energy when the machines are not fully loaded.

The option of using heat from other sources than electricity is discussed below.

4.3 More Electricity Saving Options

When protein dessolving enzymes were introduced in some detergents in Europe in 1960s, the mechanical part of the washing process was made almost unnecessary for most laundry. Soaking the clothes in such detergents and thoroughly rinsing afterwards is often sufficient. Lukewarm water is usually used to speed up the enzyme action, but hot water ruins the enzymes.

Further improvements in the detergents seems likely. A new enzyme, which dissolves fat in cold water, has been developed recently (29). This could open the way for complete cold washing.

Another way to avoid heating the water and maybe also do without detergent could be to apply ultrasonic vibrations to the laundry. This principle is used for special cleaning purposes in industry but has not yet succesfully been marketed for clothes washing. Recently, however, experiment ultrasonic washing machines have been developed in Japan (63).

One model for an efficiency advanced washing process, see Figure 4, which could be employed with presently available detergents and with only minor program changes in the machines, includes the following: The laundry is soaked in cold water with an enzyme detergent for a couple of hours with occasional rotation of the drum. Afterwards, the laundry is agitated as normally but without heating the suds. Finally, a thorough rinsing is provided. Further tests of the quality of the washing are needed to see whether this is acceptable.

If warm water is desired or required, it can be supplied from the hot water system in the house, normally much cheaper than the electrically heated water as discussed in the introduction. It is common pratice in the UK and USA to have washing machines connected to both hot and cold water, thus reducing electricity consumption to around 70 kWh/year (30).

Electricity is still needed to power the motor and the control system. As described in the chapter by Baldwin in this volume, there are several options for running motors more efficiently than presently. A 30% saving for motor and pump is anticipated in the numbers shown for advanced technology in Figure 4.

The long process described above for advanced technology should be suitable for domestic washing machines. It could fit well with efforts towards better electricity load management by using most of the electricity at night, setting the program to start for instance at 5 a.m. and finishing in the morning.

For commercial laundries and laundromates, the long process is not acceptable, but here the use of heat recovery and non-electric heat sources is an obvious option, one which is also partly used today.

4.4 Cost Effectiveness for Washing

There are no clear indications of higher sales prices for the most efficient washing machines on the market today (25,26). The 60 kWh/year saving obtained by using the best on the market instead of the average is, therefore, considered free of cost.

A washing machine for the long washing process need not be more costly to produce. Perhaps the opposite. Enzyme detergents could make a simple, robust washing machine with no electric heating system an attractive option. It could be cheaper to produce, despite more efficient motors, etc. Even with the assumption that none of the options for using cold water materialize, but with intake of warm water, the 200 kWh of electricity can be saved at a price of 4.5 cents/kWh, estimating the initial extra cost of intake of hot and cold water to be $ 40.

5 Dish Washing

For dish washing machines, many features are similar to those of clothes washing machines. Also for dish washing, approximately 80% of the electricity consumed by the machine is used to heat water. Hence, many measures to save electricity are similar to those suggested for clothes washing machines.

5.1 Behavioral Pattern

A dish washer is a less widespread appliance than the clothes washing machine. About 28% of Danish households is in possession of one. However, it was found that many used it only occasionally. This can be explained by the general family split-up when children grow up and only two persons, or even one, is left and it is not really worth using the machine. Also, most people would probably agree that to wash dishes manually is less drudgery than washing clothes by hand. For those possessing a dish washer, the average use in Denmark is about 4 times per week. This number will be used in the comparison on Figure 5.

Saving potentials by changed use of a dish washer can be substantial. A small household can fill up the machine over some days or even use it only at special occasions. As with the washing machine, the operation of the dish washer can contribute to a better electricity load management by turning on the machine during load valleys.

5.1.1 Manual versus machine dishwashing

It has been claimed that a dish washing machine could save energy compared to doing the dishes manually. This is mainly an indication of how wasteful manual dish washing can be, especially when done under a running hot water tap. In that case, 20–40 litres of hot water might be consumed, corresponding to 1–2 kWh, if heated by electricity. It is, however, also possible for an average size household to clean the dishes from one day with 3 litre of hot water in a tub

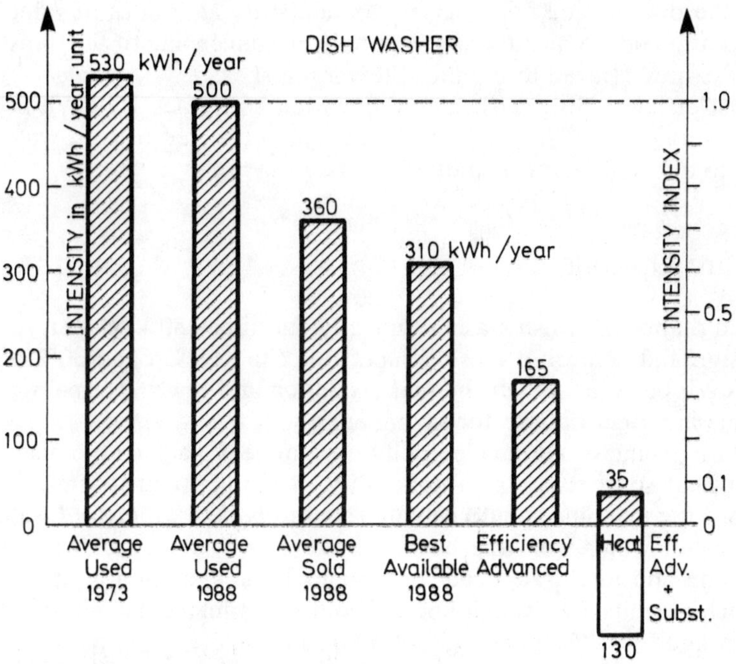

Figure 5. Annual electricity consumption for dish washing machines of various design, assuming 4 uses per week. The column to the right represents the efficiency advanced technology combined with substituting electricity used to heat water with other heat sources.

after having rinsed the dishes under cold water. This corresponds to 0.2 kWh electricity if the water is heated electrically or around one-tenth of what an average machine uses. Energywise, manual dishwashing done with care, is superior to machine washing, even if the indirect energy used to produce the machine is ignored. Dish washing by machine is a more hygienic way of cleaning and drying the dishes than manual dish washing, but this is hardly of importance in private homes.

5.2 Past Improvements

Like other appliances dish washing machines have become more energy efficient since 1973, see Figure 5. Reducing the water used has been one of the paths pursued as in the case of clothes washing machines.

Drying the dishes need not consume extra energy as it does in some of the least efficient models. Opening of a vent hole or condensing of the moisture by cold water is now utilized to dry the still warm dishes.

The most efficient dish washer on the market consumes 1.5 kWh per load, while the average uses around 1.75 kWh when connected to cold water only, according to a recent German study (7).

5.3 Future Options

Motors and pumps in a dish washer typically consumes 50 kWh per year (32). Better design and control of these components can save around 30% of this – about 15 kWh per year. Better thermal insulation and lower thermal masses in the machine are other options for saving energy.

Substituting some or all the electricity used to heat water in the dish washer with other heat sources is also a possibility. It is important to notice that for that purpose the machine should be connected to *both hot and cold* water taps. Only using hot water from the heating system will waste hot water for the initial rinsing and might also cause problems by coagulating albumin on the dishes. Such machines with both hot and cold water intake have been available some years ago (65). The hot water intake will normally be no more than 55°C warm. It could be electric heated to the 70°C typically used, but 55°C should be sufficient (48). (Manual dish washing is usually done with temperatures below 55°C).

New dish washing *detergents*, as described for clothes washing, can most likely further reduce the need for high temperatures in dish washers (61). *Ultrasonic* vibrations might be another option for washing dishes in temperate or cold water, since this is already being applied for cleaning utensiles in chemical laboratories (63).

To conclude, a future efficiency advanced dish washer could clean a load full of dishes with only 0.17 kWh electricity plus hot water corresponding to around 0.6 kWh heat. This option does not anticipate low temperature detergent or ultrasonic, which might eliminate or reduce the need for heat, see Figure 5.

5.4 Cost-effectiveness for Dish Washers

No extra cost is found to be associated with the relatively small saving of 50 kWh/year obtained by choosing the best available dish washer. Thermal insulation, better motors, etc., for the advanced model are estimated to cost $ 25 extra in production and save 145 kWh/year shown. The price of the saved electricity is 2.2 cents/kWh saved.

To reach the advanced technology level combined with heat substitution, however, would require an extra magnetic valve for the hot water connection plus installation cost, all estimated to cost $ 40. At 2 cents/kWh of hot water, the price of electricity saved is 3.5 cents/kWh.

6 Clothes Drying

The drying of clothes can occur at any temperature even below freezing-point. However, higher temperatures speed up the evaporation. About 80% of the electricity consumed by tumble clothes dryers – the common machine for the purpose – is used to heat the air. Typically, around 30% of the households in industrialized countries possess a clothes dryer (33). The rest either use commercial laundries or, more commonly, they dry the laundry on a clothes line.

Drying on a clothes line requires access to a suited place, such as a ventilated basement or loft, or an outdoor space. Hence, the design of housing can shape the need for mechanical clothes drying. Even for people who have a clothes dryer, such options are attractive and can limit the use of clothes dryer to rainy or overcast days.

Combined washing machines and dryers are now being marketed (26), but they offer no special advantages from an energy point of view.

6.1 Clothes Dryer Efficiency

Efficiency of drying in general can be measured in dimensionless terms as the ratio between the energy required to evaporate the moisture and the energy actually used. This is not a very precise and informative expression of the efficiency of dryers, since it is well known that no temperatures above ambient is required to dry the clothes. Another more practical measure of efficiency of a clothes dryer is the electricity efficiency, expressed as kWh electricity used to dry one kg laundry (dry weight) with 70% moisture. This is a reasonable measure for comparing machines, and the basis for the numbers in Figure 6. The 70% is the approximate moisture content in clothes after a normal spinning at 800 RPM in a washing machine. Higher speed spinning is used in commercial laundries and now also 1300 RPM is available in several washing machines. This saves on the drying process. With 3000 RPM, the water content, and hence the energy for drying the clothes, is cut in half compared to 800 RPM (34). This saving in electricity for drying is greater than the extra electricity used for the mechanical removal of the moisture by fast spinning.

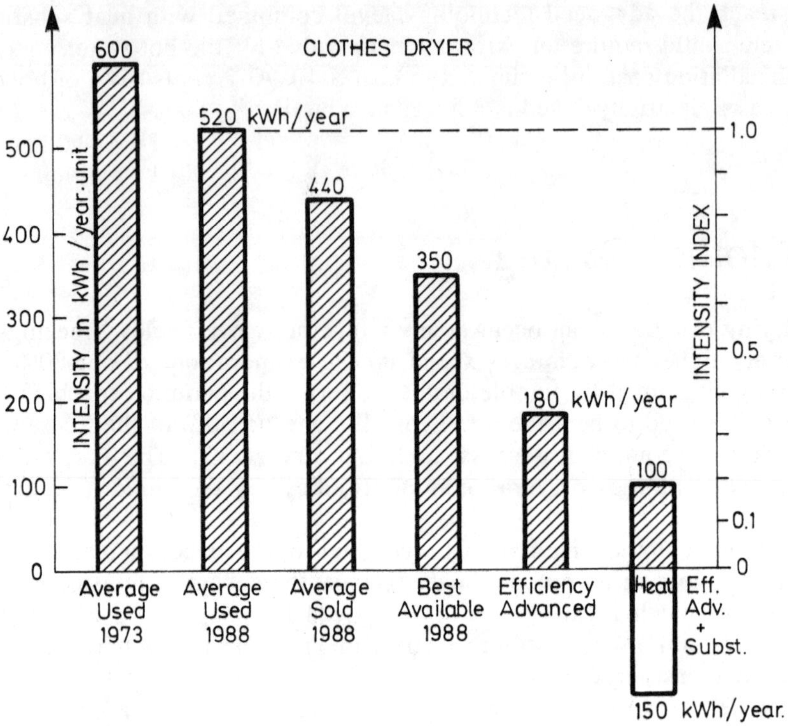

Figure 6. Annual electricity and heat consumption required for various clothes drying options, based on an average of 3.5 kg laundry dried 2.5 times a week.

6.2 Improvements in Drying Clothes

The average tumble clothes dryer in use in 1988 consumes around 4.0 kWh electricity per use, or 520 kWh per year as shown in Figure 6, if used 2.5 times per week with 3.5 kg clothes per use. Some improvements have been introduced over the years, and the best on the market today consumes 30% less electricity, mainly due to a better control system (7) and to larger volume of the drum (32).

Condensing clothes dryers use more electricity than the traditional types. However, since the air used to condense the moisture is taken from the room and vented back dry into the room again, the heat can contribute to heating the home in the cold season, rather than adding to the heat demand as in the case when air is vented to outside. Condensing clothes dryers typically cost $ 150 more than standard dryers, but saves installation cost compared to the use of outdoor air venting (34,35,36). It is assumed here that total cost of producing and installing the condensing type is similar to that of the standard type.

Heat pumps are used for several commercial and industrial drying processes. A prototype domestic condensing clothes dryer with heat pump was built already in 1972 (36). So far, however, a heat pump dryer has not been successfully marketed. In the early 1980s, heat pump clothes dryers were produced and sold for laundries (45). Electricity savings are around 60% compared to the traditional types. An efficiency advanced dryer based on this system would consume around 180 kWh per year, as shown in Figure 6.

Other options for saving electricity are available. *Heat recovery* by preheating the inflowing air with the outgoing is used in commercial laundries. The low utilization rate of domestic clothes dryers makes this investment less cost-effective. A system with a large heat exchanger could (as the heat pump) save around 60% and reach down to the 180 kWh/year shown in Figure 6.

Clothes drying by *microwaves* is another technology being developed in the USA (12), and may be marketed in 1989 (47). It is claimed to provide "substantial" energy savings.

As with clothes washers and dish washers, the substitution of electricity used for heating with cheaper energy forms seems an obvious option. Gas-heated clothes dryers are used in commercial laundries and domestically in some countries. A more energy efficient and safe solution could be to take heat for the dryer from the central heating system of the house, a principle which is used in some commercial laundries (46).

6.3 Cost-effectiveness for Clothes Dryers

No extra cost is assumed associated with making the best available clothes dryer.

Out of the options described above for the efficiency advanced dryer, the heat recovery or the heat pump condensing system is probably the most cost-effective. The extra production cost for heat pump system is estimated to be $50 as compared to the normal condensing type. With a 7% depreciation and 6% discount rate, the price of saved electricity would be 4.5 cents/kWh. However, as shown, drying clothes could be achieved in tumble dryers with only 20% of the electricity used today plus heat amounting to around 150 kWh/year. This would require a small water-to-air heat exchanger but save the electric heating system, so production cost is estimated to be only around $10 higher. However, the plumbing cost could be considerable in many cases. If done in connection with new construction or renovation of the heating system, the added installation cost is estimated to be $50, making the total additional cost $60. With a marginal price of the extra heat of 2 cents/kWh, the 200 kWh electricity saved per year cost 6 cents/kWh.

7 Cooking

In this section on cooking, only the part which consists of *heating* the food or water for tea, etc, is considered – not the cutting, mixing, etc.

Cooking is closely associated with cultural traditions, which results in significant differences in energy consumption for that purpose. If *electric* cooking is used, the variation for an average household ranges from 500 to 1200 kWh per year (33).

Today, a Danish household with electric cooking typically uses 700 kWh per year for that purpose, but only around 15% of the energy actually ends up in the food where it is wanted (37). During the process, the rest heats the hot plates, pans, cooking media like water or oil, and the surroundings. In analysing this pattern, three basic principles for saving energy in cooking appear, namely:

- better thermal insulation to the surroundings,
- lower thermal masses of cooking equipment, and
- improved control of energy input.

These principles apply to any cooking, whether using solid fuel, kerosene, gas or electricity as energy input. This section focusses on electric cooking, but with some consideration of the alternatives.

Cooking is associated with many different and strong traditions around the world. This makes international comparisons and technology transfer difficult. It is not only a matter of getting the food heated to a certain level, but equally important is the way it is heated. For saving energy in cultures with tradition for long-term cooking, thermal insulation is important, while short-term cooking could mainly gain in efficiency by lowering the thermal masses. The trends in the industrialized countries seem to have been towards short-term cooking.

As an illustration of what can be achieved in efficiency improvements a typical Danish pattern of cooking is used, see Figure 7.

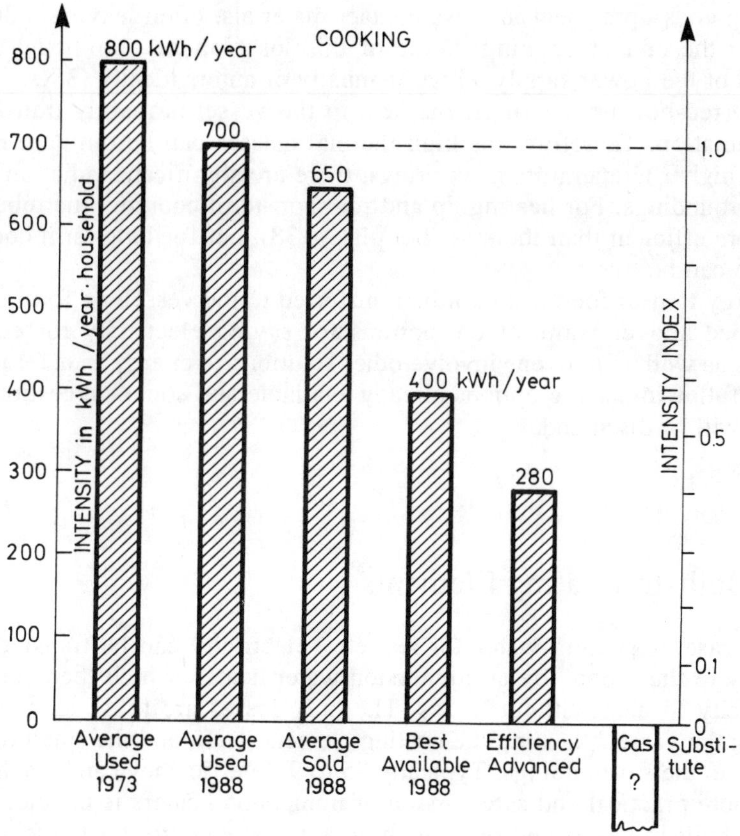

Figure 7. Annual electricity consumption for electric cooking with various equipment, when Danish cooking habits are applied.

7.1 Development in Electric Cooking

A typical electric cooking range is in its design reminiscent of gas ranges and the old coal or wood-fired kitchen ranges, where pans and kettles were put on and the heat from the range transferred to the vessels. To achieve a good heat transfer in an electric cooking range, two principles have been applied, namely: a plane mass hot plate and a red-hot tubular ring.

In the case of mass hot plate, the heat is transferred by thermal conductivity based on good thermal contact to the vessel. This requires stable, plane surfaces of the hot plate as well as the bottom of the vessel, and consequently large thermal masses. Typically 30% of the electricity used for hot plate cooking ends up as heat in these contact materials, often leaving a 300°C hot plate at the end of cooking. Some of this loss can be eliminated by better control of the power supply, either manually or automatically (38).

With red-hot tubular rings, the heat to the vessel is mainly transferred by heat radiation. Therefore, the high thermal masses can be omitted. However, due to higher temperature of the rings, there are significant radiation losses to the surroundings. For heating up and for short-term cooking, the tubular rings are more efficient than the mass hot plates (38), but for long-term cooking the reverse can be true.

Energy to heat food need not be transferred to the vessels in form of heat as described above. Some of the options for saving electricity for cooking in vessels as well as in ovens involve other solutions to energy transfer problems. In the following, some options already available and some suggestions for the future will be discussed.

7.2 Built-In Heating Elements

In the case of heating water for tea, etc., electricity can be transferred more directly to the "food" by an immersion water heater, which then converts the electricity to heat without losses. The only losses are those associated with heating the vessels, heating the heating element itself, and the small dissipation of heat to the surroundings. Typically 85 to 95% efficiency can be achieved.

A more practical and safe version of immersion heaters is the electric kettle with a built-in heater and often with an automatic controller to switch-off the power when the boiling point is reached. They have been used for a long time in the UK, but only in the last 5 years have they been marketed in Scandinavia. Efficiency of 80 to 90% has been measured, and it is lowest for small amounts of water (38). For domestic use, it is important to choose kettles which can heat as little as one or two cups of water and which also have a good water level gauge. This makes the tea or coffee brewing easier, and saves time and energy since the heating of excessive water is avoided. A 20% saving of energy has been estimated as the result of a good water level gauge (39).

When it comes to cooking food in general, a vessel with an openly exposed heating element as in the electric kettle is not practical. However, pans are being made with a built-in heating element in the bottom. They are quite commonly used in the South Pacific regions (40), and are also available in the USA. Power supply and thermostatic controller can be unplugged from these pans which can then be washed as usual. Efficiency for heating water up to boiling has been measured to be 70–80%, but total efficiency for heating vegetables or other food in water or oil is lower than that.

These electric pans are not being purchased primarily with energy savings in mind, but because they are faster. It is also cheaper and more flexible to buy a couple of such pans in different sizes and types than to install a full electric range. Such electric pans are hard to find on the European market, probably because of concern about the electric safety. For the models on the South Pacific market, some improvements could be suggested, for instance in combination with other saving options as described later.

7.3 Oven Cooking

Cooking in a traditional electric oven is often wasteful in energy because the oven box constitute a large thermal mass to be heated, and because of inappropriate thermal insulation. For most oven cooking, the normal size oven is too large. It is estimated that 90% of oven cooking only requires a 10 litre oven. Such small and well insulated ovens have been made as prototypes and shown to save around 50% electricity (38). It could be inserted into a normal big oven. This could save space and still maintain the large oven option for the Christmas turkey, etc. Similar savings have been reported by using inside walls which are highly reflective to the heat radiations (12).

The basic ideas of *microwave ovens* is that they work fast and efficiently since the heat is generated not just close to the food but actually in the food. Because the waves are reflected or transmitted almost free of losses in other materials than the food, the microwave energy is transferred close to 100% to heat in the food. However, there are losses associated with generating the microwaves. The efficiency of the microwave generator (the magnetron) and its power supply is typical below 60% today, but it can be improved.

A total efficiency for cooking a typical dish in a microwave oven is therefore similar to that of an efficient oven with electric resistance heating. The microwave oven, however, works faster. In general, efficient ovens can also save electricity as compared to cooking in pans by avoiding the cooking media like water and oil for some dishes.

7.4 New Cooking Range Developments

Induction cooking ranges generate the heat where it is wanted, like in the microwave oven, or close to, namely in the bottom of the pans. Efficiency is superior to that of a traditional hot plate during heating up, but less efficient in keeping the high temperature (13). For residential use, induction cooking ranges might not be cost-effective, but they probably will be in restaurant kitchens, etc.

A halogen lamp cooker is based on radiation from a halogen lamp, which is directed towards the bottom of the pan to heat it. It requires no heavy bottom in the pan or other heavy thermal masses. Hence, it is fast to control and rather efficient in the process of heating up.

7.5 Insulated Pans and Pressure Cookers

Heat losses to the kitchen account for about 50% of the electricity used in hot plates, depending on cooking traditions. When cooking for hours, insulated pans becomes an interesting option from an energy saving point of view. One version of this is the use of a thermally insulated box, in which certain dishes can be ready cooked after being heated to 100°C at the range. Such "hayboxes" or "Dutch ovens" have been commonly used in times of energy scarcity.

Insulated pans have been marketed recently (31,41). They have double walls and insulated lid and can be used on hot plates, tubular rings or gas. After removal from the range an extra double bottom is attached, and the pan can be placed directly on the dinner table, keeping the food warm for hours. With a special lid, the insulated pan is transferred into a *pressure cooker*. Energy consumption for pressure cooking is low because: less water is required and cooking time is reduced. In pressure cooked vegetables, the vitamin C content is found to be approximately twice as high as in conventionally cooked vegetables (42). Old style pressure cookers have often been too large for today's small families, but now pressure cookers are available in sizes down to 2 litres. The saving potentials for these insulated pans are still not fully analysed, but with appropriate adaption of the cooking habits there should be significant savings.

7.6 Future Options for Efficient Electric Cooking

Probably most of the 85% energy loss presently associated with electric cooking can be eliminated. Some of it with equipment already on the market, as described, and more by combining various known technologies. The result of some of the options is shown in Figure 7 as "efficiency advanced technology".

Electric cooking in pans is here suggested to break the traditional concept of a range, from which energy is transferred to the pans as heat. The electricity could be transferred directly into the pans and kettles as mentioned. Combining this principles with insulated pans and pressure cookers, such units could save an estimated 60% of the electricity now used for cooking on hot plates. *Faster cooking* would be another advantage of this concept.

For oven cooking, a microwave oven or a small efficient resistant heated oven could be an energy efficient solution, saving around 50% compared to common practice today.

7.7 Substitutions for Electricity in Cooking

Obviously, no electricity is needed for cooking, since most people in the world do not use electricity for this purpose. For industrialized countries, gas seems to be the most relevant alternative. For gas cooking, saving options by using insulated pans and pressure cookers are also available. Furthermore, more efficient burners, which also reduce the combustion pollutant in the kitchen, have been developed in USA (43). A combination of electric and gas cooking might be attractive, and is quite popular in Italy and France (33).

7.8 Cost-effectiveness for Cooking

Overall, no extra costs seem to be necessary for the cheapest efficiency advanced technology in cooking described above and shown in Figure 7. The options will require R&D costs, but in production a cooking system based on, say, three pans and a kettle with built-in heating, plus a couple of tubular ring plates could most likely be produced cheaper than the present range with hot plates plus various pans and kettles. It would require some changes in cooking habits, but since the savings are closely associated to other advances like faster cooking and better control, these changes could probably gradually be accepted as improvements. On this bases, the 370 kWh of electricity saved per household is free of cost, when the new equipment is introduced in the course of normal replacement as assumed in this chapter.

In countries such as Denmark, electric cooking contributes to the peak load demand for power. This makes the efficient electric cooking and the substitution options important in the energy planning.

8 Heat Distribution Systems

Most dwellings in Europe and USA are equipped with a central heating system. From a furnace, heated by oil, gas or solid fuel, or from a district heating inlet unit or a heat pump, the space heat is distributed to the various rooms. In this section the electricity used for this *distribution* and for *control* of the heat supply is considered, but not the electricity which might be used for generating the heat.

8.1 Hydronic Central Heating Systems

Central heating systems are in Europe most commonly designed as a hydronic system, i.e. a system based on water, whereas in North America air duct systems are most widespread. This section will focus on hydronic systems only, since ventilation is described in the following section.

Originally, most hydronic central heating systems in one family houses used no electricity at all but relied on the natural, thermal convection for circulating the water from a basement furnace. In order to allow for any location of the furnace and to achieve a more efficient and controllable distribution of space heat, an electric circulation pump was later added. A modern gas and oil furnace, also uses electricity for an oil pump, a blower, and various control equipment. Typically such a central heating system uses around 450 kWh/year – or more than 10% of a household's electricity consumption. Out of this, a 65 watt circulation pump running the whole Danish heating season of 225 days typically accounts for around 350 kWh (37,44). In some cases, another pump circulates the hot water, often running the whole year. This can add another 500 kWh/year.

In apartment buildings, larger central circulation pumps are used, and the electricity consumption per dwelling is usually smaller than the above mentioned for one-family houses.

8.2 Oversized, Overused and Inefficient

The 65 watt circulation pump is vastly oversized. A pumping power of the order of 1 watt is sufficient, but accounting for efficiency of pump and motor, a somewhat higher electric power is required, maybe 4 watt (37). One explanation for choosing oversized pumps has been concern about the security for proper start at the beginning of heating season. Actually, a proper sized pump for the purpose has hardly been available until recently. Today, a 20 watt circulation pump is available and thus a 65% saving can be obtained by a cheaper pump than the oversized. Pumps of this size are now being installed in new one-family houses.

Small pumps have a low efficiency. Typically, only 15% of the electric power is turned into pumping power. It is assumed here that the small motor plus pump can cost-effectively be designed to be twice as efficient as today. For a one-family house, this calls for a 5–10 watt reliable circulation pump.

On-off control by a time-clock is a simple and cheap way of turning off the circulation pump when no heat is needed in the system. This can be done during the night or by day, if the family is out. Also during heating hours the heat control system can turn the pump on and off according to the need for heat.

With the best available technology it is possible through: 1) proper sizing, 2) use of pump switch to control heat supply, and 3) better control system in the furnace, to reduce the 450 kWh per year to 100 kWh per year with equipment on the market today, as shown in Figure 8 as best available in 1988.

By technically improving the motor, pump and other components, this consumption can be cut in half, see efficiency advanced technology in Figure 8.

For large apartment-houses or commercial buildings, the pumps have a higher efficiency than the small ones described, typically 30%. The need for pump power varies with the third power of the water flow. Hence, when heat demand in part of the time is reduced to, say, half, the pump power needed is only 13% of that at full flow. By using more efficient motors and variable speed control, a saving similar to that of the small systems can be achieved.

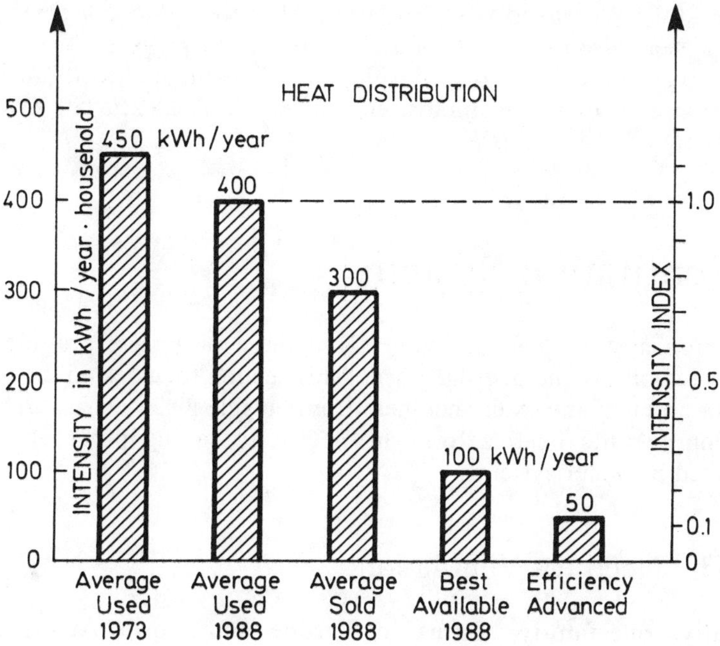

Figure 8. Electricity used per year for pumping-power and for control in a heating system in a typical Danish one-family house. No circulation of the hot water supply is included.

8.3 Superinsulated Houses

If heat conservation is carried out, the need for pumping power can be even further reduced. For a superinsulated house, the temperature will tend to be rather evenly distributed, also with simple heating systems. A furnace can be placed centrally in the house, maybe with radiators nearby in the other rooms, connected with short pipes instead of placing the radiators under windows. Also the heating season is shortened, often by weeks or even months. Although low energy houses often have been high electricity consumers with pumps and fans, it is in fact possible to run the heating system essentially without electricity use. Such heating systems will be cheaper to install than a normal system, especially in new houses.

For climatic regions where active cooling is demanded, similar efficiences in pumping can be realized, if distribution is based on a water system.

8.4 Cost-Effectiveness for Heat Distribution

Total extra cost for reaching the 100 kWh/year is estimated to be $ 40, mainly for the control systems, since the proper sizing of pump actually saves money. For the 200 kWh saved per year, the cost amounts to 2.5 cents/kWh. The control system also saves on the heating bill.

The advanced technological development is estimated to result in an extra production cost of pump, motor, etc., of $ 15. The extra 50 kWh/year can hence be saved at 4 cents/kWh.

9 Ventilation System

Ventilation systems can basically serve three purposes: provide clean air, distribute coolness and distribute heat. This section focusses on the provision of clean air and to some extent coolness from outside air. Mechanical cooling (air conditioning) and heat is from an electricity saving point of view better distributed by water systems.

9.1 The Need for Ventilation

Normally, one-family houses in Europe and North America have no mechanical ventilation systems, except maybe for an exhaust fan in the kitchen. The bathroom is typically ventilated non-mechanically by a "chimney" up through the roof. In older apartment buildings, exhaust from kitchen and bathroom is normally achieved also by "chimney"-effect through a vertical duct. In new apartment buildings, a mechanical ventilator is added in the top of the building. Most mechanical ventilation, however, occurs outside the domestic sector, mainly in the service sector, that is in offices, shops and institutions. All ventilation with the purpose of providing indoor *comfort* for humans will be considered in this chapter, while ventilation used in industrial processes will be left out.

In recent years, more elaborate and controlled mechanical ventilation systems have been installed, some also in one-family houses. In the following, the background for this will be discussed.

Insulation standard has increased and even in Nothern European climate the free heat gain from persons, windows and *electricity* in part of the year leads to indoor temperatures which are too high. The increased use of electricity has also contributed to this. The solution has been to remove the excess heat by mechanical ventilation systems. The need for ventilation for this purpose can be drastically reduced or even eliminated through the use of better window designs, and of more energy efficient lighting systems and appliances.

Clean air has been increasingly in demand, leading to installation of large ventilation systems. Even when the standards for the amount of air exchange required per person present has been followed, it has not always resulted in a satisfactory indoor climate as percieved by the occupants. This "sick building mystery" has recently been illuminated by Fanger et al. (49,50). A study of 15 randomly chosen office buildings in Copenhagen was carried out. It was found that on the average only 13% of the indoor pollutants was caused by the occupants. This is remarkable, since the number of occupants has normally been the basis for designing such ventilation systems. 25% of the pollutants was caused by smoking, and 20% by materials in the rooms. However, 42% of the pollutants originated from the ventilation system itself!

In other words, the ventilation system itself often is part of the problem rather than the solution – contributing significantly to a low air quality and to a less satisfactory indoor climate. This might explain also why 80% of the schools in a study in Switzerland were found never or only rarely to use their ventilation systems (44).

The immediate reaction to the air quality problems has often been to increase the air exchange by using even larger ventilation system. An obvious long-term solution to the indoor air quality problem would be to reduce or eliminate the causes (50). Materials which dissipate uncomfortable or even hazardous gases should be avoided. As far as the ventilation system is concerned the consequence of the above described results, could be a complete revision of where and whether mechanical systems with ducts should be used and how they should be designed.

Heat recovery from the exhaust air as a means to save heat has promoted mechanical duct ventilation systems, also in one-family houses. However, heat can be recovered without mechanical ventilation systems. A heat pump for space heating or for hot water supply can utilize the heat at the outlet of a "chimney" ventilation system (52).

All the above indicate that the future need for mechanical ventilation is not always obvious and in the least not well defined. Often the amount of air supplied in present day ventilation systems is most of the time more than necessary to remove the odours caused by the occupants, etc., (51). Ban on smoking in several places can further reduce the need for fresh air. Better ventilation inlet and outlet systems can also reduce the necessary air flow while still providing the same climatic quality. In the long-run, the average flow of air to rooms could be reduced to less than half of present practise without hampering the wanted electricity service, i.e., a satisfactory indoor air quality. Indeed, an improved air quality could probably be achieved simultaneously.

9.2 Efficiency of Ventilation Mechanics

Large ventilation systems consisting of a motor, a transmission, a radial ventilator and connection to the duct system, have a low total efficiency (53,54). Defined as the energy provided to the flowing air relative to the electric energy consumed, it is as low as 25% and often less. For large systems, a total efficiency of about 65% – or 2.5 times higher – could be achieved through more efficient and more appropriate motors, transmission and radial ventilator on the market (55). In order to suit the variable need for air, the motor must be equipped for variable speed drive.

For smaller units for one-family houses, the options for efficiency improvements are estimated to be similar.

9.3 Air Flow and Duct Systems

The power required to move the air through the duct system is proportional to the third power of the air flow. This implies that if the amount of air needed per hour is reduced to half, as suggested earlier as a typical option, the ventilator yield can in theory be lowered to only 13% of what it was before (56). In practise, it will be higher, maybe 25%. If also the mechanics are improved as described by a factor of 2.5, only 10% of the original electricity consumption is required. This is achieved without replacing the duct system, which can be a difficult and expensive undertaking.

For new systems, the dimensions of the duct and accessories provides another option. A 10% larger diameter of all ducts reduces the ventilator yield required to less than half. The shape of bendings, etc., is also important for electricity efficiency of the system. But, as mentioned earlier, the whole concept of ventilation systems might have to be revised.

9.4 Natural Ventilation

The use of natural ventilation to move the air is usually associated with manual control. In recent years, however, experiments with natural ventilation without manual control in cattle houses have shown promising results (57). Natural ventilation has here been established by the chimney effect at the rooftop, and inlets in the walls are automaticly controlled according to need. Electricity consumption is practically eliminated. There seems to be no reasons why the principle cannot be applied to dwellings and other buildings. Already a small and inexpensive ventilation inlet for use in each room of a dwelling is available with automatic control of the opening according to the humidity in the rooms (58). It is based on a humidity sensitive "bi-material" spring. The outlet can be a "chimney" duct as mentioned. This small unit is not connected to electric mains at all.

9.5 Total Efficiency Options for Ventilation

About 10% of Denmark's electricity consumption – or 500 kWh/year per capita – is used for comfort ventilation of one kind or another in dwellings, institutions, shops, industry, etc. The number of various applications, types, sizes, and conservation options for ventilation systems is too large to be reasonably analysed here. However, from the above, it seems clear that enormous efficiency improvements are available.

Studies in Denmark have shown that today's ventilation service can be provided with only 55% and 15% of the electricity currently used for, respectively, marketed and advanced technology (1,56), see Figure 9. Note, however, that the numbers for electricity consumption in Figure 9 refer to *per capita*, rather than per appliance or per household like in the other figures.

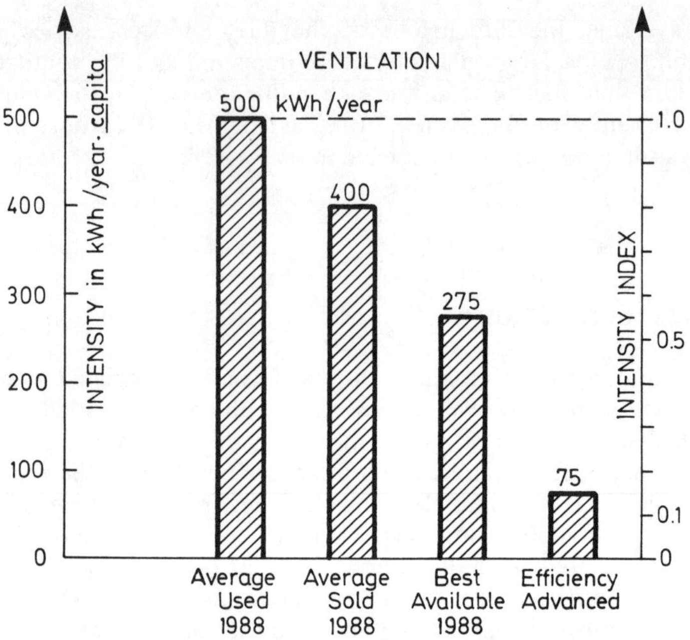

Figure 9. The columns illustrate the electricity used in Denmark *per capita* for comfort ventilation in general, not just domestically, and the estimated electricity use with more efficient technologies.

9.6 Cost-Effectiveness for Ventilation Systems

Compared to the average ventilation system installed today, no extra cost of choosing the best technology was found (1) and the 125 kWh annual savings per capita could be obtained for free.

Moving towards the advanced technology systems, the design changes would be rather radical and cost comparison extremely difficult. More control systems would be required, but smaller components could be used. However, since these motors, fans, etc., should also be more efficient, a slight extra cost is anticipated. Per capita, it is estimated the initial extra cost will be $ 50, leading to a price of 3.3 cents/kWh for the saved 200 kWh/year. As usual, the changes are anticipated to be introduced in the natural course of replacement.

10 Miscellaneous

The types of appliances already described cover an estimated 80–90% of the electricity consumed by appliances in residences as well as in the service sector. Several other types are in use (see below), but altogether these miscellaneous appliances consume in the household sector only between 300 and 500 kWh per year (7,54). Their minor role in today's electricity consumption is due to one or more of the following characteristics:

- their power rating (consumption per hour) is low,
- their normal use is for short periods only, and
- they are rather rare.

10.1 Electronics and Efficiency

Electronic equipment is an example of the low power rating. They have been introduced in the private households and similar areas, for entertainment, communication and control.

TV-sets, radios, videorecorders, and personal computers (PCs) are typical examples. Over the decades during which these appliances have been introduced, they have also become more efficient, measured, for instance, in terms of hours of entertainment per kWh. This efficiency has been pursued, not to save electricity consumption for the consumers, but to cut production costs, reduce size and otherwise improve the products. Only for equipment which run on batteries, electricity efficiency has been directly an issue.

While power ratings are small and will become even smaller in years to come for the various electronic units, their annual electricity consumption can become significant, if users get into habits of leaving radios, TVs, PCs, etc., on all day, maybe urged so by the design of the product. A recent study in New York showed TV consuming 700 kWh/year per household – 12% of total residental electricity consumption (59).

10.2 Small Household Equipments and Hand Tools

Many small appliances in the home have a rather high power rating. Some because they contain a powerful motor – vacuum cleaner, mixer, drill, etc. Others because they are equipped with electric heating element – iron, towel dryer, toaster, automobile heater, hair dryer, etc.

Motor equipped small appliances are usually not left on for a longer time than necessary due to the noise. The heating equipped devises, however, are silent and some of them can easily be left on useless for hours or days due to carelessness.

10.3 New and Rare Appliances

Humidifiers, dehumidifiers, water beds, whirlpools, and saunas are examples of appliances which are rather rare and for that reason play a minor role for the total national electricity consumption, at least so far.

Generally, the electricity saving potential of the miscellanous appliances have not been analyzed. However, their components usually contain the same efficiency potentials as described for other appliances, such as motor improvements, better thermal insulation, etc. No doubt, similar options are available here. Nonetheless, on the average, only a 30% savings with best on the market and a 50% with advanced technology are assumed here.

The cost-effectiveness of saving electricity in the miscellaneous appliances has not been evaluated. However, considering the modest estimates for saving potentials, the conservative assumption have been made that the price per saved kWh is the same as the average for all other appliances.

11 Total Saving Potentials

Electricity is used in appliances for countless purposes, which makes a thorough and detailed analysis impossible. We have illustrated some basic principles for efficiency improvement by typical examples of the major appliances.

11.1 Intensity Index Transferability

The various types of appliances have different distribution, sizes, etc., in various countries and are also used differently. This makes it impossible to transfer the per capita saving potentials in *absolute values* from one country to another. However, the *intensity index,* which is shown on the right hand scale of each previous figures in this chapter, is reasonably transferable between different countries and cultures. Actually, this index has been found to be

Table 2. Results of this analysis are here expressed in terms of the intensity indices, i.e., the intensities obtainable relative to the average of the units in use in the base year 1988. For each appliance, these indices are reasonably transferable to other countries than Denmark for which the absolute electricity consumption per capita is shown in the first column. The intensity indices for "All appliances + light" are derived from Table 3.

Type of End-use technology	All sectors consumption 1988 kWh/capita (Denmark)	Intensity index			
		Average sold 1988	Best Available 1988	Effic. Advanc.	Effic. Advanc. + heat subst.
Refrigerator	150	.75	.25	.14	.14
Freezer	235	.80	.35	.20	.20
Combined R/F	90	.80	.55	.20	.20
Clothes Washer	175	.75	.60	.30	.10
Dish Washer	90	.72	.62	.33	.07
Clothes Dryer	65	.85	.67	.35	.20
El-cooking	340	.93	.57	.40	.0
Heat distribution	270	.75	.25	.13	.13
Ventilation	500	.80	.55	.15	.15
Mischellaneous	425	.90	.70	.50	.50
Lighting	860	.85	.35	.20	.20
All appliances + light	3200	.81	.47	.26	.20

transferable from the domestic sector to corresponding appliances in other sectors. When, for instance, efficiency in small size freezers in the home can be improved to an intensity index of .20, it has been found that this index can also be applied to freezer units in shops and stores (but not large freezer storages), even though the initial intensity expressed as kWh/year per litre volume is much higher in these commercial units.

The intensity indices for the various types of appliances are shown in Table 2, together with the absolute consumption in Denmark for each type of appliance in 1988, see Table 1. In the total accounting electricity used for lighting has been included, since this completes the picture for residential electricity consumption, except for hot water and space heating. Efficiency improvements for *lighting* are analysed in the chapter by McGowan in this volume. The lighting intensity indices in Table 2 are based on that and other studies (2,56,62).

The intensity indices, can now be transferred to appliance patterns in different countries and at different times. For more aggregated analyses, the *average* intensities, shown at the bottom line in Table 2, can be transferred. However, these *average* intensity indices are derived from the Danish appliance pattern, see Table 3, and should be transferred with care.

Table 3. Annual electricity consumption in Denmark for appliances plus light is here shown for unchanged 1988 energy service level assuming the various levels of technical efficiencies implemented, based on the indices in Table 2. The average intensity index for each efficiency level is shown at the bottom.

Type of end-use technology	Annual consumption with 1988 electricity service level				
	Average used 1988 kWh/capita	Average sold 1988 kWh/capita	Best available 1988 kWh/capita	Effic. advanc. kWh/capita	Effic. advanc. + heat subst. kWh/capita
Refrigerator	150	113	38	21	21
Freezer	235	188	82	47	47
Combined R/F	90	72	50	18	18
Clothes washer	175	68	105	53	18
Dish washer	90	65	56	30	7
Clothes dryer	65	55	44	23	13
El.cooking	340	316	194	136	0
Heat distribution	270	203	68	35	35
Ventilation	500	400	275	75	75
Mischellaneous	425	383	298	213	213
Lighting	860	730	300	172	172
All appliances + light	3200	2593	1510	823	619
Intensity indices	1.00	0.81	0.47	0.26	0.20

11.2 Future Options for Saving Electricity

Table 3 and Figure 10 show what the electricity consumption in Denmark could be in the future, assuming no changes in the energy service level for each type of appliances. They illustrate the *technical options* for saving electricity.

If present trends continue and the demand for energy services increases in the future, this has to be factored into a scenario for electricity consumption. The electricity service level development for each appliance type illustrates both the stock and the utilization as stated in the section on methodology and concepts. Based on the level of 1988, the electricity service level per capita can be expressed by service index S for each appliance or aggregated for all appliances.

Future electricity consumption W_y for one type of appliance or for all types in a country can then be expressed as:

$$W_y = W_{1988}P \cdot S \cdot I \, ,$$

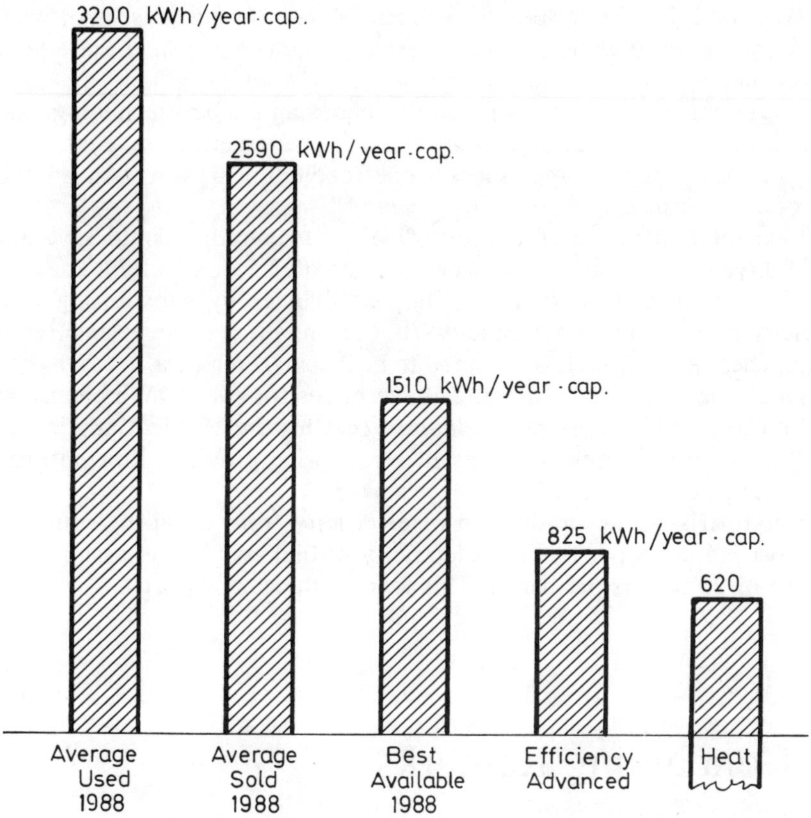

Figure 10. Electricity consumption for appliances and lighting with Danish distribution and Danish electricity service level of 1988 maintained, but with the various technical options implemented. The column to the right includes substituting heat for electricity in some appliances

where W_{1988} is the corresponding 1988 consumption of electricity for the country and P is an index for population development (6).

No scenarios will be presented in this chapter, but in the one by Meyer and Nørgård, and the one by Bodlund et al.

11.3 Cost-Effectiveness

Almost anything can be achieved in the area of technology if cost-effectiveness is ignored. In this chapter, the added costs involved in making technological improvements have been estimated for each appliance, and from that a price per saved kWh has been derived. In many cases, the price is zero, when, as here, a gradual introduction in the course of natural replacement of equipment is anticipated.

On the average, this analysis found that choosing the best technology already available on the market could over a 15 years period save about 1100 kWh/year per capita on appliances alone (excluding lighting), at an average price of *0.6 cents/kWh.*

Implementation of the efficiency advanced technology could save another 700 kWh/year per capita at an average price of *2.5 cents/kWh.* The average price for saving electricity in washing appliances by substituting heat for electricity is found to be 4.5 cents/kWh, but this price is very sensitive to the marginal heat price, which is assumed to be 2 cents/kWh heat.

It should be emphasized that the above prices for saved kWh electricity are based on the estimated extra *production* cost for the efficient appliances. As stated in the introduction, the present *retail* price provides little information for long-term energy planning.

For comparison, the production of electricity from oil at a thermal power plant cost 3.5 cents per kWh of electricity in fuel cost alone, assuming crude oil price of 20 dollars per barrel. This ignores the capital cost.

12 Concluding Remarks

There are substantial potentials for using electricity more efficiently. *The price for saving electricity in this way is per kWh superior to that of marginal production of electricity in any place in the world.* Furthermore, there are several external benefits of saving electricity as opposed for producing it, especially environmental benefits.

This should not, however, lead to the conclusion that these savings will be achieved through the market mechanism alone. Experiences have shown that this is not the case. Strong vested interests and institutional structures seems to counteract or delay the development towards a reduction in electricity demand. Hence, a strong dedication among politicians, manufactures, and – not least – the general public is required to fully realize the technically and economically feasible savings.

Finally, the technical potentials described in this chapter should not be taken as an excuse or opportunity for pursuing a policy of unlimited material growth. In the long-term, such a course would overwhelm the technical fixes, and an excellent opportunity for doing better with less electricity and less environmental impact would be wasted.

Acknowledgement

For the work presented in this chapter I owe thanks to many people. The editors of the book, the reviewers Howard Geller, Roland Bergendal, Matti Malinen and Lee Schipper as well as B. Aebischer, Wolfgang Feist, and Vitta Ebel have all provided valuable information and comments. Also my colleagues at Physics Laboratory III have been very helpful. Finally, I want to thank Ulla Christensen for patiently typing the manuscript of this chapter, for which I of course take the full responsibility.

References

1. Pedersen, P.B., Viegand, J., Nørgård, J.S. 1988. *Evaluation of electricity consumption distribution and savings potentials in the public sector* (in Danish). Energy Group, Physics Laboratory III, Technical University of Denmark, Lyngby, Denmark.
2. Johansson, M., Pedersen, T. 1988. *Technical Electricity Savings,* (in Danish), Local Governments'. Research Institute, AKF, Nyropsgade 37, Copenhagen.
3. Møller, J., 1987. *Electricity Savings in Residential Sector,* (in Danish), Report TR 258. Research Association of the Danish Electricity Supply Undertakings, DEFU, Lyngby, Denmark.
4. Møller, J. 1986. *Forecast of Residential Electricity Consumption and Power Load, 1985–2000,* (in Danish), Report. TR 240, DEFU, Lyngby, Denmark.
5. Nielsen, B. 1988. *Evaluation of Electricity Conservation Potential, Commercial and Service Sector*, (in Danish), Report TR 257, DEFU, Lyngby, Denmark.
6. Nørgård, J.S. *Methods and Concepts in Electricity Conservation Planning*. Physics Laboratory III, Technical University of Denmark, DK-2800 Lyngby, Denmark (forthcoming).
7. Ebel, W. 1987. *Die Entwicklung des zukunftigen Strombedarfs der private Haushalte in der Bundesrepublik Deutschland.* Institut Wohnen und Umwelt GmbH, Darmstadt, Federal Republik of Germany.
8. Besparelser ved lavenergifrysere, *Råd og Resultater,* 1, 1984. Statens Husholdningsråd (Danish Government Home Economic Council, Copenhagen.
9. Nørgård, J.S., Bendix Christensen, L. 1980. Debate on insulation thickness in *Ingeniøren,* (in Danish), Nos 27, 30, 33, and 35.
10. Petersen, P.H. 1988. CFC-forbrugets fordeling på anvendelser. Substitutionsmuligheder i køleindustrien, *Materialenyt*, 4.
11. Potter, T.F., Christensen, C.B. 1986. Advanced Insulation for Appliances. *Proceedings of the ACEEE 1986 Summer Study on Efficiency in Buildings* Vol. 1. American Council for an Energy-Efficient Economy, Washington, D.C.
12. Geller, H.S. 1988. *Residential Equipment Efficiency: A State-of-the-art Review.* ACEEE (American Council for an Energy Efficient Economy), Washington, D.C.
13. Personal communication with Roland Bergendal, Electrolux Major Appliances Company, Sweden, 1988.
14. Guldbrandsen, T., Nørgård, J.S. 1986 *Achieving substantially reduced energy consumption in European type refrigerator*, presented at the 37th Annual International Appliance Technical Conference, Purdue University, USA.

15. Personal communication with Toksvig Larsen, Director of R&D Dept., Brdr. Gram A/S (refrigerator manufacturer). Vojens, Denmark, 1988.
16. Pedersen, P.H., Galster, G., Guldbrandsen, T., Nørgård, J.S. 1987. *Design and Construction of an Efficient US-type Combined Refrigerator/Freezer*. Proceedings from the XVIIth International Congress of Refrigeration, Volume B, pp. 547–54. International Institute of Refrigeration, Paris.
17. Galster, G. et al. 1986. *American Type Low Energy Refrigerator/Freezer*. Final Report for Southern California Edison Company, Physics Laboratory III, Technical University of Denmark.
18. Brochures and other information material on LER200, Brdr. Gram A/S, Vojens, Denmark, 1988.
19. The Most Energy-Efficient Appliances, 1988 edition. American Council for an Energy-Efficient Economy, Washington, D.C.
20. Feist, W. 1987. *Stromsparpotentiale bei den privaten Haushalten in der Bundesrepublik Deutschland* in D. Sievert (Hrsg.): zukünftiger Strombedarf – Bedeutung von Einsparmoglichkeiten. TÜV Rheinland, Köln.
21. Aus Erfahrung gut. Prospect from AEG Aktiengesellschaft, Hausgeräte, Postfach 1036, D-8500 Nürnberg 1, Federal Republic of Germany, 1988.
22. Cronberg, T. 1988. Household Production, Technology and Time-use in the Twentieth Century, Copenhagen Business School. (Forthcoming in *Journal of Consumer Policy*).
23. Vask ved 60 grader, *Råd og Resultater*, 7, 1981. Statens Husholdningsråd (Danish Government Home Economic Council), Copenhagen.
24. *Distribution and use of electricity consuming appliances*, (in Danish). Note from Statens Husholdningsråd (Danish Government Home Economic Council), August 1, 1983.
25. Vaskemaskiner, *Råd og Resultater* and *Tekniske Meddelelser*, 3, 1988. Statens Husholdningsråd (Danish Government Home Economic Council), Copenhagen.
26. Vaskemaskiner, *Råd og Resultater*, and *Tekniske Meddelelser* 6, 1988. Statens Husholdningsråd (Danish Government Home Economic Council), Copenhagen.
27. Heebøll, J., Nørgård, J.S. 1986. *Development of energy efficient electrical household appliances, part three: Washing machines*, Report EUR 10449 EN. Commission of the European Communities, Batiment Jean Monnet, Luxemburg.
28. Abbate, G.P. 1988. *Technology Development in Home Appliances* in *Demand-Side Management and Electricity End-Use Efficiency*. Kluwer Academic Publischers, Dordrecht, Boston, London.
29. Lipolase, Information sheets on product and application, Detergent Enzyme Division, Novo Industri A/S, Copenhagen, 1988.
30. Olivier, D. 1988. *More Efficient Use of Electricity – the Megawatt Resource*. Seminar by BEEE on Energy and Climate Change: What can Europe Do?, Brussels.
31. Lovins, A.B., Bancroft, B. *The State of the Art – Appliances*. First Quarter 1988 Update. Rocky Mountain Institute, 1739 Snowmass Creek Rd, Snowmass, Colorado, USA.
32. Personal communication with Matti Malinen, Vattenfall, The Swedish State Power Board, 1988.
33. Schipper, L., et al. 1988. *Household Electricity Use in the OECD*, International Energy Studies, Lawrence Berkeley Laboratory, Berkeley, California, USA.
34. Tørretumblere, *Råd og Resultater*, 6, 1985. Statens Husholdningsråd (Danish Government Home Economic Council), Copenhagen.
35. Kondenstørretumbler, *Råd og Resultater*, 4, 1986. Statens Husholdningsråd (Danish Government Home Economic Council), Copenhagen.

36. Stoy, B., Kionka, U. 1977. Senkung des Stromverbrauchs von Electro- Grossgeräten im Haushalt, *Energiewirtschaftliche Tagensfragen*, Heft 7.
37. Nørgård, J.S. 1979. *Households and Energy*, (in Danish). Polyteknisk Forlag, Lyngby, Denmark.
38. Holck, J., Goldschmidt, L.B., Heebøll, J., Pedersen, N.H. 1986. *Development of Energy Efficient Electrical Household Appliances, part two: Cooking*. Report EUR 10449 EN. Commission of the European Communities, Batiment Jean Monnet, Luxembourg.
39. Wright. I., Baines, J. 1986. *Supply curves of concerved energy – the potential for conservation in New Zealand's houses*. Centre for Resource Management, University of Canterbury. Publ. by Ministry of Energy, Wellington, New Zealand.
40. Not cooked on the stove, *Choice*, Sidney, Dec. 1983.
41. Kuhn, H. 1988. Heinrich Kuhn Metallwarenfabrik AG, CH-8486 Rikon, Switzerland. Personal communication and brochures.
42. Afprøvning af trykkogere, *Råd og Resultater*, 5, 1967. Statens Husholdningsråd (Danish Government Home Economic Council), Copenhagen.
43. Hurley, J. 1987. *IR/Jet Rangetop Burner*. Presented at the First National Conference on Appliance Efficiency, Sponsored by American Council an Energy-Efficient Economy, Washington, D.C.
44. Brunner, C.U., Baumgartner, Müller, Stulz, Wick, 1986. *Elektrizität Sparen*, Schweizericher Nationalfonds zur Forerderung der wissenschaflichen Forschung, Zürich.
45. Brochures from Vølund Vaskerimaskiner A/S, Tommerup, Denmark, 1980, (and later communications).
46. Personal communication with J. Weinrich, sectional manager. Danish Textile Institute, Copenhagen, 1988.
47. New Products, *Newsweek*, September 19, 1988.
48. Turiel, I. 1986. *Design Options for Energy Efficiency Improvement of Residential Appliances*. Energy Analysis Program, Lawrence Berkeley Laboratory, Berkeley, California 94720, USA.
49. Fanger, P.O. 1987. *A Solution to the Sick Building Mystery*, Indoor Air '87, Congress, Berlin.
50. Fanger, P.O., Lauridsen, J., Bluyssen, P., Clausen, G. 1988. Air Pollution Sources in Offices and Assembly Halls, Quantified by the olf Unit, *Energy and Buildings*, 12, pp. 7–9.
51. Fugazza, 1988. Minska luftmängden i ventilationsanläggningen, *VVS & Energi,* 6.
52. Schipper, L., Meyers, Kelly, associates. 1985. *Coming in from the Cold – Energy-Wise Housing in Sweden*. Seven Locks Press, Cabin John, Maryland, USA.
53. Aggerholm, S., Christiansen, J. 1987. *Electricity Consumption in Schools,* (in Danish), SBI-report no. 186. Danish Building Research Institute, Hørsholm, Denmark.
54. Gabathuler, H.R. (ed). 1987. *Specialveranstaltung: Elektrizität sparen*. Impulsprogramm Haustechnik, Bundesamtes fur Konjunkturfragen, 3003 Bern.
55. Mehlsen, K. 1987. *Mulige elbesparelser ved ventilation*. Crone & Koch Consulting Engineering, Ordrupvej 60, Charlottenlund, Denmark.
56. Nørgård, J.S., Holck, J., Mehlsen, K. 1983. *Long term technical possibilities for electricity savings,* (in Danish). Physics Laboratory III, Technical University of Denmark, Lyngby, Denmark.
57. Strøm, J., Morsing, S. 1982. *Automatically Controlled Natural Ventilation*. Proceedings of the second international livestock environment symposium, Iowa. Publ. by Am.Society of Agricultural Engineers, Michigan, USA.

58. Svane, P., Egedorf, M. 1987. *Humidity Controlled Ventilation – 2,* (in Danish). The Technological Institute, Tåstrup, Denmark.
59. Personal Communication with Howard Geller, ACEEE, American Council for an Energy Efficient Economy, Washington, D.C., 1988.
60. Mills, E. 1988. *Today's Appliance Marketplace in Sweden: The Case of Residential Refrigerators, Freezers, and Combination Units.* For Vattenfall Project, Updrag 2000. Environmental and Energy Systems Studies, Lund University, Lund, Sweden.
61. Personal communication with B. Stentebjerg-Olesen, Detergent Enzyme Division, Novo Industri A/S, Copenhagen, 1988.
62. Lovins, A.B., Sardinsky, R. 1988. *The State of the Art – Lighting.* Rocky Mountain Institute, 1739 Snowmass Creek Rd, Snowmass, Colorado, USA.
63. Der Beweis, *Chancen für Leben Umvelt Technik,* 4, p. 20, 1986.
64. Personal communication with W. Feist and W. Ebel, Institut Wohnen und Umwelt GmbH, Annastrasse 15, Darmstadt, FRG, 1988.
65. Geschirrspüler, therma G11, Electrolux, *Prüf mit,* No 6, 1978. Zeitschrift des Konsumentinnenforums, Switzerland.

Heat Pumps

Thore Berntsson

Abstract

The electrically-driven heat pump is a mature technology today, with better performance, reliability and availability than a decade ago. Nevertheless, improvements can be made, for example in, their control, interaction with heat distribution systems and defrosting. Also, advanced heat pumps with better working fluids solutions, are being investigated.

Electrically-driven heat pumps for heating of buildings typically can compete with oil/gas firing and heat-driven heat pumps (internal combustion engines and absorption systems) at electricity/fuel price ratios below 2.0–2.5. At higher ratios, absorption heat pump systems would be the most competitive type due to severe environmental problems with combustion engines. Ambient and exhaust air will probably be the most interesting heat sources.

Industrial applications of heat pumps look more promising than residential ones in most countries.

Ongoing R&D indicates possibilities for introduction of new, improved technologies in the near future. For the severe environmental problems associated with today's working fluids (chlorofluorocarbons, CFCs) several promising short and long-term solutions are examined.

1 Introduction

The heat pump technology has during the last decade attracted increasing attention as one of the most promising "new" technologies for saving energy and protecting the environment. Applications of interest are in heating of all types of buildings and for energy conservation in industrial processes.

A heat pump is defined here as a device designed for the production of heat only. An enormous number of air conditioners and other units worldwide work in a cooling mode during one part of the year and in a heating mode during another. However, such systems, although having a huge impact also as heat pumping devices, have different characteristics both technically and economically than pure heat pumps and are therefore not included in this discussion.

During the last decade the number of heat pump units installed has increased rapidly in some countries. The most rapid increase has been in Sweden, where more than 100,000 units are now in operation. The reasons for this increase have been low electricity prices and the availability of suitable heat distribution systems. This development slowed considerably when oil prices decreased rapidly in 1986. In Sweden, many types of heat pumps have been used and a large number of projects for involving measurements, evaluations, component and system studies, etc., have been carried out and presented in publically available reports. Hence, many of the references in this chapter will rely on the Swedish experience.

In the following sections, heat pump applications both for heating of buildings and in industrial processes are discussed. Electrically as well as heat driven cycles are taken into account and the thermodynamic principles for various cycles are presented.

Environmental aspects are discussed only for some systems which have negative impacts. However, practically all heat pump types have an important positive environmental impact due to the use of renewable energy as the heat source and hence a decrease in emissions associated with fossil fuels. Very recently, however, some chlorofluorocarbons (CFCs), used in heat pumps of the compression type, have been shown to be partly responsible for depletion of the ozone layer and the greenhouse effect. Hence, other fluids and/or cycles must be used in the future. There exist today several suggestions for alternative solutions. The CFCs problem is discussed in the last section.

2 Basic Thermodynamics, Cycles and Components

2.1 Different Heat Pump Cycles

In a heat pump cycle, heat from a low-temperature heat source is upgraded and delivered to a high-temperature heat sink. According to the second law of thermodynamics, this is only possible if a certain amount of driving energy is used. The driving energy can be, for example, electricity, fossil fuels or industrial waste heat. The coefficient of performance (COP) is generally, regardless of the driving energy source, defined as:

$$\text{COP} = \frac{\text{Heat delivered to the heat sink}}{\text{Amount of driving energy used}}$$

The number of heat pump cycles used or suggested is very high. The most important of these are:

Closed cycles:

- Compression cycle
- Absorption cycle (Absorption heat pump type I)
- Heat transformer (Absorption heat pump type II)

Open cycles:

- Vapour recompression (steam or process gases)
- Ejector heat pump (used in special industrial applications with very small temperature lifts. Not discussed further in this chapter).

In addition, several alternative cycles are now discussed or under development, e.g., the Joule, Stirling, compression/absorption, Peltier and multi-stage absorption cycles. Most of these cycles have been well known for many years, but have not been used (except in very special cases) due to economic reasons or lack of appropriate components. Some of these cycles are discussed in Section 5.

Of the common cycles mentioned above, the heat transformer and vapour recompression ones are primarily for applications in industry. The other ones can be used both for heating and in industrial processes. The thermodynamic basics of these cycles are discussed briefly below. Detailed discussions are not possible to do here.

2.2 Compression Cycles

2.2.1 Thermodynamics

The principle of the basic closed cycle is shown in Figure 1.
The COP can be expressed as:

$$\text{COP} = \frac{\dot{Q}_1}{E} = \frac{\dot{Q}_2 + E}{E} = \eta_c \cdot \text{COP}_c \qquad (2.1)$$

where COP_c for the ideal Carnot cycle is defined as:

$$\text{COP}_c = \frac{T_3}{T_3 - T_1} \qquad (2.2)$$

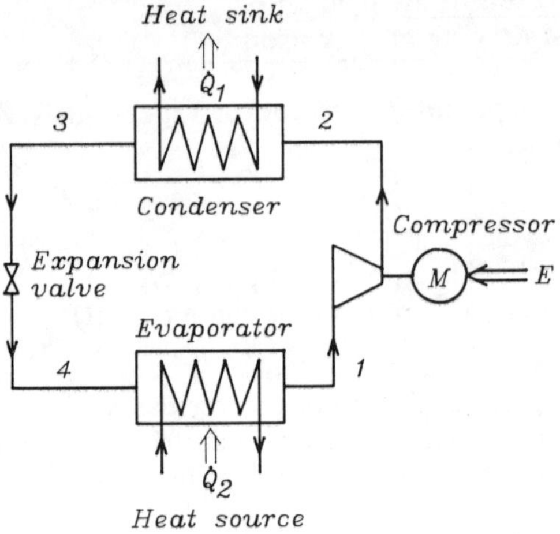

Figure 1. Basic compression heat pump cycle.

The COP may or may not include heat losses and losses in the electric motor and/or electricity consumption in auxiliary equipment such as fans and pumps. This should always be clearly stated when a COP value is given. In this chapter, all relevant losses are included in the COP unless otherwise stated. Values of η_c for the cycle are in this chapter based on the electric input to the motor, including reasonable heat losses but excluding auxiliary equipment. Real values of η_c will be given later.

It is important to note that there are several ways of improving the performance of heat pumps. The main thermodynamic losses, compared with the Carnot cycle, are associated with the isenthalpic expansion and the compression. These losses can be decreased by using multi- stage cycles, for example. When there is sensible heat both in the heat sink and the heat source, the ideal cycle to compare with is the Lorenz cycle. One means of getting closer to this cycle is the use of series-coupling of units. (Another is nonazeotropic mixtures, see Section 5.) Some examples of more advanced cycles are shown in Figure 2. The economizer cycle and the series-coupling are widely used in medium and large-scale systems. The other cycles are used only in very special cases.

The vapour recompression cycle consists only of a compressor which usually compresses steam up to a useable temperature level. If the waste heat is available as clean steam, the specific investment cost can be very low compared with the closed cycle. The compressor type normally used is the turbo one (see below), but other types can also be used.

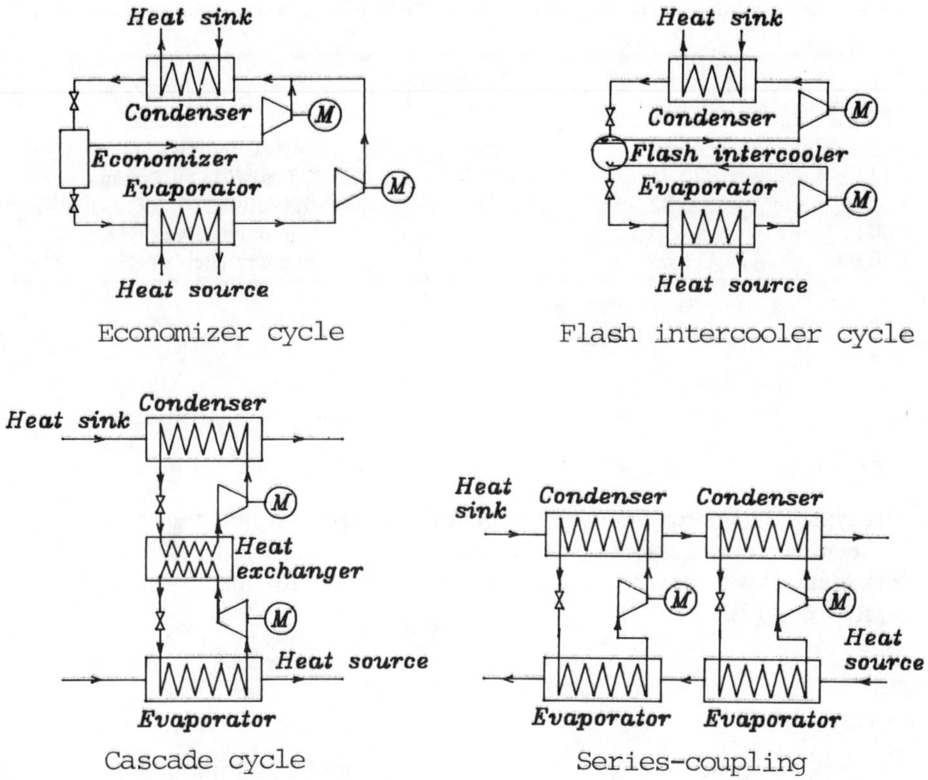

Figure 2. Various advanced heat pump cycles.

2.2.2 Components

Working fluids

The most common working fluids for heat pumps used today are shown in Table 1. Of these R22 (small domestic systems) and R12 (medium and large-scale systems) are the ones most widely used. It can be noted that there exists today no good fluid capable of working above 120–130°C. Working fluids are further discussed, in reference (1) and in Section 5. All fluids shown in Table 1 except R22 are included in the so called Montreal protocol, and their use will hence be restricted in the future, see Section 5.

Table 1. Some technical data for the most common refrigerants in compression heat pumps.

Refrigerant	Molecular weight	Normal boiling point	Maximum condensing temperature
R22	86.47	–40.8	= 60°C In turbocompr. = 80°C
R12	120.91	–29.8	75–80°C In turbocompr. = 90°C
R114	170.92	3.8	120–130°C
R500*	99.30	–33.5	approximately as R12
R502**	111.63	–45.4	approximately as R22

* R500 = R12//R152a (60.6/39.4% by mole)

** R502 = R22/R115 (63.0/37.0% by mole)

Compressors

The types of compressors used and their size ranges are shown in Table 2.

Details on compressors can be found in references (3), (4), (5), and (6). The important aspect on capacity control possibilities for the different types is discussed in (2).

Drives

The types of drive discussed here are electric motors and internal combustion engines driven by gas, diesel oil or heavy fuel oil. Other types include steam and gas turbines.

Electric motors: The electric efficiency in motors varies from approximately 65% in small units and up to at least 97% in large ones. These efficiencies are approximately valid also for part loads, down to about half the nominal load.

Table 2. Size ranges and η_c values for various types of compressors.

Compressor	Size (kW)	η_c
Reciprocating	1–5	0.42–0.56
Reciprocating	up to 500	0.45–0.58
Rotary	1–5	0.42–0.58
Screw with economizer	100–2000	0.50–0.60
Turbo	> 700	0.63–0.69

Internal combustion engines: A principle scheme of this cycle is shown in Figure 3. The efficiency and amount of recoverable waste heat for internal combustion engines varies with size, type, and manufacturer. Shaft efficiencies for diesel engines at full load conditions vary typically between 34 and 40%, whereas corresponding values for gas engines are 28–35%. Total efficiency – shaft efficiency plus recoverable waste heat amount – is typically 80–85%.

A value for the COP of a heat pump with internal combustion engine can approximately be calculated according to:

$$COP = \eta_s \cdot COP_{hp} + (0.85 - \eta_s)$$

where η_s is the shaft efficiency, $(0.85 - \eta_s)$ the heat recovered from the diesel engine, and COP_{hp} is the COP of the heat pump itself. For further details, see (2).

Heat exchangers

For evaporator and condenser types see references (7) and (8). The approach temperatures vary with the design and application but typical values are 4–8 K in the evaporator and 0–6 K in the condenser (0 K is possible due to a certain degree of superheat after the compressor).

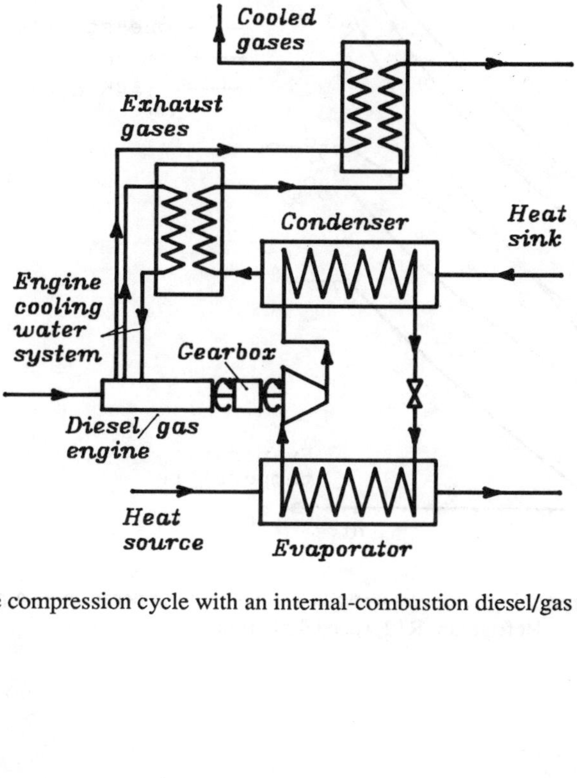

Figure 3. The compression cycle with an internal-combustion diesel/gas engine.

2.2.3 Performance

Figure 4 shows one example of COP for three different types of cycles, the basic one-stage, the economizer, and the flash intercooler. The values shown in the figure should of course only be seen as trends. Absolute values differ with the design, type of compressor, etc. With combinations of advanced cycles and series-couplings, COPs at least 60% higher than the one for the basic cycle can be achieved in extreme cases. The influence of various parameters on the COP varies of course with the conditions. Below are some examples of reasonable magnitudes for typical heat pumps for house heating.

The COP for the basic cycle is increased 4% per degree C of evaporating temperature and is decreased 2% per degree C of condensing temperature. The corresponding values for the capacity are 3% per degree C and 1% per degree C, respectively. The improvement of the COP and the capacity with subcooling of the condensate with the heat sink is about 1% per degree C of subcooled liquid. For an advanced cycle, this improvement is smaller, e.g. for the economizer cycle it is typically 0.5% per degree C. Typical COP values for the vapour recompression cycle are 5–10.

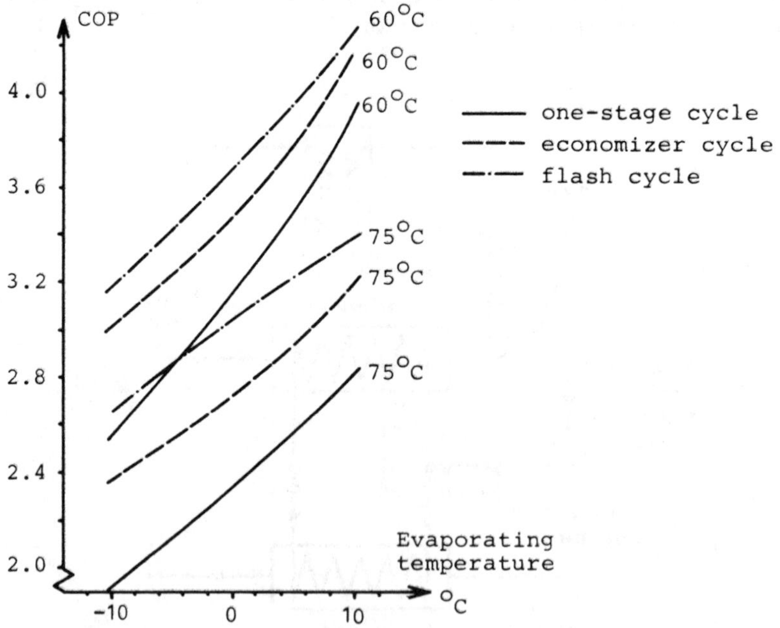

Figure 4. COP vs. evaporating temperature with condensing temperature as a parameter for large screw compressors. Refrigerant R12, speed 3000 rpm.

2.3 Absorption Cycles

2.3.1 Introduction

The basic absorption heat pump cycle, the single-stage one, has two possible configurations, see Figures 5 and 6. The cycles are displayed schematically to show the different pressure and temperature levels. Several working pairs have been suggested, but so far only two have been used in commercial units – ammonia-water and lithiumbromide-water.

The first cycle is usually referred to as the absorption heat pump cycle (absorption heat pump type I, heat boosting cycle). The second cycle is usually referred to as the absorption heat transformer cycle (absorption heat pump type II, temperature boosting cycle).

2.3.2 Performance

There are two main quantities of interest when discussing the performance of absorption heat pump cycles, namely the COP and the temperature lift. The definitions for respective cycle are:

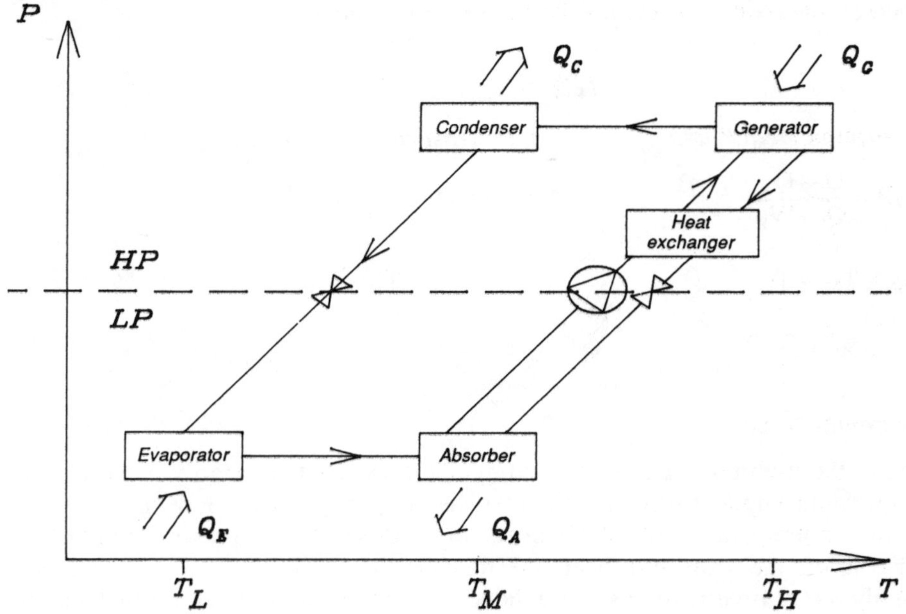

Figure 5. Principal diagram of the absorption heat pump cycle.

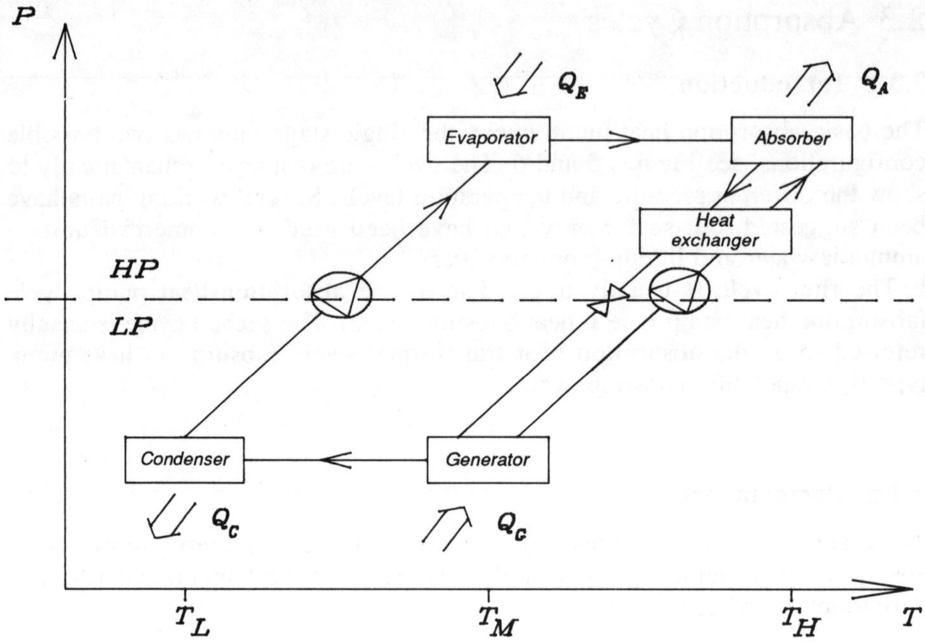

Figure 6. Principle diagram of the absorption heat transformer cycle.

Absorption Heat Pump

$$COP = \frac{Q_C+Q_A}{Q_G+W_P}$$

$$T_{lift} = T_M - T_L$$

Absorption Heat Transformer

$$COP = \frac{Q_A}{Q_E+Q_G+W_P}$$

$$T_{lift} = T_H - T_M$$

Ammonia-water

Due to the high pressure level of ammonia, this pair is primarily used for low temperature applications in residential heating. Typically, these units have a maximum temperature lift of 50–60°C. In small systems, a pressure limit of 20 bar restricts the maximum heat sink temperature to 50–55°C. These units are usually direct fired with gas or oil, hence the combustion losses are included in the COP which ranges between 1.1 to 1.4, with the lower COP levels valid at high temperature lifts.

Lithiumbromide-water

This pair is mainly used in high temperature applications because there is an absolute temperature limit imposed by the freezing point of water. Also, high viscosities and very low pressure cause problems at low temperatures. The maximum temperature lift of the heat pump cycle is approximately 45–50°C.It is 5–10°C higher for the heat transformer cycle, the risk for crystallization being the limiting factor. Because the working pressure for LiBr-H_2O is low, the energy needed for the pumps is normally negligible, which accounts for the somewhat higher COP for LiBr-H_2O than for NH_3-H_2O. Typically, an absorption heat pump with LiBr-H_2O has a COP of 1.6–1.7. This figure does not include any combustion losses. The COP of the absorption heat transformer cycle is typically in the range of 0.45–0.49. This means that nearly 50% of the waste heat can be recovered at a higher and useful temperature level with negligible consumption of primary energy.

3 Heat Pumps For Heating of Buildings

3.1 Heat Pump Applications

In principle, a heat pump can be designed to operate in three different ways in an annual load duration curve, as shown in Figure 7: monovalently, bivalently/parallell or bivalently/alternative. Monovalently designed heat pumps are rather common in new one-family houses (i.e., ones without an

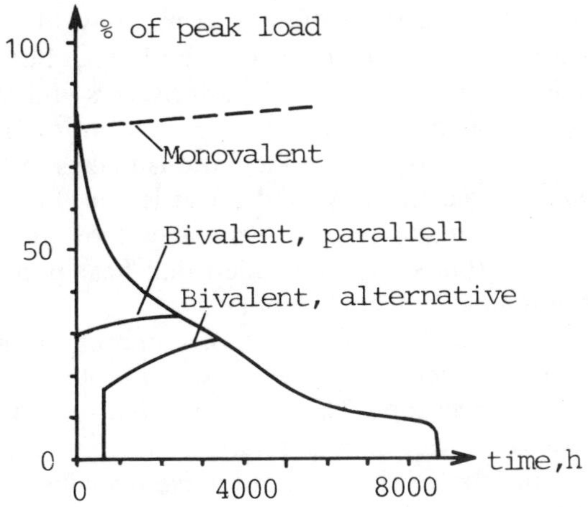

Figure 7. Three different ways of designing a heating system with a heat pump, seen in an annual duration curve.

existing oil or gas heater), especially those which have a good heat source with high evaporating temperature even in the winter. The extra investment cost for the heat pump can then be justified by the fact that the cost for a back-up system is eliminated. The choice between the two bivalent designs is dependent on the type of heat source. The optimal size, i.e., annual energy coverage, of a heat pump depends on several parameters and is discussed in Section 3.3. In very large systems such as district heating ones, several heat pumps with different heat sources can be introduced in a system.

The need for capacity control is of course dependent on how the heat pump is introduced in a heating system. Small systems normally use on/off control, while medium-sized and large ones often use more advanced techniques. In Japan and the US, rather high cycling losses during on/off operation have been reported compared to negligible losses in European systems. According to reference (9), the reason for this is that small systems designed also for cooling have a capillary tube and accumulator, whereas European systems use an expansion valve and a refrigerant receiver before this valve.

3.2 Experiences

One of the most extensive measurements on different small-scale types of heat pumps have been made by the Swedish National Testing Institute. Measurements under defined conditions have gone on in their laboratory since 1977. The results have been presented in reference (13). Some 186 heat pumps, all small-scale (5–10 kW heat output), have been tested. Comparisons between results obtained during different years show that the heat pumps' performance improves significantly up to 1983, but thereafter their improvements were marginal. Tested values for COPs and heat output are, on average, 10–12% below data given by manufacturers. Remarkably wide differences were found to exist among the various makes in forms of COP, tap water heating, defrosting efficiency, and noise levels. For instance the COP differed by 42% between the best and the worst of the liquid/water heat pumps. The authors of this study concluded that heat pumps for one-family houses are now commercially mature products.

In Table 2, η_c values for various types of compressors are shown. These values are based on reference (13) as well as many other reports. The values are valid for good compressors. The upper figures refer to relatively small temperature lifts and the lower ones to relatively large temperature lifts.

In a real application, the COP varies of course over the year. The weighted annual mean value is called the Seasonal Performance Factor (SPF). Typical values of SPFs, investment costs, auxiliary electricity consumption, and annual energy coverage for various types of heat pump systems are shown in Table 3. A striking aspect seen there is that SPF depends very little on the size of the heat pump. The reason for this is that an increasing size normally also means an increasing heat sink temperature level. Several investigations have shown that heat pump investment cost can vary considerably from installation to installation, as indicated in the table.

A few comments on the heat sources cited in Table 3 are in order.

In recent years, so called "split systems" have been installed in one-family houses. They consist of very small units of 1–3 kW heat output. The heat sink is indoor air, delivered to only one room or a small number of rooms. Systems in principle developed for air conditioning are thereby used. They can only be used as base load units.

Table 3. Technicoeconomic data for heat pumps with different heat sources.

Heat source	Size kW heat output	Investment cost US$/kW heat output	SPF	Electricity consumption in auxiliary equipment, %	Annual energy coverage, %
Ambient air	1–3 (split systems for base load)	650–800	2.1–2.3	≅ 5	15–40
Ambient air	5–300	900–1100	2.2–2.5	≅ 5	65–85 (occasionally monovalent in small systems)
Ambient air	300–5000	650–950	2.2–2.5	≅ 5	60–80
Exhaust air	All sizes	600–1500	2.6–3.2	≅ 5	35–55
Exhaust air, designed with defrosting	All sizes	600–1500	2.4–3.0	≅ 5	60–80
Soil	7–500	870–1300	2.2–2.5	6–9	Monovalent in small systems. Down to 70% in larger ones.
Ground water and energy wells	7–500	950–1500	2.4–2.7	≅ 15	Monovalent in small systems. Down to 70% in larger ones.
Sewage water	200–4000	300–600	2.5–3.0	≅ 5	30–60
Lake, sea and river water	500–>10000	160–500	≅ 2.5	Varies	Mainly base load
Industrial	>500	160–500	2.5–4.0	Varies	Mainly base load

Exhaust air units were initially designed for tap water heating only, but today they are used also for a part of the space heating. In some situations, it is even optimal to design the system to utilize outgoing exhaust air temperatures below the freezing point.

The annual energy coverage values shown in Table 3 are typically ones in installed units. They do not necessarily correspond to optimal ones, see Section 3.3.

In addition to the heat sources mentioned in Table 3, heat storage systems connected to heat pumps are also used. They have been considered mainly in seasonal storage applications, although short-term storage could also be of interest in the future, see Section 5. Possible seasonal storage systems include tube systems vertically embedeled in soil or rock and natural aquifers (12). Possible energy sources are solar heat, ambient and exhaust air, lake, sea, and river waters and industrial waste heat.

Considering that it is a relatively new and complex technology, the general experience with various types of heat pumps is astonishingly good. Some of the major problems which have arisen are:

- compressor break-downs;
- working fluid leakages;
- expansion valve control, especially in small ambient air systems;
- defrosting devices and control;
- control strategies for the interaction between the heat pump and auxiliary heating/heat distribution system, causing a decreased annual availability.

Diesel/gas-driven heat pumps

The COP for a diesel/gas-driven heat pump depends on the type of engine, degree of waste heat recovery, etc., as described in reference (2). Reasonable values can be calculated according to the discussions in Section 2.2.2.

Due to waste heat recovery, the heat pump part can be kept smaller for a given heat output than in an electrically driven system. Hence, the additional investment cost due to the engine, gear box, etc., is partly compensated by a smaller heat pump cost. The net additional total investment cost for a medium and large-scale diesel/gas system is therefore rather small, typically 10–20% compared to an electrically-driven one (2).

An interesting development going on in Japan is the mass production of small units of gas driven heat pumps, (14). So far no experiences from these have been reported, though.

One of the most important aspects with diesel/gas-driven systems is their highly negative environmental impact due to high emissions of nitric oxides (NO_x). These emissions can be up to 6 times those of a conventional oil or gas burner, calculated at the same heat output (2). In addition, many technical problems, e.g., with torque adjustment, vibrations and control have been reported from real diesel/gas-driven heat pump installations. Before technical solutions become available for the reduction of their emissions, heat pumps of this type probably will not play an important role in any national energy system.

3.2.1 Absorption heat pumps

The small absorption heat pumps are normally gas-fired in the 22–40 kW range, with NH_3-H_2O as working fluid. Over two hundred such units have been installed and successfully operated, mainly in Germany (15), (16). Their SPF is between 1.1–1.35. The variation is due to type of heat source and delivery system. For detailed information on ammonia-water systems see reference (17).

The experience so far with large LiBr-H_2O systems is that the SPF is typically around 1.6 when they are driven with steam (18). Hence, if a boiler efficiency is included, large systems would have SPFs of about 1.25–1.4, i.e., slightly higher than ammonia systems. The few installations of large absorbtion heat pumps have mainly been in waste incineration plants, where they are used for heat recovery and stack gas cleaning.

The investment cost for large absorption heat pumps, in the MW range, is 80–160 US$/kW heat output for the installed unit. The total specific cost varies of course with the heat source (the cost of which is not included in the above figure), the surrounding costs, and the application. This means that the total cost for these heat pumps is typically 70–90% of electrically-driven ones, calculated per kW heat output.

Due to the relatively few small and medium-sized ammonia- water systems installed, there is a lack of large-scale mass production, hence the investment cost situation today is very unclear for such applications. Investment costs have been estimated (16) and (19), but these costs should decrease considerably with mass production of the systems. The costs which have been estimated are clearly lower than those for comparable electrically-driven systems. Based on references (16) and (19) it can be concluded that also small and medium-sized systems could be made for a total cost of 70–90% of that of electrically-driven heat pumps.

Absorption heat pumps, having only few moving parts, are in general more reliable and have longer service lives than electrically-driven ones. The main problems with these systems today are the circulation pumps in small systems and the risk of corrosion in large systems.

3.3 Technicoeconomic System Aspects

3.3.1 Introduction

In this section, the different types of heat pumps are compared economically with each other and with conventional oil or gas heating. A rigorous such comparison involves a large number of parameters and is therefore very complex. Due to space limitations, a more generalized comparison will be done here. It will be, of necessity, approximative, focussing on the most important parameters.

3.3.2 Assumptions

For the technicoeconomic comparisons with fuel and between the various heat pump systems, the assumptions discussed below have been used. In all comparisons of this type, it is important to show the results so that a reader can make a sensitivity analysis on the most crucial parameters. Therefore, at least two levels have been used here for the most of the important parameters, except the interest rate and the service life for the heat pumps. The fuel in these calculations is always referred to as oil but it can of course also be any other type of fuel, e.g., gas. The electricity prices have been set at 1, 2, and 3 times the oil price. At ratios above 3, interest in electrical heat pumps would be very limited.

The boiler as an alternative heating system or as supplementary heat has been assumed to be already installed, therefore no investment cost for this device has been calculated. It has also been assumed that the boiler is large enough to cover the whole load, including at peak conditions when used as a supplementary heat unit. The boiler efficiency has been set at 0.80. In Figure 10 a, b, and c the boiler operating cost is assumed to be 0.028 US$/kWh; in Figure 11 a, b, and c, it is 0.056 US$/kWh.

Annual maintenance and service costs as well as service life have been set at 2%/15 years for electrically-driven heat pumps, 4%/15 years for diesel/gas-driven ones, and 1.5%/15 years for absorption systems. The figures are of course uncertain and can vary much from installation to installation. The service life for absorption systems is probably in many cases longer than 15 years, but it seems meaningless to make economic comparisons over periods longer than that.

In Figures 10 a, b, c and 11 a, b, c, three different total investment costs have been used – 323, 806, and 1290 US$/kW heat output for electrically-driven systems. For diesel/gas systems, investment costs have been assumed to be 115% of this, and the corresponding figure for absorption systems has been set at 80%, as in today's large units. Also, the SPF value used, 1.3 (including boiler efficiency), is valid for large systems.

3.3.3 Methodology

As shown in the annual duration curve in Figure 8, there are basically two types of heat pumps: the A type, which uses e.g. sewage water, soil, industrial waste as heat source, and the B type, which uses e.g. ambient air. The monovalent- designed heat pump can be seen as a special case of the A type. Although the A type mostly in reality has a certain slope in its characteristic operation curve, for these calculations, as in Figure 8, this slope has been assumed to be 0.

A technicoeconomic comparison must be able to indicate if a heat pump is economic at all and if so, the optimal size of that heat pump in an annual duration curve under various technical and economic conditions. Hence, the method presented below has been used.

Using the annuity method, the total annual cost for a heat pump and the supplementary heating can be written, see Figure 9:

$$C = \kappa(a+m) \cdot P' + \frac{b}{SPF}\int_{0}^{P'} t(P)dP \cdot \frac{b_{oil}}{\eta_b}\int_{P'}^{P_{max}} t(P)dP$$

where

C = total annual cost for heating, \$/year
κ = specific investment cost for the heat pump, \$/kW
a = annuity factor, 1/year
m = annual maintenance cost, 1/year
b = energy price for the driving energy to the heat pump, \$/kWh
b_{oil} = oil (natural gas, etc.) price, \$/kWh
η_b = boiler efficiency
P = heat load

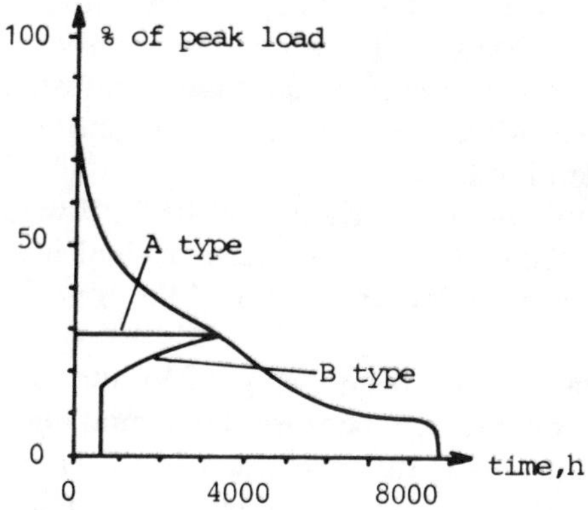

Figure 8. Two principal types of heat pumps, A and B.

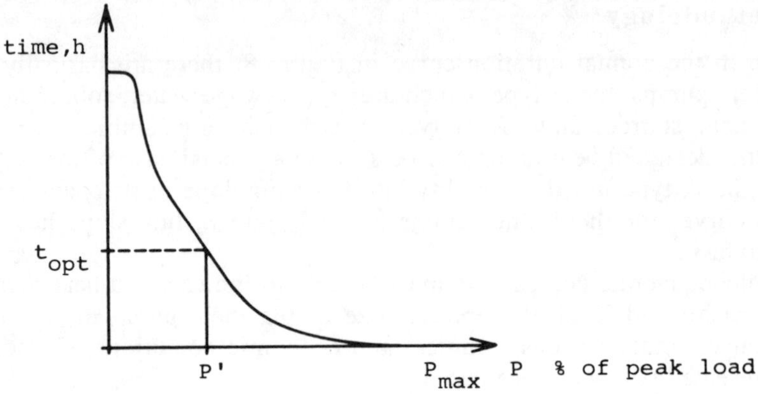

Figure 9. Annual duration curve (shown as operating time vs. heat load P) with a heat pump with heat output P' and an oil boiler for the peak loads.

The economic mix between the heat pumping and the supplementary heating is obtained when the total annual cost is at a minimum, i.e. when:

$$\frac{\partial C}{\partial P'} = 0$$

This gives:

$$t_{opt} = \frac{\kappa(a+m)}{\frac{b_{oil}}{\eta_b} - \frac{b}{SPF}}$$

Hence, by calculating values of t_{opt} one can determine if a heat pump is not economic at all (t_{opt} >~ 8000 h), if it is economic only in base load conditions (~ 4000 h < t_{opt} <~ 8000 h), or if it can cover a substantial part of the total annual energy (t_{opt} <~ 4000 h). In this way, economically interesting heat pump types can be identified.

In a thorough calculation, the variation of the SPF with t for the heat pump should be taken into account. Furthermore, marginal investment costs for the heat pump should be used. However, it would be to go too far to go into such details here.

In heat pump systems of the B type (Figure 8), the method described above is in principle not possible to use, because the annual energy coverage is lower for a given investment cost than is assumed in the method. However, it can easily be shown that the resulting error is small, at least at t values above 2000 h, and it is ignored here.

3.4 Results

The results of the calculations are shown as optimal t values versus the electricity/oil price ratio γ for three levels of investment costs, Figure 10 a, b, c and 11 a, b, c.

3.4.1 Comparisons between oil and electrically-driven heat pumps

In Table 4, the economic annual energy coverages for different types of electrically-driven heat pumps are shown. Because such heat pumps are of interest mainly at γ values definitely below 3, the results shown in the table

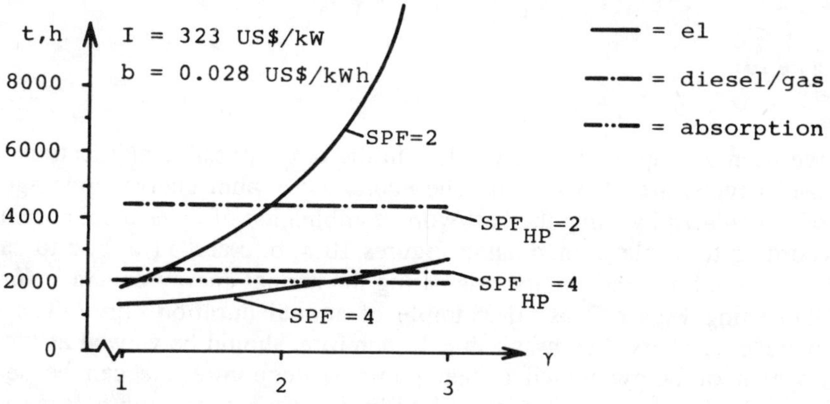

Figure 10a. Optimal annual operating time for different heat pump types compared with oil at three levels of investment cost and at the oil price 0.028 US$/kWh.

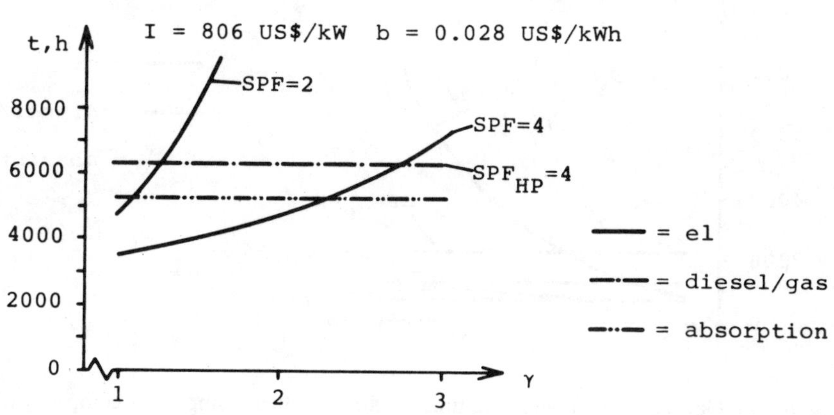

Figure 10b.

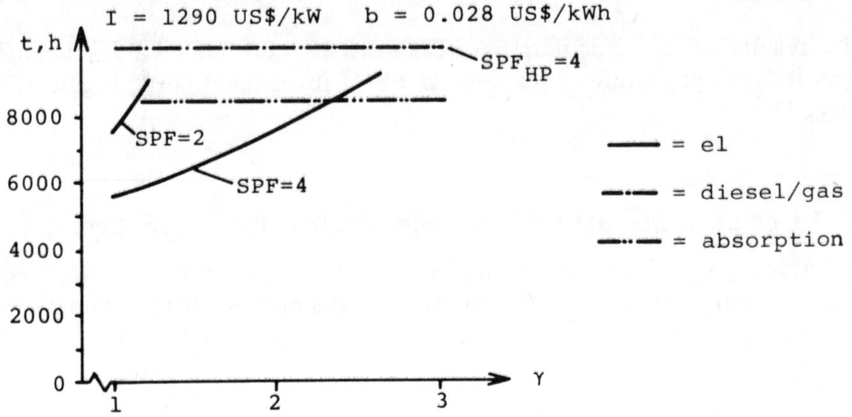

Figure 10c.

have been exemplified with $\gamma = 1.5$. In the table, possible applications for the various types are also shown. The economic annual energy coverages have been calculated by using the best/worst combination of investment cost and SPF according to Table 3, then using Figures 10 a, b, c and 11 a, b, c to calculate optimal t values, and finally calculating the annual energy coverage for each t value using Figure 7 as an example of annual duration curve. The energy coverage numbers shown in Table 4, therefore, should be viewed as the values at which or below which a heat pump is economic. As can be seen, the variations in investment cost and SPFs for each type give a large span of

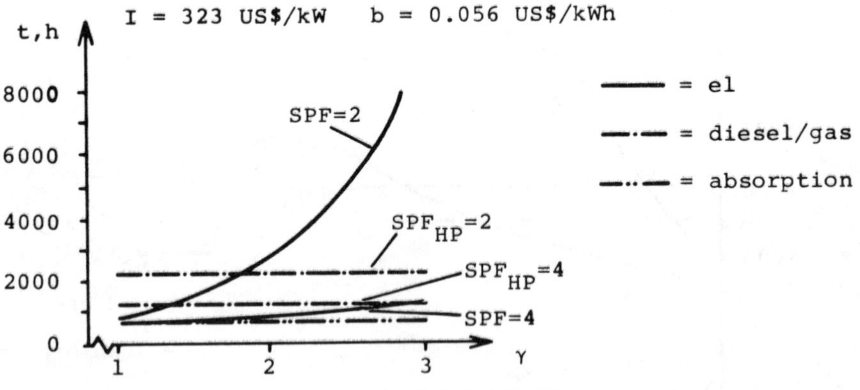

Figure 11a. Optimal annual operating time for different heat pump types compared with oil at three levels of investment cost and at the oil price 0.056 US$/kWh.

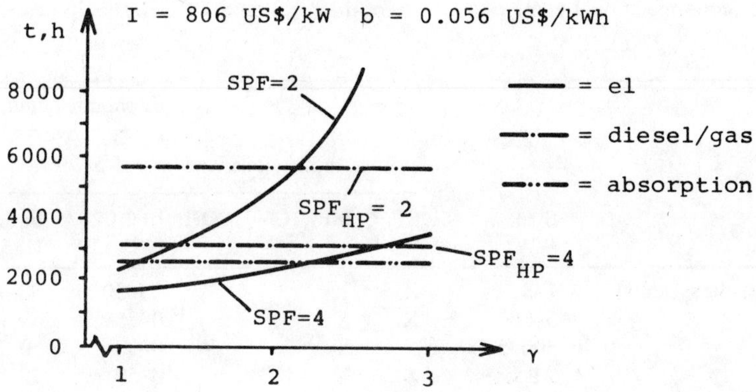

Figure 11b.

economic heat pump sizes. At a low future oil price, most heat pump types are of no economic interest or can be used only in base load situations. Exceptions are large heat pumps with water as heat source, especially industrial waste heat. At a future very high oil price, all types would be competitive. Due to the large variations in investment cost and SPF within each one, it is difficult to make any detailed comparisons between the different types. The optimum choice of system will depend to a large extent on the specific situation at each site.

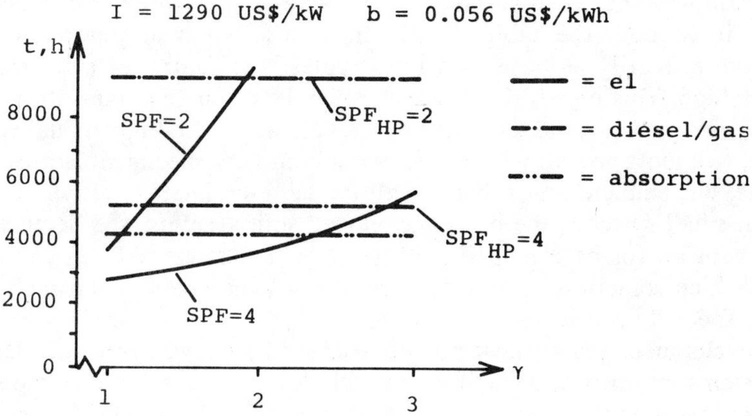

Figure 11c.

Table 4. The economic annual energy coverage for different types of electrically-driven heat pumps.

	Size (kW)	Applications SF	MF	C	DH	Economic annual energy coverages,% $\gamma = 1.5$ b=0.028 US$/kWh	b=0.056 US$/kWh
Ambient air (split systems)	1–3	X				25–30	60–75
Ambient air	5–300	X	X			0–25	40–60
Ambient air	300–5000		X	X	X	0–30	50–70
Exhaust air	All sizes	X	X	X		0–50	25–80
Exhaust air, with defrosting	All sizes	X	X	X		0–45	25–75
Soil	7–500	X	X	(X)		0–25	30–80
Ground water and energy wells	7–500	X	X	(X)		0–25	25–80
Sewage water	200–4000				X	40–75	75–85
Lake, sea, and river water	500→10000				X	50–70	75–90
Industrial waste heat	>500				X	50–80	80–95

SF = Single-family house
MF = Multi-family house
C = Commercial buildings
DH = Group centrals and district heating systems

A corresponding table like Table 4 except with $\gamma = 3$ would show that practically the only heat pump systems able to compete with oil would be those using industrial waste heat.

In addition to the three last types in Table 4, for which the technical potential is limited, the table clearly indicates that heat pumps will be competitive primarily as base load units, unless the future oil price will be extremely high. This means that more complex heat sources, like soil, ground water and energy wells, the idea with which is to achieve a high annual energy coverage, will most probably have a future only in very special situations. This means that air, ambient and exhaust, will be the most interesting type of heat source. In small systems, the newly developed split systems (see Section 3.2) with ambient air for base load should be of great interest. Also, exhaust air systems, which sometimes can have a small investment cost and a high SPF, should be specially considered.

The development for all heat pumps discussed here will probably lead to better systems in terms of reliability, control, defrosting, etc., but to moderate changes in investment cost (a decrease of less than 25%) and SPF (an increase of less than 25%) in the near future. Such changes will not alter the general economic conclusions made here. For possible new technologies, see Section 5.

A special application for heat pumps, not discussed above, is in commercial buildings (offices etc.). In such buildings, heating of the outer parts is often needed at the same time as cooling of the inner parts. Also, heating in winter and cooling in summer is needed even in climatic zones where air conditioning is normally not needed or used. Since such an installation is designed partly, or mainly, to obtain cooling, the extra investment in both a centralized or decentralized heating/cooling system to get the heat pumping effect is normally only a rather small percentage of the total cost. In such situations, heat pumping should therefore be much more economic than in most of the systems discussed above. Their actual economy depends on the amount of the extra investment, the annual operating hours when heat pumping is needed, etc., which differ from site to site.

In recent years there has also been an increasing interest in using heat pumps for the dual purpose of heat recovery and cleaning of stack gases. In such applications the stack gases are cooled below their dew point in the evaporator and are at the same time cleansed of SO_2, heavy metals, acids, etc. by one of several existing methods. Therefore, here too the investment cost is justified partly or to a high extent by other reasons than the heat pumping, making these applications economic.

Technicoeconomic studies on heat pump systems with seasonal heat storage have been made by Franck et al (20), (21) and Abel et al (22). According to these studies, the economic gain with a seasonal storage is often negative. In favourable situations, the achievable gain is normally modest. Thus, the future general market for seasonal storage systems with heat pumps looks to be very limited. However, there is one special case where such systems should be considered: in systems with cooling demands in summertime and heating demands in wintertime.

3.4.2 Comparisons between electrically and heat-driven heat pumps

Figures 10 a, b, c and 11 a, b, c indicate that diesel/gas heat pumps can compete with electrically-driven ones at γ values above between 2 (at low SPFs) and 3 (at high SPFs). These indications have been confirmed in more thorough analyses (2).

The cost and performance data for absorption heat pumps are, as mentioned, valid for large $LiBr/H_2O$ systems. In these, the maximum temperature lift is approximately 45°C, which would mean an SPF of at least 3.5 for an electrically-driven heat pump. Calculations of direct comparisons by the author have shown that such absorption systems can compete with electrically-driven ones at γ values between 2 and 3, depending on investment cost and oil price levels. Mostly, the critical γ value is around 2.5. For systems with higher temperature lifts, up to 55°C, especially small and medium- sized ones with ammonia/water, their characteristics are today more unclear, but their specific investment cost is slightly higher and their SPF slightly lower than for larger systems (see Section 3.2). This means a critical γ value closer to 3.

According to Figures 10 a, b, c and 11 a, b, c diesel/gas- driven systems can compete with oil, regardless of its price level, at today's typical SPF values, around 2.5, when the investment cost is about 900 US$/kW heat or lower. (Note that the SPF values given are related to the heat pump part only, not the engine/heat pump system, see Section 2.2). At the higher levels of these costs – 700–900 US$/kW – they are only economic in base load operation. Total investment costs for most types of heat pumps are in fact at these higher levels or higher, see Table 3. This means that also diesel/gas-driven heat pumps should be of interest mostly in base load operation in the future. However, as mentioned, the severe environmental and practical (availability, service life, etc.) problems will certainly restrict their use, at least in the near future.

Large absorption systems can compete with oil under all economic conditions used in Figures 10 a, b, c and 11 a, b, c, except at the highest investment cost and lowest oil price. Hence, systems of this type should be of interest in countries having district heating systems and γ values above approximately 2 in applications with relatively small temperature lifts. Such applications include those in which exhaust air, stack gases, or industrial waste heat are the heat sources.

3.5 General Conclusions and Discussions

Generally speaking, electrically-driven heat pumps can compete with oil firing and heat-driven heat pumps at γ values below 2.0–2.5. Unless the future oil price is extremely high, such heat pumps will normally only be economic as base load units, at least at γ values around 1.5 or above. Then the most interesting heat source is air (except in special cases, e.g. large systems with industrial waste heat). *Therefore, the most competitive types of heat pumps will probably be:*

- split systems, air-to-air, for base load in one- family houses;
- base load ambient air systems in medium-sized applications;
- exhaust air heat pumps in medium-sized applications;
- combined heating/cooling systems in commercial buildings; and
- heat pumps for both heating and cleaning of stack gases.

Most countries today have higher γ values than 2.5, and in them electrically-driven heat pumps will probably be economic only in very special situations.

In these countries in the near future, though, absorption systems should become the most competitive type of heat pump. The interesting R&D development in this field (see Sections 5 and 3.2) regarding temperature lift and investment cost may soon lead to more competitive systems also in small and medium-scale applications.

One factor not discussed above is the availability and the real number of annual operating hours achieved. Another is the oversizing of a heat pump plant, leading to part-load operation during a large part of the year. According to reference (23), such aspects can decrease the economy for heat pump systems considerably.

4 Heat Pumps in Industrial Processes

4.1 Introduction

An "industrial heat pump" is here defined as a system, which uses only industrial waste heat as heat source and delivers the heat to the industry to the processes and/or for space heating in the industry.

The technical conditions for industrial heat pumps do not vary so much among different countries as heat pumps for space heating do. However, the situation varies between industries, and industrial heat pump applications must often be tailor-made. Compared with space heating applications, industrial heat pumps often have the following advantages: high COPs (small temperature lifts and high temperature levels); long annual-operation times; small relative investment costs and acceptable pay-back periods; heat sources and heat sinks close to each other in space and time.

The number of heat pump installations in industry is very small compared with space-heating, in spite of the possibilities of the potentially favourable economics of such systems for industry in many cases. Some of the reasons for this situation are:

- lack of experimental and demonstration plants in different types of industries;
- uncertainty among the users about heat pump reliability;
- lack of good hardware in some types of applications;
- lack of the combined knowledge chemical engineering and heat pump technology in industries, consultant companies, etc.

4.2 Applications for Industrial Heat Pumps

Although the temperature levels in different types of industries vary considerably, two "typical" cases of heat source and heat sink temperatures can be identified (24):

- low-temperature case (primarily manufacturing industry):
  heat sink: 60°C – 100°C
  heat source: 20°C – 60°C
- high-temperature case (primarily process industry):
  heat sink: 120°C – 160°C
  heat source: 50°C – 110°C.

Situations where condensing temperatures much above 160°C are required are probably rather rare, except in some special types of industries.

The size range for industrial heat pumps is large, from about 100 kW to at least 10 MW heat output.

4.3 Types of Heat Pumps

The following types of heat pumps are discussed here:

- vapour recompression (also called steam compression) and open-cycle heat pumps,
- electrically driven compression cycles,
- combustion-engine driven compression cycles,
- absorption heat pumps, and - heat transformers.

Of these, combustion engine-driven and absorption heat pumps can be used only in the low-temperature case defined above, due to restrictions in engine-cooling water temperature (max. 110°C) and condensing temperature, respectively. Furthermore, for temperature reasons, heat transformers are much better suited for the high-temperature case. The vapour recompression cycle can only be used in the high-temperature case, i.e. for condensing temperatures above 120°C, when the evaporating temperature is above approximately 90°C. In some cases with very small temperature lifts, vapour recompression heat pumps can also be used in the low- temperature case. Then, it would have the same good economy as is shown later for the high-temperature case.

As pointed out in Section 2.2.2, there is a lack of good working fluids in compression systems at condensing temperatures above approximately 120°C.

Extensive work is being carried out internationally in the search for new fluids, including research into nonazeotropic mixtures (see Section 5). Due to this situation, some larger manufacturers have suggested the two-stage solution shown in Figure 12. Here the upper stage is a vapour-recompression cycle. The disadvantages of this solution seem to be its complexity and relatively high investment cost.

Most industrial applications of heat pumps, installed or discussed so far, have been in such industrial unit operations where there is a simultaneous excess of heat at a relatively low temperature and a demand of heat at a higher temperature. Unit operations of this kind involve distillation, evaporation, and drying. There are obvious advantages for heat pumps in such applications, including: agreement in time between waste heat production and heat demand; often a reasonable temperature lift and small surrounding costs for the installation. This unit-operation oriented approach is particularly applicable in industries where the unit operation in question is the one big process operation or the main energy consumer. However, in many industries, the energy system is complex and consists of many heat consumers and waste heat producers. In such plants, the introduction of a heat pump in a unit operation is often wrong from a thermodynamical point of view, and hence also from a technicoeconomic one. The reason for this is the conflict between heat exchange and heat pumping in industry, which is discussed in more detail below.

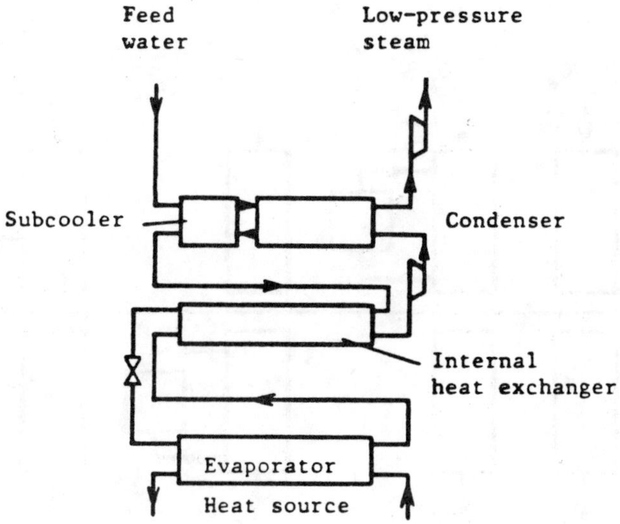

Figure 12. Two-stage electric compression heat pump for steam production.

Possibilities for improved heat exchanging in industry should be investigated first, before a heat pump is considered. A relatively new analysis method, called the Pinch Technology (25), is very useful for studies of both heat exchanger networks and appropriate process integration of heat pumps. With this method, it can be shown that most industrial systems have a "pinch" temperature, above which there is a net demand for heat and below which there is a net excess. This means that in the most intelligent heat exchanger network design, there is room for a heat pump – taking heat from below the pinch and delivering heat to another stream above it. This is shown in Figure 13, where three possible ways for process integration of a heat pump are shown. Only one of them, the last one, is thermodynamically and technically feasible, as is clearly indicated in the figure and by the discussion above. The fact that there exists a pinch temperature makes it always technically feasible to introduce a heat pump and the economic feasibility should always be investigated.

Using a combination of unit operation-based and process integration-based feasibility studies in industries, the potential for heat pumps there will most probably be shown to be greater than has so far been understood.

Another competitor to heat pumps, in addition to heat exchanging, is cogeneration. In large systems where back- pressure plants can be built in an economic way, heat pumps will normally not be able to compete in the future. Therefore, the size range of industrial heat pumps will be limited at the upper end to, say, around 10 MW heat output, except in situations with extremely low electricity prices. The competition between heat pumps and cogeneration

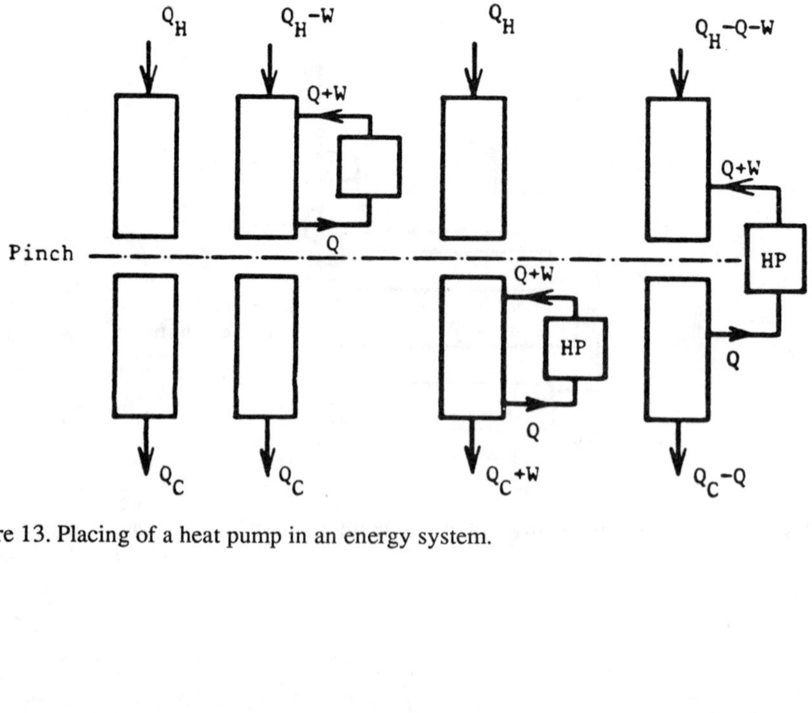

Figure 13. Placing of a heat pump in an energy system.

based on gas or diesel-driven engines or gas turbines in the small and medium-size range (1–10 MW) is discussed later. In the smaller application there is practically no limit to industrial heat pumps. In applications down to at least 100 kW, heat pumps should be competitive. In the comparisons below, data for gas turbines have been used as example.

4.4 Economic Aspects

The most crucial parameters for the future of industrial heat pumps are of course the investment cost and the future fuel and electricity prices. In this section, the impact of these and other major parameters on the economics of heat pumps is discussed. Various heat pump alternatives are compared with each other and with cogeneration.

Practically all industries today make their investment decisions based on pay-back period (PBP) calculations. Therefore, the PBP method is used here in comparing various investment alternatives. It must be pointed out here that the PBP method is not a good one for comparing systems on a life-time basis, but it does indicate the most promising types of technologies.

The investment costs used are shown in Table 5. These are reasonable in plants of about 5 MW heat output. For smaller units, the specific costs would be higher. Due to the small number of installed units, the figures used are sometimes uncertain. Anyone wanting to study the influence of this cost in the following diagrams can easily do so, as the PBP is directly proportional to the total investment cost.

Table 5. Total investment costs for various systems.

	Total investment cost/kW heat output	
System	SEK	US$
Vapour recompression	1000	143
Electric compression cycle (R12, R500)	1500	214
Electric compression cycle (R114)	1900	271
Two-stage electric compression cycle	2500	357
Dieseldriven heat pump (using heavy-fuel oil)	1750	250
Absorption heat pump	1500	214
Heat transformer	2500	357
Gas turbine (a = 0.5)	2500	357

4.4.1 Low-temperature applications

In Figures 14 and 15, the PBP is shown versus the electricity-to-fuel price ratio (γ) for two annual operating times – 8000 h (A) and 4000 h (B) – and at two different fuel price levels – 14.3 US$/MWh and 28.6 US$/MWh. Boiler efficiencies have been set at 0.85. From these figures, the following conclusions can be drawn.

Heat pumping is, in most situations, more economic than cogeneration. Only at very high γ values and low COPs can cogeneration compete. At high COPs, i.e. temperature lifts of approximately 40°C or lower, electrically-driven heat pumps are the most competitive over the whole range of γ-values. At low COPs, the internal combustion-driven systems would compete at γ-values above about 2. At these high temperature lifts, absorption systems cannot be used due to technical restrictions. The most interesting area for these is probably in situations where very cheap fuel, e.g., internally produced steam, bark, peat, can be used.

The PBP for non-electrically-driven systems can be below 3 years if the oil price and the annual operating time are high. The PBP for electrically-driven systems can be 3 years or well below that if the γ-value is low or the annual operating time high.

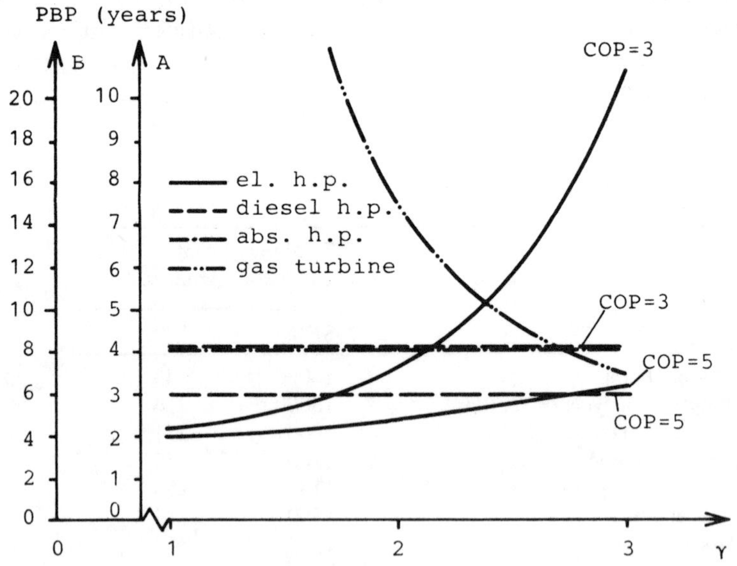

Figure 14. Pay-back period (PBP) vs. electricity/oil price ratio (γ) for low- temperature applications. Oil price 14.3 US$/MWh.

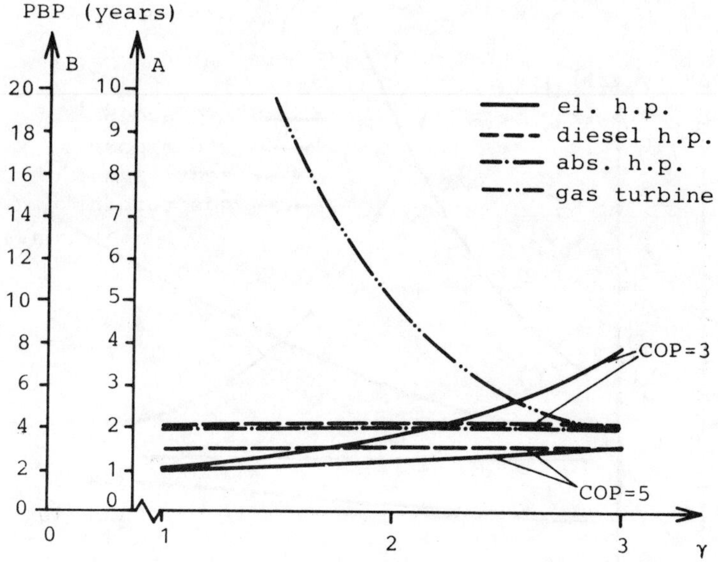

Figure 15. Pay-back period (PBP) vs. electricity/oil price ratio (γ) for low- temperature applications. Oil price 28.6 US$/MWh.

4.4.2 High-temperature applications

In Figures 16 and 17, the PBP is shown versus γ under the same conditions as in Figures 14 and 15. In situations where the evaporating temperature is above about 90°C all systems shown in the figures can be used. It is clear that vapour recompression is advantageous to all other systems at all levels of γ-values and can reach extremely low PBPs. Also, the heat transformer has a low PBP and should therefore be of interest. However, it shall be pointed out here that the calculations for the heat transformer cycle have been done under favourable technical conditions, i.e., assuming that the waste heat amount is at least twice that of the heat sink need.

At lower evaporating temperatures, where vapour recompression cannot be used, the heat transformer cycle is of interest regardless of the γ-value. The electrically- driven, closed-cycle heat pump can reach acceptable PBPs only if the γ-value is very low.

4.5 Conclusions and Discussion

From this economic comparisons, the following general conclusions can be drawn:

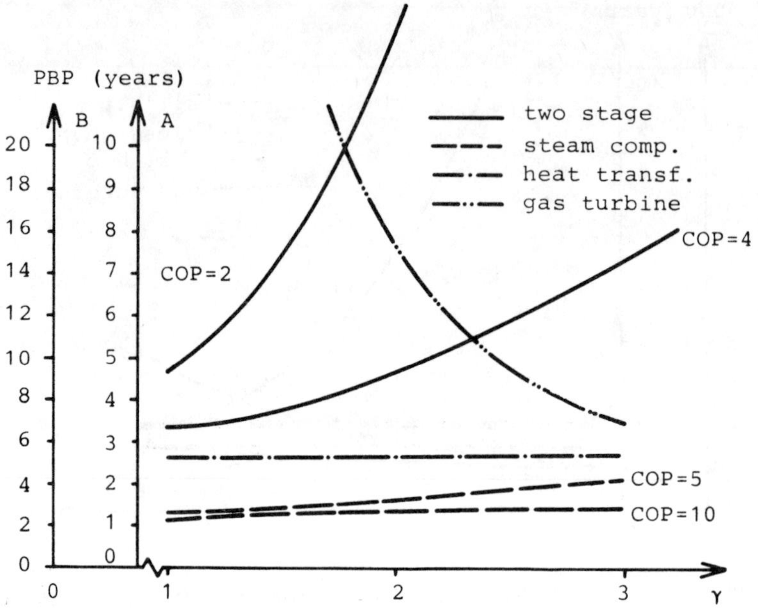

Figure 16. Pay-back period (PBP) vs. electricity/oil price ratio (γ) for high- temperature applications. Oil price 14.3 US$/MWh.

- cogeneration can compete with heat pumping only at high γ-values;
- there should be many situations where the PBP for heat pumps can become 3 years or well below;
- electrically-driven heat pumps are often of interest even at high γ-values, especially when the COP is high; and
- the vapour recompression and the heat transformer cycles should always be considered when the technical conditions are feasible.

One of the most important R&D areas for industrial heat pumps is the development of more competitive electrically- driven, closed-cycle systems for condensing temperatures above 120°C, for steam raising for instance, which can be used also in situations when vapour recompression and heat transformers are not feasible. Results so far in the R&D work with the compression/absorption cycle look promising, see Section 5.

One other important R&D area for industrial heat pumps is system oriented studies, in which suitable types of industries are defined and optimal sizes, evaporating and condensing temperatures, etc., for heat pumps are calculated. Also the influence on the process system, e.g. the heat exchanger network, flexibility, etc., must be investigated in general system studies.

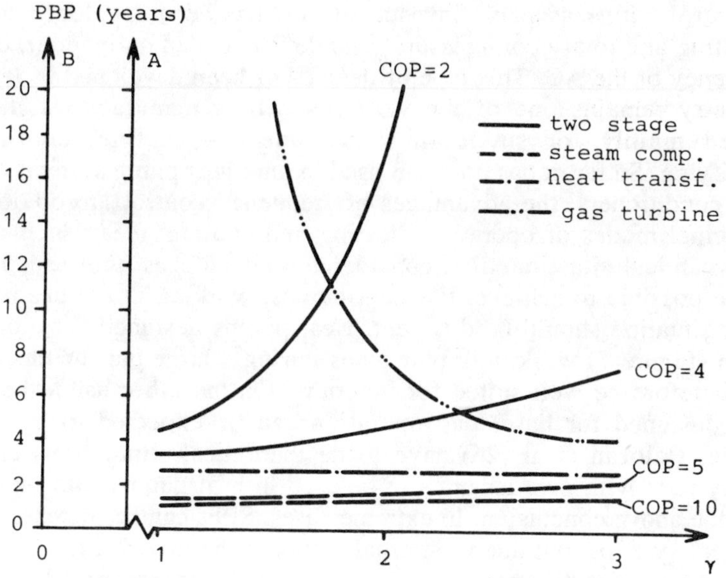

Figure 17. Pay-back period (PBP) vs. electricity/oil price ratio (γ) for high- temperature applications. Oil price 28.6 US$/MWh.

The technical and economic potential of industrial heat pumps seems to be substantial. *In fact, heat pumps should be able to compete with other energy technologies better in industrial markets than in the house-heating markets, and therefore the industrial field will probably be the most promising for the heat pump technology, at least in the near future.*

5 R&D trends

5.1 Compression Heat Pumps

Of the many ongoing R&D heat pump activities, only some of the major ones will be discussed.

5.1.1 Inverters

In Section 2.2.2, various types of capacity control have been mentioned. Speed control is the most efficient type of control. In recent years, a technology for speed control with reasonable investment cost has been developed for small reciprocating and rotary compressors. The device, called an inverter, can vary the frequency of the AC. This type of device has been developed in Japan and that country remains one of the very few which manufactures it. It was developed mainly for small air conditioners – heating, cooling, and dehumidifying. So far, it has not been used in pure heat pump systems.

In air conditioners, the advantages of frequency control are obvious. The two principal modes of operation, heating and cooling, mean such different thermodynamical situations that considerable advantages with speed control should be possible to achieve. For heat pumps, working in the heating mode only, the situation should be different. Heat pumps designed for monovalent operation (Figure 7) work with part loads during a large part of the year and should therefore be well suited for inverters. On the other hand, the gain in systems designed for base load operation can be expected to be small or negligible. Halozan et al (26) have made theoretical simulations and some laboratory tests with inverter units to be used in heat pumps, and their results confirm the above conclusion. In extreme cases, SPFs can be increased 30% in monovalent systems, but due to internal losses in the inverter, the SPF in heat pumps designed for bivalent operation is normally not improved, indeed, an even worse performance often results.

According to reference 26, other aspects to take into account are the fact that the capacity of a given compressor can be increased and that the negative operational effects of oversizing the heat pump (unfortunately a common error) can be decreased considerably.

5.1.2 Japanese super heat pump project

This project is probably the biggest R&D project ever in the heat pump field. It is scheduled to go on for 8 years, starting in 1984, and has a total budget of 80 million US $. The project organization includes many governmental agencies and private companies. The concept consists of a high- efficiency type of heat pump in connection with a chemical heat storage system. System sizes up to 30 MW are scheduled. The development aims are shown in Table 6. The aim of the heat pump-heat storage system is to use heat stored at nighttime to help meet the daytime surge in electricity demand, see Figure 18. To be able to achieve the aims of very high COPs, see Table 6, R&D on the heat pump includes multiple stage economizer systems and an expansion machine instead of an expansion valve. The work recovered in the expansion machine can be fed into the compressor.

Table 6. Development aims for the Japanese Super Heat Pump Project.

Item				Targets		
				Output temperature		Coefficient of performance COP
Sub-systems	Super high performance compression heat pump	High efficiency type	For heating only	85°C		8
			For heating and cooing	(Heating) 45°C		6
				(Cooling) 7°C		7
		High temp output type	For low temp heat source	–150°C		Over 3
			For high temp heat	–300°C		Over 3
				Output temp temperature	Heat storage capability	Heat recovery rate
	Chemical-heat storage system	High temp thermal storage		–200°C	Over 50 kcal/kg fluids	Over 75%
		Low temp thermal storage		under 10°C	Over 30 kcal/kg fluids	Over 75%
Total system		This super pump energy accumulation system for practical use must be superior to any other existing system in economy, applicability, load levelling effect etc. The targets of the pilot system should be as follows.				
		Application		Putput temperature		Energy efficiency
		Office building air Conditioning		(Heating) 45°C		4.5
				(Cooling) 7°C		5.3
		Community cooling/ heating and hot water supply		(Heating) 45°C		4.5
				(Cooling) 7°C		5.3
				85°C		6
		Industrial process heating		150–300°C		Over 2.3

5.1.3 Nonazeotropic mixtures

Nonazeotropic mixtures of refrigerants have been used for a long time in a few, very special, refrigeration plants. In recent years, there has been a growing interest for mixtures in heat pump applications. They are today also considered to be one of the most important alternatives to the CFC problem.

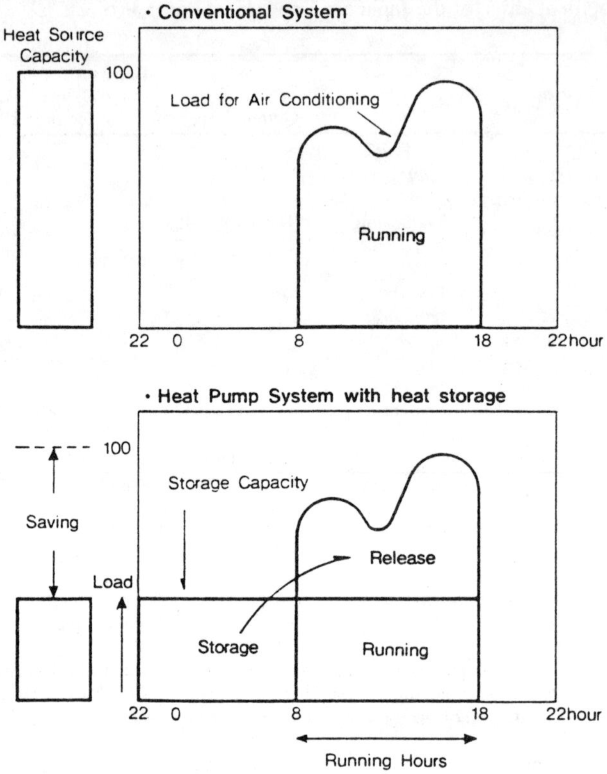

Figure 18. The effect of levelling the electric power demand by the aid of a day/night storage system.

The basic difference between a pure refrigerant and a nonazeotropic mixture is that in the mixture the temperature will increase along the evaporator and decrease along the condenser. The temperature variations in the heat exchangers are shown schematically in Figure 19. (In the figure, pressure drops and superheated parts in the heat exchangers are neglected.) In the figure, it is indicated that the necessary temperature lift in the compressor can be decreased, and hence the COP can be increased. The inlet evaporating and condensing temperatures, respectively, have been set to the same value for the pure fluid and the mixture. This would in reality correspond to a larger heat transfer area in the mixture case. This influence on the logarithmic mean temperature difference as well as lower heat transfer coefficients for mixtures means that the improvements in the COP with mixtures are in reality lower than calculated values for the ideal cycle. The possible real COP improvements

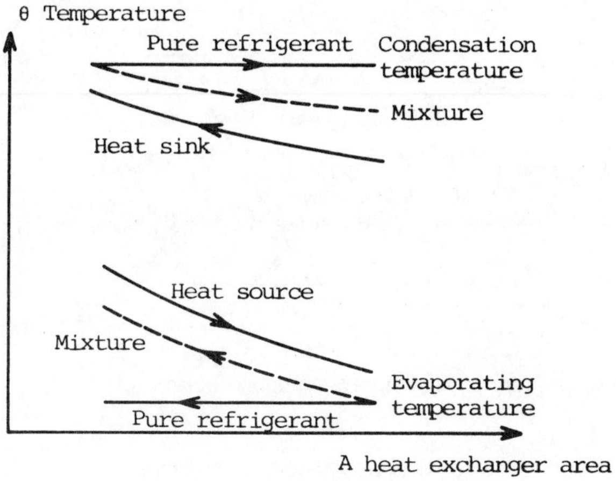

Figure 19. Temperature variations in the condenser and evaporator for a pure refrigerant and for a nonazeotropic mixture.

depend highly on the temperature gradients of the heat source and sink but range between 0% (gradients only a few degrees) and 20% (gradients 20°C or more).

In addition to the COP, several other aspects of the cycle can be improved with mixtures:

- the heating capacity,
- variation of the heating capacity with the load,
- the pressure levels in the condenser and the evaporator,
- the compressor discharge temperature,
- adjustment of series-manufactured equipments to various heating capacities,
- allowment of higher temperature gradients in the heat sink and the heat source.

These will be at least as important as the COP for the commercial introduction of mixtures.

Possible applications can range from small-scale to large- scale units, both for heating purposes and in industry. Today, nonazeotropic mixtures are being tested in some small scale plants in France and in some larger experimental plants in other countries. For further details see references (27), (28), (29).

5.1.4 Compression heat pump with solution circuit

This cycle was proposed by Altenkirch in 1950. It is also called the compression/absorption cycle. A diagram of this cycle is shown in Figure 20.

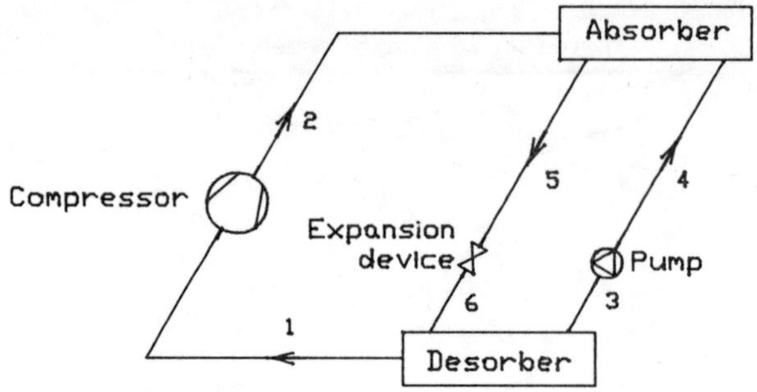

Figure 20. Principal diagram of the compression/absorption cycle.

Between the absorber and the desorber, a solution circulates in which the refrigerant is absorbed and desorbed, respectively. Several working pairs have been investigated. According to reference (30), ammonia/water is superior to all others. This is also the working pair used in the few existing full-scale plants, mainly in Germany.

As in the case of nonazeotropic mixtures, the internal heat exchanges take place at non-constant temperature conditions, so that also this cycle has the Lorenz cycle as the ideal one. The most important areas of applications are when:

- the temperature gradients of the heat sink and the heat source are large;
- a large temperature lift is required;
- a high condensing temperature (120–150°C) is required;
- the temperature gradient of the heat source is large compared to the gradient of the heat sink, e.g., for the production of low-pressure steam in industry.

For the last application above, a variation of the basic cycle (a semi two-stage one) should be used.

According to Åhlby et al (31), this cycle can be expected to give higher COPs in several types of applications than conventional heat pumps. The compressor sizes can also be kept considerably smaller. Under favourable conditions (large temperature gradients and/or high absolute temperature levels), COP improvements of up to 30% (in extreme cases higher) and compressor size reductions more than 40% can be achieved.

This cycle is more complex than the conventional one and is therefore better suited for medium and large-scale applications (above a few hundred kWs). In such sizes, the investment cost can be expected to be of the same order of magnitude as a conventional system. The reason is that the greater complexity and number of components are compensated by the smaller sizes of compressor and heat exchangers (due to higher heat transfer coefficients).

5.1.5 Measures to cope with the CFC problem

The growing concern for the influence of chlorofluorocarbons on the ozone layer and on future climatic changes has led to decisions nationally and internationally (i.e. the Montreal protocol, signed by 38 countries) about successive decreases in the use of R11, R12, R113, R114, and R115, i.e. fully substituted hydrocarbons. Although partially substituted hydrocarbons, e.g. R22, are generally believed to be much less dangerous, there are some doubts about their real environmental influence and their future use in a long-term perspective is uncertain. This situation will probably lead to the biggest changes in the modern history of the refrigeration and heat pumps, and it means a huge challenge for researchers in this field.

A large number of possible future solutions to cope with the CFC problems has been proposed and discussed. They are presented in Box 1.

As a long-term substitute for R12, R134a is usually mentioned. An extensive international R & D work on this fluid has recently begun. These efforts are focussing on R134a's (it is insoluble in normal oils), stability tests and performance in different types of systems. Furthermore, toxicity tests are not yet finished. There are also concerns on the time perspective when it can be introduced in full scale. Due to uncertainties about this R & D work is also going on in the fields of nonazeotropic mixtures and flammable fluids. Several mixtures can be used instead of R12 – e.g., R22/R142b and R22/R152a – and then the advantages for mixtures discussed above can be realized as well. One crucial question in the R & D work on alternative fluids is: can flammable fluids and ammonia be used in different types of applications, and if so, to what extent? Today, there is no answer to this question.

As indicated above, there exist several possible long-term solutions of the CFC problem. The enormous R & D work in this field, which has started or is planned, may well lead to a new generation of better designed heat pumps with smaller leakage and control problems and maybe even improved performance.

5.2 Advanced Absorption Cycles

An absorption cycle is called advanced if it yields a higher COP or a higher temperature lift than could be accomplished with a single-stage cycle using the same working fluid. A review of advanced absorption cycles has been done by Ziegler (32).

One way of enhancing the performance is by multistaging. In the multistage cycle, either the COP or the temperature lift can be enhanced, although it of course requires a more expensive and complex system. Examples of the two-stage double-lift and double-effect heat transformers can be seen in Figures 21 and 22, from reference (33). The heat pump and heat transformer cycles can

Box 1: Coping with the CFC Problem

In the presentation below is assumed that R22 can be used in the future despite recent fears that it may be forbidden. R12, R11, R113, R114 and R115, however, are presumed to be banned substances.

Measures possible in existing heat pump systems:

- Improve service and maintenance so that leakage during operation will be reduced.
- Special actions during operation in the working fluid circuit, i.e. collection of fluid when repairing.
- Exchange of R12 to R500 or R12/dimethylether (both are azeotropes) as a first step towards reduction of R12 use.
- Exchange of R12 to inflammables, i.e. hydrocarbons and R152a.
- Exchange of R12 to nonazeotropic mixtures with R22 as "main component", i.e. R22/R152a (non-inflammable) and R22/R142b.

Measures possible in heat pump systems installed during the next few years:

- Improved mechanical construction to reduce the risks for leakage.
- Improved design to reduce the amount of working fluid, i.e. indirect systems (mounted in the factory as well as at site), direct expansion instead of natural or pump circulation.
- Return to ammonia.
- Design of installations for higher pressures (40 bar) so that R22 can be used.
- Exchange environmentally damaging CFCs with nonazeotropic mixtures either of "environmental- friendly" CFC or FC-fluids or of hydrocarbons.

Measures possible in new heat pump systems installed 5–10 years from now:

- Develop new CFC and FC-fluids and possible azeotropic and nonazeotropic mixtures of these, i.e. R134a. (The reason this will take at least 5 years is because of the time-consuming but necessary process of testing the new fluids to determine their toxicity and whether they are carcinogenic. In addition, time will be required for CFC manufacturers to make the necessary changes in their production systems.)
- Develop new compression processes, i.e.:
  - The compression/absorption cycle (ammonia/water),
  - The Joule cycle or variations of it (air or some other gas), and
  - The Stirling cycle.

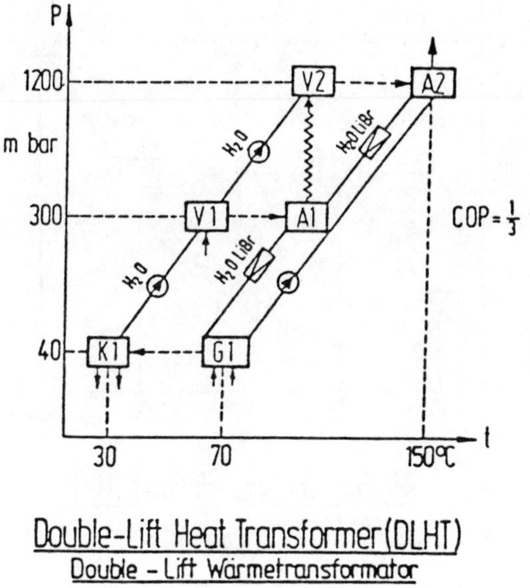

Figure 21. Double-lift heat transformer.

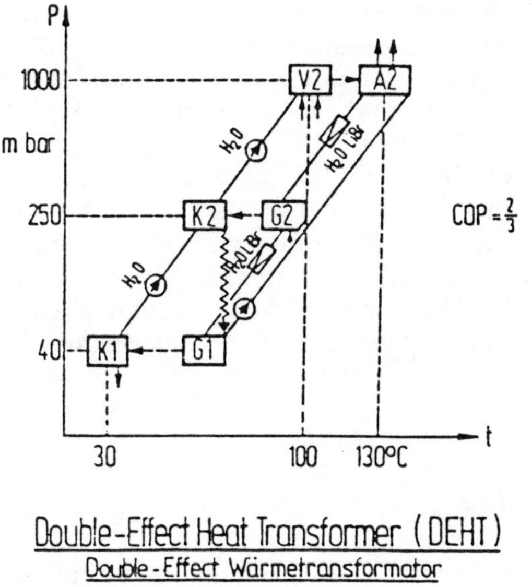

Figure 22. Double-effect heat transformer.

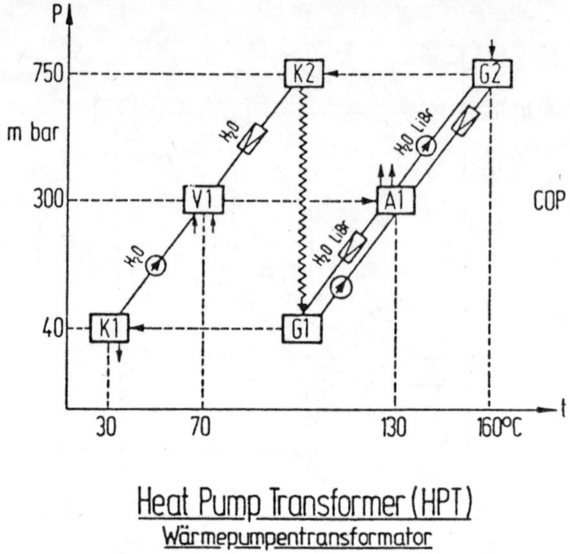

Figure 23. Heat pump transformer.

be combined, see Figure 23, to give the so-called heat pump transformer (33). In this cycle, a high COP and a large temperature lift, from the heat transformer point of view, can be achieved by substituting a part of the waste heat with a high temperature heat source.

Multistage cycles have been discussed in a number of theoretical works, see for example, reference (34), where both possible configurations and performance have been shown.

According to Bokelmann (33), a large potential for multistage absorption systems exists, which has led to a number of R&D-projects in Europe. The projects include a number of aspects, e.g., operating behaviour and control of multistage plants, heat and mass transfer equipment, material tests, new working fluids, economic studies, and optimization of manufacturing. The aim of these projects is to build two industrial demonstration plants in 1988.

Acknowledgements

The author wishes to express his gratitude to Per Holmberg, Lic. Eng., for valuable help and information concerning absorption systems, and to Karen Munch Berntsson, Ph. D., for carrying out calculations in the technicoeconomic comparisons parts.

References

1. Duffield, J., Hodgett, D. 1979. The choice of working fluids for heat pumps condensing at up to 120°C. *Proc. 15th IIR Congress, Venice*, 2:259–65.
2. Munch Berntsson, K., Berntsson, T., Hodgett, D. 1984. *Large diesel driven heat pumps for heating purposes*. Report to Swedish Council for Building Research, R129:1984 (Partly in Swedish).
3. Villardsen, V., Flemming, V. 1980. A balanced view of reciprocating and screw compressor efficiencies. *Purdue Compressor Technology Conference.*
4. Paul, J. 1978. Performance characteristics of reciprocating compressors for heat pumps of various speed considering the valve behaviour. *Wärmepumpentechnologie*, Band II, Essen.
5. Lundberg, A. 1981. Design basics-screw compressors. *Purdue Compressor Technology Conference.*
6. Dequeurce, B., Banquet, F., Denisart, J.P., Favrat, D. 1984. Use of twin screw compressor for steam compression. *Proc. 2nd BHRA International Symposium on The Large Scale Applications of Heat Pumps, York,* pp. 189–96.
7. *VDI Wärmeatlas*
8. *ASHRAE Handbooks*
9. Burtscher, J., Oberhumer, M., Streicher, W. 1986. Efficiency degradation due to cycling and refrigerant migration of an air-to-water heat pump. *Proc. Int. Workshop: Research activities on Advanced Heat Pumps, Graz.* dbv-Verlag, Verlag für die Technische Universit{t Graz.
10. Nilsson, P-E. 1985. *Wind convectors, a technical and economic study.* Report 1985:2 from the department for Building Services Engineering, Chalmers University of Technology, Sweden (in Swedish).
11. Tollin, J. 1984. *Rock heat in residential buildings – dimensioning.* Report to Swedish Council for Building Research, R183:1984 (in Swedish).
12. Berntsson, T. 1983. Subsurface heat storage – a general overview. *Proc. Int. Conf. on Subsurface Heat Storage in Theory and Practice, Stockholm.*
13. Fahlén, P. 1988. *Laboratory testing of heat pumps – Experiences 1984–1986.* Report to Swedish Council for Building Research, R1:1988 (In Swedish).
14. Yokoyama, T. 1988. Small gas engine heat pump (GHP) for commercial use in Japan. *IEA Newsletter,* 6(1).
15. Lehmann, A. 1986. Sorption Heat Pump Systems - Manufacturers, Product Information and Applications. *IEA Newsletter,* 4(1).
16. Meckel, B. 1988. Possibilities to Reduce Costs for Gas Absorption Heat Pumps in the Capacity Range of About 22 kW. *Preliminary Proc. CEC-British Gas International Workshop on Absorption Heat Pumps, London,* pp. 308–15.
17. *VDI-Berichte* 427. 1981.
18. Wimby, M., Franck, P-Å., Berntsson, T. 1987. A Large Absorption Heat Pump. *Proc. 3rd Int. Symposium on the Large Scale Applications of Heat Pumps, Oxford,* pp. 21–24.
19. von Cube, H.L. 1981. Die Absorptionswärmepumpe im Wirtschaftlichkeitsvergleich mit anderen Wärmepumpen. *VDI-Berichte* 427, pp. 15–22.
20. Franck, P-Å. 1986. *A study of heat pump systems with low-temperature seasonal heat storage in clay.* Doctor thesis. Department of Heat and Power Technology, Chalmers University of Technology, Sweden. 226 pp.

21. Franck, P-Å., Berntsson, T. 1985. Optimization of a combined heat pump/heat storage system with solar or convective heat collectors. *Proc. ENERSTOCK 85, III Int. Conf. on Energy Storage for Building Heating and Cooling, Toronto,* pp. 448–55.
22. Abel, E., Aronsson, S., Dalenbäck, J-O., Nilsson, P-E. 1987. *The economic margin for alternatives in new energy technology*. Report to Swedish Council for Building Research, D4:1987.
23. Nilsson, P-E. 1988. *Introduction of heat pumps into existing heating systems*. Document D6:1988, Department of Building Services Engineering, Chalmers University of Technology, Sweden (in Swedish).
24. Berntsson, T. 1987. Future prospects for industrial heat pumps in Europe. *Proc. 1987 IEA Heat Pump Conference, Orlando,* pp. 303–24. Lewis Publishers Inc.
25. Linnhoff, B., Turner, J.A. 1981. Heat-recovery networks: new insights yield big savings. *Chem. Engineering,* Nov. 2, pp. 56–70.
26. Halozan, H., Katona, O., Gilli, P.V. 1988. *Inverterdriven heat pumps*. Report to IEA Heat Pump Center. Graz.
27. Berntsson, T., Ljungkvist, K., Moser, F., Schnitzer, H. 1984. *High-temperature working fluids and nonazeotropic mixtures in compressor driven heat pumps - A state-of-the-art report.* Report to IEA Annex VI (only available in participating countries).
28. Munch Berntsson, K. 1986. *Nonazeotropic mixtures as working fluids in large heat pumps – Heat transfer in a falling film evaporator and system simulations*. Doctor thesis, Chalmers University of Technology, Sweden. 217 pp.
29. Kruse, H. 1987. Current status and future potential of nonazeotropic mixed refrigerants. *Proc. 1987 IEA Heat Pump Conference, Orlando,* pp. 173–94. Lewis Publishers Inc.
30. Hodgett, D., Åhlby, L. 1987. The effect of the properties of the refrigerant and solvent on the compression/absorption cycle. *Proc. XVIIth Int. Congress of Refrigeration, Vienna,* Vol. E, pp. 664–71.
31. Åhlby, L., Hodgett, D. 1987. Compression/absorption systems – Simulation of two cycles for different applications. *Proc. XVIIth Int. Congress of Refrigeration, Vienna,* Vol. E, pp. 785–92.
32. Ziegler, F. 1985. Advanced Absorption Cycles. *Proc. Int. Workshop on Heat Transformation and Storage,* Ispra, Italy, pp. 33–49.
33. Bokelmann, H. 1988. Advanced Absorption Systems. *Preliminary Proc. CEC-British Gas International Workshop on Absorption Heat Pumps, London,* pp. 234- 43.
34. Ziegler, F., Alefeld, G. 1987. Coefficient of performance of multistage absorption cycles. *Int. J. Refrig.* 10(5).

Use of Electricity in Commercial Buildings

Enno Abel

Abstract

Commercial buildings use about 10% of Sweden's electricity for their operation, excluding heating, mostly for lighting, and ventilation and air-conditioning. Savings of electricity are possible by careful planning of lighting and heating, ventilation and air conditioning (HVAC) systems. Some technical opportunities are discussed and their potential for electricity savings is estimated. Well-designed new buildings could use approximately half as much electricity per square meter as the present commercial building stock average. The potential reductions in electricity use in existing commercial buildings is difficult to estimate but should be of the order of tens of percent. These efficiency improvements are cost-effective to the consumer. A societal evaluation in comparison with the costs for new generating capacity would increase the potential for efficiency improvments.

1 Introduction

In evaluating the energy requirements in buildings, it is important to distinguish between energy needed for heating and that needed in the form of electric energy. Also, buildings are used for different purposes and have different characteristics that determine energy requirements.

The need for thermal energy is determined by different factors than those that determine the need for electrical energy. Although there may often be a number of ways in which a building's heating requirements can be met, there is generally no realistic alternative to supplying its electrical energy requirements other than to connect to the local utility. Conservation measures intended to reduce heating requirements in existing buildings will not decrease electricity demand other than to the extent that electrical energy is used directly for heating. To influence other aspects of electrical energy use, it is necessary to apply quite different conservation measures. Merely, as is unfortunately often the case, to talk about energy use or energy conservation, without specifying whether it is thermal energy or electrical energy that is meant, can all too easily result in discussions and conclusions that bear little relationship to reality.

When studying energy and electricity in buildings, it is important to apply some form of grouping into areas of similar activities and/or characteristics, and to analyse each group on its own. To maintain the discussion on a comprehensible level, we must choose a basis of divisions that allows us to divide the building sector up into only a few groups but which allows differences in energy use characteristics to be illustrated meaningfully.

For starters, we can make a rough divison of the building sector into only three groups:

- residential buildings
- commercial premises,
- industrial buildings.

In residential buildings, heating requirements are determined mainly by the buildings' structure and design. In order to reduce heating requirements, we first need to influence the structural design, e.g., by improving thermal insulation and by draught-proofing the climate screen. The actual need of electrical energy to produce a particular indoor climate is normally very small.

In commercial buildings, such as offices, hospitals, laboratories or stores, it is the use to which they are put that primarily determines the heat requirement. High ventilation air flow rates are often required in order to deal with airborne pollution and surplus heat resulting from the work or activities carried on in the buildings. The heating energy requirement for air treatment may be the major element in a building's total heat demand. By installing efficient heat recovery equipment this can be almost eliminated and thus the total need of heat decreased substantially. Improved insulation and tightness of the building envelope usually has a much smaller influence on the demand of heat. Modern commercial buildings tend to have low heating requirements and considerable surplus heat and they require cooling to maintain comfort. The electrical energy demand for powering climate control equipment may also be significant, particularly if the use of refrigeration plant for artificial cooling is required.

In industrial buildings, it is common that both the ventilation equipment and the production processes have high energy demands. In such buildings, building design often plays a relatively minor part in determining the level of energy use.

If we consider the entire building stock in a country, we find that it is residential buildings that account for the greater part of the country's building volume. In Sweden, for example, the total building stock amounts to over 2000 million m^3, of which residential buildings account for over 60%. Commercial premises of various types account for less than 25%. This proportion can well be regarded as typical for industrialised countries.

The residential buildings are also relatively easy to modify in terms of energy characteristics, although, as Elmroth notes elsewhere in this volume, retrofitting existing residential buildings with energy conserving measures can be costly. In the past, discussions of the level of energy use in buildings therefore tended to concentrate on aspects concerned with residential buildings. However, and particularly in the case of electrical energy, it is important to consider commercial buildings, which have considerable demand for electrical energy. This can be illustrated by the total electrical energy use for residential buildings and commercial premises in Sweden in 1985 (1, 2). Table 1. Electrical energy used for heating purposes in systems in which it could easily be replaced by fuels, such as in small oil-fired boilers with electric immersion heaters, and in interruptible-supply electric boilers on district heating systems, has not been included.

The difference in the characteristics of electrical energy use in residential buildings and commercial premises is also revealed very clearly if we compare the specific electrical energy use levels in the two groups. Table 2 illustrates

Table 1. Total electricity use in Swedish residential and commercial buildings, 1985.

Residential buildings (400 million m^2)	TWh
• Direct electric heating in detached houses	8
• Electric boilers in detached houses	3
• Electric heating in apartment buildings	1
Subtotal electric heating	12
• Domestic appliances	14
Total:	26
Commercial buildings (155 million m^2)	
• Electric heating	3
• Electricity for operational purposes	13[1]
Total:	16

Source: Reference 1 and 2.

[1] If electricity for operation of commercial premises interspersed in residential buildings, street lighting, sewage cleaning and water treatment plants etc., is included, the total will be 18 TWh instead of 13 TWh.

Table 2. Specific electricity use in buildings in Sweden and the U.S.

Sweden[a]	
Residential stock, 1985	65 kWh/m^2-year
Commercial stock, 1986	85 kWh/m^2-year
USA[b]	
Residential stock, 1984	55 kWh/m^2-year
Commercial stock, 1983	140 kWh/m^2-year

a Source: Reference 2.
b Source: Reference 3.

this for residential and commercial premises in Sweden and the USA. The values shown are mean values of electrical energy use (defined as total electrical energy use/total floor area) in each sector.

The difference in mean specific electrical energy use between the two sectors is considerably greater in the USA than it is in Sweden. This is due, of course, to the higher use of air-condition in commercial buildings in the U.S., and to the use of electric heating in residential building in Sweden: over half of the total use of electrical energy in residential buildings is accounted for by electric heating. In this respect Sweden, together with Norway, represents an extreme case. In the majority of industrialised countries, specific use of electrical energy in the non-residential non-industrial sector is significantly higher than in the residential sector.

2 Energy Requirements of Commercial Premises

It is relatively easy to obtain adequately detailed statistical data concerning residential buildings. Many statistical surveys of existing residential buildings and anlyses of the factors that determine heat and electricity requirements have been carried out, particularly in respect of detached houses. There is a good fund of knowledge of how existing residential buildings can be improved, and of how new residential buildings should be designed in order to ensure low heating requirements. There is also a relatively substantial basis for economic assessments of feasible measures. The use of domestic electricity, too, is well-documented and analysed, while domestic electric appliances have become (and are continuing to become) more efficient.

The same degree of attention has not been applied, in general terms, to commercial premises. It is difficult to find reliable and comprehensive statistical data. As far as the use of electricity in commercial premises is concerned, statistics are available for American conditions, published by the DOE (4), and giving details of the use of electrical energy in commercial premises in relation to various types of systems. Lawrence Berkeley Laboratory (LBL), at the University of California, Berkeley, has a detailed data base for commercial premises, including such information as detailed figures (confirmed by metering) for the use of electrical energy (5).

Statistical material available in Sweden, and updated each year, includes the total building volumes of commercial premises of various kinds, together with details of levels of use of heat and electricity in the premises. From these values, it is possible to calculate specific heat and electricity levels of use figures per unit of volume (kWh/m^3-year). However specific energy use in buildings is often related to floor areas (kWh/m^2-year), which has been done here, assuming an average room height of 4 meters.

Figure 1 shows the change in energy use in various types of premises, based on Swedish statistics.

The levels of use of electrical energy shown in the diagram refer to electricity for operational purposes only, e.g., ventilation, lights, computer,

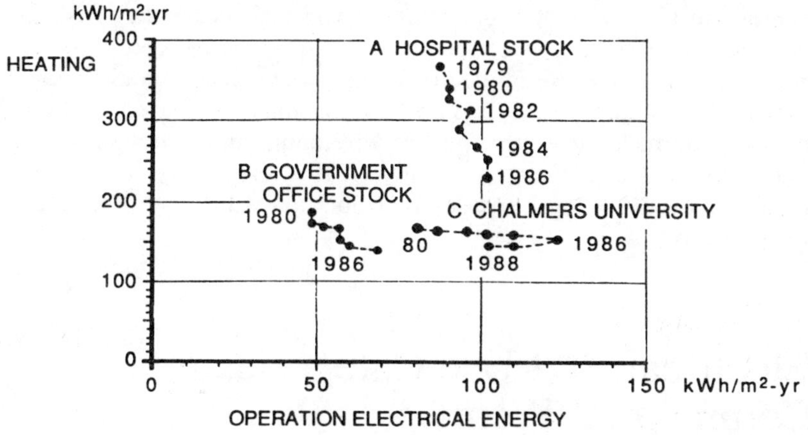

Figure 1. Measured heat and electricity use levels in three types of commercial premises in Sweden.

A: Health service and hospital buildings, 1979–1986. The data is for 45 million m^3 of building stock. About 80% of the total stock of the country's health service and hospital buildings is included (6)

B: Government office buildings, 1980–1986. The data is for 70 office buildings with a total volume of about 2.2 million m^3 (7).

C: Chalmers University of Technology 1980–1986. The data is for 30 buildings with a total volume of about 0.8 million m^3 (7).

copier, etc. Electrical energy used for heating purposes has been included in the heat use figures.

It can be seen that heating energy demand is falling, while the demand for electrical energy for operational purposes is rising, except for Chalmers. The increasing demand for electrical energy is characteristic for the entire Swedish building stock. Its total demand for thermal energy has fallen from about 130 TWh in 1980 to about 115 TWh in 1986. At the same time, total use of electrical energy, excluding that used for heating purposes, has increased from about 28 TWh to about 34.5 TWh (8, 9). One, possibly significant, part of this might be so-called "hidden" heating, where local electric heaters are used without being registered as electric heating.

The ways in which we use electrical energy differ considerably from one country to another, depending on the local cost of electrical energy, or more strictly on the relationship between the cost of electrical energy and the cost of thermal energy.

The relatively low price of electrical energy in Sweden has naturally contributed to an overall increase in its use. However, this "conscious" increase, that can be attributed directly to the low price of electrical energy, has occurred almost entirely in the use of electricity for heating of buildings, i.e. direct electric heating or electric boilers in detached houses, or electric boilers in group heating plants and district heating plants. Electrically-driven heat pumps are also another way of employing electrical energy for heating purposes, and a particularly efficient one at that. However, it is not this conscious increase in the use of electrical energy that is shown in Figure 1, but what might more suitably be called an "unconscious" increase, resulting from changes in patterns of use, changed requirements and – not least – in the physical equipment and systems employed. It is electricity for lighting and ventilation/air conditioning purposes that dominates the use of electricity in commercial premises.

3 Electrical End-Use Categories in Commercial Premises

The use of electrical energy in commercial premises can be divided up among three main types of loads:

- Apparatus and equipment being used as part of, or in connections with, the activity being carried out. The use of electricity for elevators is also included here.
- Lighting.

- Heating, ventilating and air conditioning and cooling systems (HVAC), and distribution (pumps and fans).

It is at present difficult in Sweden to obtain a satisfactory statistical breakdown of the distribuiton of electricity use between these three application areas, but an estimate is given in Table 3.

The values shown relate to premises built during the 1960s and 1970s. Electricity use in the first group also includes that for refrigerators, freezers and cookers.

Work being carried out at Chalmers University of Technology in Gothenburg includes metering of electrical energy for operation of air treatment systems. The results from this work indicate that air treatment and ventilation plant in more modern office buildings require about 30 kWh/m2-yr of electrical energy input, i.e. considerably more than shown in the table above. According to Figure 1, the real need of electricity in the health care sector is about 100 kWh/m^2-yr. There would seem to be justification for obtaining more accurate information from buildings in use and it is hoped that this will be done during 1989. Nevertheless, it is perfectly clear that lighting and air conditioning systems in commercial premises use considerable quantities of electrical energy, hence any improvements in efficiency of these applications will result in more than just marginal effects.

When discussing energy-related measures in buildings, it is important to distinguish between those applied to existing buildings and new designs and components that can be incorporated to in the new buildings. In existing buildings, it is generally not possible to do much other than to modify operating regimes and/or replace components and system elements. As the property-owner or user is generally interested only in making investments intended to reduce running costs, any improvements must be reasonably

Table 3. Theoretical calculation of electrical end-use categories in different types of existing Commercial Buildings

	Offices, kWh/m^2-yr	Hospitals kWh/m^2-yr	Department stores, kWh/m^2-yr
Apparatus and equipment, including lifts	15	15	75
Lighting	55	30	80
Air conditioning and climate control	11	30	20
Total	81	75	175

Source: Reference 1.

profitable. This means that they weigh the cost of the electricity efficiency investment against the repultant reduction in their electricity bill over the lifetime of that investment. In so doing, of course, the value the would-be investors put on money spent now (and over a loan-repayment period) versus money saved in the future over a number of years, i.e., the investors discount rate, is crucial. Figure 2 shows the permissible costs of measures intended to save electricity as a function of other parameters.

The diagram can be used to provide a first, rough assessment of how much a measure can be allowed to cost. Consider, for example, the case of changing to more electrically efficient lighting that can be expected to reduce specific electricity use by 20 kWh/m^2-yr. If the rate of interest is 8%, what is the maximum permissible cost of the investment?

From Figure 2, we can see that the intersection point indicates that N = 130. This means that, for an electricity cost of \$0.07/kWh, the total specific investment cost may not exceed 0.07 x 130 = \$9.00/m^2. For an electricity cost of 50 öre/kWh, the maximum permissible investment may not exceed SEK 65/m^2.

In existing buildings, it is often expensive to undertake improvements unless they are extremely simple to apply. However, if measures intended to reduce the use of electrical energy can be included among any other changes or modification of building services systems that must in any case be carried out,

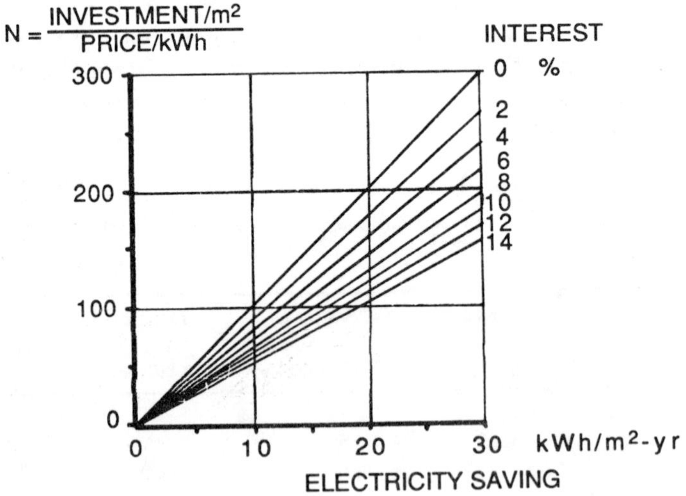

Figure 2. Break-even specific investment divided by the average price of electricity saved as a function of potential electricity saving in specific electricity use (kWh/m^2) for different interest rates, based on an operating life of ten years. This evaluation is valid for evaluating investment options from the point-of-view of the electricity customer.

a considerable improvement in economic prospects can be expected. It is important that any such opportunities for improving the efficiency of electrical energy use should be exploited. This would be facilitated if long-term plans for improving the efficiency of electrical energy use in commercial buildings with higher electrical energy demands were drawn up.

In the case of new building projects, economic justification for measures intended to improve the efficiency of electrical energy use is easier. Here, higher efficiency of electricity use can often be achieved through only marginal increases in investment. However, this does require that the parties engaged in the building process are aware of the opportunity and diligently apply themselves to achieve low levels of electricity use. Unfortunately, this is not always the case.

3.1 Apparatus and Equipment

Approxmately 15% of the electricity use in the commercial sector is for refrigeration/freezing, and food preparation, and another 25% is used for miscellaneous machines, including photo-copiers, computers, etc. An analysis of the efficiency opportunities in these areas are outside the scope of this chapter, but they are partly discussed in the chapters by Nørgård and Norford et al. in this volume.

3.2 Lighting

As described above, electrical energy for lighting constitutes a significant proportion of electricity demand in commercial premises. Besides the energy used directly in lamps and fittings, lighting also results in an indirect demand because it increases the amount of surplus heat that needs to be dealt with by the air conditioning equipment.

In premises having no natural light, or where lighting is required even during the daytime, the heat from lights often constitutes a major part of the heat load that determines the required ventilation air flow rates and cooling powers. About half of the specific electrical energy used for air conditioning of these premises may result from the heat caused by the lighting.

In premises in which only modest contributions are required from artificial lighting during daylight hours, it is often insolation rather than artificial lighting that determines the electricity demand of the climate control equipment. In this case the effect of the lighting system on electrical energy demand for air conditioning is reduced, especially if the ventilation system is designed so that mechanical cooling is not required during times of lower ambient temperatures. Here it is often doubtful whether any change in the amount of heat released by the lighting would really have any effect on the electrical energy demand for air conditioning. However, if the air conditioning operates by cooling circulating air in fan coil units, the heat from lighting represents some of the heat load that the refrigerating machinery is required to dissipate, and therefore also some of its electrical energy demand.

The electrical energy requirement for lighting in a given building is determined by:

1 Lighting requirements as dictated by the work carried out.

2 The design of the building, in terms of the extent of which daylight can be used instead of artificial lighting.

3 The design of the lighting system and selection of lamps, ballasts, fixtures and type of lighting controls.

4 Operating times as required by the use of the building.

If we are to reduce the level of electrical energy use for lighting in an existing building, the first considerations are to attempt to match operating times more closely to the work carried out, and to consider replacement of lamps, ballasts, fixtures and controls. If we are really to minimise the electricity used for lighting, it is often necessary to modify the associated electrical systems. However, even with simple measures, it is often possible to achieve significant savings, which can pay for themselves very quickly.

Replacement of lamps and lighting fixtures is a matter of cost and light quality. The new, high-efficiency types of lamps, described by McGowan in this volume, result in considerable reductions not only in electricity use but also in surplus heat. Here it is important that the cost of new lamps is such that the economic requirements as illustrated in Figure 2 can be fulfilled. It is also important that the new lamps should be designed so that they can be installed in existing lighting fittings without requiring modification to the fittings.

In the case of new buildings, there is great potential for designing extremely efficient lighting systems. For a start, the building can be designed so that maximum use is made of daylight without resulting in an excessive heat load from insolation. The lighting system can also be designed so that it results in maximum useful performance. These factors require considerable care and skill on the part not only of the designers but also of others concerned in the building process. Nevertheless, there are examples of buildings where such results have successfully been achieved. The BNB bank building in Amsterdam has been designed so that artificial light should be needed during only about 20% of the normal working hours. The building is heated with gas. Total use of electrical energy for the building, which is a large office complex, has been calculated as amounting to about 50 kWh/m^2-yr; as yet, no figures of actual metered results have been published, however.

In the case of existing buildings, it is difficult to estimate the total savings potential, because it depends on the existing lighting system, including lamps, ballasts, fixtures, controls, etc., and few measurements of obtained results have been reported. However, it should be possible to exceed the few tens of percent level, because that could be achieved with better lamps and ballasts only.

Further development of design data, design bases and design aids are required if we are to achieve a systematic design basis aimed at optimum use of daylight. We need, for example, reasonably easy-to use and reliable simulation models that allow us to optimise the design of various facade shapes and floor plans in relation to the capital and operating costs of lighting and air conditioning systems.

3.3 Air Conditioning and Ventilation

The duties of air conditioning and ventilation systems are to maintain the required indoor temperature, and also to ensure an acceptable standard of air cleanliness. Normally, it will be either the one or the other of these two main duties that determines the performance capacities of the equipment.

In residential buildings, it is generally the air cleanliness requirement that determines the ventilation performance. Normally, only very simple ventilation systems are required for this purpose. The necessary air change rate of somewhat over 0.5 changes per hour which is usually needed is often achieved by the uncontrolled ventilation through the climate screen of the building. In areas where ambient temperatures and relative humidities are high at times, it may naturally be necessary to complement the equipment with air conditioning facilities. However, in climates such as Swedens' mechanical cooling is not needed.

In commercial premises, on the other hand, it is normally the quantity of surplus heat that determines the necessary capacities of the ventilation system. Offices normally require air flow rates equivalent to at least 2.5 air changes in order to avoid occasional unacceptably high indoor temperatures. In hospitals too, the heat loads determine the necessary capacities of the air conditioning equipment, and it is only in certain special areas, such as operating theatres, treatment rooms, laboratories, etc., that high air cleanliness requirements become the controlling factor. In such areas, air flow rates of the order of 15 air changes and more may be needed.

Finally, there is one group of premises in which cleanliness requirements are paramount. This includes the above-mentioned special areas in hospitals, together with various types of laboratories and many industrial areas. A high percentage, usually 100%, of outside air is needed to maintain cleanliness for health purposes.

When discussing measures and technical modifications intended to reduce the use of electrical energy for air conditioning and ventilation purposes, it is important always to specify the design rating requirement in each particular case. If not, there is a considerable risk of applying the wrong measures.

The total energy requirement for air conditioning and ventilation consists of three main elements:

- Electrical energy for operation of fans and pumps,
- Termal energy for heating supply air,
- Electrical energy for powering refrigeration machinery. By special arrangements as absorption refrigeration equipment and gas fired engine chillers fuels could replace electricity. Still, electrically powered equipment is the normal choice.

It can be seen that it is therefore only the energy requirement for heating the supply air that is met by thermal energy. Present day systems often incorporate some form of heat recovery in air conditioning systems, thus contributing to a further reduction in the amount of additional heat normally required.

The usual means of removing excess heat from a room is to supply the room with air at a lower temperature than that of the air in the room. Cooling performance depends on the air flow rate and the temperature of the supply air, but must at all times be equal to the amount of excess heat, which can vary depending on the activities being carried on and the time of day. It is therefore necessary to control the cooling effect of the supply air, which can be done by varying either the air flow rate or the air temperature.

There are many ways of designing an air conditioning system. However, if we are concerned primarily with electricity use, we should distinguish between All-air systems and Fan-coil systems (Figure 3).

Obviously, there are also variants and combinations of these two basic system types. However, from an electricity conservation viewpoint, the important question is not the detailed design of the system, but whether or not outdoor air can be used for cooling.

In the all-air air-conditioning systems in which the supply air consists of treated outdoor air, it will be necessary when the outdoor temperature is low to heat the outdoor air to the required supply air temperature. This requires considerable heat, that however nowadays nearly always is supplied by heat recovery from the exhaust air. This means that the system is largely self-sufficient as far as heat is concerned and little external heat input will be required. Mechanical cooling of the supply air will be necessary when the temperature of the outdoor air exceeds the required supply air temperature,

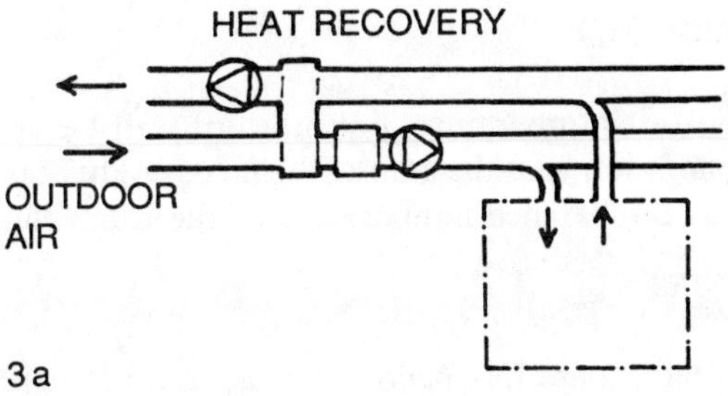

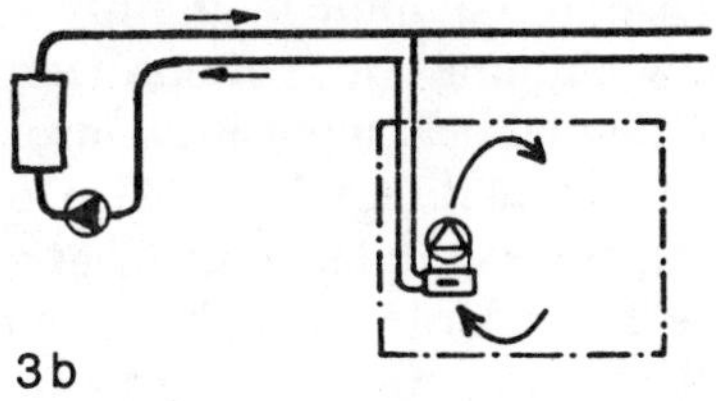

Figure 3a. All-air air conditioning systems

These are systems in which the air is treated in central units and supplied through ventilation ducts to the various rooms. The entire supply air flow can consist of outdoor air, heated to the required temperature.

Figure 3b. Fan-coil systems.

These are systems with fan-coil units in each room, cooling and possibly dehumidfying recirculated room air. The necessary cooling power is distributed by circulating chilled water in a piping system.

but at other times the cooling effect of the outdoor air can be used. However, when treating recirculated air, as is the case in a pure fan-coil system, it will be necessary to cool the circulating chilled water as soon as any excess temperature arises in any room.

The excess heat in a room, Q, consists of the difference between the thermal power input from occupants, apparatus, lighting and insolation and the thermal power loss through transmission (Q_{trans}) and uncontrolled ventilation (air infiltration) (Q_{air}). If we call the total rate of heat generation Q_{gen}, then:

$$Q = Q_{gen} - (Q_{trans} + Q_{air})$$

Losses from transmission and uncontrolled ventilation are both largely proportional to the difference between room temperature (t_r) and outdoor air temperature (t_o). We can therefore write:

$$Q = Q_{gen} - \text{Constant}(t_r - t_o)$$

At a certain outdoor air temperature, equilibrium will be established between the rate of heat generation and the rate of heat loss. This outdoor temperature can be referred to as the balance temperature of the room (t_b), so that:

$$Q_{gen} - \text{Constant}(t_r - t_b) = 0$$

Over the last ten years, there has been a steady trend towards better-insulated buildings with reduced levels of uncontrolled ventilation. At the same time, internal heat loads in offices, hospitals, department stores and other kinds of commercial premises have increased as a result of higher occupant loadings and greater use of electronic equipment, mainly in the form of computer terminals. Facades of new buildings, too, often incorporate large glazed areas, with resulting risk of substantial heat input from insolation.

All this is reflected in a reduction in the balance temperature. In modern Swedish office buildings, for example, built in accordance with the current insulation and air-tightness standards, the balance temperature during working hours is often below 0°C. During non-working hours, the rate of internal heat release is often less, although even then there is still a considerable amount of electrical equipment in operation, together with a certain minimum level of lighting, all of which releases heat. A considerable quantity of heat, too, is also stored in the structure of the building during the daytime, the effect of which is to stabilise the indoor temperature during the night and ovei weekends. Summarising, modern office buildings of this type do not need any additional heat input other than during times of extremely low ambient temperatures. During by far the greater part of the year, and virtually all working hours, there is instead a surplus of heat that must be disposed of. This is done by the air conditioning system, which requires electrical energy to drive its fans and cooling machines. As the quantity of surplus heat to be disposed of increases, so do the demands made on the air conditioning system, with greater need for electrical energy as one of the results.

As mentioned, about 30 kWh/m^2-year electricity is needed for the processing and distribution of air in new Swedish office buildings incorporating all-air air conditioning system. This can be compared with the heat required for heating the air to the appropriate temperature, approximately 20–25 kWh/m^2-year assuming efficient heat recovery.

The electricity demand of air conditioning systems can be reduced by careful design. The aim in design must be to reduce generation of surplus heat. Other significant savings can be made by correct selection and installation of fans and proper aerodynamic design of the duct systems. Savings of the order of 10–20% in electricity use for air processing and air distribution can be achieved without significantly affecting system costs.

The variable-air-volume system (VAV) provides particular potential for low electrical energy demand, as the required cooling performance in each room is controlled by varying the air flow rate. Savings of at least 30% are possible here. However, a necessary prerequisite is that the fans in the system are speed-controlled. Technically, speed control is perfectly possible, but unfortunately costs for variable-speed drives in the 5–50 kW range, as used for the majority of fan drives, are still so high that, in practice, variable speed motors are used only very seldom. A VAV system without speed-controlled fans will have much the same electrical energy demand as a conventional constant-volume system. In common all-air systems with constant airflows, variable speed drives can be used in some multi-zone air distribution systems. In general, the common solution with two speed motors is sufficient.

As mentioned, an all-air system requires mechanical cooling of the supply air only when the outdoor temepratute exceeds the required supply air temperature. A pure fan-coil system, on the other hand, requires mechanical cooling as soon as the outdoor temperature exceeds the balance temperature. This means that a fan-coil system can require mechanical cooling during much of the year. As a result, although little energy is required for driving fans in a fan-coil system, total electricity use in the cool Swedish climate can amount to about 10 kWh/m^2-year more than in a conventional all-air system.

A VAV-system with speed-controlled fans in a modern Swedish office building can reduce the electricity demand for powering the air conditioning system to about 20 kWh/m^2-year, while corresponding electricity use in a similar building with a normal fan-coil system can amount to about 40 kWh/m^2-year. However, it is not normally the electricity demand that determines the choice of system: fan-coil systems require little space, are easy to design and have low installation costs, which can be more attractive to the developer than low operating costs.

Even if the knowledge about the amount of electrical energy really consumed by HVAC systems in Sweden is insufficient, it is possible to make some estimations about possible reductions.

Of the 13 TWh electricity used yearly for operational purposes in existing commercial buildings, not more than 6 TWh can possibly have reference to operation of fans, pumps and cooling machines. If every opportunity to decrease this use of electrical energy is utilized, i.e., by installation of variable speed motors and replacement of inefficient fans and poor ducting when retrofitting buildings, maximum one third, or 2 TWh, might be saved until 2010 in the existing stock.

It is assumed that 23 million m^2 commercial buildings will be new-built and 5 million m^2 demolished until 2010 (10). The tendency today is that about 40 kWh/m^2-year or more is used for operation of the HVAC systems in new commercial buildings. If this tendency can be changed and the most electricity-efficient solutions chosen, it ought to be possible to decrease the need of electricity for HVAC to about 20 kWh/m^2-year. By that, the need of electricity 2010 would be reduced by 0.5 TWh.

The total potential for saving electrical energy by improving the HVAC systems in existing commercial buildings and choosing electricity efficient HVAC systems in new buildings would thus be maximum 2.5 TWh by the year 2010.

4 Summary Views

There is, at least in Sweden, a clear trend towards greater use of electricity in commercial premises, while heating demands fall. This applies to both existing and new buildings. There is a clear awareness in Sweden that buildings should be designed so that heat requirements are kept down. Existing buildings, too, are being improved so that heating requirements are reduced. However, when it comes to electrical energy, designers and architects are often unaware of the growing electrical requirements in commercial buildings.

Lighting and air conditioning constitute the major users of electrical energy in offices, hospitals and many business premises.

If the design of the lighting system is properly matched to the requirements dictated by use, it is possible to achieve good lighting quality with a modest mean lighting power. Suitable, well-planned grouping of the electrical supply system can assist in reducing the operating hours of the lighting system. Selection of high-efficiency lamps and efficent luminaires can provide a significant reduction in required electrical power. Yet there does not seem to be any greater use of this potential at present. One of the reasons may be that there is insufficient awareness of the new, high-efficiency lamps that are available. Initial costs may well be another reason, and greater awareness of life-cycle costs would be useful in this respect.

The quantity of surplus heat to be disposed of essentially determines air conditioning system capacity, and therefore the magnitude of the necessary electricity use. Reducing heat gains from lighting have the added benefit of reducing cooling loads.

Facade design to reduce insolation contribution is also important. It is also possible to design the air-conditioning system in different ways, more or less electrical energy consuming. There is a tendency to select systems with high electrical demand in order to save space. Again, greater awareness of life cycle cost would be beneficial.

There is a significant potential to reduce electricity demand in existing commercial buildings and to design new buildings (Table 4) so that their electricity demand is minimized. However, conscious effort and considerable work are required, and when all is said and done, it is economics that govern design decisions. In much of the Swedish stock of commercial premises, energy costs constitute only a very small proportion of total operating costs, which means that any energy conservation or efficiency improvement measures, design features or new apparatus must offer quite good profitability if they are at all considered. It is, therefore, not sufficient simply for electrical-energy-efficient technology to be available. It must also be profitable for those who apply it, which means that it must be profitable not only in terms of the country's overall energy economics, but also in terms of the simpler microeconomics. The potentials in Table 4 reflect this, rather than showing the higher potentials that might be arrived at from an analysis on the national level, when costs of new electricity generating capacity would be compared with costs for efficiency improvements.

Table 4. Approximately achievable average levels of electricity use in new commercial buildings, assuming best–practice and economically justified installations and design, based on the evaluation of the electricity customer.

Category	kWh/m^2-yr
Equipment	10–20
Lighting	15–30
HVAC	20
Total	45–70

References

1. *Driftel i lokaler. 1987. El-hushållning på 1990-talet.* (In Swedish.) Statens Offentliga Utredningar SOU 1987:69. Bilaga, pp 201–60.
2. *Energisvar*. 1987. Swedish Council for Building Research. Report G17.
3. Mills, E., Harris, J., Rosenfelt, A. 1988. Part III, Chapter 4. *Energie Internationale, 1988–1989.* (In French.) L'institut D'economie et de Politique de L'energie. Grenoble, France. Edition Economia 1988.
4. *Characteristics of Commercial Buildings 1986*. US Department of Energy. Report DOE/EIA-0246.
5. *BECA – Building Energy Compilation and Analyses, Databases on Diskettes*. 1987. Lawrence Berkeley Laboratory, Building 90H, Berkeley, California 94720.
6. Karlsson, R. 1987. *Förbrukning av energi, vatten och medicinska gaser 1986*. (In Swedish.) Sjukvårdens planerings- och rationaliseringsinstitut. Report 345 T/78.
7. *Energy use in government buildings*. 1988. Swedish State Board of Building. Unpublished data.

8. *Energy in the Built Environment.* 1988. Swedish Council for Building Research. Report G16.
9. Kraftsam. 1987. *Elkraftförsörjningen i Sverige 1987.* (In Swedish.)
10. Bodlund, B., Mills, E., Karlsson, T., Johansson, T.B. 1989. *The Challenge of Choices.* In *Electricity.*

Building Design and Electricity Use in Single-Family Houses

Arne Elmroth

Abstract

Of the 60 TWh of electricity consumed in buildings in Sweden in 1987, 18 TWh were used for space and water heating in single-family homes. With technologies and techniques available today, new all-electric homes can be built to use as little as 5,000 kWh/year for space and water heating, compared to the 16,400 kWh/year used by today's "average" home. Several options are available for retrofitting existing electrically heated homes so that they either consume less electricity or use alternative fuels.

1 Introduction

In 1987, approximately 60 TWh of electricity was consumed in buildings in Sweden (see Table 1). This figure corresponds to approximately half of the nation's total electricity consumption for that year. Of the total 60 TWh consumed, 30 TWh were used for space and water heating. The bulk of the growth in electricity consumption in buildings over the past 15 years came from electric space and water heating systems installed in single-family houses (see Figure 1). These systems consumed approximately 31 percent of the 60 TWh (18 TWh) of electricity consumed in buildings in 1987 (see Table 1); approximately 17 percent was used in homes with hydronic heat-distribution systems (10 TWh) and 14 percent (8 TWh) was used in homes with electric-resistance heating systems. Very few homes in Sweden are heated with forced-air systems.

The reasons for the substantial increases in electric space heating in single-family dwellings are diverse. There are several advantages to electric space heating. Low electricity rates and the low cost of electric-resistance heating

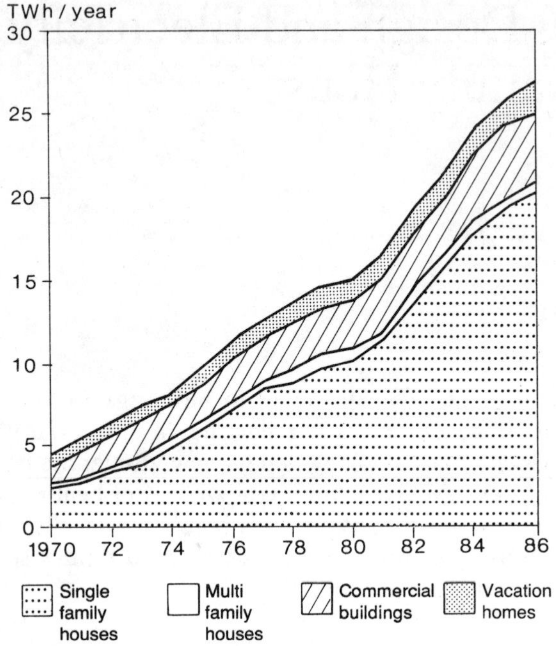

Figure 1. Climate-corrected annual electricity consumption for space and water heating classified by building type. Source: Energy in the built environment: The way forward to the 1990s, Swedish Council for Building Research, G16:1988, Stockholm, 1988.

sytems are among the most important of these[1]. Small heating loads and low electricity rates have made the use of electricity for space heating in single-family homes feasible. The low first costs associated with such systems have given electric space heating an important economic advantage over other more traditional systems (oil-fired boilers, district heat, etc). Electric systems also provide better control over indoor temperature than more traditional solutions to space heating. As a result of factors such as these, electric heating systems are installed in approximately 75 percent of the new single-family dwellings today (see Figure 2). There is a downside, however, to this trend. Although the total annual energy load for space heating in these homes is small, they are also

[1] Residential electricity was cheaper than light fuel oil on a heat-equivalent basis in Sweden between 1981 and 1985. (D. Wilson et. al., Residential energy conservation policy and programs in six OECD Countries, Lawrence Berkeley Laboratory, Draft report 1988).

Table 1. Electricity consumption in buildings, 1987.

Building type and end-use	Electricity Consumed TWh/year	Percent of Total
Electric Space and Water Heating		
Single-family Houses		
electric-resistance	8	14
hydronic distribution	10	17
forced air	0.1	–
Multi-family Buildings	4.5	8
Vacation Homes	1.7	3
Commercial Buildings	3	5
Electricity to Heat pumps	3	5
Subtotal	30.3	53
Other End-uses		
Building Services in Commercial Buildings	13	23
Appliances and Lighting in Residences		
Single-family houses	8	14
Multi-family houses	6	10
Total Electricity Consumption	57.3	100
Electric Boilers in District Heating*	5	

* This is second-class (interruptable) power and is therefore not included in the total.
Sources: Compiled from (10, 11) and personal communication with Evan Mills, Environmental and Energy Systems Studies, Lund University, Institute of Science and Technology, October 1988.

more *dependent* on electricity than less efficient homes without electrically-driven ventilation and heat-recovery systems. The load curve produced by energy-efficient houses heated with electricity can have a negative impact on utility seasonal peak loads as well. The trend toward electric space heating is likely to continue[2].

The total energy demand for heating and ventilation in new homes in Sweden, however, has been substantially reduced (by roughly one-fourth

[2] The new Swedish building regulations that came into force January 1, 1989 may even strengthen this trend: the new code will require all new single-family homes, regardless of the type of heating system they employ, to be built to the high standard that was previously applied exclusively to electrically heated homes. Although obviously the least costly way, it is questionable if electric-resistence heating will be use in Sweden in new energy-efficient single-family houses, since the favourable governmental loans for such systems will be withdrawn thus favouring other, more costly, system.

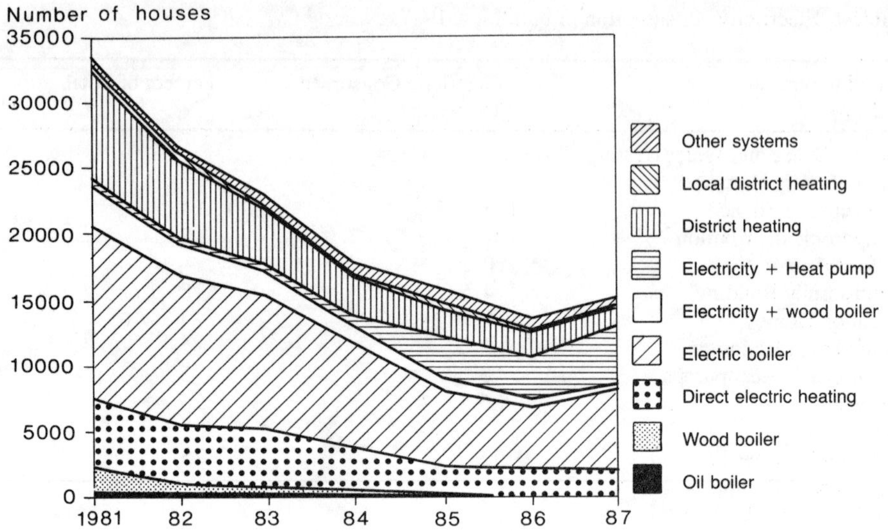

Figure 2. Number of single-family homes built in Sweden between 1981 and 1987, disaggregated by type of heating system installed.

compared to new homes in 1970) as a result of improved building design (see Figure 3). With technologies and building practices available today, single-family houses can be built that have heating loads so small that the method used for meeting the load becomes an almost insignificant issue. In a climate like Sweden's, space-heating requirements can be eliminated during the early and late winter months, thus reducing the heating season from 9 to 6 months per year. During these months, free heat from occupants, lighting, and appliances are sufficient to meet space-heating loads. The very low energy demand for space heating in new homes is achieved without reducing thermal comfort. On the contrary, energy-efficient buildings are more comfortable than non-efficient ones. Good technical design and workmanship provides uniform indoor-air temperatures, and eliminates draft problems caused by air leakages and low surface temperatures.

2 Basic Components of Household Electricity Consumption

In Table 2, the electricity requirements are given for three efficiency levels of a typical size (110 m^2) Swedish single-family house operating under Stockholm climate conditions. The figures are based both on measurements taken in new

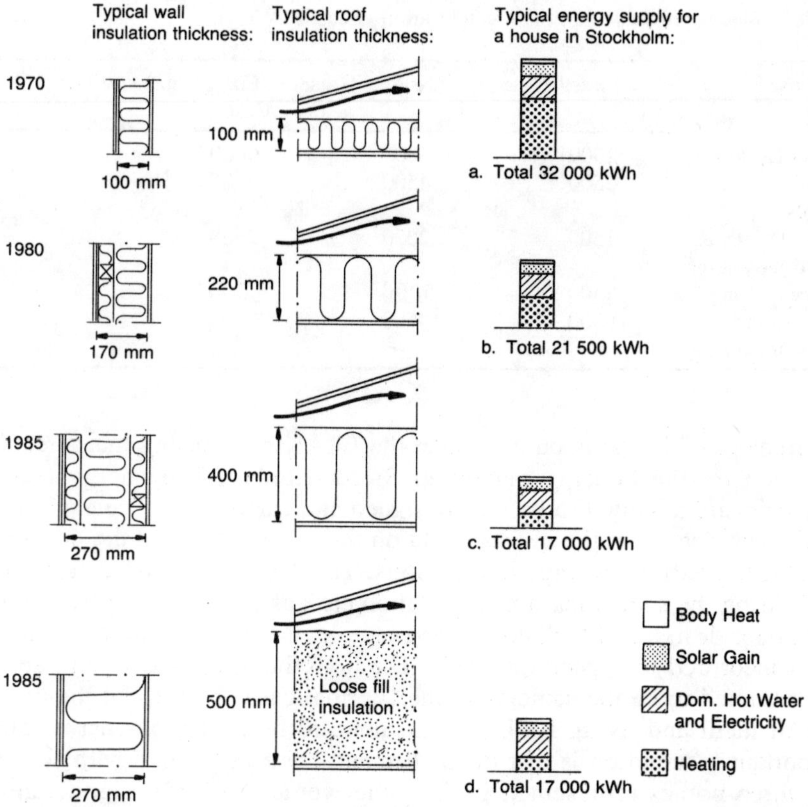

Figure 3. The evolution of typical wall and roof constructions and annual energy supplied to a 125 m^2 single-family house operating under Stockholm climate conditions.

buildings and on theoretical calculations (1,2,3). An indoor temperature of 20°C and an average ventilation rate of 0.5 air changes per hour (ACH) are assumed. The "best case" house is built with energy efficient design, technique, and equipment. The window U-values and the air-tightness of the building exceed the ELAK[3] building code requirements. The household appliances are assumed to be the most energy-efficient of those sold on the market today, and the occupants are assumed to exhibit energy concious living habits[4]. The

[3] Up to 1989 ELAK was the Swedish building code for electrically heated homes and was more strict than the requirements in SBN80 for homes with other heat sources. The building code as of 1989 for all new buildings correspond roughly to the ELAK requirements.
[4] See Nørgård in this volume.

Table 2. Electricity consumption in single-family houses kWh/year.

End-use	"Best" Case	"Normal" Case	"Energy Intensive" Case
Water Heating	2500	3000	4000
Appliances and Lights	2500	4500	6000
Fans, Pumps & Control Systems	1500	2000	2500
Space Heating	2500	5000	8000
Total Electricity Consumption	9000	14500	20500

"normal case" house is built to meet the ELAK code, using measures that are common on the housing market in Sweden today. The energy conserving measures are assumed to be well designed and carried out. The equipment and appliances represent the average sold on today's market and the occupants are assumed to exhibit average living habits. The "energy-intensive case" house is built using the same measures (insulation etc.) as the "normal case" house, but with poor design and technique (for example, the ventilation system may have been incorrectly adjusted or sized). The household equipment and appliances are assumed to be the same as in the "normal case" house, but the occupants' use of them and living habits are assumed to be energy-intensive. The most important assumption is that the investment costs associated with constructing the three homes represented in the table would not differ significantly. The differing space-heating requirements (especially between the "normal" and "worst" cases) are primarily a result of differences in knowledge and technique applied in the building design phase. In the "best" and "normal" cases, good quality control throughout the construction process is assumed to ensure that the details called for in the building design are reflected in the final product. Although additional labor costs are incurred when quality control is implemented, they are marginal with respect to most total contruction and/or retrofit investments.

The electricity requirements for each of the cases are broken down into four end-use categories: 1) water heating, 2) appliances and lights, 3) fans, pumps and control systems and 4) space heating. In the "best case" homes, the largest fraction of the electricity demand is used for appliances and lighting. Space and water heating in "best case" homes require equivalent amounts of electricity. The fans pumps and controls category is not typically shown in household electricity consumption breakdowns. Instead, the electricity consumed in this category is usually divided and hidden amongst the other columns, like space heating and appliances (or left out altogether). Some of these aggregations are not appropriate, as in the case of kitchen fans which are operated independently of other equipment and year round. As shown in the table, fans pumps and controls represent a significant portion of the electricity required in even the "normal" house. Trade-offs can occur, however, between this category and, for example, that for space heating in energy-efficient versus non-energy-efficient homes. The electricity consumed by fans pumps and controls in an energy-efficient home with a heat-recovery system, for example, might be more than that for an inefficient home relying on natural ventilation. Nevertheless, the total electrical load for the efficient home would be significantly smaller than that of the inefficient home. The electricity consumed for space heating in the energy-efficient home in Table 2 is roughly half of that for the "normal" home, and less than one-third of that for the "worst" case home.

3 Key Characteristics of Energy-Efficient Houses

The amount of energy that is required for space heating and ventilation in single-family houses is primarily dependent on the technical design and workmanship of the building. In a well-designed, well-built single-family home, occupant behavior is usually not an important determinant of the amount of energy consumed for space heating. A highly insulated building envelope and well-designed windows reduce transmission heat losses. In tightly built homes, ventilation heat losses are determined by the mechanical ventilation air-change rate chosen by the building owner/operator and on the efficiency of the building's heat-recovery equipment. The Swedish residential building regulations require a ventilation rate of 1.26 m^3/hr per m^2 of floor area (approximately 0.5 air changes per hour (ACH). Adequate ventilation is necessary, not only for comfort, but also for health reasons[5]. The heat losses associated with ventilating a building can be reduced by employing exhaust-air heat-recovery equipment. Both heat exchangers and heatpumps, used in conjunction with mechanical-ventilation systems, perform this function. The net energy loss from ventilating a house is dependent on the building's air-tightness (infiltration including open windows and doors) and on the efficiency of the heat exchanger or heatpump employed.

[5] The most important indoor-air quality problems in Sweden are toxic build-up of radon gas and mould growth. Both of these can be avoided with mechanical ventilation. The airtight building envelop prevents radon gas from the earth to enter the house.

An energy-efficient house functions as a system. When all of the parts of the system are functioning properly, a wide range of services can be provided to the building occupant(s) without producing high energy loads. Because an energy-efficient home functions as a system, however, it is not reasonable to estimate or discuss the cost-effectiveness of various system components. In a house with a leaky building envelope, neither building envelope insulation nor a well-designed and installed ventilation and heat-recovery system will function properly. Similarly, a building with a very tight envelope will not be energy-efficient without sufficient insulation of the walls, windows, floors and roof. For these reasons, component cost-effectiveness estimates are not given in this chapter. The technical measures recommended herein, however, (insulation levels, etc.) have already been accepted and are used by the building construction industry. Most of the measures are already required by the Swedish building code that is currently in force (SBN80)[6]. The most important factor in determining whether or not these measures will produce the energy-efficiency that is expected of them is the *technique* with which they are applied. This means that not only good workmanship is required, but that well thought out and detailed technical methods designed for simplifying their application are needed as well. Effective communication of design details and quality control during construction are also important. These aspects of the building process can greatly enhance the results that are achieved through the adoption of well-known and proven measures for building energy-efficient homes.

3.1 Low U-values and Minimized Thermal Bridges

The most basic characteristic of energy-efficient houses is highly insulated walls. In the Swedish climate, wall U-values of about 0.17 W/°Cm^2 are recommended. Approximately 250 mm of mineral wool insulation is required to achieve this U-value. Building thicker walls (with lower U-values) requires adapting wall construction techniques to an extent that it is not usually worthwhile. A common method of building walls is to use solid vertical studs for load bearing and additional horizontal studs on the in- or outside that will accomodate thick layers of insulation. In addition to using lots of insulation, however, it is necessary to minimize the amount of thermal bridges in walls in order to achieve maximum energy efficiency.

Several special studs have been designed which make it possible to build very thick walls with a minimum of thermal bridges. I-beams and box studs are two examples of such designs (see Figure 4). Both designs are simple and

[6] Recommendations for windows and air-tightness exceed the current building code (ELAK).

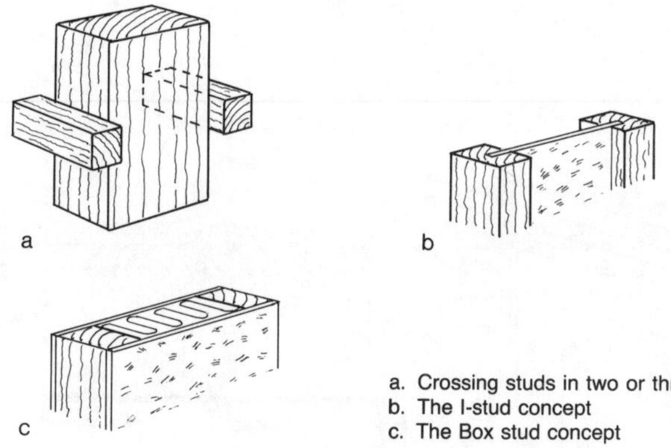

a. Crossing studs in two or three layers
b. The I-stud concept
c. The Box stud concept

Figure 4. Examples of Swedish external wall constructions designed to accomodate very thick insulation. The I-stud and Box-stud designs result in minimized thermal bridges.

easy to construct. In addition, they require less wood than solid-stud wall constructions and are therefore cheaper to build. Because they provide straight vertical passages for the insulation, proper installation of insulation is easier to achieve than in solid-stud walls. The same requirements and constraints that apply to exterior wall construction can be applied to the construction of flat roofs.

Houses with pitched roofs can easily be insulated to very high levels by applying insulation to the floor of the attic space (see Figure 5). In the average Swedish climate, U-values of 0.10 W/°Cm^2 for attic insulation are recommended. Approximately 400 mm of mineral wool slabs or 500 mm of loose fill insulation are required to achieve this U-value. Because attic spaces are usually directly accessible, attic insulation can be added to existing homes with pitched roofs as a simple and highly beneficial energy-efficiency retrofit measure. In order for attic insulation to perform properly, it must be protected from wind in the eaves which can cause air movement in the insulation. At the same time, it is necessary to maintain an air layer between the insulation and the roofing felt to ensure proper ventilation of the attic and to avoid moisture build-up in the roof structure.

The amount of insulation that is required to construct energy efficient floors varies depending on the type of foundation that is used. Prefabricated homes – which account for approximately 90 percent of all new single-family home construction in Sweden today – are most commonly built over crawl spaces. The crawl space is normally ventilated with outdoor air, and is therefore cold in the winter. In such cold crawl spaces, insulation can be added beneath the floor joists to reduce heat losses through the floor. A U-value of approximately 0.2 W/°Cm^2 is usually appropriate for floor insulation. Heat

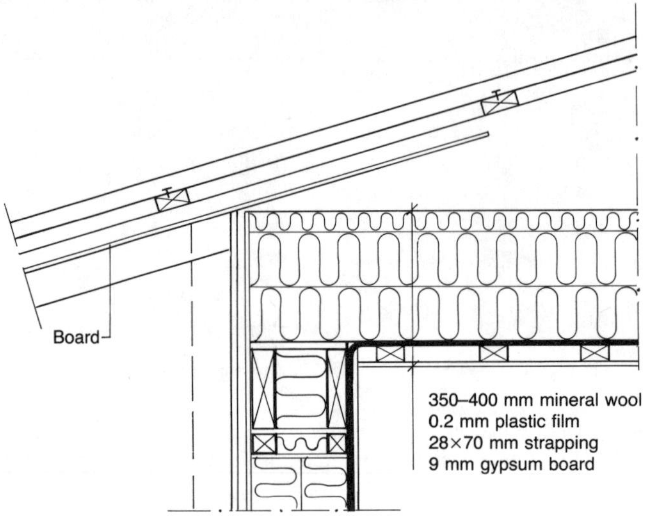

Figure 5. Example of a joint connecting an external wall to a roof. Very thick mineral wool insulation is applied in the attic. Where the roof and wall intersect, the plastic film must be overlapped and the joint must be clamped in order to ensure airtightness. A board is used to protect roof insulation from wind in the eaves. Special attention must be paid to the edges of the insulation slabs in order to avoid wind penetration.

loss through the floor, however, is not the only problem associated with cold crawl spaces. In the summer, the relative humidity in such spaces increases to a level under which mould and rot can damage the wood structure. In the winter, frost penetration can cause movement in the ground below crawl spaces and lead to shifting of the building foundation.

In order to avoid moisture problems, so called "warm crawl spaces" have been developed. In these crawl spaces, insulation is not applied beneath the floor joists. Instead, the ground and the perimeter are insulated to a U-value of approximately 0.2 W/°Cm^2. The crawl space is carefully sealed to be air-tight. In addition to reducing moisture problems, warm crawl spaces improve indoor comfort by providing warmer floor surface temperatures than cold crawl spaces. The heat losses through the ground and the perimeter are roughly equivalent to those of an insulated slab-on-grade foundation. The risk of frost damage can be reduced by applying additional insulation on the outside of the building perimeter.

Slab-on-grade foundations require special insulating techniques. Very low floor U-values can result in damage from frost penetration. Extra insulation of the building perimeter is therefore required. Most of the slab insulation is applied under the concrete, in order to protect the slab and insulation from ground moisture (see Figure 6). Warm floor-surface temperatures can be achieved by placing a thin layer of insulation on top of the slab. Equivalent average U-values of approximately 0.2 W/°Cm^2 are commonly used in Sweden.

Transmission losses through windows at night represent a major fraction of the winter heat losses in buildings. Window U-values are typically six to ten times higher than those for well-insulated walls. Because the heat losses through poor windows can be great, well-designed windows with low U-values are an important characteristic of energy-efficient buildings. Windows should have U-values lower than 1.5 W/°Cm^2. This can be achieved with a window with two linked casements and a double-glazed sealed unit with a reflective coating on the outside of the inner pane and filled with a gas such as argon. Alternatively, a coated triple-glazed unit with one reflecting coating and argon can be used. In both cases a glass coating should be chosen that does not significantly reduce the daylight transmissivity of the window unit: window transmissivity affects both the lighting levels indoors and solar heat gains. The "best on the market" windows have U-values as low as 1.0 W/°Cm^2. These windows can be used effectively for maintaining acceptable window heat loss levels while increasing the overall window area in a building design. To avoid

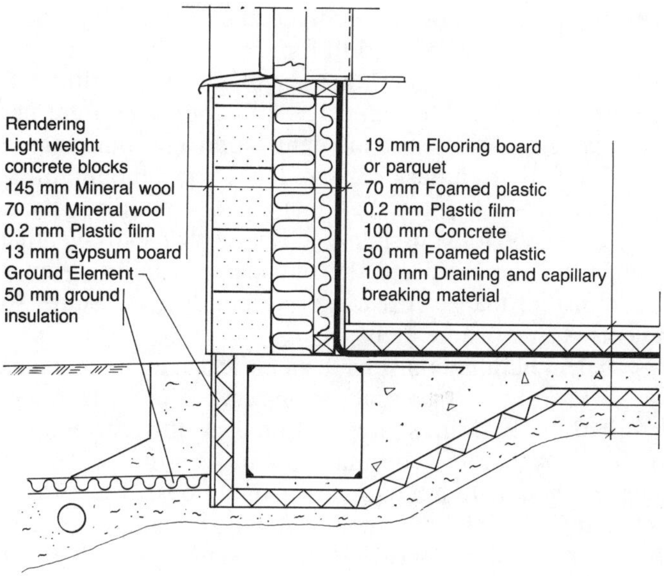

Figure 6. Slab-on-grade foundations require a special design technology. Insulation should be placed below the concrete slab in order to avoid moisture damage. To reduce the risk of damage from frost penetration, extra ground insulation on the outside of the perimeter is required.

overheating from solar heat gain in the summer, shading devices should be provided for large south, west, and east facing windows.

In addition to ensuring acceptable levels of heat loss through windows, tight windows with low U-values contribute to improving indoor comfort: as with walls and floors, lower window U-values produce higher indoor window surface temperatures, thus reducing drafts. In a draft-free building, it should be possible to lower indoor temperatures without reducing comfort. Traditionally, drafts from cold window surfaces have been avoided by placing heating elements under windows to compensate for the temperature differential. With low U-value windows this heating system design limitation is removed. The potential design and economic benefits associated with the heating system alone are significant motivations for installing energy-efficient windows.

3.2 Air-Tightness

Building technology has developed to the extent that knowledge and methods are available for building air-tight buildings. There are several important reasons for constructing buildings as air-tight as possible. First, the heat losses due to air infiltration will be reduced to a minimum. In an air-tight building, drafts caused by leaks in the building envelope are eliminated. Thermal comfort therefore increases. Mechanical ventilation system are necessary in energy-efficient houses. These systems demand airtight buildings in order to operate well. Good air-tightness also reduces the amount of sound (for example traffic noise), dust, and pollen that enters a building. By using good filters in supply-air inlets, it is possible to clean mechanically-ventilated supply air. Air leaking from the inside to the outside of a building envelope also carries with it moisture than can damage the building structure. An air-tight envelope prevents such moisture damage.

An energy-efficient building must have an air leakage rate less than 1.0 ACH (measured at 50 Pascals in a pressurization test). The costs associated with constructing air-tight buildings are marginal if the correct construction methods are used. Accurate workmanship in carrying out well-designed technical plans are the key to building an air-tight house. Measures and details related to achieving air-tightness should always be simple, and clearly and correctly shown in building drawings (see Figure 7). The most common method of achieving air-tightness in light-weight structures is to carefully seal continuous air-vapor barriers of polyethylene film around the building envelope. The film must have sealed overlappings, and all of the joints must be tightly clamped together. All necessary penetrations (from nails, pipes, ducts,

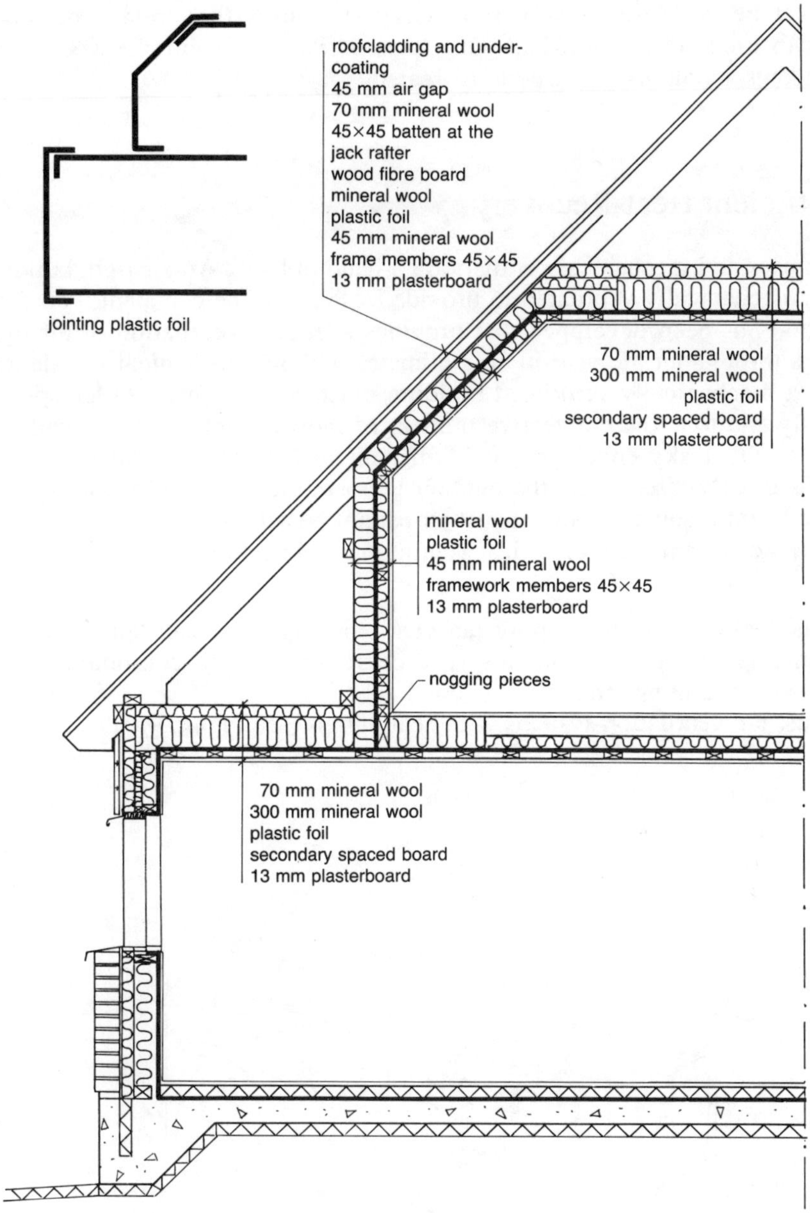

Figure 7. A well-designed house has very thick insulation. In order to make the building airtight, a continuous air-vapor barrier of plastic foil is used. Instructions for installing the plastic foil should be provided with the building plans. All of the joints must be placed in such a way that they can be carefully sealed.

etc.) must be carefully sealed. For heavy structures (concrete), the walls themselves are normally air-tight. Only penetrations and joints between walls and windows and doors etc. need to be sealed.

3.3 Efficient Heat-Recovery

Air-tightness and mechanical ventilation go hand in hand. An air-tight building must be mechanically ventilated to provide fresh air to the occupants. To date, no method has been developed that provides adequate ventilation of air-tight buildings throughout the year in cold climates without mechanical ventilation. In turn, a mechanically ventilated building must be air-tight in order operate efficiently and to avoid convective transfer of moisture into exterior walls. In buildings with leaky envelopes, it is difficult to control the ventilation rate (which is greatly affected by the outdoor temperature and wind velocity), and optimized heat recovery from exhaust air cannot be achieved.

Two basic systems of mechanical ventilation and heat recovery are available today:

- Mechanical exhaust and supply fan ventilation systems with air-to-air heat exchangers (see Figure 8). This type of system can be combined with a warm-air heating system.
- Exhaust fan ventilation systems with heat pumps (see Figure 9).

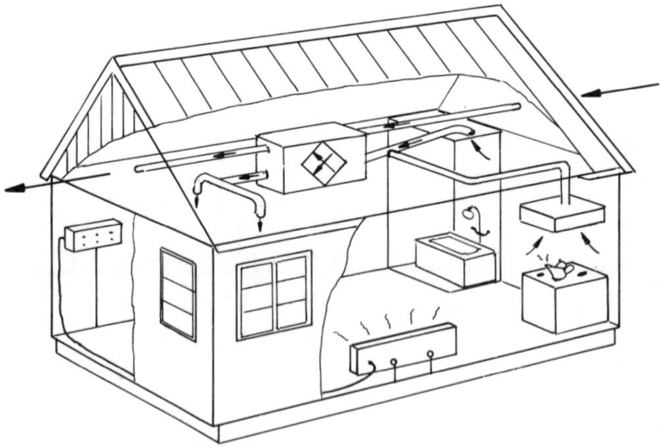

Figure 8. Example of a mechanical exhaust and supply ventilation system with an air-to-air heat exchanger. The house can have an electric-resistance heating system.

Heat exchangers and heat pumps have different mechanical qualities. The distribution of electricity consumption between space heating, water heating, and running fans and pumps in the two systems differ, but the overall electricity consumption resulting from each is roughly equivalent. Either system can be used to significantly reduce the amount of electricity required for space and/or water heating in electrically heated homes.

Heat exhangers always have supply and exhaust mechanical ventilation systems. They require, therefore, two fans. The supply air is cleaned – using good filters to remove dust and other contaminants – and often preheated. Well-designed heat exchangers supply air to the living space without causing drafts. By reducing the heat losses from ventilation, heat exchangers reduce the amount of energy required for space heating. The average efficiency of heat exchangers currently sold on the market is approximately 60 percent. The best available on the market have an efficiency rating of approximately 80 percent. The efficiency of heat exchangers drops, however, when the outdoor temperature is very low: additional electrical heat is required to keep moisture in the exhaust air pipes from condensing and freezing. Most of this energy is lost. As a result, heat exchangers tend to magnify seasonal utility peak loads by demanding the most electricity during the coldest periods of the winter. Another problem with heat exchangers is that they can become unbalanced, causing the indoor air to be under positive pressure. When this occurs, moisture can be transfered through even minute holes in the building envelope and can cause condensation inside the walls and roof.

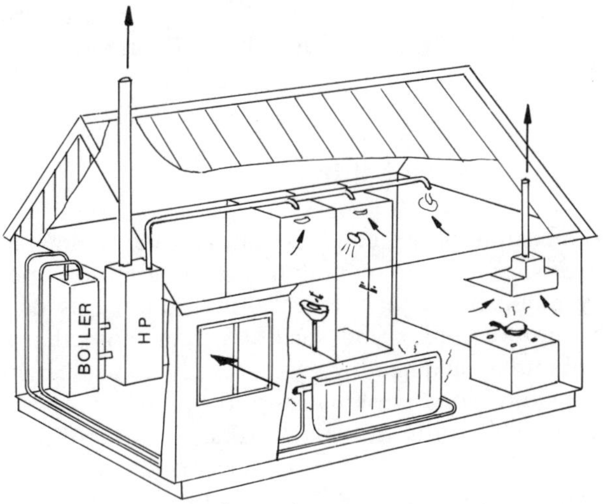

Figure 9. Example of an exhaust fan ventilation system with a heat pump. The heat pump extracts energy from the exhaust air and supplies it to the hot water tank and/or a hydronic heat-distribution system.

Slightly enlarged ventilation systems can meet the small heating loads that are achievable in very-well-designed houses. Such systems constantly circulate reheated exhaust air at a rate between one and two ACH (including 0.5 ACH of outdoor supply air). In order for this type of a system to provide adequate heat, the heat load of the building must be very small (down to 2,500 kWh/year) and the building must have an air-leakage rate well under one ACH (measured at 50 Pascals under a pressurization test). This maximum air-leakage rate must be maintained as the building ages. Because of health hazards associated with mechanical ventilation systems that recirculate exhaust air, this solution is not appropriate for multi-family buildings. Whether or not it should be applied in single-family dwellings is also questionable, because neither the air quality nor the air temperature in different rooms of a house can easily be controlled when using such a system.

Exhaust fan ventilation systems almost always cause the air inside a building to be under negative pressure. This is desirable because it reduces the risk of moisture damage in the building envelope. The heat pump in the system extracts heat from the exhaust air and supplies it to a domestic hot water tank and/or to a hydronic heating system. Cold supply air enters the building through vents in the envelope (and can cause drafts). Because heat pumps only manipulate the exhaust air, they require only one fan. Heat pumps require electricity for a compressor however, and when operating properly run almost continuously throughout the heating season. The annual coefficient of performance (COP) for exhaust-air heatpumps is in the best cases between 2.8 and 3.0. In order to maintain these COPs, very good equipment maintenance is required. Electric boilers are commonly installed as a back-up to heat pumps because they are cheap and easy to install (see Figure 9).

3.4 Accurate Workmanship

Quality workmanship during construction, and the maintenance of equipment afterward, are essential for maximizing and retaining energy efficiency in single-family homes. The techniques and technologies described herein are considerably more complex and sensitive than those that have traditionally been used for building and heating single-family homes. Highly insulated walls, for example, require good workmanship to function properly. Insulation must be connected very carefully to the vapor barrier on the inside of the wall and to the windbarrier on the outside. Air spaces or gaps compromise the thermal integrity of wall sections. There is a certain element of risk in investing in advanced measures that may not produce theoretically estimated savings due to poor workmanship. At some point, reducing the energy requirements for space heating has diminishing returns because with each increment of heat savings the energy required for running equipment increases. In addition, the ability to control indoor temperatures is degraded in homes with heating loads less than 2,500 kWh/year. Along with becoming more complex and technology dependent, high-tech energy-efficent homes have, ironically, become more electricity dependent. Occupancy of an air-tight building that lacks an effectively operating ventilation system, for example, is unhealthy because of the deterioration of indoor-air quality that can occur. On the other hand, with a well designed and properly functioning mechanical-ventilation system better air quality can be achieved than with natural ventilation alone.

3.5 Appliances and Lighting

As buildings require less and less energy for space heating, it is important to look elsewhere to find potential savings of electricity. One of the first places to look is at household appliances and lighting. In the past, some people have claimed that appliance efficiency is not important, because the waste heat from inefficient appliances contributes to meeting the heating load of the building. Appliances, however, are much less efficient space heaters than space heating equipment designed for that purpose. In addition, appliances do not provide uniform heat over time: the heat output of appliances is not coupled to outdoor temperature, so maintaining constant indoor temperatures using appliance waste heat as a heating source is not possible. Appliances do not provide spatially uniform heat either: overheating (most commonly in kitchens and laundry rooms) and underheating of specific rooms occurs as a result of heating with appliance waste heat. In non-heating periods the heat from appliances and lighting is completely lost.

3.6 Water Heating

Water heating is another large source of household energy use (see Table 2). In energy-efficient homes, hot water production can account for as much as one-third of the total household energy demand. The amount of energy consumed for water heating is largely dependent on the number of occupants in a home and their living habits. By installing thermostats on water heaters, however, the temperature of hot water can be adjusted so that it does not exceed occupant requirements. But water heaters should not be turned down too low. Measurement studies have shown that occupant consumption of hot water *increases* when water is heated to less than approximately 50°C, causing the aggregate household energy use for water heating to increase. Installing air-inducing water taps also decreases household hot water consumption, by reducing the volume of water required to produce a given flow pressure. Some of the energy used by water-heating systems is lost directly to the air in a house. Most of the energy losses associated with water heating, however, leave the house along with waste water. In addition to the energy in hot water that leaves the home in this way, energy is lost in "cold water." Cold water in a home is usually supplied at a temperature below room temperature. It is heated to room temperature by the surrounding air and material surfaces. The energy lost through warming "cold water" is approximately equal to the amount of energy that is lost to room air by the water-heating system. This heat, therefore, cannot be counted as free-heat for space heating: all of the energy used to produce hot water leaves the home through drains. Reducing the amount of energy consumed for heating water beyond the level in column one of Table 2 is difficult without reductions in comfort and/or occupant behavior. The effect of hot water production on utility load curves can be reduced, however, by delaying water heating to nighttime hours.

4 Potential Electricity Savings in Existing Single-Family Homes with Electric Heating Systems in Sweden

There are close to 500,000 single-family homes with electric-resistance heating systems in Sweden today. Together they consume 8 TWh/year of electricity for space and water heating. Most of these homes were built after 1970 and are technically well designed and are fairly well insulated. Approximately 75 percent of these homes consume less than 20,000 kWh/year (in total almost 5 TWh) (4) of electricity for space and water heating, see table 3. Most retrofits of the building envelopes of these homes are not cost-effective, even assuming a 50 percent increase in electricity prices. Less than five percent of these homes consume more than 30,000 kWh/year of electricity for space and water heating. In order to convert from direct-resistance to other heating systems, alternative heat-distribution systems (hydronic radiators or warm air) must be installed. This requires extensive modifications of the building structure, the costs of which are prohibitive. Installation of a hyronic heat-distribution system in an average existing single family dwelling (SFD) in Sweden costs approximately 23,000 SEK (roughly 3,700 USD) (5). (Together with a boiler and a chimney the total sum will be about 60–80,000 SEK.) Installing a warm-air distribution system is even more expensive. In addition, warm-air heating

Table 3. Electricity consumption for space conditioning and water heating in existing Single-family homes (SFD) with electric-resistance heating systems.

Consumption MWh/SFD,year	Number of Homes	Percentage of All Homes	Total Consumption TWh/year
<20	353500	74%	4.8
20–25	72250	15%	1.7
25–30	34250	7%	0.9
>30	20000	4 %	0.7
Total	480000	100%	8.1

Source: Reference (12)

can only efficiently meet small heating loads in air-tight buildings, so additional insulation, sealing of the building envelope, and triple-pane windows are usually required. Because the heat outputs of both of these systems are more difficult to control than that of direct-resistance heating systems, such conversions often result in *slight increases* in energy consumption for space heating.

Because the energy consumption for space and water heating in homes with electrically-fueled systems is easily metered, and because the potential energy savings from adopting conservation measures are well known, it is possible to calculate their cost-effectiveness. Several methods are available for doing these calculations (cost of conserved energy, payback period estimates, etc.). Without getting involved in detailed calculations, however, (the results of which vary significantly from building to building anyway) it is possible to make some very general statements about the value of adopting some of the retrofit measures described in this chapter. The most basic of these is that it is usually only cost-effective to make major retrofits in conjunction with other building restoration or maintenance projects. This is true even for homes with high electricity consumption for space and water heating. When major restoration or maintenance projects are undertaken, the marginal cost of the conservation measure(s) (e.g., adding insulation when building facing is being replaced) are often cost-effective. Generally speaking, buildings undergoing major restoration should have their insulation levels and air-tightness brought up to levels comparable to those in new houses.

The U-values of external walls and roofs and the number of window panes in the existing stock of electrically-heated single-family houses in Sweden are shown in Table 4. Those houses built during or after 1961 have fairly well-insulated walls and roofs. The potential savings from adding insulation in these homes is small, except in cases where major restorative work is taking place. For many of the 178,000 homes built before 1961, adding at least attic

Table 4. Thermal characteristics of existing single-family houses with electric-resistance space heating.

Year Constructed	Number of Homes	Average U-Value $W/°Cm^2$ walls	roofs	Window Area in 10 million square meters (Mm^2) 1 pane	2 pane	3 pane	total
<1940	161000	0.47	0.38	0.06	2.49	0.10	2.65
1941–1960	17000	0.36	0.28	–	0.22	0.09	0.31
1961–1975	181000	0.34	0.22	0.02	3.42	0.20	3.64
1975–1981	111000	0.33	0.20	–	0.99	1.40*	2.39
Total	470000			0.08	7.12	1.79	8.99

* Of which 0.05 million m^2 have 4 panes.
Source: personal communication, Tolstoy N. & Andersson, January 1987.

insulation would be cost-effective. A large number of these homes, however, have almost horizontal roofs with no attics. For these homes, retrofits are advisable when the felt roofing layer needs to be replaced. Insulation can be laid on top of the existing roof and cut in such a way that a roof with a steeper slope can be built above the insulation with the new roofing felt (see Figure 10). This method is widely used in both Sweden and Denmark.

Adding insulation to the external walls and floors is only cost-effective in cases where little or no insulation already exists and/or replacement of the facade material is required. From an energy-efficiency standpoint, it is always preferable to add wall insulation on the outside of the structure. Insulation on the outside both lowers the U-value of the wall and reduces the effect of thermal bridges. Insulation added on the inside of walls actually increases the effect of thermal bridges. As mentioned previously, houses with cold crawl spaces can be retrofitted with additional insulation.d

It can also be seen in Table 4 that 80 percent of the existing homes with electric-resistance space heating have windows with fewer than three panes. The bulk of them (79 percent) have double-pane windows. This is partially true even for those homes built after 1975. The addition of a third window pane is cost-effective in most cases where other window maintenance is being carried out and/or the pane can be added on the inside of the existing window (see Figure 11). If all of the double-pane windows in single-family houses in Sweden with electric-resistance heating were retrofitted with an additional pane, the resulting annual electricity savings would total approximately 0.7 TWh.[7] In addition to reducing the energy consumption of these homes, the third pane would improve thermal comfort and reduce utility peak loads.

[7] Assumptions:
total savings = $7.12*10^6 * \Delta U * Q$, where
$\Delta U = 1W/°Cm^2$, and
Q = degree hours = 100000°Ch..

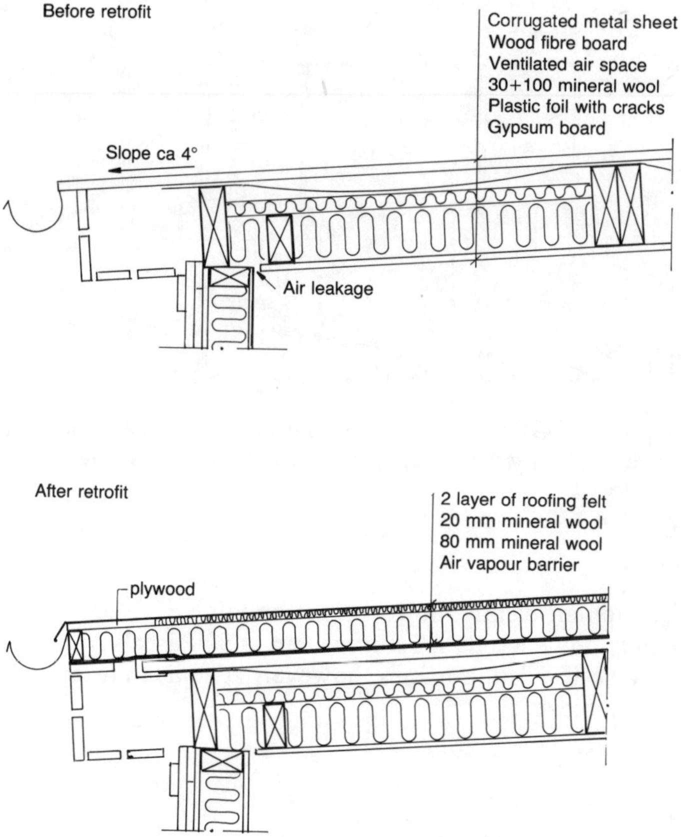

Figure 10. Examples of a low-slope roof before and after retrofit. In the after retrofit example, uncontrolled ventilation has been reduced and roofing felt laid on top of the roof cladding serves as a vapor barrier. Ten centimeters of mineral wool insulation has been added.

Many of the existing single-family homes have only natural ventilation (leaks in the building envelope) and kitchen fans to exhaust used air. The envelopes of these buildings should not be sealed at all: the negative impacts on indoor-air quality far exceed the benefits of the energy savings. In fact, the ventilation rate in many of these homes needs to be increased. Fixing the problem requires extensive and costly retrofit measures. Health problems from poor indoor-air quality can, however, motivate improving the ventilation of such buildings. First, mechanical-ventilation systems can be installed. This measure will *increase* the energy consumption of the building. The increase can be minimized by installing a heat-recovery system. In order for these

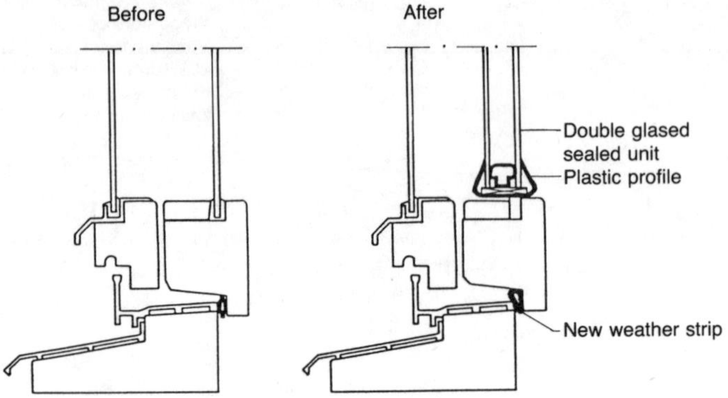

Figure 11. Example of a good method of converting double-pane windows to triple-pane. The inner pane is replaced with a sealed double-pane unit with a plastic edging strip. The weather stripping is also replaced.

systems to function efficiently, however, the building envelope must be tightly sealed (preferably down to one ACH measured at 50 Pascals in a pressurization test). In loosely built existing homes, however, adequate air leakage rates are very difficult to achieve.

5 Summary and Conclusions

New houses can be built today with very low energy demands for space heating and ventilation. In order to achieve maximum energy efficiency, they must be very well insulated, be well designed, be air-tight, and have good mechanical-ventilation systems with efficient heat-recovery equipment (heat exchangers or exhaust-air heatpumps). With the best-available technology and construction practices, single-family homes can be kept heated to 20°C throughout the winter with space-heating loads around 2,500 kWh/year. Today's "average" house uses, as a comparison, 16,400 kWh/year for heating and hot water. These heating loads are so small that the type of heating system that is installed is of minor importance. Warm-air and hydronic heating systems have the advantage of being highly fuel flexible. Although electric boilers are the simplest technology for supplying heat to these systems, gas and oil-fired boilers can also be installed. The additional expense of providing a boiler room for the latter two solutions, however, is considerable. For energy-efficient new houses, the fact that electric-resistance heating systems are much cheaper to buy and install than either warm-air or hydronic systems will ensure their continued popularity, even with high electricity prices.

Approximately one-third of the demand for electricity in energy-efficient homes comes from heating water. This demand cannot be significantly reduced without changes in occupant behavior (less hot water usage). Another third of the electricity demand, however, comes from household appliances and lights. Significant reductions in the energy requirements for these end-uses are achievable through technologically induced efficiency gains[8]. Efficient electrically-heated houses equipped with the most energy-efficient appliances available and occupied by energy concious people can consume as little as 9,000 kWh/year *without* sacrificing thermal comfort or other energy services. If 20,000 such homes were added to the Swedish building stock each year, they would only increase the national electricity requirement by 0.18 TWh/year. Their heating and ventilation loads need not exceed 0.07 TWh/year (the heating requirement from existing electrically-heated homes is currently 18 TWh/year). In addition, because of the thermal lag in well-insulated air-tight buildings, indoor temperatures can be maintained within one or two °C for a couple of hours if the power supply is interrupted.

Because most of the existing single-family homes with electric-resistance heating systems are already fairly well insulated, the potential for achieving economically profitable energy savings through building envelope improvements is relatively small. Exceptions to this general rule are: the addition of a third window pane in houses with double-pane windows, the addition of attic insulation, and the addition of insulation in crawl spaces. More extensive retrofits are not economically justifiable (from the homeowner's point of view) in most cases, unless they are carried out in conjunction with other building maintenance and restoration projects.

Although technically possible, replacement of existing electric-resistance heating systems is not likely to occur without drastic increases in electricity prices (or government incentives), because of the prohibitive investment and reconstruction costs associated with installing a heat-distribution system, building a boiler room (and in some cases an exhaust chimney), and tightening the building envelope. There are approximately 600,000 Swedish homes, however, that have hydronic heat-distribution systems fueled with electricity alone or in combination with other fuels (oil and wood). These homes could be converted to burn oil or another fuel. The cost (including parts and labor) of carrying out all of the alterations necessary for the system to burn oil is approximately 30,000 SEK (6). If the homeowner were to borrow money for the investment at an interest rate of 12 percent on a ten-year loan, the price of

8 See Nørgård in this volume.

electricity would have to nearly triple before that customer would break even[9]. Were the state to finance the conversion at 6 percent interest, the price of electricity would still have to increase by more than 2.5 times for the investment to break even.

While advanced building techniques and space conditioning technologies have drastically reduced the energy required for heating and ventilating new homes, they have brought with them an increased dependence on electricity. In addition, the electricity demand from energy efficient homes tends to get progressively more peaked as the total load diminishes. In this way, they promise to simultaneously continue to *reduce and increase the negative impacts* of single-family residential electricity customers on utilities at or near generation capacity.

Acknowlegement

The author is indebted to Ms. Deborah Wilson for constructive criticism and editing of this chapter.

References

1. Anderlind, G. 1984. *Energibehov i nya byggnader* (Energy requirements in new buildings), Stockholm: Swedish Council for Building Research R140:1984.
2. Blomsterberg, A., Stadler, C-G. 1985. *Välisolerade småhus med luftburen värme. Utvärdering av två småhus i Skultorp* (Well-insulated single-family houses with forced-air heating systems. Evaluation of Two Single-family Houses in Skultorp), Borås: National Testing Institute, technical report 1985:42.
3. Elmroth, A., Granberg., G. 1987. *SPARSAM – Fem energisnåla småhus med glasveranda och värmepump* (SPARSAM – Five Energy-efficient Single-family Houses with Sunspaces and Heatpumps), Stockholm: Swedish Council for Building Research R47:1987.
4. Isaksson, T. 1988. *Elvärme i småhus* (Electric Heating in Single-family Houses), Stockholm: Swedish State Power Board, Task 2000. Project.
5. Isaksson, T. 1988. See Ref. 4.
6. Isaksson, T. 1988. See Ref. 4.
7. Isaksson, T. 1988. See Ref. 4.
8. Isaksson, T. 1988. See Ref. 4.

[9] Other assumptions: efficiency of oil boiler 70 percent (7), constant oil price of 31öre/kWh (8), annual heating load 13,360 kWh/yr based on 1986 stock average (9).

9. Schipper, L., Ketoff, A., Meyers, S., Hawk, D. 1987. Residential Electricity Consumption in Industrialized Countries: Changes Since 1973. *Energy*. no. 12.
10. Isaksson, T. 1988. See Ref. 4.
11. Elmroth, A. et. al. 1988. *Energy Answers '87 Questions and answers on energy conservation and management in buildings and the built environment*, Stockholm: Swedish Council for Building Research G17:1988.
12. Isaksson, T. 1988. See Ref 4.
13. Engström, L. ed. 1988. *Energy in the Built Environment: The Way Forward to the 1990s*, Stockholm: Swedish Council for Building Research G16:1988.

Electrotechnologies and Steelmaking

Sven Eketorp

Abstract

Primary electricity represents 7% out of total energy requirements for integrated steel plants starting from ore and 65% for plants based on scrap. Great energy and electricity savings can still be made. Electricity can with different methods be replaced by coke, coal or gas, but the contrary is also true. Converters operating with injection of coal and oxygen are versatile metallurgical tools. New methods such as casting to near final dimension open the way for small-scale steelmaking plants characterized by high productivity, short process chains, low capital cost, drastically reduced energy consumption per ton of output, and improved environmental conditions. The alternatives are many from plants using 100% coal and making steel from ore as well as scrap and producing all electricity in house, to the challenging possibility of a futuristic steel plant based on 100% electricity and hydrogen reduction.

1 Introduction

Iron and steelmaking is one of the oldest technical branches and, at least in Sweden, the other industries came into existence in order to serve mines and ironworks. Today, this old and traditional industry is in a turbulent, not to say convulsive, period with great changes in technology and structure. All this naturally affects energy: how much energy and which sources get consumed.

1.1 Production Facts

In this changing world of steelmaking, the bold dreams of the 60's have not come true. The experts then just extrapolated the production of that time, Figure 1, ignoring the factors that decreased consumption: better yield in steel plants, higher steel grades with improved properties and better constructions, competition from other materials, and above all the steel saturation in

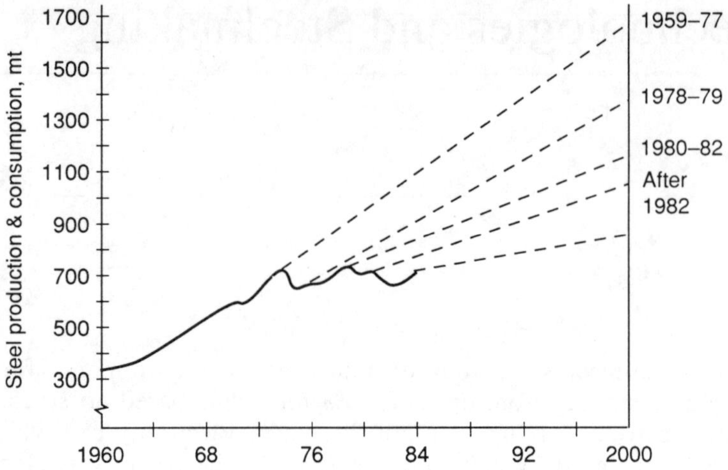

Figure 1. Trend lines of forecasts of world steel production.

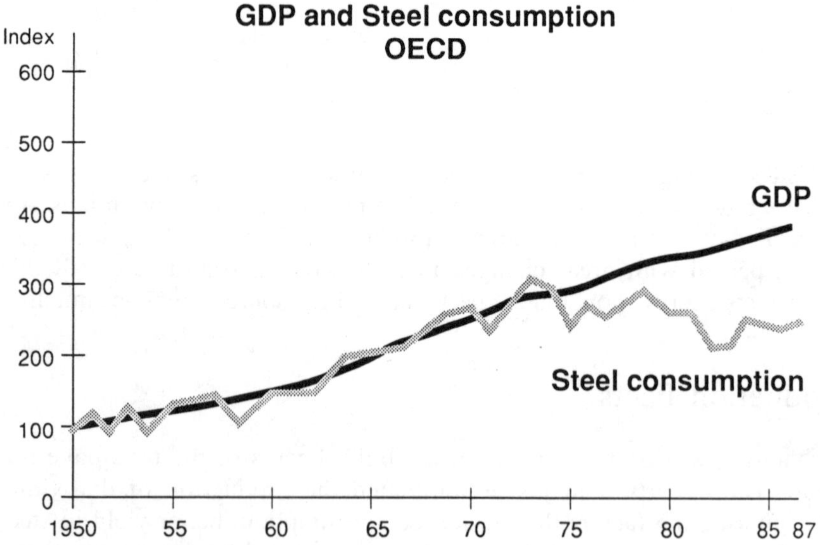

Figure 2. GDP and steel consumption. OECD.

Table 1. Crude steel production (thousand metric tons)

	1977	1987	+ change %
Latin America	21730	39547	+ 82
Africa (excl. S. Africa)	2069	4044	+ 95
Middle East	1948	3423	+ 75
Asia (excl. Japan)	17868	40059	+ 124
Total developing countries	43615	87073	+ 99
Eastern Europe	204169	223600	+ 9
Total CPE:s*	232211	289545	+ 25
Western Europe	155129	151354	− 2
Total industrial countries	399716	360840	− 10
World total	675542	737458	+ 9

* including Cuba, China, DDR, Korea
Source: IISI Statistics 1988–03–16

industrialized countries. The same fan-shaped projections had been made for energy as well. Steel consumption, however, no longer parallels GDP-growth, and has not since 1975. (See Figure 2) Neither does energy demand.

It is important to know where in the world future steel production will increase. Table 1 shows clearly that growth is high in developing countries (production doubled 1977–87), while decreasing 10% in industrialized countries. It is, therefore, important that the "new" steelmaking countries apply energy saving technology.

1.2 Energy Facts

The International Iron & Steel Institute (IISI) has spent much effort in promoting lower energy consumption. In 1982 (1) they published energy data for model plants which could act as reference values possible to obtain if all possible means, using currently available cost effective technology, were used to save energy. Figure 3 shows that for an integrated plant starting from iron ore, 76% of the energy used is for ironmaking and 12% for hot rolling; while in the scrap melting plant, 67% is used in the electric arc melting furnace (EAF). As is evident from Figure 3, the best way to decrease energy is through improved circulation of scrap in the community. But this will require that constructions easy to scrap be developed and that the problem of impurities in scrap be solved.

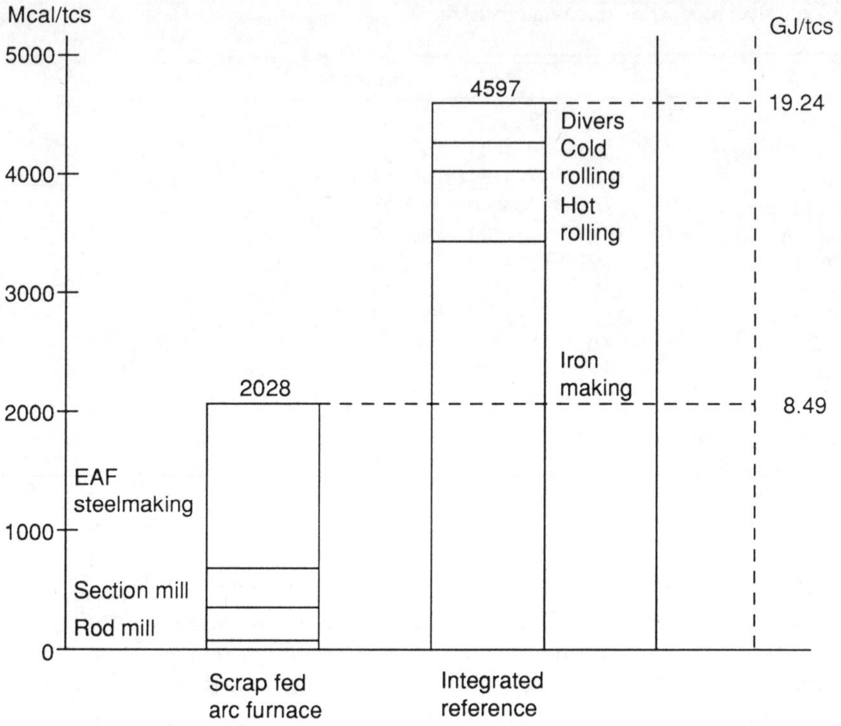

Figure 3. Energy consumption in IISI reference plants (1).

Energy structures for these two different model plants are shown in Table 2. The importance of coal for the integrated plant and electricity for the scrap based plant is obvious.

Table 2. Purchased energy structures for IISI model plants

	% of total energy input	
	Integrated plant	Scrap plant
Coal	90.9	–
Electricity	4.9	74.7
Pellets	2.5	–
Natural gas	–	15.5
LPG	0.2	–
Oxygen	1.6	2.4
Carbon	–	3.3
Electrodes	–	1.6
Unspecified	–	2.5
	100.0	100.0

In Swedish Steel AB (SSAB) Sweden where hot rolled sheet is produced from ore in one plant and from scrap in another, total energy consumption is 5150 kWh/t from ore and 1330 kWh/t from scrap, of which electric energy is 385 kWh/t and 702 kWh/t, respectively (1 kWh = 3.6 MJ). Out of a total of 4315 kWh/ton finished steel in Sweden in 1986, electric energy represented 838 kWh/t. Integrated steel plants with blast furnaces and coke oven plants can be almost self-supporting with electricity produced from process gases.

The iron and steel industry is highly energy intensive. In Sweden annual energy use is 500 MWh per employee, which is 10 times the energy use per employee of the heavy machinery and construction industries. The *specific* use of electric energy mesasured as kWh per SEK (Swedish kr) production value is surpassed only by the pulp and paper industry, Figure 4.

The cost of electrical energy for iron and steelmaking in Sweden represents 4% of the steel sales value and 11% of its processing value. For scrap based plants and some special steel plants, the cost of electrical energy is 6–9% of the sale value. It is evident that the cost of electrical energy is an important factor in steelmaking and intensive discussions are underway at present in Sweden on how to save electrical energy, especially since Sweden is the only country to have decided to abandon nuclear power. Sweden, though, has, a relatively low price of electricity compared to other countries, Table 3. The differences, between countries, however, might be smaller for the different steel industries.

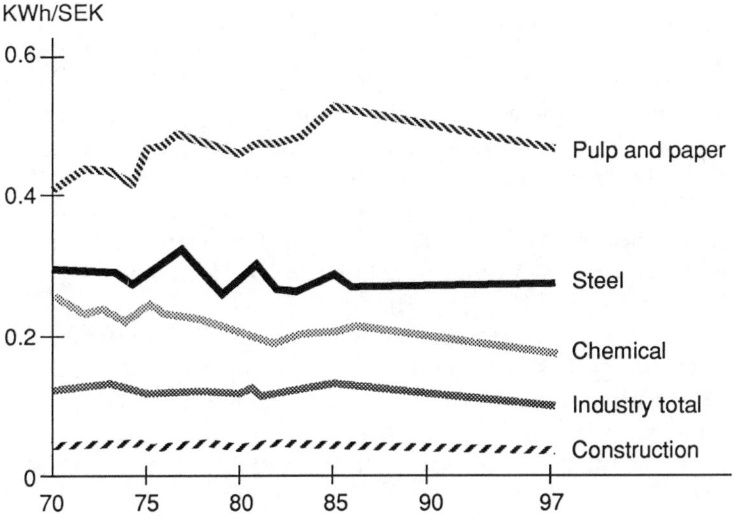

Figure 4. kWh per SEK production value for some Swedish industries.

Table 3. Cost of electrical energy for some countries in 1985

	Industry $/kWh	Households $/kWh
Australia	0.031	0.052
Denmark	0.046	0.086
France	0.037	0.085
Great Britain	0.046	0.067
Italy	0.059	0.077
Japan	0.091	0.123
Norway	0.014	0.039
Sweden	0.029	0.040
USA	0.052	0.078
West Germany	0.047	0.082

Source: IEA, Unipede

2 Decreasing Electrical Energy in Present Processes

2.1 Overall Savings

There are two ways of attacking the energy problem. One is to try to make innumerable and perhaps small changes in conventional technology, the other is to find entirely new process routes. Both ways are and must be used.

Even the most energy efficient steelworks in the world operate at a total energy consumption which is 3 to 4 times the theoretical figure, due to different types of heat losses. This is true for ironmaking from ore as well as from scrap. Naturally, a decrease in energy use is more difficult the closer to the theoretical minimum energy level that the process is operating. Even in today's most highly efficient steel industry, however, it is thought that a 5–10% decrease is possible within 10 years. (2).

That it is possible to lower energy consumption substantially by efficient use of conventional technology has been admirably shown by Swedish Steel AB (SSAB)(3). In 1979 a drive to lower energy consumption was started because of the enormous increases in cost of fuel oil, LPG (liquid petroleum gas) and even coal. The goal was set to reduce total energy consumption by 7% in each of the years 1980, 1981 and 1982. As shown in Figure 5 this goal was achieved: energy consumption was 6422 kWhr/t (23.12 GJ/t) in 1979 and 5166 kWhr/t (18.60 GJ/t) in 1982, i.e. 20% lower.

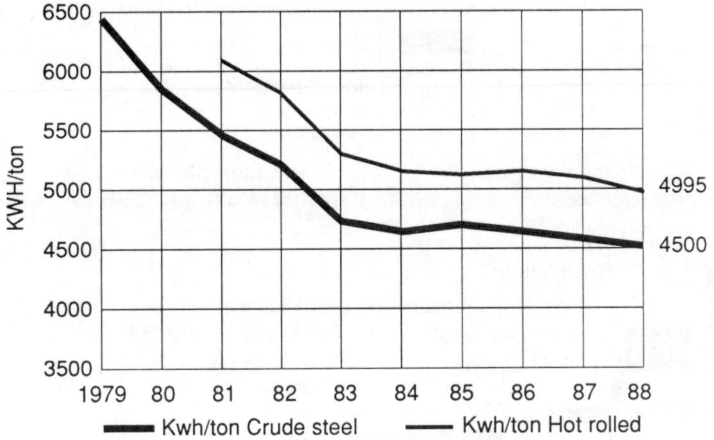

Figure 5. Energy consumption SSAB, Sweden. 1988 estimated.

Conservation has since then been continuing even if, as noted above, saving is more difficult at lower levels. The target figure for 1988 – 4500 kWh/t – will apparently be reached. Similar extraordinarily good results have been achieved in many other countries.

Contributing to SSAB's energy efficiency achievements are a number of different measures: closure of small furnaces; more scrap-based steel; increased share of continuous casting (from 65% in 1979 to 100% in 1982); elimination of primary ingot mills; improved blast-furnace operations; increased use of burnable by-product gases (internally and in the community heating system); energy recovery from hot-rolled materials; and the introduction of control systems based on microprocessors.

Key energy losses to fight are heat losses in off gases and in slags and metal. S. Toyoda (4) showed that the total waste heat is 2280 Mcal/ton (9.54 GJ/t) for a typical integrated steelworks. In Figure 6 each type of heat, recovery technology, either already in use or under development, are cited. Some 300 Mcal/t (1.26 GJ/t) of crude steel can be recovered and used with present technology and 380 Mcal/t (1.59 GJ/t) with technology under development. The total 680 Mcal/t (2.85 GJ/t) is 30% of the heat now wasted. This is 12% of total energy consumption. High-temperature energy (to the left in Figure 6) is much easier to recover than the low-grade cooling water energy (to the right).

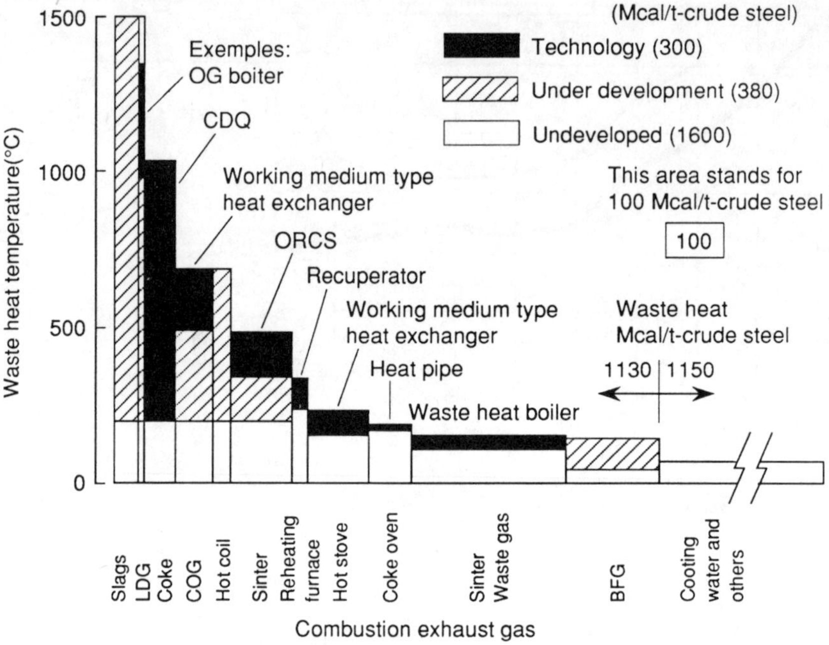

Figure 6. The main sources and quantities of waste heat and status of recovery technology. CDQ = coke dry quenching, LDG =LD gas, COG = coke oven gas, BFG = blast furnace gas, OG boiler = gas boiler (4).

The steel industry has been successful in replacing oil with coal and electricity. Figure 7 shows Swedish figures for 1986. Electrical energy was used in this way:

Scrap melting	31%
Heating and heat treatment	11%
Motors and lighting	57%
Steam generation	1%

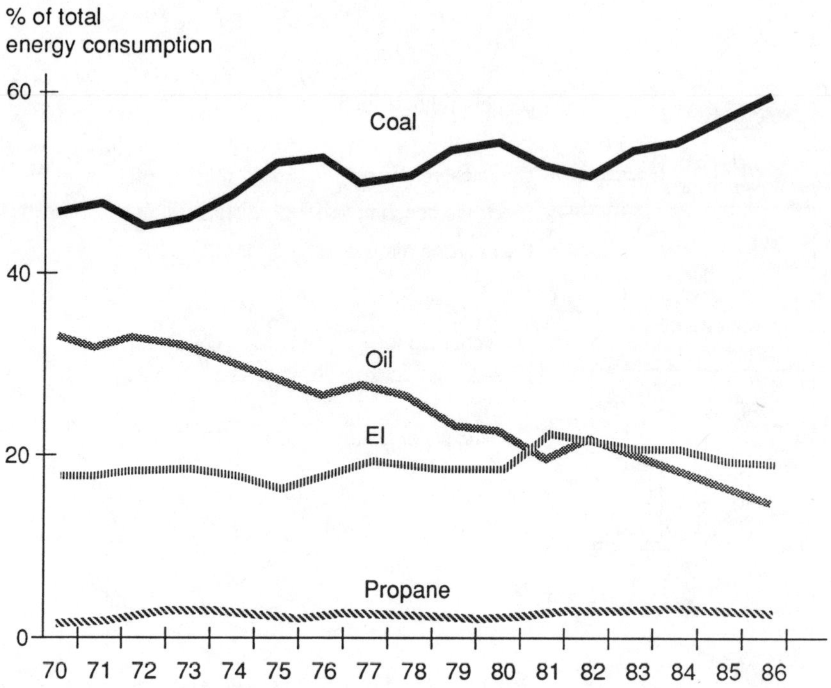

Figure 7. Use of different energy in Swedish steel industry 1986.

From a process metallurgical standpoint, the time is close at hand for attacking energy consumption in the scrap melting process, but the electricity used for motors and lighting is far larger. When dealing with such problems it is often difficult to distinguish between the cost of electric energy and the amount used, because cost often has no simple relation to amount. Different tariffs are used, nevertheless the cost per kWh can often be decreased by production planning.

Great care should be taken to choose the correct type of motor and power in kW. Electrical efficiency can vary from 85% for smaller sizes to 96% for larger. The usual motor size is 10–15 kW and these motors represent most of the electricity consumption. Every saving of 1% means a lot of energy. It is also important to discuss the need for speed regulation on motors, pumps and fans.

All the above points regarding motors are necessary to consider, but it is through changes in the processes themselves – or the shortening of the process chain – that drastic changes (decreases or increases) in electric energy consumption will really be made. Therefore, I will deal in some detail here with the units for electric melting and their efficiency.

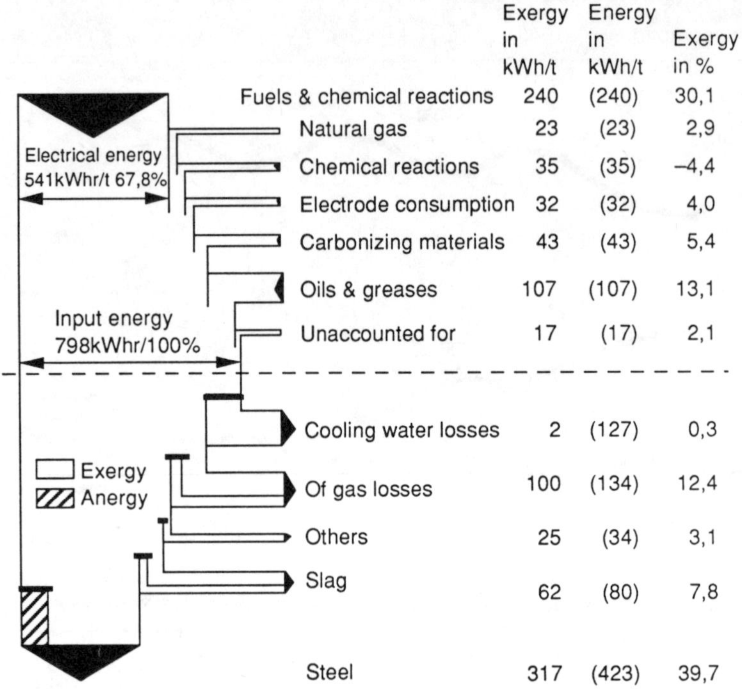

Figure 8. Energy – anergy flow in a UHP (Ultra High Power) arc furnace. 1 kWh = 3.6 MJ (5).

2.2 The Electric Arc Furnace (EAF)

The second law of thermodynamics deals with the energy "quality". Two forms of energy are relevant, namely exergy, which is transformable, and anergy which is not transformable (5). In Figure 8 the energy flow of an arc furnace for scrap melting is shown.

The cooling water losses are shown to be 100% anergetic, while 77% of the slag heat should be recoverable. In this connection, heat pumps should be mentioned. By adding some high quality energy, electrical or fossil, it should be possible to raise the fraction of exergy (i.e. of useful transformable energy) in waste heat.

Improvement of the electric arc furnace can be achieved in many ways. Figure 9 shows the specific changes in electric energy consumption (kWh/t) that result from changes in operating conditions. For instance, supplying 1 m^3 of natural gas per ton of steel via oxy-fuel burners can save 7.8 kWh/t but the time when the burners are used is very important and influences the energy yield.

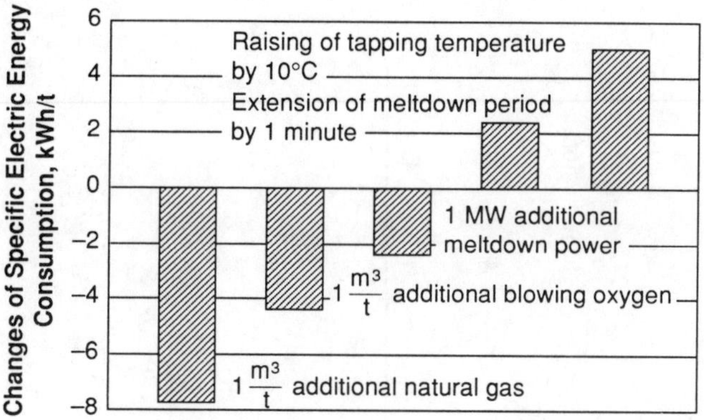

Figure 9. The effect of various EAF equipment and operating factors on power consumption.

By using a number of means to improve productivity, results which were unbelievable 20 years ago can now be achieved. Table 4 gives some results from some efficient arc furnaces in Europe 1986 (6). In Japan average electric energy consumption per ton of good ingots in 1987 was 396.7 kWh/t, compered to 470 as the average for arc funaces shown in table 4. The arc furnace has in a short time gone through a change from a furnace with 3 electrodes to a sophistcated melting machine using water-cooled walls, oxygen and fuel, etc., Figure 10.

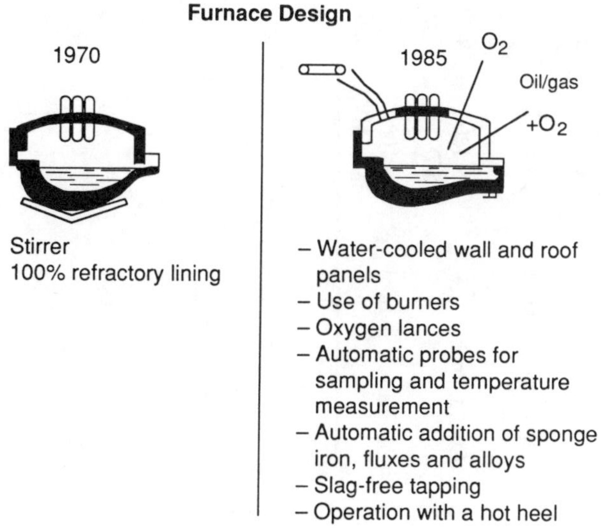

Figure 10. A comparison of furnace designs.

Table 4. Technicalk data for some prominent electric arc furnaces in Europe 1987 (6)

Company	Furnaces Power	Production kt/year	Ladle furnace	Scrap preheat	O_2 Nm^3/h	Electrodes kg/t	Energy kWh/t Oil	Electricity	Total
Metas, Turkey	2 x 35 T 20/24 MVA	400	–	–	35		60	420	480
BSW, West Germany	2 x 60 T 60 MVA	820	+	+	22	2.7	34	410	444
DDS, Denmark	2 x 110 T 70 MVA	620	+	–	24		33	465	498
Thyssen, West Germany	1 x 110 T	550	–	–	–	3.6	–	565	565
Chelsa, Spain	1 x 130 T 65/78 MVA	600	+	–	20	2.9	–	490	490
Domnarvet, Sweden	1 x 100 T 55/66 MVA	410	–	+	27	2.2	–	470	470

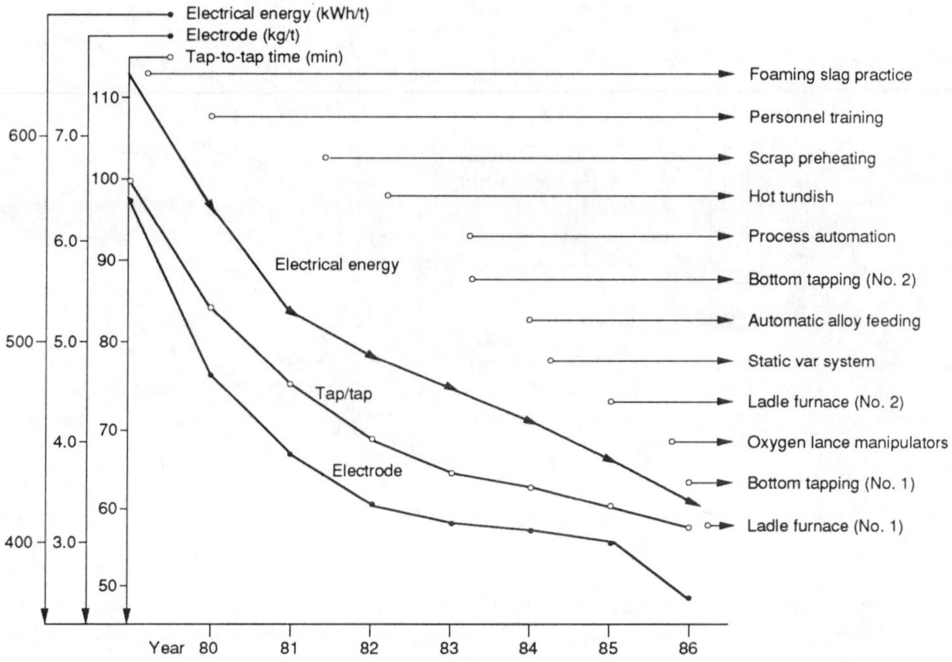

Figure 11. Evolution of productivity, electrical energy and electrode consumption at BSW, Germany.

The spectacular evolution of electrical energy productivity and electrode consumption is illustrated in Figure 11 from one of the foremost furnaces in Europe – BSW, W. Germany. The introduction of a number of changes is marked in figure.11.

The most direct way to save electrical energy is to preheat the scrap to 200–300°C by using the off-gas heat, Figure 12. As seen in Figure 8, exergy in off gases is 75% of total and scrap preheating can lower electrical energy with about 40 kWh/t, save electrodes and refractory and shorten heat time. New processes using preheating to much higher scrap temperatures achieve much larger reduction in kWh/t.

Most arc furnaces substitute part of the electrical energy by using oxy-fuel burners, see Table 4, to increase the melting rate. The burners are aimed at the furnaces "cold spots" where the additional thermal energy is most effective. Power savings can be 25–40 kWh/ton and productivity increase 5–10%.

Most furnaces today are today operated with long arcs with the highest available voltage and submerged in a deep foamy slag. This results in higher power factors, lower electrode consumption because of lower current and

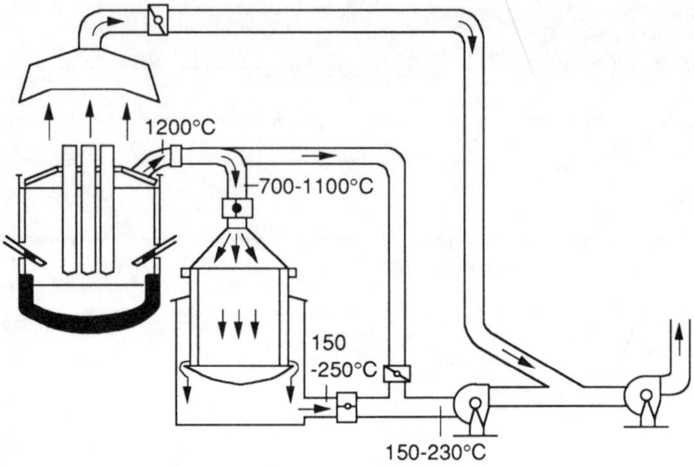

Figure 12. Conventional scrap preheater.

shorter heat time and above all in overall lower thermal stress on the refractory lining. This is true also for water-cooled furnace linings which present furnaces normally are equipped with.

The EAF is today used as a melting machine into which a maximum of electrical and thermal energy is introduced in a minimum of time. Almost no metallurgical work is done in the furnace itself. Control of temperature and composition is performed in a secondary unit – the ladle furnace – which usually is equipped with electrical heating. A typical ladle furnace has a power rating of 12 MVA and consumes 30–40 kWh/t but relieves the primary furnaces of 50 kWh/t per 100°C temperature increase.

All the improvements described above and many more such as excentric bottom tapping, automatic charging of additives, detailed process control etc., have resulted in drastical cuts in tap-to-tap times (time between two heats), Figure 13.

3 Replacing Electrical Energy by Coal and Coke

A radical way of diminishing the use of electrical energy is to replace it by other energy sources. The high cost of coke in blast furnaces and of electrical energy in scrap melting arc furnaces have been driving forces for new

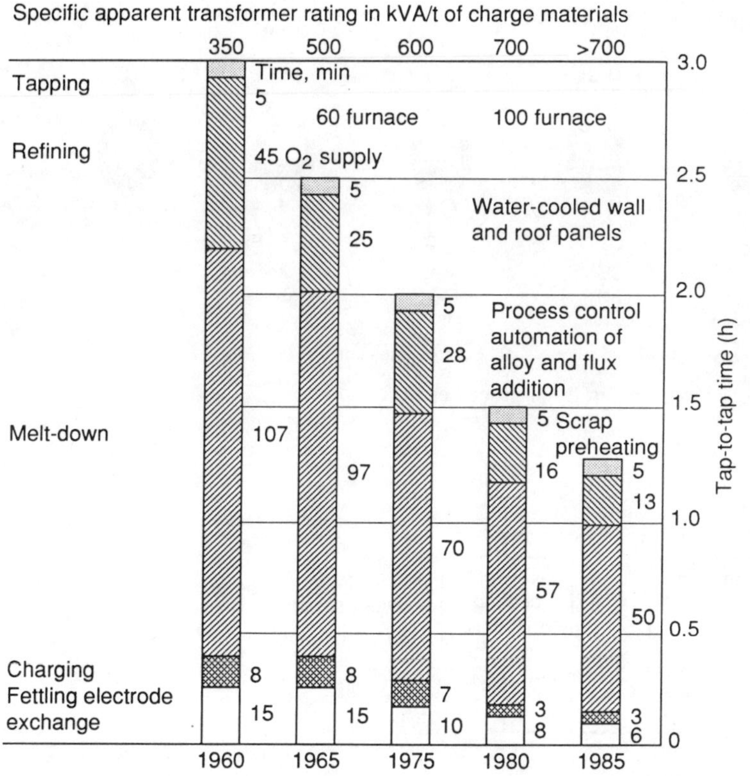

Figure 13. Development of tap-to-tap times.

processes in iron and steelmaking. Modern bottom-blowing oxygen steelmaking processes have often been the starting point for these developments. The converter, known as the basic-oxygen furnace, is a very versatile tool not only for steelmaking but also for another family of processes, which are called smelting reduction. A number of processes that are proposed or under development will now be discussed.

The liquid iron bath can be used as a rapid reaction medium into which coal, oxygen and different ores can be injected. Figure 14 shows how the (steel) converter can be used for many purposes (7).

There can be many main products as well as by-products, and the converter can be charged with different materials. To the left in Figure 14 is the usual bottom-blowing steelmaking converter, labelled OBM (or Q-BOP). By injecting coal through nozzles in the bottom, KMS, the amount of scrap can be increased. With more coal injected, KS, it is possible to melt 100% scrap. If coal as well as iron ore is injected together with the oxygen, KSG, the amount

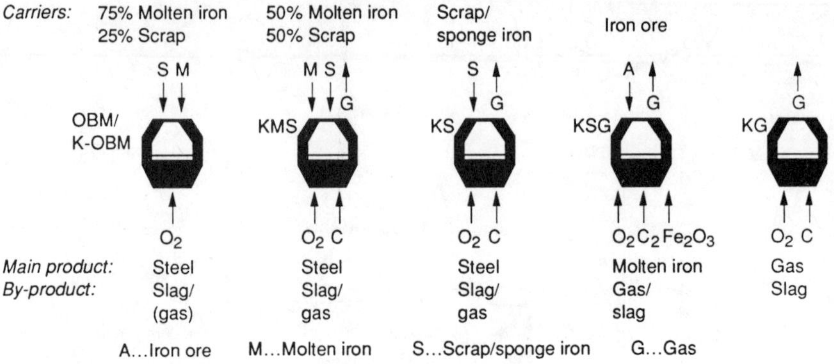

Figure 14. The Klöckner-CRA family of processes (7).

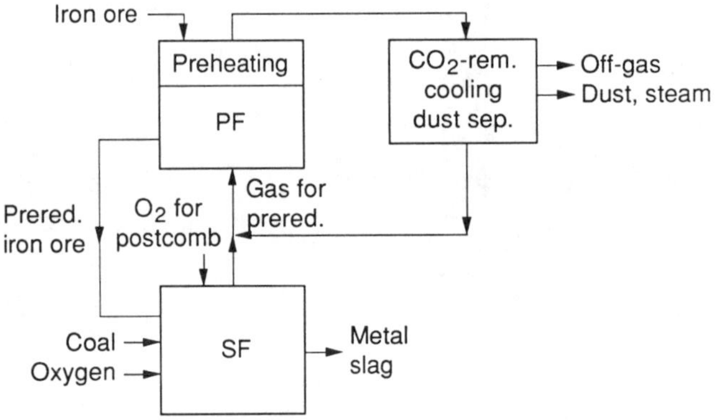

Figure 15. Smelting reduction with one high-temperature reactor (8).

of gas will be large due to the requirements of the endothermic reaction and the fact that only CO and H_2 are produced. The converter can also be utilized simply as a coal gasifier, KG. This will be discussed later on.

Based on converter operation, a whole family of processes have been proposed and are under development in Germany, Japan, Australia, Sweden, etc. Figure 15 shows a system where the smelting reduction reactor (SF) generates gas for the prereduction furnace (PF). The gas is circulated via a CO_2-remover and some oxygen is used for postcombustion of the CO + H_2 in the converter. Figure 16 shows a system (8) where the SF is used with a high degree of postcombustion and a pressurized coal gasification reactor supplies the gas for prereduction.

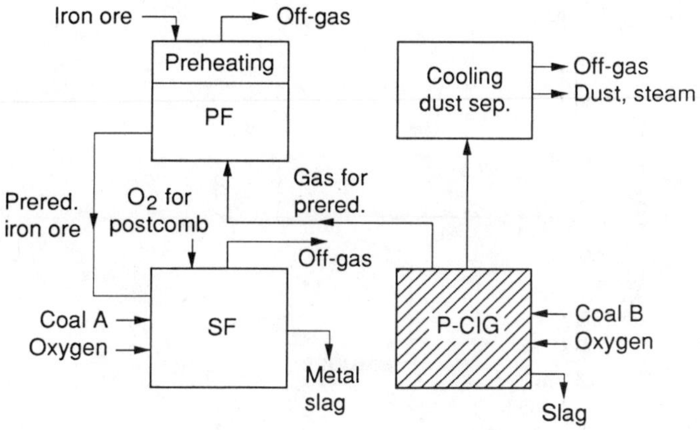

Figure 16. Smelting reduction with two high-temperature reactors (8).

It is obvious that these process systems are very complicated, with many reactors, CO_2-removal, gas circulation and heat exchangers. It would be very attractive if a *single-stage* coal-oxygen process could be developed. If, however, ore, coal and oxygen react in a converter, at 1650°C the gases produced will be CO and H_2. This results in an immense coal consumption per ton of iron produced being 1.9 tons/ton Fe with dry carbon and up to 5.2 tons/ton Fe with bituminous wet coal (9).

In order to bring down the coal consumption per ton Fe combustion of the gases in the converter itself is necessary. The problem is to burn these gases without too much oxidation of the iron bath and also to transfer the heat of combustion to the iron bath. This so called *postcombustion* has proved to be possible and much work is at present directed to the mechanism and possibilities of postcombustion. Postcombustion is defined as the percentage $CO + H_2$ combusted to $CO_2 + H_2O$.

Figure 17 shows a flow sheet for the HI smelt (High Intensity Smelting) developed by CRA, Australia and Kloeckner in Germany (10). Coal, prereduced material and air are fed to the horizontal converter.

Of special interest are the results with high postcombustion of gases. At the combination of 45% postcombustion and prereduction of the ore to FeO, 650 kg coal/t is possible. The technical and economical balance between prereduction and postcombustion must be studied further. 500 kg coal/t might be reached which should be compared to the blast furnace figure of around 500 kg coke/t (which is produced by 750 kg/t high grade coal). Heat transfer to the bath is high, being above 80% and independent of postcombustion degree.

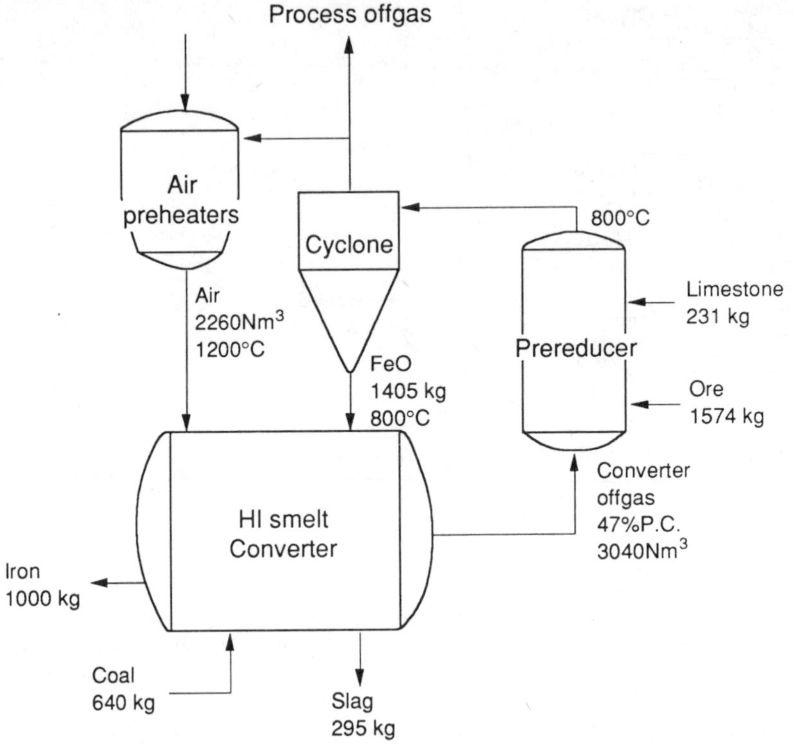

Figure 17. A flowsheet for the HI smelt process for the case with 47% postcombustion and prereduction to FeO (10).

As indicated in Figure 14, this way of operation can also be used for *scrap melting without direct use of electric energy*. A direct application of this is the Asluc process being developed by Sumitomo Metal Ind., Japan (11). Asluc = All Scrap Melting with Lumpy Coal and Coke. The converter, Figure 18, is equipped with top, side and bottom blowing of oxygen. The heat is started by charging coke at the bottom of the vessel. Before charging the scrap, the coke is ignited with top and bottom oxygen. The combined oxygen blowing oxidizes coke and hot scrap and burns the CO. The bath is stirred by bottom blowing. As seen in Figure 19, it is possible to arrive at a consumption of 130 kg coke/t and 130 Nm^3O_2/t. The postcombustion degree then is 60% and heat transfer efficiency is 70%.

Thirty years ago many tests were made with the FOS (fuel-oxygen-scrap) process using a bath similar to EAF. It was found impossible to avoid oxidation of the iron bath. Now, with stirring from the bottom and side, these

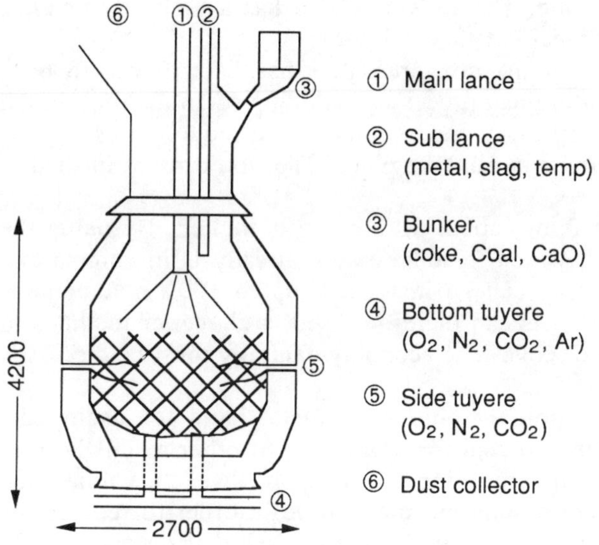

Figure 18. 15 ton test converter for scrap melting (11).

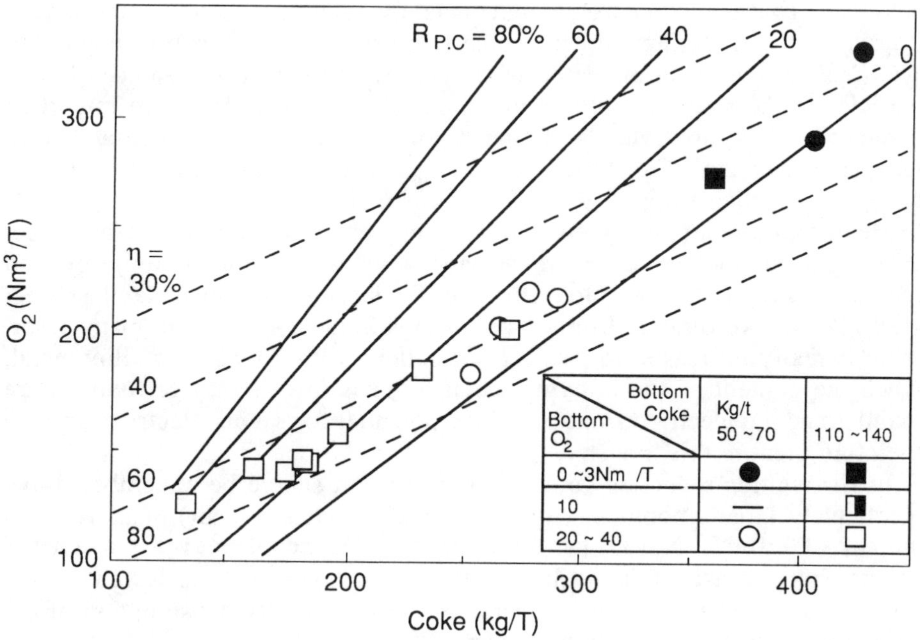

Figure 19. Effect of bottom coke and O_2 on postcombustion and heat efficiency (11).

drawbacks have been overcome. The converter also has a high productivity due to high temperature and ideal kinetic conditions.

It is interesting to note that the converter thus has been found to be a universal tool for many metallurgical processes including scrap melting. In the iron bath, reactions like the strongly endothermic oxide reduction as well as scrap melting and coal gasification can take place. The iron can be finished to steel in the same converter.

In normal oxygen steelmaking, about 30% scrap is melted. By using the above described technology, which to some extent already is in commercial operation, the amount of scrap can be increased (up to 100%). A country facing increasing electricity costs can therefore use the converter methods to replace electricity with coal or coke. The economy naturally must be discussed in each special case.

Having such a universal tool, the time is close at hand for useing the converter blown with oxygen and coal for production of other metals besides iron. Large scale investigations are going on in Japan, Sweden, China, etc., testing this possibility for chromium and manganese in order to replace the submerged electric arc furnace ordinarily used for producing feroralloys. In an all Japanese effort directed by MITI, it has been decided to investigate these methods thoroughly.

In tests (12) with an oxygen converter blown with oxygen from above and below and charged with chrome ore pellets, flux and coke using a rather heavy slag layer, the chromium content in the slag decreased as shown in Figure 20. The Cr/Fe in the metal increased with time. Strong stirring, existence of coke particles in the slag and appropriate composition of the slag were important parameters. Very low values of final Cr in the slag were reached and were independent of the $CO_2/CO+CO_2$ in the gas phase. This indicates that also in chromium reaction a high degree of postcombustion is possible.

Metal oxide reduction in converters is interesting for many metals. Direct alloying from the oxide or ore is easy with nickel, cobalt, tungsten and molybdenum (Ni, Co, W, Mo). One can, for instance, produce stainless steel (18% Cr, 8% Ni) directly from oxides or ores in the converter. In many cases, there is really no reason to produce ferroalloys with 70–80% of alloy metal, when the content of the alloying metal often is low. Today ferroalloys are mainly made in electric furnaces and the potential for saving electric energy is therefore great in this industry.

In all smelting reduction converter processes, it should be remembered that a relatively large amount of oxygen is used (in HI smelt air), usually in an amount of 130–150 Nm^3O_2 per ton of product. Oxygen is not really an energy source, but the use of it results in other advantages such as lower off gas volumes and higher reaction temperatures. An electricity consumption of 0.5 kWh/Nm^3O_2 is normal for large modern oxygen plants.

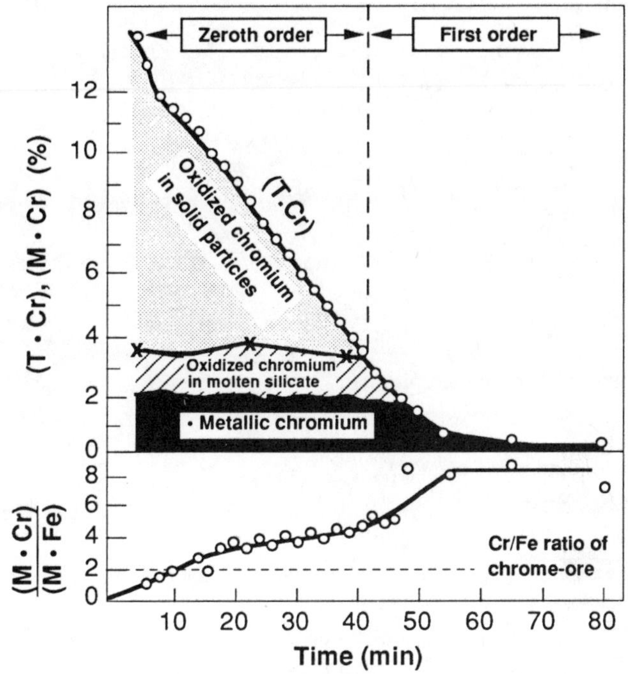

Figure 20. Morphology of chromium in slag during reduction (12).

If all the possibilities of replacing electricity by fuel are applied and waste gases are used to generate electricity, then, in fact, purchase of electricity for steelplants based on scrap or iron ore could be avoided.

4 Replacing Coke and Coal with Electricity

Electricity has many advantages in the process industry. It

- is a clean energy source
- is easy to control
- can be turned on and off easily
- has high efficiency
- gives no waste gases, and
- is especially suitable for reaching high temperatures

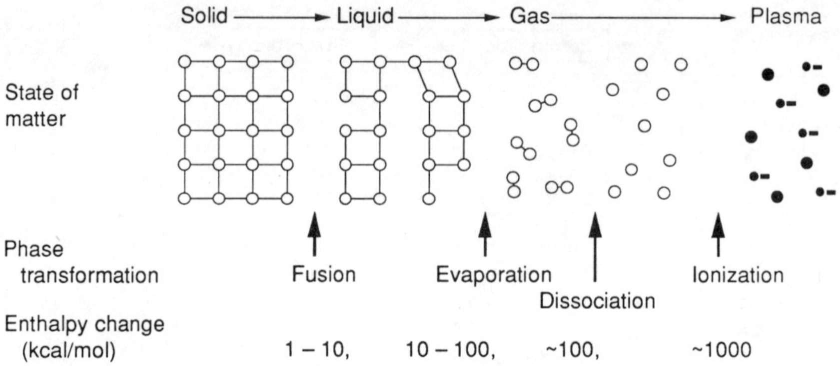

Figure 21. The fourth state of matter, plasma.

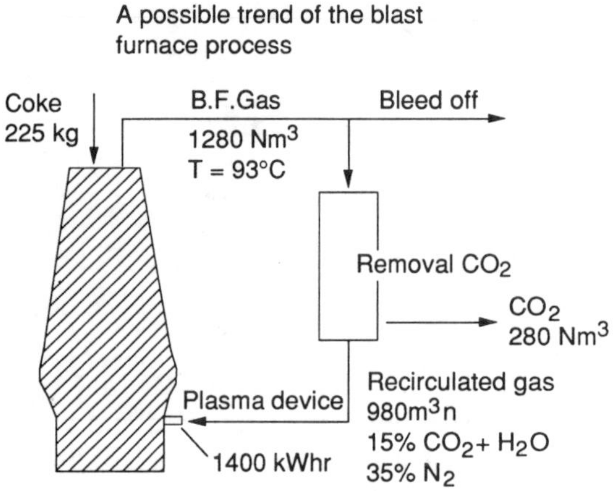

Figure 22. A possible trend of the blast furnace (13).

There are fewer examples of electricity replacing coke or coal than the contrary, because of the high cost of electrical energy in many countries. In almost all attempts to introduce electricity in primary ironmaking processes, plasma has been used.

Plasma is called the fourth state of matter, Figure 21. To transport electrical energy through plasma is especially suitable when gas can or has to be injected. The plasma has an extraordinary high enthalpy.

With smelting reduction processes being developed much effort has been spent on improving the classical blast furnace. In the Piro-gas process (13),the consumption of expensive coke has been halved (to 225 kg/t) by circulating the gas, removing CO_2 and replacing coke energy with 1400 kWh/t of electrical energy by superheating the air with plasma, Figure 22.

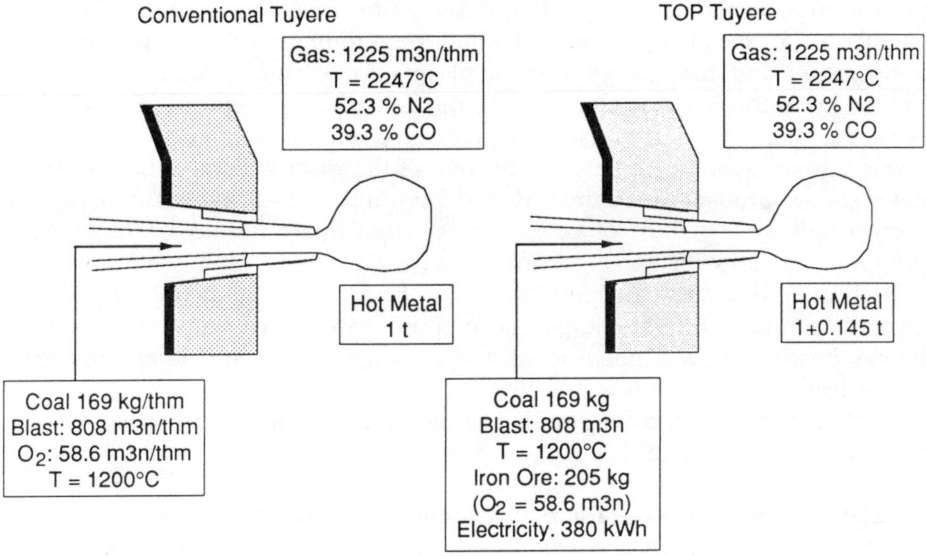

Figure 23. Principle of the TOP process (14).

Many attempts are being made to replace air in the blast furnace with oxygen, which means a consumption of oxygen of 250–450 Nm^3/t and thus 125–225 kWh/t. In the TOP (Tuyere Ore Plasma) process (14), it is proposed to increase the production of an existing blast furnace by injection of iron ore, i.e. "solid" oxygen and coal with a plasma gas, Figure 23.

A further step in this direction is the Plasmasmelt process, Figure 24, in which prereduced ore and coal powder is injected with a plasma gas carrying 1210 kWh/t of iron into the coke filled shaft. The same principle is used in the Plasmachrome process where chrome ore fines and coal are injected with plasma into the shaft for production of ferrochrome. With only ore fines, which are much cheaper on the market than lump ore, the electric energy consumption is 10% higher than in the normal submerged arc furnace, but with 70% fines the energy need is 10% lower than conventional (15).

A plant with a nominal capacity of 80,000 tons/year of FeCr has been in operation for more than a year in Malmö, Sweden. At full production Swedechrome will, from its off gases, produce 15% of the district heating energy needed in Malmö, a city of 250,000 inhabitants.

The conventional furnace for melting scrap to a high-carbon metal has been the cupola, mostly used in foundries but also as scrap melter in steelworks prior to oxygen converters. The Plasmascrap process (16) is different from the cupola. It uses the energy from fuel combustion to preheat the scrap to melting temperature and the energy from a plasma for melting and superheating, Figure 25. In this way the electrical energy is used for the purpose for which it is best suited. The shaft is used for preheating and air is injected at different levels for partial and complete combustion of the gases generated below. Tests have shown production figures of 170 kWh/ton and 45 kg coal/ton, which corresponds to 1/3 of the electrical energy used in arc furnaces. The process produces a metal with 1.5% carbon.

Preheating of ladles (for liquid steel) can be achieved by metallic resistance heaters, resulting in total energy consumption half of that with traditional oil or gas heating. Lower noise level and no waste gases are other important advantages.

Heating furnaces in rolling mills use metallic or ceramic resistance heating for temperatures up to 1200°C. Again, lower energy consumption, better environment and better possibilities for control are obtained. In SSAB Luleå, a 37 MW induction furnace for slab heating is in production. It is 60 m long, consumes 410 kWh/ton and has an efficiency of 58%. Cooling water has a temperature of 85°C.

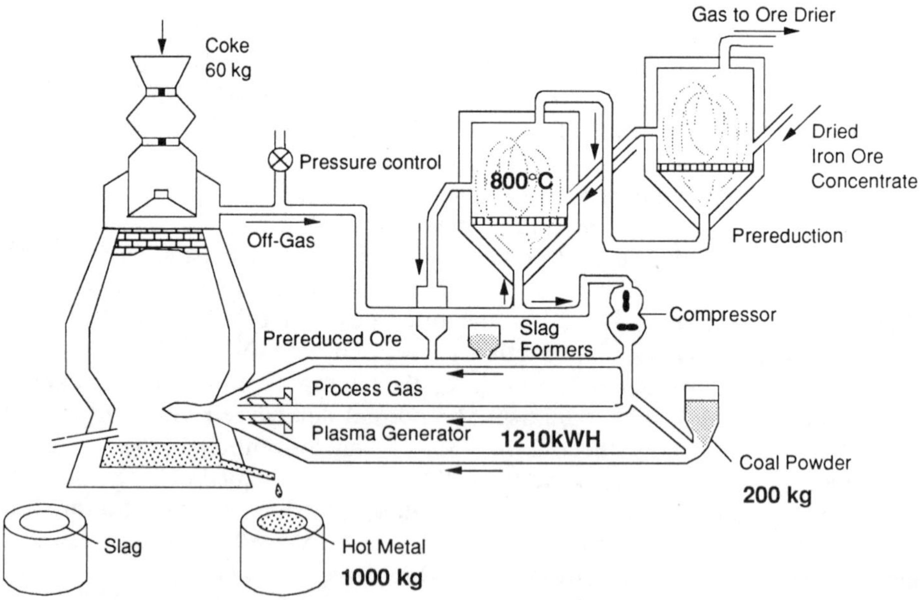

Figure 24. SKF Plasmasmelt process.

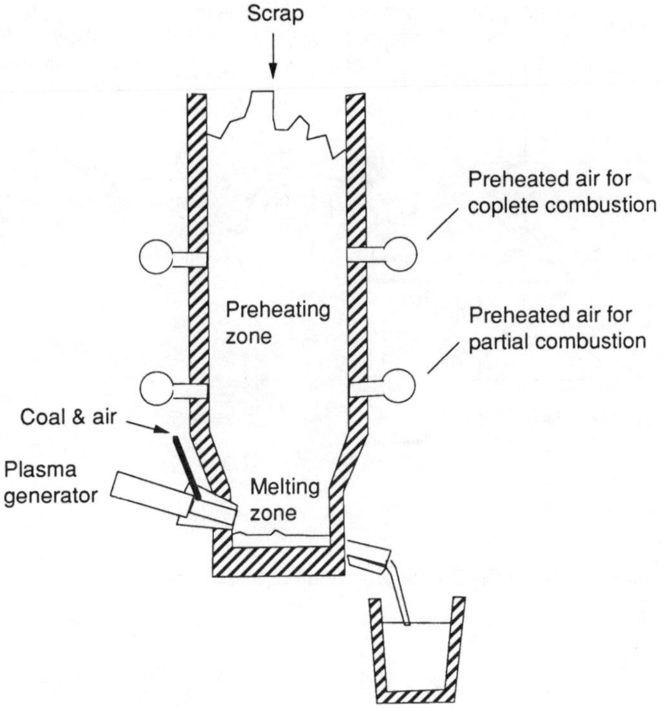

Figure 25. Straight-shaft Plasmascrap furnace (16).

5 A Combination of Metallurgy and an Electric Power Plant

As discussed earlier, the steelmaking converter can be used as a coal gasifier by injecting coal powder and oxygen into the iron bath. The P-CIG (Pressurized Coal Iron Gasification) is a process especially suitable for producing gas for further use in gas turbines. A test plant has been run at MEFOS, Sweden as a joint project between Interproject Service AB of Sweden and Nippon Steel Corporation of Japan (17).

The converter-reactor was charged with 5–6 tons of iron and operated at 1400–1500°C. Coal powder, particle size <1 mm, dried to about 2%, was injected together with slag formers and oxygen, Figure 26. The reactor was normally operated at 3–3.5 bar and the feeding rate of coal was about 500 kg/ton, h, which resulted in an off-gas flow of 3800 Nm^3/h of a gas with 70% CO, 30% H_2 and <1% CO_2.

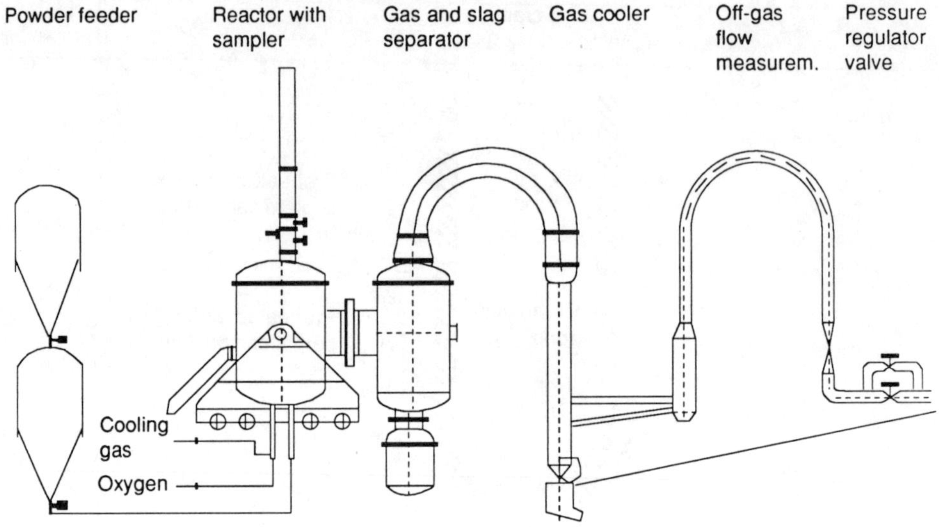

Figure 26. Schematic figure of the P-CIG pilot plant process (17).

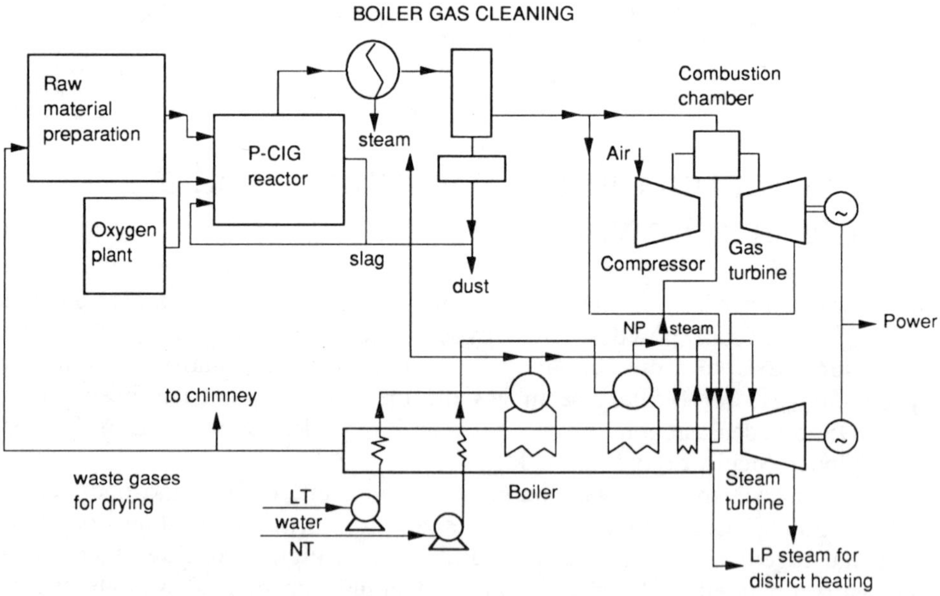

Figure 27. The P-CIG process for coal gasification in a combined cycle power plant (17).

The slag – with lime and dolomite additions – absorbs sulphur and furthermore the dust leaving the process consists mainly of iron which will also absorb sulphur. The final gas has a sulphur content of 20–25 mgS/MJ, a value well below the stipulated Swedish standard of 50 mgS/MJ.

The P-CIG process is highly suitable for integration with combined cycle electric power generation. Figure 27 shows the concept of the integrated P-CIG combined cycle power plant. Some conditions will be somewhat different compared to natural gas as a fuel in the power plant. The heat radiation from the CO/H_2 gas will be higher compared to natural gas and the high temperature will put high demands on the material in the combustion chamber. Also the NOx formation must be taken into consideration. The gas should not hold dust with a melting point below 1000°C because of sticking to the turbine blades.

A demonstration gasification plant has been designed and cost estimated. A reactor holding 40 ton iron will gasify 22 ton/h of new coal. The coal is injected together with 5 ton/h of flux and 9000 Nm^3/h O_2. Steam is used for cooling the melt and 40,000 Nm^3/h gas is produced at a pressure of 4 bar. In a second step, the reactor will be prepared for operation at 20 bar.

6 Radically New Methods Influencing Energy Consumption

At the first Soviet-American workshop on energy consumption and energy aspects of the production and utilization of materials, it was decided to give these two areas high priority:

- Net shape and near net shape of continuously cast strip and sheet
- Rapid solidification and powder metallurgy.

Most energy savings are realized by painstaking small steps in order to improve efficiency and productivity, decrease losses and recuperate heat. The two areas just mentioned belong to another class. They represent a real leap in the steel making methods, and save energy by radically shortening the process chain.

Sir Henry Bessemer patented as early as 1865 a method of direct casting of thin sheet, Figure 28. Efforts are presently being made in many countries to cast slabs 50 mm and less thick and 1500–2000 mm wide. Several methods are proposed for strip casting and direct cold rolling. A Swedish method produces wire directly from the melt. Spray casting with the Osprey process, where the metal body is built up by spraying, is showing promising results. Figure 29 depicts some of these new methods (2).

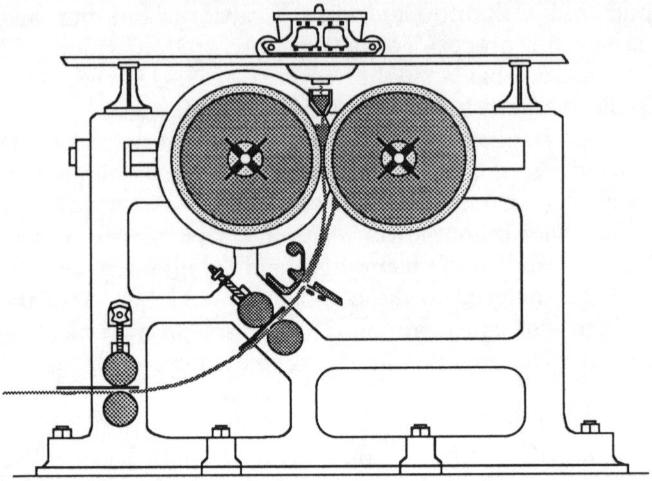

Figure 28. Henry Bessemer patent 1865.

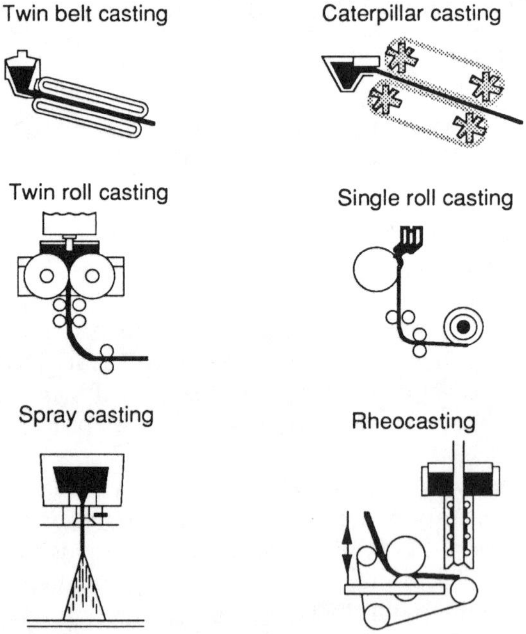

Figure 29. New casting processes (2).

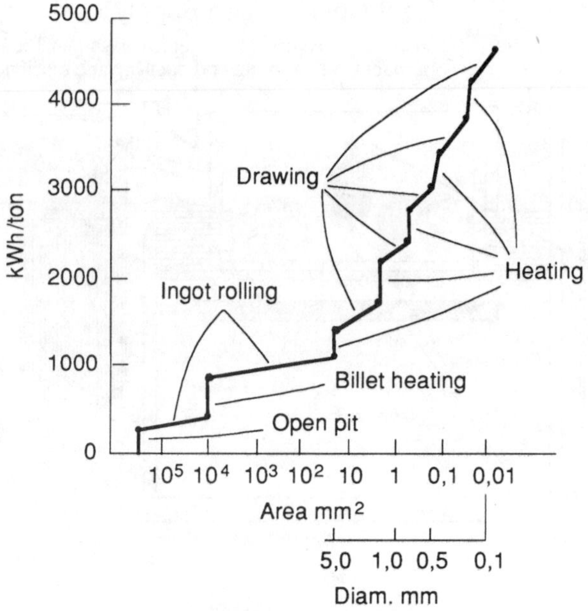

Figure 30. Energy consumption in conventional rolling and drawing from ingot to wire (stainless 2333).

More direct casting will eventually suplant hot rolling and hot working because of its many advantages:

- lower capital cost
- lower man-power cost
- increased yield
- less material being worked – or stored – at a given time
- possibilities for quite new alloys, and
- great potential for energy savings.

Figure 30 shows the energy needed for rolling and drawing in the production of wire of different sizes. For extremely small sizes, a saving of almost 5000 kWh/ton can be obtained if this wire can be cast directly from the melt.

A normal figure for energy consumption from molten steel to hot rolled strip or sheet is 650 kWh/t. With more direct methods the following savings can be anticipated:

	Savings %
Casting to 25–50 mm slab	50
Casting to 5–10 mm strip	60
Casting to 3 mm strip (no hot rolling)	75

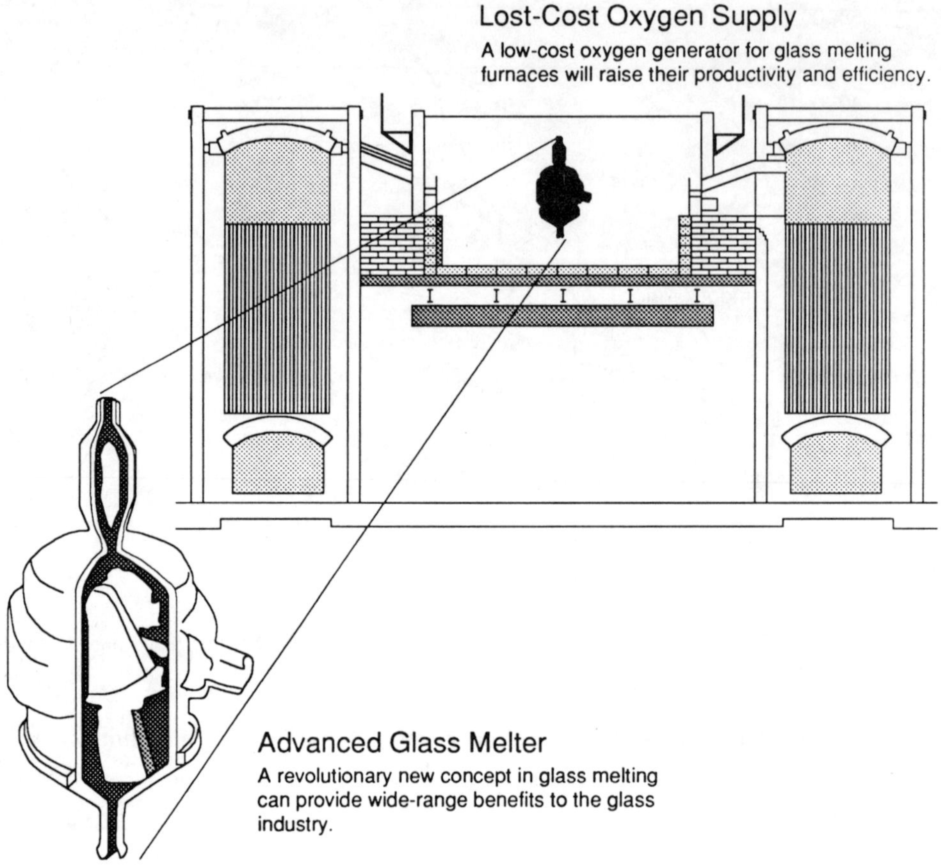

Figure 31. Advanced glass melter.

All of these new processes necessitate very sophisticated methods for control of steel cleanliness, metal flow and temperature. In my opinion, these methods open the way for small-scale, precision steel industries with low energy consumption and much improved environmental conditions (18).

In the glass, industry, the concept of a gas – fired "advanced glass melter" linked to a low-cost oxygen generator has been developed, see Figure 31. In due time the steel industry will go in this same direction.

7 A Scenario. A Steel Plant with 100% Electricity and H_2O as Waste Gas

In this time of computers, process controls and automation, it seems evident that the steel plant of the future will be completely different from the present. The steel industry is characterized as a large scale "heavy" industry with often deplorable environmental conditions, high capital cost and high energy consumption.

In a book such as this it seems appropriate to discuss a steelplant with electricity as the only primary energy source. This not completely utopian proposal can be coupled with other radical process suggestions. In the following, a scenario for such a steel plant is sketched. If these ideas will be realized, though, a great amount of research and development work will have to be done.

The starting point must be to formulate some absolute basic demands. Two of them are general:

1. 100% electric energy produced from hydro, wind, tide, waves, solar cells etc. In Sweden no nuclear power is allowed after the year 2010. If other primary energy sources are necessary, distant power plants should have zero emission, except for CO_2.
2. Hydrogen (H_2) used as reducing agent in order to produce waste gases of only water as vapour or liquid and with no NO_x-formation.

Some specific process metallurgy demands are:

3. A continuous process from ore to cast product.
4. Direct steelmaking from ore to steel.
5. No reduction of elements other than iron in the primary step from ore. No Si, S, P, C, H or N in this product.
6. Removal of a refractory lining by using water cooled self-lining reactors when slag is present.
7. Deoxidation without the formation of slag inclusions.
8. Heating without contamination from electrodes.
9. Use of only finely shredded scrap.
10. Direct casting to final size and shape.
11. Sophisticated control of flow of solids, liquids and gases.

The realization of these demands results in an unconventional process line where the first step is precipitation of iron, saturated with oxygen from a bath of molten oxide (molten ore) (19). As FeO forms low temperature eutectica with all known oxides used in refractories it, is necessary to use a self-lining reactor, where the oxide slag freezes on to a water cooled wall, figure 32. As

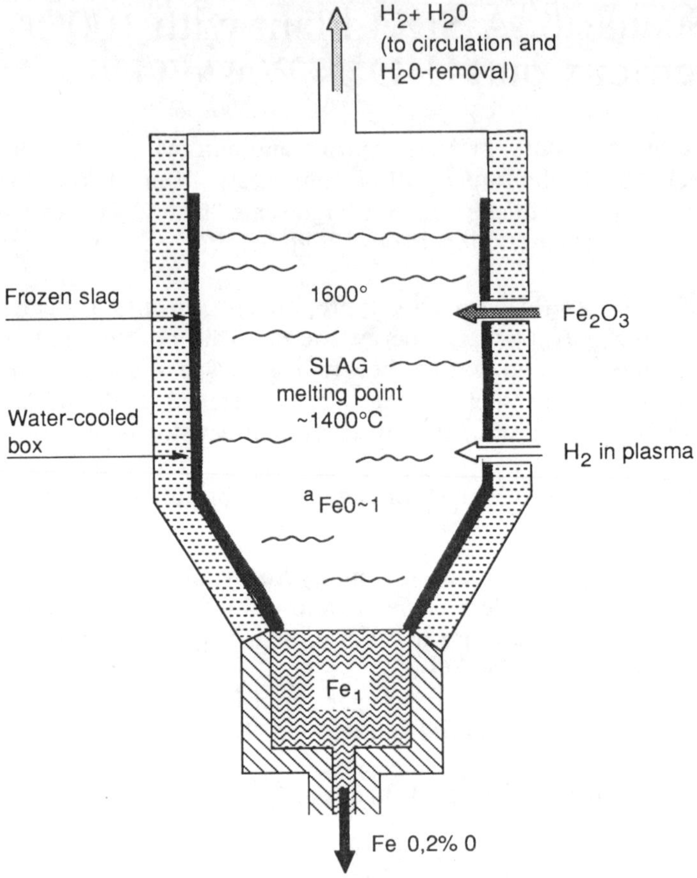

Figure 32. Self-lining reactor for producing iron by precipitation from an oxide slag with hydrogen (19).

rich an oxide concentrate as possible is used. Hydrogen for reduction is injected and is superheated in a plasma to compensate for the endothermic reaction, heating of constituents and heat losses.

Laboratory experiments by Nakamura et al. (20) using hydrogen plasma and oxide in a water-cooled crucible have shown that the efficiency of the hydrogen reduction is high because H_2 always is in contact with iron oxide. The rate of reduction was constant (at an extremely high rate) throughout the whole reaction period. The metal formed had:

Carbon	Silicon	Aluminium	Oxygen	Phosphorous	Sulfur
0.002%	<0.03%	0.007%	0.152%	0.003%	0.005%

The produced metal from the process, according to Figure 32, is saturated with oxygen and due to the strongly oxidizing conditions (in sharp contrast to the strongly reducing conditions in the blast furnace) almost no carbon silicon, phosphorous, sulfur, nitrogen or hydrogen are introduced into the metal. Steel is made *directly* from iron ore.

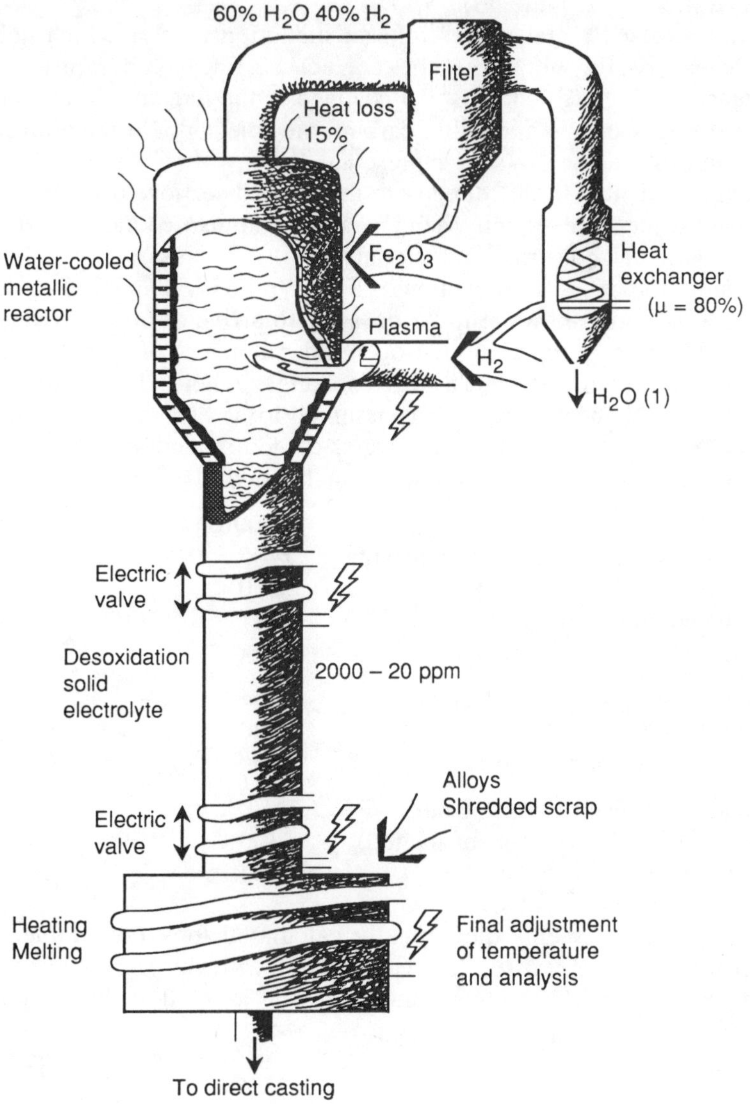

Figure 33. A process line with 100% electrical energy (Scetch: B. Zethraeus).

The iron (steel) must be deoxidized from 2000 ppm to 10–20 ppm. Consistent with demand 7, this is done by electrolytic deoxidation, using an oxygen cell for diffusion of oxygen through a solid electrolyte of ZrO_2, ThO_2 or similar material by applying a certain voltage across the cell.

Between the different process steps are placed electric valves consisting of an induction coil where the metal can be frozen or melted and furthermore regulated in speed (Figure 33).

The deoxidized metal passes into an induction reactor which delivers the heat necessary for obtaining the correct casting temperature. A certain percentage of finely shredded scrap is added to the induction unit and powders of all necessary alloys, including carbon, are injected. No metallurgical work is done in this reactor. No slag is present.

The finished steel is cast directly to the desired sections of wire or strip or it is sprayed to heavier sections. Thin sections can also be elements in different types of sandwich construction materials.

Consumption of H_2 for reduction of Fe_2O_3 to Fe is 600 Nm^3 H_2/ton Fe. Tests in Japan (20) have shown it to be possible to arrive at 60% efficiency (H_2 to H_2O) in this high temperature hydrogen reduction. The off gas with 60 % H_2O and 40% H_2 is circulated to a heat exchanger where H_2O is removed. The efficiency of this heat exchanger is assumed to be 80% and yield of Fe in the process 98%. Heat losses from the water-cooled reduction vessel are assumed to be high, 15%. The energy requirements then will be (21):

	kWh/ton Fe
Heat of reaction and heating of reactants	1405
Heat losses	210
Recirculated energy in gas	-571
To be introduced by plasma	1044
Production of H_2 by electrolysis of water. 4.2 kWh/Nm^3 H_2	2520
Total energy for primary metal	3564
Deoxidation in electrolytic cell and heating in induction reactor (estimated)	300
Grand total	3864

The energy requirements theoretically calculated look favourable. If such a steel plant could be realized, the capital cost with this very compact process would be very low. The environment in such a small scale plant should be exellent.

8 Conclusion

Modern metallurgical developments offer possibilities for great freedom of choice of energy sources. Electricity can be replaced by different fuels, but electricity can also replace these fuels if local conditions or economics demand such changes. New technologies will make it possible to operate efficient small scale plants characterized by high productivity, short process chains, low capital cost, improved energy consumption and better environment.

A steel plant to 100% based on electricity and hydrogen reduction is a challenging possibility.

References

1. *Energy and the steel industry*. International Iron & Steel Institute, Brussels 1982.
2. Springorum, D. 1988. *The Changing World of Steelmaking – the European View*. Metallurgical Plant and Technology, pp. 6–13.
3. Personal communication from Mr. B. Westin, Swedish Steel AB, May 1988.
4. Toyoda, S. 1983. Trans ISIJ 23, pp. 1–13.
5. Fett, F., Pfeiffer, H., Siegert, H. 1982. Stahl und Eisen 102:461–5, (in German).
6. Personal communication from Mr. L.E. Carlsson, Swedish Steel AB, May 1988.
7. von Bogdandy, L., Innes, J. 1984. Stahl und Eisen 104, 1025–1030.
8. Axelsson, C.L., Kaufmann, D., Torssell, K.1988. *The CIG process for smelting reduction and coal gasification*, Scand. J. Metallurgy 17, pp. 30–37.
9. Smith, R.B., Crockett, M.J. 1987. *Coal based ironmaking*. Ironmaking and Steelmaking 14, pp. 49–75.
10. Innes, J.A., Moodie, J.P., Webb, I.D., Brotzmann, K. April 1988. *Direct smelting of iron ore in a liquid iron bath – the HI smelt process*. 7th Process Technology Division Conference AIME Toronto.
11. Ishida, H., Marukawa, K., Anezaki, S., Hirata, T., Shiode, S. April 1988. *Development of scrap melting process by STB Test converter*. 7th Process Technology Div. Conference AIME Toronto.
12. Kajioka, H., Ishikawa, H., Katayama, H., Kuwabara, M., Fujita, M. 1986. *Fundamental study on production of ferro-chrome by the smelting reduction*. Trans. ISIJ, 26, pp. B 194–95.
13. Michard, J.A. et al. Nov 1983. Rev. Metallurgie, pp. 851–67.
14. de Lassat de Pressigny, Y., Michard, J. April 1988. *The TOP Process – a new ironmaking process in the blast furnace*. 7th Process Technology Division Conference, AIME Toronto.
15. Private communication from Mr. S. Santén, SKF Plasma Technologies AB, Sweden.
16. Santén, S., Feinman, J. April 1988. *Pilot plant studies of the Plasmascrap Process* 7th Process Technology Div. Conference, AIME Toronto.
17. Axelsson, C-L., Sato, K., Torssell, K., Törneman, B. *The P-CIG process for coal gasification*. Coal Powder 87 Conference, Nov. 27–29, Düsseldorf
18. Eketorp, S. April 1988. *My wishes for the process metallurgy future*. 7th Process Technology Div. Conference, AIME Toronto.

19. Eketorp, S. 1981. *Gedanken zu Stahlerzeugung im Jahre 2000*. Stahl und Eisen 101, pp. 852–859. In German.
20. Nakamura, Y., Ito, M., Ishikawa, H. 1981. *Reduction and dephosphorization of molten iron oxide with hydrogen-argon plasma*. Plasma Chemistry and Plasma Processing 1, nr. 2, pp. 149–160.
21. Calculations by Pei Wen Guo, Royal Institute of Technology, Stockholm.

New Electrothermal Processes

Helmut Schaefer and Manfred Rudolph

Abstract

Conversion from electricity to heat and the energy transfer to the object to be heated can be based on quite different principles. There are a multiplicity of electrothermal technologies with special heat distribution, energy transfer densities and operational characteristics. Electrothermal technologies are constantly being further developed in detail, and are penetrating into new fields of application, bringing about various technical and economic advantages.

1 Introduction

1.1 Systematic Overview of Electrothermal Processes

Electrothermal technology basically includes all processes using electrical energy for conversion into useful energy.

The type of heating is determined by the point at which electrical energy is turned into heat. A distinction is made between:

- *direct heating* where electrical energy is applied in the form of an electromagnetic field to the object to be heated. Conversion into heat occurs within the object to be heated.
- *mediate heating* where electrical energy is converted into heat outside the object to be heated. It is then transferred to the object through convection, thermal radiation or conduction; sometimes these phenomena are combined.

There are also mixed forms of direct and indirect heating such as in arc heating.

The physical mechanisms involved in energy conversion and the energy input into the object may vary considerably. A distinction can be made between the following technologies:

- *Resistance heating*: generation of Joule's heat in a resistance being part of an electric circuit. The resistance may be either the object to be heated (direct resistance heating, also referred to as ohmic heating) or a resistive heating conductor transmitting heat through thermal conduction, convection and/or radiation to the object to be heated (mediate resistance heating). Therefore, electric infrared heating also counts as mediate resistance heating. In many cases, the mechanisms of heat transfer are combined.

- *Inductive heating*: generation of Joule's heat through induced eddy currents (inductive heating without iron core) or circular currents (inductive heating with iron core). Inductive heating is usually direct, that is, whenever the electric current is induced almost exclusively in the electrically conducting object to be heated. If, however, the object resides in an electrically conducting vessel or container the current is largely or wholly (in the case of a non-conducting charge) induced in the container. Heat is then transferred to the mostly liquid charge through convection.

- *Dielectric heating*: generation of heat in dielectric material by agitation of polarized molecules, caused by an electromagnetic field oscillating at high frequency. Depending on the type of field and on the frequency range, radio frequency heating in a capacitor field and microwave heating in standing or travelling electromagnetic waves are distinguished. Dielectric heating is always of the direct type.

- *arc heating*: generation of heat through current flowing in an ionized gas channel (arc discharge). If the gas is identical with the object to be heated (gas reaction furnace), the type of heating is called direct. In all other cases heat is transferred indirectly from the arc to the object to be heated, primarily by radiation, or convection if there is a direct contact. If the arc flows onto the object to be heated ohmic heat is generated inside the object.

- *Plasma beam heating*: generation of a very hot beam of ionised plasma through arc discharge or inductive energy coupling. In striking the object to be heated, the plasma beam transfers heat mainly through convection.

- *Electron beam heating*: heat generation caused by accelerated free electrons striking the object to be heated.

- *Laser beam heating*: generation of monochromatic light and/or infrared radiation through electrically excited coherent quantum emission.

There are a wide range of *operating frequencies* and/or frequencies at which electromagnetic energy is transferred to the object to be heated; see Figure 1 (1).

1.2 Technical and Operating Properties

Physical peculiarities of electrical energy bestow on electrothermal technologies a number of features that distinguish them from other heating processes based on burning fuels for generating heat.

The most important features are:

- High temperatures can easily be obtained without a major decrease of efficiency.
- Power input can freely and simply be controlled and easily be metered.
- Local temperature distribution in the object to be heated can be achieved according to needs (see Figure 2).
- There is a wide variation of energy transfer density and effective cross-section (see Figure 3).
- Fast processes at high heating rates lend themselves to chained production processes; they can be adapted to the entire work cycle. In metal processing, high heating rates have the added advantage of producing less scale due to shorter heating periods, thus saving material cost and avoiding problems in subsequent processing.
- A high degree of freedom relating to process atmosphere and pressure is permitted.

1.3 Economic Criteria

In general, the costs and benefits of a plant must be analysed if the economic viability of a plant is to be judged.

As an example of the case of heating metals for hot forming, Figures 4 and 5 (4) show the cost structures for an induction heating plant and for a gas-fired rotary-hearth furnace, as a function of their annual production. Production capacity for each of the plants, when double-shifted, is 6,000 tonnes per year. The comparison is on a 1984 price basis, with a conversion rate of 3.14 DM per $ (1.58 for 1987).

In Figures 6 and 7, the influence of independent variations of energy price levels on the result of the above cost comparison is illustrated. It has to be noted that every such relationship is only valid for one particular case in terms of operation mode and annual production.

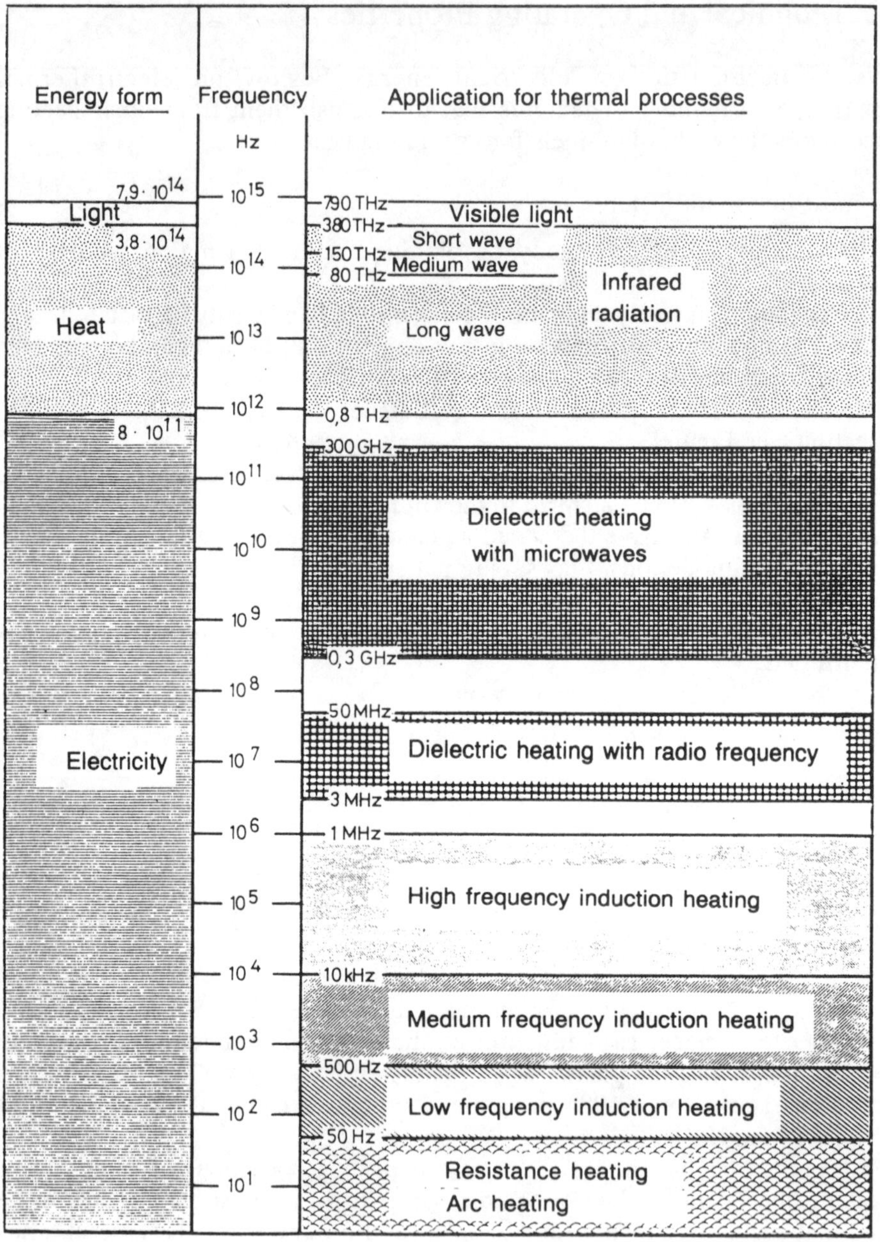
Energy form
Frequency
Hz
Application for thermal processes
7,9 · 10^14
Light
3,8 · 10^14
Heat
8 · 10^11
Electricity
10^15
10^14
10^13
10^12
10^11
10^10
10^9
10^8
10^7
10^6
10^5
10^4
10^3
10^2
10^1
790 THz
Visible light
380 THz
Short wave
150 THz
Medium wave
80 THz
Infrared radiation
Long wave
0,8 THz
300 GHz
Dielectric heating with microwaves
0,3 GHz
50 MHz
Dielectric heating with radio frequency
3 MHz
1 MHz
High frequency induction heating
10 kHz
Medium frequency induction heating
500 Hz
Low frequency induction heating
50 Hz
Resistance heating
Arc heating

Figure 1. Range of frequencies for electrothermal technologies: Resistance heating and arc heating are possible at an operating frequency of zero (DC), because the generation of heat is not based on electromagnetic interactions in this case. Nevertheless, the mains frequency is often applied because it requires no frequency conversion. In certain applications care must be taken to avoid electrochemical processes.

Due to the effect of the induction law being dependent on frequency, inductive heating only works with alternating current. The frequencies used cover a range from the main's frequency to about 1 MHz. The current frequency for the individual application is mainly determined by the desired degree of field displacement of electromagnetic waves in the object to be heated.

A significant amount of heat can also be generated in the frequency range above approximately 3 MHz in electrically non-conducting substances containing electrically polarised molecules. This field of dielectric heating goes up to near the frequency limit at which electromagnetic waves are referred to as electrical energy.

Beyond a frequency of 8×10^{11} Hz, however, (corresponding to an atmospheric wavelength of 375 micrometer) electromagnetic waves assume the character of "temperature radiation". For this reason, electric infrared heating with its spectrum reaching partly into the visible light range, may be referred to as mediate resistance heating. But also in other cases of mediate resistance heating, a larger or smaller amount of heat generated in the heating conductor is radiated to the object to be heated. Depending on the laser-active medium used, the frequency of laser radiation also lies in the infrared or the visible spectral range.

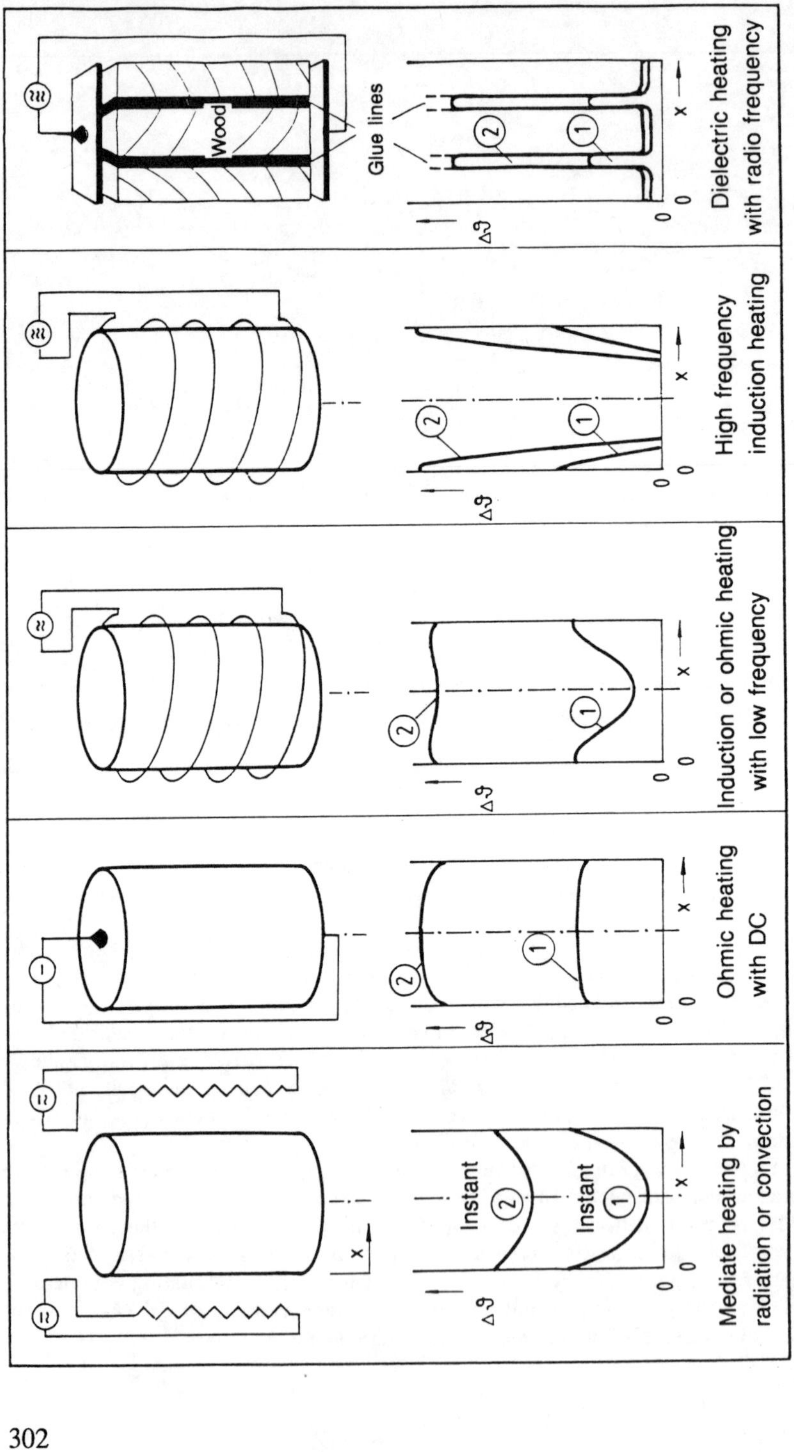
Mediate heating by radiation or convection
Instant 2
Instant 1
Ohmic heating with DC
Induction or ohmic heating with low frequency
High frequency induction heating
Dielectric heating with radio frequency
Wood
Glue lines

Figure 2. Temperature distribution for various electrothermal technologies at two different stages.

In mediate resistance heating, like in types of heating using heat media or combustion gases, heat is always being transferred from the outside to the surface of the object to be heated, usually in the form of radiation and convection. If the object is a solid, the energy is transmitted to the inside of the solid through heat conduction. For this reason the temperature distribution always drops from the surface to the center of the object, so ideal heating to reach an even and exact temperature throughout the object is impossible.

In the case of ohmic DC-heating, even heat generation occurs throughout the entire volume of the object, so that the core temperature can exceed the surface temperature.

If the frequency of induction or ohmic heating is adjusted to the thickness of the target, a very even temperature profile can be obtained at the end of the heating process.

If, however, high frequencies are applied, together with high energy transfer densities, energy absorption and hence heating is concentrated at the extreme edges of the object.

One particular type of selective heating can be obtained with direct electroheat when the object to be heated is inhomogeneous and when individual sections of the object have different material properties, i.e. widely differing energy absorption. One case in point is glue drying in a capacitor field in wood processing. Because of the high dielectric losses of solvents, these absorb most of the energy, while the wood itself is hardly heated.

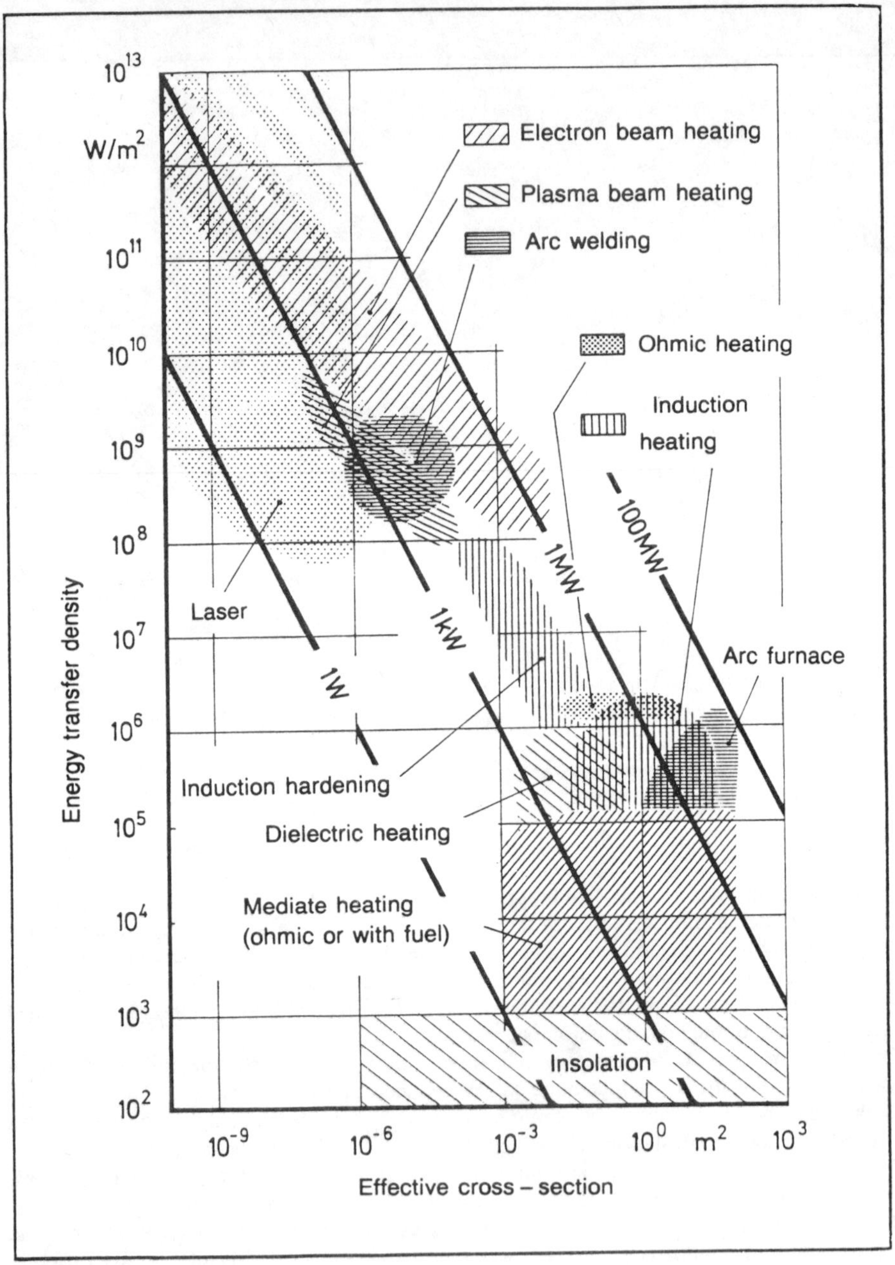
Electron beam heating
Plasma beam heating
Arc welding
Ohmic heating
Induction heating
W/m²
Energy transfer density
Laser
1W
1kW
1MW
100MW
Arc furnace
Induction hardening
Dielectric heating
Mediate heating
(ohmic or with fuel)
Insolation
10¹³
10¹¹
10¹⁰
10⁹
10⁸
10⁷
10⁶
10⁵
10⁴
10³
10²
10⁻⁹
10⁻⁶
10⁻³
10⁰
m²
10³
Effective cross – section

Figure 3. Energy transfer densities of electrothermal technologies:

Energy transfer densities that can be achieved with the various technologies are plotted against the corresponding range of effective cross-sections. For orientation purposes, the heat input is represented as a family of straight lines. The effective cross-section is that part of the surface through which the heating power passes to the object to be heated, be it in the form of heat transmission, electromagnetic fields or the application of electrically charged particles. The heat input per unit of surface of the effective cross-section is the energy transfer density.

An illustrative comparative value for energy transfer density is the maximum solar radiation on a clear summer's day which is about 1 kW per m^2 of vertically irradiated surface area.

Conventional heating processes based on fuel and mediate resistance heating can have maximum densities of energy transfer by means of radiation and convection, which are higher by around two powers of ten. However, values of 10^5 W/m^2 can only be reached in high-temperature processes.

About another factor-of-ten higher is the energy transfer density achievable in dielectric, ohmic and inductive heating or in large arc furnaces. In the latter, the high value is due to the high temperature of the radiating arc. In the first three processes, electrical energy is transferred to the object; compared with heat transmission this is possible at a much higher density.

Inductive hardening can be performed with energy transfer densities up to 10^8 W/m^2. Combined with the small penetration depth of power, this leads to extremely high heating-up velocities which are usually between 10^2 and 10^3 K/s. In the extreme case about 10^4 K/s is possible, which means that the desired temperature is reached after one tenth of a second.

High power densities are often due to the energy being transmitted to very small effective cross-sections. This effect is desirable in electrothermal joining and cutting processes; an attempt is sometimes made to focus the effective cross-section to as small a point as possible in order to enhance energy transfer density. Electron-beam and laser-beam heating, in particular, are methods permitting the minutest effective cross-sections down to about 10 (mic.)^2. Due to their extremely high energy transfer densities, all types of material can be evaporated at the point of action within split seconds; such processes are used for very precise removal of material as in cutting and drilling.

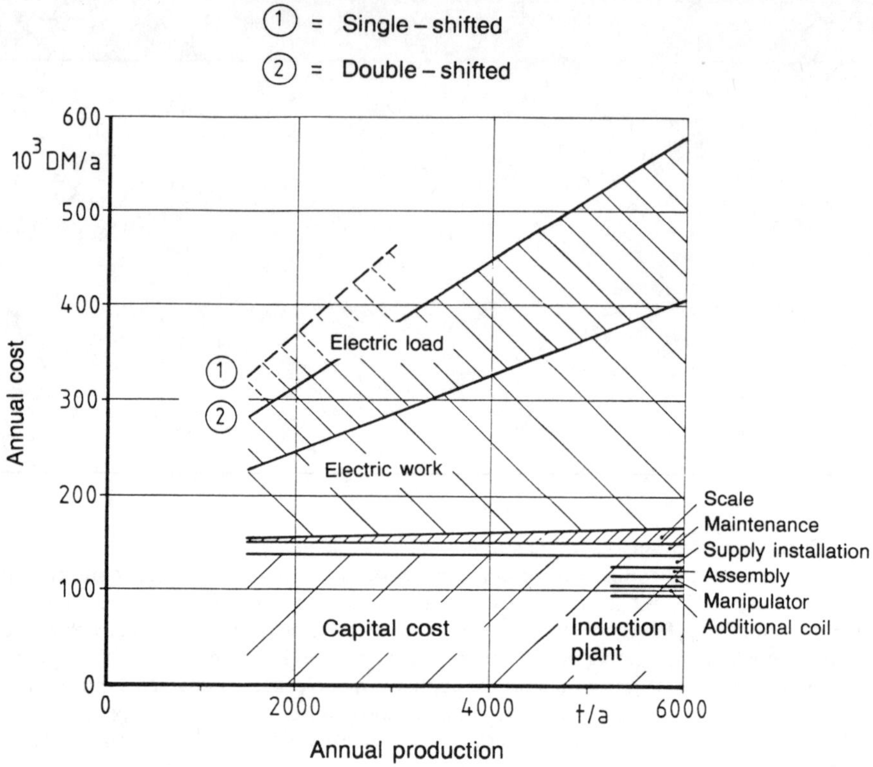

Figure 4. Annual cost of an induction plant in a forge shop.

The question of profitability must always be analysed for every individual case. A premature generalization of a comparative analysis may lead to serious misinterpretations.

In any economic analysis there are also aspects which are difficult to quantify in a cost accounting system, which, however, may be of decisive importance in the corporation's purchase decision and selection between the various technologies available.

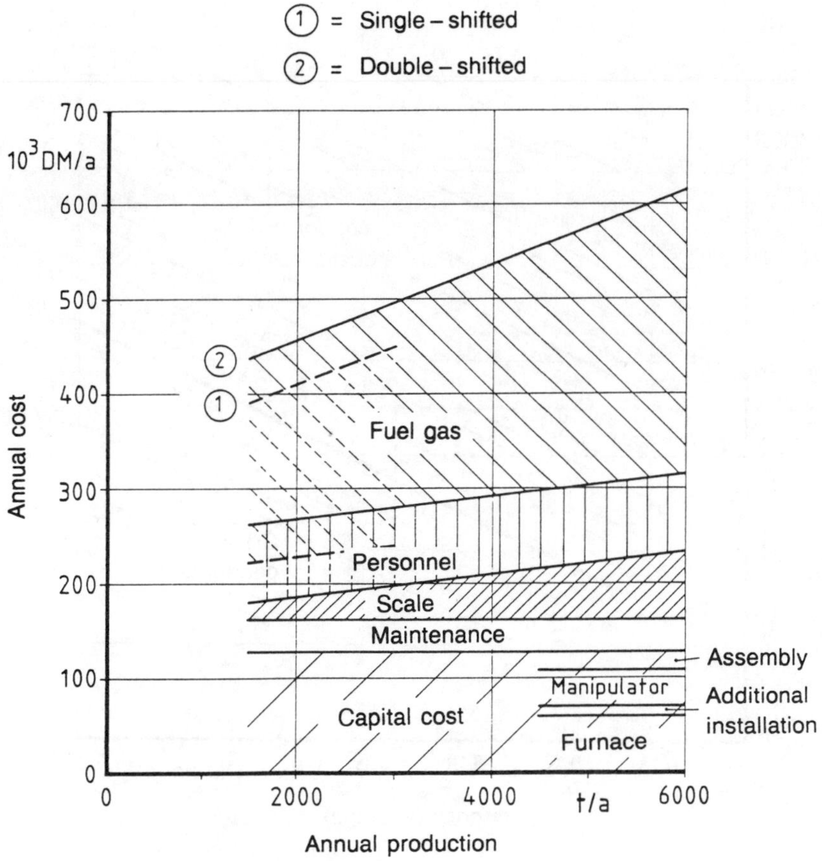

Figure 5. Annual cost of a rotary-hearth gas furnace in a forge shop.

1.4 Chains of Utilization of Primary Energy Requirement

As a rule, an electrothermal process requires considerably less final energy than a fuel-based process. Looking at the input of primary energy, the picture is completely different for countries where most of the electrical energy is generated in thermal power plants. Even then, however, the use of electricity instead of fuels for heating processes does not necessarily mean a waste of primary energy. There are important sources of primary energy (like nuclear or lignite fuels) that up to now can only serve this purpose via electricity.

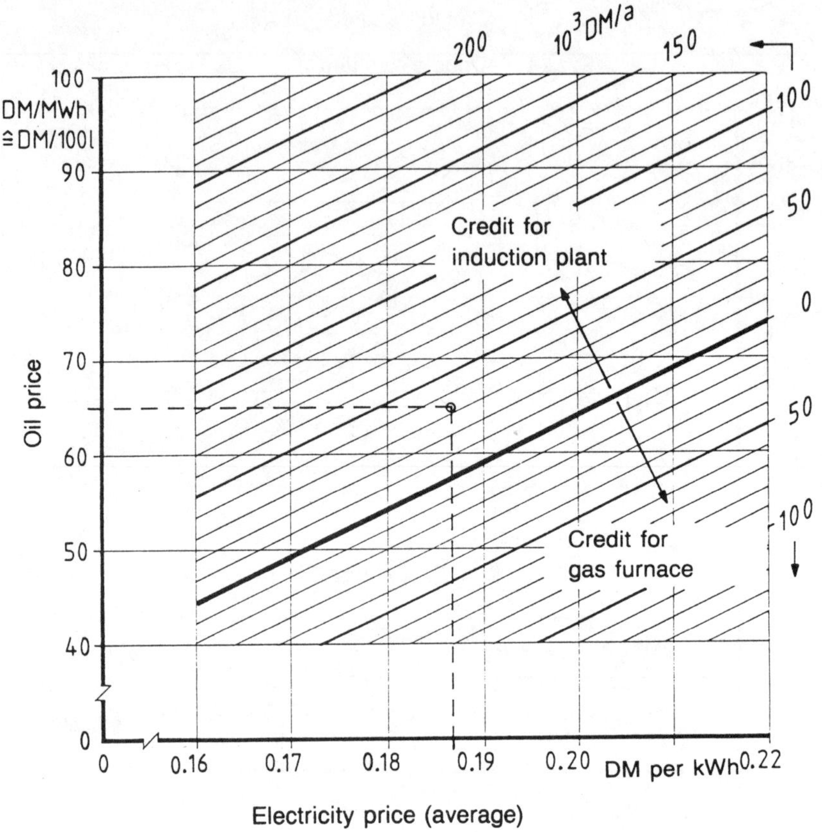

Figure 6. Variation of energy price levels for comparison of heating costs in a forge shop. Annual production: 6,000 tonnes with double-shifted operation. The thick, solid diagonal line represents economic break–even conditions for heating costs in the prototypical forge shop.

When comparing the qualities and the quantities of primary energy required in different technologies, all accessory process chains must be analysed to the extent that they are of relevance in energy terms. This concerns in particular losses and own-consumption at the various stages of converting, transporting and distributingenergy between the levels of primary energy and final energy supplied to the consumer. If there are differences in the qualities or quantities of the material input required for the process, this can lead to important differences in the energy requirement for the production of these materials (see Figure 8).

A number of other process comparisons are quantified in (3).

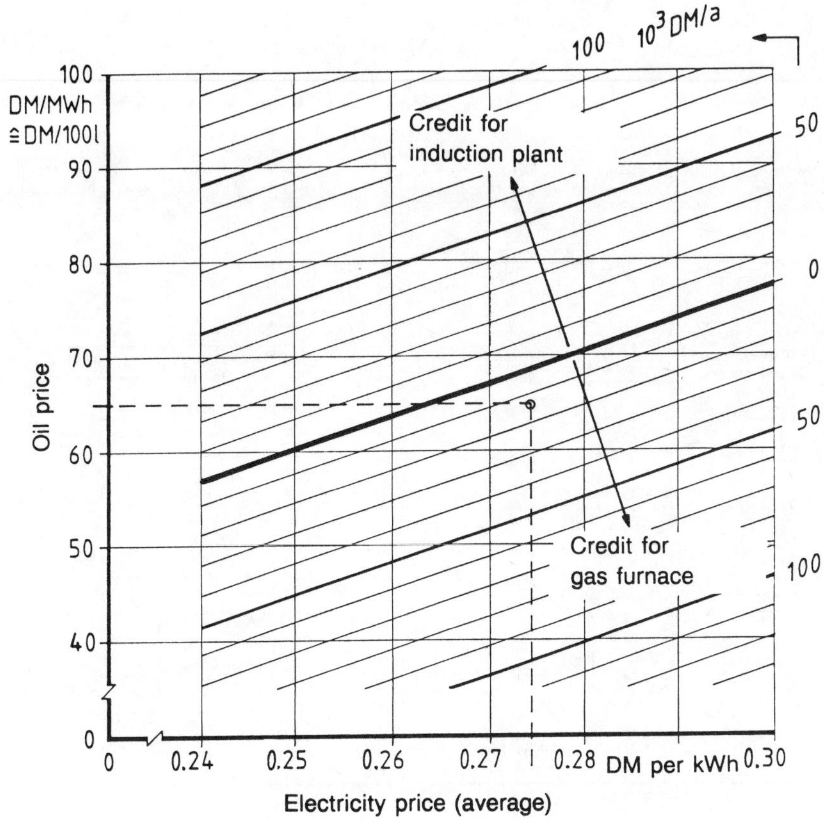

Figure 7. Variation of energy price levels for comparison of heating costs in a forge shop. Annual production: 3,000 tonnes with single-shifted operation.

2 New Electrothermal Processes and Technologies

None of the electrothermal technologies can rightly claim to be new in principle. Even the laser, the most recent technology applied in this field, was developed almost 30 years ago. Nevertheless, almost all the electrothermal technologies involved are constantly undergoing further development in detail, either with a view to reducing energy consumption, enhancing process quality, or adapting it to new functions. Despite some mostly cost-related difficulties

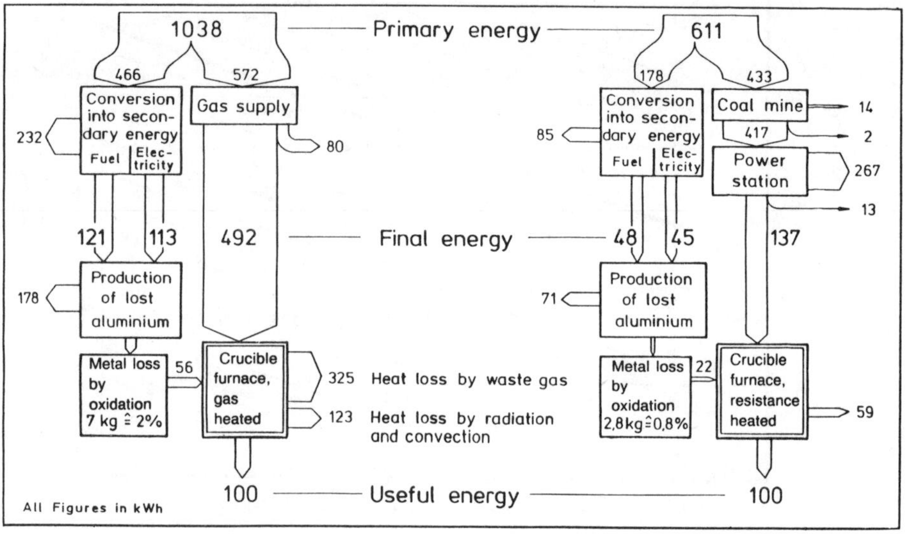

Figure 8. Energy flow diagram for 350 kg aluminium, melted in small crucible furnaces of 150 kg capacity.

Based on the same useful energy and the same quantity of molten metal, the electric furnace has only a fraction of the final energy requirement at 137 kWh because of the absence of waste gas losses and the lower surface losses; this compares with 492 kWh in the gas-heated furnace. This difference is partly compensated for by losses in the field of conversion, which are naturally much higher in thermal power generation than in gas supply. Taking into account the energy process chain up to the stage of primary energy, then metal losses due to scale exert a considerable influence. Because of the more even temperature distribution and the lower temperature at the surface of the bath, burn-up in the electric furnace is much below that of the gas-heated furnace. Considering the consumption of electricity and fuel to produce the burnt-up aluminium and/or the primary energy required, then the resulting advantage for the primary energy balance of the electric furnace is even greater than that coming from the actual energy required for melting.

and obstacles, there is a growing trend for these electrothermal technologies to penetrate into new fields of application, thus substituting fuels in final consumption by electrical energy.

In the following section, an attempt is made at outlining the present state of the art in electrothermal technologies and their applications and discussing the most interesting development perspectives. Reference (4) describes these processes in detail.

2.1 Melting and Holding Metals

Cast iron is made in different types of melting furnaces based on either fossil fuel (cupola) or electrical energy (induction furnace).

Interest in electric melting systems has grown as a consequence of increased quality requirements and the need to melt at low raw material input and low environmental pollution. At roughly 30% of primary melted iron, electromelting is a genuine alternative mainly to medium and small cupola melting furnaces and in the field of duplicate castings.

The heating principle involved in induction furnaces varies according to the furnace design; they are either channel furnaces (inductive heating with iron core) or crucible furnaces (inductive heating without iron core). Both designs are sketched in Figure 9.

The induction crucible furnace works either at main's frequency (50 or 60 Hz) or at medium frequency (up to 1000 Hz). For technological and economic reasons, the crucible furnace operating at main's frequency is advantageous as a melting unit for medium to high melting rates, in uniform melt composition and in a multi-shift operation. Moreover, it can be used for storing, overheating, holding and casting, and it can be combined with other melting units such as the cupola.

The medium frequency induction crucible furnace is the unit of choice if the greater investment for frequency conversion is compensated for by one or several of the following advantages:

- higher specific output at a given furnace size and shorter melting times,
- frequent change of alloys, combined with complete emptying of the furnace,
- start-up without residual sump and with small lumps of charging material, or
- avoiding of scale losses, especially with high-quality alloys.

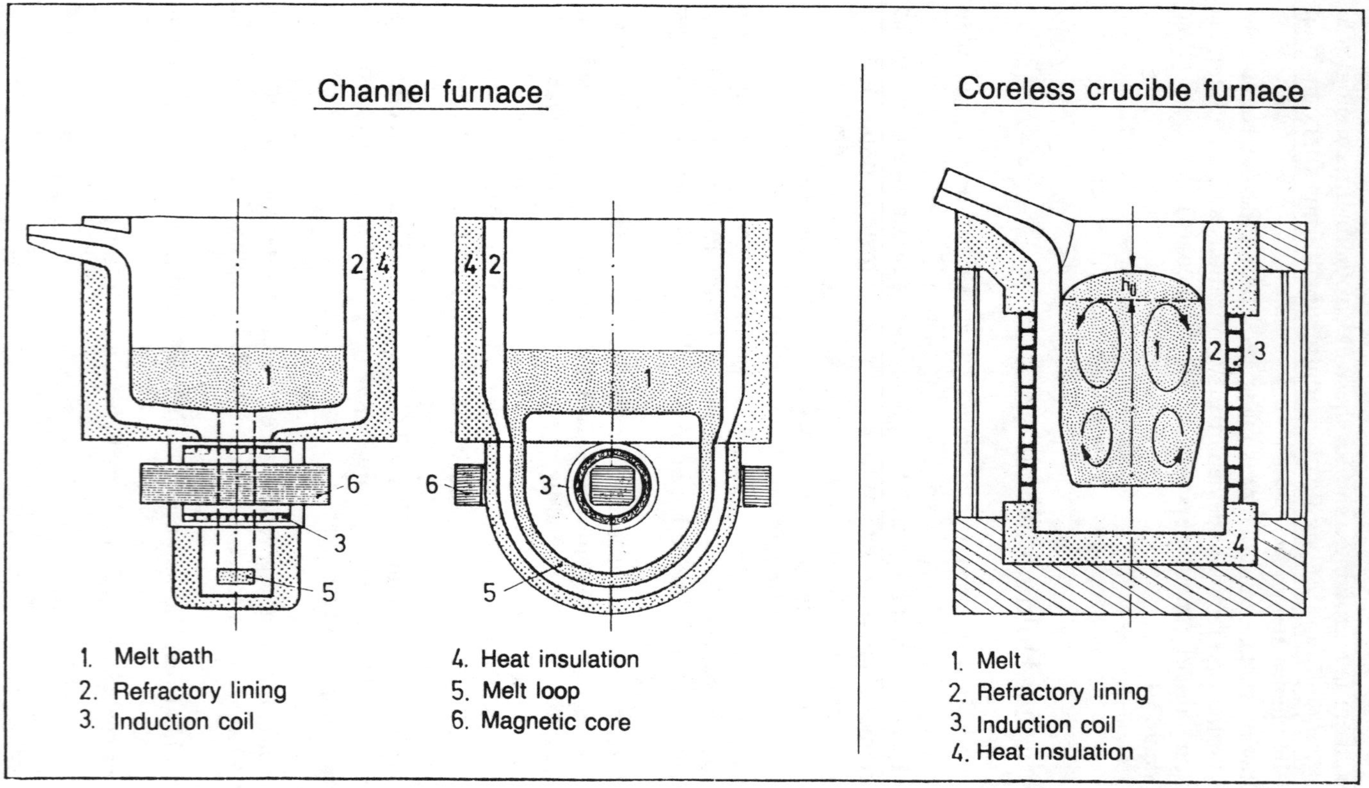

Figure 9. Basic design of induction furnaces.

Starting up the furnace without problem has a favourable effect on energy consumption particularly in single-shift operation.

As an additional unit, the induction channel furnace is used for overheating and holding the melt, provided the analysis or composition of the melt or the type of alloy does not have to be changed.

The main reasons for the induction crucible furnace taking the lead over the cupola are as follows:

- flexibility in terms of the charging material, because the composition of the alloy can be corrected at any time. Connected with that is the capability of very fine tuning to the desired analysis. There is a very low loss of material in the form of scale and consequently very little change in the alloy composition. Alloys can be changed quickly and without much problem;

- homogeneity of the melt in terms of composition and temperature due to the movement of the bath. In this way, the required casting temperature can be accurately maintained;

- great variety of operating modes, depending on the requirements of the casting line and type of furnace: either batchwise operation with complete emptying after every batch or quasi-continuous operation with frequent discharge of small quantities;

- simplicity of operation, hence little demand for manpower and relatively comfortable working conditions;

- environmental protection in the foundry: there are no combustion exhaust gases, and the waste air and dust produced is much less than in the cupola furnace.

For the melting of non-ferrous metals, the induction channel furnace is also used as a melting unit. In smaller foundries, resistance-heated crucible and/or bath furnaces are also used for melting and holding. The main competitors of electric furnaces in the case of non-ferrous metals are oil and/or gas-heated furnaces.

2.2 Heating Metals for Hot Forming

One of the most important applications of inductive heating is heating rods, blocks, arbors or slabs to the temperature required for hot forming (mainly die forging, pressing or rolling).

For heating in an axial passage, the individual items are slowly passed through one or several induction coils in the direction of their longitudinal axis; during this pass they absorb energy. Figure 10 shows the temperature range for the surface and the core of a steel bar over time during induction heating.

In order to obtain an even temperature distribution across the entire workpiece, which is needed for hot forming, the operating frequency must match the dimensions of the cross-section of the items to be heated. Needless to say, material-specific properties have to be taken into account. Figure 11 illustrates the most favourable range for the diameter (or edge length) of the item and the frequency required for heating steel bars to a temperature of 1,200°C (5).

The most common application is induction heating of small ingots, i.e., heating rough forgings cut to length in mass production.

Rods are completely heated in combination with a hot press or hot shears for cutting the rods into forging grade ingots. One of the largest plants of this type is designed for a nominal throughput of 25 tonnes per hour (t/h), with the heating line being 25 m long and consisting of 20 individual coils and a nominal power rating of around 10 MW (6).

If only one section of the rod is to be forged, only this section needs to be heated. Depending on the throughput and the degree of automation, either each rod comprising the section to be heated is introduced into an individual coil; or up to six rods lie side by side in an oval pancake coil. If the sections to be heated are short and the throughputs high, special tunnel inductors are used enabling a continuous transverse passage of the rods and minimum cycle times of 1 second (7).

Because of its process and operating advantages, inductive heating is leaving its competitors behind. Fuel-heated (oil or gas-fired) continuous discharge furnaces, rotating hearth furnaces, bogie hearth furnaces, or chamber furnaces can be favourable when the dimensions of workpieces are unsuitable for induction heating or when small process lots with different dimensions change frequently.

One advantage of induction systems which is often of decisive importance is the cost of material lost during heating due to surface scaling and scrap forgings. Induction heating takes only about one-quarter of the time required in fuel-heated furnaces. Combined with the absence of overheating of edges, this adds up to a dramatic reduction of scaling rates and scrappage.

The specific electricity consumption of an induction plant at nominal operation is mostly between 380 and 460 kWh per tonne of heated material, while the specific final energy consumption (in relation to the calorific value) of a fuel-heated furnace at nominal throughput is higher by at least half;

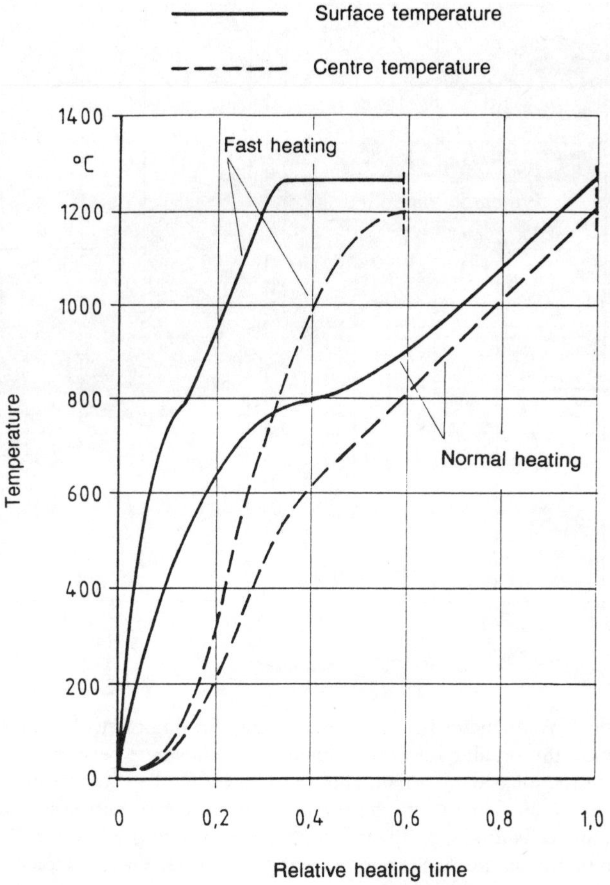

Figure 10. Temperature of a steel billet during induction heating.

If a uniform coil system is used and a suitable frequency selected then the temperature curve is the same as plotted under 'normal heating'.

In large plants, the system is designed to handle an increased power transmission per unit of length; this is done either by increasing the number of turns per unit length in the induction coil or by raising the coil voltage. This 'fast heating' process causes a change in the temperature curve. In the second section of the heating system, the power input must be adjusted such that the surface temperature is maintained at a target value. In fast heating, the heating period is reduced to about 50 to 60% of the corresponding reading for normal heating. With about the same sweep rate, the length required for the heating system is reduced by the same amount. Specific current consumption, too, is slightly reduced because of the shorter heat radiation period of the workpiece. On the other hand, fast heating units are less flexible; if partial utilization drops below 80% of the nominal throughput, it is difficult to obtain an adequately uniform temperature profile. There is growing danger of the temperature rising excessively high, thus causing a possible melt.

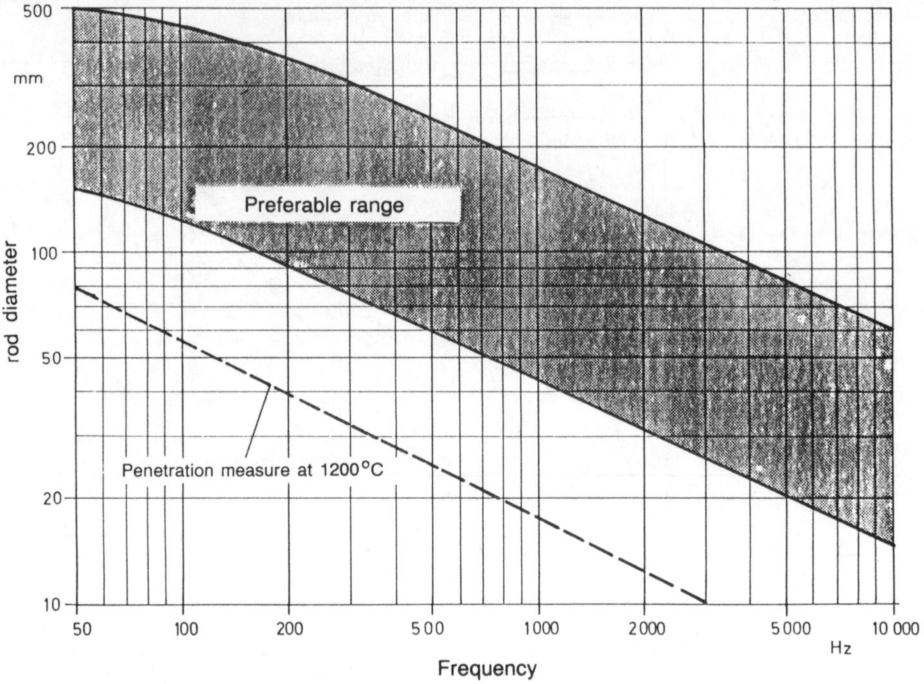

Figure 11. Selection of frequencies for induction through heating of steel rods:
For a given frequency, the suitable range of diameters is somewhere between 2.5 times and 7 to 10 times the penetration measure at a temperature of 1,200°C. If the diameter is too small, the electric efficiency is too low, therefore the specific current consumption is excessively high. However, if the diameter is too large, the centre area remains cold while the surface layer has reached hot forming temperature. Since overheating of the surface layer is tolerable only within narrow limits, the power transmitted to the workpiece has to be reduced dramatically once the overheating point is reached. The ensuing temperature compensation by means of thermal conduction takes relatively long; it is accompanied by intense heat radiation across the hot surface. This again leads to an increased specific current consumption.

unfavorable conditions, like inadequately preheated air, poor burner tuning, excessive secondary air, etc., may raise consumption up to 1,000 kWh/t and more.

Basically, the same process advantages of induction heating also apply to ohmic heating of steel rods. This heating process lends itself to elongated items with a uniform cross-section along the total length. The workpiece to be heated is moved into the device on rolls or by means of a grapple transverse to the longitudinal axis; it is then clamped into contact units. One of the contact units must be movable in axial direction in order to adjust to different lengths of workpieces and to compensate for thermal expansion. Depending on the dimensions and the required current, each set of contacts consists of a number of die-shaped individual contacts (up to four at the face and up to eight spread over the circumference). A maximum current of 10 kA can be transmitted over each contact, provided the contact is water-cooled and hydraulically pressed on the surface of the workpiece. In order to prevent the contact surface from denting, the contact pressure must be reduced before the material starts to soften; in the case of steel softening begins at about 500°C.

Forges apply ohmic heating to diameters between 10 and 50 mm and lengths between 150 and 750 mm. Ohmic bar heating units are used upstream milling units for dimensions of 80 to 120 mm square and lengths up to 10 m. Units with up to 12 MVA are already in operation (8).

Most steel grades require about 200 to 230 kWh/t of specific active energy for heating from cold to 1,200°C. In the relevant range of dimensions, the specific energy consumption from the mains is normally between 270 and 320 kWh per tonne of heated material; this is markedly below the values typical of induction heating.

Ohmic heating is also applied in heating round wire in a continuous-flow system. The wire has a diameter between one and 30 mm. It is heated to the temperature required for hot forming in the downstream structuring mill. Figure 12 (9) illustrates the layout of such a plant. The largest plants have a power rating of 1500 kW.

2.3 Inductive Hardening

Inductive hardening is mainly used for hardening surfaces or limited areas of the total surface as well as for through-hardening sections of a workpiece. It is rather seldom, however, to through-harden complete workpieces. In selecting appropriate power, frequency and inductor shapes, the heat development can be confined precisely on the zones to be hardened.

The centrepiece of an induction hardening plant is the inductor. In general, it consists of a hollow copper profile through which water flows. It is fed by an AC source of suitable frequency and power. The shape of the inductor follows the desired heating pattern and/or geometry of the part to be hardened. Figure 13 shows the basic shapes of inductors. The inductor is usually directly connected with a quenching device in the form of a water shower that rapidly cools down the workpiece just heated, thus causing the formation of martensites required for hardening.

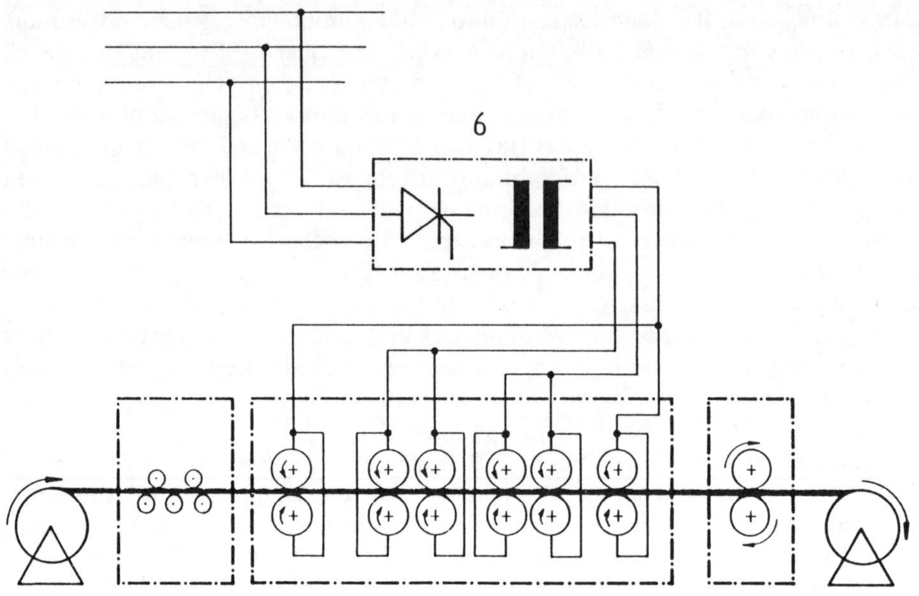

Figure 12. Plant scheme for continuous ohmic wire heating:

The wire is wound off the coil (1), straightened and mechanically descaled (2), following which it is guided through a conductive heating system (3). The current transmitted via roller contacts heats the wire to a temperature of 1,000 to 1,050°C. The wire is then taken through two to three shaping passes in a stretcher-reducing mill (4), either cut to length or wound on coils again (5). The three sections of the overall heating system permits a three-phase connection via power controller and transformer (6).

Figure 14 illustrates the interaction between the most important process parameters:

- frequency,
- power density, (i.e. the power induced per cm^2 surface of the workpiece,
- surface temperature at the end of the heating period, and
- heating depth (determined by the minimum temperature of 800°C required for austenitic formation).

This is the example of a cylinder with a very large radius in relation to the effective penetration measures. For comparison, the penetration measure of the electromagnetic field sizes is shown for the frequencies in question. The higher

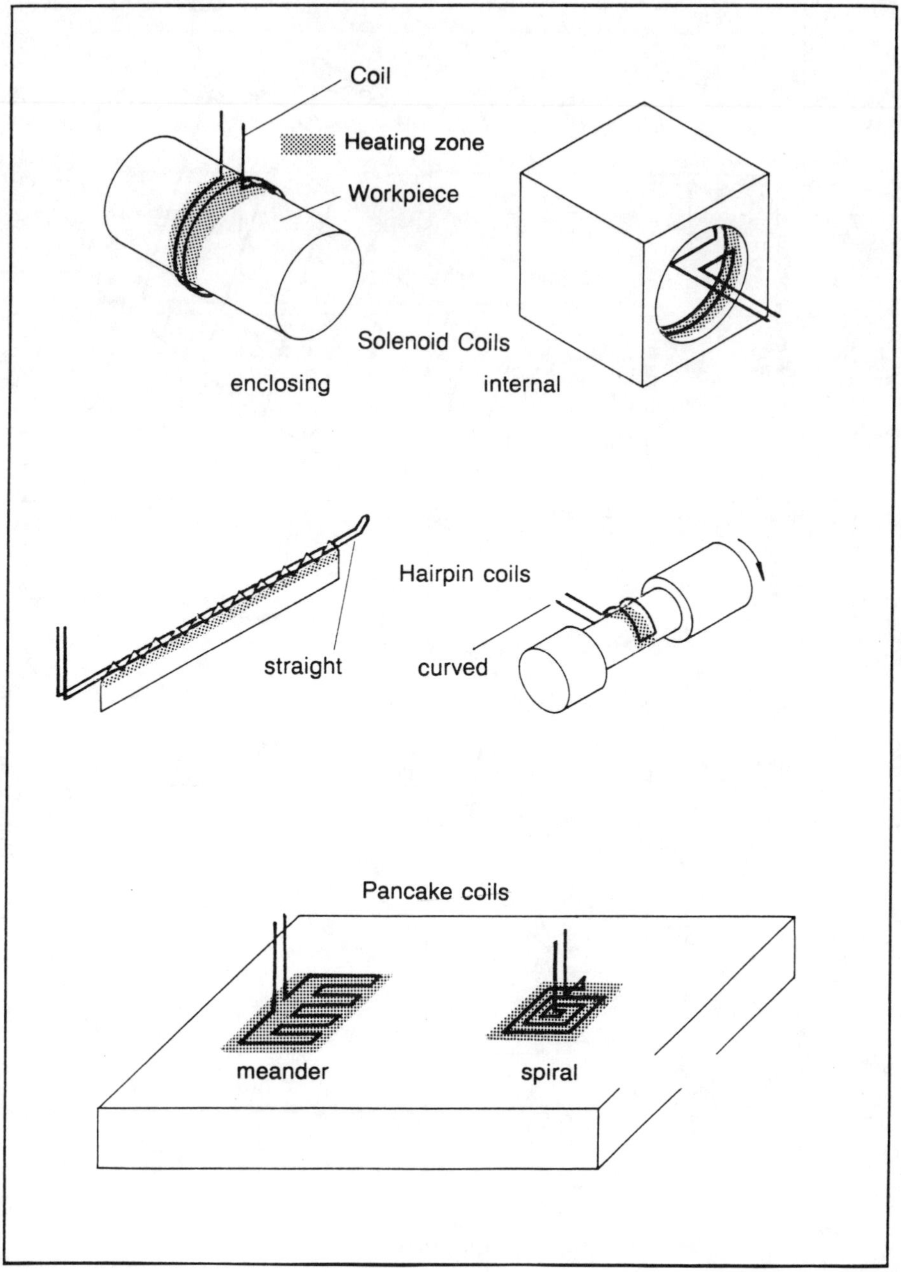

Figure 13. Basic shapes of coils for induction surface heating.

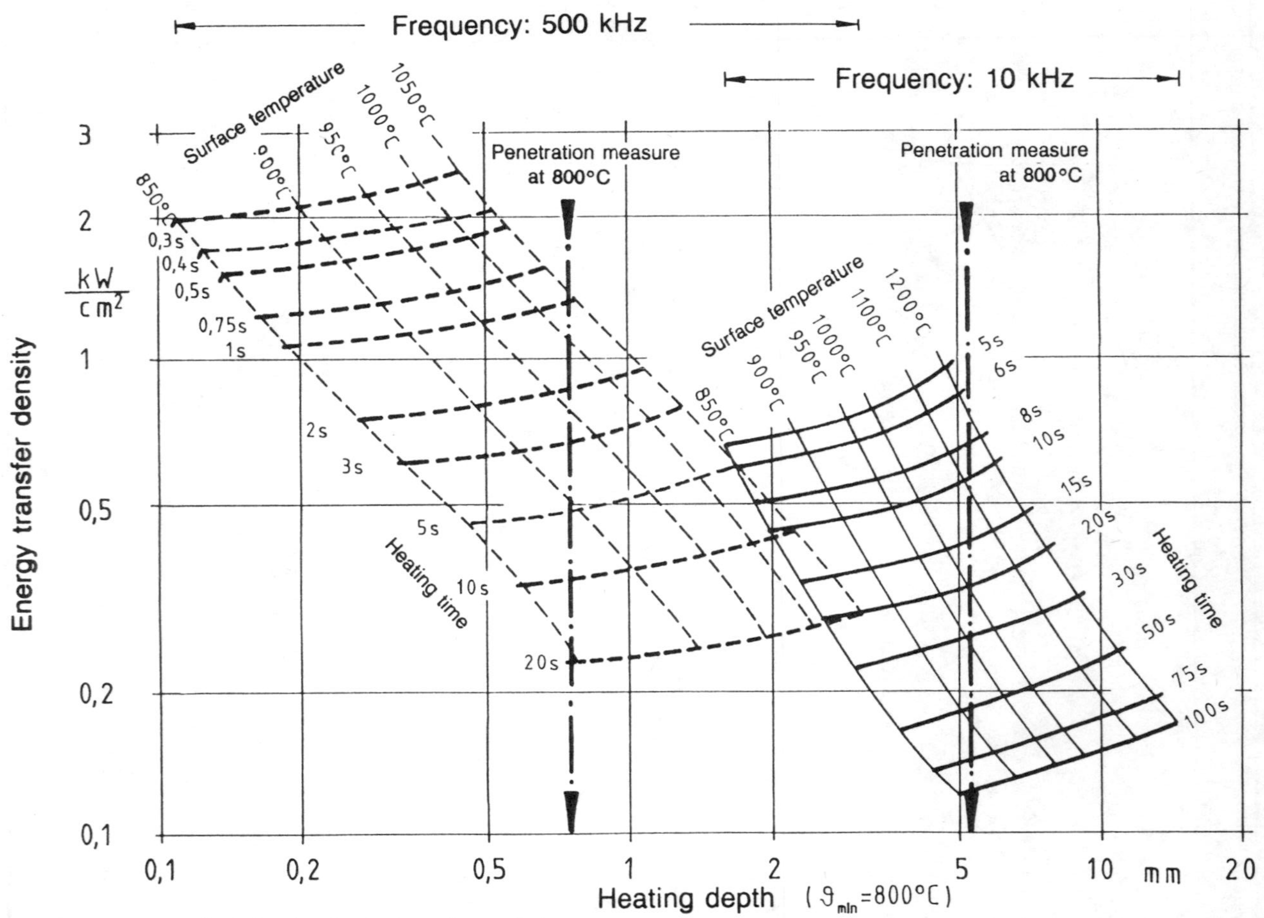

Figure 14. Process parameters for induction hardening of a steel piece.

the frequency used, the thinner are the surface layers on which the heating is concentrated. Basically, greater power densities can be induced at higher frequencies.

Heating can be locally limited not only in depth but also in the surface zones of the workpiece where hardening is intended. This is done by adjusting the shape and size of the inductor to the task. If the inductor or the workpiece is moved forward in the course of heating, the process is of the scanning type, the remainder being single-shot hardening. In both cases the workpiece can be rotated around its own axis during the hardening process. Compared with scanning-type hardening plants, single-shot hardening plants have a greater rated power at shorter processing times per piece. This explains why single-shot plants are usually preferred to the scanning type whenever large quantities and a high throughput are involved, provided, of course, that the method is suitable for technological reasons.

Compared with other technologies of surface hardening (such as carbonizing, nitriding and flame-hardening), the main advantage of inductive hardening is that the desired profile can be achieved with the appropriate choice of method (scanning or single shot), the inductor shape and the process parameter; it is also precisely reproducible. Partial hardening, that is hardening of a predefined hardness area only, is also generally possible for complex shapes of workpieces. Due to the short heating period, scaling is low and reworking is often superfluous. Undesired coarse grain formation can be avoided reliably. Distortion is comparatively low. In some applications, expensive alloyed case hardened steel can be substituted by cheap heat treatable steel when an inductive hardening process is applied. Since usually only minor areas of the entire workpiece are heated, the efficiency of this process is higher in terms of primary energy than that of comparable non-electric processes; this holds true although there are relatively high losses incurred along the way from primary energy to the induced thermal power.

An inductive hardening plant reqires little space, it is simple to operate, and it can be easily automated. Together with the absence of almost any emission from the plant, it lends itself to integration into production lines.

One disadvantage is the relatively high purchase price of an induction plant. High utilization is therefore one of the conditions for its profitable use.

Inductive hardening is applied for many different types of workpieces such as:

- crankshafts and camshafts,
- valves and their drive elements,
- axles, drive shafts and steering components,
- driving gear and gearwheels,
- track links,
- machine chain guides,

- machine parts exposed to high surface loads, such as cold rolls, bending rollers, and paper cylinders.

2.4 Electric Glass Melting

Plate glass, e.g., window panes, or glass containers, e.g., bottles, are produced from raw material which is melted together with a share of recycled cullet material, and chemical reactions occur along the way.

This process is operated mostly in continuous tank furnaces which are fired directly with fuel oil or natural gas. Figure 15a shows a simplified energy flow chart for a current design of this type of furnace. For standard grades used in container production, the amount of active energy contained in the glass melt is about 0.6 kWh/kg.

Since molten glass contains moving ions, an application of AC can provide for direct resistance heating. The furnace illustrated in Figure 16 has a hexagonal layout with two levels of electrodes superimposed, each consisting of six molybdenum electrodes horizontally protruding from the side walls. Because of the vertical design of this electric furnace it is also referred to as shaft furnace.

In a furnace heated exclusively with electricity, all the heat is generated within the glass melt; the heat insulating effect of the enclosed batch makes it possible, therefore, to maintain the temperature in the vault above the melt at a significantly lower level than in a fuel-fired furnace. The external heat losses of the electric furnace are reduced accordingly. Moreover, there is a total absence of heat losses in the form of combustion exhaust gases which gives the all-electric furnace a very high energy efficiency. Figure 15c shows a simplified energy balance of such a furnace.

For technological and economic reasons, the use of an all-electric shaft furnace is particularly interesting under two aspects:

- Because of the favorable flow conditions homogeneity and viscosity are improved when melting coloured and opaque glass; this is combined with a considerable saving of expensive colourings and plaining additives.
- In small size furnaces, i.e., melting capacities under about 30 tonnes per day, the gap in energy consumption between electric and fuel-fired furnace widens to an extent that gives the electric furnace a competitive edge even in view of the current prices of electricity and fuel. In this connection, it is mainly the compact design of the electric furnace and its small volume in relation to a given nominal melting capacity that are of particular significance. This has a positive effect also when there is a need to change glass grades to be molten.

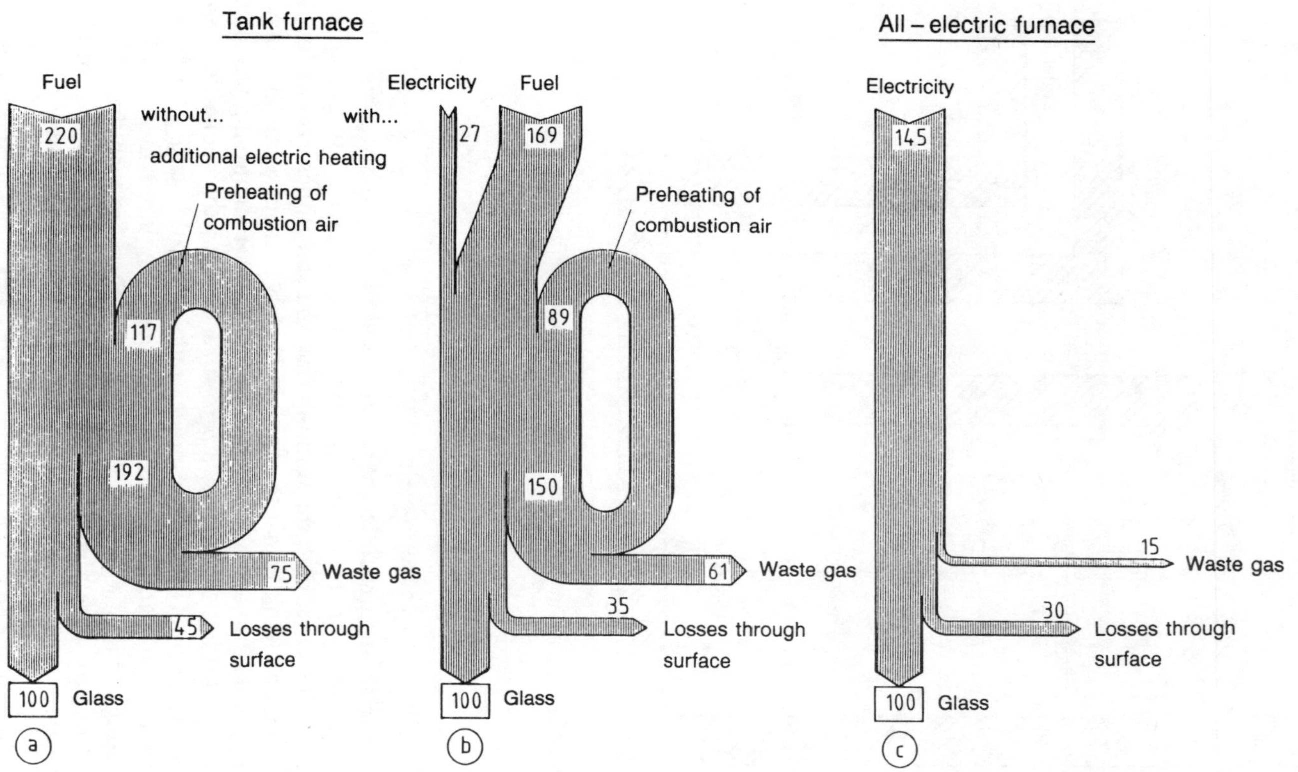

Figure 15. Energy flow for glass melting furnaces.

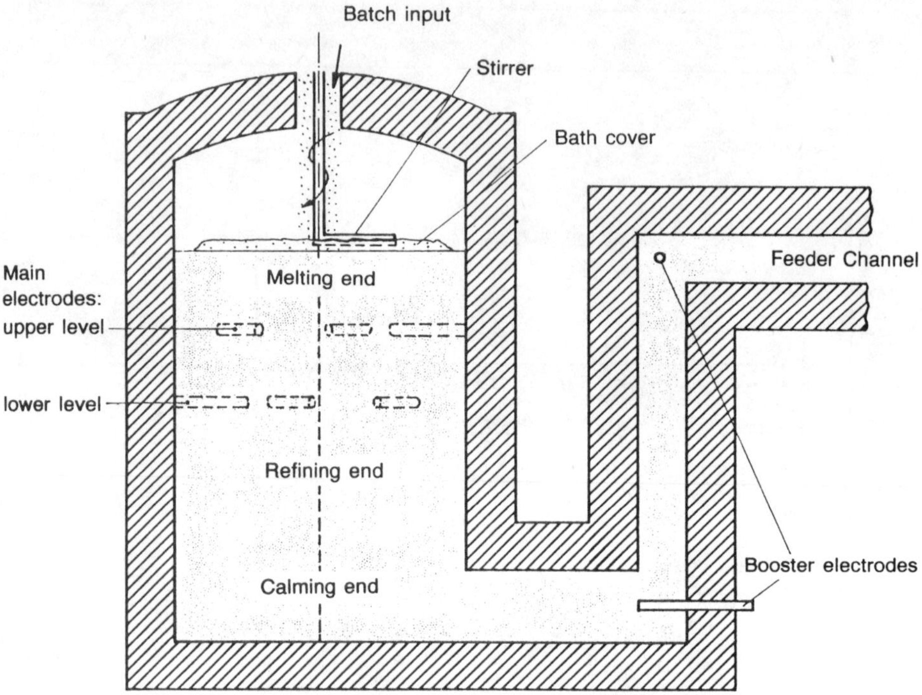

Figure 16. All-electric glass melting furnace (longitudinal section).

The all-electric shaft furnace is not widely used in the FRG, whereas electric heating as an addition to oil or gas-fired tank furnaces has gained wide acceptance. In this case, both the refining end and sometimes also the melting end of conventional tank furnaces come with molybdenum electrodes that are usually fitted in the furnace base. This is instrumental in increasing the melting performance of a tank furnace by up to 40% beyond the fuel-side thermal rating of the furnace. The electric power required in this case is only about 20% of the thermal rating. The specific energy required in an additional electric heating for melting is therefore only slightly more than the theoretical value. A simplified energy balance of a tank furnace is shown in Figure 15b.

Other advantages of an additional electric heating are as follows:

- Flexible adjustment to different melting capacities is possible by controlling the electric power.

- The share of recycled material (cullet) can be raised; this may create problems in tank furnaces without additional electric heating because the material tends to sink to the bottom of the doghouse.

2.5 Electric Infrared Heating

The infrared oven is mainly used for drying thin coats of lacquer and varnish, textile and paper webs, for baking and roasting as well as for heating plastics.

The heat generating elements are infrared radiators; the material to be heated, therefore, absorbs energy in the form of radiated heat. The permissible radiation intensity is between about 1 and 100 kW/m^2, depending on the type of heating task.

The quality of the energy flows in an infrared oven are shown in Figure 17.

Infra-red (IR) radiators are subdivided into three main groups that differ according to wave length and radiator temperature and/or their characteristic visible radiation:

- short-wave IR radiators with a radiator temperature of over 2,000°C (see Figure 18),
- medium-wave IR radiators with a radiator temperature up to about 1,100°C (see Figure 19),
- long-wave IR radiators with a radiator temperature up to about 700°C (see Figure 20).

The radiation spectrum (see Figure 21) is primarily a function of the radiator temperature. The higher it is, the smaller is the wavelength of maximum radiation and the more marked is this maximum radiation. The spectral absorptivity of the material to be heated, is one of the most important criteria applied to the selection of the radiator type. In order to achieve a maximum energy absorption, the maximum emission of the radiator should be in the wave length at which the absorptivity of the target surface is particularly great. It should be noted that the spectral absorptivity does not only depend on the material to be heated but also on the condition of its surface.

In IR-drying of moisture or solvents, the partial radiolucency of substances causes the radiation to penetrate beneath the surface, with short-wave radiation having a deeper penetration effect than long-wave radiation. For the choice of IR-radiators for tasks like curing varnishes it is therefore preferable that the maximum emission of the radiator is at wavelengths slightly smaller than the maximum absorption of the solvent to be evaporated so that the major share of radiation penetrates into the varnish base. This causes progressive drying

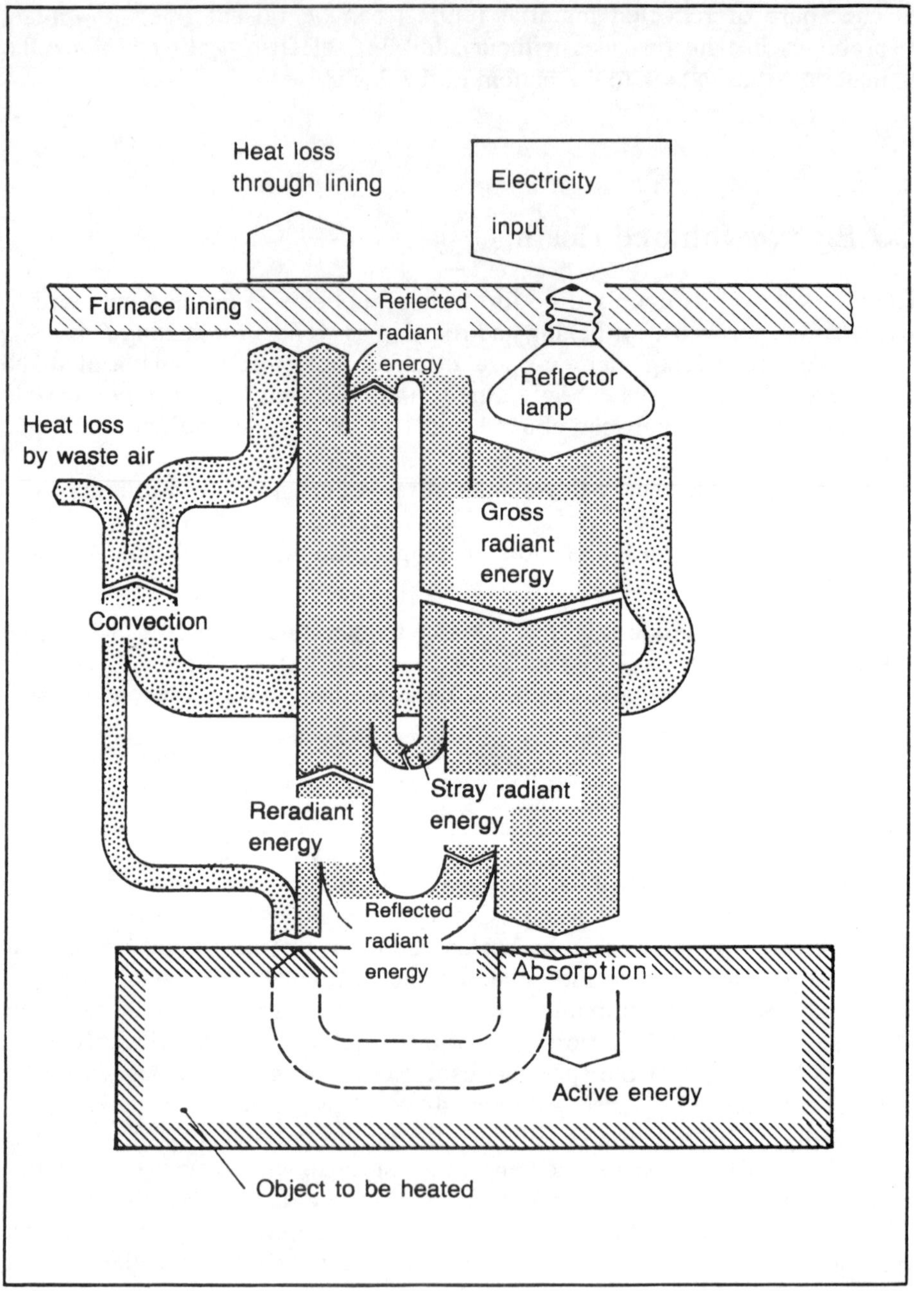

Heat loss
through lining
Electricity
input
Furnace lining
Reflected
radiant
energy
Reflector
lamp
Heat loss
by waste air
Gross
radiant
energy
Convection
Stray radiant
energy
Reradiant
energy
Reflected
radiant
energy
Absorption
Active energy
Object to be heated

from the inside to the outside, thus avoiding bubbles and encrustination of the coat. In addition, the curing rate is significantly increased, thus shortening not only curing times but also reducing the energy consumption, because the substrate is not fully heated.

There are a wealth of applications for IR-heating and new ones are being added constantly. A few examples are in:

- drying and fixing textile webs,
- gelling polyvinyle chloride pastes on textile substrates,
- drying and curing primers and varnishes on wood and metal,
- drying paper webs and curing coatings,
- plastifying thermoplasts in the production of mouldings, and
- drying and/or heat treating food.

Figure. 17. Energy flow in an IR furnace.

A share of Joule's heat produced by the radiators is dissipated by convection through the hot surface of the radiator (partly also by heat conduction through the radiator mount). The remainder is radiation energy that adds up with the radiation reflected from the interior wall of the furnace to form the gross radiation inside the furnace. Even with an optimized design of furnace chamber and radiator layout, some of this gross radiation is scattered and passes by the material to be heated. Of the radiation striking the material, a share is reflected at the surface, the remainder being absorbed and converted into heat. Of the absorbed energy again only a share can be used as active energy in the material itself; because the object is normally hotter than the ambient atmosphere inside the furnace, there is also convective heat emission from the object's surface. In addition, the heated surface of the object emits radiation to the mostly cooler interior walls of the furnace. Together with the reflected share this causes the reradiation from the object into the furnace chamber. As much as possible of the overall radiation striking the furnace walls is intended to be reflected and thus to benefit the gross radiation. This explains why the interior walls of furnaces are usually clad with aluminium or other reflective material. Nevertheless a certain amount of radiation is inevitably absorbed by the furnace wall. If the atmosphere inside the furnace is hotter than the wall surfaces, the latter also absorb heat through convection. In a stationary state, both heat shares together are taken to the outer side of the wall where they appear as surface loss.

As a rule, the convection phenomena in the atmosphere of the furnace do not provide any useful heat transmission to the object in an IR-furnace. For this reason air circulation and air exchange are basically kept at a minimum. In drying processes, certain exchange rates may be necessary, however, for the moisture or solvent vapours not to exceed a certain concentration.

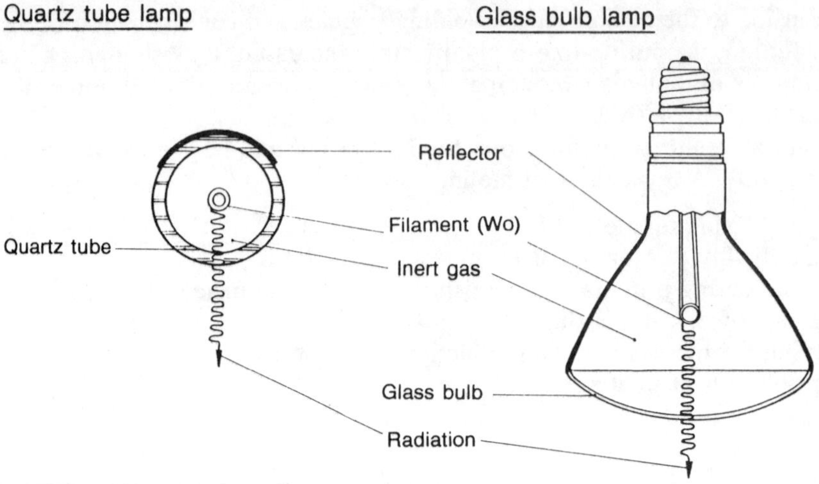

Figure 18. Types of IR radiators: Short-wave reflector lamps.

The design of these radiators resembles that of incandescent lamps. The reverse of the spherical, parabolic or cylindrical glass bulb is interior-reflected by means of evaporated silver or highest grade aluminium. The bulb is either evacuated or filled with inert gas. High-output radiators are also available as tungsten-halogen lamps.

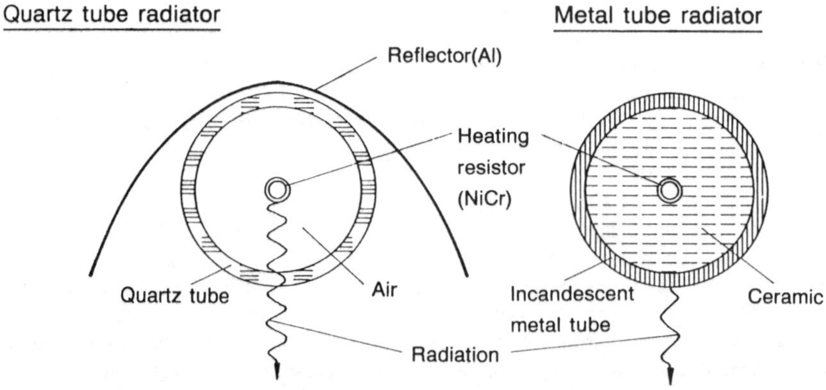

Figure 19. Types of IR radiators: Medium-wave radiators.

The quartz-tube radiator has a heating resistor in a quartz glass envelope that may be internally mirrored at the reverse for the benefit of a directed radiation emission; it may also come with an outer reflector. The major share of the power emitted by the resistor is transmitted by the envelope because quartz glass starts to have a major absorbing effect at wavelengths beyond about 4 micrometers. Accordingly, the quartz envelope does not get as hot as corresponds to the temperature of radiation. Usually, the surface temperature is below that of far-infrared radiation elements. For this reason, the convective heat emission of quartz tube radiators is very low which is of great advantage in IR-heating. In contrast, the metal tube radiator has the outer incandescent tube at radiation temperature. Metal tube radiators can be operated at such low temperatures that their radiation is in the far-infrared range.

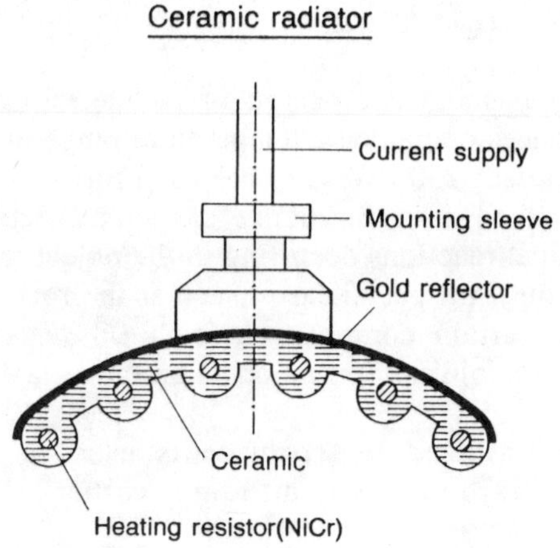

Figure 20. Types of radiators: Long-wave radiator.

Since this type of radiator emits from its surface hardly any visible light, it is also referred to as black-light lamp or far-infrared radiation element. As a rule, it is designed as a large-area ceramic radiator with NiCR heating resistors embedded into the ceramic structure. The latter has a high emission quotient at the surface. It is mirrored at the reverse.

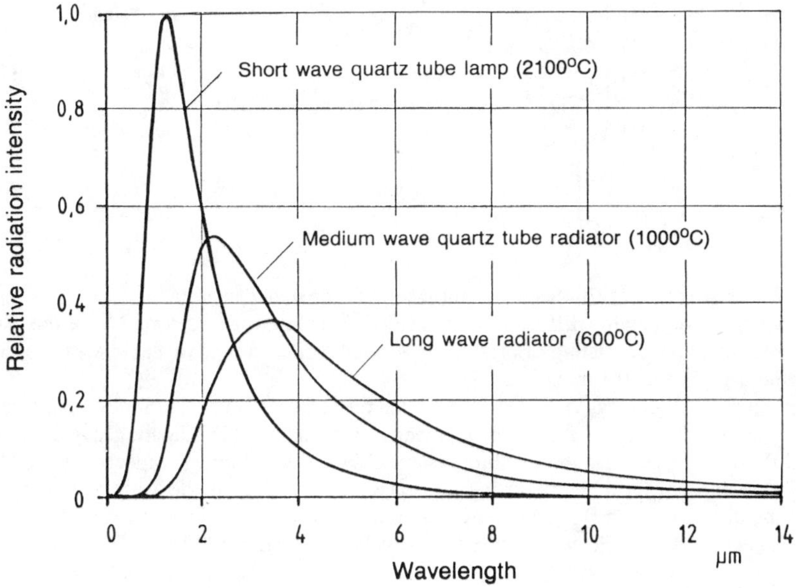

Figure 21. Spectral emission of different IR-radiators of equal total radiation power.

2.6 Fluidized-Bed Heating

Figure 22 illustrates the basic design of an electric fluidized-bed furnace. Fluidized-bed furnaces operate at a temperature range up to about 1,300°C and a power rating of about 250 kW; they are used for various types of metal heat treatment. Air is used as a fluidizing gas in mere thermal processes and without any chemical reactions occurring in the object (as in annealing, neutral hardening, tempering). In special cases, such as in process annealing of copper, an inert gas of a certain composition (e.g., 98% N_2 and 2% H_2) may be required in order to avoid the formation of an oxide layer at the surface of the material to be annealed.

Moreover, the fluidized bed furnace is used in thermochemical heat treatment such as carburization, nitriding, carbonitriding, etc., where the

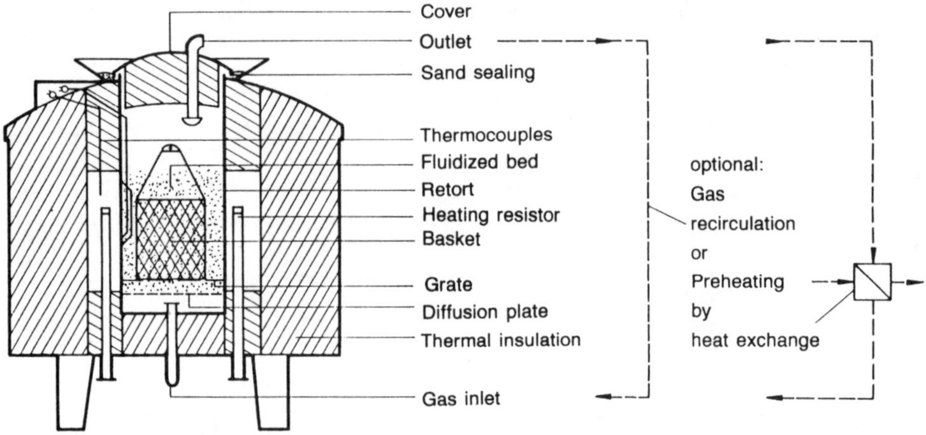

Figure 22. Construction of fluidized-bed furnace with resistance heating.

A gas-permeable metallic diffusion plate is fitted horizontally in a retort. Above the plate is the fuidizing medium, consisting of high-melting-point oxide ceramic (usually $Al_2 O_3$ or SiO_2) with a graining of about 0.1 mm.

In operation, a fluidizing gas is passed through the retort from bottom to top. Beyond a certain flow velocity, each particle of the fluidizing medium is maintained in a floating stage. This state is called fluidization because the medium behaves like a fluid in many respects. This refers in particular to the convective properties of heat compensation in the fluidized bed (local temperature differences $< \pm 0.5$ K) and the heat transfer to an object situated within the fluidzed bed. The heat transfer coefficient practically achievable is between 300 and 800 W/(m^2 K).

Retort ovens may be either electric or fuel-fired. As a rule, electric ovens have the heating element fitted on the outside of the retort. The hot fluidizing gas leaving the oven through an outlet may be either recirculated directly or used for preheating the fresh fluidizing gas.

carbon and/or nitrogen content of layers near the surface of a workpiece is intendedly changed. This reaction is initiated through the appropriate process gas.

Due to the high heat transfer coefficients in a fluidized-bed furnace, the temperature can be varied relatively fast within a very wide range. For instance, a cylinder with a diameter of 50 mm of aluminium-wrought alloy takes 11 minutes in a fluidized bed to reach the core temperature required for annealing, which is 495°C. In a normal chamber furnace, it takes 75 minutes (10). The time course of temperature changes can also be exactly reproduced for holding, cooling and quenching processes. The maximum cooling intensity is somewhere between that obtainable with air and oil.

Fluidized-bed furnaces are not only available as stationary units for discontinuous operation but also as continuous units for tempering, for instance, oil-hardened wire. The inlets and outlets of the unit are below the surface of the fluidized bed. They are isolated by a defluidized filling of the medium, through which the wire is drawn (11).

2.7 Dielectric Heating

Two different technologies exist in dielectric heating. They are:

- radio-frequency (RF) heating in the electric field of a capacitor, and
- microwave heating in the electromagnetic field of travelling or static waves.

2.7.1 RF heating in a capacitor field

In RF heating in a capacitor field, a voltage is applied with a frequency from 2 to 100 MHz between two or more electrodes (see Figure 23). The shape and layout of electrodes vary according to the heating task, with the aim always being to create an electric field that can best fulfill the given heating task (e.g., fast, even, partial, selective heating of a homogeneous or inhomogeneous material of a given size or shape). Very often a heating capacitor with parallel faces is used, but form electrodes and stray-field electrodes are also common. The use of stray-field electrodes (see Figure 24) has advantages over plate electrodes in cases where thin and elongated pieces or sections are exposed to dielectric heating.

If a water-containing material is exposed to a radio-frequency electric field, this is referred to as radio-frequency drying; it produces the following typical phenomena which distinguish RF-drying from conventional thermal drying in a convection drier, radiation or contact drier:

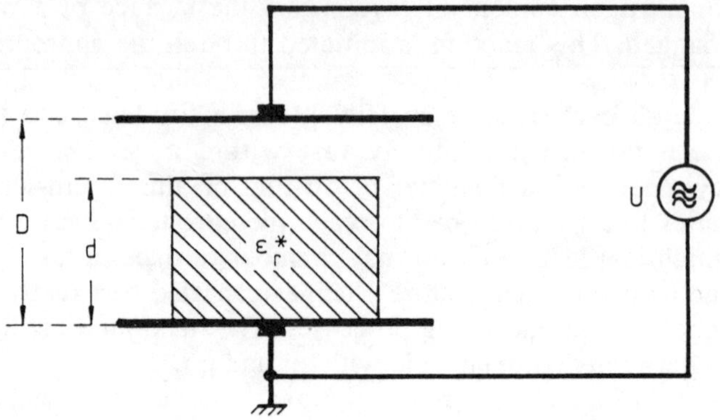

Figure 23. Dielectric RF heating in a capacitor.

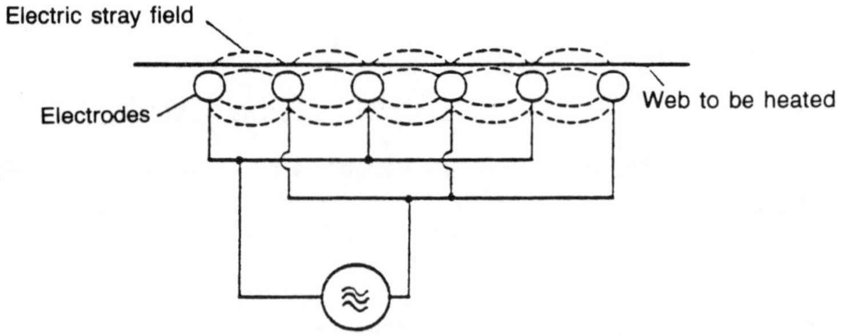

Figure 24. Dielectric RF heating with stray-field electrodes.

Stray-field electrodes are used for drying webs or filaments The layout consists of tubular electrodes which are fitted transverse to the direction of the web. The electrodes may be fitted either on one side of the web only (as shown in this figure) or on both sides (festooned field electrodes).

Electrodes of the same polarity are always parallelled and connected to one pole each of the RF generator.

- Thorough drying effect: the heat required for evaporating the moisture is generated inside the material to be dried from the energy contained in the RF electric field. This means that contrary to the common thermal drying process moving from outside to inside, RF drying permits a largely even reduction of humidity across the entire volume.

- Selective energy input of water particles: due to the marked dipolar character of water molecules, they are particularly capable of absorbing electrical energy and converting it into heat. Moreover, Joule's heat is developed because the molecules of the moist material are partly dissociated in ions and therefore exhibit ohmic conductivity. However, the degree of selectivity of energy absorption between water particles and dry matter at each point of the material to be dried depends on local moisture. At a moisture content above around 10% (in relation to the mass of dry matter), energy is almost exclusively absorbed by the water particles, and the dry matter is heated only indirectly; this selectivity of energy absorption in the water particles, however, is widley lost if the moisture content is reduced to just a few per cent.
- Decreasing overall energy absorption of the material to be dried as the moisture content goes down.

The effect of these phenomena on the drying process itself and its result is illustrated in the chart in Figure 25.

Capacitor-field heating can be used for a great number of very diverse applications, and potential applications are far from being exploited. In many cases, however, RF heating competes with conventional thermal processes. There are two main factors that prevent capacitor field heating from being widely accepted:

- The investment cost of such a system is relatively high;
- In applications where little practical experience is available, a capacitor field heating process must be carefully designed and tested so as to exclude the risk of local overheating of the material and to achieve an optimum heating result.

Compared with conventional alternatives, a capacitor field process is not always profitable. There are good reasons, however, for this heating technology to gain a more important position in the marketplace. Here are some:

- The system is compact and accelerates the drying process because heat is generated inside the material to be dried and/or because of the selective energy conversion in drying processes. This could also have a positive effect on the specific energy consumption, in spite of the relatively low efficiency involved in RF generation.
- The capacitor field heating offers decisive advantages in terms of quality, evenness and reproducibility of heating results. Such advantages become most obvious when RF heating is combined with a conventional upstream process.

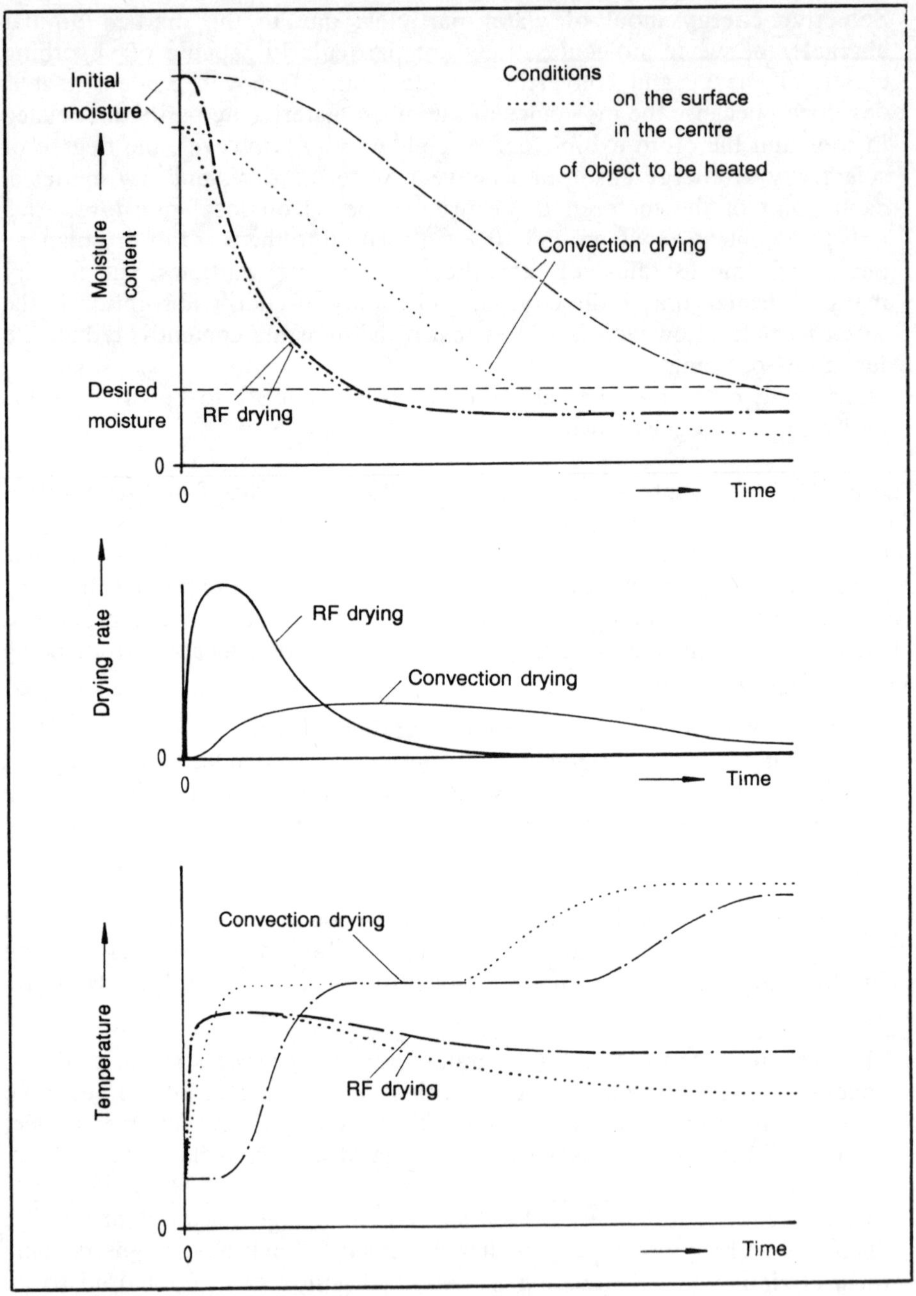

Initial
moisture
Conditions
on the surface
in the centre
of object to be heated
Moisture content
Convection drying
Desired
moisture
RF drying
0
0
Time
Drying rate
RF drying
Convection drying
0
0
Time
Temperature
Convection drying
RF drying
0
0
Time

The most important applications of RF heating are in:

- fusing thermoplasts,
- preheating hot-curing resins,
- thawing, preserving or baking food,
- drying chemical fibres, fibre glass, textile fibres, paper, food, wood, and casting cores, and
- bonding and joining wood and curing fibre board and chip board.

2.7.2 Microwave heating

The main components of microwave heating comprise the mains supply units such as high-voltage transformer and rectifier, the magnetron to generate an operating frequency (usually 2.45 GHz), transmission elements (waveguides) and the actual microwave heater. Depending on the type of wave duct, a distinction is made between:

- cavity resonator (see Figure 26),
- travelling wave system (see Figure 27), and
- aerial radiator.

When the microwave radiation strikes the surface of the material to be heated, some of the radiation is reflected, with the remainder penetrating into the

Figure 25. Comparison of thermal and electric drying (convection vs. RF).

The object to be dried (reels in this example) lost some moisture at its surface at the beginning of the drying period.

A comparison between RF drying in a plate heating condenser and convection drying in a hot-air furnace shows the following advantages in favour of the RF drying method:

- Faster drying: the desired moisture is reached much faster due to the higher drying rate. The maximum drying rate is limited by the vapour permeability of the object to be dried because there may be an excessive rise in pressure at high evaporation rates and consequently internal destruction of the structure of the matter.
- More uniform drying: even if several pieces of different initial humidity and/or size are dried simultaneously, different degrees of energy absorption at progressive drying causes a self-regulating effect that leads to a remarkably uniform drying result, without there being a need for separate measurements of temperature or humidity.
- Lower thermal load of the object: the object's temperature remains lower even during the phase of maximum drying rate, then drops again so that there is no risk of overheating as exists in convection drying, especially for the surface of an object.

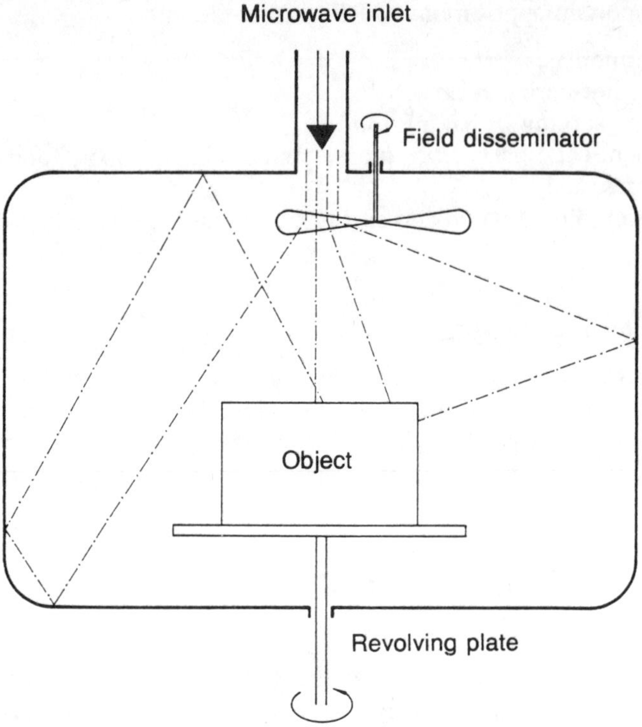

Figure 26. Microwave heating in a cavity resonator.

The cavity resonator is a square-shaped space with enclosing metal surfaces and one or several energy coupling points. A multitude of reflections form a system of standing waves. To ensure uniform heating of the object the aim is to reach a uniform energy density distribution within the resonating space, with a maximum number of different types of resonance being desirable. For this reason, a large resonance space with rounded corners is advantageous. Field disseminators and/or rotary plates are used to excite all resonances; this causes a continual variation of the system impedance. The best known and most widely spread example of a microwave heating device with cavity resonator is the domestic microwave oven. For industrial use, there are also systems available for flow-type heating, where the object to be heated is transported on a belt through the cavity resonator. The required inlets and outlets in the walls of the cavity must be fitted with suitable filter designs so as to keep leakage radiation below the maximum permissible value (in normal operation this is 5 mW/cm^2 at a distance of 5 cm from the unit).

material where it is absorbed at a constant rate (if it is a homogeneous material) and converted into heat. Contrary to RF heating in a capacitor field, therefore, the heating intensity in the material is always uneven, as is illustrated in Figure 28 in the example of microwave radiation on both sides of a 5-cm-thick board.

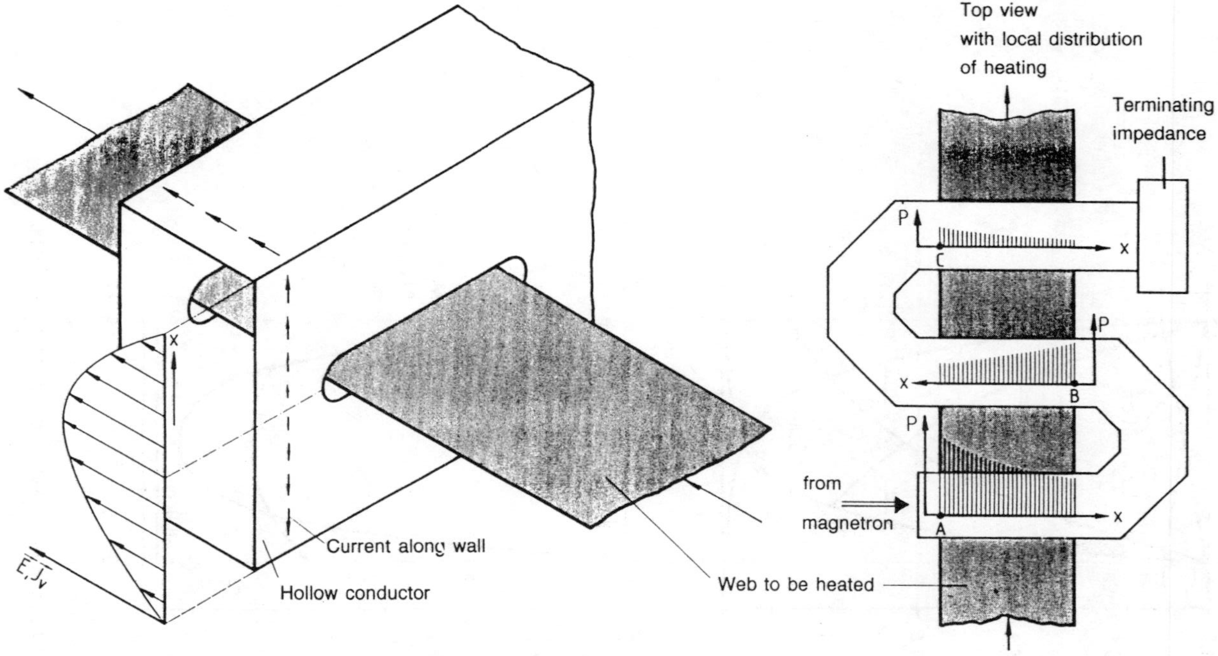

Figure 27. Microwave heating in travelling wave system.

Travelling wave systems usually come with a meander-type layout of square waveguides with transversal electric wave duct. The object to be heated has either the shape of plates or webs. It is pulled through slot openings in the center plane of the waveguide's broadside. In this plane the effective values of electric field strength and displacement current reach their maximum. Since the wall currents of the waveguide are virtually undisturbed in this slot position, there is only a very slight lateral radiation. Heating is caused by the displacement currents moving in longitudinal direction of the web. Power absorption drops exponentially from point A across the width of the web; the same applies to the next meander train of the waveguide from point B, etc. The microwave power remaining in the waveguide at the end of the meander is absorbed in a terminating impedance.

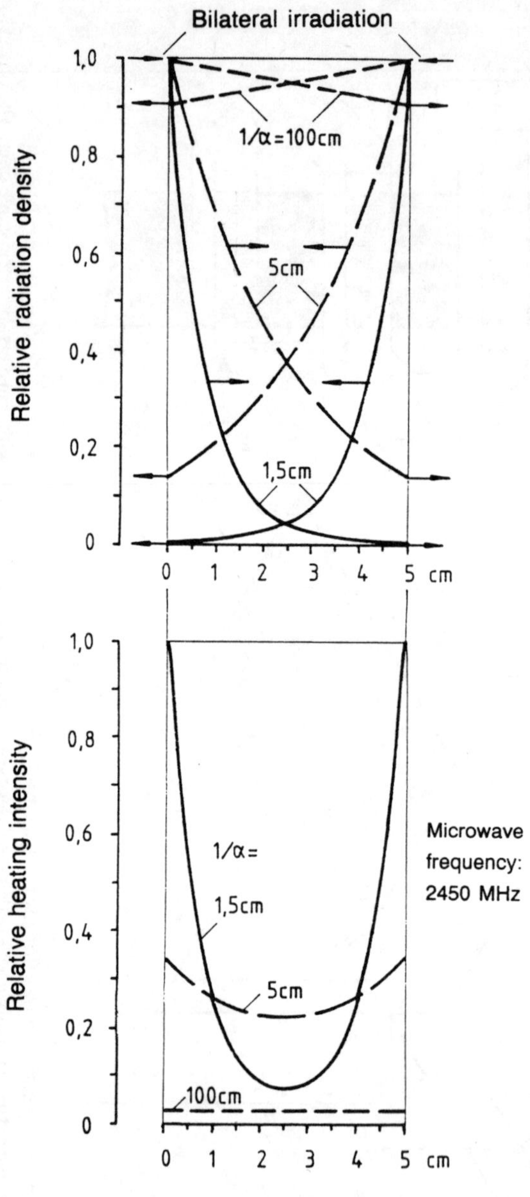

Figure 28. Bilateral penetration of microwaves into 5 cm thick plates.

Ovens with a microwave rating of up to 850 W have become a generally accepted domestic appliance. Most modern appliances have stepless power control by clocking the magnetron at variable pulse widths. Main applications comprise:

- heating fast food, drinks, etc.,
- cooking meals, and
- thawing deep-frozen food.

Compared with the conventional heating methods, microwave ovens significantly cut down cooking times as well as energy consumption. These advantages are more marked when small lots are involved, as is illustrated in the example of boiling potatoes in Figure 29.

In catering, the tasks are very similar to those in home cooking. In addition, there are a large number of applications which have also gained an important position; among others there are applications in the following fields:

- rubber industry: heating rubber to about 200°C for vulcanization;
- plastics industry: heating for curing epoxy resin parts, fusing thermoplasts, heating extruders for thermoplasts;
- food industry: heating for preserving (pasteurization, sterilization), drying (pastry, for instance), baking; and
- medical application: diathermy, warming fresh blood packs, thawing deep-frozen donor organs.

2.8 Conductive Heating of Fluid Food

In recent years, industrial preservation of fluid food (e.g., milk, vegetable soups, custards, and many more) has concentrated more and more on heat treatment in continuous flow-type heaters. They operate with plate or tubular heat exchangers heated with steam or hot water. Since heat is transferred to the food to be heated by means of convection, an excess temperature at the inner surfaces of the heat exchanger cannot be avoided. This temperature must be higher if the material is to be heated more rapidly.

Due to different reaction velocities of the individual conversion processes and their temperature behaviour, effective conserving with an optimum preservation of nutrients is achieved by heating the food for a few seconds at a high temperature (up to 140°C). This requires high heating temperatures which can hardly be realised in conventional heat exchangers, especially in the upper temperature range, apart from the fact that many types of food are sensitive to even slight overheating.

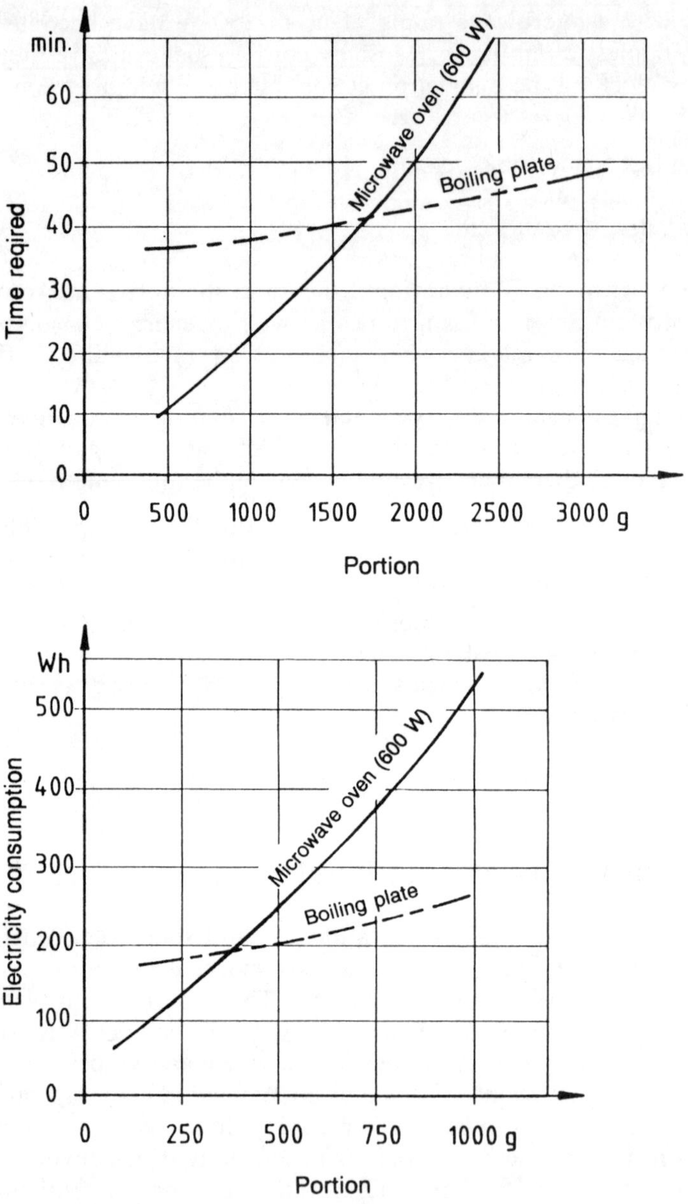

Figure 29. Load characteristics for boiling potatoes.

Contrary to cooking plates, microwave ovens have a very low load-independent basic requirement. However, because of considerable conversion losses, the load-dependent characteristics are relatively steep compared with the corresponding characteristics of an automatic cooking plate.

One way of solving these problems is to include in the heating section a mains-fed ohmic heater. The electric current flowing through the food produces extremely fast and very even heat up to the desired temperature. Unfavourable overheating can be avoided because the inner surfaces of the ohmic heater do not require excess temperature.

Other advantages of this technology lie in their technical simplicity and compact design, as well as their high energy efficiency. These features also raise this method above the competing microwave heating method.

2.9 Plasma Beam Heating

A plasma beam is formed when two factors are combined: creation of a gas flow and a concentrated energy input.

The gas flow is created by guiding a certain amount of gas through a nozzle system, thus creating a defined flow pattern. The parameter and characteristics of the gas flow can be selected to suit the requirements of the application. The main working gases comprise Ar, H_2, N_2 and air.

The input of electrical energy causes the flowing gas to turn into a state of plasma. Wherever the plasma beam comes into contact with the object to be heated, there is an intense heat transfer by convection. However, plasma can also fulfill other functions. In a space filled with the plasma beam, ambient air is displaced, with the plasma beam being scattered to a certain degree. This may lead to the formation of an inert gas cloud (as in plasma welding or in a plasma melting furnace). If the gas flow strikes the object at an adequately high velocity, the molten droplets at the surface of the workpiece can be blown off (as in plasma cutting). Because of its state of excitation, plasma gas is reactive. This may have a deleterious effect (e.g., increased formation of nitric oxides), but it can also be harnessed in plasma chemistry for producing certain chemical compounds.

The best established and most widely spread application of plasma-beam heating is plasma cutting. Usually, plasma torches with direct arc are used for cutting steel and non-ferrous metals. In rare cases, plasma torches with indirect arc are used for fabrics, plastic films and metal foils. The main features of the plasma beam comprise not only its high temperature but also the pressure applied to the material to be cut and the inert gas effect.

Compared with the main competing processes of mechanical sawing and acetylene cutting, plasma cutting excels in higher cutting speeds and equal or better cutting quality. In spite of the high investment cost, therefore, plasma cutting has proved to be profitable in many different applications.

In plasma injection, a powdered material is blown by feeder gas flow into the plasma beam (usually argon) emitted from the generator nozzle. Due to the very high plasma temperature of 15,000 to 20,000 K, the individual injection particles, that are 10 to 100 micron in size, are molten and partly evaporated while being transported to the surface of the workpiece. In addition, these particles are accelerated to velocities of between 100 and 500 m/s. In striking the surface of the workpiece, the material passes extremely rapidly through the various stages of phase reconversion and cooling. In so doing, the injection material clutches mechanically to any unevenness in the surface of the workpiece (14).

Vacuum plasma injection is applied in all those cases where the coating has to comply with the most exacting requirements of density, purity and absence of oxides. In this variation, the injection material particles strike the surface at an even higher velocity. There is no undesired interaction with the ambient atmosphere. This process is mainly used for coating gas turbine blades with special alloys in order to achieve a reliable protection against hot gas corrosion (15).

The main advantages of plasma welding over competing arc welding are the welding speed which is about double and a smaller heat affected area. Disadvantages are the increased process and material cost (argon consumption!), the larger and more expensive burner design (that may also create space problems at the point of welding), and greater dimensional accuracy required at the weld itself. For these reasons, plasma welding lends itself primarily to applications where high levels of productivity and excellent welding quality are demanded.

An interesting option also exists in the form of plasma generators for heating gas that is used as a low-loss energy supply in high-temperature processes. The high specific enthalpy contained in hot plasma gases enables an energy transfer at a process temperature of, for instance, 1500°C and an efficiency of over 80%. This corresponds to more than three times the thermal efficiency of fuel-based processes, and even if electricity for the plasma generator is produced in thermal power plants the expense of primary energy is not higher. Another advantage is that if the appropriate plasma gas is chosen it is easy to ensure an atmosphere free of oxygen or carbon for the process (16).

2.10 Electron-Beam Heating

In electron-beam heating, free electrons are generated at the thermionic cathode of an electron gun; these are accelerated in a high-voltage system.

The resulting electron beam can be focussed, widened or guided when exposed to electric or magnetic fields. A high vacuum (10^{-2} to 10^{-4} Pa) is necessary for both generating the electron beam and ensuring its sufficiently

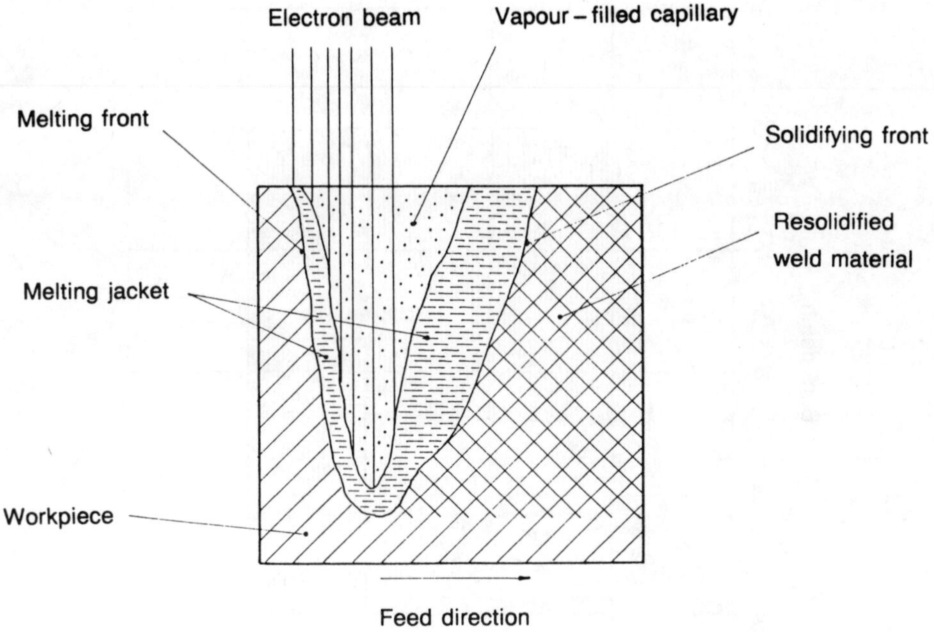

Figure 30. Deep-welding effect of electron beam.

At a power density of 10^9 to 10^{11} W/m^2, the electron beam penetrates deeply into the workpiece forming a metal vapour- filled capillary which in turn is enveloped in a jacket of molten metal. Due to the relative movement between electron beam and workpiece, the electron beam pushes a melt front. Material is constantly becoming molten, and due to the effect of vapour pressure and surface tension, it is transported in tangential direction around the vapour capillary into the rear part of the melt jacket. While forming the weld at the solidification front, the material solidifies again.

unimpeded passage to the object to be heated; failing this, there would be excessive energy losses and incorrectable beam scattering due to the interaction between electrons and gas molecules.

The efficiency of an electron gun, expressed as the ratio between radiant power and total electric power (for acceleration voltage, control voltage and cathode heating), is around 90%.

Electron-beam heating is applied in vacuum remelting processes for extremely pure special steel grades needed for making cutting and punching tools, huge steel blocks for turbine rotors and semi-finished metal products of Ti, Zr, Ta, V, Hf, W and/or alloys.

Another group of electron-beam-heating processes aims at improving the surface properties of metals by means of evaporation, melting or hardening.

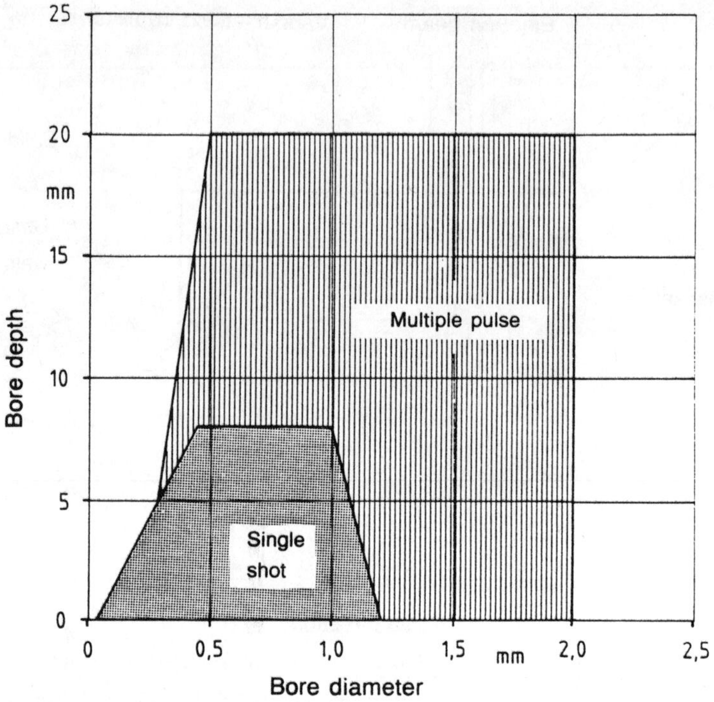

Figure 31. Electron-beam drilling parameters for steel.

In electron beam welding, the deep-welding effect is characteristic (see Figure 30). At more than 200 mm, the maximum welding depth in electron beam welding is about twenty times higher than that achievable by other methods. Electron-beam welding is also superior in terms of the depth-to-width ratio of the weld that can be obtained (up to 50:1 in great welding depths). Comparatively very high welding speeds allow great productivity. Since the melting flanks are nearly parallel and the heat affected areas are very small, electron-beam welds are virtually shrinkproof and distortion-proof, thus avoiding reworking.

Electron-beam welding can also be applied to joining different types of metals. This opens up new perspectives for material-saving designs while offering the choice of the best material (17).

Electron beams can also be used for thermal drilling (see Figure 31) without the material hardness being of any significance. Due to the extreme speed of the process and the inertia-free beam control, drilling frequencies in small diameters and depths can be several kHz. This makes electron-beam drilling an interesting proposition for perforating, i.e., drilling many small holes in a single workpiece.

The most important application of electron-beam perforation is in the construction of turbine components (combustor casings, turbine blades). It has also come to be used in other fields such as fibreglass spinning pots, filter and screen processes, etc. (18).

2.11 Laser-Beam Heating

The use of the laser as a tool in thermal processing is based on the following properties of laser beams:

- high power output,
- an extremely narrow spectrum (near-monochromatic),
- focus diameter under 0.1 mm,
- highest energy transfer densities achievable (in continuous mode up to 10^{11} to 10^{12} W/m^2, and much more during short pulses),
- generation of very short flashes at high repetition rates.

Some of the characteristic data of different laser designs are compiled in Table 1. With the exception of the wavelength data, the figures stated are approximate values that can be obtained with commercially available systems.

Laser-beam heating uses mainly CO_2-lasers and Nd-YAG-lasers. The excimer-laser is of particular interest for its high pulse power at very short pulse times and because of its short-wave radiation.

Laser beams are applied in many different sectors, such as in mechanical engineering for thermal finishing (see Figure 32). The main applications include cutting, welding and drilling.

Table 1. Laser data for thermal material processing.

Design	wavelength µm	max. power W CW*/pulse	pulse duration seconds	max. pulse frequency kHz	efficiency %
solid state laser:					
Nd-YAG	1.064	$10^3/10^6$	10^{-4}–10^{-2}	0.3	0.1–4
ruby	0.694	$10^2/10^5$	10^{-4}–10^{-2}	0.1	0.1–1
gas laser:					
CO_2	10.6	$10^4/10^4$	10^{-4}–10^{+1}	2.5 (15)	5–10
excimer	0.19–0.35	–/10^7	10^{-8}–10^{-7}	0.5	1–5

* CW means continuous wave operation; the pulse figure refers to the maximum power in the pulse.

Laser welding, in particular, is one of the growing applications in mechanical engineering and automotive engineering. The advantage of lasers over electron beam processes is that it does not require a vacuum for beam guidance. It can be expected, therefore, that in the long-term laser welding will be preferred for welding depths up to 30 mm.

Laser technology has also opened up new applications in the field of welding and joining where conventional methods proved either impossible or difficult in view of technical and quality problems. Here are some examples:

- Butt welding of coated sheet metal has become possible with laser, for instance in can production where the stiffness of the wall can be maintained while the wall thickness was reduced; welded parts are profiled after joining.
- Welding is possible in heat critical areas without any significant material distortion.
- The laser beam can reach locations inaccessible to conventional welding processes.

The width of overlapping material in lap-welding is a considerable cost factor in large-scale production. Since lasers produce much narrower welds than conventional methods, the overlap can be significantly reduced.

As in electron-beam hardening, lasers also afford processes based on heating the surface of a workpiece in order to change its physical, chemical or optical properties: transformation hardening, surface coating and/or surface alloying, evaporation of films for lettering, marking, applying decorative patterns on ceramics, for instance, or for wire stripping.

The criteria applied to judging the potential usage and the limits of laser processes do not only pertain to the process technology itself but also its profitability in relation to competing technologies. Cost advantages arise in favour of laser processes wherever greater varieties and smaller batches are involved because the benefits of flexibility inherent in the use of laser combined with computer-numeric-controlled tools become evident in this case; provided, of course, that the expensive unit is fully utilized.

3 Electrothermal Processes – Current State of the Art and Outlooks

Figure 33 gives a breakdown of the estimated power consumption in individual final consumer sectors in the FRG for the most important applications in the year 1986 (4). Roughly 28% of the entire power consumption goes into

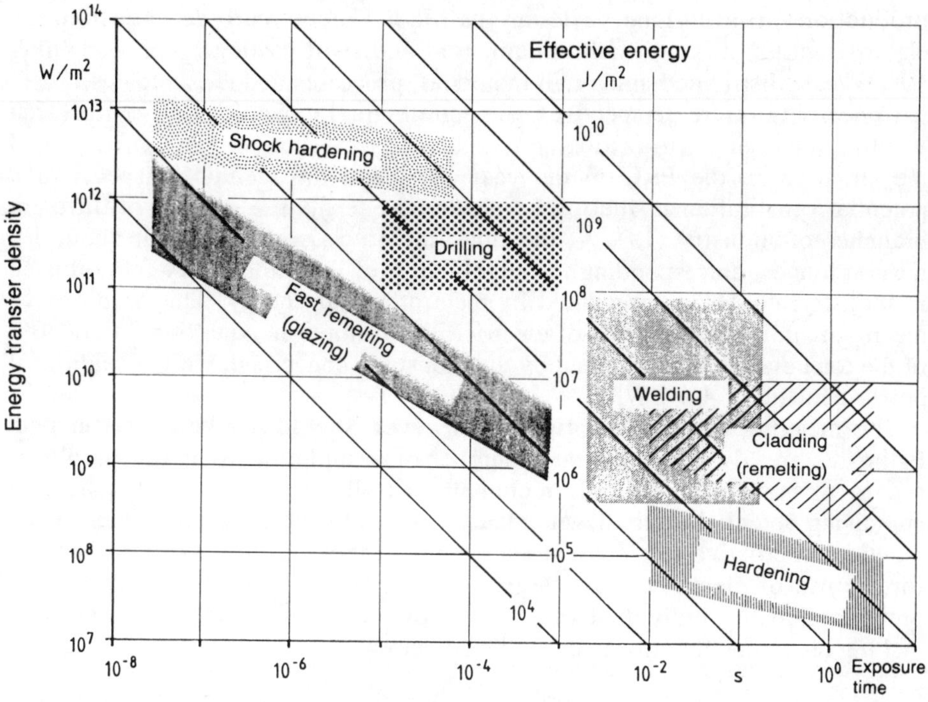

Figure 32. Processing characteristics for laser-beam heating.

Energy transfer density is one of the characteristics determining the effect that a laser beam of a given wavelength has on the spot to be processed in the surface of a workpiece. Another important parameter is the laser's dwell time on that spot. This follows from the pulse duration of a pulsed laser or the travel speed of the laser beam and the size of its focus in CW-lasers. Depending on the application, very different combinations of effective parameters of energy transfer density and dwell time may be required. The product of both is the effective energy applied per surface unit.

process heating. The quantitative significance of electrothermal processes therefore is much greater than consumption for electric lighting (10%) and electric space heating (around 11%).

At 45 TWh, one of the largest shares of all is electroheat in industry. It should be noted, however, that this figure includes very energy-intensive electrolytic processes in aluminium and chlorine production which together account for about half of this share. Around 4 TWh are used for producing steel in electric arc furnaces and between 0.5 and 1 TWh each go into the production of acetylene, calcium carbide, silicon carbide, phosphorus, electrographite, zinc, ferrochromium, iron and steel castings. The remaining 10 TWh are distributed among all industries, processes and technologies.

Concerning future perspectives of electrothermal technologies, the Research Institute for Energy Utilization and Technology in Munich, FRG, investigated the situation in the FRG in the year 1981 and the technical aspects of a potential substitution of fuel oil by electrical energy in a number of different branches of industry (19). At the time, the study concluded that about 1.9 million tonnes, corresponding to roughly 60% of light and heavy fuel oil used in industry, could be substituted by electrothermal energy. This would have meant an additional 15 TWh of electricity consumption, equalling barely 30% of the total electricity consumed by the industries concerned, with an additional power demand of 4.4 GW.

The meaning behind such figures, however, should not be overestimated. Estimates are always based on a number of assumptions that can hardly be regarded as safe. After all, technical criteria alone are inadequate for appraising potential future developments. Apart from cases where there is no practical alternative to electrical energy for generating process heat, the odds for the use of electroheat are largely a function of the economic conditions prevailing in the individual case. Some of the criteria applied in decision-making on competing processes are listed below:

- Electrothermal plants are often considerably more expensive to buy than competing fuel-fired plants.

- Electrothermal plants usually exhibit a much more favourable energy behaviour at partial capacity utilization.

- Since both these factors run counter each other when it comes to characteristic costs incurred, capacity utilization can influence the profitability of a plant to varying degrees.

- When comparing energy costs, which is one of the main issues for the question of profitability, the relationship between the prices of electricity and fuel comes into play. By definition, such price forecasts must be tentative.

- Cost advantages inherent in electrothermal processes are sometimes hidden among those items that cannot be precisely calculated in advance but are estimated based on practical experience. (Examples include service and maintenance, low scale loss, less scrap.) Whether or not such cost advantages

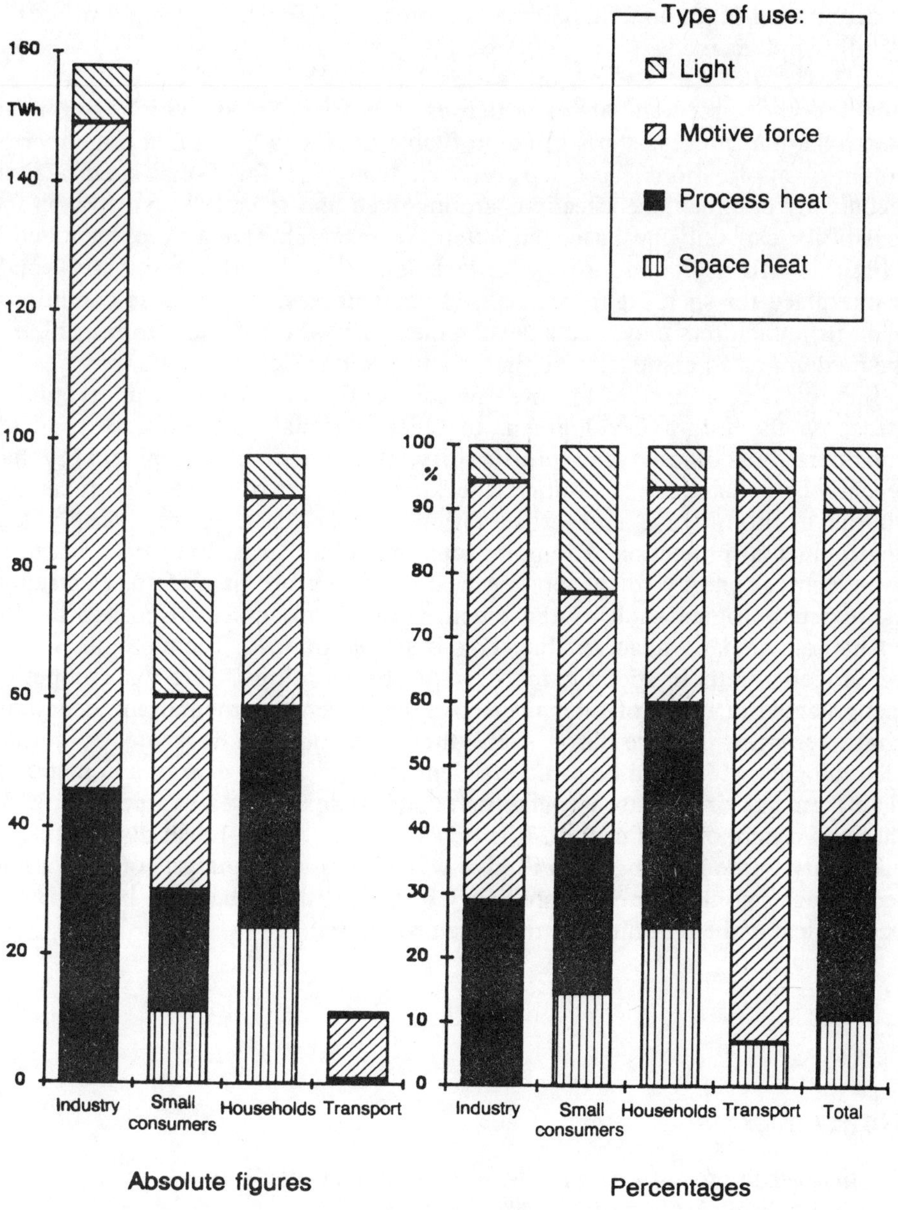

Figure 33. Breakdown of electricity consumption, FRG, 1986.

materialize in the individual case is normally established once the project is fully implemented.

Besides the problem of establishing whether or at which point an electrothermal process starts to be profitable, it is also a fact that a number of potential applications have hardly been tested so far. Very specific and technically complex specifications are involved and reliable statements on the feasibility can only be based on extensive research and development work which is often claimed to be unjustified in view of the mostly small marketplace for such highly specialised technologies. Under these conditions, plant manufacturers may face a development risk which is quite formidable for the medium-sized companies normally active in this field.

Finally, it is important to note one factor that may have an unfavourable effect on the future development of electrothermal processes. Beyond all considerations of cost and profitability, the use of electrical energy has recently come under some political attack in the course of discussions focussing on the economic use and conservation of resources, energy supply and environmental protection. The point often made is that it is always a waste to use electrical energy for generating heat and therefore ipso facto, electricity-using technologies such as the electrothermal processes discussed in this chapter are socially unacceptable. This is simple-minded. Technologies should be judged on their specific merits and shortcomings vis-a-vis competing technologies in terms of their respective social, environmental and economic characteristics. As we have seen, there are quite a few electrothermal technologies which can be more efficient in their use of energy and materials than their gas or oil-based counterparts and in some cases can perform tasks that other technologies cannot. It is hoped that the broad-brush discrediting of electricity in the heat market will give way to a rational analysis of the various technologies so that for every specific job at hand, the technology best suited in economical and environmental terms can be chosen.

References

1. Rudolph, M., Schaefer, H. 1986. *Wärme*,92(3): 73–78.
2. Decker, E. 1983. *Elektr.wirtsch.*, 82(17–18): 660–67.
3. Decker, E., Dickopp, A. 1980.*Elektrowärme int.*, 38(B6): 280–85.
4. Rudolph, M., Schaefer, H. 1989. *Elektrothermische Verfahren*, Berlin, Heidelberg, New York: Springer.
5. Fasholz, J. 1984. *Induktive Erwärmung*, Heidelberg: Energie, 84pp. 3rd ed.
6. Thelin, C. 1983. *Elektrowärme int.*, 41(B2): 92–97.

7. RWE, ed. 1979. *Induktives Erwärmen zum Warmformen von Stahl*, Essen, 12pp.
8. RWE, ed. 1979. Konduktives *Erwärmen zum Warmformen von Stahl*, Essen, 8pp.
9. Decker, E. 1983. *Proc.RWE Conf.*, Nuremberg, 1985(5): 1–23, Essen, RWE.
10. Sommer, P. 1981. *Fachber.Hüttenprax.Met.weiterverarb.*19(9): 670–75.
11. Sommer, P. 1983. *Drahtwelt*, 69(1): 16–19.
12. Reitler, W., Rudolph, M. 1986. *Elektrowärme int.*, 44(B6): 275–79.
13. Conrad, H., Krampitz, R. 1983. *Elektrotechnologie*, Berlin: VEB Technik.
14. Steffens, H.-D., Busse, K.-H. 1987. *Elektrowärme int.*, 45(B3–4): 183–89.
15. Lugscheider, E. Weber, Th. 1987. *Elektrowärme int.*, 45(B3–4): 190–95.
16. Müller, R. 1987. *Elektrowärme int.*, 45(B3–4): 146–54
17. Hiller, W., Steigerwald, K.H. 1981. *Trennen und fügen*, 8, Messer Griesheim, ed.
18. Meyer, E. 1980. *DVS Bull.*, 63. Düsseldorf: Dt. Verl. f. Schweisstech.
19. Krabiell, H. 1983. *Elektr.wirtsch.*, 82(18): 668–70.

Keeping Score in Electricity Conservation Programs

Gautam S. Dutt and Margaret F. Fels

Abstract

Conservation measures are chosen on the basis of their expected savings. This requires analysis of electricity use and calculation of the savings which can or have been achieved. General issues relating to data needs of electricity-use scorekeeping are presented. The widely used PRISM techique for analyzing energy end use in buildings is described, with illustrative examples of its application in different kinds of buildings. Unresolved problems in scorekeeping peak demand are considered.

1 Fundamentals of Scorekeeping

1.1 Introduction

The success of electricity conservation programs requires knowledge of the electricity use and savings potential in new and existing end uses. Although projections of electricity use and savings can be made from engineering and other analysis, experience has shown that often the projected values do not correspond to the actual measured values, due to uncertainties and inaccuracies in the assumptions implicit in the projections. For this reason, there has been growing recognition that evaluation of energy conservation programs should rely only on measured energy use. For the evaluation to be feasible, it should be based as much as possible on existing reliable data sources. Analysis of energy consumption and demand from measured data has been called *scorekeeping*.

Scorekeeping can perform several important functions. Scorekeeping of existing energy use can point to areas of opportunity for energy conservation. It can be used to calibrate models that project energy use and conservation and evaluate conservation programs. Scorekeeping can provide feedback to users who may then evaluate and improve their efficiency of energy use. Knowing the patterns of existing energy use and success of conservation programs is also useful in forecasting future demand. In its ability to evaluate conservation programs and forecast demand, scorekeeping can be an important planning tool.

In this chapter, we will first elaborate on the principles and issues involved in scorekeeping electricity use (Section 1). Second, we will examine one sector–buildings–in detail (Section 2). Finally, we will address special problems in scorekeeping electricity use, focusing on peak demand (Section 3).

1.2 Principles

1.2.1 Data from existing meters

To be widely applicable, scorekeeping methods should use existing reliable sources of data wherever possible. Existing sources of data include electric meter readings and other data that may be needed to disaggregate and normalize electricity consumption. (Disaggregation and normalization are discussed below.) Existing data sources greatly reduce the cost of the data, and also permit historical comparisons. Reliability of data is of course important if the analyses are to be accurate. Electric meters used for billing purposes are carefully calibrated and widely available. They constitute the primary source of energy data in scorekeeping. Recent developments in electric meter design will expand the range of easily accessible data sources for scorekeeping. Some of these are discussed below.

1.2.2 Disaggregation of energy consumption

Electric meters used for billing indicate electricity use totaled over a variety of end uses. For instance, the monthly electric bill in a house represents the total of electricity uses that may be for space and water heating, ventilation, lights, and a range of household appliances. In evaluating the success of energy conservation programs, it is often desirable to know the distribution of the total electricity use among the different end uses. In some cases, especially for large users, there may already be several meters, perhaps because different rates apply to different end uses. In these cases, the separate meters may provide a useful disaggregation. The electricity use measured by a single meter may sometimes be disaggregated into a number of sub-components using various analytical tools. For instance, household electricity use can be separated into components depending on its weather dependence (e.g. space heating and non-heating consumption). In some cases, the required level of disaggregation may not be possible on the basis of existing meter reading data alone. In those cases, submetering may be necessary. Already, submetering is used by electric companies to measure components of electricity use in a sample of users within a group as a forecasting tool. This type of sample submetering can be used to establish rules by which the existing meter readings of much larger numbers of users may be disaggregated.

Remote reading meters are likely to be widely available in the near future and will permit easy access to more frequent energy consumption data. High frequency data acquisition from electric meters to measure changes in power and power factor may permit accurate disaggregation of the total consumption into its components (1,2,3).

1.2.3 Normalization of energy consumption

An important use of scorekeeping in determining trends in energy use and in evaluating conservation programs is through comparisons. Energy use data are usually not sufficient for making useful comparisons. Two users such as office buildings may differ in size, two plants may differ in production levels, etc. Even the same user over time may alter his/her energy use for a variety of factors: the production level of a plant changes over time, the heating season may be more severe in one period than in another, energy saving retrofits may have been installed, etc. To make useful comparisons, the energy use data must be adequately normalized. The process of normalization is best understood in terms of the following general formula (4):

$$\text{Energy use} = \text{energy intensity} \times \text{level of activity}. \tag{1}$$

This equation may be applied to the entire economy, to a sector, to a plant or other end user, and in fact to individual end uses as well, provided the "energy intensity" and "level of activity" are defined appropriately.

Table 1 illustrates levels of activity and energy intensities for the economy, various sectors, users, and end uses. Scorekeeping is carried out primarily at the level of the user, so that the energy use is normalized by the level of activity to calculate an energy intensity, sometimes referred to as an "energy consumption index", that is used in comparisons. The level of activity for some users may be confounded by the weather where this influences energy use, as in buildings. In these cases, the normalization process must include a weather adjustment. The weather adjustment of building energy use is examined in detail in Section 2 of this chapter. In Table 1, levels of activity and energy intensity at the more aggregated "sector" level and the more disaggregated "end use" level illustrate the relationship of scorekeeping to analysis at these other levels. The end-use level analysis is usually the basis for specifying conservation measures and for making projections of energy use.

Table 1. Level of activity and energy intensity indices, for economy, sector, user, and end use.

	Level of Activity	Energy Intensity
Economy	gross national product	MJ / $
Sectors		
transportation	passenger-km;	MJ per passenger-km
	tonne-km	MJ per tonne-km
industry	value added ($)	MJ / $ value added
buildings	m^2 floor area	W / m^2
User (typical of scorekeeping application)		
industrial		
plants	tonne of output	GJ / tonne
automobile	vehicle-km	litres/100 vehicle-km
office building	m^2 floor area (normal year of weather)	W / m^2 (normal year)
house	dwelling unit (normal year)	GJ / year per unit (normal year)
End use (illustrative)		
lighting	lumens	(lumens / watt)$^{-1}$
auto travel	vehicle-km	litres / 100 km
auto travel	work trip; pleasure trip	litres / trip
air travel	seat-km	litres / 100 seat-km
refrigerator	unit, or	kWh / year per unit, or
	litres interior vol.	kWh / year per 100 litres
space heating	heating degree-days (normal year of weather)	W / °C

Scorekeeping permits the evaluation of the effect of these conservation measures as aggregated in the analysis of meter readings. This "downward" linkage of scorekeeping permits the calibration of models for predicting energy use and savings. With improved models, conservation measures can be specified with greater certainty of energy savings and thus can be more cost effective. On the other hand, scorekeeping analysis of the energy intensity of individual users and their levels of activity may be summed to determine the effect of changes at the user level on the sector and on the economy. This "upward" linkage of scorekeeping makes it useful in forecasting future demand and in planning between alternatives (5).

1.2.4 Statistics in scorekeeping

Since making comparisons of the energy consumption index is an important use of scorekeeping, some measure of the reliability of the estimated index is essential. There are two sources of errors in the estimation of the consumption index: (a) improper assumptions used to disaggregate or normalize raw consumption data, and (b) fluctuations in energy use introduced by factors that are not controlled for in the disaggregation or normalization. An example of the first kind of error is the exclusion of solar energy in the weather normalization of energy use in houses, or the assumption that the energy intensity of steelmaking is weather independent. An example of the second kind of error is the effect of occupancy changes on the energy consumption of an office building. In any real analysis process, these errors cannot be avoided entirely, but the analyst can determine, with statistical techniques, the possible impact of these errors on the estimates.

One approach to reduce the first type of error is to select the best energy consumption index. For instance, energy consumption in a group of similar users may be normalized in a number of different ways. Then the index with the smallest coefficient of variation could be selected as the most desirable one. If the most reliable normalization is unusually time consuming, a compromise index must be selected, depending on the objectives of the scorekeeping exercise. For example, consider energy used for domestic water heating in an apartment building. The best index may be energy use per resident, but the number of residents may not be known or may vary. Energy use per bathroom may be a better index than the more common energy use per housing unit, without requiring much additional data.

Another way to reduce the first type of error is through regression techniques. If the energy use is represented as a function (usually linear) of one or more independent variables as the basis for normalization, then the R^2 of the regression may be used to determine how well the normalization function fits the observed data. A high R^2 indicates that variables not considered are not important. For instance, in the weather normalization of energy use data in houses using the Princeton Scorekeeping Method (PRISM), described in Section 2 below, outdoor temperature explains virtually all of the month-to-month variation so that the non- inclusion of solar radiation is not seen to reduce significantly the validity of the weather normalization. The effect of errors of the first kind are usually expressed as standard errors on the estimated energy consumption indices (6).

The second type of error may be reduced by eliminating subsets of data for which there were changes believed to be significant but which cannot be adjusted for in the normalization procedure. For instance, a change of occupants in an office building is expected to alter energy consumption drastically, by altering the internal loads from computers and other office equipment, by altering the operating schedules of the building, etc. The effect of these changes cannot be easily quantified. A straightforward way of avoiding errors introduced by them is to exclude the relevant subsets of data from the analysis. However, in such cases the increased accuracy may be eroded by the reduced reliability from data attrition.

Other changes, difficult to model *a priori*, are caused by changes in energy prices, incentives for energy conservation, changes in tax laws, etc. The effects of these may be corrected for in scorekeeping exercises by including a control group, whenever possible. So, for example, the evaluation of a particular residential energy conservation program on a group of participating houses should include a comparison with other "control" houses that did not participate in the conservation program.

Nevertheless, there will be unknown or unquantifiable changes that will introduce unavoidable errors in the results. The possibilities of such errors will remain for the scorekeeper to grapple with.

1.2.5 Comparisons

An objective of scorekeeping is to produce energy consumption indices that can be used for comparison. Comparisons may be with:

(a) other users, to determine the level of energy use relative to typical values;
(b) the same user at another time period, to evaluate the effects of some changes introduced between the two time periods;
(c) a target value determined as the theoretical limit, the best available, next year's goal, etc.; or
(d) a design or predicted value, for determining how the measured performance compares with that predicted, and for calibrating the procedure used for prediction.

1.2.6 Discrepancies and diagnostics

It is the usual case in scorekeeping that the measured energy consumption index differs significantly from the engineering estimate. While the scorekeeping estimate, based on measured energy use, is the "real" value, much can be learned by understanding the discrepancy between the two values. This is particularly important because conservation measures are chosen on the basis of their expected savings. If the actual energy consumption, as determined by scorekeeping, is substantially "higher" than expected, as is often the case, finding out the reasons for the shortfall may point the way to substantial new opportunities in energy savings. For instance, the lower-than-expected energy savings achieved by attic insulation in U.S. single family houses led to the discovery of anomalous "attic bypass" heat losses (7). These bypass heat losses represented about 10% of energy use in U.S. houses, and their elimination, at relatively low cost, represented a new conservation opportunity. Many years later, the internal loads in an office building were found to be much larger than "handbook" values, and were discovered to be the result of increased office use of microcomputers. Energy efficiency improvements in microcomputers were found to play a major role in reducing energy use in office buildings (8). The recent scorekeeping of the large-scale Hood River Conservation Project in Oregon showed that only half of the predicted savings were realized. Whether this shortfall is due to inappropriate retrofits or whether the participating houses were more efficient than expected, with fewer energy-saving opportunities, such scorekeeping results are an important consideration in designing future conservation programs (9).

The failure to predict energy consumption and savings, discovered through scorekeeping, has led to the development of diagnostic techniques that are increasingly a part of procedures for specifying energy-saving improvements in processes and buildings (10,11).

2 Scorekeeping Buildings

2.1 Introduction

Scorekeeping techniques have perhaps been best developed for the analysis of energy consumption data in buildings. The techniques can be divided into physical models intended primarily to adjust consumption for changes in the weather, and socio-economic models. Some models combine both physical and socio-economic parameters in a single model. Experience suggests, however, that the interpretation of physically meaningful parameters from such models is difficult because of non-linear interactions among the parameters. Instead, a two-stage approach in which the physical model is first used to obtain weather-adjusted consumption indices seems preferable. These indices then can be analyzed for socio-economic determinants in a second stage, as discussed below (12,13).

2.2 Weather Adjustment of Building Energy Consumption

2.2.1 Uses of weather-adjusted energy consumption

Weather adjustment of energy consumption serves many functions. The energy consumption of different buildings may be compared, leading to the identification of buildings that are less energy efficient and therefore targets for conservation activities. Actual consumption may be compared against a target, useful in verifying compliance with performance-oriented standards. Comparison of the measured consumption with design or predicted value can be used to calibrate computer models of building energy use. Such comparisons are part of a process of improved building design, based on measured performance. In a retrofit program, pre and post-retrofit energy use can be used to determine energy savings and cost effectiveness of different alternatives. This evaluation of retrofit energy savings provides feedback to identify successes and failures–essential to the support of energy management activities, both at the individual building level and as part of a larger program.

2.2.2 PRISM as an example

Our basic tool for analyzing the energy consumption history is derived from the PRInceton Scorekeeping Method (PRISM) (6). The PRISM method is used by over 100 groups in the U.S., Canada, and Western Europe (and recently by one group in the U.S.S.R.). Of various versions, the "heating-only" PRISM-HO model has been the most widely used and is applicable to the situation when the energy use includes space heating and other end uses, but not cooling. Thus, it applies to electricity used for space heating, water heating, cooking, and appliance use, but not for air conditioning. (Analogous PRISM models for cooling have been developed, but the emphasis in this discussion will be on PRISM-HO.) Using monthly energy consumption and daily-average outside temperature data, PRISM-HO uses a statistical procedure to estimate three parameters that characterize the building's overall energy consumption pattern. (The monthly data requirement is obviously most convenient when monthly billing data are available, as they commonly are in the U.S. In other countries, in Western Europe and elsewhere, in which billing is quarterly and meter reading may be even less frequent, scorekeeping will require special data collection.)

The assumptions of the model are that a building's heating system operates only when the outdoor temperature T_{out} drops below a certain level (called the heating reference temperature τ), and that for each degree drop in outdoor temperature there is a constant rate of increase in heating fuel use rate (the heating slope β). In addition, the house may use a fixed amount of heating fuel

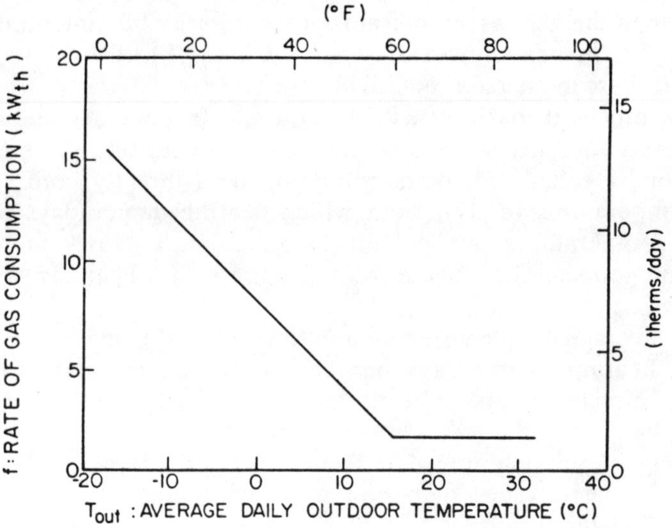

Figure 1. Daily energy consumption (f) as a function of outdoor temperature (T_{out}), for a single idealized house. Source: Ref. 6.

per day (the base level α) independent of T_{out}. The daily consumption rate of the heating fuel, therefore, depends on the outdoor temperature as shown in Figure 1. The relationship is non-linear because the heating system needs no energy input when $T_{out} > \tau$. Thus, the heating energy consumption rate, f, is:

$$f = \alpha + \beta\,(\tau - T_{out}) \qquad \text{for } T_{out} < \tau \qquad (1)$$

$$\text{and } f = \alpha \qquad \text{for } T_{out} \geq \tau$$

For an interval of N days, the average daily heating fuel consumption rate, F, is given by:

$$F = \alpha + (\beta/N) \sum_{j=1}^{N} (\tau - T_j)_+ \qquad (2)$$

where T_j is the daily-average outdoor temperature for day j in an interval of N days, and the "+" indicates that the summation needs to be carried out only for days when $\tau - T_j$ is positive, or $\tau > T_j$.

The quantity $\sum_{j=1}^{N} (\tau - T_j)+$ is the number of heating degree-days for the period, calculated to base temperature τ. Although heating degree-days in common use in the U.S. as an indicator of the severity of winter are calculated to a standard reference temperature τ = 18.3°C (65°F), PRISM allows for the variability of base temperatures across houses.

On this physical basis, PRISM requires (a) actual meter readings (approximately monthly) and corresponding meter reading dates from which consumption is calculated, or consumption data directly, and (b) average outdoor temperatures (daily), from which heating degree-days to different reference temperatures are calculated in exact correspondence to the consumption periods. The input to the procedure is thus F_i and H_i where:

F_i = average daily consumption in time interval i, and
$H_i(\tau)$ = heating degree-days per day in time interval i, computed to reference temperature τ.

PRISM uses a regression procedure to fit the set of data points $\{F_i\}$ and $\{H_i\}$ for an approximately year-long period to a linear model:

$$F_i = \alpha + \beta\, H_i\,(\tau) + \varepsilon_i \tag{3}$$

where ε_i is the random error term. For any value of the reference temperature τ, the $\{H_i\}$ set is calculated, and the α and β parameters are determined by ordinary least-squares linear regression. Using an iterative procedure based on Newton's method (14), the "best τ" is found as the value of τ for which a plot of F_i vs. $H_i\,(\tau)$ is most nearly a straight line, i.e., the regression has the highest R^2 value. The corresponding values of α and β are the best estimates of these parameters. The parameters normally have the following significance:

α = non-temperature-dependent use, called base-level consumption;
β = heating slope, or rate of increase in fuel use per degree drop in outside temperature;
τ = reference temperature, or outside temperature below which the space heating system needs to operate to maintain interior temperature.

The heating slope can be additionally interpreted as equal to L/η where L is the "lossiness" of the building, i.e., the rate of heat loss per degree of inside-outside temperature difference, and η is the heating system efficiency.

The reference temperature has another physical significance:

$$\tau = T_{in} - Q_{int}/L \tag{4}$$

where T_{in} is the interior temperature, Q_{int} is the rate of heat provided by lights, appliances and people in the building and solar energy entering the building. When the outside is at this temperature τ, the intrinsic heat Q_{int} is just adequate to maintain the interior at the thermostated temperature T_{int}. For this reason, τ is also called the break-even or balance-point temperature.

PRISM thus permits the statistical analysis of consumption data in estimating the three parameters α, β and τ. PRISM also calculates an index of the building's consumption of the heating fuel by applying the estimated parameters α, β, and τ to a long-term (say 10-year) annual average of heating degree-days, calculated to the reference temperature τ estimated for that building. This index, called the Normalized Annual Consumption (NAC), is calculated as follows:

$$NAC = 365\alpha + \beta H_0(\tau) \tag{5}$$

where $H_0(\tau)$ is the average heating degree-days (calculated to reference temperature τ) for the long-term normalization period. Once a normalization period has been established, finding H_0 over the range of possible integer τ values requires only a one-time calculation for each weather station used.

The physical basis for PRISM assumes that the energy use can be divided into a "heating" part that increases linearly with decreasing outside temperature below a reference value, and a base-level consumption that is temperature independent. Some end uses other than space heating can also be related to the outside temperature: water heating energy use is higher in the winter because hot water demand is higher and because the water has to be heated from a lower cold water temperature; winter days are shorter and darker so that lighting requirements are also higher. In a study of houses with submetered heating energy use, the non-heating consumption showed (on average over many houses) sinusoidal seasonal variations both for gas and electricity as the primary heating energy source; see Figure 2. The effect of this seasonal variation on the PRISM estimates, based on total energy consumption, is seen as follows (15): the PRISM base-level estimate α is close to the summer consumption while the increased base-level consumption becomes included within the PRISM heating estimate, which means that heating is overestimated. The NAC index representing total energy consumption is unaffected and serves as a reliable scorekeeping index. It should also be noted that the division between heating and non-heating consumption is somewhat arbitrary in that all energy input to the house, including the part of the base-level consumption that remains within the house, contributes to space heating when it is cold enough. Thus, the energy input to the heating system itself is not the best index for the heating energy needs.

In spite of its summer bias, the base-level consumption, α, is a useful index. If the end uses making up the base-level consumption are known, then the α value for one household may be compared with other households with a similar mix of end uses (11), as well as against a standard or target value. The American Council for an Energy-Efficient Economy (ACEEE) periodically

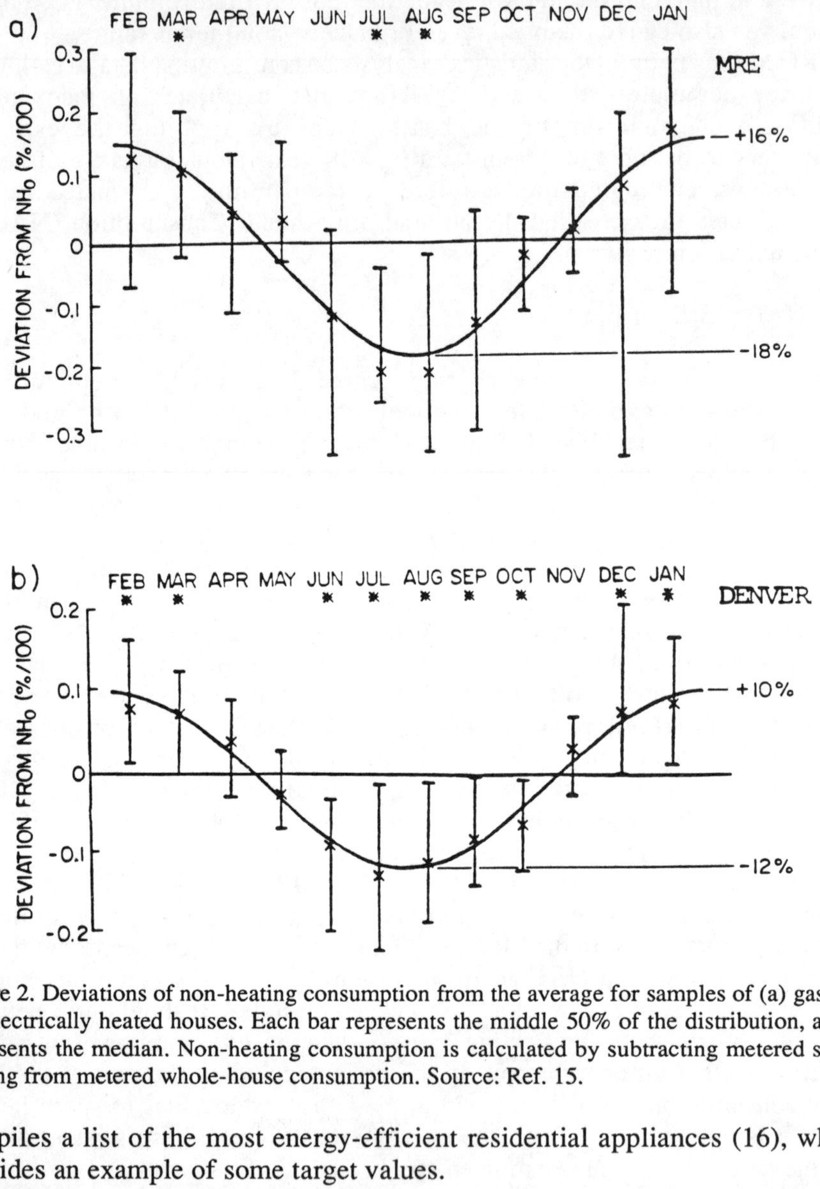

Figure 2. Deviations of non-heating consumption from the average for samples of (a) gas and (b) electrically heated houses. Each bar represents the middle 50% of the distribution, and x represents the median. Non-heating consumption is calculated by subtracting metered space heating from metered whole-house consumption. Source: Ref. 15.

compiles a list of the most energy-efficient residential appliances (16), which provides an example of some target values.

PRISM's great advantage is that it calculates standard errors for all estimated parameters. The temperature dependence of energy consumption is non-linear, such that these error estimates cannot be accurately determined from standard statistical procedures. Goldberg developed the statistical theory that permits the calculation of appropriate error bounds on the estimates (14).

As noted before, there are other versions of PRISM: PRISM-CO when the energy use is for space cooling and other uses, excluding space heating; and PRISM-HC, an experimental (for research) heating-plus-cooling version, for use when the energy use is for space heating and cooling, as well as other uses (17).

2.3 Applications of PRISM

2.3.1 Single-family houses

PRISM was originally developed for single-family houses and is best adapted to these buildings. For electric heat pumps that operate from outside air temperatures, the amount of heat delivered is not directly proportional to the energy consumed by the heat pump, since its coefficient of performance depends on the outside temperature. When PRISM is applied to this class of buildings, systematic distortions are introduced in some of the parameter estimates (18). In particular, the reference temperature estimate is low, as can be seen from a comparison with houses heated by electric resistance, in Figure 3. The bias on the estimate of NAC, the weather-adjusted index of total energy consumption, was affected very little (around 1 or 2%). Thus, PRISM is satisfactory for evaluating total energy use and savings but care should be taken in interpreting the individual estimates of α, β, and especially τ. Exhaust-air heat pumps work off a much more constant interior air temperature and should have no estimation bias.

In district heating systems, scorekeeping at the individual-house level should be based on heating energy delivered, as measured by flows and temperature differences. PRISM should work well.

The principal application of PRISM in single-family houses has been in the evaluation of retrofit energy savings. The Modular Retrofit Experiment saw the first application of PRISM to a large number of houses (19,20). Median results, based on PRISM analyses on individual houses in the pre- and post-retrofit periods, are summarized in Table 2. The standard errors of the estimates are shown in parentheses. It is noteworthy that the NAC values are extremely well determined, i.e., their standard errors divided by the estimated values are very small. This has been found in all studies using PRISM, for electricity as well as other energy sources (21): the NAC is on average determined to within a few percent, and serves as the best scorekeeping index for comparing pre and post-retrofit consumption levels. The β estimate and the heating component of the NAC, βH_0, are generally less well determined, with α and τ the least well determined. A comparison of PRISM results for electricity, natural gas and fuel oil is shown in Table 3.

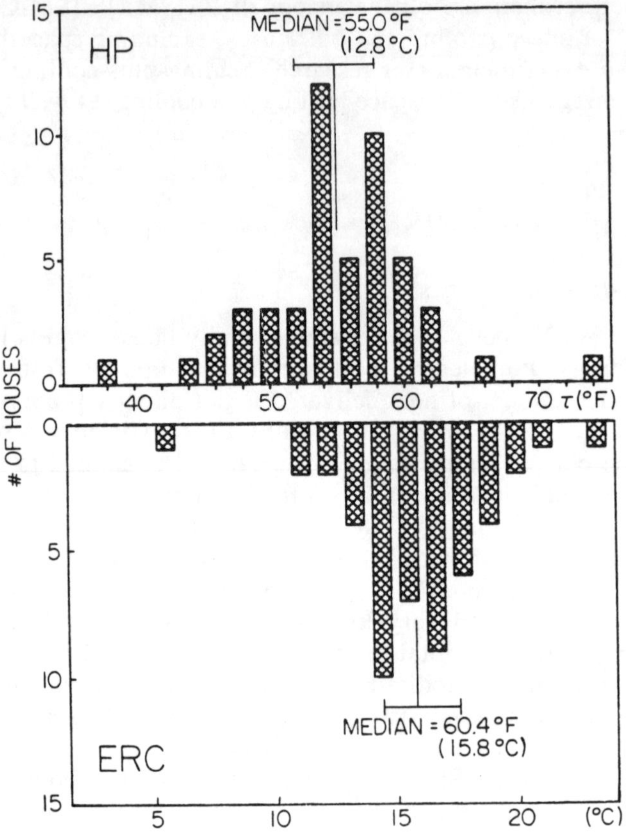

Figure 3. Comparison of PRISM estimates of reference temperature, τ, for 50-house samples of heat-pump (HP) and resistively heated (ERC) houses. Source: Ref. 18.

PRISM has been used to compare actual savings, based on changes in NAC, with savings predicted by engineering calculations. Such comparisions are problematic, in view of the difficulty of isolating individual conservation measures, from each other and from behavioral changes. One study, illustrated in Figure 4, ended up with small sample sizes but nevertheless meaningful results (22).

Instead of individual PRISM runs based on 12 months of data, it can also be run with 12-month sliding periods, moving forward a month at a time. The estimates, plotted against estimation period, can be an excellent graphical representation of energy use over extended periods of time, as shown in Figure 5 for an electrically heated house with a retrofit between years one and two.

Table 2. Summary of median PRISM results for Modular Retrofit Experiment.*

	Control (C)	House Doctor (HD)	Major Retrofit (MR)
# houses	40	58	40
R^2 pre	0.991	0.990	0.992
R^2 post	0.989	0.989	0.987
τ pre (°C)	18.0 (1.3)	17.5 (1.4)	18.2 (1.4)
τ post (°C)	17.8 (1.1)	17.2 (1.2)	16.9 (1.2)
$\Delta\tau$ (°C)	0.4 [–]	0.5 [–]	1.2 [–]
p-value	0.188 [–]	0.013 [–]	0.0001 [–]
α pre (kW_{th})	1.54 (0.24)	1.41 (0.25)	1.58 (0.25)
α post (kW_{th})	1.43 (0.22)	1.30 (0.19)	1.58 (0.18)
$\Delta\alpha$ (kW_{th})	0.01 [0.35%]	0.16 [11.5%]	0.07 [5.0%]
p-value	0.788 [0.662]	0.002 [0.0001]	0.141 [0.255]
β pre (W_{th}/°C)	478 (29)	428 (27)	462 (28)
β post (W_{th}/°C)	415 (26)	381 (24)	362 (25)
$\Delta\beta$ (W_{th}/°C)	41 [8.8%]	54 [12.3%]	90 [20.0%]
p-value	0.0001 [0.0001]	0.0001 [0.0001]	0.0001 [0.0001]
(βH_o) pre (GJ_{th}/year)	105.6 (6.19)	101.3 (6.65)	99.8 (6.76)
(βH_o) post (GJ_{th}/year)	97.9 (5.99)	75.7 (4.87)	70.1 (4.93)
$\Delta(\beta H_o)$ (GJ_{th}/year)	12.6 [12.7%]	18.6 [19.3%]	30.7 [30.4%]
p-value	0.0001 [0.0001]	0.0001 [0.0001]	0.0001 [0.0001]
NAC pre (GJ_{th}/year)	152.6 (4.31)	143.4 (3.69)	158.0 (3.61)
NAC post (GJ_{th}/year)	139.3 (3.72)	116.0 (3.15)	119.3 (3.19)
ΔNAC (GJ_{th}/year)	14.3 [9.8%]	21.5 [15.3%]	35.6 [21.9%]
p-value	0.0001 [0.0001]	0.0001 [0.0001]	0.0001 [0.0001]

* For each set of houses, the following four lines of results are shown for each parameter θ (=α, β, τ, βH_o, NAC):
medians:

median for θpre (median for s.e. of θpre)
median for θpost (median for s.e. of θpost)
median for change $\Delta\theta = \theta post - \theta pre$ [median for relative change $\Delta\theta/\theta pre$]

significance: p-value for total change [p-value for relative change].

Here s.e. = standard error. Note that a positive value of $\Delta\theta$ indicates savings. The p-value is calculated from the Wilcoxon signed rank test of the hypothesis that the median change was zero; a p-value of 0.01 indicates significance at the $1-p = 99\%$ confidence level.

Because of the arbitrariness of the zero in the °C temperature scale, no relative (%) change is given for $\Delta\tau$.
Source: Ref. 20.

Table 3. Comparison of median accuracy measures of PRISM results across fuels.[a]

Heating fuel:	Electricity (i)	Gas (ii)	Oil (iii)
# houses	50	276	207
R^2	0.972	0.990	0.971
se (α)/α	0.123	0.160	0.476[b]
se (β)/β	0.097	0.076	0.105
se (τ) (°C)	1.7	1.2	3.0
se (βH_o)/βH_o	0.091	0.067	0.191
se (NAC)/NAC	0.030	0.027	0.073

a Median R^2 values and standard errors are shown for three New Jersey samples:

i) a set of 50 houses which were heated but not cooled by electricity (Source: Ref. 17)
ii) gas-heated houses in the Modular Retrofit Experiment (a pre and a post-retrofit period for each of 138 houses) (Source: Ref. 20)
iii) a sample of oil-heated houses (Source: Ref. 21)

Standard errors of α, β, βH_o, and NAC are expressed as fractions of estimate.

b Since all houses in the other two samples have their water heated by the space heating fuel, the median of se (α)/α for the oil sample corresponds to the subset of 133 houses with oil-fired water heaters.

A variant of PRISM has been used to analyze energy consumption trends in aggregates of houses, from data on total monthly energy sales to all residential heating customers (e.g., electricity customers with electric heat) in a utility's service area. Here, in essence, the entire set of customers is treated as a single house whose α, β, and τ parameters are estimated by the same statistical procedure (23). The NAC per customer is a weather-adjusted index of energy consumption in the sector. Fig. 6 shows trends in gas and electricity consumption per customer, based on NACs and their standard errors, from 1970 to 1987, for the aggregate of all utilities in one state. Also shown is average raw (unadjusted) consumption per year. The trends evident in the weather- adjusted NACs are masked in the raw data.

2.3.2 Multifamily buildings

We have applied PRISM to multifamily buildings heated by oil and gas in the U.S. The results for a 60-unit multifamily building are shown in Figure 7, in terms of a sliding analysis of each parameter (24). In another multifamily building of the same size, PRISM-HO was applied to gas consumption, and PRISM-CO to the electricity used for lights, appliances and air conditioning

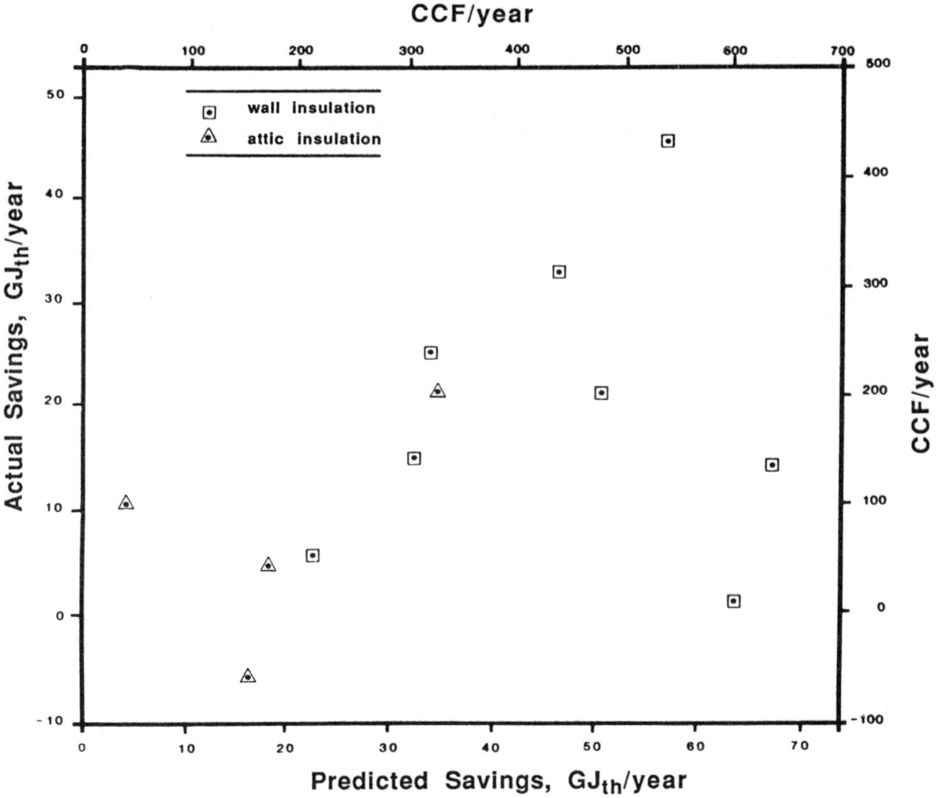

Figure 4. Actual versus predicted savings from wall and attic insulation in a sample of Minneapolis houses. Source: Ref. 22.

(window units in each apartment). The results are summarized in Figure 8. In electrically heated buildings, for which individual metering is fairly common, PRISM-HO can often be applied at the individual apartment or the whole-building level. For any fuel, analysis of the whole-building consumption over time can be confounded by variable vacancy rates. In applying PRISM to individual apartments, problems such as heat transfer between neighboring apartments, common-area usage, and location of apartment within building and solar orientation need to be considered.

2.3.3 Commercial buildings

There are several features of office and other commercial buildings which may lead to problems in interpretation for a simple procedure such as PRISM. The

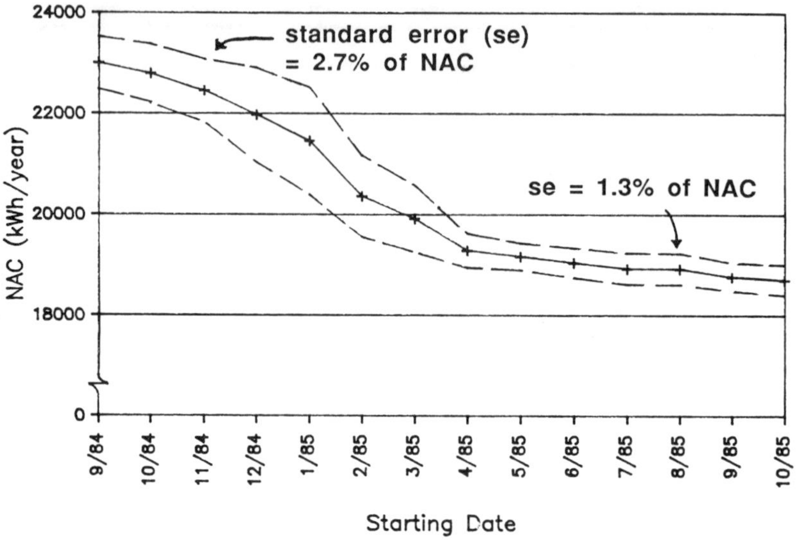

Figure 5. PRISM-HO estimates for sample electrically heated house over sliding estimation period. Shown is Normalized Annual Consumption (NAC) vs. estimation period: each NAC is estimated from one year of monthly meter readings, ending on the date shown. Corresponding error bounds (one standard error on each side of the estimate) are indicated by the dashed line. Estimated savings for this house, retrofitted in the middle of the two-year period, are 18 (±3)%. Source: ongoing research at Princeton University.

energy consumed by lights, office equipment, etc., contribute substantially both to internal loads and to base-level energy use, reducing the relative magnitude of heating energy use. These buildings cover a wide range of uses, differing both in their schedules and in their special features. Office and other buildings have set operating hours, which can strongly influence energy use of such occupancy-dependent components as lights, mechanical ventilation, and office equipment. Whether it is a weekday or weekend, and other vacation schedules, e.g., academic year, also affects energy use and must be taken into account in scorekeeping. One useful approach is to analyze separately periods corresponding to workdays and non-workdays, later combining the two in order to obtain annual consumption totals.

Among special features influencing building energy use is the operation of refrigeration equipment, restaurant, laundry, swimming pool, large numbers of computers, etc., all of which make comparisons across buildings difficult. A

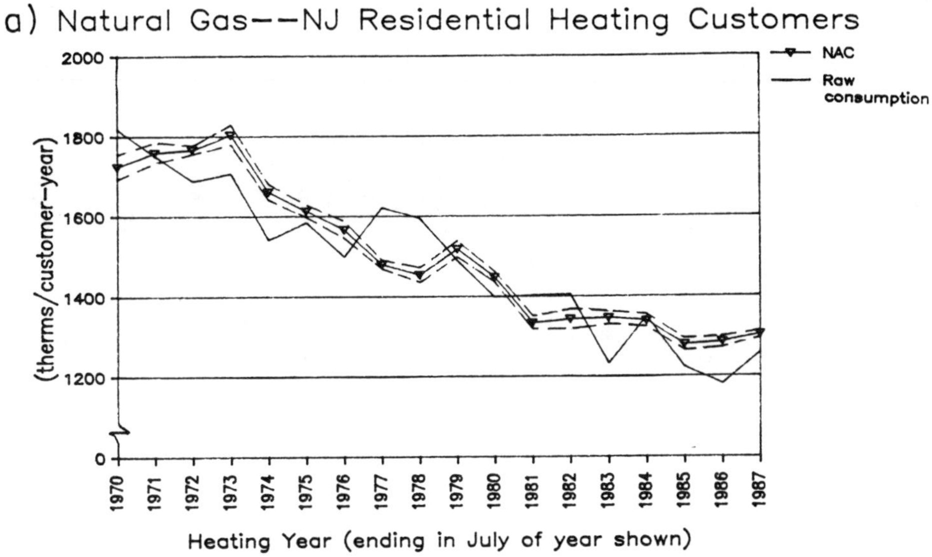

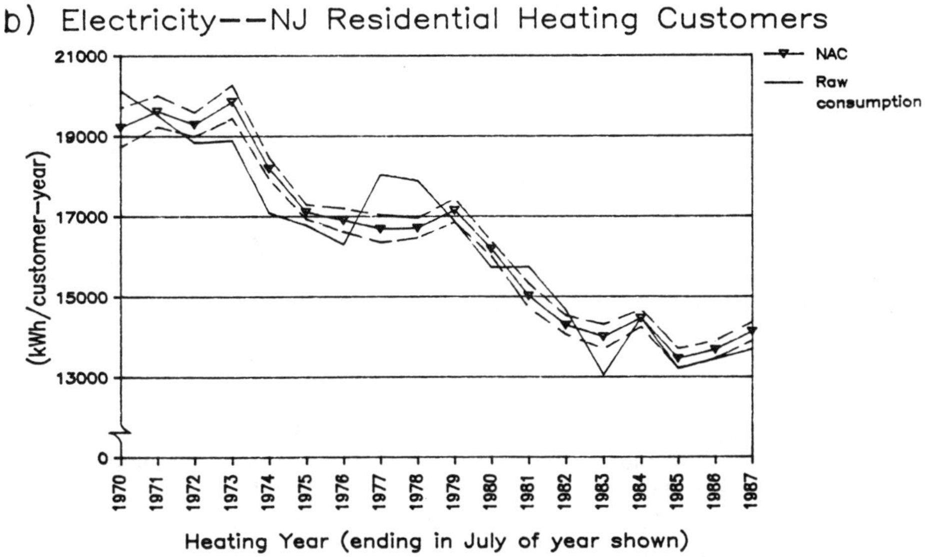

Figure 6. Annual per-household gas and electricity consumption for New Jersey residential heating customers as a function of heating year. NAC and its standard error bounds, derived using the aggregate PRISM model, are shown. For comparison, the raw (unadjusted) average per-customer consumption for each year is indicated. Source: Ref. 23.

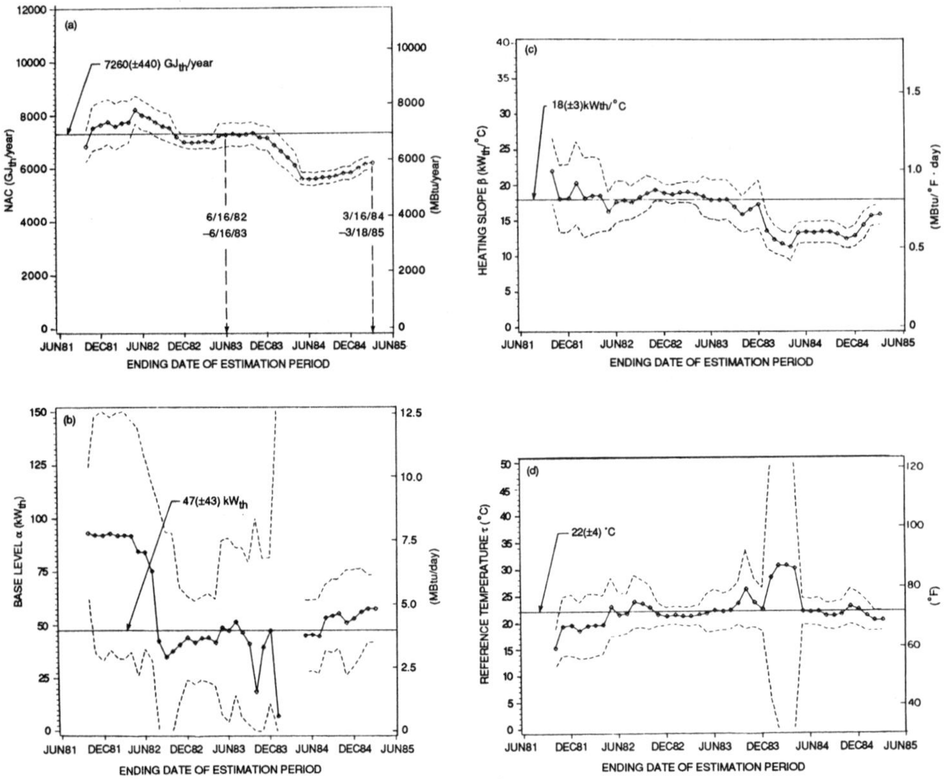

Figure 7. PRISM-HO estimates for gas-heated apartment building over sliding estimation periods: (a) Normalized Annual Consumption (NAC); (b) base-level consumption α; (c) heating slope β; (d) reference temperature τ. Dashed lines show error bounds. Horizontal lines show PRISM results for June 1982 to June 1983 estimation period. Source: Ref. 24.

conventional approach is to normalize consumption by floor area to obtain energy intensity, and then to be careful in comparing buildings being put to very different uses. The average 1982 commercial building stock in the U.S. consumed 1.04 GJ/m^2, two-thirds of which was as fuel (25). Energy-efficient office buildings are expected to consume only a third as much.

The "heating and cooling" version PRISM-HC was applied to Enerplex, a pair of well-instrumented, energy-efficient office buildings in New Jersey (26). Separate PRISM runs were carried out for weekdays and weekends in each building. Predictions of total monthly electric consumption based on the separate PRISM runs agree well with actual data (Figure 9), and the corresponding NAC is well determined. The predictions of the individual

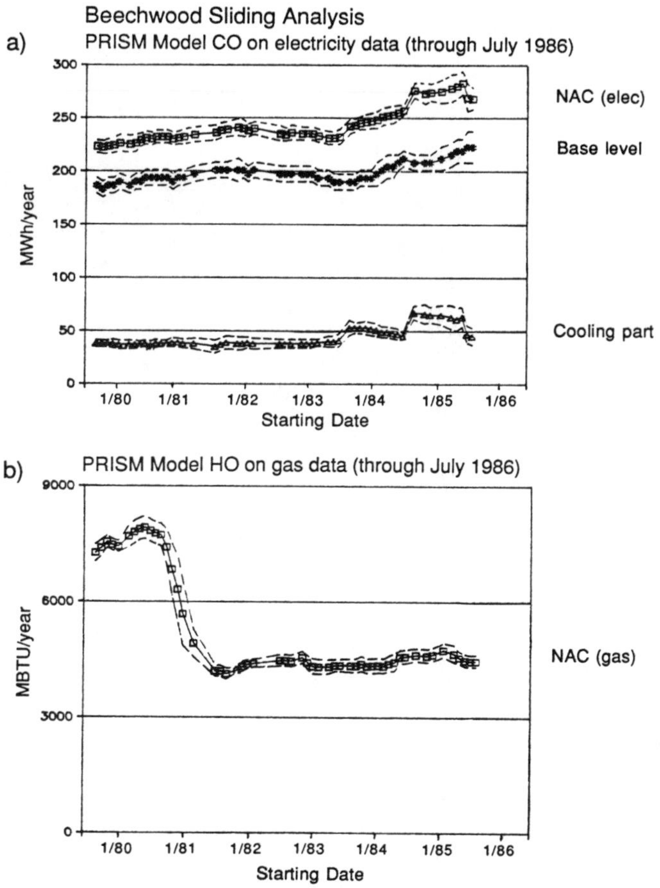

Figure 8. PRISM results for a multifamily building heated by natural gas and cooled (with window units) by electricity: (a) PRISM-CO applied to electricity data, with NAC broken down into base-level and cooling components; and (b) PRISM-HO applied to gas data. Dashed lines show error bounds. Source: ongoing research at Princeton University.

heating and cooling components are underestimated compared to their values obtained from submeter data. The reason is that there is both cooling and heating energy use (in different periods) within some months which cannot be distinguished from PRISM's elevated base-level estimate. Separate PRISM-HO (heating only) and PRISM-CO (cooling only) runs on submetered heating and cooling data model well, and are the preferred alternative when such data are available. With only total building consumption data, the NAC remains an accurate and appropriate index of energy use, as in single-family houses.

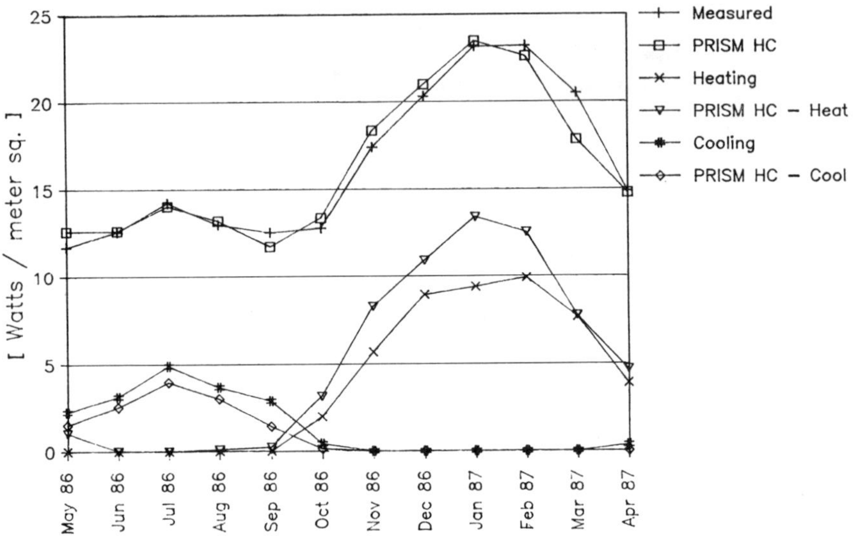

Figure 9. PRISM-HC (heating and cooling) model applied to total electricity consumption at Enerplex South, an office building in New Jersey. Measured and PRISM estimates of total energy use are compared on the top two curves. Heating and cooling estimates from PRISM-HC are compared with submetered heating and cooling data in lower curves (see labels in figure). Source: ongoing research at Princeton University.

2.4 Socio-Economic Analysis

After adjusting for differences in weather using a physical model such as PRISM, the consumption data may be analyzed to determine the effects of socio-economic determinants, in a two-stage approach. Hirst has used this approach to determine electricity savings in a large number of single-family houses (13). Cross-sectional models were used to analyze variations in NAC reduction, as a function of household demographic and structure characteristics, and participation in a Bonneville Power Administration (BPA) energy conservation program. He also compared savings one, two, and three years following participation in the program, as shown in Figure 10 for the three levels of program participation.

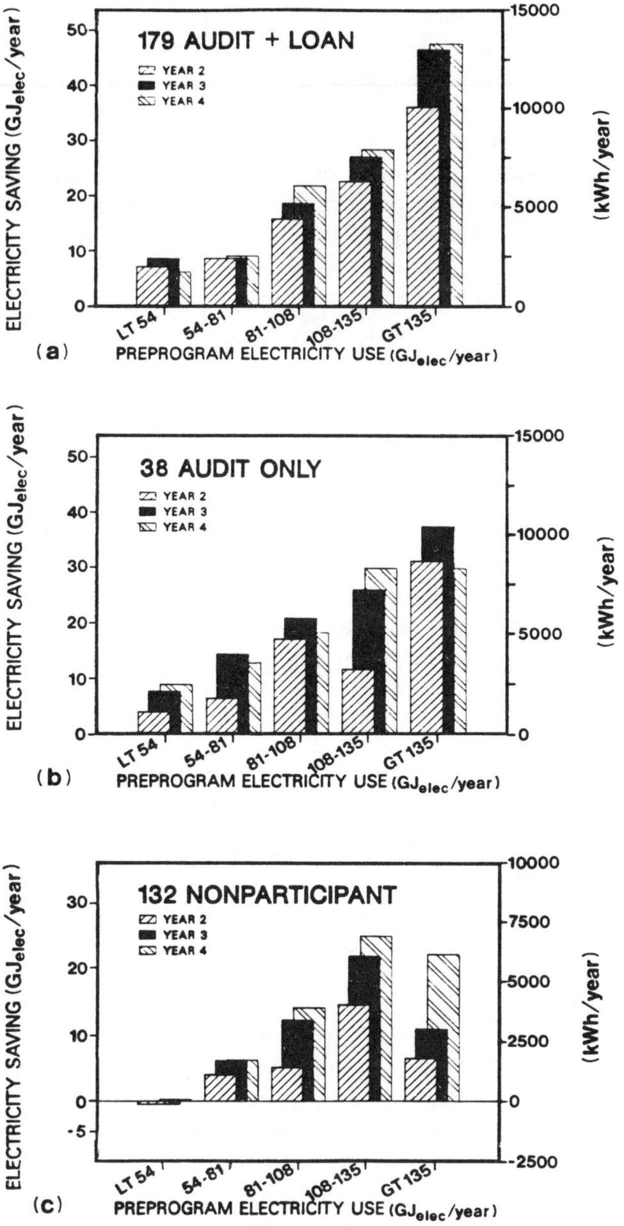

Figure 10. Reductions in annual electricity use in years 2, 3, and 4 (relative to year 1) as a function of pre-program electricity use for (a) audit + loan, (b) audit only, and (c) non-participant households in a Bonneville Power Administration energy conservation program. Source: Ref. 13.

Econometric models have also been developed and used to disaggregate total electricity consumption in terms of its appliance-specific components (27). While these conditional demand models are useful in estimating average consumption for different appliances, they are not useful in identifying the houses or the appliances which are inefficient. For such determination, the energy consumption data must be combined with nameplate data on the appliances involved, their performance in the laboratory and in the field, obtained through submetering. Combined with data from individual household appliances, conditional demand models can be used to understand the sources of hourly variations in electricity use (28).

3 Special Problems in Scorekeeping Electricity Use

Techniques for scorekeeping energy use in buildings are well developed, as discussed in Section 2. There are, however, a number of difficulties that need to be addressed. We have divided these into two groups – those relating to electrical *energy use*, and those relating to *peak demand.*

3.1 Electrical Energy Use

3.1.1 Efficiency of electricity generation, transmission, and distribution

The generation of electricity, especially from thermal power plants, involves large losses. In addition, there are losses in the transmission and distribution of electricity. Hence, electricity used by the consumer (*site* energy use) is derived from a much larger quantity of *primary* energy use.

Scorekeeping results are often presented both in terms of site energy and primary energy. For calculating primary energy equivalents, it is conventional to use the average efficiency of generation, transmission, and distribution of all power generation within the country or region, instead of trying to determine which source provided electricity to which end user at what overall conversion efficiency. The site energy is adequate for comparison among all-electric users, but does not allow a reasonable comparison in the case of mixed use of fuels, since fuel energy and electrical energy may be very different in cost, and since ultimately it is conservation of fuel resource energy that is of interest. For over two decades, there has been little improvement in electricity generation efficiency (29) and primary energy use has remained at roughly three times site electricity use. This ratio is expected to decrease in time as much more efficient technologies for electricity generation are installed (30). And if the recent breakthroughs in the laboratory in the discovery of high-temperature superconducting materials lead to the development of usable superconducting wire, then transmission losses will be reduced as well.

3.1.2 Power factor

Many end uses, especially those with motors, have inductive loads for which the power factor is less than unity, i.e., the alternating current is not in phase with the voltage. The result is that a certain power consumption corresponds to a larger volt-amperage. The larger current flows necessary to meet the demand incur larger resistive losses in transmission and distribution (losses that are proportional to the square of the current). At present there is no metering of the power factor, but some large users are assessed a surcharge based on an estimate of the power factor and others are required to install capacitors to balance their inductive loads and increase their power factor. Since it is the current and not the power factor itself that leads to transmission loads, future meters might accumulate the square of the user's current draw to determine the additional resistive losses.

3.1.3 Thermodynamic quality

An analysis of the thermodynamic quality of energy use considers whether the end use is work or heat, and if it is heat, at what temperature, in keeping with the second law of thermodynamics (31). The second-law analysis of energy use sets theoretical limits on conversion efficiencies and serves as targets for conservation design. Scorekeeping based on existing meter readings generally analyzes consumption data from many different end uses of varying thermodynamic quality. Thermodynamic quality is thus not considered in scorekeeping where normalization by "level of output" provides a more useful basis for comparison of metered electricity use data.

3.1.4 Non-congruent metering

One problem encountered in comparisons of energy indices from different users is due to non-congruent metering, i.e., the meters do not record consumption for the same combination of end uses. Hence, electricity used for water heating in one office building may be on a separate electric meter from all other uses, while space heating and ventilation fans may be on a different meter from other uses in another building. In these cases, it is usually possible to combine meter readings to get total electricity consumption, but the benefits from disaggregation by multiple meters are then lost. Moreover, substantial savings in a single end use may be statistically indistinguishable in an examination of total energy use. The problem is exacerbated when multiple fuels are used. For example, annual total gas use in an apartment building showed no significant reductions after a retrofit measure that halved summertime gas use for water heating, verified through other data (32). Meaningful comparisons of engineering estimates and measured data in general require adequate disaggregation. The advantages of scorekeeping using disaggregated energy use data may warrant more careful design of metering configurations in future installations.

3.1.5 Fuel shifting and fuel choice

Change in the energy source for some or all end uses creates problems for scorekeeping, in the same way as non-congruent metering does. It also presents opportunities for insight if data on multiple fuels are available. In the evaluation of the Modular Retrofit Experiment (20), for example, the house with the largest gas savings (40%) was a control house that had shifted to wood for heating during the course of the experiment. This house was eliminated from the statistical analyses of savings. The conclusion of the experiment might have been different in the absence of knowledge of fuel switching. Change in energy consumption following fuel shifting can give information on the relative efficiency of different end uses, where the level of activity is unchanged. The performance of an oil (or gas) furnace in a house cannot be determined from a laboratory test alone, because of interactions between the house and the furnace. For a house in which the heating fuel was changed from electricity to oil or vice versa, comparison of energy use with the two energy sources gives information about their relative efficiencies as installed. Because of the usefulness of this type of information, scorekeeping should not be limited to electricity use at any site, but, rather, it should consider the energy consumption of all fuels, wherever possible, especially if fuel shifting is occurring.

Fuel choice is an important policy issue. Some processes are more energy efficient (using less site energy per unit of output) but require more electricity. For instance, the Plasmamelt process for raw steel production requires about a third less total energy per tonne of steel than the Elred process but over three times as much electricity (33). In addition to site energy use and primary energy use totals of alternative processes, the planner may wish to multiply energy use components of site energy use by appropriate shadow prices (that take into account market prices plus externalities such as environmental costs, energy independence and security, etc.) for each fuel type to arrive at an imputed cost of the site energy use. This will permit a better comparison of alternatives that involve greatly different mixes of energy carriers.

3.1.6 Self generation, cogeneration, and total energy systems

Some users, notably large industries, produce some or all of their own electricity. They may use the waste heat to meet process heating requirements in the form of steam (cogeneration) or hot water (total energy system). In these cases, scorekeeping electricity use based on purchased power alone will underestimate actual consumption, since the electric company does not meter the amount of self-generated electricity. Here again, a scorekeeping analysis of all energy sources at the site is warranted. Four energy consumption measures may be proposed: purchased site electricity use and primary energy equivalent, as before, and total site energy use and total primary energy use equivalent. In the latter two indices, the site is treated as the control volume and all energy flows into the site are recorded, while in the former, only the electricity purchased from the grid is recorded. Additionally, the amount of independently generated electricity should be recorded.

3.1.7 Annual-cycle storage

A scorekeeping problem arises in the case of systems involving annual-cycle storage, e.g., hot ponds or ice ponds, in which the energy is collected and stored in one part of the year to meet demand in another part of the year. Electric meter readings alone cannot be used to relate measured consumption to month of use, or to separate weather-dependent from weather-independent energy use, or to normalize consumption to a year of average weather. Until additional scorekeeping techniques are developed, we may only record and compare annual energy use for annual-cycle storage systems.

3.2 Peak Demand

3.2.1 The nature of the peaking problem

For many electric utilities, meeting peak demand is a major problem. For all practical purposes, electricity cannot be stored so the instantaneous demand must be matched by a corresponding supply. Generating capacity, and transmission facilities, must be capable of meeting the peak demand. Generating capacity is usually divided into three categories: base load, load following, and peaking. Base-load equipment is best operated at constant output, while load-following units can alter their output in response to demand. Peaking generating equipment is operated for short periods when the base-load and load-following equipment cannot meet the demand. In general, base-load generation has higher capital and lower operating costs, the reverse being true for peaking equipment.

Variations in demand are the result of operating schedules (offices and factories), fluctuations in daylight availability (lighting), and climate (heating and cooling). They can be seen in terms of daily cycles and seasonal cycles. Peak-shaving strategies (to reduce peak kW, and thus consumption during particular hours) are often of greater interest to utilities than energy conservation strategies (to reduce total kWh). Electric meters that only record energy consumption provide no information about demand distribution or peaks, and current scorekeeping techniques are deficient in their handling of demand patterns.

3.2.2 Scorekeeping peak demand

Existing data on a user's demand are very limited, which limits scorekeeping on this basis. Large consumers in the U.S. have a separate demand meter, which records the highest 15-minute or half-hour electric consumption in the billing period. They are billed according to this "demand." (Small users–houses, in particular–are not metered or billed for their contribution to this demand.) The only scorekeeping index from available data for large users is the load factor, based on the total energy consumption and the consumption in the peak interval for the billing period, typically a month:

Load factor = average power / peak power.

A load factor can be calculated for each month, but it lacks information important in determining the contribution of the user's peak demand on generating capacity needs: the time when the peak occurred and the distribution of the demand. The first is important in knowing if the peak was coincident with the peak demand for other users. The second is important in knowing the sharpness of the peak. To get this additional information, and to collect baseline data for residential customers, utilities frequently submeter and record individual end uses. Figure 11 shows 24-hour load profiles for two water heaters in two houses. The profiles are similar but not identical so that the combined demand of the two is less than twice the individual demand. The non-coincidence of demand of a large number of users (N) of the same end use is expressed by the diversity factor, defined as follows:

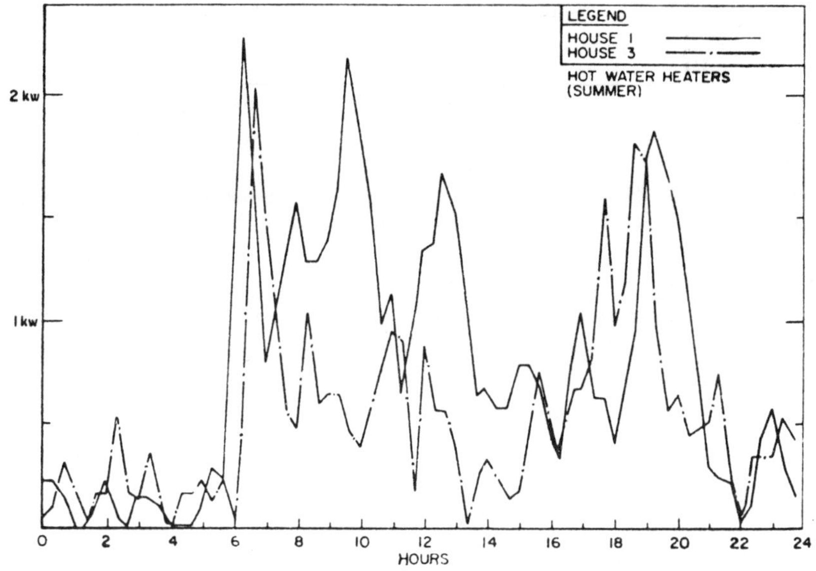

Figure 11. Summer load profiles for water heater use in two Twin Rivers, NJ houses. Source: Ref. 49, p.292.

$$\text{Diversity factor} = \frac{\text{combined demand of N users}}{\text{N x demand of the median user}}.$$

The combined effect of all users, or all of a set of users whose combined demand is being continuously measured, may be represented by a *load duration curve*. A load duration curve is a frequency distribution of the occurrence of different levels of demand, and projection of this curve is the basis for planning generating capacity of different types–base load, load following, and peaking.

Changes in the load duration curve for individual users, or individual end uses, will require submetering and extensive data analysis that fall outside the scope of scorekeeping. Evaluating the peak demand effect of conservation and load management programs will thus continue to be based on submetering samples of users (34,35). New technical and institutional arrangements for peak load management offer the possibility of scorekeeping of *all* users through limited amount of additional data collection. Unlike energy scorekeeping, there is no universal scorekeeping index for load management but rather one designed to suit a particular load management strategy.

3.2.3 Alternative peak load management strategies and scorekeeping possibilities

There are various technical options and institutional arrangements for managing peak loads through peak shaving or peak shifting (36,37). The institutional arrangements involve incentives for reduced on-peak consumption and/or shifting demand to off-peak periods, and usually transfer the management responsibility to the user. Examples include the Peak Activated Rate and Real Time Pricing experiments of Orange and Rockland Utilities (38). Users may choose between technical options (e.g., energy management systems, thermal storage), scheduling changes (carrying out electricity-intensive processing in off-peak periods), or behavioral changes (not using air conditioning on certain hot days). It is in the electric company's interaction with the user that the main potential for scorekeeping exists.

Interruptible service

Traditional peak shaving is through users with previously arranged interruptible service, and of course load shedding, i.e., shutting down service to groups of users without previous arrangements when there is a shortfall in generating capacity. There is little opportunity for scorekeeping here. An important issue in these cases is knowing the amount of generating capacity installed by the user, and the extent of its use. Thus, apart from an analysis of the electric company's supply and demand picture, it is important to evaluate the electricity supply and demand for the whole economy, including the independent producers.

Peak shifting

Traditional peak shifting has been through off-peak metering and billing–time-of-day rates in which users pay a lower rate for consumption in pre-designated off-peak periods. In residences, one or more appliances are wired through the off-peak meter and can only operate during off-peak periods. The appliance chosen involves thermal storage (such as storage- type electric water heaters) or discretionary uses (e.g., dishwashers). Here, scorekeeping is straightforward–energy scorekeeping for the main and the off-peak meters.

Utility direct control

Recently, the direct control of user devices by the utility has become a popular idea (36,37,39,40). In this approach, the utility sends a signal via radio, power line, or telephone to a receiver at the user location, which controls the device. There is no communication from the user to the utility. If there is a separate meter for devices subject to direct utility control, e.g., for billing purposes, then scorekeeping can be based on the separate user meters; if no other meter is in use, then there is no opportunity at present for scorekeeping load management in terms of peak (kW) reduction by time of day, as opposed to total energy (kWh). In future, an electric meter could be built into the receiver/controller for additional data.

Local control

Another possibility is a controller at the user's location that regulates one or more devices, based on sensors for total demand, temperature, and/or time. The controller takes advantage of preferential off-peak rates. There is no utility-user link and no data can be directly collected by the utility. Depending on the sophistication of the controller, energy consumption and demand may be available to the user, but only indirectly to the utility.

Distributed control

Here, the control is influenced by conditions at the user site but with a communication link to the utility. Distributed control reduces user inconvenience while meeting utility load-management objectives better than completely local control would do. Recent advances in electronics have made this type of control less expensive while increasing its capability (41,42). More advances may be expected. With two-way communication between utility and user, not only is more sophisticated control possible, but site energy use data may be collected by the utility. One may anticipate, for example, power line communication from sensors in various end-use devices and the electric meter to a controller which is equipped with a (radio or telephone) transceiver. The utility would relay information such as spot electricity prices to the controller, which decides how to control the various devices, based on measurements such as refrigerator and water heater loads, or room temperature, time, and pre-programmed price elasticities. With few additional sensors, the controller could relay data on the on/off status of the devices and electric consumption to the utility, which can use the data for energy and demand scorekeeping.

3.2.4 Special issues in scorekeeping peak demand

Annual cycle storage

The above discussion on scorekeeping peak demand has largely been based on considerations of shifting peaks over a 24-hour period. Annual cycle thermal storage, in the form of hot water, ice, etc., may reduce seasonal variations in demand resulting from weather-dependent variations in heating and cooling energy requirements. Analysis of the weather dependence of peak demand will need to consider the nature of the thermal storage, on a case-by-case basis.

Photovoltaics and wind power

Analysis of demand in terms of load duration curves presupposes that the demand may be divided according to frequency of occurrence for division among base-load, load-following, and peaking power plants. This assumes that the economics and choice of generation depend on frequency of demand alone, and not on the time of demand. In the near term, solar photovoltaic and wind power may play a role in providing peaking power (43,44). In such a case, a simple load duration curve is no longer useful, but must be separated into periods where solar and wind energy may or may not contribute. Analytic techniques taking into account the expected availability of photovoltaic generation, analogous to those proposed for wind systems, will need to be developed (45,46,47).

Advanced gas turbines

Another assumption, implicit in many load management programs and consequently in their scorekeeping, is that base-load plants (typically large coal and nuclear plants) are more capital intensive and have lower operating costs, with the reverse being true for peaking power plants, typically gas turbines. The assumption is questioned by the increasing unreliability of some nuclear plants, which could lead in some regions to load-following power plants (dominated by older, smaller, less capital-intensive fossil fuel plants) becoming *less* expensive to operate on average than base-load plants (48). The assumed hierarchy of costs has come seriously in doubt with the development of advanced gas turbines that are so inexpensive and efficient that they have been proposed for base-load and load-following applications as well (29,30). The large-scale introduction of advanced gas turbines that are well suited for all frequencies of operation could have the effect of greatly diminishing the urgency and need for peak demand management programs.

4 Conclusions

Scorekeeping the energy use of buildings involves well developed techniques that are in the process of being extended to include new developments in metering technology. Increasingly, these techniques are being used to evaluate energy conservation programs on the basis of actual energy saved. A number of scorekeeping issues remain unresolved, especially in the area of peak demand. Creative solutions are needed in order to produce the required data and appropriate analytical tools so that demand scorekeeping of future peak-shaving programs is feasible, as energy scorekeeping of current conservation programs is today.

Acknowledgements

This report results from research performed at the Center for Energy and Environmental Studies, with funding from the U.S. Department of Energy, the Electric Power Research Institute, and the New Jersey gas and electric utilities and the New Jersey Department of Commerce, Energy and Economic Development through the New Jersey Energy Conservation Laboratory. The authors thank Cathy Reynolds for help in preparation of this draft.

References

1. Robinton, M.A., Starkie, A. 1983. *End-use meter development and design of the electric ARM*, Final Report EPRI EA-3247, Electric Power Research Institute, Palo Alto, Calif., USA.
2. McCarthy, P.M., Barnes, H.H., Harris, L.R. 1984. *An examination of methods for providing residential customers with estimates of end-use*. Report EPRI EA-3576, Electric Power Research Institute, Palo Alto, Calif., USA.
3. Hart, G. 1984. *Progress report, non-intrusive appliance load data acquisition method.* MIT Energy Laboratory Technical Report, Cambridge, Mass., USA.
4. Socolow, R.H. 1977. The coming age of conservation, *Annual Review of Energy*, 2:239–89.
5. Williams, R.H. 1989. *Innovative Approaches to Marketing Electric Efficiency*, this volume.
6. Fels, M.F. ed. 1986. *Measuring Energy Savings: the Scorekeeping Approach*, Special Issue, *Energy and Buildings, 9*, Nos. 1 and 2.
7. Dutt, G.S., Beyea, J., Sinden, F.W. 1978. *Attic heat loss and conservation policy*, Proc. Energy Technology Conf., American Society of Mechanical Engineers.

8. Norford, L.K., Rabl, A., Harris, J.P., Roturier, J. 1988. Electronic Office Equipment: The Impact of Market Trends and Technology on End-Use Demand for Electricity, this volume.
9. Keating, K. ed. 1988. Special issue of *Energy and Buildings* on evaluation of the Hood River Conservation Project (forthcoming). See also E. Hirst, R. Goeltz, D. Trimble, 1987. *Electricity use and savings in the Hood River Conservation Project,* Oak Ridge National Laboratory Report ORNL/CON 231, Oak Ridge, Tennessee, USA.
10. Baldwin, S.F. 1988. *Energy-efficient electric motor drive systems: motor and system diagnostics*, Princeton University Center for Energy and Environmental Studies, Working Paper 93.
11. Lyberg, M. D. ed. 1987. *Source Book for Energy Auditors*, International Energy Agency, 2 Vols.
12. Fels, M.F. 1986. *Some thoughts on two-stage PRISM*. Princeton University Center for Energy and Environmental Studies, unpublished memo, Princeton, N.J., USA.
13. Hirst, E. 1986. *Electricity savings one, two, and three years after participation in the BPA residential weatherization pilot program*, in Ref. 6, pp.45–53.
14. Goldberg, M. 1982. *A Geometrical Approach to Nondifferentiable Regression Models as Related to Methods for Assessing Residential Energy Conservation*, Ph.D. Thesis, Dept. of Statistics, Princeton University; also, Princeton University Center for Energy and Environmental Studies Report 142, Princeton, N.J., USA.
15. Fels, M.F., Rachlin, J., Socolow, R.H. 1986. *Seasonality of non-heating consumption and its effect on PRISM results*, in Ref. 6, pp.139–48.
16. ACEEE, 1988. *The Most Energy-Efficient Appliances 1988 Edition*, American Council for an Energy-Efficient Economy, Washington, D.C., USA.
17. Stram, D., Fels, M.F. 1986. *The applicability of PRISM to electric heating and cooling*, in Ref. 6, pp.101–10.
18. Fels, M.F., Stram, D.O. 1986. *Does PRISM distort the energy signature of heat-pump houses?*, in Ref. 6, pp.111–18.
19. Dutt, G.S., Lavine, M.L., Levi, B.G., Socolow, R.H. 1982. *The Modular Retrofit Experiment: exploring the house doctor concept*. Princeton University Center for Energy and Environmental Studies Report 130, Princeton, N.J., USA.
20. Dutt, G.S., Lavine, M.L., Levi, B.G., Socolow, R.H. 1986. *The Modular Retrofit Experiment: Design, Scorekeeping, and Evaluation*, in Ref. 6, pp.21–33.
21. Fels, M.F., Goldberg, M.L., Lavine, M.L. 1986. *Exploratory scorekeeping for oil-heated houses*, in Ref. 6, pp.127–36.
22. Hewett, M.J., Dunsworth, T.S., Miller, T.A., Koehler, M.J. 1986. *Measured versus predicted savings from single retrofits: a sample study*, in Ref. 6, pp.65–73.
23. Fels, M.F., Goldberg, M.L. 1986. *Using the scorekeeping approach to monitor aggregate energy conservation*, in Ref. 6, pp. 161–68.
24. DeCicco, J.M., Dutt, G.S., Harrje, D.T., Socolow, R.H. 1986. *PRISM applied to a multifamily building: the Lumley Homes case study*, in Ref. 6, pp.77–88.
25. Goldemberg, J., Johansson, T.B., Reddy, A.K.N., Williams, R.H. 1988. *Energy for a Sustainable World.* New Delhi: Wiley Eastern.
26. Rabl, A., Norford, L., Spadaro, J. 1986. *Steady state models for analysis of commercial building data*, Proc. American Council for an Energy-Efficient Economy, Vol. 9, Summer Study on Energy Efficiency in Buildings, Santa Cruz, Calif., USA.
27. Parti, M., Parti, C. 1980. *The total and appliance-specific conditional demand for electricity in the household sector. Bell Journal of Economics*, 11(1): 309–21.

28. Parti, M., Sebald, A., Chaudhury, I.B. 1986. *Integrated load and time of day models for electricity in residences*. Proc. American Council for an Energy-Efficient Economy, Summer Study on Energy Efficiency Buildings, Vol. 9, Santa Cruz, Calif., USA.
29. Williams, R.H., Larson, E.D. 1988. *Aeroderivative turbines for stationary power*. Princeton University Center for Energy and Environmental Studies Report 226, Princeton, N.J., USA.
30. Williams, R.H. Larson, E.D. 1989. *Expanding Roles for Gas Turbines in Power Generation*, this volume.
31. APS, 1975. *Efficient Use of Energy. A Physics Perspective*. A Report of the American Physical Society Summer Study on Technical Aspects of Efficient Energy Utilization, Vol. 25 of the American Institute of Physics Conference Proceedings.
32. Englander, S.L., Dutt, G.S. 1987. *Domestic hot water savings in multifamily buildings*. Princeton University Center for Energy and Environmental Studies Report 218, Princeton, N.J., USA.
33. Eketorp, S. 1989. *Electrotechnologies and steelmaking*, this volume.
34. Limaye, D.R., Whitmore, C. 1984. *Selected statistical methods for analysis of load research data*. Report EPRI EA-3467, Electric Power Research Institute, Palo Alto, Calif., USA.
35. EPRI, 1981. *Load data management and analysis*, Stone and Webster Management Consultants and Applied Energy Research, Final Report EPRI EA-2178, Electric Power Research Institute, Palo Alto, Calif., USA.
36. Rabl, A. 1985. *Technology for load management*, in *Meeting Energy Challenges*, Proc. 2nd Pacific Gas and Electric Energy Expo, Oakland, California, (Smith, C.B., Davis, T., Turnbull, P.W., Eds.). Pergamon Press, Vol. 1, pp. 169–78.
37. Isaksen, L. 1985. *Residential load management technology review*. Report EPRI EM-3861, Electric Power Research Institute, Palo Alto, Calif., USA.
38. Orange and Rockland Utilities, Inc. 1987. *Electric Master Energy Plan Supplement: Demand-Side Management Plan Peak Load Reduction*. Pearl River, N.Y., USA.
39. Blevins, R.P. 1986. *1985 survey of utility residential end-use projects*. Report EPRI EM-4578, Electric Power Research Institute, Palo Alto, Calif., USA.
40. Butterfield, K. 1987. *Air-Conditioner Cycling Programs*. Energy Auditor and Retrofitter, July/August, pp. 24–28.
41. Rosenfeld, A.H., Bulleit, D.A., Peddie, R.A., 1986. *Smart meters and spot pricing. IEEE Technology and Society Magazine*, pp. 23–28.
42. Peddie, R.A., Bulleit, D.A., Rosenfeld, A.H. 1985. Managing electricity demand through dynamic pricing, Chapter 20 in *Energy Sources: Conservation and Renewables*, Hafemeister, D., Kelly, H., Levi, B. eds. American Institute of Physics Conf. Proc. 135, pp. 382–401.
43. Korsun, V., Stranix, A.J. 1984. *Improving the match between the daily solar insolation peak and electrical demand peak, Solar Energy, 33*(2): 171–74.
44. Martin, B., Diesendorf, M. 1982. Optimal thermal mix in electricity grids containing wind power, *Electrical Power and Energy Systems,* 4(3): 155–61.
45. Haslett, J., Diesendorf, M. 1981. The capacity credit of wind power. *Solar Energy*, 26: 391–401.
46. Moretti, P.M., Jones, B.W. 1982. Analysis method for non- schedulable generation in electrical systems. *Solar Energy*, 28(6): 499–508.
47. Sethuram, R.B., Moretti, P.M. 1984. Predicting utility peaks with and without wind generators. *Solar Energy*, 33(1): 89–90.

48. Sachs, H.M. 1988. New Jersey Dept. of Commerce, Energy and Economic Development, Newark, N.J., USA, private communication.
49. Socolow, R.H. ed. 1978. *Saving Energy in the Home: Princeton's Experiments in Twin Rivers*, Ballinger, Cambridge, Mass., USA.

Superconductivity and the Efficient Use of Electricity

Nils-Johan Bergsjö and Lars Gertmar

Abstract

The potential applications of superconductivity are listed and examined. Wide space use of room temperature superconductors is estimated to reduce electricity use by less than 7 percent from the 1986 level for total three different cases: Sweden, OECD Europe and North America. This reduction will not be feasible for 30–40 years at the earliest. A general transition to more efficient superconducting (S.C.) motors and to S.C. power transmission accounts for most of the estimated saving in electricity, together about 5%. A transition to S.C. generators could save an additional 0.5% and this is also the estimate for a transition to S.C. power transformers. No electricity savings worth mentioning seem feasible from S.C. magnet energy storage and distribution or from a general use of liquid nitrogen temperature superconductors.

1 Background

1987 was the year when Solid State Physics came into our living-rooms. We have never seen such an impact in scientific discoveries (1,2) as the interest for high temperature superconductivity. While thousands of scientists all over the world are competing on the possibilities to beat the latest temperature records, the man in the street is fed by the mass media with levitation trains, no loss energy conversion, no loss transmission systems and many other things that stimulate one's imagination.

But engineers and scientists working on applications have a more realistic view (3–7) of the possibilities and challenges with the new superconducting (S.C.) technology. For everyone working in the energy field, the new S.C. discoveries raise certain questions:

– what are the possible applications of the new S.C. materials?
– what are the realistic applications to be?
– what is the time-scale for converting into the S.C. era?
– what are the impacts on the global energy supply?

The following paper is intended to give some of the answers to these questions.

2 About Superconducting Materials

The first material to show superconductivity was Mercury (Hg) and this was discovered already in 1911 by the Dutch scientist Kamerlingh Onnes. The reason for the discovery was the invention of a machine for the production of liquid helium. Many of the pure metals were found to be superconductors. It was very soon clear that the S.C. effect was destroyed by even very weak magnetic fields. The current density was also low. About 1950, superconductivity was found in some intermetallic compounds, and at the beginning of the 60's, compounds like NbTi and Nb_3Sn (often referred to as type II compounds) were developed and showed very promising results concerning critical magnetic fields (H_c) and current density (J_c). Figure 1 shows the relation between critical temperature, magnetic field and current density for the best S.C. materials used today, including indications for the new high transition temperature (T_c) materials.

The most commonly used S.C. material is NbTi, even though its critical electromagnetical parameters are not as good as those for Nb_3Sn. NbTi is, however, very ductile and it is easy to produce conductors from this material. Nb_3Sn is very brittle and can only be used in film structure or by doing heat treatment after winding if used in a magnet.

The earlier superconducting wires had problems with so called "flux jumps" which is a phenomenon producing heat in the conductor. They also showed a hysteresis behaviour. This in turn results in transition to normal state in magnets at much lower current densities than expected. It was also impossible to use these conductors for a.c. applications. During the last 20 years, the material has been constantly improved. By drawing the wire down to diameters as small as 1 µm or less, the critical current density has been improved almost by a factor of 4 (see Figure 2) and the ac-losses have been eliminated. From the view of the electrical energy applications, the R&D work on the metallic low-temperature S.C. materials is well summarized in books by Foner & Schwartz (8) and by Bumby (9) and in reviews by others (10,11).

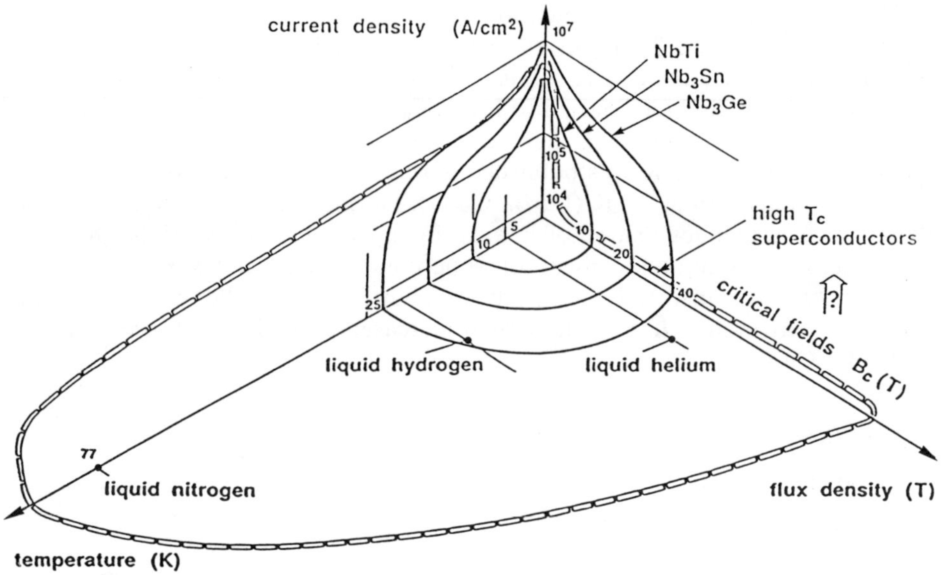

Figure 1. The relation between critical temperature, magnetic field and current density. To be superconducting, operational stresses must be inside the "onion"-sloped area. The outermost chain-like envelope is a projection of the new high-temperature super conductors.

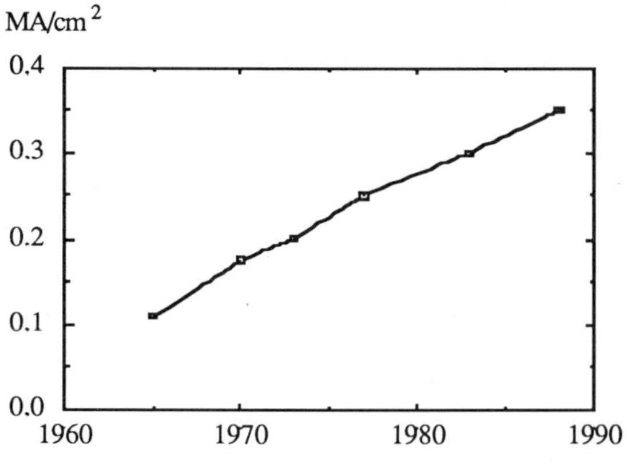

Figure 2. The improvement of the critical current density for NbTi-conductors.

The development of the transition temperature (T_c) has been constantly improved (see Figure 3). From 1973 to 1986, the best material was Nb_3Ge with superconductivity at 23 K. As is well-known to everyone interested in the field, Müller and Bednorz started to investigate the possible superconductivity in different oxide compounds during this period. Their first discovery of a compound with T_c>23 K was published during 1986 (1) with a first indication of superconductivity at 31 K (La-Ba-Cu-O). During the first months of 1987, there was a tremendous surge of interest in superconductivity, highlighted by Chu's and Wu's announcement of superconductivity at 93 K (2). These discoveries have inspired thousands of people around the world and the transition temperature has recently been increased up to 125 K for an alloy of Tl-Ca-Ba-Cu-0.

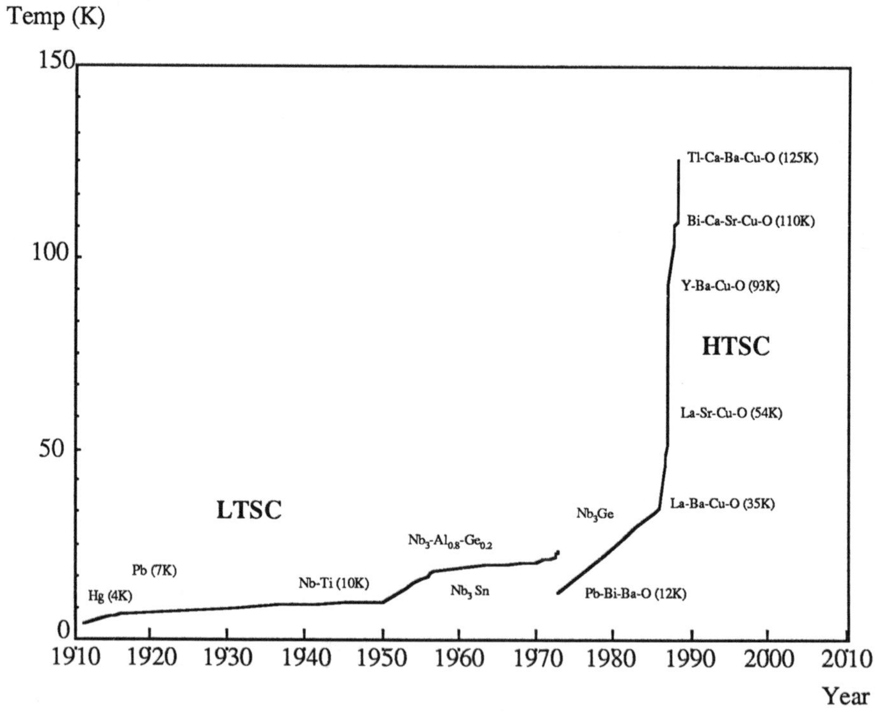

Figure 3. The development of the critical temperature for low and high temperature S.C. compounds.

Some efforts have already been made to produce wires and films from the new S.C. compounds for their practical uses, but there are a lot of hurdles to be overcome. A main hurdle for the electrical energy applications is the low values of the S.C. materials current density (12). A generator needs, for example, about 10^4 A/cm^2 at 2 Tesla. We are several orders of magnitude below that value in bulk S.C. materials, especially when the superconductor is located in a magnetic field. Some good results have, however, been reported for single crystal films, but it is not a practical way of using these for energy applications.

Another hurdle to be cleared is how to make wires out of S.C. ceramic materials, reducing the high contact-resistivity and the ageing problems. However, thousands of people are working on these problems today (12–15) so we are confident some very rapid improvements will be forthcoming. In the following discussions about applications, we have assumed that good S.C. materials are available.

3 Electrical Generators

3.1 General on Electrical Generators

Independent of the primary energy source, most of today's electricity is produced by synchronous generators. These are equipped with normally conducting copper windings and are working with a very high efficiency – approximately 98%. Before going into the S.C. alternatives, we will summarize the basics of electromechanical power conversion (16).

Electrical machinery is a collective notation for generators, motors and transformers. The first two machines are equipped with a rotating shaft and are used for conversion between mechanical and electrical power. The sizes of all three machines are essentially characterized by the active power (W) which they are rated to convert. A more correct measure is often the apparent rated power (VA) based on the product of the rated voltage and the rated current. The difference is not more than 10–20% and accounts for the losses and the power factor. In our study, we do not need to make a distinction between the two variables. In order to simplify matters, we always interpret apparent power (VA) as active power (W).

The power conversion of the rotating machines is based on an interaction between the air-gap flux density, B (Vs/m^2) and the armature winding current conductors, positioned close to the air-gap. The converted power (P) principally follows the relation:

$$P = \text{const}\ A\ B\ L\ D^2\ n$$

A is the current density (A/m) along the surface of the air-gap, D is the air-gap diameter, L is the length, and n the rotor speed (1/s).

In a superconducting machine, A and B could be considerably increased compared to normally conducting windings. Hence, for a given size, the rated power can be increased considerably by using the S.C. alternative. This is achieved at the same time as conductor losses are practically eliminated.

4.2 Earlier R&D-Activities on S.C. Generators

A large, normally conducting a.c. generator has losses which are about 1.5% of the rated power. The development of superconducting machines has so far been concentrated on eliminating the rotor losses by introducing a superconducting d.c. field winding. This will reduce the losses to about 1%. Once an a.c. superconducting armature winding can be introduced, the losses can be reduced by another 0.25% of the rated power, see Figure 4 for typical efficiency characteristics.

Rotating electrical machines are by far the biggest research area for large-scale electrical applications of superconductivity. The target is to reduce the losses thereby compensating for the increased cost of the S.C. solution.

The development of S.C. generators began in the middle of the 60's when the first superconductors of type II had been developed. Projects have been running in Europe, U.S. and Japan for about twenty years now (17–20). During the last ten years, the sizes of the prototypes have been increased from 10 MVA to hundreds of MVA (see Table 1). Studies of S.C. components especially rotors intended for prototypes at the level of 1000 MVA are also

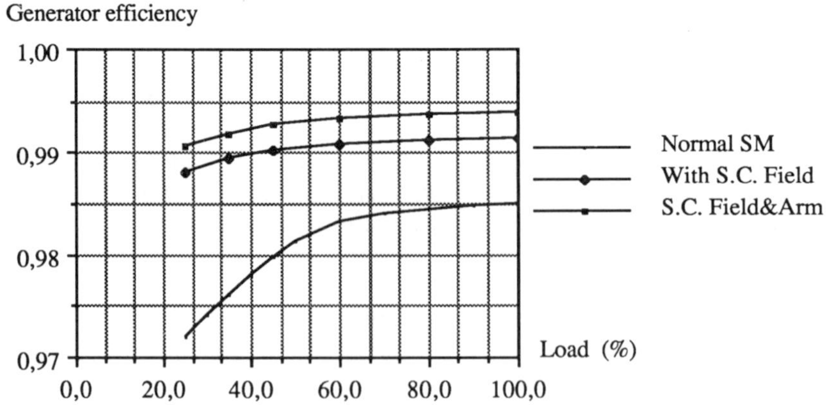

Figure 4. Typical values of synchronous generator efficiencies.

Table 1. List of experimental S.C. generator items with ratings above 10 MVA. The Super-GM project is run by Central Research Institute of Electic Power Industry, Japan.

Experimental generators

Manufacturer	Country	Generator rating	Status
MIT	USA	10 MVA	Gen. tests going on from 1984
All-Union Institute	USSR	20 MVA	Gen. tests up to 1981
GE	USA	20 MVA	Gen. tests up to 1982
Fuji & Mitsubishi	Japan	30 MVA	Gen. tests up to 1983
Hitachi	Japan	50 MVA	Gen. tests up to 1984
Kansai	Japan	200 MVA	Designed 1987
Super-GM project	Japan	70 MVA	To be built and tested 1992–5
Super-GM project	Japan	200 MVA	To be designed 1995
Electrosila	USSR	300 MVA	Gen. tests starting 1989
Westinghouse	USA	300 MVA	Terminated 1986, no completion
Int Res & Dev (NEI)	UK	1300 MVA	Design study
Asea Brown Boveri	Switzerland	Various	Design studies

Experimental rotors

Manufacturer	Country	Rotor diameter	Status
Toshiba	Japan	1000 mm	Rotor tests going on from 1985
Alsthom	France	1060 mm	Rotor tests going on from 1980
KraftWerksUnion	Germany	1170 mm	Rotor tests going on from 1986
Ansaldo	Italy	1200 mm	Rotor tests going on from 1984

performed. Figure 5 shows a proposed design of a 1000 MVA machine being a result from the Japanese "Moon Light Project" (21). These R&D programmes provide a good basis for cost evaluations of the S.C. solutions.

The cost evaluations give the results that we under different assumptions will obtain a lower system-cost (life-cycle-cost) for helium-cooled turbo generators when the ratings are at least above the 700–2000 MVA range. Very few generators are given such a high rating. The range of variation depends on the fact that the system availability is estimated differently and that the value of the lost energy production at shut-downs are based upon widely differing assumptions by the power engineering companies all over the world.

The results obtained so far show that there are no fundamental restrictions to develop and manufacture a.c. generators with 4 K superconducting windings. But they will give technical and economical benefits only at high power ratings and hence will become key components in very large power stations. It is therefore necessary to build full scale systems and run key

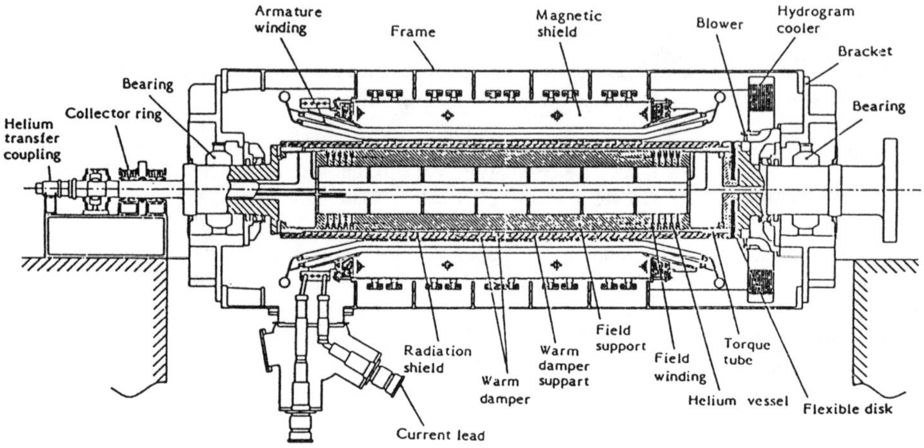

Figure 5. A proposed design of a 1000 MVA S.C. generator (21).

component projects in order to demonstrate their reliability. These activities are taking place in Europe, U.S. and Japan (17–21).

4.3 Future S.C. Generator Design and its Implementation

The new high temperature S.C. (HTSC) material breakthrough in attaining superconductivity well above 77 K indicates that the power rating cross-over range of 700–2000 MVA earlier mentioned will be reduced to 1/3, i.e. to above 250–750 MVA range. This means that it will be possible to work with liquid nitrogen and that S.C. generators will appear in the top of the power rating range where there is the greatest demand.

In order to develop these generators, national and international R&D-projects must be created and completed. As an example, Super-GM was established in Oct. 1987. It is a Japanese 500 million US dollar (MUSD) R&D-program with many industrial partners, such as Mitsubishi, Hitachi and Toshiba. It is aiming at a complete design of a 200 MW pilot generator by 1995. It is planned to be constructed and tested by 1999. An interim evaluation is planned by 1991–1992 before manufacturing and field testing a 70 MW model generator. MITI is applying 200 MUSD and the industrial participants budgets are 300 MUSD. The project includes a large portion of superconductor technology – both metallic conductors and oxide compounds.

The technical challenges of creating the breakthrough of superconducting machines are formidable. They cover a large number of strongly interconnected areas, such as:

- New superconductor winding materials and their:
  Cooling system design and thermal insulation
  Electrical insulation
- Electrical machine design and characteristics including:
  3-dimensional design of new electromagnetic circuits
  Transient electrical circuit analyzes
  Power system interaction
- Structural mechanics construction details
  Rotating or non-rotating items
  3-dimensional modal analyzes
  Vibration levels
- Electronic relays for system protection
- Materials science and testing

Low temperature physics is of course also essential as long as we are working with liquid helium or nitrogen.

The goal is to quarantee at least 25 years lifetime for a S.C. generator. Computer simulations and practical experiments on the different construction details in a full or reduced scale are promising and attractive development methods.

S.C. design will result in considerably reduced sizes and improved power system stability. It may also be possible to eliminate today's use of generator transformers.

Assuming that we have got the room temperature S.C. generator, what will the energy benefit then be? Estimations show that we will halve the losses of today's generators. There will still be the rotational losses and probably also a shielding iron circuit with minor iron losses. The sum of these today accounts for approximately 45% of the losses.

When studying the impact on generation, it is reasonable to use a loss reduction variation with primarily 40% and finally 60% of today's generator losses. The transition period will naturally depend on first the development period and then on the pay-back from the loss reduction and on the expected lifetime. Achieving 40 – 60% generation loss reduction will be twenty to thirty years late after the appearance of the S.C. generator. In section 8, we will use 40% loss reduction from a typical efficiency of 98.5% resulting in a generation loss reduction of 0.6% of the total electrical energy production.

4 Electrical Power Transmission and Distribution

4.1 Power Cables

The possibility of making superconducting cables was studied already during the 1960's. Practical experiments have been made in Austria, Germany and USA. There are some facts that have to be taken into consideration when discussing this application:

- Ninety-nine percent of power transmitted today is a.c. Transmitting d.c. requires converter stations at each end of the line which have a negative influence on the price. Some circumstances do justify such costs for d.c. transmission and this is when the lines are very long (lower losses), when connecting together systems with different frequencies, or for stabilization purposes.

- Over 90% of power transmitted is by overhead cables. Underground cables, necessary for S.C. transmission, are several times as expensive as overhead cables.

- The driving force for considering S.C. a.c. power transmission is the knowledge that about 8% of power generated is lost in transmission and distribution.

Work on superconducting power transmission lines began in the mid 1960's with a number of paper studies on economics of superconducting power transmission lines (SPTL). A typical study was made by IBM and published in 1967 (22). This presented a conceptual design of a very large (100 GW) line carrying d.c. current over a distance of 1000 km. 100 GW represented about 50% of the peak generating capacity of the US at that time. Also in Sweden, an unpublished study was made by ASEA comparing a 100 km long 500 MW S.C. cable with the existing "Kontiskan-cable" (between Sweden and Denmark). The S.C. cable appeared to be about 10 times as expensive.

In the early 1970's, there were several significant SPTL projects funded by different authorities in the US. The most well-known is a project at Brookhaven National Laboratory (BNL) studying a flexible a.c. line using Nb_3Sn conductor.

A summary of this project (lasting until 1986) gives a good overview of the state of SPTL (23). The work started 1972 and the technical choices which were made remain valid today for lines using conventional superconductors. These are:

- A conductor incorporating a Nb_3Sn superconductor,
- A flexible cable in a rigid insulated enclosure,
- Cooling by means of helium, and
- An electric insulation of lapped polymeric tape which is permeated by the coolant. The use of lapped tape insulation permits the cable to bend on a reel.

From 1972 through 1981, BNL developed all the necessary components of a SPTL. In 1982, they began testing of two short (115 m) cables. The testing and evaluation continued through 1986 at which time the project was completed.

Another very important study (24) was done by Philadelphia Electric Company (PECO) and funded by US Department of Energy (DOE). It involved a theoretical expansion of the Pennsylvania-New Jersey-Maryland interconnection to approximately three times its present size. The route was 106 km from an assumed Energy Park to a location inside Philadelphia and the transmitted power was assumed to be 10 GW. Fourteen different technologies were evaluated including different voltage levels and different cooling systems. The study was made in 1977 and was updated by BNL during 1987 including a 77 K superconductor. Since knowledge about the behavior of the HTSC conductors is poor, a number of assumptions were necessary:

1 The current density is equal to that of the PECO study which included Nb_3Sn, and the tape conductor can be made thick enough without fracturing.

2 The a.c. loss is the same as the loss measured in the BNL test line

3 A superinsulation is used as in the helium-cooled system.

4 The total refrigerator efficiency at 77 K is 15 times that at 7 K.

5 The conductor-price is the same as for Nb_3Sn.

6 The refrigerator cost remains unchanged and the cost of helium is eliminated.

Table 2 compares a conventional high-pressure-oil-filled pipe type, a combined aerial/underground installation and the 77 K superconducting case. The main conclusions that can be drawn from this analysis are:

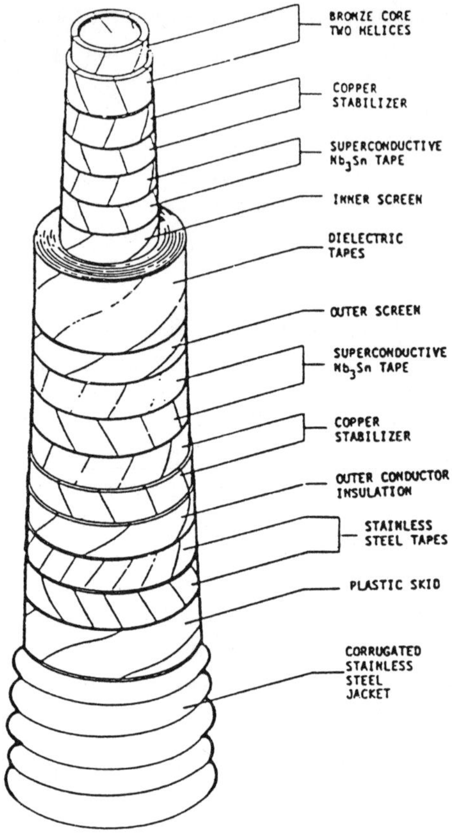

Figure 6. The Brookhaven National Laboratory cable (23).

1 The superconducting cable is 1.5 times the cost of the combined aerial/underground system when using HTSC cables.

2 Nb_3Sn cable was also studied giving a cost of 1.9 times the cost of system B. This indicates that HTSC cables may be about 25% cheaper than (LTSC) cables.

3 S.C. power transmission lines are economic alternatives to conventional underground lines only in large capacity installations (5000 MW or more).

4 The widespread use of S.C. cables for distribution requires room temperature superconductors.

Table 2 Comparison of conventional and S.C. 10 000 MVA cables (24).
System A is a conventional high-pressure-oil-filled pipe type.
System B is a combined aerial/underground installation.
System C is the 77 K superconducting case.

	System A		System B		System C	
Item	kUSD	%	kUSD	%	kUSD	%
Right of way	37,989	1	139,230	11	8,910	0
Clearing and roads	2,249	0	7,923	1	628	0
Foundations			27,000	2		
Towers			74,358	6		
Pipe	310,217	9				
Enclosures					415,048	21
Manholes	40,819	1			176	0
Excavation and backfill	311,604	9			150,661	8
Cable or conductors and devices	940,061	27	86,400	7	638,741	33
Oil	55,049	2				
Terminations	29,261	1			17,820	1
Monitoring systems					6,372	0
Pressure systems	4,565	0				
Engineering	15,246	0	13,500	1	9,450	0
Underground section			186,480	15		
Capitalized cable maintenance.	46,584	1	58,202	5	23,273	1
Refrigerators					263,772	14
Substations	256,585	7	229,500	18	193,606	10
Shunt or series compensation	495,497	14	29,322	2	30,780	2
Capitalized aerial trans. losses			189,425	15		
Capitalized cable losses	445,498	13	12,455	1	26,889	1
Capitalized substation losses	216,218	6	216,218	17	162,648	8
Capitalized compensation losses	324,270	9	1,339	0	407	0
Total	3,531,712	100	1,271,352	100	1,949,181	100

Source: Refrence 24.

4.2 Power Transformers

The possibility of making superconducting transformers was also studied in the early sixties. In the first such study, Mc Fee examined the feasibility of a S.C. power transformer in 1961 (25). It was already clear at that time that the a.c. losses in the winding was the problem. Mc Fee concluded that to minimize the "flux jumps" in the winding, the magnetic field intensity must be kept low. Several studies were performed between 1961 and 1977, but all concepts turned out to be non-economical because of excessive losses. In 1981,

Riemersma et. al. published "Application of superconducting technology to power transformers" (26). The conclusions of this study were:

1 The technology exists today to construct a S.C. transformer with regard to conductors, insulation and mechanical structure.

2 S.C. transformers above certain ratings are economically feasible in that their life-cycle-costs are smaller than those of conventional transformers. This advantage is derived from increased efficiency.

3 S.C. transformers have a weight advantage but the size is about the same as that for conventional transformers of the same rating.

4 Under fault condition the multi-winding S.C. transformer concept presents a thermal problem which requires additional work in order to find an acceptable solution.

Another problem which still exists is the reliability of the cooling equipment for liquid helium. It is not reliable enough for applications like power transformers.

In the beginning of the 1980´s, different publications indicated the possibility of reducing the a.c.-losses by producing ultra-fine filament superconductors from NbTi. J.H. Murphy at Westinghouse R&D Center published a paper (27) which indicated that the critical filament diameter was as low as 0.1 μm.

During the last few years, it has been possible to produce fine filament NbTi wires. A. Fevrier at Laboratoires de Marcoussis has published papers (28,29) about tests on S.C. power transformers and a.c. loss measurements on wires. The wire diameter is as small as 0.2 μm. The latest transformer which has been tested was designed for 80 kVA and the efficiency was 97.8%. The main contribution to the losses was from the current leads and the vessel. By recalculating the results for a 220 kVA transformer, an efficiency of 99.2% was obtained. According to these publications, the existing helium cooled ultra-fine filament superconductors result in loss saving and weight savings by at least a factor of 3.

A hurdle to overcome is how to avoid quenching in case of faults. In order to make an economic dimensioning of the transformer, the overcurrent can only be in the order of twice the operating current. This means that the fault current in the transformer would be dramatically reduced by the quenched windings. This is considered to be an advantage for the transformer but a fast switch would be required in series with such a transformer in order to avoid excessive heating of the winding after a quench.

It is to early to know whether the new htsc-material is usable for transformers. In a study by Fevrier (30), the losses for htsc-material have been calculated. The assumption is a filament-diameter of 10 μm and a value of the current density of 2x104 A/cm2 at 2 T. If it is possible to use such a material, it would be rather attractive to make a transformer. In table 3, a comparison is made between two designs of 1000 MVA three phase transformers (31–33).

The a.c. losses and the critical current densities are assumed in this study to be the same as presently available with the ultra-fine multifilament NbTi-wires. The production costs are presumed to be the same as for conventional transformers.

As can be seen from the table, the conductor losses are very small. This means that it is possible to increase the conducting area to avoid the quenching problems of the transformer. Of course it is very uncertain today to give a price indication of the superconducting materials so these parameters depends very much on the future behaviour of htsc materials.

The following conclusions can be made about S.C. transformers today:

1 The total weight of the transformer can be reduced by a factor of 6.

2 The total costs will be reduced by about 30%. The price of the superconducting material is still very uncertain.

3 The transformer has to be combined with a fast breaker and eventually a current limiter.

Table 3. Comparison of 1000 MVA three phase transformers (31–33).

		Conventional	HTSC–transformer
Investment cost	(M$)	9	9
Iron losses	(kW)	400	200
Current lead no-load losses	"	–	4.7[a]
Heat leak, dielectric losses	"	–	1[a]
Overall no-load losses	"	400	206
Cost of no-load losses[b]	(M$)	3.7	1.9
Conductor losses	(kW)	1600	6[a]
Current lead load losses	"	–	2.3[a]
Cost of load losses[c]	(M$)	5	0.03
Refrigerator cost	"	–	0.06
Fast power switch	"	–	0.6
Conductor weight	(tonne)	150	80×10^{-3}
Total weight	"	600	110

a Refrigerator efficiency 10 W/W included
b Presently worth 9000 $/kW
c Presently worth 3000 $/kW

The efficiency increase is about 0.17% for a 1000 MVA transformer and more for a smaller one.

When studying the impact on power transmission and distribution losses, it is reasonable to assume several transformers for transmission of power. The transition period will naturally depend on first the development period and then on the pay-back from the loss reduction and on the expected lifetime. Achieving mentionable loss reductions will be tens of years after the appearance of the S.C. power transformer. In section 8, we will use up to twice the value mentioned above resulting in a loss reduction of 0.15 – 0.3% of the total electrical energy production.

4.3 Magnet Energy Storage

Energy storage in huge subterranean coils is a potential application of superconductivity.

When there is no energy storage in use, the electrical power production must fully balance the consumption. As an example, the production must follow the day-and-night's and season's variations in demand as shown in Figure 7, for Sweden. The nuclear power plants are operating with only a seasonal variation. During the summer, the fossil fuel plants are very little used because there is a low demand for heat production. The hydro power is used for load-following.

For a country with a high amount of fossil fuel power plants, the introduction of an energy storage system gives the freedom to produce and to transmit a superfluous amount of energy during low-load periods. It can be stored diurnally for a later use at peak-load. The storage can be localized close to the big consumers of the power grid, e.g., large cities. This gives a transmission advantage as the power is made available close to the consumer. It ought to be noted that the building of any energy storage is an investment of tens of percent of a large-scale production unit but without any increase of the total energy production capacity of the power system.

A S.C. magnet energy storage (SMES) will have a high efficiency (>90%) and a high energy density compared to competing systems being built (<≈70%) or imaginable (35–39). In comparison with its competitors, it will have a very rapid response (<20 ms) to demands of storing or discharging. So, SMES can be used not only for load-leveling but also for load-following, for spinning reserve, for transient stabilization and for synchronous resonance damping. As an example, we are treating stabilization in section 5.4. SMES provides more freedom of power system operation than energy savings.

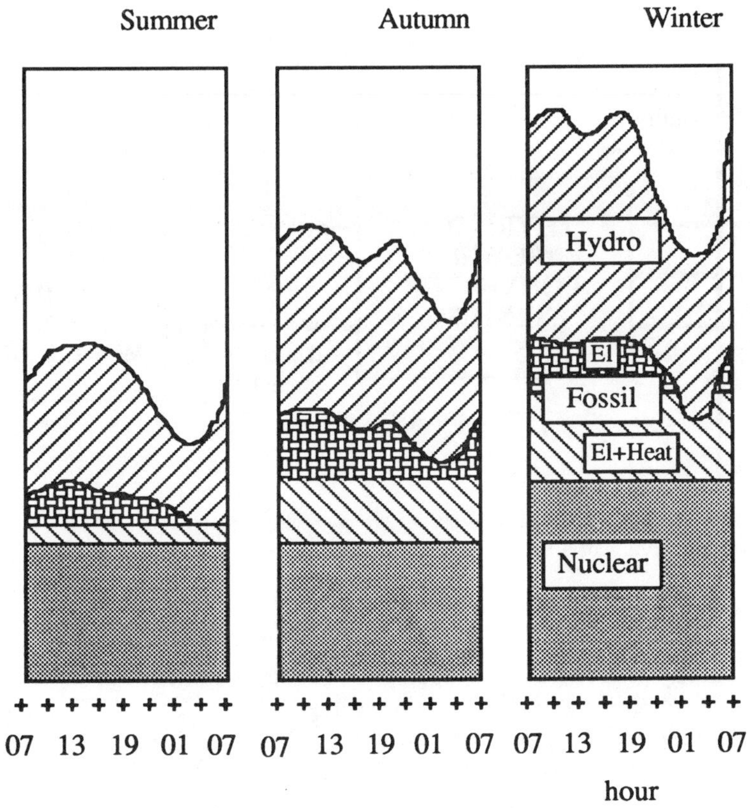

Figure 7. Power production seasonal variation during day and night in Sweden.

It is necessary to work with d.c. energy for the system realization. See Figure 8 for an overview. The electrotechnical relation is well-known:

$$W = L * I^2 /2$$

Thyristor convertors are transforming the a.c. power into d.c. and they are very efficient tools with small losses. They are well established in power electronics (34) with power ratings up to those of the HVDC transmissions. The thyristors are controlled so that a d.c. voltage across the coil is obtained. There is an increase or a decrease of the d.c. current following:

$$U = L * dI/dt$$

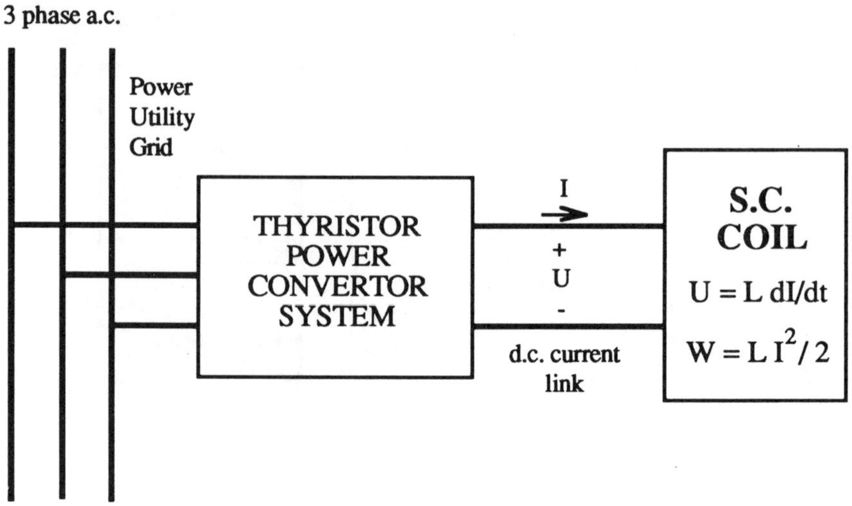

Figure 8. System overview of a S.C. magnet energy storage.

The stored energy will increase or decrease as long as the voltage is not equal to zero but will remain constant when the voltage is zero. The SMES system is almost lossless and very rapid compared to the competing systems in operation today. They are mainly based on pumped-hydro and compressed-air (35–38). Studies of battery storage are also made (36,39).

When using LTSC coils, the SMES must be rather large to be economical. An artist's view showing an example of a 5000 MWh toroidal coil design with a subterranean localization is displayed in Figure 9. In contrast to the solenoid coil, the toroidal design provides almost zero magnetic flux density on the earth's surface and in the air above the unit.

Research on SMES is mainly done in the US and in Japan (40–45). Programmes have been running for more than ten years. The work is aiming at a realization at the turn of the century. Some small plants, sized 10–20 MWh, are planned to be built as the first steps to the commercialization (40,43). As an example, two US R&D-teams have been selected by the US Government to develop bedrock NbTi designs for 30 MWh, operating with superfluid helium at 1.9 K and commissioning by 1992. The use will be for load-levelling, spinning reserve and transient stabilization but also for pulse power supply to Strategic Defense Initiative (SDI) lasers (45). The solenoidal coil will be 10 m high and 100 m in diameter, carry currents of 50–100 kA, and use 10 tonnes of NbTi.

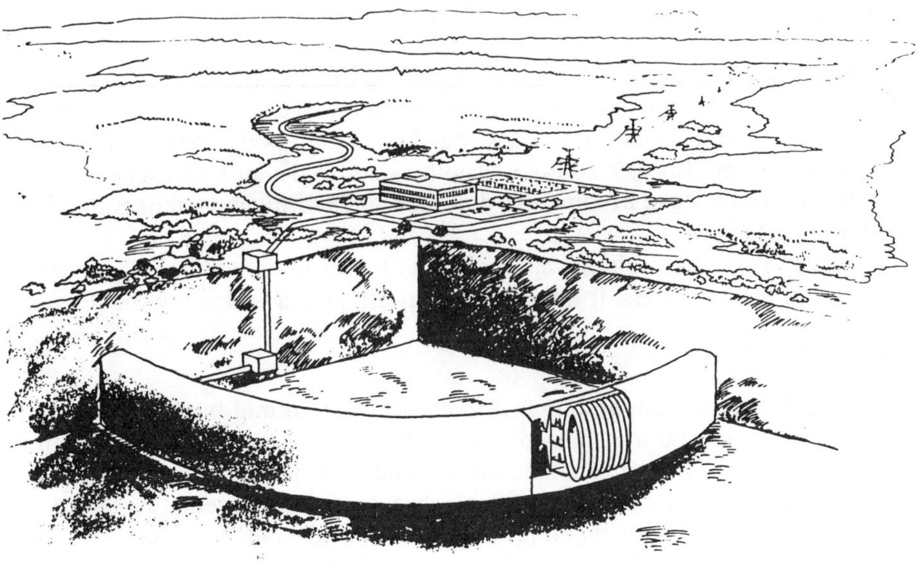

Figure 9. An artist's view of a S.C. magnet energy storage.

HTSC materials ought to reduce the cross-over energy rating to more reasonable values such as 500 MWh. This will open some new possibilities to those countries which have a demand for large energy storage. These countries are today producing the main part of their electrical energy from fossil fuel.

Sweden is a partner of the Nordel power system. In this system, power production is easily adjusted to meet the demand fluctuation by varying the hydro-power as can be seen in Figure 7.

The conclusion on large scale S.C. magnet energy storage is that it is feasible and highly efficient. We have no great demand for it in the Nordic countries, but there may be a demand in some other countries, e.g., the US and Japan. It will only marginally affect the power transmission losses. However, it will give electric utilities greater freedom to meet their peak load demand and thus reduce the mechanical spinning reserve and optimize the power plant operations.

The energy savings, due to a power system operation with magnet energy storage units, is 5 – 15% of the oil and gas used to generate peaking energy (6). The transition period will be tens of years and will depend on the pay-back from the loss reduction and from reduced investments in normal power production units.

4.4 Magnetic Power-System Stabilization

A much smaller amount of energy than in a diurnal SMES can be used to stabilize power transmission systems. The natural frequencies of the generator transmission system are only a few hertz. The system is damped by adding and extracting power. Energy storage systems sized only tens of megajoules are now being used instead of hundreds and thousands of megawatthours in the last section.

Let us analyze a simple power system with two lines in parallel, Figure 10. The system has no losses. It transmits a constant power following the relation:

$$P = E_1 * E_2 \sin \delta / X$$

A line fault appears. Lightning strikes close to a terminal bar at the generator end, for instance. The transmission line can no longer transmit power from the generator. The turbine accelerates the generator until the fault is cleared, i.e., the faulty line is disconnected after some tenths of a second. The power transmission can now be taken up again but the reactance is increased to 1.5 times its initial value, X. There is no longer a full parallel connection between the two terminals. In order to keep the power level, the angle δ between the two voltages E_1 and E_2 must now be increased. The system has to have a design so that it can reach the new working point in a stable way.

A S.C. coil can be an efficient tool from the fault appearance to the new stable condition. It can temporarily store a part of the surplus energy and dampen oscillations.

In this application, as with the diurnal SMES, the S.C. coil is fed via a thyristor convertor. The response time is again very short. Electric energy can be stored and discharged immediately if some base energy is kept in the coil. Thus, there exists a fast possibility to change the system impedance in an electronic way.

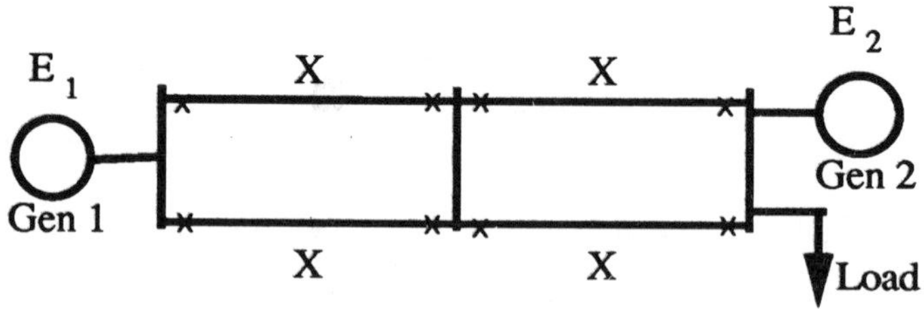

Figure 10. A power system with two lines in parallel. S.C. coils can be located either at the busbar close to generator 1 (Gen 1) and/or at the busbar in the middle (a) of the transmission system is the line reactance, and E_1, and E_2 are the electromotive force of the generators.

However, the tools other than S.C. coils for securing the power system stability after a fault are many. A short fault time, a rapid reconnection or short transmission line segments are the cheapest tools. The serial impedance is sometimes compensated by series capacitors in the middle of the line segment.

Damping of oscillations after a fault have already been mentioned. In general, stationary oscillations may appear without a prior fault in power transmission systems due to a low damping. This, too, is also a feasible application for S.C. coils. Bonneville Power Administration (BPA) has installed a S.C. coil – a pulse magnet – sized 30 MJ as an R&D project (47), but it was never taken into commercial use. It damps 0.35 Hz power oscillations in the high voltage AC transmission link called Pacific AC Intertie between Seattle and Los Angeles. These oscillations appear when the parallel high voltage DC-link is out of use. Japanese researchers are also working on S.C. power system stabilization (48).

Damping is today primarily achieved by so-called power system stabilizers acting on the field windings of the synchronous generators affecting the voltage and the power. As a second alternative, thyristor controlled shunt capacitors are used to vary the voltage and thus the power.

Damping by using S.C. coils is not a general solution, but if such a tool is developed, it will be used and be a companion-piece to the capacitor batteries now found in power system design. S.C. coils, however, will not reduce the annual transmission losses to any extent worth mentioning. The pay-off is obtained by reducing the allowable stability restriction and thus increasing the output power of the generators.

5 Electrical Motors

5.1 General on Electrical Motors

The asynchronous motor (AM) is the dominant industrial motor. It is often also called the induction motor. The alternating three-phase current is supplied to its stator winding directly and to its rotor winding by induction from the stator. Its characteristics are well described in textbooks, see reference 16. Here, we will focus on losses and efficiency. Typical values for asynchronous motor efficiencies are shown in Figure 11; the upper curve is valid for two-pole motors and the lower one for eight-pole motors. Four- and six-pole motors fall in between the two curves.

The motor losses are divided into three major parts: 1) stator copper losses, 2) rotor copper losses and 3) iron and rotational losses. A rough distribution of these is one-third for each group.

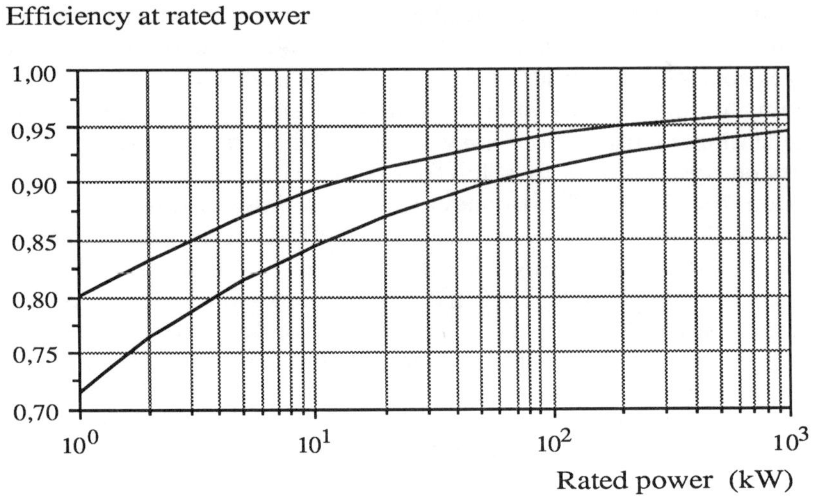

Figure 11. Typical values for total motor and drives efficiencies.

For household machines and hand-held tools, mainly rated less than a few hundreds of watts, we want to use a single-phase supply. Although they always have very low efficiencies 10–70%, they do not contribute more than 2–3% to the total motor energy consumption because they are used non-continuously, which can be derived from Table 8. We do not deal in details with these motors due to their small energy savings potential.

For industrial or tractional variable speed drives, the d.c. commutator motor fed from a thyristor convertor is often used, but the inverter-fed asynchronous motor is increasing its share. The d.c. motor drive typically has an efficiency curve like the lower curve in Figure 11. The inverter-fed AM drive has an efficiency like the constant speed AMs but reduced by 1.5 – 2.5%. All these values are given for rated power and rated speed.

5.2 Earlier R&D-Activities on S.C. Motors

R&D-activities on S.C. d.c. motors started contemporarily with the R&D on liquid helium cooled generators. It has been mainly directed towards the development of compact electrical motor drives, e.g., for ship propulsion. These R&D-programmes have not been as huge as those of a.c. generators. Notwithstanding the goal of developing much smaller machines, these R&D-activities have not reached a technical breakthrough either.

The largest technical difficulty is to design the current collector of the d.c. homopolar machine. The duty of this collector is to transmit the high direct current to and from the rotor.

Based on his own experience from the British S.C. motor R&D-programme, J.R. Bumby has authored a useful textbook (9) on the subject of S.C. rotating machines.

S.C. motors can so far provide a size reduction of 3/4, excellent speed control dynamics and higher efficiencies. A great number of d.c. homopolar motors have been successfully completed and tested. To our knowledge, there have been no extensive R&D-activities on S.C. a.c. motors due to the problems with a.c. hysteresis losses in superconductors.

5.3 A Future S.C. Motor Design and its Implementation

The first two S.C. motors – based on the Meissner effect in ceramic superconductors operating at liquid nitrogen temperature – have been built by Sanyo Electric and by Argonne National Laboratory, according to newspapers (49–51)! But as long as is needed a cryostat around the S.C. winding, there will never be a large number of S.C. motors and thus no essential reduction of the energy consumption.

Much of the material earlier written about establishing the future S.C. generator applies also to developing the S.C. motors. The observant might already have noticed that we partly used the notation "machines" in section 4. Keeping in mind that motors are sized from watts to megawatts, therefore, a lot more of experimental work than calculations will be applied to develop the small-sized S.C. motors. The expected lifetime of the motor can be accepted to be shorter than that of a S.C. generator. Due to the fact, that the asynchronous motor's starting characteristics are heavily dependent on its rotor's resistance losses, it is worth noting that the old asynchronous workhorse cannot simply be substituted by the same basic design but now equipped with room temperature S.C. wires. The fully S.C. motors must have, for example, a power electronic controller to take care of the start-up (52). By adding some cheap features to the electronics, this controller will also be capable of an energy optimization during the running operation.

Here, Baldwin's chapter in this book slightly overlaps our chapter. His, others and our opinion (53–55) is that the use of power electronic variable speed drives will increase with the increasing energy prices. Non-S.C. workhorses will be the a.c. motor and the d.c. motor fed from inverters and convertors respectively. The energy gain will be mainly achieved by reducing the process losses due to the elimination of the loss-suffering process control equipment, e.g., pump throttling valves. In order to take a full advantage of the power electronic evolution, the mature motor technology will primarily take into account increased maximal speed and/or the use of high-performance permanent magnet motors for further improving the drive's efficiency before going into a use of S.C. motors.

Assuming that we have got the room temperature S.C. motor, what then will be the energy saving for both the constant speed and the variable speed drives? Estimations show that the fully S.C. motor will reduce losses down to one-third compared to today's motors. There will still be the rotational losses and also some iron losses. The sum of these are today causing about 30% of the losses. Although the power electronic controllers are very efficient – 98% is a common value – their losses have to be accounted for in this analysis. A non-S.C. rotor winding makes for a simpler but not so energy-efficient motor that does not need a power electronic controller to take care of the start-up. This motor, i.e., a S.C. motor with a non-S.C. rotor winding, will have an estimated loss reduction of 1/3.

When studying the impact on the end-use of electricity of S.C. motors, it is therefore reasonable to use a loss reduction variation with three steps 33, 50 and 67% of today's motor losses. 67% is an estimation for the high-speed drives. The transition period is very difficult to estimate, but it will be in the range of ten to twenty years after the appearance of the S.C. motors.

6 Magnet Applications

Already the helium cooled superconductors have found a lot of magnet applications. In this chapter, we will comment on the different applications, specially those with some future impact on the energy production or consumption.

6.1 Magneto-Hydrodynamic Generators

Hot gases flowing at high temperatures and close to the speed of sound through powerful magnets can produce electricity in large quantities. The hot gases are produced by burning fossil fuels. It is a complicated process but does provide a clean and efficient way of burning coal and natural gas (56). See Figure 12 for a drawing of how a MHD generator works.

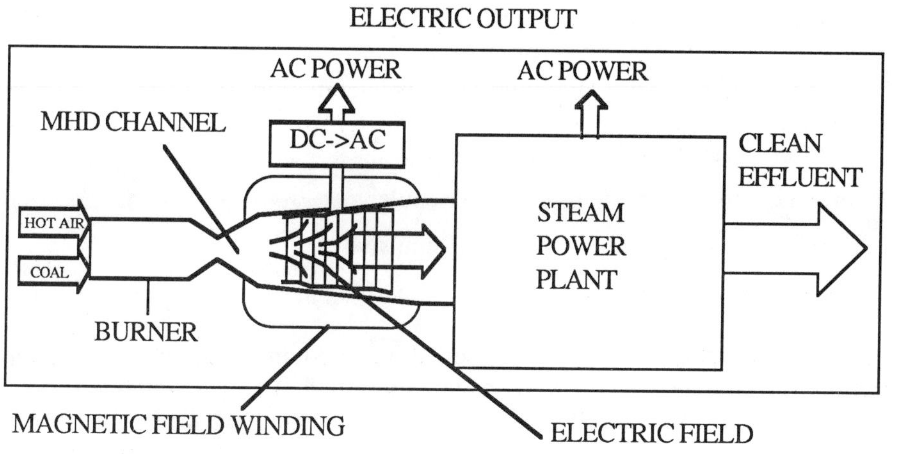

Figure 12. A principal drawing of a MHD-power plant.

Coal is burned in a combustor which produces electrically conducting gas and the gas passes through the magnetic field to generate electric power. After passing through the magnetic field, the remaining heat is used to operate steam turbine which produces additional electricity in a conventional way.

The countries that first began to develop MHD during 1960´s were USSR, US and UK. The aim was to build a MHD power station burning coal in preheated or oxygen-enriched air. Specially in the Soviet Union, there is a very ambitious program going on. The final goal is to produce 2000 MW MHD-generators. For the time being, the USSR is building a MHD rated at 270 MW and this is fed to a conventional steam turbine generator rated at 312 MW. The power station is scheduled to be completed 1993.

It is known, however, that there are a lot of problems with the materials needed to work at the high temperature. The MHD channel must stand more than 3000 K.

The reason for trying this technology is the possibility to increase the efficiency of converting heat to electricity from about 35 to about 50% while at the same reducing the pollution. Yet, it would be foolish to become to optimistic about the technology. The engineering difficulties are extreme. The hurdle is not to get superconducting material or to make the superconducting magnet. The problem is to develop the high temperature material needed.

Williams and Larson's chapter on electricity generation in this volume give details on competing systems which are much more easy to actualize.

6.2 Magnetic Levitation (Maglev)

The concept of magnetic levitation is to use magnetic forces to raise a train or other vehicle from the ground like a hovercraft. There are three different versions of magnetic levitations. These are:

- Permanent magnet systems (Repulsive)
- Electrodynamic system (Repulsive)
- Electromagnetic systems (Attractive)

The electrodynamic system has been developed in Japan by the Japanese National Railways (57). They have developed a very advanced system prototype which has reached a speed of more than 500 km/h. In Germany however, the development of the electrodynamic system was terminated after a very successful operation of a prototype. They have instead investigated a system called Transrapid which uses attractive forces for the levitation. For the time being, there seems to be a very limited use for magnetic levitation. The programs will continue in Japan and also in Germany, but it seems to have an economic interest only for the transportation in and out of the centers of the very big cities.

We do not foresee any large energy savings due to Maglev.

6.3 Fusion Magnets

The only energy sources which can meet the very long term worldwide needs for energy are solar, fission breeder and fusion. All these technologies have their drawbacks and they are all very complicated. The fusion technique seems to offer some long term advantages. Particularly the radioactivity of the fusion plant is much shorter lived than that of a breeder fission plant and there is no possibility of nuclear run away. Hence, there have been work on the fusion technique since the early 1950´s.

Even though it is a very long time to any commercial products, big superconducting magnets are already being used in for the different fusion experiments (58). There are two magnetic configurations which are thought to be capable of commercialization – Tokamak and Tandem Mirror. All commercial reactor designs involves magnets with very big diameters more than 10 meters. Test magnets with bores from 3.5 to 5 meters have been built and tested for Tokamak in the Large Coil Task at Oak Ridge and for Mirror Fusion systems in the Mirror Fusion Test Facility in Livermore. See Figure 13 for a conceptual design of the Tokamak reactor.

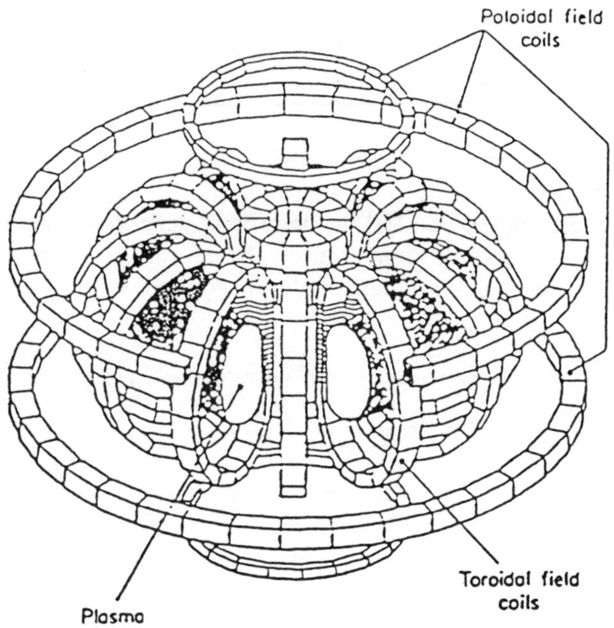

Figure 13. A conceptual design of a Tokamak reactor.

6.4 Other Magnet Applications

The most interesting application today is MRI (Magnetic Resonance Imaging) (59). It is a competitor to the conventional X-ray-technique for studying the inside of the human being. Several hundreds of these low temperature S.C. MRI-systems have already been built.

Other applications for high superconducting magnetic fields are:

- High energy physics
  where the magnets are used in very big accelerators.
- Magnetic separation
  which is a system where it is possible to separate magnetic from nonmagnetic materials.
- Laboratory magnets
  which have been build for several purposes. It is possible to reach very high magnetic fields with a combination of superconducting and normal water cooled magnets.

We do not foresee any large changes of the energy consumption due to these magnets.

7 Impact on Generation, Transmission and End-Use

The energy benefits from S.C. electrical equipment designs of two temperature levels can generally be listed as in Tables 4 and 5.

The saving due to a use of MHD-generator is very little depending on the S.C. magnet and its winding temperature. Therefore, we have omitted MHD from the impact discussion and treat it only in section 6.1.

Table 4. Estimated long term percentage energy yields due to a possible future use of room temperature superconductors.

Generation	
S.C. generators	0.6% of the total electrical energy production
MHD generators	Treated only in chapter 6.1
Transmission and distribution	
S.C. cables	<≈5% of the total electrical energy production
S.C. transformers	0.15–0.3% of today's transformer ratings
Diurnal SMES	5–15% of the oil&gas peak power energy production
S.C. power system stab.	Negligible
End-use	
S.C. motors	1/3 – 2/3 of today's motor losses
Magnet applications	
Medical and R&D	Negligible
Maglev	Negligible
Others	
Electronics	Not treated in this work
Computers	Not treated in this work
Processes	Only in cables, transformers and motors as above

When forecasting the long-term energy savings, we have chosen as a basis the existing energy production and consumption (60) shown in Tables 6 and 7 and we do not take into account other technology improvements in end-use.

Table 5. Estimated long term percentage energy yields due to a possible future use of liquid nitrogen cooled superconductors.

Generation	
S.C. generators	Large turbo generators as for room temperature superconductors i.e. 0.6% of rating, small turbos and all hydros negligible
MHD generators	Treated only in chapter 6.1
Transmission and distribution	
S.C. cables	Negligible
S.C. transformers	0.15–0.3% of today's largest transformer ratings
Diurnal SMES	5–15% of the oil&gas peak power energy production
S.C. power system stabilization	Negligible
End-use	
S.C. motors	Negligible
Magnet applications	
Medical and R&D	Negligible
Maglev	Negligible
Others	
Electronics	Not treated in this work
Computers	Not treated in this work
Processes	Only in cables, transformers and motors as above

When working on savings, we have found, like many others (53,54,61–66), that there is an imbalance between the knowledge of the electrical power production (Table 6) and distribution (Table 7) and final consumption. The high-quality statistics end with the utility's power energy meters in the consuming sectors, see Table 7. The differences between the totals of Table 6 and 7 are due to imports and exports of energy.

As an example of the lack of statistics, we have not found many modern lists of motor population by size and use. There is an American one from 1977, Table 8 (66) and a Finnish one written in Swedish and based on statistics from the late 70's (67).

Table 6. Electrical power production 1986 by energy sources (GWh) (60).

	Sweden	OECD Europe	North America
Hydro	61,250	429,484	604,454
Nuclear	70,243	634,530	510,147
Conventional thermal	7,080	975,860	1,982,750
Geothermal and other	6	3,183	10,944
Total (GWh)	138,579	2,043,057	3,108,295

Table 7. 1986's electrical energy consumption (GWh) by sectors (60).

	Sweden	OECD Europe	North America
ENERGY SECTOR	18,897	334,024	459,484
(of which:			
Transmission & distribution losses	10,647	136,026	211,619)
INDUSTRY SECTOR	48,148	766,066	895,975
(of which:			
Iron and steel	5,035	101,971	64,021
Chemical	6,312	170,620	184,499
Non-ferrous metals	2,371	87,652	103,266
Paper, pulp & print	17,681	82,327	153,036
Machinery	3,923	64,968	108,133)
TRANSPORT SECTOR	2,557	42,025	6,799
All to railways			
OTHER SECTORS			
Agriculture	1,300	239,411	9,069
Commerce and public	21,300	357,770	812,158
Residential	40,008	517,850	926,730
Non specified	1,737	3,620	–
Total (GWh)	133,947	2,050,766	3,110,215

We have estimated the energy savings for HTSC-motors on the basis of the US list, and we found the savings to be 3.6–6.8% of today's motor energy consumption, see Table 9. For motors, we will finally use 5% in Table 10. This is a little lower than the average of 3.6 and 6.8%.

Table 8. Electric motor population by sizes, numbers, etc (66).

	Electric motor size, horsepower					
	<1	1–5	5.1–20	21–50	51–125	>126
Total number, millions	660	55	10.5	3.3	1.7	1.0
Average life, years	12.9	17.1	19.4	21.8	28.5	29.3
Weighted average size, hp	0.26	2.1	11.9	32.5	86.7	212
Average efficiency,%	65	77	82.5	87.5	91	94
Average load, % of full load	70	50	60	70	85	90
Average capacity factor, %	3.5	7	17.4	27.7	37.5	43.2
Total annual use, TWh	30.5	33.9	103.4	155.2	337.7	573.3
Total annual motor energy 1234 TWh						

Table 9. Estimations of electric motor energy savings with high temperature superconductors, based on 1977 U.S. Electric motor population.

	Electric motor size, horsepower						Total energy savings in % of 1977's 1234 TWh
	<1	1–5	5.1–20	21–50	51–125	>126	
Average efficiency, %	65	77	82.5	87.5	91	94	
Total annual use, TWh	30.5	33.9	103.4	155.2	337.7	573.3	
Total annual use, TWh if the losses are reduced							
by 33%	25.9	30.8	96.6	148.1	326.9	561.4	3.6%
by 50%	24.0	29.5	93.5	144.9	321.8	555.6	5.2%
by 67%	22.4	28.3	90.6	141.7	316.8	549.9	6.8%

We have estimated energy reductions in Sweden, OECD Europe and North America due to room temperature S.C. equipment and liquid nitrogen temperature S.C. equipment. The results from the calculations are displayed in Table 10 and 11.

For room temperature S.C. equipment, and the results in Table 10, we estimate:

1) a reduction of the (transmission and) distribution losses to be 5% of the total electrical energy production in Table 6;
2) the generation losses to be reduced by 0.6% of the same amount; and
3) the energy consumption reduced by 5% of the 2/3 of the industry sector energy due to S.C. industrial motors.

Table 10. An optimistic estimation of the energy reduction due to room temperature S.C. electrical equipment by sectors. As examples, the distribution losses for Sweden are reduced by 5% of the power production, i.e. 0.05*138579 GWh = 6.9 TWh. The generator losses for OECD Europe are reduced by 0.6% of the power production, i.e. 0.006*2043057 GWh = 12.3 TWh. The motor losses for the US are reduced by 5% of 2/3 of the industry sector consumption, i.e. 0.05*2/3*895975 GWh = 29.9 Twh. The transport sector's savings are negligible, etc.

TWh (1986)	Sweden	OECD Europe	North America
ENERGY SECTOR	7.7	114.8	174.2
(of which:			
Distribution losses	6.9	102.5	155.5
Generation losses	0.8	12.3	18.6)
INDUSTRY SECTOR	1.6	25.5	29.9
TRANSPORT SECTOR	0.0	0.0	0.0
OTHER SECTORS	0.0	0.0	0.0
Total (TWh)	9.3	140.3	204.1
Total (% of 1986's value)	7	7	7

For nitrogen temperature S.C. equipment, and the results in Table 11, we estimate:

1) no reduction of distribution losses,
2) 1/10 of 0.6% for the generation losses, and
3) no S.C. motors in industry.

When looking at the energy reduction due only to a transition to liquid nitrogen S.C. equipment, we find very little benefits here even in the long term. The estimation is in the region of a fraction of a percent for each of generation losses and distribution losses.

A total generation, transmission and distribution loss reduction of 1.0% requires a full transition to S.C. generators and the establishing of S.C. power transformers based upon a.c. S.C. windings and at least a partial transition to them. SMES will only cause minor national savings as long liquid nitrogen S.C. windings are used.

Table 11. Estimations of the energy reduction due to liquid nitrogen S.C. electrical equipment by sectors.

TWh (1986)	Sweden	OECD Europe	North America
ENERGY SECTOR	0.1	1.2	1.9
(of which:			
Distribution losses	0.0	0.0	0.0
Generation losses	0.1	1.2	1.9)
INDUSTRY SECTOR	0.0	0.0	0.0
TRANSPORT SECTOR	0.0	0.0	0.0
OTHER SECTORS	0.0	0.0	0.0
Total (TWh)	0.1	1.2	1.9
Total (% of 1986's value)	0.06	0.06	0.06

It is noteworthy that today's distribution losses are given above as a fixed percentage (5%) and that room temperature S.C. cables account for the main part of the total savings of 7% shown in Table 10.

The time-aspects of superconductivity's penetration of electricity production and end-use are not encouraging. Superconductivity has now been a scientific area for 75 years. Yet, there are no large-scale energy applications in general use. It will take an estimated 20 years or more of R&D to attain a high-quality S.C. room temperature conductor suitable for electrical energy applications. And after that, additional years of R&D will be needed to develop the desired applications. Our opinion of a realistic time measure for developing these applications is 10 years. As a comparison, a Japanese survey on the prospects of superconductivity (68) shows a 40% probability for an electric power product development within 10 years after the discovery of a superconducting material operating at liquid nitrogen temperatures.

The S.C. materials supply situation will depend on what compounds will constitute the room temperature conductor. While working on this paper, a couple of new elements, such as Bismuth and Thallium have been brought to the scene. Today, there is no clear opinion on which materials will really be used.

The requirements of other construction materials are obvious. To work with liquid nitrogen temperature calls for insulation materials operating at this temperature. Most of the applications calls for a semiconductor convertor, but these and their control equipment are in no way critical parts.

8 Conclusions

Our conclusions are:

• S.C. cables and S.C. motors account for roughly equal amounts of energy savings – a few percents – but long-distance S.C. cables are not economical below thousand(s) of MW as long as room temperature superconductors are not available. Furthermore, a large use of S.C. motors is not feasible without room temperature superconductors.

• S.C generators and S.C. power transformers have potential energy savings of only about one percent.

• S.C. magnet energy streagh and S.C. power system stabilization offers more freedom in operating the power production units, but it does not result in any large energy savings.

• Magnetic levitation and other S.C. magnet applications are feasible but do not contribute to any large savings.

• S.C. motors offer the only significant opportunity for a more efficient end-use of electrical energy through the use of superconductors, although some of these savings can more easily be achieved by using high-efficiency constant-speed motors or by using variable speed drives based on power electronics.

• Use of room temperature superconductors could reduce electricity production in Sweden, OECD Europe and North America by 7% of their 1986 levels at most, but probably not earlier than 30–40 years ahead. With liquid nitrogen temperature superconductors, no national energy saving worth mentioning seems feasible.

References

1. Bednorz, J G., Müller, K A. 1986. Possible high T_c superconductivity in the Ba-La-Cu-O system, *Z. Phys. B 64*, pp. 189–93.
2. Wu, M.K., Chu, C.W., et al. 1987. Superconductivity at 93 K in a new mixed phase Y-Ba-Cu-O compound at ambient pressure, *Phys Rev Lett 58*, p. 908–.
3. Clarke, J. 1988. Small-scale analog applications of high-transition-temperature superconductors – a review article. *Nature*, p. 29–35.
4. Malozemoff, A.P., Gallagher, W.J., Schwall, R.E. 1987. *Applications of high temperature superconductivity*, IBM, a part of *Chemistry of high Tc superconductors*. Am. Chem. Soc. Symposium Series, Washington D.C.
5. Data management. *Management Today*, 1987, p. 1–31.

6. Superconductors could affect demand for oil and gas worldwide by end of the century. *Oil & Gas Journal*, 1987, p. 48–50.
7. Warm superconductors could have big impact on many industries. *Research & Development*, 1987, p. 60.
8. Foner, S., Schwartz, B.B (ed), 1974. *Superconducting Machines and Devices*. New Your: Plenum Press.
9. Bumby, J.R. 1983. *Superconducting rotating electrical machines*. Oxford: Clarendon Press.
10. Laumond, Y. 1985. *Superconductivity in electrical technology*. Alsthom Review: 31–40.
11. Sabrié, J-L. 1986. Cryoelectricity. Alsthom Review: 3–50
12. Küpfer, H, et al. 1986. Critical current and upper critical field of sintered and powdered superconducting YBa2Cu3O7. *Z. Phys. B - Condensed Matter 69:* 159–66.
13. *High temperature superconducting materials. The current situation.* ASM International, Metals Park, Ohio 1987.
14. Flükiger, R. 1987. *The material aspect in advanced superconducting wires*. Magnet Technology Conference MT-10 Boston.
15. Larbalestier, D.C. 1987. *Implication of the new superconductors for magnetic technology*. Magnet Technology Conference MT-10 Boston.
16. Fitzgerald, A. et al. 1983. *Electric machinery*, New York: McGraw Hill.
17. Lambrecht, D. 1985. *Overview of the development of superconducting generators*, Magnet Technology Conference MT-9 Zürich, pp. 273-76.
18. Kumagai, M. et al. 1986. *Development of superconducting a.c. generator*. IEEE Trans on Energy Conv, Vol EC-1, p. 122–29.
19. Fevrier, A. 1987. *Latest news about superconducting a.c. machines*. Magnet Technology Conference MT-10 Boston.
20. Maki, N. et al. 1987. *Development of superconducting a.c. generator*, Magnet Technology Conference MT-10 Boston.
21. *Feasibility study on S.C. generator and related technologies*. MITI Moon Light Report, Japan, 1987.
22. Garwin, R.L., Matisoo J. 1967. *Superconducting Lines for the Transmission of Large Amounts of Electrical Power over Great Distances*. Proceedings of the IEEE, 55(4): 538.
23. Forsyth, E.B., Thomas, R.A. 1986. Performance of the Brookhaven superconducting power transmission system. *Cryogenics*, 26: 599.
24. Philadelphia Electric Company (PECO) and US Dept of Energy (DOE).
25. Mc Fee, R. 1961. Superconducting Power Transformers – A Feasibility study. *Electrical Engineering*, 80: 754–60.
26. Riemersma. et al. 1981. *Application of superconducting technology to power transformers*. IEEE Trans. on Power Apparatus and Systems, PAS-100, 10: 3398–405.
27. Murphy, J.H. 1984. *Advanced applications of superconductors*. Advance in cryogenic engineering, 29: 133–40.
28. Fevrier, A. et al. 1987. *Preliminary tests on a superconducting power transformer*. Magnet Technology Conference MT-10 Boston.
29. Fevrier, A. et al. 1986. *220 kVA superconducting transformer*. Proc. Eleventh Int. Cryog. Eng. Conf., Berlin, West Germany 474.
30. Fevrier, A. 1986. *Latest news about superconducting a.c. machines*. Magnet Technology Conference MT-10 Boston.
31. Schauer, F. 1986. *Supraleitende Transformatoren*. Habilitation at the Technische Universität Graz.

32. Aichholzer, G., Schauer, F. 1984. *Coaxial turn transformer.* Proc Int Conf on Elec Mach (ICEM), 1: 58–.
33. Schauer, F. et al. 1987. *Assessment of potential advantages of high Tc-superconductors for technical applications of superconductivity.* Kernforschungszentrum Karlsruhe, Germany, KfK4308.
34. Thorborg, K. 1988. *Power electronics.* Prentice Hall Int'l Ltd UK.
35. Swisher, et al. 1983 *Energy storage technology.* Energy Systems Handbook, New York: Wiley.
36. 1985. *New electric power technologies: Problems and prospects for the 1990's.* Washington D.C.: US Congress, Office of Technology Assessment, OTA-E-246.
37. Schainker, R.B. 1985. *Compressed-air energy storage (CAES): overview, performance and cost data for 25 MW and 220 MW plants.* IEEE Trans on Power Appar. & Syst., Vol PAS-104: 791–95.
38. Smith, D.J. 1986. Utilities employ energy storage for load management. *Power Engineering*, pp. 28–34.
39. Rowan, J.W. et al. 1985. Zinc-chloride energy storage is approaching maturity. *Energy Storage*, pp. 45–47.
40. Loyd, R.J. et al. 1987. *Design advances in superconducting magnetic energy storage for electric utility load leveling.* IEEE Trans on Magnetics, Vol MAG-23(2): 1323–29.
41. Boom, R.W. et al. 1987. *Operational aspects of superconductive magnetic energy storage (SMES).* IEEE Trans on Magnetics, Vol MAG-23(5): 3275–77.
42. Yoshihara, T. et al. 1987. *Design study of SMES system using high temperature superconductors.* Magnet Technology Conference MT-10 Boston.
43. Hayakawa, K. et al. 1987. *Test plant as the first step towards commercialization of SMES for utilities.* Magnet Technology Conference MT-10 Boston.
44. Masuda, M. 1984. *A brief report of Japanese SMES.* Proc. of the Tenth Int'l Cryo. Eng. Conf., Helsinki, Finland, 31 July-3 Aug. Guildford, Surrey, England: Butterworth, pp. 641–44.
45. Chaffee, C.D. 1987. *Strategic Defence Initiative, Public utilities to benefit from SMES system.* Superconductor Week.
46. Christensen, G.S. et al. 1987. *Optimal control applications in electric power systems.* New York: Plenum.
47. Rogers, J.D. et al. 1985. *Operation of the 30 MJ superconducting magnet energy storage in the Bonneville Power Administration electrical grid.* IEEE Trans on Magnetics, Vol MAG-21: 752–55.
48. Mitani. Y, et al. 1987. *Fundamental study on electrical power system stabilization using superconducting magnet energy storage.* Magnet Technology Conference MT-9 Zürich pp. 353–56.
49. *Sanyo Electric develops superconductor motor using Meissner effect.* Comline Electronics, 1987, p. 1.
50. Experimental motor makes use of power of superconductors – Sanyo Electric. *Asian Wall Street Journal*, 1987, p. 8.
51. Force of superconductor drives an electric motor – Argonne Nat Lab. *Wall Street Journal 3 Star*, Eastern edition, 1987, p. 32.
52. Lipo, T.A. 1987. The potential for high temperature superconducting a.c. and d.c. motors. *Electrical Machines and Power Systems*, 13:373–85.
53. Baldwin, S.F. 1988. *The materials revolution and energy-efficient electric motor drive systems.* Annual Review of Energy, Vol 13.
54. Baldwin, S.F. *Energy-efficient electric motor drive systems*, a chapter of this book.

55. Brown, J.E. et al. 1984. *Electrical variable-speed drives*. IEE Proceedings, 131A: 516–58.
56. Carabetta, R.A. et al. 1987. *Status of the DOE open-cycle, coal-fired, MHD power generation program*. IEEE Trans. on Energy Conv. Vol EC-2(4): 549–55.
57. Kyutani, Y. 1987. *Recent progress by Japan Nat. Railways on MAGLEV*. Magnet Technology Conference MT-10 Boston.
58. *NET – Next European Torus – Status Report*. NET Report 51: 97–102.
59. Vetter, J. et.al. 1987. *A 4 Tesla Superconducting Whole-body Magnet for Imaging and Spectroscopy*. Magnet Technology Conference MT-10 Boston.
60. Energy statistics. *International Energy Agency – IEA*, Paris 1988.
61. Comnes, G.A. et al. 1987. *Efficient alternatives for electric drives*. Oak Ridge National Lab, ORNL/TM-10415.
62. Höltkemeier, P. 1988. *Rationelle Energieverwendung*. BWK Bd. 40: 131–35.
63. *Industrial electric motors and controllers*. Worley Consultants Ltd, Aukland, New Zealand, Report No PB-88-139589-XAB, 1987.
64. *Energi- och elanvändning (Use of energy and electricity) 1985-1997-2010*. Statens Energiverk, Stockholm, Sweden, 1987.
65. *Industrial energy use*. Washington D.C.: US Congress, Office of Technology Assessment, OTA-E-198, 1983.
66. *Classification and evaluation of electric motors and pumps*. US Dept of Energy – DOE, Arthur D Little Inc & Argonne National Lab, DOE/TIC-11339 or DOE/CS-0147, 1980.
67. Boije, H. 1982. *Kartläggning av de elektriska motorernas energiförbrukning och förluster inom finsk industri (A 'map' of Finland's industrial motor energy consumption and losses)*. Helsinki Univ. of Techn., Lab. of Electromechanics, Otaniemi, Finland.
68. Nakajima, A. 1988. *Preparing for the age of superconductors: Experts judge the prospects*. Nikkei High Tech Report, Vol III: 18.

Electronic Office Equipment: The Impact of Market Trends and Technology on End-Use Demand for Electricity

Les Norford, Ari Rabl, Jeffrey Harris and Jacques Roturier

Abstract

Electronic office equipment represents a fast-growing but poorly understood component of the "other" end-use in commercial buildings. We have analyzed available data and market projections, to characterize electricity usage and efficiency potential. Measured power use was typically 20–40% of nameplate ratings. Power use varies for similar equipment; in particular, desktop PCs use about ten times the power of equivalent laptop models. Typical daytime loads are about 10–20 W/m^2, roughly equal to lighting loads in a well-designed new office. Future load growth depends on many market and technical factors; U.S. office equipment electricity use in 1995 could range from 130 TWh ("market saturation," current technology, expanded use of computerized printing), to about 25 TWh if today's most efficient hardware and operating systems became the market norm.

1 Introduction

Since the 1973 energy crisis, there have been continued improvements in the efficiency of energy use in buildings, as well as steady gains in our ability to quantify the details of energy end-use. Yet, virtually all energy use studies and energy conservation programs leave an unexamined residual, called "miscellaneous" or "other." This may include commercial buildings with diverse or mixed occupancy, structures that are unusual or highly specialized, and equally diverse energy-using equipment. This is an important omission, because the "other" category can represent 20–25% (or more) of total energy use in a commercial building (1). Of special interest within the "other" energy uses in office buildings is electronic equipment related to the production, use, storage, or transmission of information. These information functions represent the principal energy-using "industrial process" in an office building, and the major use of energy for other than occupant comfort and lighting.[1]

One of the ultimate objectives of our research is to establish office electronics as a distinct, recognized end-use within the commercial sector. We are compiling and analyzing information from diverse sources on the market trends, characteristics, loads and usage patterns, and options for improved efficiency in both electronic equipment hardware, software, and operations. We rely, wherever possible, on measured data for both items of equipment and complete buildings. This analysis is limited to the office building subsector, and focuses on one of the fastest-growing components of office electronics: personal computers (PCs) and their immediate peripherals, such as printers and display terminals.

One recent report, based on a study for the Electric Power Research Institute (EPRI), identified computer loads as a potential source of about 125 TWh (1 TWh = 1 billion kWh) in added electricity consumption between the early 1980's and the 1990's (3, 4). This represents a rather dramatic 24% increase over the estimated 526 TWh of electricity for total commercial electricity use in 1982[2]; it is also over half of today's consumption for all commercial lighting (about 200 TWh). The same study cites estimates of power requirements for plug circuits in new, New York City commercial buildings (for all plug-in equipment, not just office computers) that range from 55–130 W/m^2 for offices and banks, to 215 W/m^2 for large computer service centers. These levels substantially exceed lighting power intensities of about 15–20 W/m^2 in a well-designed new office building. At present, there are no definitive, measured data on electricity loads and operating profiles for computers and other electronic office equipment, based on a statistically meaningful sample of U.S. commercial buildings. However, the fragmentary data we have seen suggest that these EPRI load-growth cases are high, for two main reasons: (1) the calculations rely on nameplate ratings rather than actual power requirements, and (2) the report does not account explicitly for potential efficiency improvements. We discuss both of these issues below.

[1] Exceptions to this include office buildings with elevators, and mixed-use buildings that combine offices and non-office uses such as restaurants, laundries, food stores, or light manufacturing. A recent analysis of several hundred California buildings shows that, with the exception of elevators and these specialized process loads, it is "information equipment" which represents the largest single component of the "other" commercial energy end-use (2).

[2] This EPRI estimate, extrapolated to 1982 from earlier data, is lower than the U.S. Deptartment of Energy data from its 1983 national survey of commercial buildings: total commercial sector electricity use of 656 TWh, of which 153 TWh was in office buildings (5).

The EPRI study, already widely quoted in the utility industry, has sparked debate among demand forecasters in utilities and regulatory agencies. The California Energy Commission recently compared its own staff's forecasts for "other" commercial sector electricity use with those made by the major California electric utilities. In a recent comparison, the two largest California electric utilities were expecting 1985–1999 growth rates for the commercial "other" end-use (not limited to office electronics) that were 1.5 times and 3 times those predicted by the California Energy Commission staff (6). In addition to utility demand forecasters, trends in energy use and load profiles for computers and electronic equipment are of growing concern to: standard-setting and enforcement organizations, facility managers and "shared-savings" contractors, and energy analysts and policymakers.

2 "Office Electronics" as a Commercial End-Use

One difficulty in writing about office electronic equipment is that there is no single, widely-accepted definition of what is included. This paper concentrates on general-purpose office electronics, such as PCs, workstations, and minicomputers with associated terminals, along with the computer peripherals used for data storage, inter- and intra-office communications, and other business functions. We discuss briefly some energy characteristics of other office equipment related to the processing and storage of information on paper – mainly copiers and telefax machines. In terms of energy use, it may be less important to determine whether the actual computing and data storage functions occur in a central computer or at the desktop, than to understand the factors affecting hardware efficiency for all types of computers, as well as the future intensity of usage for major types of input/output equipment such as terminals, printers, and digital scanners.

2.1 Market Status and Trends

The overall U.S. market for data processing equipment is about $80 billion per year (7). Peripheral equipment and copiers account for over one-third of total dollar sales, with mainframes, single-user PCs, and multi-user micros and minis each representing about 20%. Table 1 presents some published market-research estimates for 1986 stock and 1986 annual sales (including

Table 1. Estimates of U.S. stock and annual sales for selected electronic office equipment[a] (in thousands of units).

Equipment Type	1986 Stock[b]	Annual Sales[c]	
		1986 (est.)	1996 (proj.)
Single-User PCs	32,500	7,800[d]	
Business Use	(13,000)	2,600[d]	16,900
Home Use	(17,500)	–	
Schools	(2,000)[e]	–	
Text Processors	–[d]	660[d]	2,330[d]
Minicomputers[f]	375	90	220
Mainframes	2.4	9.9	13.8
VDT Monitors	23,000	7,600[g]	
Mono	(15,000)	–	
Color	(8,000)	–	
Printers and Plotters	21,000	6,100	17,500
Impact[h]	(18,000)	(5,000)[i]	
Non-impact	(2,000)	(900)[i]	
Plotters	(1,000)	–	
Office copiers	–	1,300	3,200
Fax machines	590[j]	130	320
Electronic typewriters		1,300	1,150

a Sales and stock estimates from different sources (see below) may use inconsistent definitions and assumptions; they should be compared with caution.
b Except as noted, 1986 stock estimates are from Ref. 11, citing unpublished data from Future Computing, Inc., Dallas, Texas. The stock data for PCs exclude multi-user PCs.
c Except as noted, estimated 1986 sales and (projected) 1996 sales are from Ref. 10.
d There are widely conflicting estimates of 1986 PC sales, due in part to differing definitions. Ref. 38 estimates sales of 7,100 K units, citing Dataquest, Inc., San Jose, Calif. A lower figure for 1986 PC sales, 4,800 K units, was reported in Ref. 39. Other estimates of 1986 PC sales reported in Ref. 12 are even lower: 2,000 K and 2,700 K units (pp 103, 104); these studies focus on market-share and thus may not include all sales. These last numbers are closer to the Computer and Business Equipment Manufacturers Association (CBEMA) estimate of 2,600 K units, which we show on the next line as PC sales for "business use." The CBEMA definition excludes PC's under $1,000, which leaves out many units purchased for home (recreational) or school use. CBEMA also has a separate category for dedicated text processors, (shown separately in the Table) which may be lumped with PCs in other stock and sales estimates.
e Ref. 41 estimates only 300 K units "in use" in schools during 1985–86, citing the National Center for Educational Statistics. However, the CBEMA data may exclude many units costing under $1000.
f Includes minicomputers and super-minis (costing up to $1.5 M), technical workstations, and multi-user microcomputers (costing less than $25,000).

g This estimate for monitors is for "PC terminals" (i.e., may exclude terminals used for multi-user minis and mainframes), as reported in Ref. 42. Note that Ref. 10 has a much lower estimate for 1986 sales of "general-purpose terminals," 1,890 K units, dropping to 1,480 K by 1996 due to "competition from low-end PCs."

h Including dot-matrix printers.

i Ref. 43 estimated sales by printer type; the total for 1986 was close to the total shown from Ref. 10: 5,900 K vs.6,100. Laser printers are the fastest-growing component of non-impact printer sales. Laser printer sales predictions for 1987 range from 80–400 K units, based on different market studies cited in Ref. 44. The authors of this compilation, predicted 1987 laser-printer sales at 160–200 K units, based on their review of other studies. However, this same firm was quoted in a May 1987 issue of *Publish* as predicting 1987 laser printer sales at only 81 K units.

j Fax stock estimates are extrapolated from 1984 estimates in Ref. 45.

replacements), for selected office equipment. According to Table 1, about 13 million out of the total 32.5 million single-user PCs were in business use in the 1986 stock, as well as 375,000 minicomputers and over 2,000 mainframes. (A single mini or mainframe may support a few to several dozen users.) Some business and technical users have both a PC and shared access to a mini or mainframe; the PC may be used either for separate applications or to log-in to a larger machine. A different source estimated that about 11 million PCs were in business use in 1987, representing about half of all 21.6 million "computer screens" used (7, 8). Table 1 shows that monochrome monitors still outnumbered color monitors by nearly two to one. Non-impact printers, including inkjet and laser printers, were under 10% of the stock but over 15% of new sales in 1986.

The projected annual sales for each type of equipment in Table 1 are significant – 25–45% of the 1986 stock. One recent market study reported that, of 2.8 million PCs planned as business purchases in 1987–88, about 10% of these were to replace existing PCs (9). A separate study projected business sales of 3.6 million PCs in 1986, with 22% for replacements and 78% for new or expanded use (8).[3] Annual PC sales were expected to rise, to 5.3 million in 1990 and 9.8 million in 1994, but with the fraction of new units dropping to 64% in 1990 and 55% in 1994 – indicating a "maturing" and increasingly saturated market for business PCs.

Another important trend, not obvious from Table 1, has been the tendency for PCs and other smaller computers to displace mainframes in many new and established markets. The number of PCs sold in 1986 was 2.5 times larger than in 1982. Over the same period, the number of mainframe units sold grew only slightly, from 2,100 to 2,400. The total stock of single-user PCs in the workplace increased more than tenfold, from 1.2 million to 13 million,

3 For comparison, an estimated 1.2 million office typewriters were sold in 1986 (10).

between 1981 and 1986 (11). Stromer (8) cites a forecast by Comtec that PCs, which in 1986 represented 51% of all computer screens in business use, would have 70% of the (much larger) business market by 1994.

Most analysts predict a continuation of the trend toward smaller computers, along with other trends that would both increase energy use (growth of laser printing) and decrease it (popularity of laptop PCs). One source estimates the following annual growth rates in dollar sales, from 1987 to 1990, based on a review and comparison of forecasts by other industry analysts (12):

6% for large computers (> $1 M),
11% for medium computers ($100 K – $1 M, 15–128 users),
27% for multi-user micros, workstations ($6 K – $25 K),[4]
17% for desktop PCs,
28% for laptops,
31% for non-impact (laser) printers (increasing from one-third to two-thirds of all printer sales), and
–7% for impact printers.

The same report identifies several near-term trends that may affect both computing power and energy intensity.

A significant shift toward laser printers, which require 5–10 times the power of impact printers (see Table 2 and the discussion, below) is perhaps the most dramatic trend. Some single-user PCs and workstations are configured with their own dedicated desktop laser-printer.

Color printers and plotters are increasing in popularity; estimated 1987 sales were 650,000 units, projected to increase to 1.1 million by 1990. Most of these are inkjet, dot-matrix, or pen-plotters, with relatively low energy intensities – whether monochrome or color. However, industry sources expect color laser-printers to be introduced soon, but targeted initially to mainframe applications because of the expected $30–40 K price tags (13).[5]

High-resolution and color monitors are gaining in popularity; both tend to be more energy-intensive than conventional monochrome units. High-resolution monitors may double their market share, to 4 million out of 14 million units sold in 1990. Color monitors use about 1.5–2 times the power of monochrome displays (see Table 2).

[4] No estimates were given for computers in the $25 K – $100 K range.

[5] Color copiers, priced at around $4 K, will be a necessary complement to most applications of color computer print-out. Both color laser printers and color copiers are three or four times slower than equivalent monochrome equipment (depending on whether they use a three- or four-color process). Energy requirements per page would increase by nearly this same factor, but the increase in overall energy use for color printing may be less, since many printers already have significant standby time (for laser printers, power requirements drop by about 50% in standby mode – see Table 2).

Conversely, low-energy-use laptops are expected to increase their share of new PC sales, from about 30% in 1986 to 40% in 1990, but there is uncertainty over how much their usage (for travel and take-home work) might displace vs. add to usage and connect-time by desktop PCs.

Desktop publishing systems are relatively new products that typically combine a powerful PC workstation or minicomputer, a high-resolution display, and a laser printer, as well as such other energy-intensive devices as optical scanners. The rapid growth in popularity of desktop publishing systems is expected to continue, with sales jumping from an estimated 84,000 in 1986 to nearly 550,000 by 1990. The market for desktop optical scanners themselves, only 10–15,000 units in 1986, may increase nearly tenfold by 1990.

2.2 Equipment Saturations in Commercial Buildings

Studies conducted by California utilities and the California Energy Commission collected detailed inventory data and nameplate ratings on computers and other miscellaneous equipment (R. Alereza, California Energy Commission, personal communication, and reference 2). The initial results for 855 commercial buildings in one California utility's service area showed an average electricity connected load for all "miscellaneous equipment" in large and small offices of 24 W/m^2. Computers (excluding other office equipment, such as copiers) averaged about half of this, or 11 W/m^2 (R. Alereza, personal communication). Newer (post-1975) large office buildings tended to have significantly larger computer equipment loads, on average; these loads were typically 16–22 W/m^2 (2). For other building types, the differences between pre-1975 and post-1975 buildings were mixed, showing no clear pattern.

The analysis by Alereza, Nguyen, and their colleagues included a set of "usage factors," based in part on an earlier American Society of Heating, Refrigerating and Air-Conditioning Engineers (ASHRAE) study (14). The usage factor was defined as the ratio of power consumption under "normal operating conditions" to nameplate rating; estimated values were 0.85 for computers and related equipment, and 0.8 to 1.0 for most other office equipment (except copiers, at 0.5). For computer equipment, these estimates appear high by at least a factor of two, based on the measured data to be presented in Table 2.

2.3 Design Requirements for New Buildings

The latest draft of ASHRAE Standard 90.1P, in discussing sizing criteria for space-conditioning equipment, recommends a value of 8 W/m^2 for receptacle loads in new offices (15). The latest California energy codes for commercial buildings allow 11 W/m^2 for miscellaneous plug loads, or more if special occupancy requirements are documented (16).

The limited amount of data available on new offices – and numerous anecdotes – suggest that current guidelines of roughly 10 W/m^2 for miscellaneous loads may understate actual daytime loads for at least some new office buildings, and for many existing or renovated buildings, as well. For example, in nine new office/mixed-use buildings in Northern California of 1,300 to 32,500 m^2 the (weighted) average load for miscellaneous plug-in equipment (not just electronic equipment) was 17 W/m^2, with a range of 9 to 24 W/m^2 (S. Taylor, Linford Engr., Oakland, CA, 1987, personal communication).[6] This was 85% of the (weighted) average lighting load of 19 W/m^2; equipment loads actually exceeded lighting loads in two of the buildings. For three of these buildings, totaling nearly 46,000 m^2, equipment load data were available for each individual tenant's space, as shown in Figure 1. For these buildings, miscellaneous electric equipment loads were higher than the values assumed in California's statewide building standard. On the other hand, we have found few examples of office buildings that require 55+ W/m^2 for computers only, as cited in the EPRI article for computer loads in new, large New York office buildings (4).

2.4 Typical Equipment – Actual vs. Rated Loads

We used a portable power meter to make short-term measurements of the actual power requirements for selected PCs, components, and peripheral equipment. We measured power in "active" operation and standby mode, and compared both measurements with nameplate ratings. Measured data are presented in Table 2 and Figures 2 and 3. Our principal findings are:

Nameplate ratings typically overstate actual measured power by factors of 2–4 for PCs and by factors of 4–5 for printers (see Figure 2). This is especially significant since most electricity demand forecasts and building design (HVAC sizing) decisions are based on rated rather than measured equipment loads.

6 Maximum daytime loads, including an assumed diversity factor.

Table 2. Rated vs. measured loads for selected equipment.

Equip. Type and Model	Features[a]	Rated[b] Power	Measured Power[c] Average	Peak	Avg. as% of Rated	Peak	Screen Only Rated	Meas'd
Desktop PCs								
IBM PC #1	2 FD	240	94.5	99.5	39	95	40	28.5
IBM PC #2	2 FD, 3 SB	240	91.2	97.8	38	93	40	26.0
IBM PC #3	2 FD, 2 SB	240	93.9	98.1	39	96	40	26.0
IBM XT #1	FD, 20Mb HD	440	111.0	114.5	25	97	40	30.0
IBM XT #2	FD, 10Mb HD	440	117.0	123.0	27	95	40	27.0
IBM XT #3	FD, 30Mb HD, 3 SB	440	115.0	120.0	26	96	40	27.0
IBM XT #4	FD, 10Mb HD, 2 SB	440	117.0	121.5	27	96	40	27.0
IBM AT #1	2 FD, 30Mb HD	500	166.0	169.5	33	98	40	30.5
IBM AT #2	2 FD, 30Mb HD, 3 SB	500	165.0	167.0	33	99	40	30.0
IBM PS-2/30	FD, 20Mb HD	12	76.0	78.0	36	97	74	32.0
IBM PS-2/50	FD, 20Mb HD, SB, Color	322	109.0	109.0	34	100	74	43.0
IBM PS-2/60	FD, 70Mb HD, SB, Color	580	172.0	172.0	30	100	92	48.0
IBM PS-2/80	FD, 70Mb HD, SB, Color	672	209.0	209.0	31	100	184	61.0
Mac-512	FD		39.5	43.0	92			
Mac-Plus	FD	60	31.4	34.3	52	92		
Mac-SE #1	FD, 20Mb HD	100	40.0	44.5	40	90		
Mac-SE #2	FD, 20Mb HD	100	45.0[d]	45	45	100		
Mac-2	FD, 40Mb HD, Color	358	130.0	133.0	36	98	128	66.0
Compaq Port.I #1	2 FD, 3 SB	192	97.0	100.5	51	97		
Compaq Port.I #2	2 FD, 2 SB	192	117.0	122.0	61	96		
Compaq Port.I #3	2 FD, 2 SB	288	80.3	82.5	28	97		
Compaq Port.III	2 FD, 20Mb HD, 2 SB	288	85.3	89.5	30	95		
Atari 1040ST	FD	118	46.2	48.6	39	95	43	31.7
AT&T Unix PC	FD, 20Mb HD	240	94.0	95.2	39	99		
Laptop PCs								
Zenith 181	2 FD	21	11.6	16.1	55	72		
Data General 1	2 FD	–	6.2	9.5	–	65		
Sharp 7000		–	29.4	30.5	–	96		
Toshiba 1100-Plus		–	10.8	12.1	–	89		

Equip. Type and Model	Features[a]	Rated[b] Power	Measured Power[c] Average	Peak	Avg. as% of Rated	Peak	Screen Only Rated	Meas'd
Impact and Inkjet Printers[e]								
HP Thinkjet		17	3.1	9.8	18	32		
Epson RX-80		50	9.7	24	19	40		
Epson MX-100		100	19.1	31	19	62		
Imagewriter II #1		20–180	13.0	23.0	–	57		
Imagewriter II # 2		20–180	11.2	41.0	–	27		
Okidata 83A		76	19.7	44.3	26	44		
Okidata 92		76	18.3	46.0	24	40		
IBM Proprinter		84	26.3	46.0	31	57		
Diablo 630		336	46.0	94.0	14	49		
Laser Printers								
HP 2686A		850	140.0	315.0	16	44		
Laserwriter Plus		531	129.0	233.0	24	55		

a Nameplate ratings are sometimes given in voltage and amperes; we assumed a power factor of 0.8 for computers and screens, and 0.7 for printers, plotters, and copiers (rated watts = volts x amperes x PF).
b Abbreviations used to describe equipment features: FD = floppy disk drive (5–15 W for a 360 Kb drive and controller card); HD = hard disk drive (25–55 W for a 10–30 Mb hard drive and controller card); SB = special boards for expanded memory, graphics, expanded calculation, printer port, or modem (5–10 W for most boards); Color = color monitor (monochrome is standard except as noted).
c Includes computer and monitor screen. Except as noted, measured power is an average for several short-term measurements using a Dranetz Model 808 Electric Power emand Analyzer. There was little variation across the short-term readings; accuracy of the averages reported in the Table is about $\pm$ 2 W.
d Measured with a conventional watt-hour meter, by clocking 1 revolution (1.8 Wh).
e For printers, the "average" column refers to the power draw when in standby mode (not printing); "peak" refers to typical usage when printing, typically 2–3 times the standby power. In addition, printers in use have pronounced (up to 80–700 W) short-term power oscillations, on the order of seconds, probably due to drum heaters and paper feeds. These are not included in our "peak" values. The overall average power required while the equipment is turned on will thus be between these "average" and "peak" numbers, but is difficult to estimate since it depends on the printing demand. This last number, if available, would be more directly comparable to the "average" shown in this Table for other equipment.

The Thinkjet is an inkjet printer, with notably low power requirements. The Diablo is a daisy-wheel impact printer; daisy-wheels are fast being replaced by laser printers. The rest are dot-matrix impact printers.

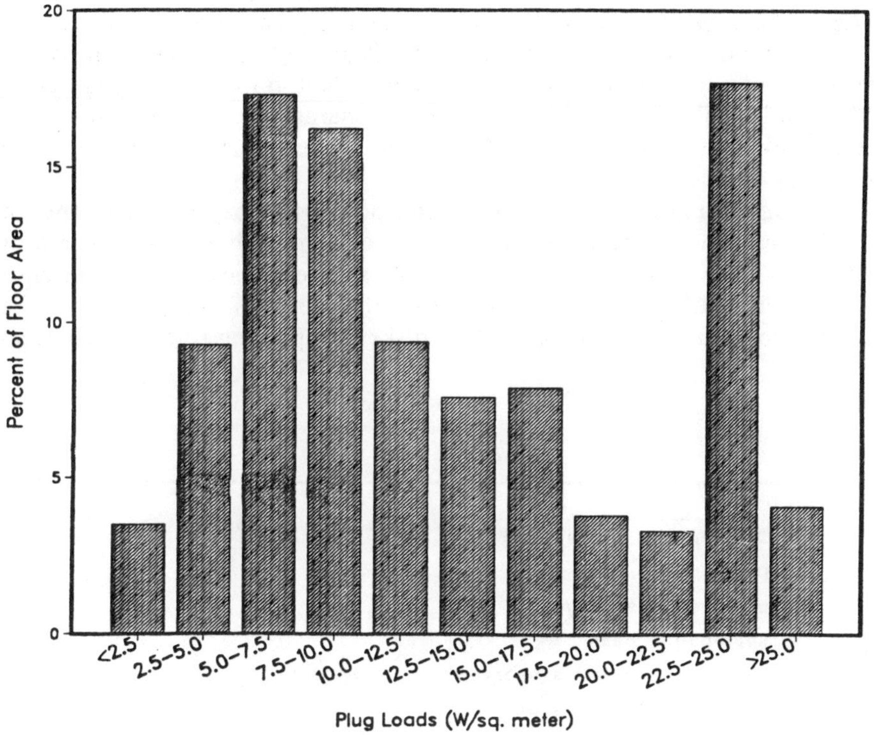

Figure 1. Distribution of tenant equipment load intensities in 3 new mixed-use commercial buildings in Northern California. The mean value is 14.1 W/m^2, the median, by floor area is 11.2 W/m^2. These maximum daily equipment loads, for all plug-in equipment (not just office electronics) include an assumed diversity factor. Source: S. Taylor, Linford Engr., Oakland, CA, 1987, personal communication.

There are significant variations in power requirements for equipment in the same general category: laptop vs. desktop PCs (Figures 2 and 3) or impact vs. laser printers (Figure 2).

For PCs, in particular, the most recent models tend to provide more computing and storage per watt of power.

Once turned on, PCs are essentially a constant load; power requirements for printers vary by a factor of about 2 between standby and printing modes.

Here are further observations for each type of equipment reported in Table 2:

Desktop and laptop PCs. Nameplate power ratings typically overestimate actual power consumption by a factor of 2–4. There is also a large range in measured power use: over 20:1 between the lowest-power laptop models (incorporating liquid-crystal displays and CMOS type logic boards) and the most powerful desktop workstations. Newer models with equivalent performance are often more energy-efficient.

PC components. Figure 3 shows measured power for major components of four IBM desktop PCs and one laptop. Use of floppy disk drives increases power by 0–7 W. This is the only noticeable source of power variation once the computer is turned on; it is so small that today's computers can be considered constant power devices. Hard disks are spinning (drawing constant

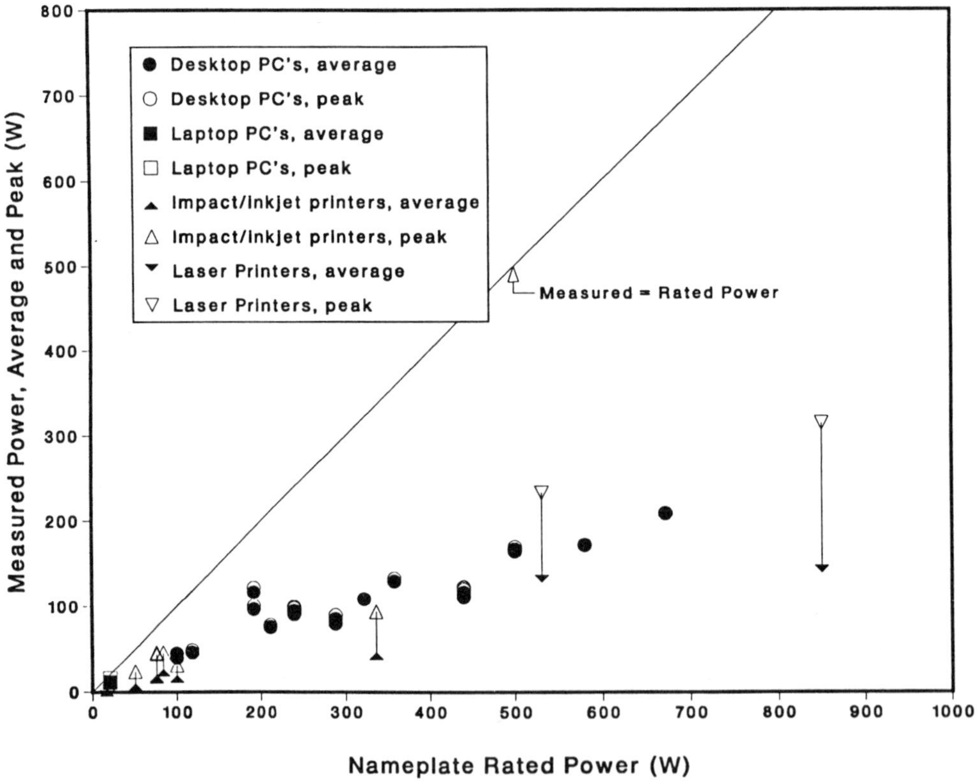

Figure 2. Measured vs rated power for PCs and printers. Data plotted from Table 2 show the range in power requirements for similar types of equipment and the large discrepancy between nameplate ratings and actual measured power (average while in use).

power) whenever the machine is on. Power for hard disks does not appear to scale with storage capability, at least in typical PC sizes. Manufacturer's specifications for a current-generation hard disk drive (17) rate a 65 MB half-height drive at only 11 W, less than the 15–54 W we measured for several older 10–30 MB, full-height drives. Power supplies for the five desktop units we measured required 11–25 W. The power supply for the Toshiba laptop used about 4 W (one-third of the total). Integrated circuit boards (including the disk controller cards) accounted for 28–42% (32–71 W) of the total PC power requirements for our IBM models. In a later section, we discuss two major

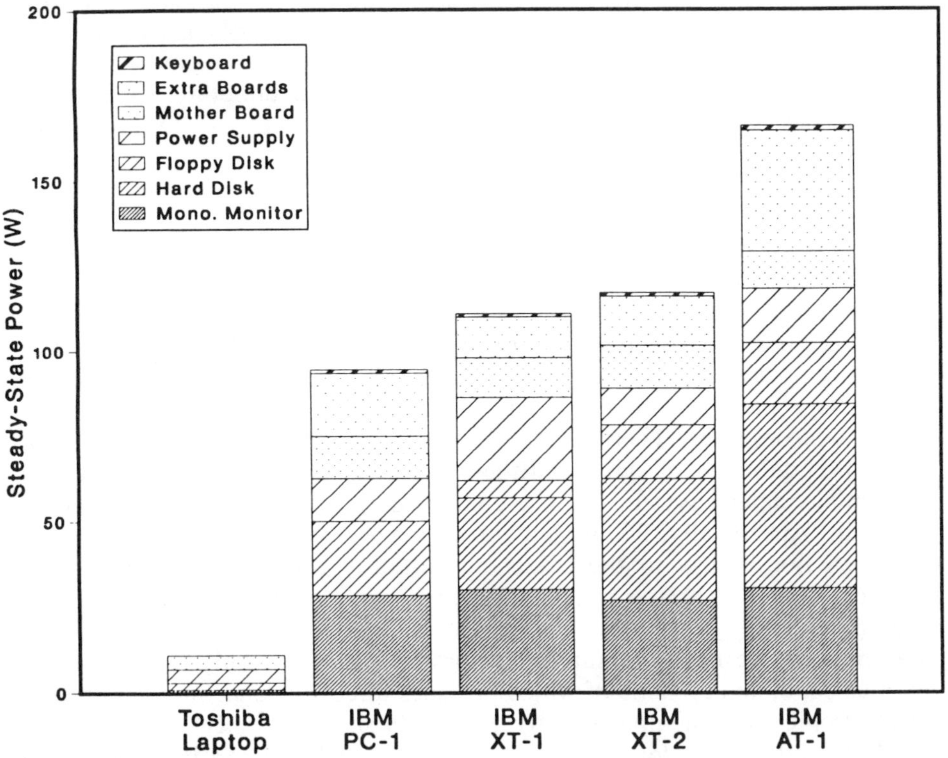

Figure 3. Measured component power requirements for five PCs. Measurements were made using a Dranetz Model 808 Electric Power emand Analyzer. Component power was determined by disconnecting each component and noting the power drop. These averages of several short-term measurements are accurate to about +/– 2 W. Comparisons with other PCs are shown in Table 2. Source: Field measurements by Princeton Univ. CEES.

technological trends affecting energy efficiency of computer logic boards: continued miniaturization and the shift to low-power transistors.

Monitors. Monochrome monitors for IBMs and the Atari required 26–32 W. Color displays are much more energy intensive, requiring 43–66 W. Both monochrome and color screens show essentially no power reduction when the screen intensity is turned down, or when a "screen-saver" feature is enabled. The typical monochrome monitor using video-type (CRT) technology requires two to three times the power of a complete laptop computer and as much power as an entire Mac-Plus, which also has a CRT display. On the other hand, the LCD screen for our Toshiba laptop used a negligible amount of power; we could not detect a change when disconnecting the screen.

Printers and plotters. Table 2 shows a variation of about 2:1 between average and peak power for printers – much larger than for computers.[7] As with computers, the nameplate power ratings for printers are very conservative (by a factor of 4:1 or 5:1), and there is a wide range of measured power levels: a 10:1 ratio for inkjet vs. dot-matrix printers, a further doubling for a daisy-wheel impact printer (Diablo 630), and another increase by two to three times for small laser printers.

2.5 Usage Profiles and Energy Consumption

Electric utilities are primarily concerned with the growth of on-peak loads due to electronic office equipment. This makes it important to understand the actual usage patterns that determine daily load profiles, as well as annual energy consumption. There is much speculation about daily profiles for electronic equipment usage, but little measured data. Most mainframe and many minicomputer systems operate continuously; some may be shut down on weekends. Anecdotally, PC usage patterns range from those consistently shut off when not in use to those left on virtually all the time. Control of peripherals, such as a laser printer shared by several workstations, is probably much more lax.

At present, we have measured load profile data from only a few sources (Figures 4a, b, and c), none of which provides direct evidence on load profiles for electronic office equipment only. Figure 4a shows a typical weekday load profile for all "tenant loads" (including both plug loads and lighting), for a

[7] For printers, "average" power is defined as standby (non-printing) mode; "peak" power is measured during a print cycle. Many printers experience brief (10 second) power peaks during warm-up: about 80 W for the inkjet and dot-matrix printers, and 500–700 W for the laser printers.

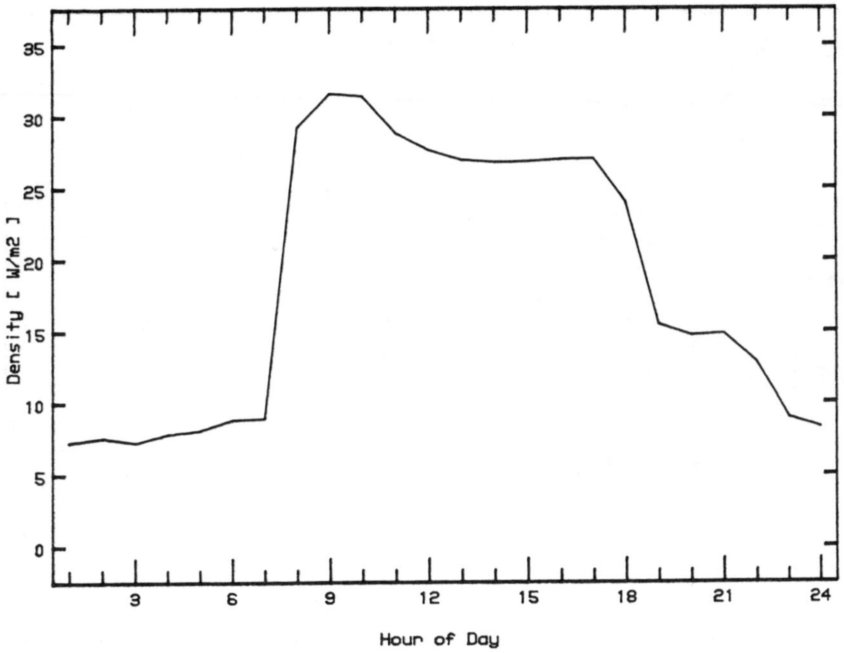

Figure 4a. Typical daily load profile – miscellaneous plug loads in a New Jersey leased office building (12,100 m^2). Hourly submetered tenant lighting and general-purpose office equipment loads for a typical spring weekday, 1986. Lights account for about 16 W/m^2, and are almost all turned off when the cleaning crew leaves at 10 p.m. Metered data exclude the central computer facility (40 kW, or about 3.3 W/m^2) and kitchen equipment. Miscellaneous office equipment appears to be a nearly constant load, at 9–11 W/m^2. About 40% of the office equipment is running at night. Source: Field measurements by Princeton Univ. CEES.

medium-size leased office building in New Jersey. Lights were not metered separately, but power levels for individual fixtures, combined with a count of the fixtures, gave a power level of 15 W/m^2 for general purpose lighting and 1 W/m^2 for task lighting. From on-site visits, the lighting levels at night are estimated to be 1–2 W/m^2. The remaining tenant load due to office equipment is thus about 15 W/m^2 at 8 and 9 a.m.[8] At night, office equipment power drops to about 5–6 W/m^2. We conclude that approximately 40% of the office

[8] A central computer facility, with a total load of about 40 kW, would be equivalent to about 3.3 W/m^2 if it were allocated to the total floor area.

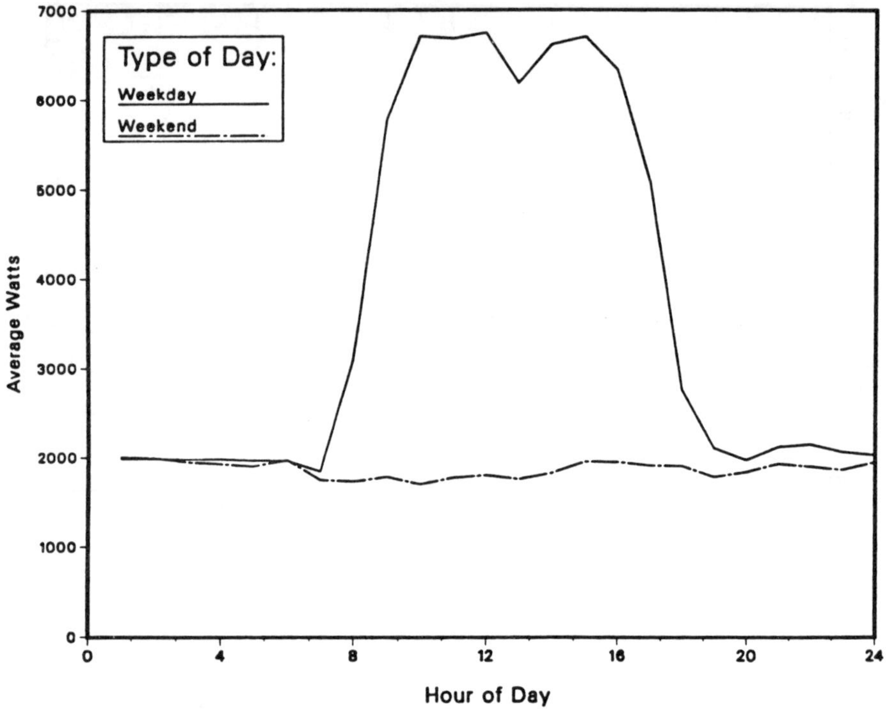

Figure 4b. Average weekday "plugs" load profile, small office building. These data show weekday average hourly loads for April 1988 for a recently-constructed, two-story building of about 850 m^2. Computers and other office electronic equipment represent most of this load; not included is a separately metered minicomputer, which would add about 10% to the loads shown. Data come from monitoring equipment installed by Battelle Pacific Northwest Labs.

equipment is left running at night. Some of this equipment is needed for remote communications, but the majority, including word processors and copying machines, appeared to be powered up unnecessarily.

The second data source (Figure 4b) comes from end-use monitoring of a small office building (about 850 m^2) in the Pacific Northwest. Based on a detailed equipment inventory, about 90% of the plug loads shown are due to electronic office equipment.[9] Maximum load, about 7.9 W/m^2, occurs from

9 Not included in Figure 4b is power for a separately metered minicomputer, with a daily load profile similar to the plug loads. If included, it would add about 10% to the plug loads shown.

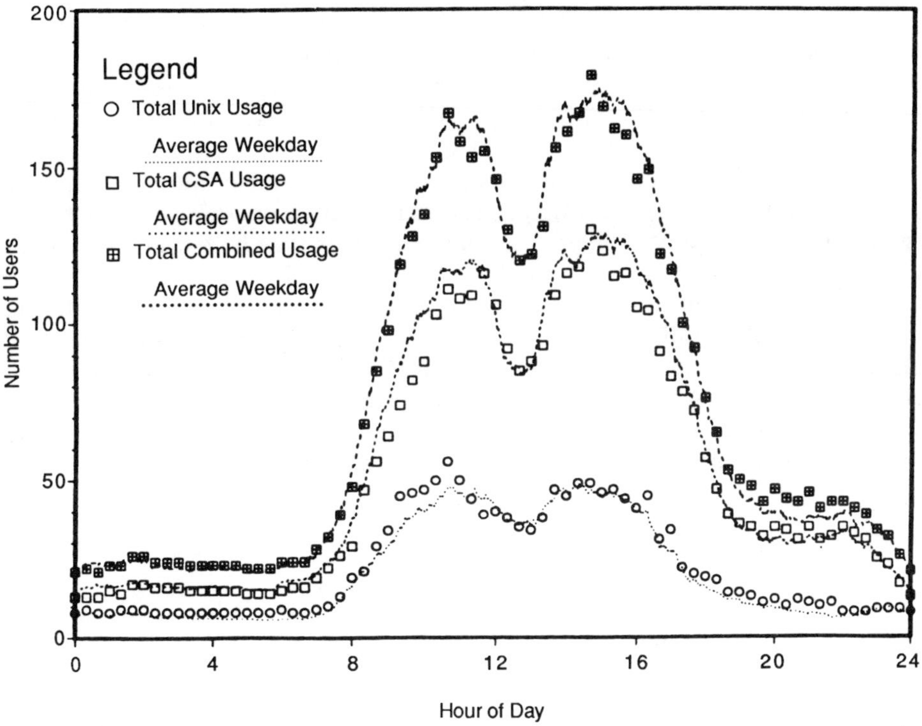

Figure 4c. Typical weekday profile of logged-in users at Lawrence Berkeley Laboratory, for networked Unix and VMS computers. Note the lunchtime drop in usage. (Some terminals may be left on even when the user is not logged in.) Usage patterns vary little over the five weekdays; weekend use on the two machines is typically 40% of the weekday usage on Saturdays, and 20% on Sundays. Source: System accounting data from LBL.

9:00 a.m. to 4:30 on weekdays; night and weekend power is fairly constant at about 30% of the daily peak.

A third data source comes from log-in records for a bank of eight minicomputers at a DOE National Laboratory (LBL), connected through a local network to users with "dumb" terminals. Once again, the data provide only an indirect indicator, since they show the number of logged-in users rather than electricity load. Figure 4c shows a typical weekday profile of logged-in users. There is a drop at lunchtime and well-defined morning and afternoon peaks, in contrast to the "block" profile of total tenant loads in

Figures 4a and 4b, which characterizes some other occupancy-driven (scheduled) office end-uses such as lighting. Usage patterns are quite stable for the five weekdays; Saturday usage is typically 40% of the weekday average, and Sunday about 20%. This usage profile measures computer use more directly than Figures 4a and 4b, but does not include peripherals such as printers and disk drives, which (along with the central minicomputers) are constantly on. At LBL, user charges for connect-time provide some incentive to log off the central computers, but some terminals may be left on even when the user is not logged in.

In the future, improved data will come from continuous, on-site monitoring of commercial buildings, such as the forthcoming data from the ELCAP project (Electric Loads and Conservation Assessment Project) sponsored by the Bonneville Power Administration. Even with careful submetering, however, it will be difficult to keep track of rapid turnover in office equipment stock, and to isolate electronic equipment from other office plug loads.

3 Energy Efficiency Opportunities

Major energy-efficiency opportunities exist both in hardware and system operations to reduce expected load growth for office equipment. Some of these improvements may occur naturally, in response to technology advances or such market pressures as user demands for increased PC power without use of a fan, reduced desktop "footprints," and improved portability). Others may need encouragement.

3.1 Potential Hardware Improvements

Advances in microelectronics will continue to reduce electrical power requirements for a given level of computing or storage capability. The potential reductions in power are so great that, at the least, forecasts of increased computing power should be decoupled from electrical load growth.

The power consumed by an electronic chip depends mainly on the type of transistors used. Laptop computers which must be capable of extended battery operation rely on "complementary metal-oxide semiconductor" (CMOS) technology, rather than the "n-channel metal-oxide semiconductors" (NMOS) used in desktop PCs. Both types draw similar amounts of power when a transistor is switched, but CMOS requires almost no power when inactive (18). Memory chips, in particular, are inactive most of the time. Widespread use of CMOS chips in desktop systems could eliminate a large fraction of the power now consumed by chips – about 30–40% of the total for the IBM PCs we measured.

Reduced electrical power means less thermal energy to remove from a chip; the thermal benefits of CMOS become more important to chip designers as more transistors are packed on a chip. CMOS has thus become today's dominant technology for more powerful integrated circuits (19, 20). One chip designer (20) predicted in 1986 that within a decade most digital circuit designs, including microprocessors and memory, would shift from NMOS to CMOS. Other analysts have predicted that CMOS would account for nearly half the integrated circuit market by 1990 (21)[10]. While CMOS is slightly slower than NMOS, cost is the main factor limiting its usage to date: there is at present a premium of about 50–100% for CMOS, at the retail level.

The shift from NMOS to CMOS will be encouraged by the need to reduce heat generation as chip densities increase, with chip voltages remaining at the industry standard 5 V. Beyond some point, however, further density increases will require lower voltages, to prevent excessive electrical gradients. Some developers of new 4 or 16 MByte memory chips are already using 3.3 V (22). If the voltage is lowered in proportion to the decreasing device length, the energy per switching operation will shrink with the cube of the device length. The number of devices per chip has been quadrupling every three years, a trend that should persist until the early 1990's. The speed of personal computers, measured in millions of instructions per second (MIPS), is projected to increase by a factor of ten between 1985 and 1995 (23), a rate slower than the increase in chip density. If these projections hold true, by 1995 the typical business computer will perform many more calculations in the same period of time, but consume less energy than today's machines.

For many applications, conventional video-type screens (CRTs) can be replaced by "flat screen" displays (liquid-crystal, gas plasma or electroluminescent). The newer laptop LCD screens use backlighting for improved visibility; even with this addition, power for a flat LCD screen is currently 7–10 watts, compared to 25–50 W for a CRT. Color LCD screens of acceptable quality appear to be technically feasible (and slightly more power-intensive) but will be more expensive. Technical advances in materials and recording heads for disk storage are expected to increase storage density by a factor of five within five years (24). Since the drive motors do not have to work any harder, the gain in data storage should come at no increase in

10 J. Sturm, Princeton University, and J. Meindel, RPI, personal communications

electrical power. Data storage on optical disks, initially applied to larger systems, offers even greater gains in the ratio of power use to megabyte of data stored.

Finally, recent innovations in laser printing and photocopying could significantly reduce power requirements. Lovins (25) reports that some Canon and Siemens photocopiers use a cold-compression roller rather than a heated drum to fuse toner onto the paper, with a tenfold savings for this component of the copier. Copy quality is a concern to some users we talked to, since the surface of the copies is glossier than with a plain-paper copier using conventional heat-and-pressure fusing. With much lower standby power requirements (12 W, according to Lovins), there is no warmup time as with conventional copiers, and quieter operation since no fan is needed to cool off the paper. Also, cold-fusion copiers may reduce cooling loads and improve comfort where occupants share a small office with a desktop copier. A new series of high-end computer printers (6000 lines/minute, or about 1–2 million pages/month) uses a magnetic rather than electrostatic method to attract the toner to the rotating print drum, saving an estimated two-thirds of the energy required by an equivalent laser-printer (26). The system uses a magnetic rather than electrostatic method to attract the toner to the rotating print drum. Pressure, rather than a static charge, transfers the toner to the paper (27). Toner fusing involves heat and pressure, rather than cold-fusion. The manufacturer claims that this new process is more reliable (with 100 moving parts rather than 4000) and lower in first-cost than an equivalent conventional machine.

Two companies have recently introduced an alternative to the laser printer, based on ion-deposition technology (28)[11]. Both the Miltope Series 75 (rated at 75 pages/minute) and the C. Itoh Megaline (30 pages/minute) use a Delphax printer engine. The printer drum is much harder than the light-sensitive drums used in laser printers; this makes it possible to fuse the toner to the paper by means of pressure rather than heat-and-pressure, which is standard for most laser printers (and plain-paper copiers). C. Itoh Electronics is considering application of the same technology to mid-range, 15–20 ppm printers. In a cost comparison prepared for C. Itoh, the ion-deposition printers had total per-page costs at the low end of the range for comparable laser printers, but this range is quite large (about 1.5–3.8 cents/page) and electricity costs represent under 5% of the total.

Recently introduced "laser copiers" combine a digital scanner with a laser printer. The digitizing of images allows some copy-editing to be done at the machine, but as yet such copiers are not standardized to accept input from other computer files, remote copiers, or other electronic sources (29). Once this occurs, we are likely to see systems that combine scanners, printers, and

[11] J. Ashak and M. Le, C. Itoh Electronics Corp., personal communication

copiers take over a large share of the market for separate devices. The initial effect will be to increase equipment saturations and energy intensities, while improving the quality, convenience, and flexibility of hard-copy output. Eventually, however, we may see a decrease in total energy use for printing, as better-integrated systems reduce the need for printing drafts and extra copies, while also using hard-copy input and output devices more efficiently (i.e., less idle-time at full power).

3.2 Software and Operations

The most effective way of reducing an office machine's energy use is to turn it off (or to a "standby" power setting) when not in use. Even a computer with low-power CMOS chips uses energy for hard-disk drives, the monitor, and the power supply; together these represent at least two-thirds of the total (see Figure 2). The time (and extra steps) required to "boot up" a computer when it is first turned on will influence user decisions to manually turn off the power for varying periods of expected idle time. Booting times varied from 32 to 55 seconds, for several PCs we measured. Most of the problem lies in the memory check performed at the beginning of each startup, a diagnostic which is optional with some machines.

Most video monitors require only about ten seconds to show a picture once the power is switched on. Thus, a separate, automatic power switch for monitors would be a logical step. At present, some video monitors have manual power switches only, while some have no switch at all. Ideally, screens and other components would automatically switch to a dormant, low-power mode when not used for a (user-specified) time interval. They would then be reactivated at the touch of any key (or, for equipment tied to a communications network, by receipt of an incoming message). The necessary switching technology is already used in some laptop PCs, in order to stretch battery operating times (30); it could now be readily applied to desktop machines.

Relatively little is known about the percentage of time that major energy-using components of a computer or system are actually required, during normal periods of use. However, some suggestive data on "idle times" for terminals were obtained from system accounting files on the LBL minicomputer network (Table 3). These data show that, as of any given moment, about half of the logged-in users had been "idle" for at least ten minutes during on-peak hours, and more than 60% of users had been idle longer than ten minutes during evening and weekend off-peak hours.[12] In

[12] "Idle time" is defined as a period with no keystroke; this may include times when the user is waiting for a batch command to be executed. "On-peak" periods were 9 a.m. to 7 p.m.,

Table 3. Idle-time for LBL computer users[a].

Range of Idle Time[b]	Percent of Users Idle			
	On-peak[c]		Off-Peak	
	Bin %	Cum. %	Bin %	Cum. %
0 minutes	29.3	29.3	25.3	25.3
> 0 and <= 4	11.2	40.5	9.1	34.4
> 4 and <= 10	10.0	50.5	4.0	38.4
> 10 and <= 30	15.1	65.6	7.4	45.8
> 30 and <= 60	11.6	77.2	5.7	51.5
> 60 and <= 120	10.6	87.8	6.7	58.2
>120 minutes	12.2	100.0	41.8	100.0

a Observations were made for one week in June, 1987, for networked users of the LBL Unix 8 minicomputer. The data exclude idle-time for the microcomputer used to display the queue for a networked (multi-user) laser-printer.
b Idle time is defined as a period with no keystroke; this may include periods when the user is waiting for a batch command to be executed. In many cases, idle times of less than 10 minutes do not indicate that the user has "wastefully" left the terminal logged in, while engaged in unrelated activities. Conversely, power to individual terminals may be left on, even after they are logged off the LBL network.
c On-peak is defined as 9 a.m. to 7 p.m., weekdays only.

many cases, idle times of less than 10 minutes may not indicate that the user has "wastefully" left the terminal logged in while engaged in unrelated activities. Conversely, power to individual terminals may be left on, even after they are logged off the LBL network.

Manually turning equipment on and off is an inconvenience, but encouraging office workers to switch off unused PCs (like lights) may be a reasonable option for the near term, until hardware and software vendors introduce automated power switching as a standard feature. For a PC, screen, and small printer that use 150 W and are left on continuously, the annual energy charge will be about $92, and the demand charge $15, for a total of $107.[13] Indirect effects on space conditioning loads are quite site specific; some of the cooling load may be met with "free cooling," using an outside-air economizer which requires fan energy only. Turning off the PC for nights and weekends would

weekdays. Although the off-peak period represented the most hours (about 6150 hours/year) it had a much smaller user base, averaging 2.8 users compared with 11.7 average users during the on-peak period. Thus, nearly two-thirds of the usage (user-hours) was on-peak.

[13] We assume typical commercial building electricity costs include an energy charge, of 7 cents/kWh, and an on-peak demand charge, roughly $8/kW per monthly billing period, or $100/kW for the year.

cut the energy charge by nearly a factor of four, while leaving the demand charge unchanged. Total costs (excluding effects on space conditioning energy) would thus decrease from $107 to about $40 per year.

Since it is unlikely that all the PCs in a given office (i.e., on the same utility meter) would be in use at the same time, automated standby switching could realize additional savings of perhaps 30% in diversified on-peak usage, thus reducing demand charges. With effective automatic switching, typical electricity operating costs would thus be reduced to about $25–30 per year. (Of course, both the baseline numbers and potential savings would be reduced by future technological improvements, just as they would be increased by the trend to larger PCs and more energy-intensive screens and printers). Other significant savings may be possible through automated switching of such energy-intensive equipment as laser printers and copy machines.[14]

3.3 Energy Labeling and Other "Market-Enhancement"

Some opportunities for more energy-efficient hardware and software will occur through technology evolution and market pressures, often as side-effects of non-energy objectives. Others may require special efforts to accelerate their introduction. One way to draw more attention to energy issues and opportunities, from hardware manufacturers, software developers, and purchasers, is to introduce an energy performance label, similar to the labels on other appliances. This could be done either through a voluntary industry testing and rating program, or through existing appliance energy labeling programs, either at the Federal level or in states such as California, with their own labeling programs. To reflect hardware efficiency, an energy label would only need to include information on steady-state power. To give credit to efficient operating systems (as in laptops) that automatically shut down components until they are needed, the energy label should include energy use over a "typical operating cycle."

How much impact could energy labels have? Buyers of the first few generations of PCs and other equipment are mainly concerned with cost, performance, compatibility, and reliability – but not power requirements. In the past, most PCs were so expensive that their energy cost seemed irrelevant. But when one can buy the equivalent of a basic IBM PC for about $600, annual energy costs of $50–100 may begin to be noticeable – both the purchase price and annual energy costs of a basic PC are now roughly comparable to those of

14 For example, a laser printer typically requires at least 30 seconds for the compiler to compose and format a page; this may allow enough time for the toner-fusing drum to reheat, if if it were normally left in a cool(er) standby mode.

a large refrigerator. Of course, most business PCs are more powerful and more expensive, and total usage costs are still dominated by the user's own salary and benefits. Added to this, the person who decides which computer equipment to purchase is often not the one who pays the utility bills (a problem not unique to computers). Still, one might expect some market segments, such as larger corporate and governmental purchasers, to respond if better information were available on relative energy efficiency, especially for the factor-of-ten differences in energy costs such as those for laptop vs. desktop PCs. Alternatively, utilities could distribute to their customers comparative information (brand and model specific) that makes it easier to consider energy efficiency along with other features. We favor labeling and information strategies over mandatory energy efficiency requirements for computers and electronic equipment. Mandating efficiency standards for office equipment appears too risky for now given the rapid rate of change in both electronics technology and new applications.

4 Future Trends in Office Equipment Electricity Demand

This section draws on the preceding discussions of current energy use patterns, market trends, and potential efficiency improvements to outline, in broad terms, some alternative scenarios of possible future energy use by electronic office equipment in the U.S. The purpose of this exercise is not to present and defend a single estimate of future demand, but rather to use order-of-magnitude numbers to illustrate some of the major factors at work. Such estimates are, by necessity, highly judgmental, given the rapid rates of technological change and the large number of unknowns, even for today's stock and energy usage of office electronics.

Our starting-point is a review of some key assumptions in the EPRI report referred to earlier (3, 4). That report presented, with appropriate disclaimers, a possible scenario for U.S. electricity load growth in computers and related equipment, from 1982 to the early 1990's. EPRI projected an added 124 TWh of electricity consumption (31 GW increase in connected load) for selected sectors involved in "information activities": manufacturing (headquarters and office functions only), education, retail trade, finance, health care, office services (other than major manufacturing firms), and telecommunications "infrastructure" serving both homes and businesses.[15]

[15] The office electronics we consider in the present report are concentrated in only three of the sectors addressed by EPRI: manufacturing, finance, and office services, which represent

The EPRI scenario appears to us to be at the high end of a range of possibilities. Our estimates differ from those in the EPRI report on two main issues: actual vs. nameplate-rated power and the relative impacts of efficient technology vs. expanded levels of service.

The EPRI estimates of average power per unit of equipment exceed the measured data we presented earlier. For example, electricity use by PCs (representing about 17% of the EPRI total) is calculated at 340 W/unit, high by at least a factor of two.[16] Similarly, energy use for terminals (about 5% of the EPRI total) is based on an average rating of 300 W, or over five times the actual power consumption we measured, i.e., 40–50 W for a terminal plus 10 W for a modem or other networking electronics. These differences in power per unit affect the final numbers even if we accept the EPRI estimates of added equipment stocks, typical operating hours, and diversity factors.

On the other hand, minicomputers, which contribute 27% of EPRI's total growth in electricity use, were assumed to average 8500 W. This may be slightly high for future technology, but is close to the 6500 W we measured for a VAX 8700 (with hard disks and tape drive). That machine, in principle, can support 100 users, but typically averages only 10. In this case, the difference between 65 W/user and 650 W/user is a very significant one – even when we add the 40–50 W/user for a dumb terminal and modem or, in some cases, the added 100–150 W for a PC that functions as a terminal while to the central computer. These numbers suggest that, for minis and mainframes, it may be inherently more difficult to achieve energy-efficient (full) utilization, compared with the potential for "on-demand operation" of individual PCs and some other equipment.

In addition to overestimating unit power requirements, the EPRI report does not explicitly account for future efficiency improvements. Instead, these are "... assumed to be offset by the increased power needs resulting from increased requirements for computer processing capacity and peripherals." (31) In contrast, we believe that the potential gains in both hardware and operating efficiency are important enough to be analyzed separately from trends toward increased equipment saturation, usage intensities, and changes in the "levels of service" for electronic equipment.

It is hard to find definitive, published data showing how close we are to market saturation, but anecdotes can be suggestive. One analyst reports that at the Xerox Corporation in 1983, there was roughly one VDT monitor for every ten employees; by 1986 this ratio was one in four and by 1988 it was

slightly over half the total estimated load growth (or 66 TWh). The remainder of EPRI's load forecast involves mainly non-office computer services in health care, retail sales (point-of-sale terminals), and extensive use of home videotext. This makes a direct comparison difficult.

16 The EPRI report cites a range from 60 W for a Macintosh, to 614 W for an IBM PC/XT with color screen and dot-matrix printer. Our corresponding estimates, from Table 2, would be 40 W to 175 W.

expected to be one to one (32). Saturation of terminals may be one of the simplest cases to estimate – few office workers, managers, or professionals really want to use two or more display screens at the same time.

A number of published market surveys cite trends toward a maturing market for business PCs, as well as for larger computers. Increasing fractions of new sales are devoted to replacements, upgrades, and a transfer of market share from larger central computers to smaller, networked ones. The market study by Stromer (8), cited earlier, forecasts a continued growth in PC sales, but a decline in the annual growth rate, from 33% in 1987 to 12% in 1994. Another annual survey of 7000 business users found a steady decline in respondents' "expected growth in hardware budgets over the next 3–5 years," from a high of 11.8%/year growth in the 1982 survey, to 7.8%/year in 1987 (33).

For purposes of broad estimates, it is convenient to think of computing power in units roughly equivalent to a PC workstation (or the corresponding share of a larger machine). From an energy perspective, we have noted that the choice of a mainframe or mini vs. a PC may affect energy requirements, but the trend seems to be from larger machines to smaller ones, which seem to have more potential for high utilization factors.

Our base-case scenario for 1988 (see Figure 5) begins with an estimate of 25 million "computer screens" (workstation-equivalents) in business use, up from 21.6 million in 1986 (8). Of these, business PCs represented 60% of the total, or 15 million, up from an estimated 11–13 million in 1986 (see Table 1 and 8). These 25 million computer screens were distributed among more than 60 million white-collar workers,[17] not all of whom are potential computer users. We then assume average power requirements of 200 W per "screen." Of this, 150 W is for computing and data storage (either a PC workstation or a shared central computer); the remaining 50 W represents printers and other peripherals (either individual impact printers or shared laserprinters). Assuming an average of 5000 annual hours of operation (probably high for workstations but low for central computers/peripherals) annual energy use is about 25 TWh for today's 25 million users. The corresponding daytime peak power is about 5 GW (assuming little load diversity for office equipment at the peak daytime hour).[18]

To translate today's use into the "saturation" case for 1995, we begin with an estimate that the white-collar workforce is expected to grow by at least 30%,

[17] The Bureau of Labor Statistics (BLS) estimated 1986 white-collar employment, including management, professionals, technicians, sales and administrative personnel, at 61 million (34). The EPRI report cited a 1983 BLS estimate of 55 million.

[18] Both these numbers, and the EPRI estimates, exclude the impact of office equipment loads on heating and cooling energy. Added cooling and fan energy could represent another 20–40% of indirect energy use, depending on HVAC system efficiencies, for those buildings (and hours of the year) that require cooling; similarly, heating energy would be reduced in some cases.

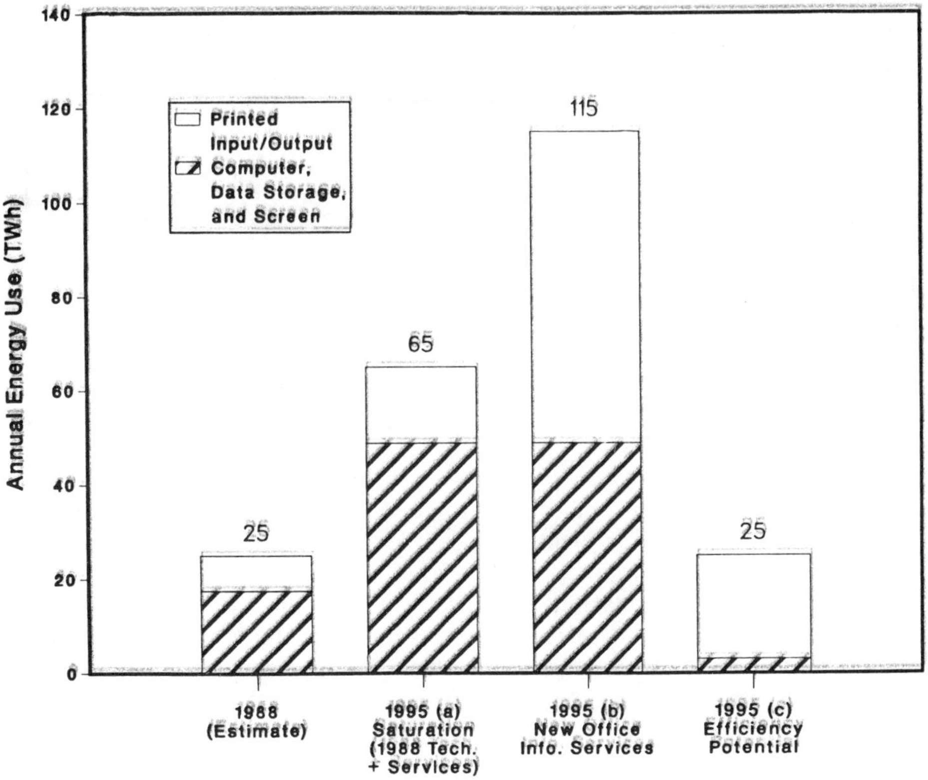

Figure 5. Alternative load growth scenarios for office electronic equipment in U.S. buildings. The current (1988) estimate is based on 25 million "screens" serving 60 million white-collar workers, at an average of 200 W/user (equivalent to a PC workstation or shared mini- or mainframe, plus screen and peripherals) and 5000 hours of operation per year. The 1995(a) scenario assumes today's hardware efficiencies and levels of usage, but shows full "market saturation" at 65 million screens. The 1995(b) scenario shows the impact of new office information services, mainly for increased input/output (I/O) using computer-generated or – scanned print media. Increased power requirements for I/O, from 50 W/user to 200 W/user nearly doubles total energy use, assuming the same technical efficiency and operating hours. The 1995(c) scenario ("technology potential") shows the impact of applying today's most efficient hardware and operating systems universally, reducing the per-user power from 150 W to 15 W for computers, and from 200 W to 100 W for I/O. Better control of idle-time would reduce average annual operating hours to 3000, bringing total U.S. electricity use for office electronics back to the range of today's 25 TWh.

adding about 15 million potential users.[19] We will assume that "saturation" of computer usage would correspond to 65 million, for an expected white collar xworkforce of at least 75 million in 1990. Thus, today's stock of electronic equipment would grow by about 40 million screens (2.6 times) to saturate the mid-1990's market – not yet allowing for any significant changes in office automation technology.[20] The "market saturation" scenario – 1995 (a) in Figure 5 – would thus increase office equipment energy use to about 65 TWh. At first glance, this number appears to coincide closely with the 66 TWh estimated by EPRI for three office-related sectors, however, the EPRI number represented energy demand growth since 1982, not total energy use. Moreover, there is some office equipment use in the other sectors analyzed by EPRI.

This first scenario does not yet account for potential increases in other types of electronic office equipment – mostly related to high-quality printed output and to electronic scanning, digitizing, and transmission of information in hard-copy form. Advances in laser printing are making high-quality computer output easier and cheaper; this is likely to translate into a potentially large source of electricity demand growth. Other recent product innovations with large market potential include scanners, optical character recognition, and presentation graphics systems. For these devices, we envision continued technical advances, sales growth, and powerful new applications – most likely as components of integrated office automation systems. For example, scanners, laser printers, and copiers may soon be integrated as components of a single desktop machine, perhaps shared by at least 3–4 users.

We think that such changes in office automation technology will affect energy use for hard-copy input/output devices more than the energy needs for computing and data storage. Thus, the second scenario – 1995 (b) in Figure 5 – includes an added 150 W/user, beyond the 200 W already estimated for computers, screens, and printers. In this scenario, increased energy use by advanced copiers, scanners, fax machines, telecommunications, and other electronic office equipment would nearly double our annual energy estimate, from 65 TWh to about 115 TWh. This is, again, a market saturation scenario for 1995, which allows for a significant change in office automation services,

19 The BLS predicted a 30% workforce growth for several sectors where white-collar employment is concentrated (retail trade, finance and insurance, business services, education, and legal services); together, these sectors account for about half of all white collar workers (35). The Computer and Business Equipment Manufacturers Association (CBEMA) (36) projected the number of "office workers" increasing from 58 million in 1984 to 81 million by 1995, a growth of 40%.

20 This is reasonably consistent with the market forecast cited by Stromer (8) which envisions roughly a tripling of PCs in use between now and the mid-1990's, some of which would replace larger machines.

but still does not reflect any improvements in hardware or operating system efficiency.

The last scenario – 1995 (c) in Figure 5 – attempts to account for the hardware and operating system efficiency opportunities discussed above, most of which are already on the market. *It illustrates the technical potential for reducing projected electricity demand by a factor of about five*, to about 25 TWh, or the same level as today, but with virtually full saturation of the office sector, and major changes in the level of office automation. This scenario assumes that market trends, reinforced by aggressive demand-side policies and programs, would allow today's "best" (most energy-efficient) technology to become the market average by the mid-1990's.

For example, the 150 W/user represented by the computer, data storage, and screen, could be reduced to 15 W without sacrificing computing power or capabilities, by using today's best laptop technology. The 200 W/user that we projected for input/output devices could probably be cut in half, with the use of more efficient printing and scanning technology. Thus, the scenario assumes a reduction from 350 W to 115 W/user, on average. Better automated control of equipment usage and idling time could reduce average annual usage from 5000 to 3000 hours. This is still more time than the average user would require for computing or information management, thereby allowing some off-peak usage hours for non-interactive computing, printing, or data transfer jobs. These alternative assumptions, for 60 million users, give us a final order-of-magnitude estimate of roughly 20–25 TWh/year as the "technical efficiency potential" for energy use by office electronics by the mid-1990's.

In the longer term, there are other powerful forces that may slow energy demand growth for office electronics. Continuation of today's trends toward increased system integration and inter-system compatibility could ultimately allow electronic media to replace much of the paper now being generated. This is in contrast to our present, hybrid approach, which in many cases uses office electronics to produce more paper output, encode and transmit the visual image with a fax, print it again, and then convert it back to electronic storage using a scanner. Projects for a "paperless office" have been discussed (on paper, of course!) for several years, but the time may finally be approaching when major vendors introduce practical hardware and software products aimed at this goal.[21] In addition to gains in office worker productivity and

[21] For example, IBM has just announced a major new office automation system called "Image Plus," which includes image-processing as well as conventional functions such as text-processing and electronic communications. The objective of this system is "to move from a paper-based to a computer-based system for information storage and retrieval." The system is targeted initially to medium or large-scale firms or government agencies, especially those involved in "case-work" activities which produce a high volume of non-coded document information in diverse formats, such as insurance, banking, real estate, health care, social services, and criminal justice (37).

office support costs, the aggregate energy and cost savings for paper manufacturing, printing, distribution, file storage, and waste disposal (or recycling) could be substantial.

5 Summary and Conclusion

As we have seen, the "other" end-use category represents up to 25% of electricity use in new U.S. office buildings. Electronic office equipment merits recognition as a distinct end-use, based on its aggregate energy consumption, contribution to peak demand growth, and potential for improved energy efficiency. Some of our major findings include:

- Actual electric loads for the equipment we measured, when in operation, were typically 20–40% of nameplate ratings. Energy analyses based on nameplate ratings thus overstate actual power requirements.
- Daily load profiles and on-peak diversity factors seem to vary by type of equipment and by office scheduling practices. More detailed, measured data are needed.
- Smaller-scale, dispersed computers and other equipment (small copiers, etc.) have been gaining market share compared with large, centralized systems. This trend, if it continues, may affect market penetration among smaller buildings and firms and management strategies for thermal loads that are dispersed throughout the occupied space, but have little impact on overall energy use for computing.
- Input/output devices (terminals, printers, scanners, fax machines and other communications equipment), not computing and data storage equipment itself, are major users of power and the largest source of load growth and uncertainty.
- There is significant potential for improved energy efficiency in both hardware and software, often as a by-product of other objectives such as miniaturization. Several policy options could accelerate these market and technical trends.
- Absent adequate data, both building designers and energy demand forecasters may tend to overestimate office electronics loads. Extrapolating from limited data, we estimate that typical daytime loads for office electronics (where present) in today's newly built U.S. offices are closer to 10–20 W/m^2, rather than the 30–50 W/m^2 sometimes cited.

- Future load growth depends on: (1) increased saturation of equipment and functions (more users of computers and related equipment); (2) changes in usage and "levels of service" (increased computing power and data storage, higher-resolution screens, laser printing, color output, etc.); and (3) improved energy efficiency for a given level of service (more efficient chips and screens, better software control of idling equipment, etc.).
- A recent EPRI report estimates 124 TWh of added electricity use for electronic office equipment in U.S. buildings, from 1982 to the mid-1990's. We believe this estimate to be at the high end of a range of possibilities. Our alternative order-of-magnitude estimate, first assuming saturation of the market with office automation equipment based on today's state-of-the-art, suggests total electricity consumption of 115 TWh by 1995.
- However, technical improvements in hardware and operating systems, based on technology already available but not widely used, could reduce this by a factor of about 5, to 20–25 TWh – roughly the same as today's energy use.
- In the longer term, better integrated inter and intra-office electronic communications may bring us to the reality of a "paperless office," with major impacts on both energy use and office workforce productivity.

5.1 Data Needs and Future Work

A number of steps are needed to develop better information on trends in energy use by electronic office equipment and on opportunities for efficiency. As part of our ongoing research, we plan to work, in cooperation with other organizations, on:

- development and dissemination of standard definitions for miscellaneous end-use loads, equipment types, etc., to allow better comparability of results from on-site surveys and end-use load metering projects;
- analysis of data from several completed or planned commercial on-site surveys, as well as from end-use metering of plug loads from the ELCAP project and other sources;
- examination of energy use (rated vs. measured) and the underlying technology trends for other types of electronic equipment (telefax machines, copiers, scanners) and for other non-residential building types (schools, retail, etc.); and
- a more extensive, independent assessment of available market-research studies on market trends, including important but seldom-examined issues such as retirement and replacement rates, and the longer-term energy implications of a "paperless" office.

Acknowledgements

The authors wish to express their appreciation to the following individuals who provided us with data and other assistance in preparing this report: S. Taylor, Linford Engineering; E. Westman and F. Peterson, Bonneville Power Administration; R. Alereza, California Energy Commission; P. Black, H. Otten, and S. Greenberg, Univ. of Calif.; N. Mueller, Wisconsin Power and Light; R. Lucas, Battelle Pacific Northwest Labs; and W. Thomas, IBM. Special thanks are due to J. Spadaro and P. Curtiss of the Princeton Center for Energy and Environmental Studies, and to B. Nordman of LBL, for their help with detailed measurements of equipment power requirements and computer network usage.

The work described in this paper was funded in part by the Assistant Secretary for Conservation and Renewable Energy, Office of Buildings and Community Systems, of the U.S. Department of Energy under Contract No. DE-AC0376SF00098. Research at Princeton University's Center for Energy and Environmental Studies was supported in part by funding from the New Jersey gas and electric utilities through the New Jersey Energy Conservation Laboratory, from the Prudential Insurance Co., and from the New Jersey Department of Commerce and Economic Development. Partial support for this research was also provided by the French Energy Agency (AFME) under contract #6–09–0118.

References

1. Piette, M.A. 1986. A comparison of measured end-use consumption for 12 energy-efficient, new commercial buildings. *Proceedings of the 1986 ACEEE Summer Study on Energy Efficiency in Buildings*, Santa Cruz, Calif., 3:176–92.
2. Nguyen, D., Alereza, R., Hamzawi, E. 1988. Energy consumption by computers and miscellaneous equipment in commercial buildings. *Proc. of the 1988 ACEEE Summer Study on Energy Efficiency in Buildings*, Asilomar, Calif., 3:169–80.
3. Roach, C. 1987. *Productivity tools for the information economy: possible effects on electricity consumption*, final prepublication draft report to the Electric Power Research Institute, Palo Alto, Calif.
4. Squitieri, R., Yu, O., Roach, C. 1986. The coming boom in computer loads *Public Utilities Fortnightly*, December 25, pp. 30–34.
5. Energy Information Administration (EIA), U.S. Dept. of Energy. *1986 Commercial buildings consumption and expenditures*, 1983, Report DOE/EIA-0318 (83).
6. California Energy Commission (CEC). 1987. Demand forecasting issues: CFM 7, P300–87–015 : 3–18–25. Sacramento, Calif.
7. 1987 U.S. market report. *Electronics*, January 8, 1987 pp. 51–74.
8. Stromer, R. 1988. New service forecasts future of microcomputer market through 1994 *PC Week* vol. 5, March 1, p. 130.

9. Business computing in a flux. 1987. *Personal Computing*, December, p. 369, citing as a source Comtec Market Services.
10. Computer and Business Equipment Manufacturers Association (CBEMA). 1987. The computer, business equipment, software & services, and telecommunications industry, 1960–1996, Washington, DC.
11. Statistical Abstract of the United States. 1988. 108th Edition, Table 1286. U.S. Dept. of Commerce, Bureau of the Census.
12. Computer Industry Abstracts. 1987. Data Analysis Group, Inc., La Mesa, Calif., Second Quarter.
13. Nihei, W. 1987. The color of the future. *PC World*, February, pp. 278–83.
14. Alereza, T., Breen, J.P. 1984. Estimates of recommended heat gains due to commercial appliances and equipment. *ASHRAE Trans.* 90:2A, pp. 25–58.
15. American Society of Heating, Refrigerating, and Air-Conditioning Engineers, Inc. (ASHRAE). 1987. Proposed standard for the energy efficient design of new buildings Third Public Review Draft, Sept: 29, pp. 8–21 and 9–2,-3, Atlanta, Georgia.
16. California Energy Commission (CEC). 1987. Notice to building officials on alternative compliance methods for Title 24 code requirements, Sacramento, Calif., Nov. 23.
17. Manufacturers' advertisement for hard-disk drives. 1988 *BYTE*, April, p.53.
18. Horowitz, P., Hill W. 1980. *The Art of Electronics*, p.328. Cambridge University Press: Cambridge.
19. Meindl, J.D. 1987. Chips for advanced computing. *Scientific American,* 257(4):78–88.
20. Chen, J.Y. 1986. CMOS – the emerging VLSI technology. *IEEE Circuits and Devices Magazine,* March pp. 16–31.
21. Davies, R.D. 1983. The case for CMOS. *IEEE Spectrum*, October, pp. 26–31.
22. Chou, S., Simonsen, C. 1987. Chip voltage: why less is better. *IEEE Spectrum*, April, pp 39–43.
23. Peled, A. 1987. The next computer revolution. *Scientific American,* 257(4):57–64.
24. Kryder, M.H. Data storage technologies for advanced computing. *Scientific American,* 257(4):117–25.
25. Lovins, A. 1987. *Advanced electricity-saving technologies and the South Texas project.* Report to the Electric Utility Department, Austin, Texas, Rocky Mountain Institute, p. 19.
26. Bull. 1988. Product literature, Bull MP 6090, and personal communication from Bull sales representative.
27. Eltgen, J-J.P., Magnenet, J.G. 1980. Magnetic printer using perpendicular recording. *IEEE transactions on magnetics*. Vol. 16, September, No. 5.
28. Cook, R. Page printers. *BYTE*, pp. 187–97.
29. Keer, K. 1986. Technology is redefining copying-machine functions. Office Systems pp. 34–44.
30. Hogan, M. 1988. Product development. *PC World*, May, p 118.
31. Roach, C. 1987. See Ref. 3, p. 7.
32. Clevinger, M. 1987. comments on "smart office buildings," in Technology and the future of the U.S. construction industry. U.S. Congress, Office of Technology Assessment, pp. 34–38.
33. Carlyle, R.E. 1987. Midrange shootout: mini/micro survey. *Datamation*, 33(22):60–76. (Citing a 1987 market survey by Datamation/Cowen & Cowen.)
34. Statistical Abstract of the United States. 1988. See Ref. 11, pp. 376–77.
35. Statistical Abstract of the United States. 1988. See Ref. 11, p. 380.

36. Computer and Business Equipment Manufacturers Association (CBEMA). 1985. Computer and business equipment marketing and forecast data book Washington, DC.
37. IBM Internal product information and personal communications from W. Thomas, IBM Southwest Marketing Division, Oakland, Calif.
38. Statistical Abstract of the United States. 1988. See Ref. 11. Table 1285.
39. Computer Industry Abstracts 1987. See Ref. 12. p. 101. Original source is Hambrecht and Qust, Infoworld, Jan. 12, 1987, p. 15.
40. Computer Industry Abstracts. 1987. See Ref. 12. pp. 103–4.
41. CBEMA. 1987. See Ref. 10, p. 111.
42. Computer Industry Abstracts 1987. See Ref. 12, p. 123. Original source is *The shifting images of the monitor market* in Computer Reseller, 1986, p. 44.
43. Dower, J.W., Luft, N., Puopolo, S. 1987. *Printout Annual*, p. 4.
44. Computer Industry Abstracts.1987. See Ref. 12, pp. 119–20.
45. Becker, S. 1986. *Fax meets high-tech objective of speed and cost-effectiveness Office Systems*, pp. 60–68, citing a survey by Venture Development Corp., Natick, Mass.

The Role of Sectoral Shift in Trends in Electricity Use in United States and Swedish Manufacturing and in Comparing Forecasts

Gale A. Boyd and Marc H. Ross

Abstract

Changes in electricity intensity in each industrial sector and shifts in the mix of production by these industries determine the trends in aggregate electricity intensity. Moreover, the trends in real electricity intensity and output mix shifts have different economic driving forces. For these reasons, trends in Divisia indices of shift and real electricity intensities are examined for U.S. and Swedish manufacturing. We find that the shift away from electricity-intensive manufacturing plays a large role in offsetting real electrification in the U.S. but the shift effect is much smaller in Sweden. We find that the index of shift for the U.S. is driven less by price and income effects and more by other economic trends such as maturation of product markets, more efficient use of materials, and increased import of energy-intensive goods.

1 Introduction

Industrial production, output mix, and electricity intensity play an important role in determining future electricity use patterns. For this paper we construct Divisia index numbers to measure the effect of output mix on aggregate measures of U.S. and Swedish manufacturing electricity use. The U.S. analysis draws on our previous work in the area, but is based on a more detailed data base. We discuss how the trends in both the U.S. and Swedish indices are related to trends in the materials manufacturing sector. The historical index number analysis is descriptive, not predictive. To relate the historical Divisia index number analysis to electricity forecasting, we review some previous

Divisia index analysis of two forecasts. These forecasts embody different underlying output mix assumptions that are revealed by the Divisia decomposition.

In the U.S., the ratio of energy use per dollar of output (or aggregate energy intensity) of manufacturing declined 33% 1972–1985, or an average of 2.6% per year.[1] This decline was due in large part to improvements in the efficiency of energy use in making each kind of product. These efficiency improvements were motivated by shortages of natural gas in the early 1970s, rapid increases in the real prices of petroleum and gas to the early '80s, and less rapid, but substantial increases in the real price of electricity (Figure 1). In addition, some energy efficiency increases were caused by on-going technical change, e.g., by general-purpose innovations in manufacturing technology.

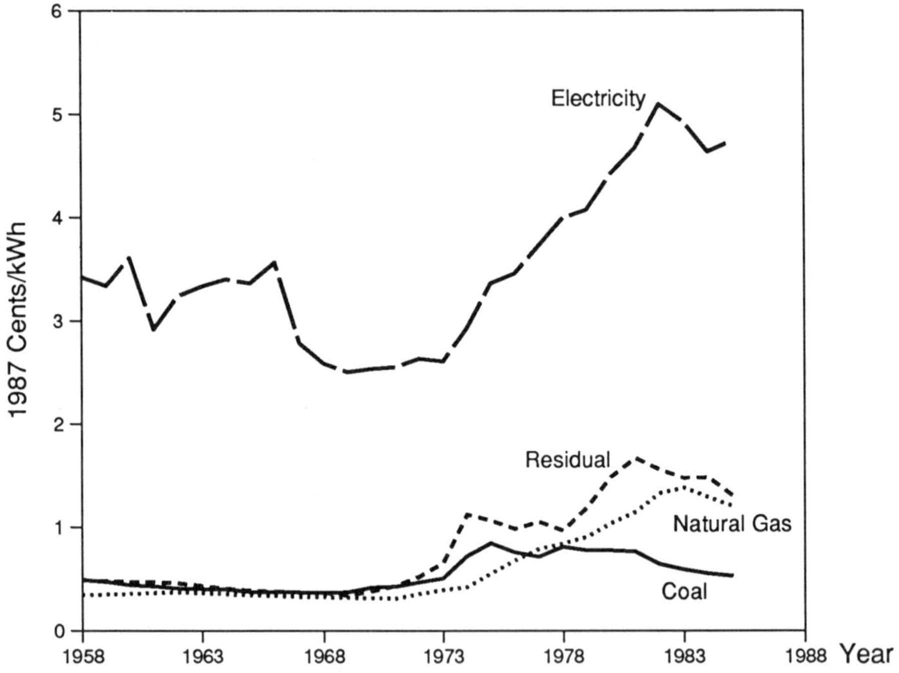

Figure 1. National average real energy prices to industrial customers in the U.S. deflated with producers' price index.

[1] In this computation, fuel and electrical energies are combined to obtain total primary energy using a ratio of 3.3, characteristic of fossil-fuel-powered systems. I.e., 1 kWh of electricity is counted as 3.3 kWh of fuel. Fuel used for heat and power is counted, while feedstock and other non-energy uses of fuel are not. Production is real output (inventory- corrected shipments).

Aggregate fossil fuel intensity declined in the U.S. from 1958 to 1985, electricity intensity increased until 1970–71 and has been declining gradually ever since 1975 (Figure 2). One factor in the increased use of electricity per unit of production has been price. As Figure 1 shows, the price of electricity relative to fuel decreased until the early 1980's. The real price of electricity, nevertheless, increased from the early 1970s to the early '80s. Electricity now costs 4.5 times as much as natural gas in terms of its heating value. This means that electricity is not used by manufacturers as a simple substitute for fuel in the U.S., but to perform functions that absolutely require electricity or to substitute for fuel in processes where the overall efficiency of electricity use is much higher than that of fuel. Electricity intensity also is affected by technical change, i.e., general purpose innovation often involves increased electricity use.

As in the U.S., the aggregate electricity intensity of Swedish manufacturing rose rapidly before the early 1970s, but unlike the U.S., the intensity has continued to rise, after dipping during the peak production years of 1974–75. (See Figure 3.) Meanwhile, the pattern of aggregate fossil fuel intensity was similar for the two countries.

Electricity use by Swedish manufacturing has been subject to many of the same general factors as in the U.S., but there are major differences. One is that natural gas is not an important fuel in Sweden. Swedish industry, depending heavily on oil, had a stronger impetus to reduce its fuel dependence when the oil shocks of the 1970s hit. A second important difference is that the price of electricity is relatively low in Sweden. In 1985 the average price to industry was 2.9¢/kWh (based on the 1985 exchange rate for kroner), compared to 5.2¢/kWh in the U.S. The low price of electricity in Sweden, as well as and the lack of natural gas which is cheap and convenient, has meant that much electricity is used in roles done by fuel in the U.S. In the U.S. the 1985 ratio of electrical energy to fuel energy (including wood-based fuel) used for heat and power by manufacturers is 0.20, in Sweden the ratio is 0.42. Tables 1 and 2 show in detail the U.S. and Swedish energy use patterns for the energy intensive U.S. Standard Industrial Classification (SIC) and International Standard Industrial Classifications (ISIC) codes. If only fossil fuels are counted the difference is even more striking.

The effects of technical change and energy prices on energy demand of individual industries have been quantified in studies such as Jorgensen for the U.S. and Walfridson for Sweden, to name only two (1,2). While it is not our goal to summarize their results, since the main thrust of those studies is to estimate these effects for selected industries, we note that the electricity-using bias of technical change in energy-intensive industries is generally supported by these studies.

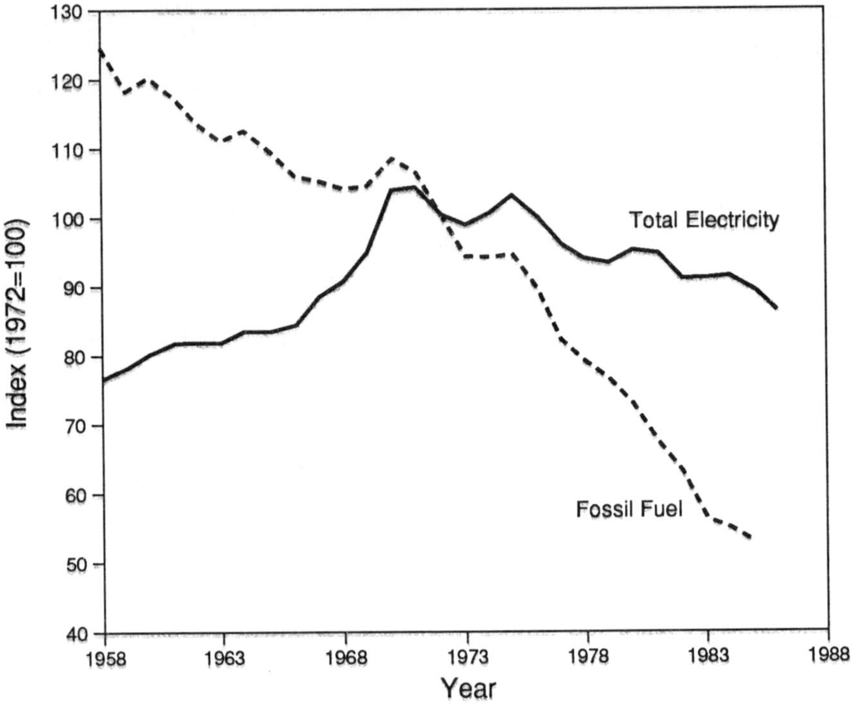

Figure 2. The aggregate fossil fuel and electricity intensities of U.S. manufacturing, indexed to 1972.

Interestingly, when Walfridson estimates the coefficients of his economic models for the aggregate manufacturing sector he has trouble reconciling his aggregate energy price and technical change coefficient with those obtained in the disaggregate industry sectors. Similarly, Solow in reviewing aggregate energy factor demand studies summarizes the difficulties in reconciling aggregate energy coefficient estimates (3). One reason for this is that the decline in aggregate energy intensity was also associated with a shift in the mix of production. In recent years, there has been a pattern of decline in the relative production of energy-intensive products, especially bulk-materials such as steel and cement. As with energy generally, the aggregate electricity intensity depends in part on changes in the mix of production. The effect of the decline in relative production of energy-intensive products is similar for electricity and for fuels. Solow illustrates the importance of this production shift on energy modeling. He shows that even in the absence of substitution possibilities of energy for other factors of production at the micro level, the economy may appear to exhibit energy substitution at the macro level.

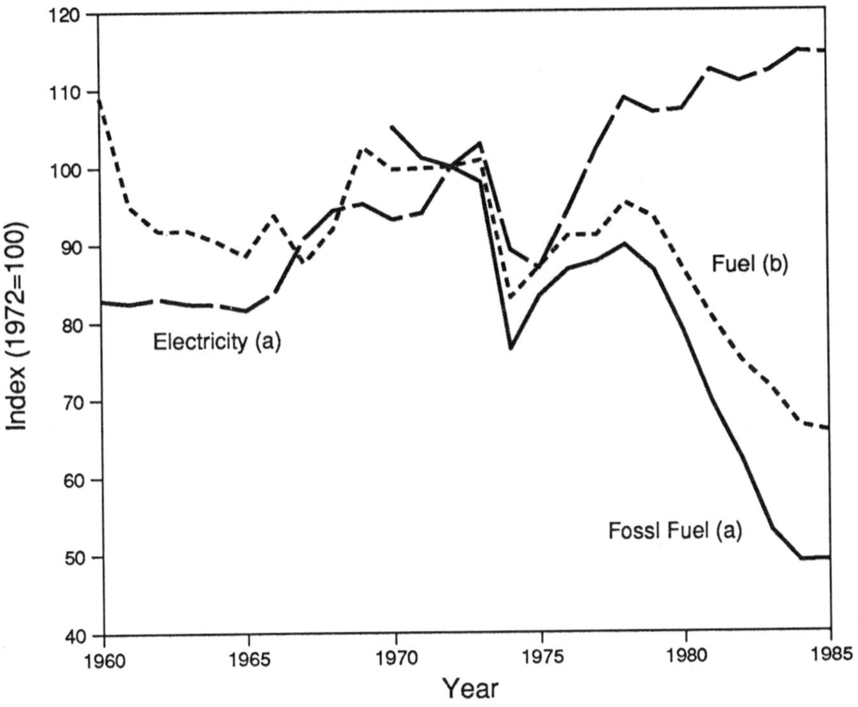

Figure 3. The aggregate fossil fuel and electricity intensities of Swedish manufacturing, indexed to 1972. Energy series a are for manufacturing excluding petroleum refining, and excluding biomass fuels and fuel used as feedstock. Energy series b is for all industry and includes biomass fuel and fuel used as feedstock.

This suggests that factor demand modeling generally and energy demand modeling specifically need to be done at a disaggregate level. However, it may also be useful to introduce a measurement concept that directly accounts for the effect of the shifting output mix in aggregate energy intensity. Quantifying and understanding the shift effect at a macro level can lead to a more basic understanding of underlying trends in the economy, since changes in energy efficiency and output mix probably have different driving forces. This understanding of basic trends is also useful for forecasting and, in particular, in scenario design.

The measurement and effect of the shift in the mix of production on electricity consumption is the focus of this report. The trend in electricity consumption will be decomposed into: 1) a trend due to the shift and 2) a trend due to changes in the electricity intensity (electricity use per unit of production) in the separate sectors. The developments underlying the second trend are discussed in the chapter by Kahane in this volume.

Table 1. Energy used for heat and power in manufacturing in the U.S., 1985.

Sector	SIC[a]	ISIC[b] (approx.)	electricity[c] (billion kWh)	fuel[d] (billion kWh)	Total[e] (billion kWh)
Food processing, tobacco	20,21	31	50.0	236	403
Textiles, apparel	22,23,31	32	31.6	55	162
Lumber, furniture	24,25	33	20.6	96	166
Pulp and paper, printing	26,27	34	69.2	650	884
Chemicals	28	351,352	130.5	590	1031
Petroleum refining	29	353,354	35.3	676	795
Rubber and plastic products	30	355,356	26.6	38	128
Cement, clay, glass	32	36	34.1	238	353
Primary metals	33	37	134.4	558	1012
Metal Products, and machines	34–39	38.39	138.0	223	689
TOTAL			670.1	3359	5624

[a] Standard Industrial Classification.

[b] International Standard Industrial Classification.

[c] Does not include electricity generated from fuel at the site because the fuels are included in the next column.

[d] Includes biomass fuels (about 0.5 trillion kWh in SICs 26, 24 and 20), but it does not include fuel used as feedstock, about 0.9 trillion kWh in 28. Includes petroleum process by-products used as fuels, especially in 29 and 28.

[e] Primary energy. This includes losses in generation and transmission of electricity, using a typical fossil-fuel heat rate: losses = 2.34 x delivered energy. Thus to obtain the total for food: 3.34 x 50 + 236 = 403.

Source: Manufacturing Energy Consumption Survey.

2 Sectoral Analysis of Electricity Use

2.1 The Divisia Decomposition

In our analysis of trends in energy use we start with: the trend in total manufacturing production, then determine, by the Divisia technique, the effect of shifts in the mix of production among the sectors within manufacturing, and finally obtain a measure of the trend in the efficiency of energy use. This third quantity we call the real energy intensity. The following presents a discussion of the Divisia decomposition. For more details see Boyd, et al.(4).

Table 2. Energy used for heat and power in manufacturing in Sweden, 1985.

Sector	ISIC	Electricity (Billion kWh)	Fuel (Billion kWh)[a]	Total (Billion kWh)[a,b]
Food processing, tobacco	31	2.03	3.9	10
Textiles, apparel, leather	32	0.42	~0.8	~2
Lumber, furniture	33	1.80	6.7	12
Pulp & paper, printing	34	17.21	43	91
Chemicals	351,352	5.40	10.8	26
Petroleum refining & products	353,354	0.60	10.6	12
Rubber & plastic products	355,356	0.69	~1.4	~3
Cement, clay & glass	36	1.24	5.6	9
Primary metals	37	7.99	15	38
Metal products, machines	38,39	6.39	~8	~26
TOTAL		43.76	105	229

[a] Includes biomass fuels; does not include fuel used in non-energy uses such as feedstock and lubricants.

[b] Includes losses in generation and transmission of electricity with losses = 1.83 x delivered energy. Thus to obtain total for food: 2.83 x 2.03 + 3.9 = 10.

Source: United Nations Industrial Statistics Yearbook, Vol I: General Industrial Statistics, and Energy Balances of OECD countries 1975–85.

If e_i is the energy use in the ith sector, then aggregate energy use is $E = \Sigma_i e_i$ in any time period. It is an identity that

$$E = Q \, \Sigma_i \, (e_i/q_i) \, (q_i/Q)$$

where q_i is a measure of production in the ith sector and Q is a measure of total production. For example, we could define electricity use in kWh and production of steel in tons of steel mill products. Then the sector intensity would be in kWh/ton. Two aspects of this identity may be important: 1) the level of sectoral disaggregation and 2) the measures selected to describe production. We return to these issues below.

We define the (compound) growth rate, G, of any quantity X(t) by

$$G_X(t) = \ln \, [X(t)/X(t-1)].$$

The growth rate for aggregate energy can be approximately decomposed into shift, g_S, and real g_R, components:

$$G_{E/Q} \approx g_S + g_R$$

where

$$g_S \equiv \overline{G}_{q/Q} = \overline{G}_q - \overline{G}_Q$$
$$g_R \equiv G_{e/q} = G_e - G_q$$
$$G_X \equiv \Sigma W_i(t) \ln [X_i(t)/X_i(t\text{-}1)]$$

and the weight, W_i, is the (moving) energy share of the ith sector:

$$W_i(t) = 1/2\,[(e_i/E)_t + (e_i/E)_{t-1}].$$

This is the Divisia decomposition.

In other words, the growth rate in energy use due to shifts in the mix of production, g_S, is the energy-weighted average of the growth in sectoral production shares q_i/Q, and the growth rate in energy use associated with changes in energy efficiency, g_R, is the energy-weighted average of the growth in sectoral energy intensities e_i/q_i. Thus, the growth in the aggregate energy intensity differs from the weighted-average growth in sectoral energy intensity, g_R, by the weighted-average growth in production shares. For example, if production in the energy-intensive sectors declines relative to that in the other sectors, g_S will be negative. Note that the measures of production in each sector, $q_i(t)$, do not need to be commensurate from one sector to the next since only the ratios $q_i(t)/q_i(t\text{-}1)$ enter. Thus, for example, production indices can be used, or completely different measures of production can be used in the different sectors.

In the coming Figures 6 and 7 indices will be shown where

$$I_{E/Q} \approx I_S \cdot I_R$$

and $I_x(T) = \exp\,[\sum_{t=1}^{T} g_x(t)]$

2.2 Disaggregation of Energy Use by Sector

Energy use in both U.S. and Swedish manufacturing tends to be concentrated in a few sectors, the sectors where bulk materials are manufactured (Tables 1 and 2). In the U.S., over 3/4 of all energy used in manufacturing occurs in the bulk-materials sectors: pulp & paper, chemicals, petroleum refining, stone, clay & glass, and primary metals. Some 58% of the electricity used in manufacturing occurs in these sectors. In the following, we will use the term materials to refer to the bulk products of these sectors.

Swedish manufacturing is even more concentrated in the materials sectors, with the predominance of pulp and paper and iron and steel. Over 50% of the electricity used in manufacturing occurs in these two sectors and about 75% in the 5 materials sectors listed above.

The concentration of energy use in the materials sectors explains the high energy intensity of these sectors compared to the rest of manufacturing. This concentration of energy use and energy intensity is responsible for the sensitivity of energy use in manufacturing to the mix of products.

Energy intensity can be evaluated in many ways; in Table 3 it is the ratio of the cost of energy to value added in 1985. (The ratio cost of energy to total cost, or value of shipments, is also shown, but this ratio is skewed in cases such as petroleum refining, because input materials are relatively expensive.) Table 3 shows that the electricity cost to value added ratio is roughly five to ten times greater in materials manufacturing than in the rest of manufacturing.

Care is warranted in categorizing manufacturing sectors, i.e. in the level of

Table 3. Ratio of the cost of energy to value added and to value of shipments, various sectors, United States 1985, percent.

Sector	SIC	Electricity[a] Cost to Value Added	Energy[b] Cost to Value Added	Electricity[a] Cost to Shipments	Energy[b] Cost to Shipments
Pulp and paper mills	261–263	11	29	5	12
Basic chemicals	2812–2816, 286	11	26	4	10
Petroleum refining	291	12	36	1	3
Cement	324	20	45	11	24
Basic iron and steel	331	13	30	4	10
Primary non-ferrous metals	333	94	112	16	19
Other manufacturing[c]		2.0	3.1	1.0	1.5

[a] Not including electricity generated and used on site.

[b] Includes purchased electricity and fuels, but not by-products of the process such as biomass fuels in pulp and paper and petroleum refinery by-products nor organic feedstocks.

[c] SICs 20–25, 27, 30, 31, 34–39.

Source: "Annual Survey of Manufactures, 1985", U.S. Bureau of the Census.

disaggregation, for analysis of trends. To obtain a clear analysis, major sectors should be characterized by fairly uniform energy intensities. The two-digit Standard Industrial Classifications are a product of historical development and are not fully satisfactory for energy analysis. A general rule is that upstream manufacturing activities are energy intensive while downstream activities are energy non-intensive. For metals and allied products, the 2-digit SICs enable a reasonably clean separation between energy-intensive upstream manufacturing and non-intensive downstream manufacturing. The existing sectors for paper and allied products (SIC 26, ISIC 341) and chemicals and allied products (SIC 28, ISICs 351 and 352 combined) are unsatisfactory from this perspective, however.

Because of these considerations, we have adopted for our analysis of U.S. manufacturing the sectoral categorization of the National Energy Accounts (5), with additional disaggregation in paper and chemicals making a total of 71 sectors (Table A.1 in the Appendix). This categorization is roughly at the 3-digit SIC level. For example, it separates converting operations in paper and board manufacturing (e.g. box making) from pulp and paper mills, and it separates various basic chemicals manufacturing activities from downstream product manufacture like pesticides and herbicides, drugs, toiletries and paints. This level of disaggregation is more than adequate for analysis of most product mix issues. We believe that expansion of the analysis to the 360 4-digit sectors would not lead to significant changes in our results. There are very few major sectors in our analysis which incorporate 4-digit sectors which have both very different energy intensities and divergent trends in production. (Possible problem areas are the booming industrial gases sector within industrial inorganic chemicals and pulp mills within pulp and paper mills.) Moreover, earlier work suggests that the effect of the shift can be largely captured with only two well-chosen sectors (4), although some aggregation effects are observed.

Let us be clear what the level of disaggregation means for our analysis. All steel mill activities are, for example, in one sector. Thus, integrated production of steel from ore is in the same sector as production of steel from fabricators' and consumers' scrap. The effect of a shift in materials flow toward the scrap-based process will show up as a decline in real energy intensity. The energy embodied in scrap enables lower-energy steel production using the scrap-based route. If, instead, we could carry through an analysis beyond the 4-digit level of disaggregation, to where the two steel processes were in separate sectors, such a shift would show up as a shift effect, with the real energy intensity essentially unchanged.

The same arguments apply to the categorization of Swedish manufacturing. Here, we have adopted a 28-sector disaggregation (Table A.2). This 3-digit ISIC categorization disaggregates chemicals into largely energy-intensive and non-intensive groups and primary metals into ferrous and non-ferrous metals, but unfortunately does not separate pulp and paper mills from converted product manufacture. We briefly explore the latter issue in Section 3.1.

2.3 Production Data

Selection of production data for analysis poses a serious issue for the energy forecaster, because different descriptions of production lead to substantial differences in trends. Two problems in particular can be identified: 1) growth in sectors downstream from primary production and 2) product heterogeneity. The first problem is, in principle, easily resolved. For broad sectors, such as paper and allied products (SIC 26, ISIC 341), the relative growth in downstream subsectors leads to higher growth trends for the broad sector than in the energy-intensive manufacturing itself. This problem can be overcome by selecting sectors narrowly enough so that they correspond to the energy-intensive part of the industry or by using an indicator of production tied to the energy-intensive activity.

The second problem arises in describing an energy-intensive sector with many heterogeneous products, such as organic chemicals. For some of these sectors tonnage data is not available or not convenient for analysis. This lack of data causes the Federal Reserve Board to use such awkward measures as kWh or employee-hours as proxies to construct their production indices. Such measures are unsatisfactory for energy analysis.

A different primary source of data for production trend analysis is value of shipments and purchases. This economic data is routinely converted to real output or value added, correcting for changing prices. Price change is not a major source of error in periods of low inflation, but in periods of high inflation major errors in trends can be introduced. This problem is illustrated by the most electricity-intensive sector in manufacturing: primary aluminum (SIC 3334, part of ISIC 3720), a single homogeneous product. For the U.S., we compare three different primary sources of production information: real output, physical production and electricity use (Figure 4). The real output measure has diverged significantly from both physical production and electricity use. In the late 60's and early 70's electricity use in primary aluminum was under reported by the Census, explaining much of the discrepancy in that period.

In the following analysis, two data series on production trends will be used in order to provide some indication of the variability from this source of ambiguity. For the U. S., the economic series is the Bureau of Labor Statistics (BLS) real output (inventory corrected shipments) and for Sweden it is based on indices of real value added from Statistics Sweden.

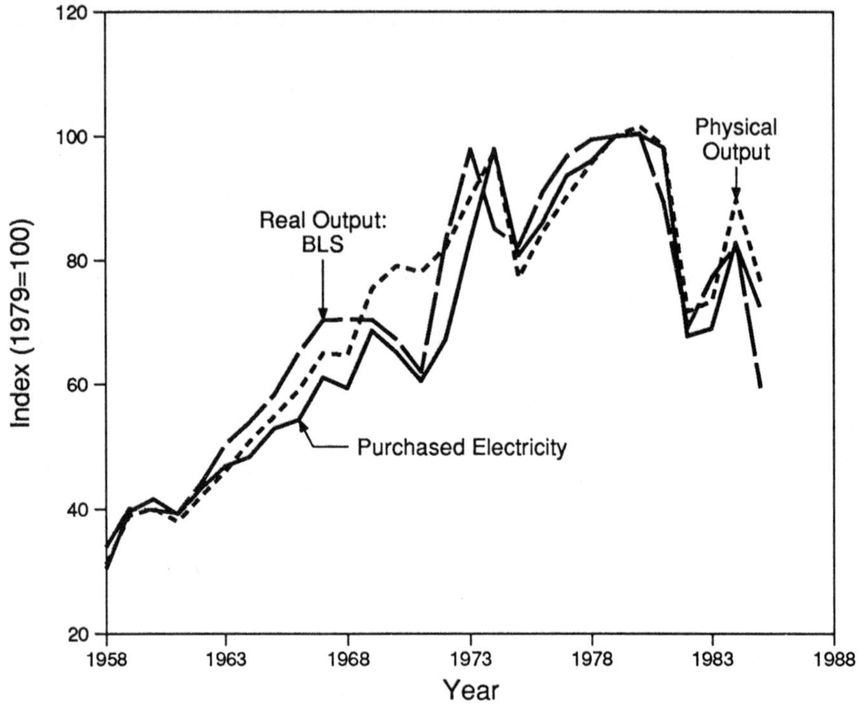

Figure 4. Different measures for production of primary aluminum in the U.S.

For the Swedish physical series, we use tonnage data for paper and board obtained from Sweden's paper trade association, the SCPF, to represent ISIC 341. The economic series mentioned above is used for the remaining sectors. The physical series for the U.S. uses tonnage data for paper and board, steel mill products, industrial chemicals, petroleum refining input, cement and aluminum (6).

2.4 Electricity Data

In the use of electricity data, only one major issue arises: the counting of self-generated electricity at the manufacturing site where it is used. In the U.S., data on such on-site electricity is collected by the Census, but it is usually not included in analysis of electricity demand and is not included in the National Energy Accounts. We calculate our Divisia indices for both "purchased" electricity from off-site and hydropower generated at the site, as well as "total" electricity which includes on-site generation with fuel. For the Swedish data the calculations are based on total electricity.

The inclusion of on-site electricity in the U.S. data is important because on-site generation was rapidly declining as a fraction of total electricity until the early 1980s (Figure 5). Analysis of purchased electricity thus suggested faster growth in the use of electricity than occurred.

3. Results and Interpretation

3.1 Analysis of Historical Trends

The Divisia decomposition for electricity in U.S. manufacturing is summarized in Table 4 and illustrated in Figure 6. It shows that there was a small positive shift, i.e. a shift toward more electricity-intensive manufacturing, in the period before the early '70s and a substantial negative shift thereafter. From 1972 to

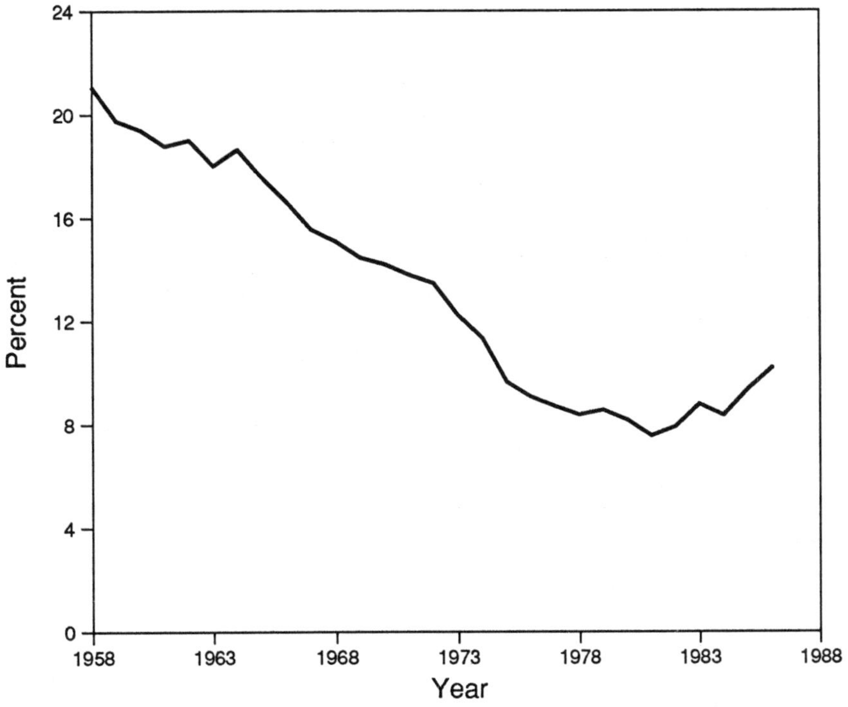

Figure 5. Percentage of total electricity generated on site, U.S. manufacturing sector.

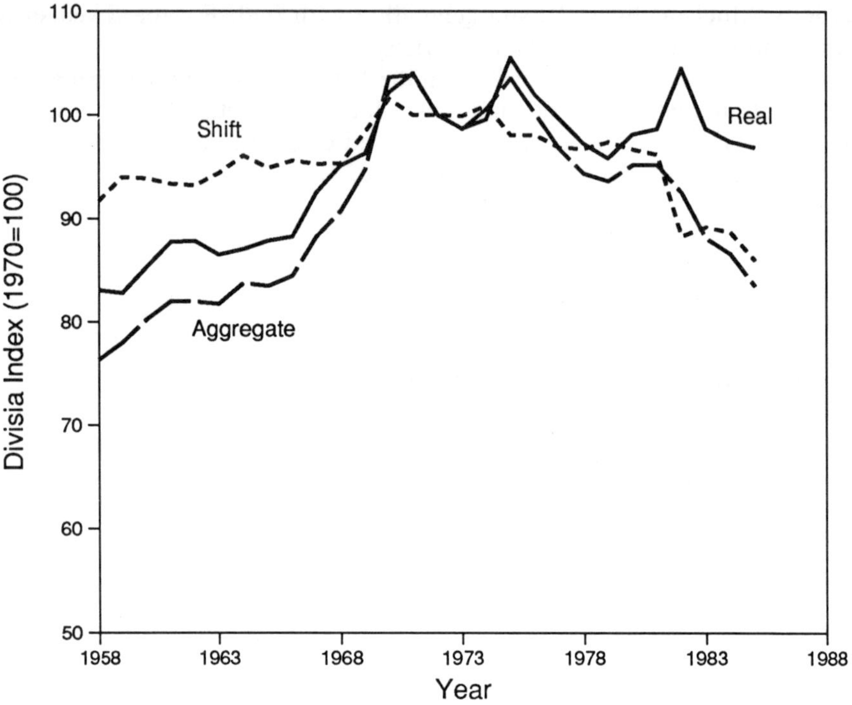

Figure 6. Divisia analysis of electricity consumption by U.S. manufacturers total electricity intensity using physical output.

1985 the change in the shift index was -14%. That is, 14% less electricity was used in 1985 due to the decline in bulk-materials manufacturing in the U.S. economy since 1972. The shift is similar for purchased and total electricity. Moreover, averages for the two production series are presented in Table 4.

Table 4. Summary of Divisia analysis for electricity, U.S. manufacturing* (regression growth rates, percent per year).

	1958–72	1972–85	1978–85
g (shift)	0.3	–1.4	–2.3
g (real)			
purchased	2.5	0.3	0.4
total	1.8	0.0	0.5

* Results for the two output series have been averaged.

The results for electricity intensity calculated with the physical production series, shown in Figure 6, indicate the downward shift occurring almost entirely in the 1975 and 1982 recessions. In effect these recessions, rather than being temporary declines, constituted major adjustments in the role of energy-intensive manufacturing in the U.S. economy. The real output series, in contrast, leads to a shift (not illustrated) which declines somewhat more steadily over the period 1975–1985.

In Sweden, there is slow relative growth in electricity-intensive manufacturing from 1966–85, but since then it changes little (Figure 7). The growth of pulp and paper production in Sweden and, in contrast to the U.S., the relative stability of metals production, have offset the growth in production in less electricity-intensive sectors. Both the Swedish and U.S. shifts in electricity intensive manufacturing drop with the severe 1975 recession in materials production, but the Swedish index recovers by 1978. The U.S. index does not, however – indicating more of a structural change in the U.S. There is no downward shift in 1982 in Sweden corresponding to the dramatic 8% decline

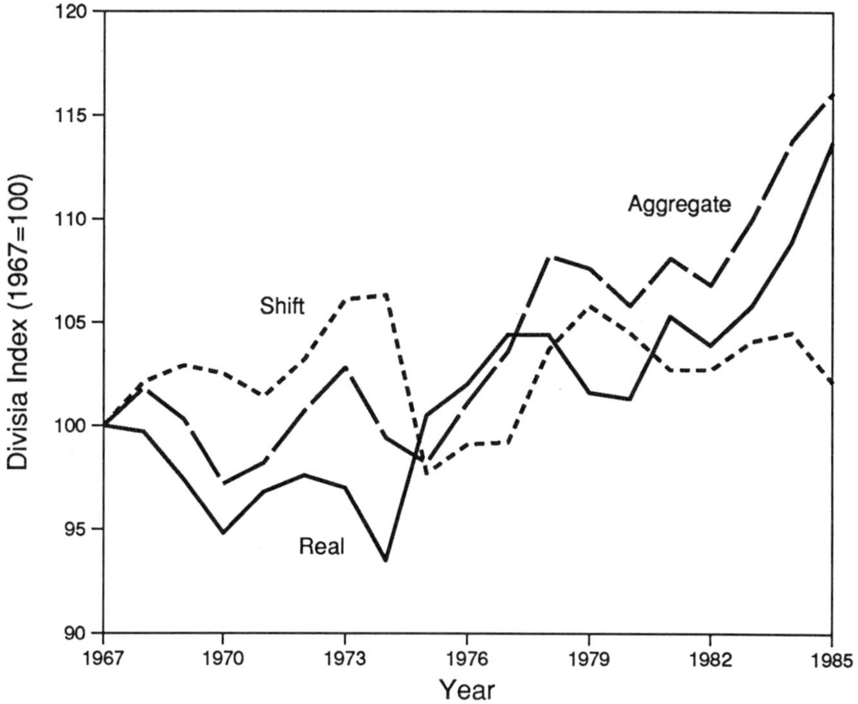

Figure 7. Divisia analysis of electricity consumption by Swedish manufacturers: Total electricity intensity using value added production index.

in that year in the U.S. (Figures 6 and 7). While electricity- intensive production in the U.S. declined in 1982, both because of a shift away from consumption of electricity-intensive products and a shift toward imported bulk-materials, Sweden maintained its production of electricity- intensive products.

The aggregate electricity intensity has been falling in the U.S. and rising in Sweden (Figures 2 and 3). The growth rates were -1.3 percent per year and +1.1 percent per year, respectively, for the period 1972–85. Almost all of this difference is due to the shift effect – i.e., the relative decline of electricity-intensive production in the U.S., a decline which has not occurred in Sweden (Table 5).

The real electricity intensity of U.S. manufacturing rose throughout the period to 1971. This was a period of rapid electrification. The rise was particularly rapid in the years from 1964 to '71. Since the early '70s, it has grown little, if at all, although there has been considerable structure. The structure is interesting: the real intensity rises sharply in the recession years 1975 and 1982, and, perhaps, in the 1970–71 recession. This is the result, at least in part, of less-efficient electricity use during periods of low production as shown both by 1) analysis of electricity intensity with respect to capacity utilization for the separate industrial sectors and 2) by records from paper mills and petroleum refineries (7). The size of the effect is consistent with most motors running the same number of hours annually independent of production levels, and indicates an important electricity savings opportunity (see chapter by Baldwin in this volume).

The recent pattern for real electricity intensity in the U.S. depends on which electricity series one examines. The trend for total electricity is lower growth than for purchased electricity for the entire 1972–85 period, but total electricity grows more strongly than purchased in the most recent years (in accord with Figure 5). For example, for total electricity using the physical output series (Figure 6) there is a slight decline in real intensity in the mid and late '70s and a slight rise in the '80s. During this period, some industries are

Table 5. Summary of Divisia analysis for electricity, Swedish manufacturing* (regression growth rates, percent per year).

	1966–85	1978–85
g (shift)	0.5	–0.1
g (real)	0.3	1.1

* Results from the two output series have been averaged.

becoming more electricity intensive, and others less intensive. There is a rough overall balance between electrification and efficiency improvement, as discussed in the chapter by Kahane in this volume.

The real electricity intensity has risen slightly more in Sweden than in the U.S. during the period since 1966. It rose 14% in Sweden and 10% in the U.S. (1966–85). During recent years the growth rate has been about 1% per year in Sweden, compared to about 0.5% per year in the U.S. (Tables 4 and 5).

As discussed in section 2.3, industrial output measurement issues are very important in determining trends in real energy intensity and sectoral shift. For the U.S., there are significant differences between the results using the BLS real output series and physical series for the materials sectors (Table 6). The trend differences range from about 0.5 to 1.0 percent per year, with the physical output series indicating lower growth in real intensity and less of a shift from the energy-intensive industries. These results are somewhat surprising. It means that the relative production of materials in tons has been declining less rapidly than the real value ascribed to them. Examination of trends, such as changes in product mix, in the separate sectors has not revealed an explanation.

For Sweden, we replace the value added production index for sector ISIC 341 with data on tons of paper and paperboard production in order to create a second output series. The results using these two series are shown in Table 7. The mixed production series shows the same effect as for the U.S., the change in shift is more positive and that of the real intensity is more negative. Little weight should be given to this result, however, because we could have used, for example, the sum of pulp production in tons and paper and board production in tons as our measure of output for ISIC 341 for the mixed series. Unlike paper and board tonnage, this measure has behaved very similarly to the value added production index.

Table 6. The effect of different production series on the Divisia analysis of total electricity use, United States (regression growth rates, percent per year).

	Output Series	1971–85	1978–85
Shift			
	BLS output	–1.6	–2.7
	Physical output	–1.1	–1.9
Real			
	BLS output	0.2	0.9
	Physical output	–0.2	0.2

Table 7. The effect of different production series on the Divisia analysis of total electricity use, Sweden (regression growth rates, percent per year).

	Output series	1966–85	1979–85
Shift			
	Production Index	0.1	–0.2
	Mixed	0.9	0.1
Real			
	Production Index	0.7	1.3
	Mixed	0.0	1.0

3.2 Materials Consumption and Production: Interpretation of Trends

Not only is materials manufacture energy and electricity intensive, but the role of materials is changing decisively in industrial societies, ie., bulk materials' share of gross national product (GNP) is declining. This combination of factors is having a powerful effect on energy use. What is behind the relative decline in materials consumption? Is it a temporary phenomenon or is it part of a long-term change in the structure of advanced industrial economies? We present a regression analysis for the U.S. and anecdotal evidence for the U.S. and Sweden on the possible causes of the shift.

The shift that occurred in the U.S. is correlated with the energy price shocks of the seventies. This leads to the hypothesis that the relatively higher energy content of the materials industries has led to increases in relative prices of these intermediate goods, causing consumers to substitute less energy-intensive goods. The effect of consumer income and tastes can also play a role in the composition of output. To estimate the relative effects of prices, income, and changing tastes and technology a simple regression model is used:

$$I_s = a + b\ \text{GNP} + c\ \text{RELPR} + d\ \text{TIME}.$$

The Divisia index of shift is regressed against GNP in billions of 1982 dollars, the lagged relative price of materials vs. all commodities as measured by the ratio of the Producer Price Index of intermediate goods to the Index for all commodities (RELPR), and a time trend (TIME) in years.

We would expect the price effect to be negative, reflecting substitution away from energy-intensive products. The sign of the coefficient of GNP is difficult to predict. Capital intensive growth would require relatively more materials while affluence may suggest a shift away from primary products to luxuries and services. Meanwhile, in association with recessions, the GNP coefficient should be positive, reflecting the stronger effect of recessions on the materials industries. The time trend is for unexplained changes in tastes and technology. The results of the regression are presented in Table 8.[2]

The relative price variable has the expected sign, but is not significant. The GNP coefficient is positive and only marginally significant, suggesting that, over the historical period growth, was materials intensive, hence electricity intensive. The regression also shows a significant negative time trend. It is noteworthy that most of the sectoral shift cannot be explained directly by price and income effects. Rather, it appears to be related to other trends in the economy: trends such as maturation of product markets, changing consumer preferences, technical change, and the import-export picture. The following presents some anecdotal evidence of these trends in the U.S. and Sweden.

Materials appear to have a life cycle as industrial economies mature (8). A century of steel consumption in the U.S., Figure 8, illustrates the cycle. Steel consumption (shipments plus net imports) in tons peaked in 1973–4, but steel consumption in tons per real dollar of GNP peaked during the First World War and has been in decline, although with wide swings, for the past seven decades! The consumption of steel and cement in Sweden show a similar pattern, although the peaks have been much more recent (9).

There are five major reasons for this declining role for steel: 1) The markets for heavy consumer products such as cars and refrigerators have become saturated i.e. consumption of these products now tends to follow population growth rather than economic growth because most purchasing units (drivers, households) have the equipment. 2) Innovative consumer products such as electronic devices tend to have low weight per dollar of cost. Moreover, consumer preferences, hence spending, are shifting from goods to

Table 8. Regression results for the determinants of U.S. sectoral shift.

Coefficient	Estimate	t-ratio
a	0.764	5.43
b	0.000197	1.39
c	–0.300	–0.74
d	–0.0146	–2.61

2 The Durbin Watson test indicated first order autocorrelation, a two-stage procedure was used to correct this in the final reported coefficients.

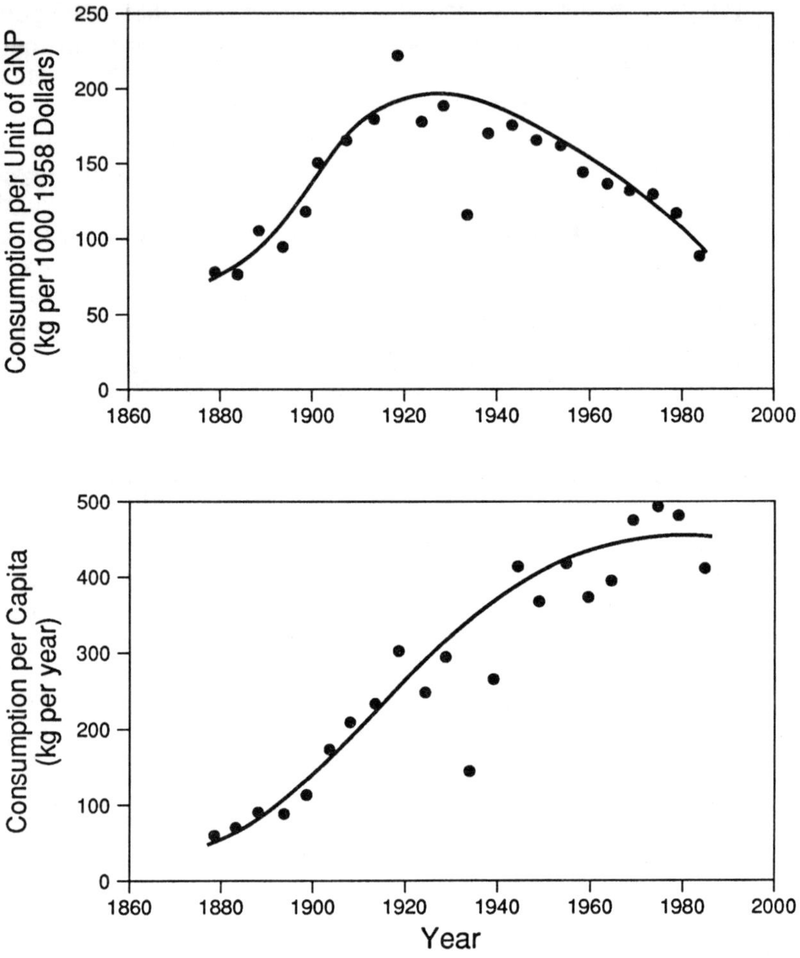

Figure 8. A century of steel consumption in the U.S. Apparent consumption, shipment of steel mill products plus net imports.
Reference 8.

services. More of the expenditure is thus for value added and less for materials. 3) As any given product is redesigned and improved, its weight tends to be reduced. For example, less steel is being used in new cars, in part because of materials prices but also technical change, e.g., higher-strength steels and vehicle redesign. This is a wide-spread phenomenon, observed in pipelines, structures and machinery as well as in consumer products. 4) Competition between materials is resulting in increasing efficiency in materials use. Newer materials such as plastics, which are more effective for certain applications, are being substituted for old materials such as steel. 5) Materials-intensive products are becoming more durable due to technological improvements. For example, the average age of cars in the U.S. has increased 2 years in the past decade, largely due to improved reliability of parts and improved coatings to inhibit rust.

What is true for materials such as steel and cement is rapidly becoming true for other materials in the U.S. as well (Figure 9). Consumption of paper and aluminum, although enjoying strong growth in some new markets, are no longer keeping pace (in tonnage terms) with GNP. Many major markets have saturated or are in decline. For paper and paperboard, for example, while consumption of office papers is booming, due in large part to photocopiers and computers, the markets for corrugated boxes and tissue have become saturated, and packaging, aside from corrugated containers, is in decline because of competition from plastics.

Plastics is the only major category of material whose consumption in tons is still growing faster than GNP in the U.S. Plastics production (which is roughly the same as consumption in the U.S.) grew 4% per year in the 1976–86 period.

Of course, there *are* rapidly growing markets for high-technology materials such as new ceramics, composites and engineering plastics. The cost per ton of these materials is predicted to remain well above the cost of bulk materials like steel, aluminum and today's commodity plastics, and consumption in tonnage terms is predicated to remain much smaller. Although important in economic and engineering terms, they will not be significant in aggregate industrial energy analysis.

Production in contrast with consumption in a nation depends of course on international trade. The U.S. and Sweden are both major importers and exporters of materials. Although the U.S. began to import heavily some important manufactured materials in the 1970s, the balance of trade from an energy-of-manufacture perspective did not shift very much. While steel imports increased, exports of chemicals increased (the latter because feedstock prices were controlled and low in the U.S.). There was also a major increase in pulp and paper trade.

Starting about 1980, the U.S. began to become a heavy net importer of manufactured materials. Steel imports have been strong. Aluminum imports are increasing. Exports of bulk chemicals are weakening. In the U.S. there are no longer any energy resources remote from broad markets, so electricity in the Pacific Northwest for aluminum production and natural gas in the Gulf states for chemicals are no longer cheap. Paper is the only major manufactured material where U.S. producers are in a position of strength. The net effect of international trade in manufactured products was that the U.S. imported about 3% of its total energy consumption in the form of embodied energy in 1984 (10). We are not aware of a comparable analysis of the energy impacts of Swedish international trade.

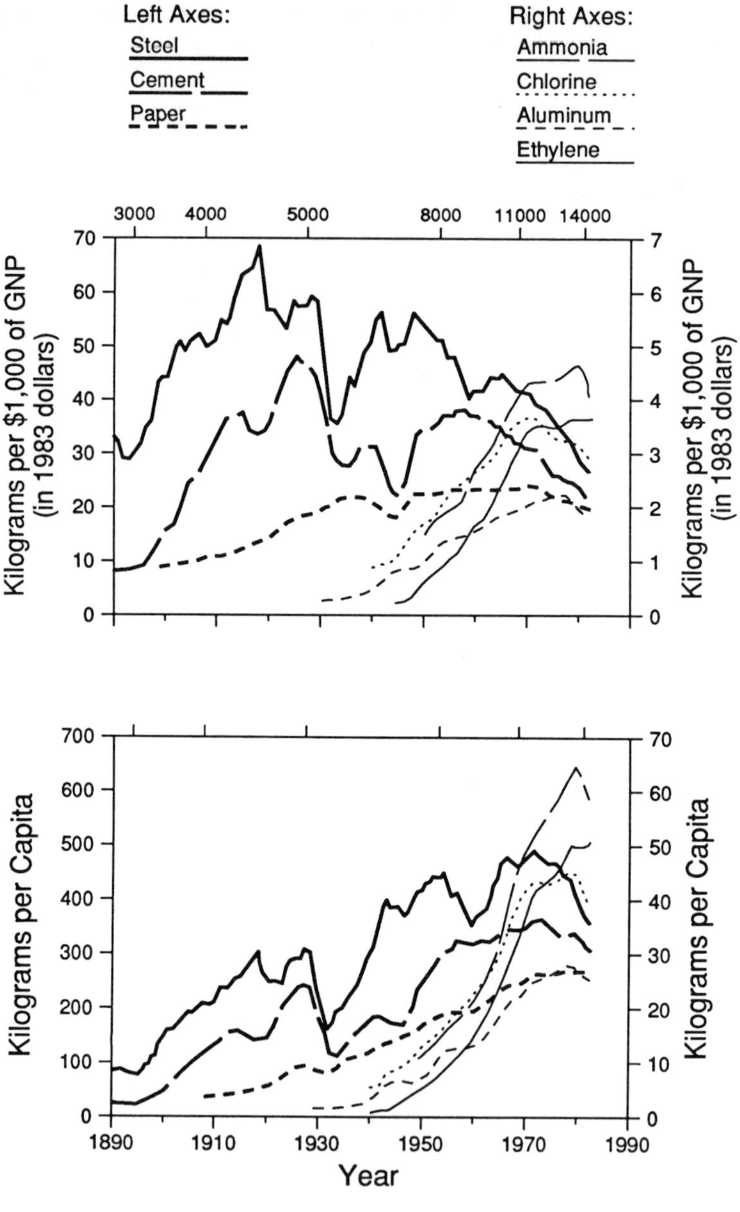

Figure 9. Trends in apparent consumption of Materials per capita and per $1000 GNP. Reference 8.

4 Analysis of Forecasts

The decomposition of growth of aggregate energy intensity into growth in real energy intensity and growth from a shift in production shares is also a useful tool for energy forecasters. Without it, one can only compare forecast growth in energy use and forecast growth in overall production. Design of energy forecast scenarios often focuses on the role of GNP or manufacturing growth, but not always on the composition of output, or the structure of the economy. This tool enables comparison of the effect of composition shifts in terms of a single index.

One such comparison is made in Figure 10 (for more details see Boyd (11)). The same energy model for U.S. industry, the Purchased Heat and Power System (PURHAPS), is used to generate two forecasts of electricity use. The forecasts differ primarily in the inputs for the mix of production. One, from the Energy Modeling Forum Study 8, EMF-8, is a forecast in which the materials industries thrive. This type of forecast assumes that capital

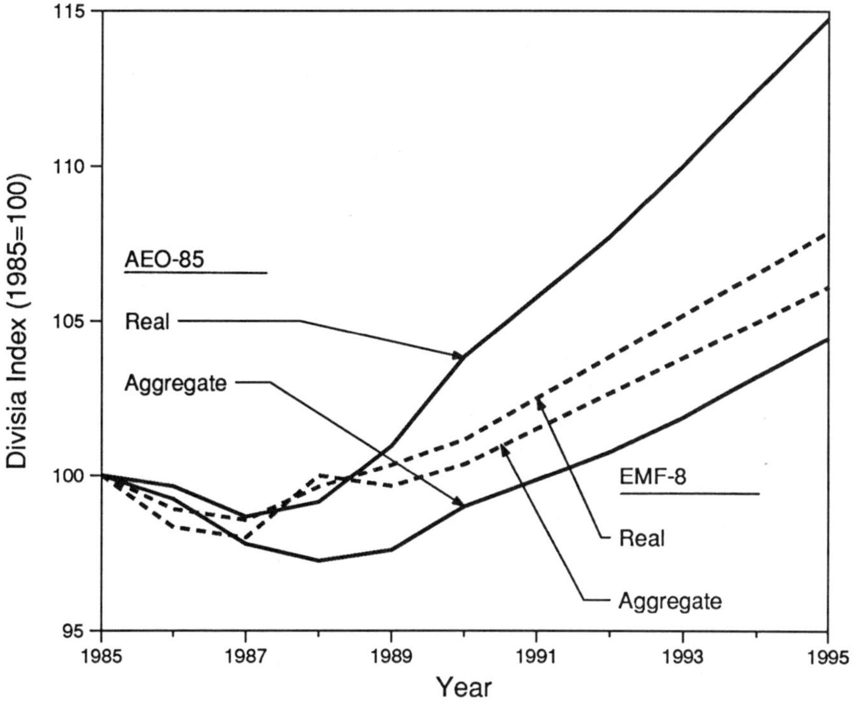

Figure 10. Comparison of the 1985 Annual Energy Outlook (AEO-85–85) and the Energy Modeling Forum Study 8 (EMF-8) forecasts using the Divisia decomposition.

investment and growth will be materials intensive, as it was in the 1960s and early 1970s, rather than reflecting the 1980s pattern of investments emphasizing electronics and other high-technology products. This forecast exhibits very little difference between the index for aggregate intensity and that for real intensity. In other words, in this forecast the production in electricity- intensive sectors keeps pace with that of other sectors. In the second forecast, taken from the 1985 Energy Information Administration Annual Energy Outlook, AEO-85, the aggregate electricity intensity grows much more slowly than the real because production in electricity-intensive sectors is assumed to continue declining with respect to total production. This forecast was created in part to reflect the concept that the change which has been occurring in the mix of products is structural, i.e. long term rather than temporary. It was the first of the major energy forecasts to adopt this perspective.

It so happens that the AEO-85 forecast also incorporates assumptions leading to rapid growth in real energy intensity, in particular the electricity intensity of the separate sectors grows rapidly. While the bottom line, the energy forecast, is similar for the two forecasts illustrated, the economic forces that would give rise to these forecasts are different. The outlook that energy planners have for structural changes in their economy plays a key role in evaluating forecasts such as these. An index of shift makes these trends explicit and easy to compare.

5 Conclusion

The identification of the impact of shifts in the mix of production on energy consumption has been shown to be a useful tool in developing a simple or summary understanding of trends in energy consumption. In the recent past, the relative decline in production of bulk materials has been responsible for the decline of electricity use per unit of overall manufacturing production in the U.S. A similar process has not occurred in Sweden, largely because electricity-intensive manufacturing is dominated by the pulp and paper industry. In mature industrial societies, consumption of paper and paperboard in tons is still keeping pace with consumption of manufactured products overall.

Similarly, one should ask of a forecast or a scenario: What is the impact of the changing relative role of electricity-intensive production? How does the forecast shift compare with recent history?

Appendix

Table A.1 List of manufacturing sectors used for the U.S.

NEA Sector	Description
13990	Ordnance and accessories
13010	Guided missiles and space vehicles
13030	Tanks and tank components
14000	Food and kindred products
15000	Tobacco manufacturers
16000	Broad and narrow fabrics, yarn and thread mills
17000	Miscellaneous textile goods and floor coverings
18990	Knitting mills
18040	Apparel made from purchased materials
19000	Miscellaneous fabricated textile products
20990A	(20990 & 20070) Lumber and wood products and Prefabricated wood buildings
21000	Wood containers
22000	Household furniture
23000	Other furniture and fixtures
24020	Paper mills, except building paper
24990A	Pulp, board and building paper mills
24990B	Converted paper products
25000	Paperboard containers and boxes
26000	Printing and publishing
27010A	Industrial organic chemicals
27010B	Industrial inorganic chemicals*
27020	Fertilizers
27030	Pesticides and agricultural chemicals
27040	Miscellaneous chemical products
28010	Plastic materials and resins
28020	Synthetic rubber
28990	Cellulosic man made fibers and organic fibers
29000	Drugs, cleaning and toilet preparations
30000	Paints and allied products
31011	Petroleum refining
31012	Miscellaneous products of petroleum and coal
31990	Paving mixtures, blocks, asphalt felts and coatings
32000	Rubber and miscellaneous plastics products
33000	Leather tanning and finishing
34000	Footwear and other leather products
35000	Glass and glass products
36010	Cement, hydraulic
36990	Stone and clay products, ex. cement hydraulic
37011A	(37011 & 37012) Coke oven products and blast furnaces and basic steel
37990	Primary iron and steel ex. blast furnaces and basic steel

Table A.1 List of industry sectors used (Contd.).

37999	Forgings
38040	Primary aluminum
38990	Primary nonferrous metals, ex. primary aluminum
39000	Metal containers
40000	Heating, plumbing, and fabricated structural metal products
41010	Screw machine products
41990	Metal stampings
42000	Other fabricated metal products
43000	Engines and turbines
44000	Farm and garden machinery
45000	Construction and mining machinery
46000	Materials handling machinery and equipment
47000A	(47000 & 48000) Metalworking machinery and equipment and special industry machinery and equipment
49000	General industrial machinery and equipment
50000	Miscellaneous machinery except electrical
51000	Office, computing, and accounting machines
52000	Service industry machines
53990	Electrical transmission and distribution equipment
53010	Instruments to measure electricity and industrial apparatus
54000	Household appliances
55000	Electric lighting and wiring equipment
56000	Radio, television and communication equipment
57000	Electronic components and accessories
58000	Miscellaneous electrical machinery, equipment and supplies
59000	Motor vehicles and equipment
60000	Aircraft and parts
61990	Other transportation equipment
61060	Mobile homes
62000	Professional, scientific and controlling instruments and supplies
63000	Optical, ophthalmic, and photographic equipment and supplies
64000	Miscellaneous manufacturing

*Electricity used for uranium enrichment is omitted.

Table A.2 List of manufacturing sectors used for Sweden.

ISIC	Description
311/2	Food products
313	Beverages
314	Tobacco
321	Textiles
322	Wearing apparel
323	Leather and products
324	Footwear
331	Wood products
332	Furniture and fixtures
341	Paper and products
343	Printing, publishing
351	Industrial chemicals
352	Other chemical products
353	Petroleum refineries
354	Petroleum, coal products
355	Rubber products
356	Plastic products NEC
361	Pottery, china, etc.
362	Glass and products
369	Non-Metal products NEC
371	Iron and steel
372	Non-ferrous metals
381	Metal products
382	Machinery NEC
383	Electrical machinery
384	Transport equipment
385	Professional goods
390	Other industries

The following sources were used to compile data for the figures and to perform the Divisia analysis of U.S. and Swedish manufacturing trends.

Energy Information Administration. 1987. State Energy Price and Expenditure Report 1985, U.S. Department of Energy, DOE/EIA-0376(85).

U.S. Department of Commerce, Annual Survey of Manufactures. various years.

U.S. Department of Energy. 1987. Manufacturing Energy Consumption Survey. Monthly Energy Review, January.

U.S. Bureau of Labor Statistics, Productivity Measures for Selected Industries, and unpublished supporting data.

Energy Balances for OECD Countries 1970–1985. 1987. International Energy Agency, Paris.

United Nations, Industrial Statistics Yearbook, Vol. I, various years.

Ross, Marc. 1984. Measuring Trends in the Productivity of Chemicals. University of Michigan, Department of Physics, discussion paper.

Acknowledgments

We would like to thank the U.S. Department of Energy, Assistant Secretary for Fossil Energy, Office of Planning and Environment for their support of the analysis of sectoral shift issues in the U.S. In particular we wish to thank H. W. Hochheiser for his valuable assistance. The research contained within this paper would not have been possible without the assistance of Ron Fisher, who provided substantial programming support, Vickie Lazar, who did the graphics, and Anita Jandula, who did the typing. Finally we are indebted to Adam Kahane, Gunnar Larsson, Thomas B. Johansson, Robert Williams, and an anonymous referee for their helpful comments on an earlier draft. Any remaining omissions or errors are, of course, our responsibility.

References

1. Jorgensen, D.W. 1984. The Role of Energy in Productivity Growth. EPRI Annual Report, EA-3482, Electric Power Research Institute, Palo Alto, CA.
2. Walfridson, B. 1987. Dynamic Models of Factor Demand: An Application to Swedish Industry, Ph.D. thesis, Department of Economics, University of Gotesborg, Sweden.
3. Solow, J. 1987. The Capital Energy Complementarity Debate Revisited. *The American Economic Review*, Vol. 1. 77(4): 605–14.
4. Boyd, G., et al. 1987. Separating the Changing Composition of U.S. Manufacturing Production from Energy Efficiency Improvements: A Divisia Index Approach. *Energy Journal*, 8(2): 77.
5. Faucett Associates. 1984. National Energy Accounts 1958–1981. U.S. Department of Commerce. PB85–142024.
6. Ross, M. 1984. Materials Trends in the U.S. Economy. University of Michigan, Department of Physics, discussion paper.
7. Boyd, G., et al. 1986. Disaggregation of Industrial Fossil Fuel Use in the 1985 National Energy Policy Plant, Methodology and Results. Argonne National Laboratory Report, ANL/EES-TM-303.
8. Williams, R.H., Larson, E.D., Ross, M.H. 1987. Materials, Affluence and Industrial Energy Use. *Annual Review of Energy*, 12: 99–144.
9. Steen, P., et al. 1986. Energy – For What and How Much?. Liber Förlag, Stockholm, (in Swedish).
10. Roop, Joseph M. 1986. The Trade Effects of Energy Use in the U.S. Economy: An Input-Output Analysis. Proc. 8th Annual North American Conf. of the International Association of Energy Economists, Cambridge, MA.
11. Boyd, G., et al. 1986. Energy Intensity in Manufacturing: A Comparison of Historical Results with Forecasts. Proc. 8th Annual North American Conf. of the International Association of Energy Economists, Cambridge, MA.

Technological Change and Industrial Electricity Use

Towards Better Generalizations

Adam Kahane

Abstract

Changes in industrial electricity consumption are fundamentally the result of changes in production mix and technology. Decisions about technology are driven by very diverse market pressures, and other pressures, not simply by the relative price of electricity or the characteristics of electricity-using equipment. Electricity demand analysts and forecasters must therefore examine the pressures on individual industries in particular market areas, and cannot rely on simplistic views about general trends.

Examples to support this argument are taken from case studies of the steel, paper and cement industries in France, Japan, Sweden and the U.S. It is possible to identify the key pressures driving electricity consumption, but they differ across industries and countries. General conclusions are drawn about the links between technological change and electricity consumption. No general conclusions can be drawn, however, about whether consumption will increase or decrease.

1 Introduction

Trends in the use of electricity in industrial production can be measured by the change in electricity consumption per unit of output (the real intensity), correcting for changes in the composition of output (the shift effect). In the U.S., the average real intensity has been essentially flat since 1972, which means that overall production technology has become neither more electrified nor more electricity-efficient.(1) Over the same period, however, the real electricity intensity has increased markedly in Sweden and decreased markedly in Japan. (2)

These divergent trends raise a basic question about electricity demand for analysts and forecasters: What determines the real electricity intensity in different industries and countries? If the real intensity is measured in physical units (eg, kilowatt-hours per tonne of product) and if the effect of changes in product mix can be separated out – Boyd and Ross attempt both in this volume – then changes in the real intensity are essentially driven by changes in the production technology employed.

This chapter examines the impact of technological change: Will the introduction of new technologies into factories result in increased or decreased industrial electricity consumption? Two general views have been put forward on this question. One view, articulated from different engineering and economic perspectives, is that the particular productivity-enhancing properties of electricity-driven technologies (flexibility, controllability, cleanliness at point of use, etc) will encourage the adoption of these technologies and will push up industrial electricity consumption, as well as consumption per unit of output (real intensity) (3).

Another view is that the overall rationalization of production achieved by new technologies will generally keep unit consumptions flat or push them down, in spite of the introduction of certain electricity-driven technologies. (4)

These two propositions are *not* mutually exclusive. In fact, most of the authors cited recognize that both technology effects exist, although they differ in which is emphasized and thought to dominate.

Figure 1 illustrates a framework for considering the links between technological change and electricity consumption. The amount of electricity consumed in the industrial sector is determined, by definition, by how much and what kinds of goods the sector produces and how the goods are produced. Market and other external pressures, such as the demand for and profitability of different products, determine how much and what kinds of goods are produced. These in turn determine what kind of technology is used, with different technologies having different intensities or unit consumptions (electricity consumption divided by output, expressed either in real money or physical terms). Choices between alternative technologies are also driven by market and other pressures, such as environmental regulations and the prices of different inputs. Technology choice is therefore driven, indirectly as well as directly, by various external pressures.[1]

[1] This model can be refined. For example, the output level has an impact on unit consumption, even disregarding changes in technology, because unit consumptions go up at low levels of capacity utilization. Another point is that technology choice takes place over several different time scales, from daily or monthly decisions about which installed machines to run, to annual decisions about what kind of new machines to buy among those already on the market, to longer-term decisions about R&D; all of these are driven by market and other external pressures. In more general economic terms, the question is how the demand for electricity is affected on balance by changes in the level and composition of output, the prices of inputs (labour, capital, energy, materials, etc) and autonomous technological change.

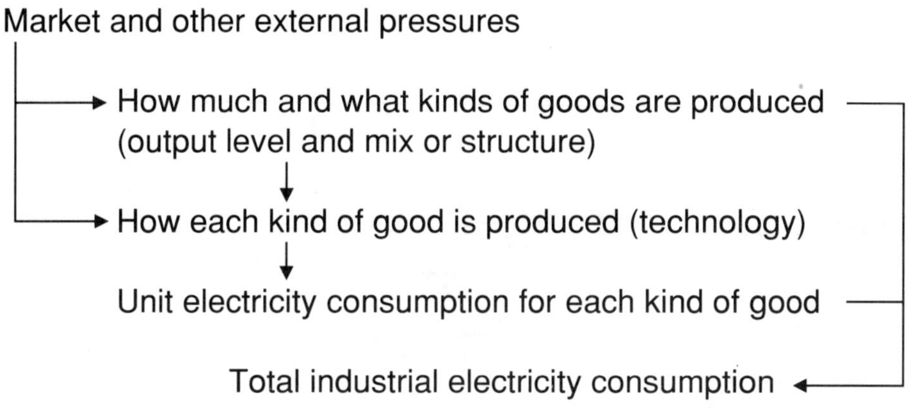

Figure 1. Framework for analyzing industrial electricity consumption.

What general impact have these pressures had on technology and on electricity use? The historical changes in unit consumptions of individual industries are very mixed. For the period 1976–86, for example, unit consumptions generally increased in Sweden, decreased in Japan and varied in the U.S., although they increased in some industries and decreased in others in all three countries (Figure 2).[2] Although it is possible to calculate the average of the changes in real intensities (as Boyd and Ross have done), it is difficult to find any general trends across industries or countries.

One external factor has been generally important: the price of electricity relative to other fuels (Figure 3). Relative electricity prices have been highest in Japan and lowest in Sweden, with the U.S. and France in between; this correlates well with the average changes in real intensity. To go beyond this simple correlation, it is necessary to study individual industries in more detail.

2 Industry Case Studies

Case studies were undertaken of iron and steel (ISIC 371), pulp and paper (3411) and cement (3692) in France, Japan, Sweden and the U.S.[3] The general

[2] Movements in unit consumption are particularly sensitive to the method of calculation (e.g. choice of end year, industry definitions and output measure). See Kahane and Squitieri, Reference 4.

[3] Motor vehicle manufacturing in Japan and the U.S. was also reported in Kahane, Industrial Electrification: Case Studies of Four Industries, Reference 4, but is not discussed here.

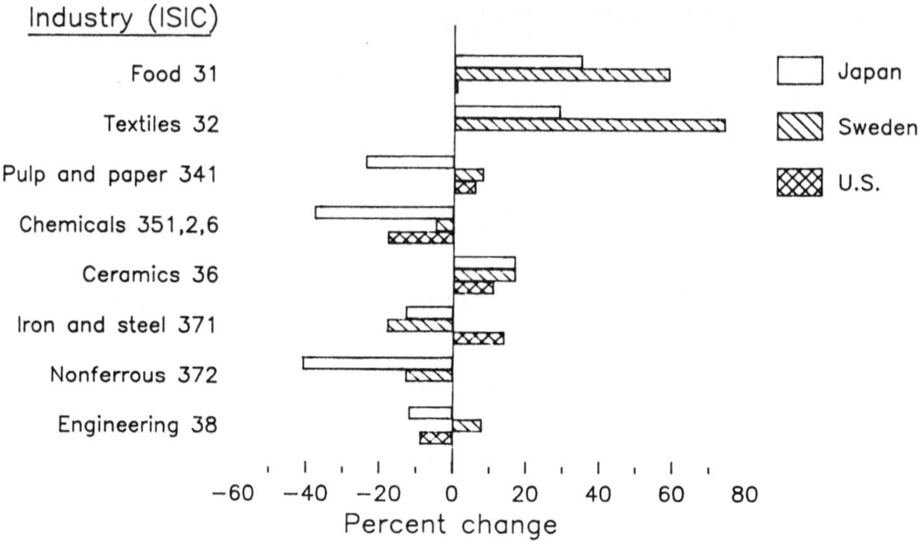

Figure 2. Changes in electricity intensity, 1976–86
Calculated from total electricity consumption, purchased plus self- generated, and indices of real manufacturing production (weighted by base year added value) for Japan, and real value of production for Sweden (in 1980 krona) and the U.S. (in 1984 dollars). For the U.S., the industry definitions are approximate, the change is for 1976–85, and the figure shown for iron and steel includes non-ferrous metals.

Sources: The Institute of Energy Economics, "Energy Balances in Japan", Tokyo, 1987; Petsala, B., National Energy Administration, Stockholm, private communication, 1988; and Holte, J., Energy Information Administration, Department of Energy, Washington, personal communication, 1988.

approach was to examine national-level government and industry association statistics, as well as data from industry researchers, consultants and individual companies.

2.1 Iron and Steel

Electricity consumption per tonne of raw steel produced rose between 1973 and 1984 in France, Japan and the U.S., and fell in Sweden (Figure 4). For the most part, these changes (as well as the absolute consumption levels) can be explained by changes in steel industry technology, with the largest factor being

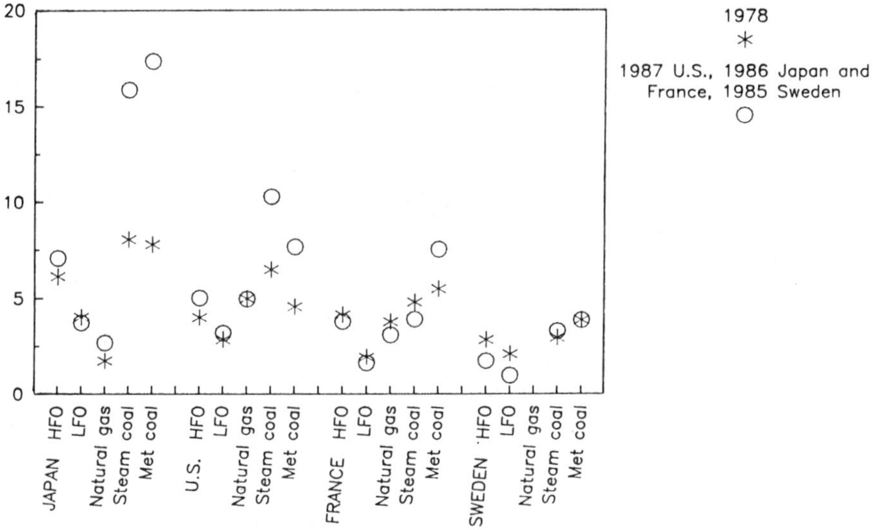

Figure 3. Prices of electricity relative to other fuels
Calculated from average prices (including taxes) in net calorific value thermal equivalents. Electricity is counted at end use, not primary energy value.

Source: International Energy Agency, "Energy Prices and Taxes", Paris, 1988.

increased electric arc furnace (EAF) steel production for scrap.[4] Even though the mix of final steel products generally shifted towards cold-rolled, specialty and other higher value-added products (which all require somewhat more final processing), this shift accounted for less than 10% of the total change in unit consumption. (5) Therefore, the changes shown in Figure 4 can be interpreted (approximately) as the net result of the different changes in technology.

The share of EAFs in total raw steel production increased in all four countries (Figure 4), pushing up unit consumption. For many kinds of steel, production by EAFs is less expensive, with respect to both operating and capital costs, and is economic at smaller capacities, making siting and expansion more flexible. (6) In addition, production of higher value-added alloy and other specialty steels favours EAF use; in this way, product mix indirectly affects unit electricity consumption. (The Swedish share of specialty steel in total final products was 33% in 1984, compared to 14% in the U.S. and

4 Production of a tonne of raw steel from scrap using an EAF typically requires 500–600 kWh, compared to less than 100 kWh via the coke oven-blast furnace-basic oxygen/open hearth furnace route. (See Kahane, Industrial Electrification: Case Studies of Four Industries, Reference 4.)

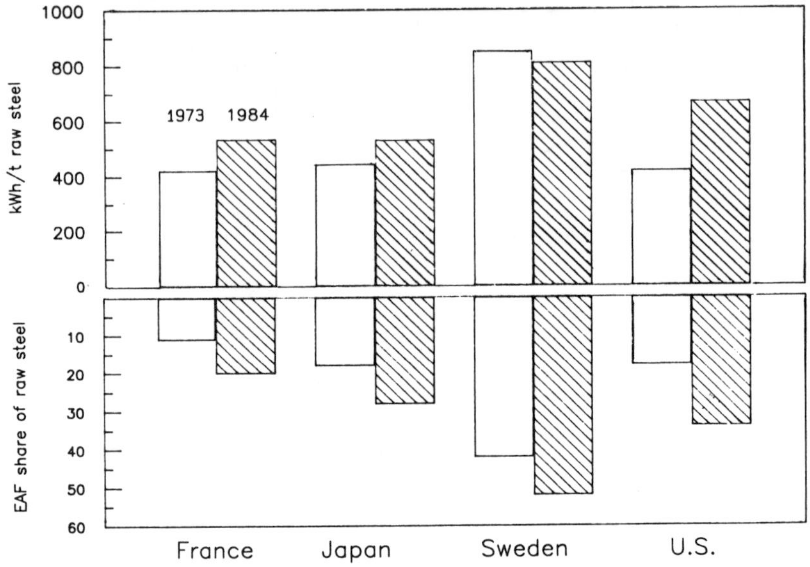

Figure 4. Steel industry electricity use, kWh/t
Calculated from total electricity consumption, purchased plus self generated, and metric tonnes of raw steel produced.

Sources: Chambre Syndicale de la Siderurgie Francaise, "L'Energie dans la Siderurgie", Paris, 1985; Japan Iron and Steel Federation, "Japanese Steel Statistics Yearbook", Tokyo, various years; Lindblad, B., The Swedish Ironmasters Association, Stockholm, personal communication, 1988; American Iron and Steel Institute, "Annual Statistical Report", Washington, various years.

18% in Japan, which partly explains the much higher use of EAFs shown in Figure 4.)

Increased use of EAFs does not, however, entirely explain the changes in industry unit consumption. In France, Japan and the U.S., the increased EAF share accounted for only about half of the total increase in unit consumption (7), and in Sweden other changes resulted in a fall in unit consumption in spite of the increase in EAF share (Figure 4).

Other important technological changes that pushed electricity consumption *up* were:

- Additional pollution control equipment, primarily dust collecting blowers and water recirculation pumps, installed in response to stricter limits on air and water emissions,

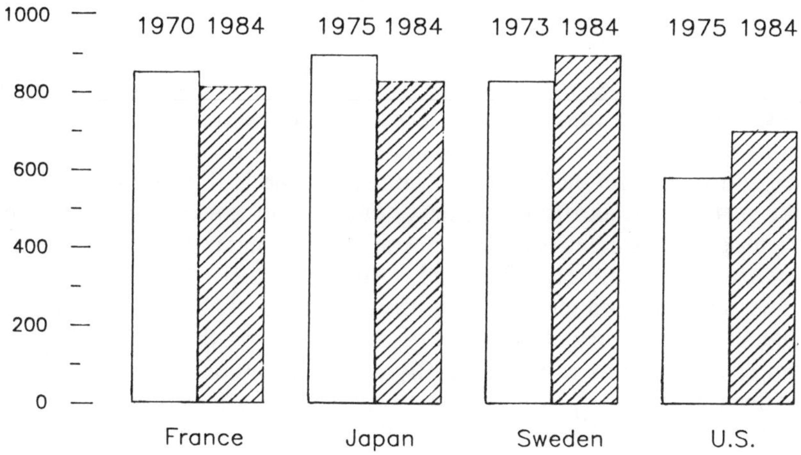

Figure 5. Changes in pulp and paper industry unit consumption, kWh/t
Calculated from total electricity consumption, purchased plus self- generated, and metric tonnes of pulp and paper produced.

Sources: Pinsac, J-L., Centre d'Etudes et de Recherches Economiques sur l'Energie, Paris, personal communication, 1986; Japan Paper Association, "Statistical Bulletin", Tokyo, 1985; Ministry of International Trade and Industry, "Yearbook of Pulp and Paper Statistics", Tokyo, 1985; Axelsson, L-E., Swedish Pulp and Paper Association, Stockholm, personal communication, 1988; American Paper Institute, "1985 Statistics of Paper, Paperboard and Wood Pulp" and "Patterns of Fuel and Energy Consumption in the U.S. Pulp and Paper Industry", 1985.

- Replacement of steam turbine with electric blowers on large blast furnaces, and
- Additional pumps and fans for waste heat and gas recovery in all process stages, primarily to conserve fuel.

Changes that pushed consumption *down* included:

- Technologies to accelerate EAF melting and hence conserve electricity, including ultra-high power operation, oxygen injection, water-cooled linings, scrap preheating and oxy-fuel burners,
- Variable speed drives on pumps and fans, and
- Factory consolidation and rationalization.

Electricity-using technologies (notably EAFs) are, in many cases, chosen to respond to external pressures. Nevertheless, electricity-conserving technologies (notably technologies that accelerate EAF melting) are also chosen. As a result, the net change in steel industry electricity consumption can be either positive or negative, as shown in Figure 4.

2.2 Pulp and Paper

In the pulp and paper industry, electricity is used in making pulp as well as making paper and board products. Electricity consumption divided by tonnes of pulp, paper and board[5] decreased between the mid-1970s and 1984 in France and Japan and increased in Sweden and the U.S. (Figure 5).

Like in the iron and steel industry, changes in the mix of pulp and paper goods produced is relatively unimportant in understanding the movements in unit consumption, although (also like iron and steel) the distinction between changes due to product mix and those due to technology change is not completely satisfactory. The effect of changes in product mix, i.e. pulp/paper mix and paper type mix, was generally relatively small. (8)

Changes in pulping processes used had a larger effect on unit consumption: negative in France, Japan and the U.S. and positive in Sweden. In the first three countries, pulping shifted towards less electricity-intensive processes with the replacement of sulfite, semi-chemical and mechanical pulping by sulfate, and the replacement of wood pulp by pulp from recycled fibers.[6] In Sweden, by contrast, mechanical replaced chemical pulping, with the growth in exports of mechanically-produced newsprint.[7] This illustrates again the complex interaction between changes in unit consumption, product mix and technology choice.

The residual changes in unit consumptions can generally be attributed to the net effect of technology changes in individual processes, i.e. other than technology change through the choice of pulping process. This was essentially zero in France, negative in Japan, and positive in Sweden and the U.S.[8] These technology changes include additional pumps and fans for heat recovery

[5] This ratio is the most useful general measure of unit consumption, instead of consumption divided by tonnes of paper and board (as might be appropriate for some individual countries) or divided by tonnes of pulp (the analogy to consumption per tonne of raw steel), because some countries with large forest industries (notably Sweden) export pulp as a final product.

[6] The share of thermo-mechanical pulping, a particularly electricity-intensive process, increased quickly, but from a small base and generally replacing other mechanical pulping, and therefore did not have a large net effect on unit consumptions.

[7] The share of mechanical in total pulping explains pulp and paper electricity intensity much less adequately than the share of electric arc furnaces explains steel electricity intensity.

[8] The positive movements in this variable for Sweden and the U.S. are exaggerated in this calculation, for two different reasons. In Sweden, the underlying calculation included shifts between different types of mechanical pulping (e.g. thermo-mechanical) and between different types of chemical pulping (e.g. sulfate) in the residual component, not in the pulping mix component as for the other countries. In the U.S., the survey which produced the underlying consumption statistics was changed in 1982, and so the 1975–84 unit consumption increase (included entirely in the residual) may be overstated. For these reasons, the real net effect of technological changes is probably less of an increase than indicated. See references to Figure 5.

Figure 6. Cement industry electricity use, kWh/t
Calculated from total electricity consumption, purchased plus self- generated, and metric tonnes of cement.

Sources: Japan Cement Association, "Selected Statistics of the Japanese Cement Industry", Tokyo, 1985; Portland Cement Association, "Energy Report of the Portland Cement Industry", Skokie, 1985; Petsala, B., National Energy Administration, Stockholm, personal communication, 1988.

systems to conserve fuel, and additional pollution control equipment, which all push consumption up. They also include changes in motor sizing and number of motors and the installation of variable-speed motors, which push consumption down.

2.3 Cement

The cement industry provides a tidy case study because production material flows are straightforward. The industry's output is also fairly homogeneous, so changes in product mix can be neglected for purposes of analyzing unit electricity consumption. Given this simplicity, the 1972–84 changes in unit consumption, up in the U.S. and Sweden and down in Japan (Figure 6), seem surprising.

The key technological change in the cement industry since the early 1970s has been the gradual replacement of wet process clinker kilns with dry process kilns. This replacement, which has occurred in all three countries, has been driven by the much lower unit fuel consumption of the dry process, which avoids the need to evaporate water from a clinker slurry. However, even though the dry process has a lower unit fuel consumption, it has a higher unit electricity consumption because it requires large suction fans and additional drier drives and dust control equipment. (9)

The increase in Swedish and U.S. unit electricity consumption is explained by this replacement of wet with dry process kilns, along with some additional motor drive for pollution control. The decrease in Japan, even though the same process conversion was being implemented, points to additional changes in technology in Japan. These were directed towards electricity conservation, and included the replacement of roller with ball mills in raw material grinding, and improved cement classifiers to reduce re-grinding.

The introduction of different technologies in Japan is partly due to the differences between Japanese and U.S. input prices, raw materials and final product specifications, and illustrates the importance of local market conditions on technology choice. In addition, some of the technologies already installed in Japan are also economic in other countries; in these cases, the Japanese example suggests that further equipment modernization in Sweden and the U.S. will drive down unit consumption, in contrast to their experience to date.

3 Implications and Conclusions

The three industries studied are not completely representative of the industrial sector as a whole. The other important primary materials industries (aluminium and chemicals) have generally shown declining unit electricity consumptions (10), and unit consumptions in the light industries (such as motor vehicle manufacturing) are driven more by a collection of small technology changes than by a few dominant process changes. (11) Nevertheless, the very varied connections between technological change and electricity use in these three industries alone suggest several general implications and conclusions.

3.1 Electricity and Technological Change

Many of the new technologies installed in response to external pressures use electricity. The most common examples are technologies for automation, pollution control, water, waste and heat recovery, additional product finishing, and specialized heating. In general terms, electricity contributes to the solution of many industry "problems".

If all other aspects of production were unchanged, then, we could expect that technological change would generally push up unit electricity consumption. The reason why unit consumptions actually decreased in some cases (after correcting, as far as possible, for changes due to product mix) is that other aspects of production did change, in response to these same or other external pressures.

3.2 Add-On Technologies vs. Process Change

A key distinction here is between "add-on" technological change – one which doesn't make a basic change in the type of process used or the way it is organized – and process change. Introduction of an electricity-driven add-on technology – such as electric instead of fuel boilers in paper mills or induction heating in steel finishing – will usually drive up unit consumption. On the other hand, introduction of an electricity-driven process technology – such as mechanical pulping or an automated assembly line – will not necessarily push up unit consumption because process changes are often accompanied by overall rationalization. This rationalization, by reducing material throughput (high consistency pulp) or waste (continuous casting), or by eliminating process steps (EAFs, automation), can indirectly reduce unit consumption and offset the direct increase from the new electricity-using equipment.

In the three industries studied, this indirect effect of process rationalization has been important. The process changes that have been central to technological change in these industries and that dominate changes in electricity use have generally been electricity-using. Nevertheless, they have not always pushed up unit consumptions. In steel, greater use of EAFs pushed up unit consumptions, but continuous casting and direct rolling pushed them down. In paper, more mechanical pulping and paper finishing pushed up unit consumptions, but greater waste paper use and reduced water throughput pushed them down. In cement, dry process kilns pushed consumptions up, but modifications to the grinding processes pushed them down. In other words, a new process can use electricity but still result in decreased unit consumptions.

In addition to the indirect effect of process change on electricity use through overall process rationalization, there is another reason why technological change is not always accompanied by increased unit consumption. In industries where electricity costs are significant, the pressure to reduce these costs through electricity conservation is itself an important pressure for technological change. This is important both in electricity-intensive processes such as EAF steelmaking, where improvements in furnace design and operation have significantly reduced unit consumptions, and in distributed operations, where general factory "energy awareness" has resulted in electricity savings. Since technological change includes the introduction of these electricity-saving as well as electricity-consumptive technologies, unit consumptions have not always increased.

3.3 Truly New Technologies

The technologies examined in the case studies and emphasized above are generally conventional, i.e. have been used in these industries in some form for several decades. The case studies left largely unanswered questions about the impact of "truly new" technologies on electricity use.

Two groups of new technologies can be considered here. First, technologies that build on the important, recent research frontiers – e.g. new materials, biotechnology, catalysis and information technology – have not yet been widely applied to industrial processes. At this point, we have little indication of the overall impact of these technologies on industrial electricity consumption. A study of automation in motor vehicles manufacturing – a high-profile but relatively simple application of information technology – showed no clear net trend. (12) This examination did, however, suggest one important point: the downward indirect effect of the full application of a frontier technology (through improving the efficiency of, or perhaps modifying, an entire production process) is likely to be larger than the direct effect (through the electricity consumption of the new machine).

The second group of new technologies are "electrotechnologies", those relying on electromagnetic, electromechanical, electrophysical or electrochemical effects. (13) Many of these technologies are available for commercial application and are enthusiastically promoted by the electric utility industry. Collectively, they have the potential to significantly increase total industrial electricity consumption. (14) In the industries studied, however, new electrotechnologies (i.e. excepting EAFs) have not had a significant impact on electricity consumption to date.

3.4 Conclusions

The case studies suggest that some approaches to understanding industrial electricity demand will be more helpful than others to analysts and forecasters.

"Electrification" and "electrotechnologies" are very poor organizing principles for studying the impact of technological change on electricity use. In other words, the two views mentioned at the beginning of this paper are not helpful in generalizing the nature of this impact.

The important trends in technologies adopted can only be understood by considering the external pressures on each industry. These pressures, although often similar across industries – cost reduction, environmental regulation, etc. – lead to completely different technology solutions in each industry and hence to different movements in unit electricity consumptions. For this reason, it is unreasonable to expect general trends in unit electricity consumptions.

International comparisons are useful in broadening our understanding about technological possibilities. Nevertheless, the comparisons reveal little about the prospects for individual countries, since industries in each country must ultimately respond to local market forces. On the other hand, the comparisons can be very helpful if the domestic forces in one country are expected to change to resemble those in another (e.g., much higher electricity prices in Sweden) or if market forces are expected to converge among countries (e.g., Europe in 1992).

In summary, forecasters cannot rely on electricity-oriented conceptual models, general trends in consumption, or international comparisons. This leaves us with the familiar tools: econometric and engineering models. It is imperative, however, that these tools be applied based on information about particular markets (not other industries or regions) and on information about market pressures in general, not simply the price of electricity.

References

1. Boyd, G.A., Ross, M.H. 1987. *The role of sectoral shift in trends in electricity use in United States and Swedish manufacturing and in comparing forecasts*, in this volume. The trend was different from 1958 to 1972, when the real intensity increased. Similar results are reported by Marlay, B., *Industrial Electricity Consumption and Changing Economic Conditions* in The Changing Structure of American Industry and Energy Use Patterns: Issues, Scenarios and Forecasting Models, ed. Faruqui, A., Electric Power Research Institute EM-5075-SR, Palo Alto.
2. Swedish National Board for Technical Development, *Industrial Use of Energy, 1970–84*, Energy Technology No.2, Stockholm, 1987. Kahane, A. *A Bottom-Up Analysis of Electricity Use in Member Countries of the IEA*, Electricity in IEA Countries: Issues and Outlook, IEA/OECD, Paris, 1985.
3. Schurr, S. 1986. *Electricity Use, Technological Change and Productive Efficiency. An Economic-Historical Perspective*; Schmidt, P. *The Form Value of Electricity: Some Observations and Cases*; Burwell, C. *Industrial Electrification: Current Trends and Case Studies*; Jorgenson, D. *The Role of Energy in Productivity Growth*, all in Electricity Use, Productive Efficiency and Economic Growth, ed. Schurr, S., Sonenblum, S., Electric Power Research Institute, Palo Alto.
4. Kahane, A. 1987. *Electricity Use in U.S. and Japanese Manufacturing*. Papers and Proceedings of the Eighth Annual North American Conference, ed Wood, D., International

Association of Energy Economists. Kahane, A. 1986. *Industrial Electrification: Case Studies of Four Industries*, Report LBID-1261, Lawrence Berkeley Laboratory, Berkeley. Kahane, A., Squitieri, R. 1987. *Electricity Use in Manufacturing. Annual Review of Energy 12*. Walker, W. 1985. Information Technology and the Use of Energy. *Energy Policy* October 1985.

5. Kahane, A. Industrial Electrification: Case Studies of Four Industries, Reference 4.
6. Schmidt, P. Reference 3.
7. Kahane, A. Industrial Electrification: Case Studies of Four Industries, Reference 4.
8. Kahane, A. 1988. Industrial Electrification: Case Studies of Four Industries, Reference 4, for France, Japan and the U.S., Larsson, G., Swedish State Power Board, Stockholm, private communication.
9. Garrett-Price, B.1985. *Potential for Energy Conservation in the Cement Industry*, Pacific Northwest Laboratory.
10. Boyd, G.A., Ross, M.H., in this volume.
11. Kahane, A. Industrial Electrification: Case Studies of Four Industries, Reference 4.
12. Kahane, A. 1988. Industrial Electrification: Case Studies of Four Industries, Reference 4, Uekusa, Y. *Information-Intensive Society and Energy Problems*, Energy in Japan, Institute of Energy Economics, Tokyo.
13. Resource Dynamics Corporation. Industrial Technology Assessment Guide, Electric Power Research Institute, Palo Alto, October 1987, and supporting publications.
14. Resource Dynamics Corporation. Electrotechnology Reference Guide, Electric Power Research Institute, EM-4527, Palo Alto, 1986.

Expanding Roles for Gas Turbines in Power Generation

Robert H. Williams and Eric D. Larson

Abstract

For the next two to four decades the gas turbine is likely to assume major new roles in power generation–in central station and cogeneration applications, in industrialized and developing countries. The gas turbine will be much more widely used both because the changing circumstances of the electric power industry are especially conducive to the gas turbine and because improvements in gas turbine technology are making gas turbines more competitive. In a wide range of circumstances, new, highly-efficient, gas turbine-based power plants will be able to provide electricity at lower cost and with less adverse environmental impacts or safety problems than coal or nuclear steam-electric plants.

Already gas turbine/steam turbine combined cycles are attracting the interest of power planners throughout the world. While combined cycles offer many advantages compared to conventional steam-electric power plants, this class of energy-efficient gas turbine-based power systems does not exhaust the possibilities. In addition, advanced gas turbine cycles involving steam-injection for power and efficiency augmentation have great promise for some important market applications. The various steam-injected gas turbine cycles that are available or under development are based on turbines derived directly from aircraft engines. Both types of turbines have major roles to play in power generation; hence, in decades ahead a more balanced mix of heavy-duty industrial and aeroderivative turbines than is presently anticipated will most likely evolve in power markets.

1 Introduction

A revolution is underway in electricity generating technology. It may soon radically transform the power industry, in both industrial and developing countries. This revolution involves not an exotic new technology, but rather an upgrading of the familiar but little-used gas turbine–the neglected step-sister of the steam turbine in power generation.

Though it has always offered a low unit capital cost, the gas turbine has traditionally been restricted in utility applications to peaking plants, because of its low efficiency and requirement for high-quality fuel. Its peripheral role in power generation also fostered less-than-rigorous maintenance practices, helping to give the gas turbine a reputation of poor reliability among utility planners.

Although in many parts of the world its use for cogeneration has been inhibited by institutional constraints, the simple cycle gas turbine's high thermodynamic efficiency for combined heat and power production as well as its low capital cost have long made it an economically attractive option for cogeneration applications. But the simple cycle gas turbine has been used mainly in cogeneration applications characterized by steady steam loads due to its poor part-load performance.

Innovations, though, are making gas turbines competitive in cogeneration markets characterized by variable heat loads and in central-station applications with conventional baseload and load-following technologies, using low-quality as well as high-quality fuels.

This chapter will show how the changing environment for power generation tends to favor a technology like the gas turbine and how changing gas turbine technology is making the gas turbine an increasingly strong competitor as a power source. A distinction is made here between heavy-duty industrial turbines (designed specifically for stationary applications) and aeroderivative turbines (derived from jet engines). The heavy-duty industrial turbine is quite familiar in the power sector and is the technology of choice for the gas turbine/steam turbine combined cycle systems that are beginning to come into wide use in various parts of the world. The various configurations of aeroderivative turbines of interest–e.g., the steam-injected gas turbine, the intercooled steam-injected gas turbine, the chemically recuperated, steam-injected gas turbine–are much less well-known. In this paper, which has a future-oriented perspective, emphasis is on aeroderivative turbines, because their most interesting cycle modifications represent relatively new developments, and because this new technology offers advantages in some important markets. Both types of turbines have major roles to play in power generation; hence in the decades ahead a more balanced mix of heavy-duty industrial and aeroderivative turbines than is anticipated will most likely evolve in power markets.

2 The Policy Context

The electric power industry needs a technological revolution. Business-as-usual is becoming increasingly untenable. Public concerns about nuclear power risks and the environmental problems posed by fossil fuel power plants have made electric utility planning more and more difficult. And compounding these problems are the escalating costs of electricity that are reducing its competitiveness as an energy carrier.

The Chernobyl accident has led to a considerable stiffening of public opposition to nuclear power and to sharp limitations on nuclear power programs in various parts of the world. As we will show, gas turbines would be economical alternatives to nuclear power in many instances.

The large quantities of SO_2 and NO_x emitted by existing fossil fuel plants, especially coal-fired plants, are among the most significant pollutant emissions leading to acid deposition, which has become a serious international problem because of the transnational air transport of these pollutants. The need to control acid deposition is leading to proposals to reduce emissions from existing power plants in some areas. Gas turbine systems would be an attractive alternative in such situations.

Global warming from the greenhouse effect, associated with the buildup of CO_2 in the atmosphere, has also become a major environmental concern. Again, as we will show, gas turbines would make significant contributions in efforts to cope with the greenhouse problem.

The electric utility industry has also been experiencing rising electricity costs–a marked departure from the long-term historical trend, as illustrated by the situation in the U.S. Up until 1970 the outlook for electric power looked bright in the U.S. Between 1950 and 1970, the real cost of fossil-fueled steam plants fell by more than a factor of two (Figure 1). And cost-cutting innovations in the U.S. power industry persistently led to a reduction in the average real price of electricity by about 25% for each doubling of cumulative electricity production throughout the period 1926 to 1970 (Figure 2). If the trend up to 1970 had persisted to the present, the average price of electricity in the U.S. today would be 1/3 lower than in 1970. Instead, the average electricity price in the U.S. today is 40% higher than in 1970 or more than double what it would have been if the long-term, cost-cutting trend continued (Figure 3).

Escalating capital costs for new central station power plants have contributed greatly to these rising electricity prices. A nuclear plant ordered in the U.S. in 1970 cost about $900 per kW (1).[1] In a 1985 study by the International Energy Agency (IEA), it is estimated that the cost of a new nuclear plant that would be

[1] Unless explicitly indicated otherwise, prices are presented here in 1987 dollars. Conversions to 1987 dollars were made using the U.S. gross national product deflator, if original data were in the nominal dollars of other years.

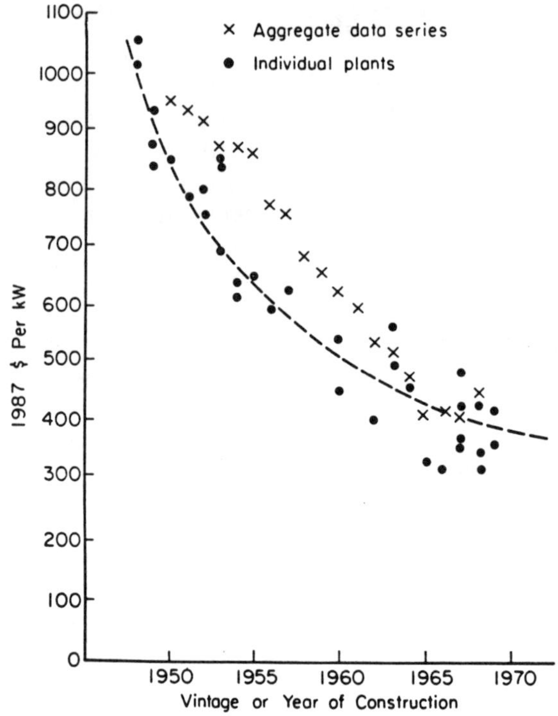

Figure 1. The cost of capacity additions of fossil-fueled steam-electric plants in the U.S., in 1987 dollars per kW.

Source: Reference 1.

started up in 1990 in Europe would cost $1900 per kW if construction took ten years or $1700 per kW if construction could be completed in six years (3). The Electric Power Research Institute (EPRI) has estimated that the cost of a new nuclear plant ordered today in the U.S. would be $3060 per kW, but that this cost could be reduced to $1670 per kW in a "reborn" nuclear industry, i.e., one featuring a streamlined nuclear licensing process, a shorter construction period (six instead of eight years), and improved labor productivity (4). The costs of coal-fired steam-electric plants with flue gas desulfurization (FGD), which cost about $500 per kW in the U.S in 1970 (1), have also escalated markedly. The IEA has estimated that the installed cost of a plant with two 600 MW units would be about $1230 per kW in Europe (3); EPRI has estimated a slightly higher cost ($1340 per kW) for a plant with two 500 MW units in the U.S. (4).

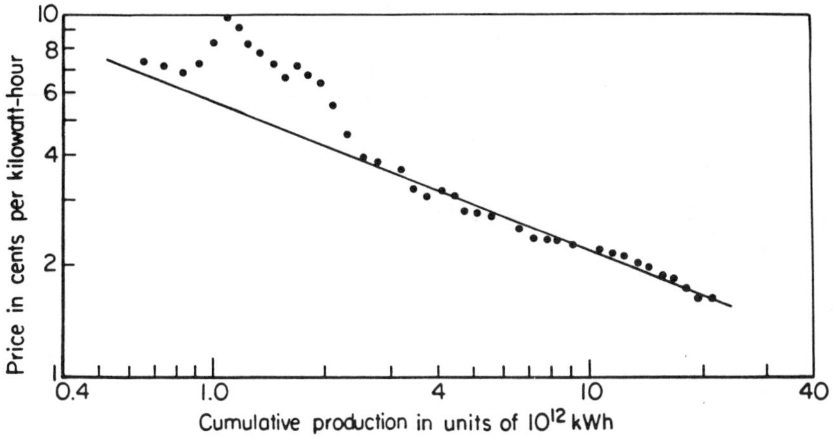

Figure 2. The average price of electricity (in 1970 cents/kWh) vs. cumulative production of electricity in the U.S., 1926 to 1970. The trend line corresponds to a 25 percent reduction in price for each doubling of cumulative production. (The price rose above the trend line during the Great Depression in large part because of the high and relatively inflexible fixed charges associated with electricity generation that continued in the face of depression-diminished demand.)

Note: $1 of 1970 = $2.80 of 1987.

Source: Reference 2.

In a 1987 World Energy Conference (WEC) study, it is projected that the ongoing escalation in power generation costs will continue (Table 1). For industrial countries, it is projected that the share of GDP that will have to be spent on electricity supply expansion will be in the range 2.0% to 2.7% of GDP in 2000, compared to 2.2% in 1980, even though electricity demand is projected to grow only 0.65 to 0.83 times as fast as GDP, 1980–2000. For developing countries, where electricity demand growth is expected to be more rapid, it is projected that the share of GDP that will have to be spent on electricity supply expansion will increase from 1.5% of GDP in 1980 to the range of 2.6% to 5.5% in 2000. For the already capital-constrained developing countries, this means that supporting electricity demand growth rates that are modest by historical standards will be extremely difficult.[2]

2 The WEC capital expenditure estimates are for electricity demand growth rates for developng countries in the range of 4.5% to 6.8% per year, 1980–2000. For comparison, the growth rate has averaged about 7% since 1980, and the long-term historical average growth rate has been about 9% per year.

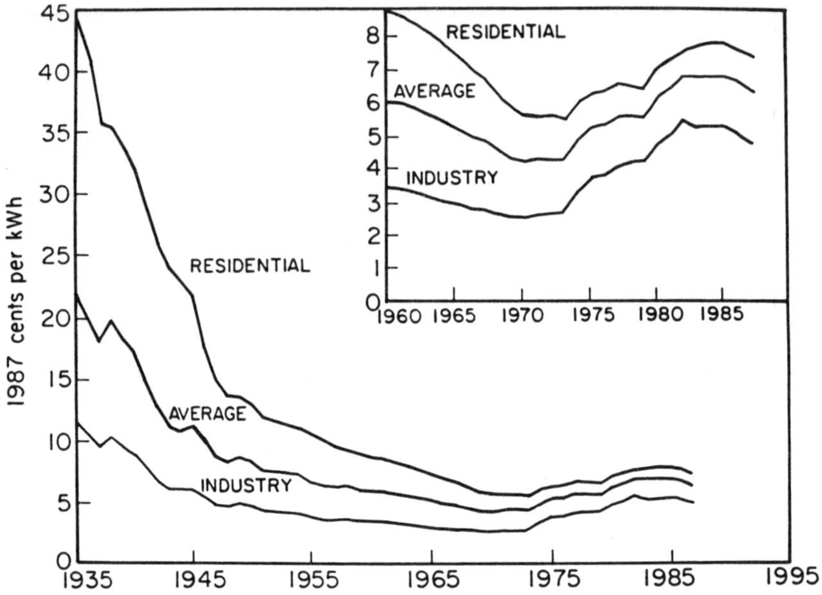

Figure 3. Long-term and recent (insert) electricity price trends in the U.S. The prices shown are the total revenues divided by electricity sales, expressed in 1987 cents per kWh.

Source: Reference 5.

3 The Prospects for Improving Steam-Electric Power Technology

Before discussing the opportunities for innovation afforded by advanced gas turbine technologies, let us examine the prospects for improving the technology for nuclear and fossil fuel-based steam-electric power generation.

The cost escalations plaguing steam-electric power generation are due in part to tightening environmental and safety rules. Other important factors include inadequate quality control in equipment manufacture and construction, bottlenecks that have arisen because each big project has been in many ways unusual, and escalating labor costs arising from shortages of qualified manpower and declines in labor productivity. Many such problems result in not only direct cost increases, but also indirect cost increases associated with the accumulated interest charges from extended construction periods.

Table 1. Capital requirements and growth rates for electricity projected by the World Energy Conference.[a]

	1980	2000L[b]	2000H[b]
Cost for New Generating Capacity ($/kW)			
Hydroelectric	2740	3360	4110
Nuclear	2060	2540	3080
Fossil Fuel, Thermal	1030	1230	1510
Average Capital Requirements by Region ($/kW)			
Industrial Countries			
Generation	1480	2000	2390
T & D	2740	2770	3030
Developing Countries			
Generation	1690	2070	2480
T & D	810	1700	2480
Centrally Planned Economies			
Generation	1370	1810	2320
T & D	1370	1960	2620
Overall Capital Requirements for Electricity [10^9 $/year (% of GDP)]			
Industrial countries	226 (2.2)	302 (2.0)	488 (2.7)
Developing countries	44 (1.5)	148 (2.6)	381 (5.5)
Centrally planned economies	60	147	233

Average Growth Rates (%/year)	*1980–2000L*[b]	*1980–2000H*[c]
For GDP		
Industrial countries	2.0	3.0
Developing countries	3.5	4.5
For Primary Energy Consumption		
Industrial countries	0.15	1.3
Developing countries	2.5	4.7
Centrally planned economies	1.8	2.3
For Electricity Generation		
Industrial countries	1.3	2.5
Developing countries	4.5	6.8
Centrally planned economies	2.7	3.2

[a] According to a 1987 World Energy Conference (WEC) study (6).

[b] 2000L (2000H) is for the WEC low (high) growth scenario.

A 1974 analysis by John Fisher of the escalation in nuclear power costs in the decade leading up to the first oil crisis provides an important insight associated with these construction-related problems that seems relevant for power cost escalation generally since 1970 (2):

"When measured in constant dollars per kilowatt of capacity, the cost of constructing a nuclear power plant increased by perhaps 50 percent in the past decade... When power plant costs rise an explanation is required, as we expect all power plant costs to decline through the economies of scale and new technology. The environmental movement was responsible for part of the rise in nuclear plant costs, by causing various procedural delays and by requiring additional expensive safeguards to protect against hypothetical accidents. But there appears to be another cause for increasing construction costs, associated with a growing portion of high-cost field construction and a shrinking proportion of low-cost factory construction for the very large power plants now being built... the costs associated with a shift to field from factory can more than offset anticipated economies of scale..."

Fisher pointed out that historically, as electric utility plant capacity doubled every decade, factory capacity also doubled, as did field construction at each site. Manufacturing and construction costs per kW declined in the factory and in the field, since each of these increased its scale of operations. As long as both activities grew in proportion, the economies of scale produced similar cost reductions in each, and therefore an overall cost reduction, even though the unit cost of field construction was always higher than the unit cost of factory construction. This pattern held until plant size reached about 200 MW. Then, because design engineers felt that scale economies would be much more important for nuclear than for fossil fuel plants, nuclear power plant capacities were built in sizes of the order of 1000 MW–shifting a greater portion of the construction from the factory to the field, upsetting the pattern of the past, with the result that a much larger fraction of the construction was carried out at smaller, less-efficient field locations. Fisher's important insight is that the widely touted economies of scale in power plant construction are illusory because: (*a*) field construction is inherently more costly than factory construction, and (*b*) with field construction it is never possible to get very far down the "learning curve," in contrast to the situation with factory production. This "diseconomy of scale" problem has persisted to the present.

The increase in scale for coal-fired power plants to the present range of 500 to 600 MW was in part a competitive response to large-scale nuclear plant construction. Also, the pursuit of scale economies was seen as an opportunity for continuing the historical reductions in the cost of electricity (Figures 2 and 3), after the long-term trend toward increased thermodynamic efficiency of power generation ceased in the late 1950s (Figure 4). But as in the case of nuclear power, the point of diminishing returns to scale appears to have been reached or exceeded for coal plants. Even where scale economy gains might be realized in construction, these gains tend to be offset by losses in reliability for the larger units (7).

What are the prospects for improving thermodynamic efficiency? Since the 1920s, most gains in efficiency in steam-electric power plants have been due largely to increases in maximum steam temperatures and pressures. By the 1950s, peak temperatures had reached 565 °C and peak pressures 165 bar for subcritical steam units and 241 bar for supercritical steam units.

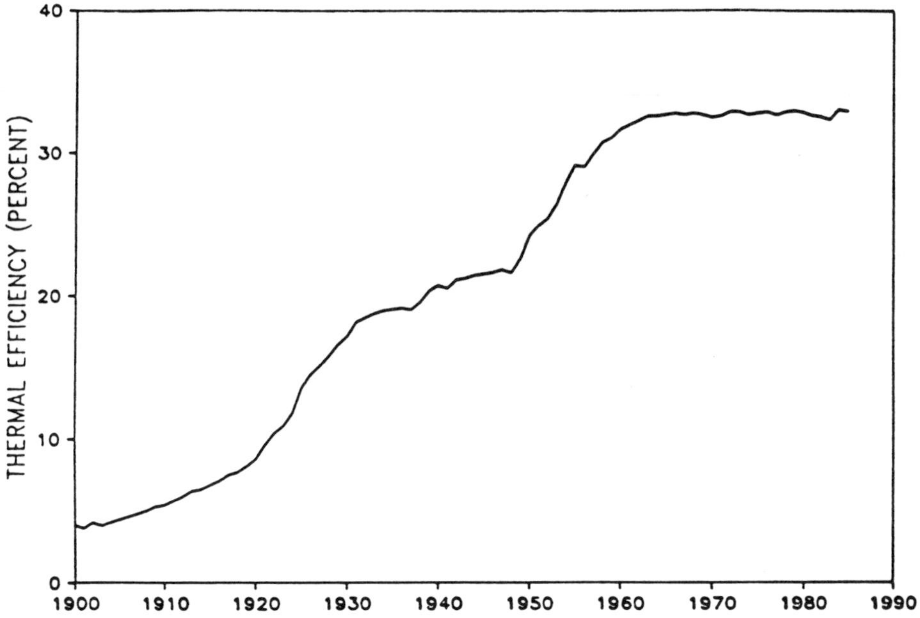

Figure 4. The historical trend in the average efficiency (higher or gross heating value basis) of electricity generation in central station thermal power plants in the U.S.

Source: Reference 5.

There are ongoing developmental efforts aimed at improving steam-electric power plant efficiency by increasing peak steam temperatures (8,9). But as peak steam temperatures are increased, problems of materials strength, oxidation, and corrosion rapidly become more serious, dictating shifts to more costly high-strength, oxidation and corrosion-resistant alloys for the large steam-tubing heat exchangers that transfer heat from the combustor to steam at high temperature and pressure. (See, for example, Figure 5, which shows, for a number of alloys used in steam tubing exposed to high temperatures, that the maximum allowable stress declines rapidly beyond a critical threshold.)

Peak steam temperatures have not increased since the 1950s, and in fact utilities today tend to choose a slightly lower peak temperature of 540 °C in coal plants. They do so not only because of the lower capital cost, but also because, even with judicious choice of better tubing materials, higher temperature operating conditions have led to more forced outages, owing to tubing damage from problems such as coal-ash corrosion (10).

A 1976 Westinghouse study–the results of which are consistent with many other studies carried out since the 1950s (11)–indicates the magnitude of the

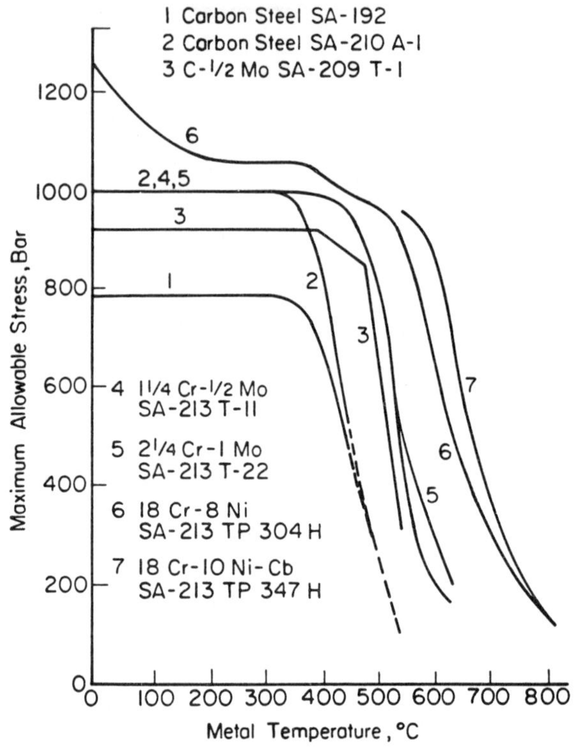

Figure 5. Effect of temperature on the maximum allowable stress for different steel alloys used for steam tubing in high temperature service, according to the boiler code of the American Society of Mechanical Engineers (ASME) (10). 1 = low strength carbon steel; 2 = intermediate strength carbon steel; 3 = a ferritic alloy containing 0.5 percent molybdenum; 4 = ferritic alloy containing 1.25 percent chromium and 0.5 percent molybdenum; 5 = ferritic alloy containing 2.25 percent chromium and 1.0 percent molybdenum; 6 = austenitic stainless alloy containing 18 percent chromium and 8 percent nickel; 7 = austenitic stainless alloy containing 18 percent chromium and 10 percent nickel.

tradeoff involved in increasing the maximum steam temperature of a 500 MW steam plant; an increase from 540 to 650 °C would increase the plant efficiency by 6%, but at the cost of a 26% increase in capital cost (12). On a lifecycle cost basis, the price of coal would have to increase from $50 to $200 per tonne before it would be worthwhile to shift to the higher peak steam temperature.

While the outlook for major improvements in steam-electric power technology is not auspicious, it may be feasible to increase efficiency without pushing peak working fluid temperatures further, through development of the recently proposed Kalina cycle (13,14). The Kalina cycle is a novel modified Rankine cycle that uses as a working fluid a mixture of ammonia and water, the relative proportions of which vary throughout the cycle. A 3 MW demonstration plant, to be built at the U.S. Department of Energy's Engineering Center in Canoga Park, California, is expected to be operating by 1989 (15). The big uncertainties regarding the Kalina cycle are the complicated "plumbing" and possible difficulties associated with managing the binary working fluid at high temperatures and pressures, which might lead to significant capital and operating and maintenance cost penalties. Also, the performance estimates for the Kalina cycle are for very small assumed pressure drops and tight temperature differences–conditions that are difficult to achieve in practice (14).

4 The Outlook for Stationary Power Applications of Gas Turbines

Rising costs of steam-electric power plants and slow, uncertain electrical load growth have led to dramatic reductions in the construction of new central station power plants in many countries; in the U.S., for example, there were virtually no orders for central station power plants between 1982 and 1986 (Figure 6). These conditions are leading utility planners to give more attention to the gas turbine for meeting future power needs. On the basis of contacts made with the majority of the large electric utilities, Gluckman at the Electric Power Research Institute estimated that *some 40 GW of new gas turbine-based generating capacity is on order or being planned in the U.S for installation by 1995* (16).

Emerging utility interest in gas turbines is complemented by a boom in gas turbine sales to cogenerators in the U.S. Electric utility rate increases and the Public Utility Regulatory Policies Act of 1978 (PURPA), along with the 1982 and 1983 Supreme Court decisions upholding its provisions, have led to a competitive challenge for utilities from independent cogenerators and small power producers in the U.S. PURPA encourages cogeneration and the production of electricity from renewable energy sources in small installations by requiring utilities: (*a*) to purchase the electricity from qualifying producers at a price equal to the cost the utility can avoid by not having to otherwise supply that power and (*b*) to provide back-up power at reasonable rates. Between 1980 and 1987, 62 GW of electrical generating capacity was certified as qualifying for PURPA benefits by the Federal Energy Regulatory

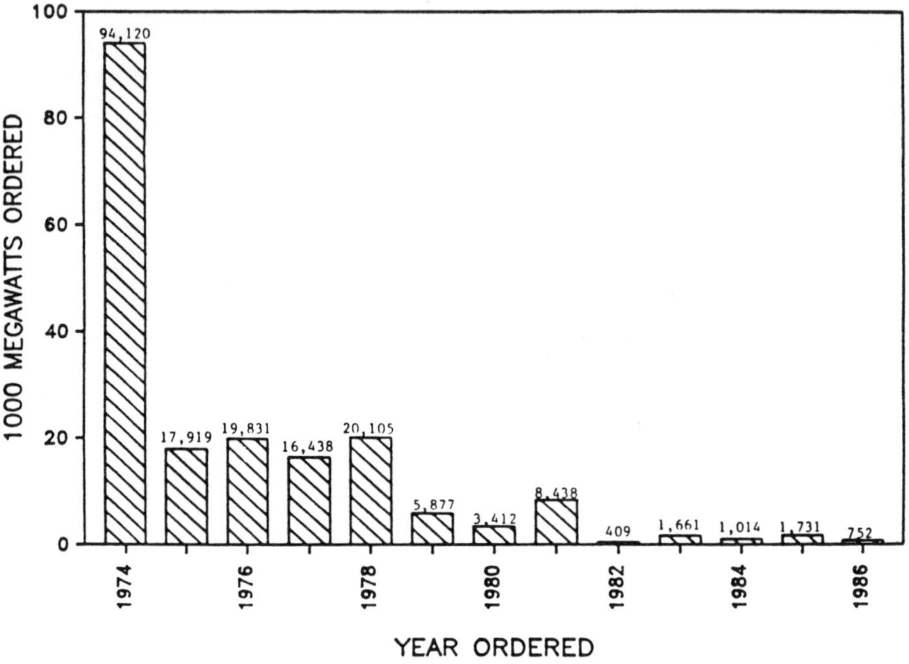

Figure 6. New orders for central station power plants in the U.S. (17). The total in 1986–752 MW–was for 11 gas turbines. This was the first year since World War II that no steam or hydroelectric turbine units were ordered by U.S. electric utilities.

Commission (FERC), nearly three-fourths of which is due to cogeneration (Figure 7). *Some 20 GW of the cogeneration capacity certified in this period–more than the sum of all utility orders for all kinds of central station power plants, 1980–1986* (Figure 6)–*was based on the gas turbine* (Figure 7).

Recent interest in the gas turbine for stationary power reflects both long-standing attractions of this technology and recent improvements that make it possible for the gas turbine to compete in a wider range of markets.

4.1 Traditional Roles for Gas Turbines

The historical attraction of the gas turbine for utilities has been its low cost–$300 per kW (4) or less–a small fraction of the cost of coal or nuclear power plants. This low cost reflects the utter simplicity of the simple cycle gas turbine. While costly heat exchangers are required in a steam turbine power

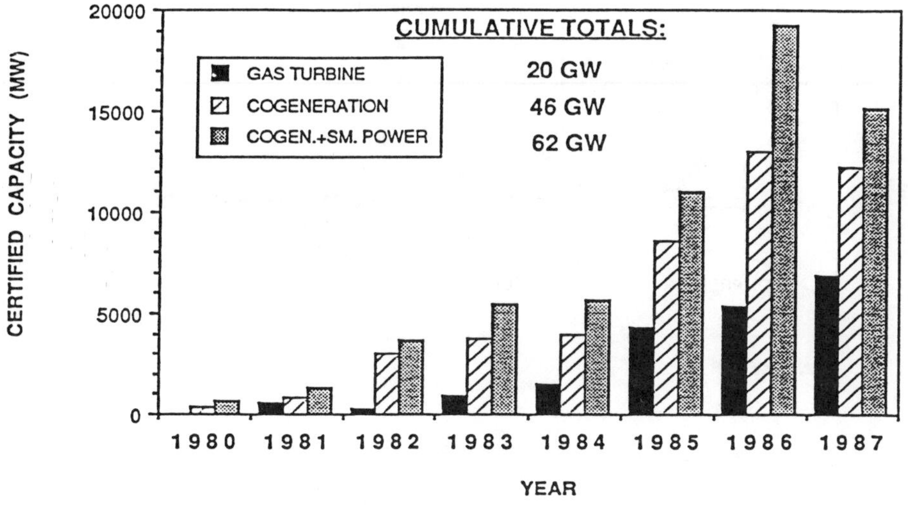

Figure 7. Annual cogeneration and small power production capacity for facilities certified in the U.S. by the Federal Energy Regulatory Commission to be eligible for the benefits allowed under the Public Utility Regulatory Policies Act (19).

plant to transfer heat from the combustor to the steam working fluid that drives the turbine, in a gas turbine power plant the hot fuel combustion products drive the turbine directly (Figure 8a). Also, while large condensers and often cooling towers are required to condense a steam turbine's exhaust steam, the exhaust from a gas turbine is discharged directly to the atmosphere.

But simplicity has been a mixed blessing for the simple cycle gas turbine. It has meant a low efficiency–the average efficiency of utility peaking units in the U.S. in 1985 was only 29%[3] (18). Also, clean fuels have been required to avoid damaging the turbine blades with the combustion products–a constraint that has limited the use of the gas turbine mainly to liquid or gaseous fuels that have been costly or whose long-term availability is uncertain. Because of these constraints, utilities have used gas turbines mainly for peaking service.

Its low unit capital cost has also helped make the gas turbine attractive for cogeneration applications. Because of the relative insensitivity of gas turbine unit costs to scale (Figure 9), the gas turbine tends to be favored over the steam turbine for all but the largest cogeneration installations. The use of the high-temperature (425 to 540 °C) turbine exhaust to raise steam in a heat

3 Efficiencies are given here in terms of the fuel's higher (gross) heating value.

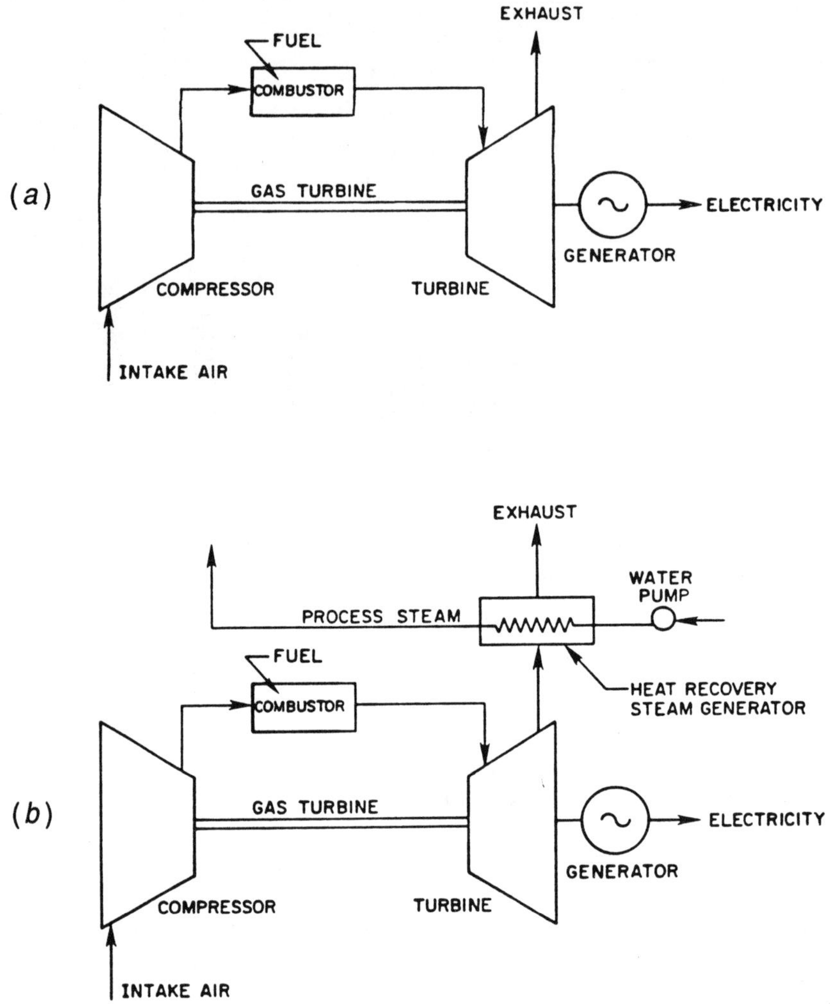

Figure 8. (*a*) *Simple power cycle*: fuel burns in air pressurized by compressor, combustion products drive turbine, and hot turbine exhaust gases are discharged to atmosphere. (*b*) *Simple cogeneration cycle*: like simple power cycle, except that hot turbine exhaust gases are used to raise steam in HRSG for heating.

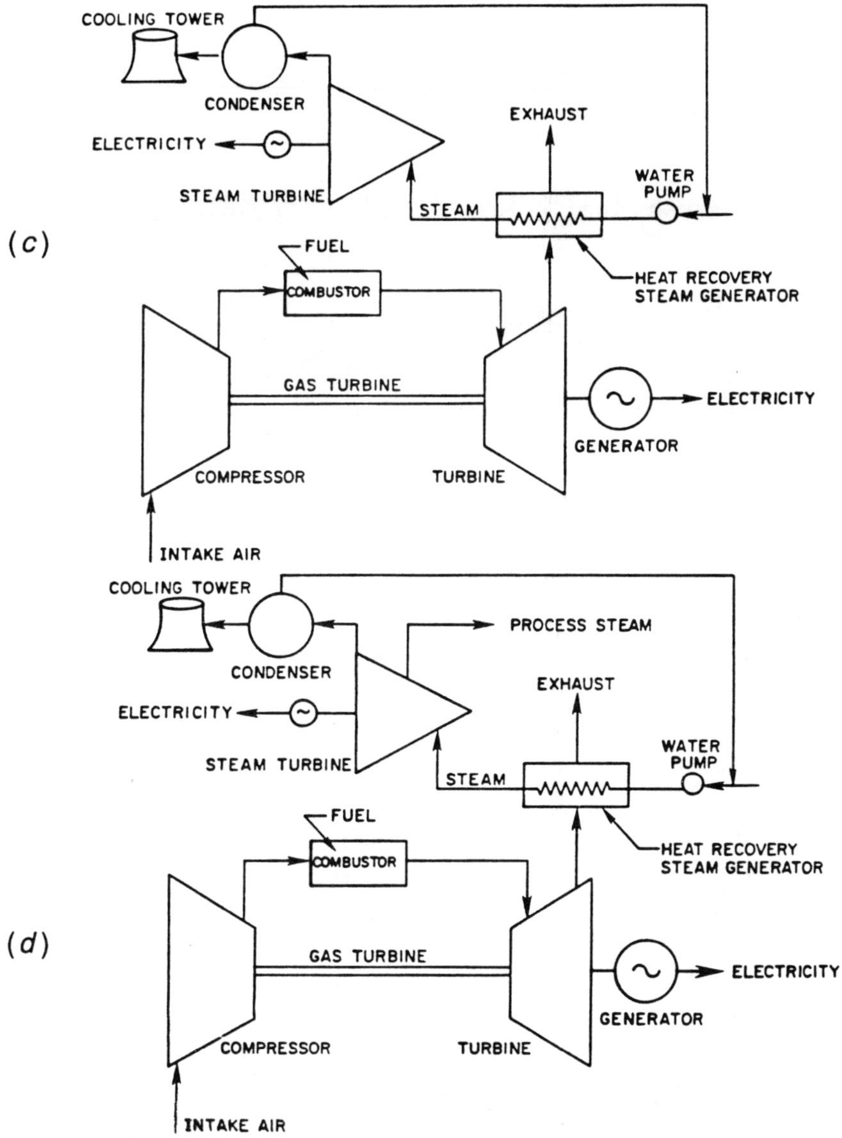

Figure 8. (*c*) *Combined cycle for power*: like simple cogeneration cycle, except that steam from HRSG is used to produce extra power in condensing steam turbine. (*d*) *Combined cycle for cogeneration*: like combined cycle for power, except that some steam is bled from steam turbine for heating.

(*e*)

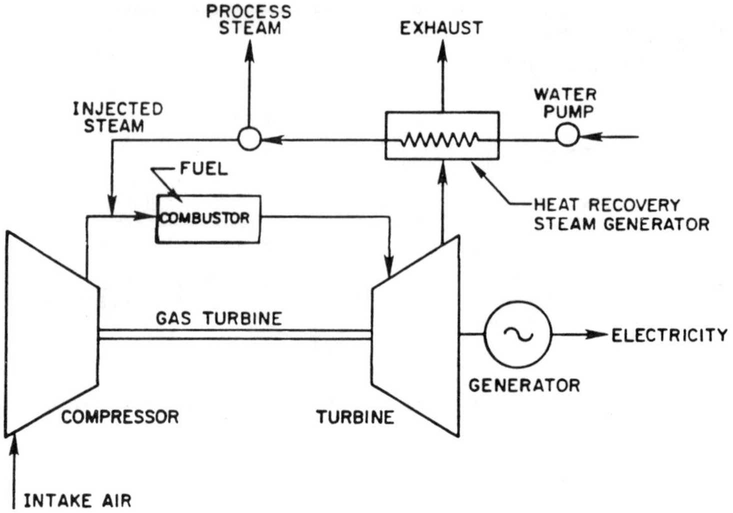

(*f*)

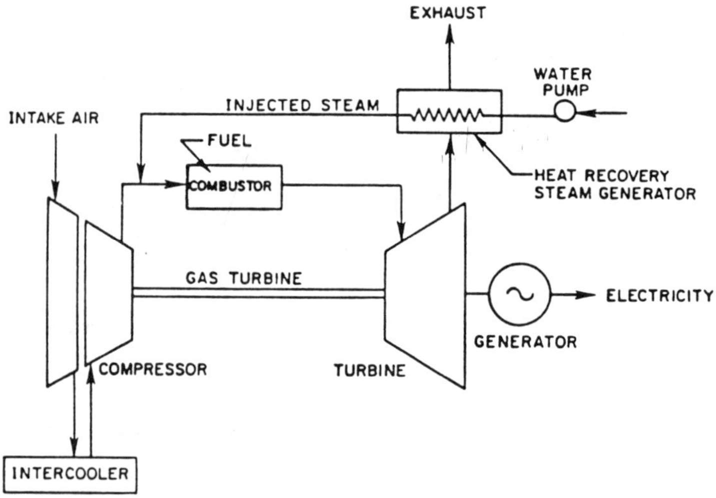

Figure 8. (*e*) *Steam-injected gas turbine* (STIG): like simple cogeneration cycle, except that steam not needed for heating is injected into combustor for increased power output and higher electrical efficiency. (*f*) *Intercooled steam-injected gas turbine* (ISTIG): like STIG with full steam injection except that intercooler between compressor stages allows for operation at much higher turbine inlet temperature because of improved cooling of turbine blades.

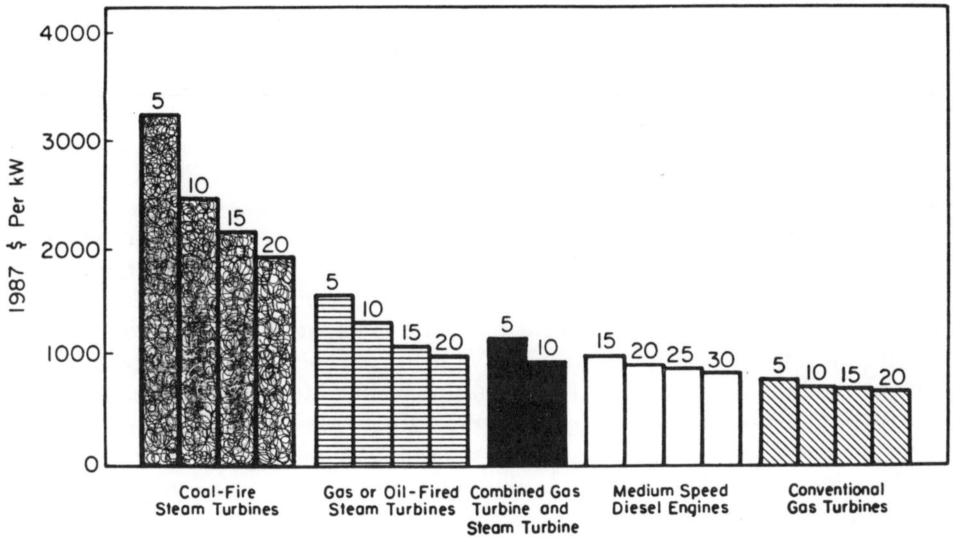

Figure 9. Unit installed costs for small-scale cogeneration systems. The numbers at the tops of the bars are the installed electrical generating capacity in MW (20).

recovery steam generator (HRSG) for heating applications (Figure 8b) makes the gas turbine a thermodynamically efficient cogeneration device, even if the efficiency of the turbine for producing power only is relatively poor.

A major shortcoming of the simple cycle gas turbine in cogeneration is that it is often uneconomical in applications involving highly variable steam loads, because achieving a high thermodynamic efficiency depends on being able to make use of the full electrical and thermal output capacities.

It is now possible to overcome the constraints which have confined gas turbines to peaking service for utilities and baseload service for cogeneration because: (*a*) the performance of the basic gas turbine cycle is improving steadily, and (*b*) various simple cycle modifications offer opportunities for both improving efficiency and reducing capital cost. A brief review of the history of the gas turbine is helpful in understanding these possibilities.

4.2 A Brief History

An early major milestone in the history of the gas turbine was the initiation of German and British programs in the mid-1930s to explore the use of gas turbines for aircraft propulsion. The success of these initial efforts led the U.S. to launch major jet engine development programs during and following World War II: the cost of these programs between 1940 and 1980 totalled about $10 billion (11). These efforts have been successful, both in improving jet engine reliability and thrust-to-weight ratios and in increasing efficiency by increasing turbine inlet temperatures, at an average rate of more than 20 °C per year, between 1950 and 1980 (Figure 10).

Improvements in jet engine technology and electricity demand growth that was more rapid than expected stimulated considerable interest in the use of short lead-time gas turbines for stationary power applications in the late 1960s. Between 1965 and 1975, installed gas turbine capacity in the U.S. electric utility industry increased from 1.3 GW to 43.5 GW (11). But subsequently commercial interest in stationary gas turbines ground to a halt as a result of the sharp rise in oil and gas prices, concerns about gas scarcity and oil import dependency, and a sharp reduction in electricity demand growth; the installed gas turbine capacity of the U.S. utility industry in 1985 was no greater than in 1975 (18).

The end of commercial interest in gas turbines for stationary power in the U.S. did not slow fundamental progress in improving gas turbine technology, however. One reason is that commercial airlines pressed vendors to improve the efficiency of jet engines. The rising world oil price increased the fuel costs of air passenger travel from 11 to 32% of the total cost of air passenger service in the 1970s (21)! U.S. Department of Defense support for research and development on jet engines for military applications also continued at a high level, averaging about $450 million per year in the decade ending in 1986 (22). Continuing R&D in this area is expected to bring significant further increases in turbine inlet temperatures by the turn of the century, as a result of major improvements in blade materials (Figure 10) and more effective blade-metal cooling technologies.[4]

A paradoxical aspect of the development of the stationary gas turbine is that most of the well-known "low-technology" cycle modifications available for improving performance–e.g., reheat, intercooling, regeneration, evaporative regeneration, steam injection, and steam-reforming of fuel–remain largely unexploited, even though enormous "high-technology" advances have been made in turbine blade materials, design, and fabrication. This is because such

[4] A significant difference between the steam turbine and the gas turbine is that the maximum temperatures of the metal and the working fluid are oppositely related for these technologies. In steam turbine systems the maximum temperature of the metal is always higher than that of the working fluid; thus metallurgical contraints that limit the maximum metal temperature make it difficult to increase efficiency by increasing the working fluid temperature. But in gas turbine-based cycles the metal temperature is lower than that of the working fluid, and the difference between these temperatures is increasing continually with improvements in turbine blade cooling technology, leading to continual improvements in efficiency.

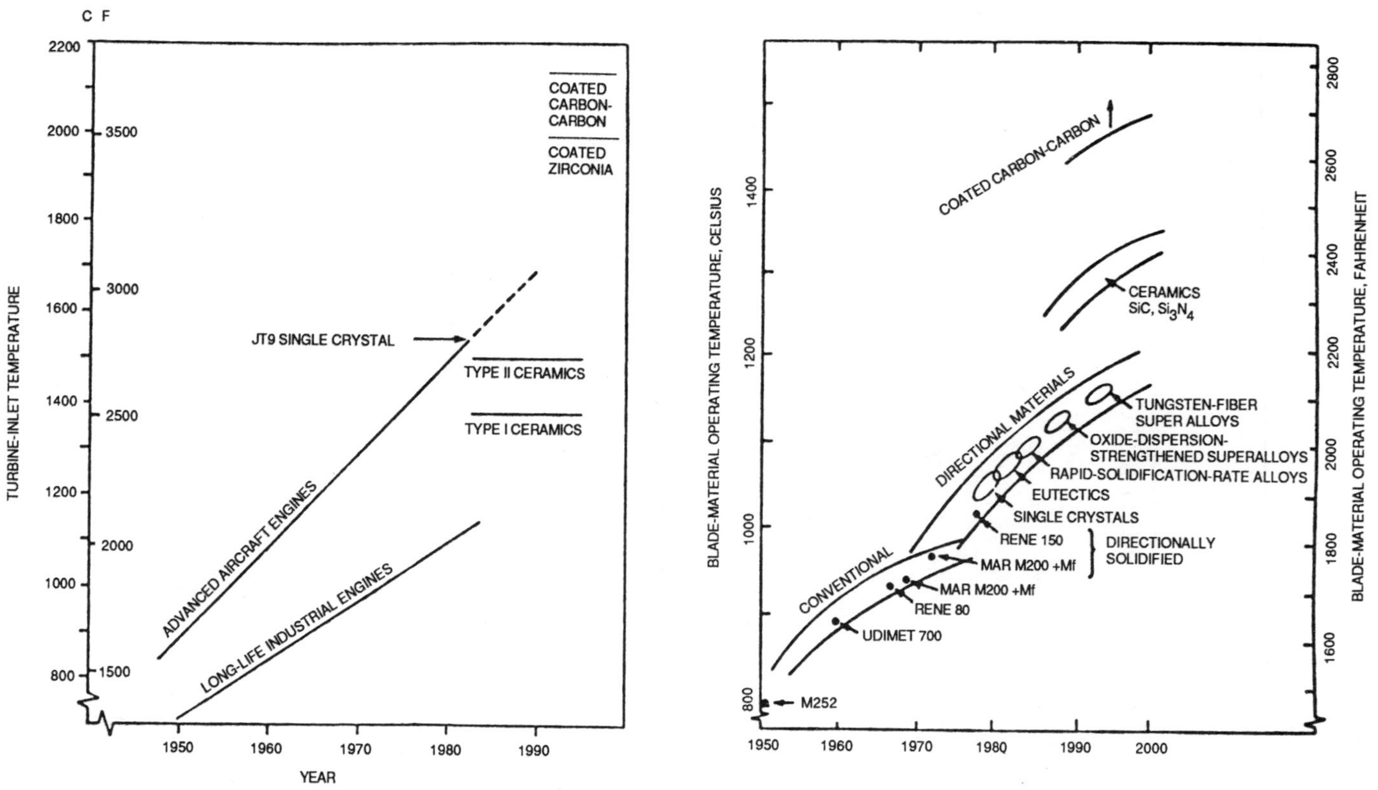

Figure 10. The trend in turbine inlet temperatures for advanced aircraft jet engines and long-life industrial turbines (left) and turbine blade material operating temperature (right) (23).

Note: When an aircraft engine is modified for stationary applications the rated turbine inlet temperature is reduced about 110 °C to promote long-life operation.

cycle modifications involve the introduction of heavy or bulky heat exchangers or the use of large quantities of steam or water, neither of which is relevant to aircraft applications. This situation presents an enormous opportunity because it means that major improvements can be made in the performance of gas turbines for stationary power applications with relatively modest R&D efforts.

4.3 Progress in Combined Cycle Technology

One gas turbine cycle modification is familiar to the electric utility industry: the gas turbine/steam turbine combined cycle (Figure 8c), which accounted for some 4.7 GW of utility generating capacity in the U.S. in 1985 (24). With advanced gas turbines now commercially available for stationary applications (25), a combined cycle efficiency of 45% can be realized in 200 MW plants costing $520 per kW. For comparison, a 36% efficient natural gas-fired steam-electric plant consisting of two 500 MW units would cost $760 per kW (Table 2). These combined cycle plants could produce electricity at a busbar cost only three-fifths of that for large new coal plants at the 1987 average natural gas price in the U.S. (Table 2).

Of course electric utilities cannot be certain that natural gas prices will remain low over the entire expected lives of combined cycle facilities. However, recent developments in coal-gas turbine technology offer a means by which utilities can minimize the risks.

In 1979 a major private-sector effort was launched in the U.S. to demonstrate the ability to operate gas turbines on gasified coal.[5] This effort led to the construction of a 94 MW combined cycle power plant coupled to a

5 A gas turbine can also be fired directly with coal using a pressurized fluidized bed combustor (PFBC) – a technology that is is inherently simpler than a gas turbine/gasification system. With this technology, sulfur is removed during combustion by a reaction with dolomite that forms an inert magnesium oxide-calcium sulfate complex, and particulate matter is removed from the combustion products in cyclones before they are directed to the turbine. This technology could be a serious competitor to the first generation of gasifier/gas turbine power plants (33). But with the PFBC it would not be possible to exploit expected continuing improvements in gas turbine technology, because the temperature of the combustor must be limited. The optimal temperature for sulfur capture is a bed temperature of about 850 °C, with a rapid decline in sulfur removal for either higher or lower temperatures (10). Also, operation above about 950 °C can lead to ash agglomeration in the bed. The agglomerates thus formed can lead to segregation of the bed and a cessation of fluidization (34). These problems do not arise with gasification, where sulfur and particulates are removed from the gas stream before the gas enters the combustor. These problems could also be overcome using designs being considered for advanced PFBC systems, involving the partial gasification of coal in a carbonizer, followed by burning the char/lime residue from the carbonizer in a PFBC unit in high excess air, followed by the burning of the fuel gas mixed with the exhaust gas from the PFBC unit in a topping combustor (35). However, such systems are not simpler than gasification systems–they are more complicated.

Texaco coal gasifier, at Cool Water, California, operated by the Southern California Edison Company as part of a joint industrial effort involving the Electric Power Research Institute (EPRI), the Bechtel Corporation, the General Electric Company, Texaco, and a Japanese consortium under the rubric the Japan Cool Water Program. Plant operation began in June 1984, and the demonstration is expected to run till June 1989. The effort has been a technical success. The plant was built on time; the actual capital cost ($263 million) did not exceed the initial target; and the plant has operated reliably, with low pollutant emissions (see below) (32).

Cool Water technology could not provide power competitively at the scale of this demonstration plant; the busbar cost would be about two thirds higher than for a conventional steam plant with flue gas desulfurization (Table 2). However, there are substantial scale economies to be exploited. First, there are scale economies to be gained in the combined cycle unit, arising from the scale-sensitivity of the steam turbine sub-unit. Second, there are scale economies to be realized in gasification. With the Texaco gasifier, gasification takes place in oxygen, provided by a scale-sensitive oxygen plant. Based on EPRI capital and operating cost estimates, 600 MW units using advanced gas turbines that have recently become available (25) would probably be competitive (Table 2).

The success of the Cool Water demonstration is leading to the formulation of proposals for utility capacity expansion strategies that offer flexibility in the face of continuing uncertainty about future electricity demand and fuel prices (36). It is proposed that while natural gas prices are relatively low, natural gas-fired gas turbines be installed to support demand growth, thereby avoiding financial commitments to larger amounts of more costly capacity that might not be needed. These units could be expanded to combined cycle units as demand grows and gas prices rise; and ultimately they could be modified to operate on gasified coal if necessary.[6]

4.4 A Comparison of Industrial and Aeroderivative Turbines

The combined cycle is a good technology for beginning a transition to greater use of gas turbines in stationary power applications. It marries the new gas turbine technology to the familiar steam turbine. But it would be a mistake to

6 A variation on this proposal involves "repowering" existing steam plants with gas turbine topping cycles. The repowered plants would be fired initially with natural gas, with the flexibility to shift later to gasified coal (36,37).

Table 2. Cost and performance characteristics for U.S. central-station power plants.[a]

1. *Coal and Nuclear Steam-Electric Plants*

Type	*Coal*[b,c]			*Light Water Reactor*[d] *Current*	*Light Water Reactor*[d] *Target*
Unit Size (MW)	2 x 500	500	200	1100	1100
Efficiency (%)[e]	34.6	34.6	34.6	33.4	33.4
Unit Cost ($/kW)	1340	1410	1880	3060	1670
Levelized Busbar Cost (cents/kWh)					
Capital[f]	1.61	1.69	2.25	3.66	2.00
Fuel	1.86	1.86	1.86	0.91	0.91
O&M	0.89	0.99	1.37	1.11	1.11
TOTAL	4.36	4.54	5.48	5.68	4.02

2. *Natural Gas-Fired Plants*[g]

	1987 Natural Gas Price[h]				*2 X 1987 Natural Gas Price*			
	Steam	*ACC*	*STIG*	*ISTIG*	*Steam*	*ACC*	*STIG*	*ISTIG*
TIT (°C)	540	1260	1200	1370	540	1260	1200	1370
Unit Size (MW)	2 x 500	205	4 x 51	114	2 x 500	205	4 x 51	114
Efficiency (%)[e]	36.3	45.0	40.0	47.0	36.3	45.0	40.0	47.0
Unit Cost ($/kW)	760	520	410	400	760	520	410	400
Levelized Busbar Cost (cents/kWh)								
Capital[f]	0.9	0.63	0.49	0.48	0.91	0.63	0.49	0.48
Fuel	2.08	1.68	1.89	1.61	4.17	3.36	3.78	3.22
O&M	0.49	0.29	0.29	0.29	0.49	0.29	0.29	0.29
TOTAL	3.48	2.60	2.67	2.38	5.57	4.28	4.56	3.99

3. *Coal-Gas-Fired GasTurbine Systems*[c,i]

	<——Oxygen-Blown Gasifier——>				<——Air-Blown Gasifier——>		
	<——Cold Gas Clean-Up——>				<——Hot Gas Clean-Up——>		
	<—— CCC——>			<——ACC——>		<–STIG–>	<–ISTIG–>
TIT (°C)	1090	1090	1090	1200	1200	1200	1370
Unit Size (MW)	100	250	500	600	520	2 x 50	110
Efficiency (%)[e]	34.3	35.7	36.0	37.9	37.6	35.6	42.1
Unit Cost ($/kW)	2720	2000	1680	1550	1160	1300	1030
Levelized Busbar Cost (cents/kWh)							
Capital[f]	3.26	2.39	2.01	1.85	1.39	1.56	1.23
Fuel	1.88	1.81	1.79	1.70	1.71	1.81	1.53
O&M	2.11	1.19	0.89	0.80	0.45	0.71	0.60
TOTAL	7.25	5.39	4.69	4.35	3.55	4.08	3.36

——————>

Notes for Table 2

a All costs are 1987 U.S. dollars.

b Capital costs, efficiencies, and O&M costs are EPRI estimates, for a bituminous coal-fired subcritical steam plant with flue gas desulfurization (4).

c The assumed coal price is $1.79/GJ, the average U.S. utility price projected for 2000 by the U.S. Department of Energy (24).

d Reactor plant size, unit capital costs, and efficiencies are EPRI estimates (4). The two sets of capital costs are the current cost and an EPRI target for "improved" conditions–resulting from higher construction labor productivity, shorter construction period, streamlined licensing process, etc. The assumed nuclear fuel cycle cost is $0.84/GJ, EPRI's projection for the period 1990–2000 (4). The assumed O&M cost is the 1985 U.S. average for nuclear power plants (18), twice as large as the EPRI estimate for new plants (4).

e Based on the fuel's higher heating value and for operation at 100% load.

f For a 6.1% real discount rate [recommended by EPRI (4)], a 30-year plant life, and a 70% capacity factor. No taxes or tax incentives are included.

g The steam plant involves 165 bar, 540 °C steam (with single reheat to 540 °C) driving a conventional turbine/electric generator (4). The advanced combined cycle (ACC) is a recently commercialized 135 MW GE Frame 7F gas turbine plus a 70 MW steam turbine; the indicated performance is a General Electric estimate (25). The STIG unit is a commercial steam-injected gas turbine based on the GE LM 5000 (L. Gelfand, Manager, Advanced Programs and Ventures, General Electric Marine and Industrial Division, Cincinnati, Ohio, personal communication, February 1987). The ISTIG unit is a proposed intercooled steam-injected gas turbine, based on the LM 8000 (26,27,28). The unit capital costs for the steam-electric and ACC plants are EPRI estimates (4); that for STIG units is a Bechtel estimate (29); that for the ISTIG is based on estimates made by GE and the staff of the California Energy Commission (28,30). The O&M costs are EPRI estimates (4) for all but STIG and ISTIG units. For the latter the values are those estimated by EPRI for combined cycles (4), even though a Bechtel analysis indicates that steam-injected gas turbine systems offer inherent O&M cost savings compared to combined cycle units (29).

h The average gas price for U.S. electric utilities was $2.10/GJ in 1987.

i The performance/cost values for current combined cycles (CCC) and advanced combined cycles (ACC) fired with gas derived in oxygen-blown gasifiers are EPRI estimates for the Texaco gasifier (4). The corresponding numbers for systems using an air-blown Lurgi fixed bed gasifier are from a GE study exploring less costly, more energy-efficient alternatives to the Texaco gasifier (31), using EPRI cost analysis guidelines (4), to be consistent with the EPRI cost estimates for oxygen-blown systems.

limit utility use of advanced gas turbines to this option, because alternatives may offer advantages in some applications.

In exploring alternative gas turbine strategies, it is useful to distinguish between the characteristic features of heavy-duty industrial turbines and aeroderivative units. Various vendors offer heavy-duty industrial units in sizes ranging up to 70 to 135 MW. The tendency has been to design them with modest compression ratios (8 to 16). They are thus well-suited for combined cycle operations because the turbine exhaust gases are thereby relatively hot–593 °C for the most recently offered advanced industrial unit (25)–making it possible to produce high quality steam in the heat recovery steam generator (HRSG).

In contrast, aeroderivative units are lightweight and compact, with relatively small capacities–30 to 35 MW at the high end of available capacities–and the trend is toward high compression ratios (18 to 30); all such characteristics reflect jet engine design requirements. Though simple cycle aeroderivatives are relatively efficient as electricity producers, such engines tend to be poor candidates for combined cycle applications, since the turbine exhaust gases are not nearly as hot as in heavy-duty industrial units. Until recently this attribute led to the neglect of aeroderivatives as a serious candidate for central station power generation. However, recent developments have shown that aeroderivatives are good candidates for other efficiency and output-augmenting cycle modifications such as steam injection, discussed below.

While aeroderivative turbines are not nearly as familiar as heavy-duty industrial turbines for stationary applications, they warrant greater attention not only because, with appropriate cycle modifications, they can perform as well as or better than industrial units (see below), but also because aeroderivatives have some other important attributes.

First, aeroderivatives can bring back to power generation the advantages of cost-cutting mass production. Moreover, when an aeroderivative is introduced for power generation, it is already well-advanced on the learning curve, because large-volume production of jet engines for aircraft applications has preceded it.

Second, the aeroderivatives are expected to benefit from continuing advances in jet engine technology, which can be transfered quickly and at low incremental cost to stationary applications. While there are only modest ongoing development efforts to improve industrial turbines in the U.S., there is continuing heavy U.S. government support for jet engine R&D. This includes, for example, the new $3.4 billion, 13-year Integrated High Performance Engine Technology program supported by the Department of Defense and the National Aeronautics and Space Administration (38). Such R&D efforts are expected to lead to major improvements in aircraft engine technology, including substantial further increases in turbine inlet temperatures (Figure 10).

Many utility managers are reluctant to consider aeroderivatives in capacity expansion plans. One concern is that, because in their manufacture emphasis is given to the use of special materials to meet the low weight and compactness requirements of jet engines, aeroderivative engines are inherently more costly per kW than industrial turbines, where such constraints are not relevant. While the use of more costly materials does tend to raise the cost of aeroderivatives, a compensating factor is that a greater proportion of the aeroderivative power plant can be built at the factory, where costs are easier to control than in the field. Moreover, the various cycle modifications that would be employed for stationary applications of aeroderivatives tend to lower unit costs. For example, when a simple cycle gas turbine is modified for both steam injection and intercooling, its output can be tripled, resulting in a lower unit capital cost than that of a combined cycle based on an industrial turbine (Table 2).[7]

Another concern is that because aeroderivative engines are more delicate than heavy-duty industrial turbines, they are less reliable. This might be true if aeroderivative turbines were maintained like industrial units; instead they are maintained like jet engines. Their compact, modular construction makes it easy to remove and replace failed parts quickly.[8] In fact the entire basic engine can be removed and replaced with a spare (flown in, if necessary) from a lease-engine pool, resulting in short downtime.[9] With aeroderivative units, it is not necessary to schedule downtime for major maintenance, as is the case with heavy-duty industrial units. Also, statistical data on utility use of industrial turbines, combined cycles, and aeroderivative turbines compiled by the North American Electric Reliability Council show no significant differences in the availabilities of the three types of engines (42).

A closely related concern is the cost of maintenance. It is widely believed that maintenance costs of gas turbines, heavy-duty industrial as well as aeroderivative, are much higher than those of steam-electric plants. This belief is supported by utility experience; between 1982 and 1985 maintenance costs for utility gas turbines averaged 0.76 cents per kWh, compared to 0.26 cents per kWh for coal-fired steam plants (18). Some utilities report maintenance costs for gas turbines as high as 1.0 to 1.5 cents per kWh (42). These statistics should be interpreted with care, though, because the data for coal-fired plants are for carefully maintained baseload units, while the gas turbine data are for peaking plants that typically operate at an average capacity factor of only 5 to 7% and are often not carefully maintained. In considering gas turbines for baseload or load-following utility service, a more appropriate historical record is that for gas turbines operated in baseload cogeneration configurations at

[7] General Electric's LM-5000 is the 33 MW aeroderivative gas turbine which provided the basis for the first design of a 110 MW intercooled steam-injected gas turbine (ISTIG) (27). The LM-5000 in turn is derived from the CF6 turbofan jet engine which sells for $6 million (39). As the CF6 weighs 4770 kg, it is indeed costly on a per unit mass basis–$1260 per kg. [For comparison, at the time of this writing, gold and silver were selling on the world market for $14,500 and $225 per kg, respectively.] But this "high technology" part of ISTIG contributes only $53 per kW or 13% of its estimated installed cost of $400 per kW (Table 2).

[8] Complete inspection (with any necessary replacements) of the hot section of a GE LM-2500 aeroderivative turbine requires a crew of five working 100 person-hours (40). The same job requires a six-person crew working 480 person-hours for a GE Series 5000 industrial turbine that has a comparable output (41).

[9] This possibility arises because the gas generator, the "high technology" part of an aeroderivative engine, for which maintenance is most crucial, is easily transported. The gas generator for the largest aeroderivative turbine available, General Electric's LM-5000, weighs just 4770 kilograms and measures only 1.8 m x 2.1 m x 4.6 m.

industrial plants. Preventive maintenance programs carried out over the last 20 years for aeroderivative gas turbines used for cogeneration at the Dow Chemical Company resulted in maintenance costs of 0.2 to 0.3 cents per kWh (42).

Another concern often expressed about aeroderivative turbines is that utilities will not be interested in them because of their small unit capacities. However, pressed by the financial risks of building large power plants, many utilities are already beginning to shift the focus of their planning efforts to smaller units. Moreover, utilities would be able to improve overall reliability with multiple small units on the same site. The ongoing trend toward more competition in power generation is also making market conditions more favorable for introducing these smaller-scale power-generating technologies.

5 Steam-Injected Gas Turbines

The most significant development to date relating to stationary power applications of aeroderivative gas turbines was the introduction in the early 1980s of the steam-injected gas turbine (STIG), a variant of the simple gas turbine in which high pressure steam recovered in the HRSG is injected into the combustor, where it is heated to the turbine inlet temperature and then expanded in the turbine (Figure 8e).[10] Steam injection can give rise to large increases in power output and electrical efficiency. The only extra work required with steam injection, compared to a simple cycle gas turbine, is that needed to pump the feedwater to boiler pressure, which is negligible compared to the work required to compress the main flow air. This and the fact that the specific heat of steam is double that of air account for the large increases in efficiency and power output that arise with steam injection (26,43). Aeroderivative engines are chosen for steam injection, because, unlike heavy-duty industrial engines, these units are designed to accommodate mass flows considerably in excess of their nominal ratings, so that only minor modifications are required to operate them as baseload STIGs (44).

10 The injected water must be treated to avoid turbine blade corrosion problems. Because the minimum water treatment level required is not yet known (42), present practice is to be conservative. Even so, water treatment costs are minor. For STIG cogeneration units based on the Allison 501-KH, water treatment costs have been estimated to be 0.09 cents per kWh (for 1.63 liters per kWh and water treatment costs of 0.05 cents per liter, personal communication from C. Koloseus, International Power Technology, Inc., April 1985). For central-station STIG units based on the GE LM-5000, the capital cost for make-up and waste-water treatment (based on typical river water quality in the Eastern U.S.) has been estimated to be less than $20 per kW, some 5% of the total installed cost (29).

Injecting small amounts of steam (or water) in stationary gas turbines (heavy-duty industrial as well as aeroderivative) for the control of NO_x emissions is a well-established practice (45,46). Only recently has injecting large amounts of steam attracted serious commercial interest as a means of increasing efficiency and power output in stationary applications. Yet the concept is not new. The idea of using steam injection to increase power and efficiency is discussed in textbooks (47,48), in various articles dating from the mid-1970s (44, 49–55), and in a 1951 Swedish patent application (56) that was rejected in 1953. The injection of water into gas turbines dates to the earliest use of jet engines, when water was often injected to increase thrust during takeoff.

5.1 STIG for Cogeneration

The commercialization of STIG for cogeneration applications grew out of the post-PURPA flurry of interest in gas turbine cogeneration in the U.S. The STIG concept was introduced to cope with the most troublesome problem for simple cycle gas turbines in cogeneration applications: their poor part-load performance. With a STIG unit, steam not needed for process applications can be injected back into the combustor to produce more electric power; the provisions of PURPA often make it attractive to sell this extra power to the utility, thus extending the economic viability of gas turbine cogeneration to a wide range of variable-load applications (26).[11]

The first commercially operated STIG cogeneration units involved the use of the Cheng cycle, a patented version of STIG introduced by International Power Technology, Inc. (57,58). Cheng cycle units have been marketed using the Detroit Diesel Allison 501-KH turbine. Without steam injection, this turbine is rated to produce about 3.5 MW of electric power at 24% efficiency when producing power only. With full steam injection, it will produce about 6 MW at 34% efficiency (26). At the time of this writing six units based on the Allison 501-KH had been installed and two more ordered. More recently, three larger STIG units based on General Electric's LM-5000 have been installed at industrial sites [the first involving an in-the-field modification of a simple cycle cogeneration unit (59)] and fourteen more are either under construction or on order; and seven STIG units based on GE's LM-2500 are being planned

[11] A combined cycle with a condensing steam turbine using steam extraction to provide steam for process (Figure 8d) can also be used economically in variable steam-load applications: steam not needed for process is expanded through the lower turbine stages and condensed to produce more power. But the scale economies associated with steam turbines limit the economical use of the combined cycle to relatively large installations. STIG technology allows the gas turbine to be used in small-scale, variable steam-load applications.

(personal communication from M. Horner, Marine and Industrial Turbine Division, General Electric Company, Cincinnati, Ohio, December 13, 1988, and William Flye, Stewart & Stevenson Services, Inc., Houston, December 7, 1988). The LM-5000–derived from the CF6–50 high-bypass-ratio turbofan engine used in wide- body commercial airplanes (e.g., the DC-10 Series 30, the Boeing 747, and the Airbus A300)–is a 33.1 MW unit with a compression ratio of 25:1 and an efficiency of 33% when operated as a simple cycle on natural gas; with full steam injection the output and efficiency of the LM-5000 increase to 51.4 MW and 40% respectively (60). The LM-2500 is a 21.4 MW unit with a compression ratio of 18.5:1 and an efficiency of 33% in the simple cycle mode; with full steam injection its output and efficiency are 26.8 MW and 36%, respectively (60).

5.2 STIG and ISTIG for Central-Station Power

The use of steam injection for cogeneration has stimulated interest for central station applications, in which all the steam raised in the HRSG is injected for power and efficiency augmentation. A paper by a Bechtel analyst indicates that STIG plants based on the LM-5000 and using once-through steam generators would have several advantages over combined cycle units with cooling towers–including a unit capital cost lower by one-sixth, water requirements less by one-third, a 6% higher availability, and the possibility of remote operation without operators in continuous attendance (29).

A major drawback of STIG is that it is less energy-efficient than combined cycle technology now on the market (25). Accordingly, despite a modest capital cost advantage for STIG, the busbar cost would usually be lower for combined cycles (Table 2).

A more interesting candidate for central-station applications is a proposed modified STIG using intercooling between the two compressor stages (Figure 8f).[12] One result of intercooling is that less power is needed to run the compressor. The addition of an intercooler to a simple gas turbine increases the power output but decreases the efficiency; the reduced compressor work

[12] The intercooled STIG could also be used for cogeneration. An ISTIG unit based on the LM-5000 would produce about twice as much steam in the heat recovery steam generator as a STIG unit, largely because the turbine exhaust flow would be increased 26%, and its temperature would be increased from 395 to 441 °C (27). About 44% of the produced steam would have to be injected for NOx control and cooling of the combustor. The rest – some 27,200 kg/hour at 41.3 bar plus 15,400 kg/hour at 13.8 bar – could be used for process. In the full cogeneration mode the electrical output of the ISTIG unit would be 97 MW, and electricity production would represent 42.3% of the higher heating value of the fuel input (personal communication from M. Horner, General Electric Company, September 1988).

requirements would be more than offset by the extra fuel requirements for heating the cooled air exiting the compressor up to the turbine inlet temperature. But modern aeroderivative turbines use air bled from the high-pressure compressor to cool the turbine blades, so that intercooling in this instance leads to an efficiency gain as well. Because of the lower temperature of the air used to cool the blades, the metal temperatures can be kept acceptably low, while the turbine inlet temperature is raised. Detailed design work carried out at General Electric indicates that an intercooled STIG (ISTIG) based on the LM-8000[13] will be able to operate at a turbine inlet temperature of 1370 °C and produce about 114 MW with an average efficiency of 48.3% and a guaranteed efficiency of 47.0%, at an installed capital cost of $400 per kW (28).

The projected ISTIG efficiency is somewhat higher than that for an advanced combined cycle and its estimated capital cost is somewhat less, leading to a lower busbar cost (Table 2). The busbar cost would probably be less than for a large coal-fired steam-electric plant with flue gas desulfurization even if the natural gas price is double the average for 1986 (Table 2).

Despite the indicated efficiency advantage of the ISTIG compared to the combined cycle, this is not the result of a systematic comparison of steam-injected and combined cycle designs. And the estimated performance difference is too small to declare unequivocally that steam-injected designs are more efficient. In looking to the future, the balance could tip in favor of combined cycles, for example, if the Kalina cycle were successfully developed and used instead of the steam Rankine cycle in combined cycles. But there are also many possible modifications to the STIG cycle.

One such cycle modification is the chemically recuperated gas turbine, which involves using some of the turbine exhaust heat to reform the fuel with steam in the presence of an appropriate catalyst (61,62). For example, methane fuel could be converted into a mixture of hydrogen, carbon monoxide and carbon dioxide by reacting it with steam. As the steam-reforming reaction is highly endothermic, the chemical energy content of the products is greater than that of the fuel from which it is derived; thus through steam reforming, low-quality heat can be converted into high-quality chemical energy. To the extent that some of the turbine exhaust heat can be used for chemical recuperation as an alternative to heat recuperation through steam injection, there would be a net cycle efficiency improvement because of the reduction of the latent heat loss to the stack. (More than half the heat used to raise steam in the HRSG is the latent heat needed to evaporate water, which is lost to the stack in a STIG cycle.) Steam injection plus steam reforming has been shown to

[13] The LM-8000 is based on an upgraded version of the CF6 jet engine, the CF6–80C2, for which production totalled 110 engines in 1987 and is expected to be 260 engines in 1988 (39).

increase the efficiency of a simple cycle gas turbine from 30% to 47% when operated on methanol, while reducing the unit capital cost of the turbine by 15% (63). Further gains would be possible if intercooling were combined with steam injection and chemical recuperation, as shown in an evaluation of chemically recuperated, natural gas-fueled ISTIG units carried out by General Electric analysts for the California Energy Commission (64).

Because of the uncertainties relating to an efficiency comparison of advanced STIG and combined cycle designs, decisions to commercialize advanced STIG technologies should be made on grounds other than efficiency alone.

An ISTIG has several advantages over a combined cycle unit: it is simpler, requiring no steam turbine, condenser, or cooling tower; pollution controls would be less costly than with combined cycle units (see below); the small unit capacities of ISTIG units implies flexibility in capacity planning, improved reliability, and ease of maintenance through lease-pool arrangements; their small size also makes them good candidates for cost-cutting innovations and the economies of mass production; and steam-injected gas turbines will continue to benefit quickly from expected continuing improvements in jet engine technology.

A minus with STIG cycles is that they exhaust significant quantities of steam to the atmosphere. While the absolute level of makeup water requirements actually favors steam-injected cycles–some 0.6 kg per kWh for an ISTIG unit, compared to 4.1, 1.8, and 0.8 kg per kWh, for a nuclear plant, a large fossil-fuel fired steam-electric plant, and an advanced combined cycle, respectively (27)–all the water required for STIG cycles must be processed to boiler quality. In fact, STIG cycles require about five times the high-quality water and demineralization processing capacity required for combined cycles (65). Although these water processing requirements typically would not give rise to significant economic penalties (footnote 10), it could be a source of concern in water-scarce situations. But in such instances makeup water requirements could be reduced to zero by condensing water vapor out of the exhaust stream of the heat recovery steam generator for reuse, using a condensing heat exchanger. For STIG units, it has been estimated that complete water recovery could be achieved for an 11% increase in capital cost and a 1.2% increase in fuel requirements per kWh (65).[14]

Such considerations, collectively considered, suggest that advanced STIG cycles warrant development. It has been estimated that to develop ISTIG would take four to five years and cost $100 million, including the cost ($40 million)

[14] For an ISTIG unit operated on natural gas about two-thirds of the water vapor in the exhaust stream would have to be recovered to reduce makeup water requirements to zero, at zero relative humidity. The rest of the water vapor in the exhaust is a product of fuel combustion.

for the first unit (31). As no proof-of-concept is involved, only good engineering design, the technological risk associated with development is small. Accordingly, bringing the technology to market requires only the sale of a few units to pay for the relatively modest development costs.

5.3 Coal-Fired STIG/ISTIG

The uncertainty about the long-term availability of natural gas at affordable prices is a major obstacle to a broad shift by utilities to advanced gas turbines. The Cool Water project did demonstrate that gas turbines can be shifted to operate successfully on gas derived from coal, should natural gas become scarce and expensive. But a shortcoming of the "backstop technology" demonstrated at Cool Water is that, even when an advanced gas turbine is used and scale economies are exploited, the overall cost of electricity would be no less than that of power from a conventional coal-fired steam-electric plant with flue gas desulfurization (Table 2). In the present utility investment climate, it will be hard to persuade many utilities to take a chance on a new technology that offers no economic benefits. Accordingly, most utility planners willing to emphasize gas turbines in their capacity expansion plans are those who think that over the expected lives of these plants the chances of having to shift to gas derived from coal are remote.

This situation would be different if the coal-gas backstop technology offered significant economic advantages over conventional coal-fired steam plants. A 1986 analysis for the U.S. Department of Energy by the General Electric Corporate R&D Center offers one approach to addressing this challenge (31). This study, which considered several gasifiers and gas turbine technologies, identified three strategies as the most promising for reducing capital costs and improving the coal-to-electricity conversion efficiency, thereby reducing overall costs: (*a*) replacing the oxygen-blown Texaco gasifier with an air-blown Lurgi fixed-bed gasifier, thus eliminating the need for the costly and scale-sensitive oxygen plant; (*b*) employing hot-gas desulfurization instead of a scrubber (cold-gas cleanup) for sulfur removal, thus improving efficiency; (*c*) using an ISTIG instead of a combined cycle, further improving efficiency and eliminating the scale-sensitive steam-turbine bottoming cycle for the combined cycle. With these strategies, the GE study estimated that the installed capital cost would be about $1000 per kW in a 110 MW unit with an overall coal-to-busbar efficiency of 42.1%. The resulting busbar cost would be one-fifth less than that of either a coal-fired steam plant with flue gas desulfurization or combined cycle plant fired by coal gas derived from a Texaco gasifier, and one-sixth less than EPRI's target for nuclear power in a "reborn" nuclear power industry (Table 2). The Lurgi/ISTIG technology would offer the environmental benefits of coal gasification (see below) in a power plant one-sixth or one-fifth as large as a commercial-scale version of the technology demonstrated at Cool Water, thus bringing to coal-based power technology the advantages of small-scale units.

The Lurgi gasifier is a proven, commercially available system. ISTIG is not commercially available, but it involves no technological proof-of-concept. The unproven part of the proposed system is the hot-gas clean-up, which would involve the use of iron and zinc oxide catalysts to remove the sulfur from the gases exiting the gasifier (in the form of H_2S and COS) before these gases are delivered to the combustor. The proposed sulfur removal technology has been proven technically in bench and pilot scale investigations, and significant progress is being made in this area (35). Nevertheless, field experience is needed to demonstrate the long-term performance of the catalysts involved (31).

The attractiveness of the Lurgi/ISTIG concept led the U.S. Department of Energy to select for funding a pilot/demonstration project proposed by General Electric to develop essential features of this technology under the Department's Clean Coal Technology Demonstration Program. The $156 million project was to include testing the hot-gas cleanup concept at an existing General Electric gasifier test facility, followed by the construction of a 5 MW pilot plant and a 50 MW commercial demonstration project based on the use of a STIG unit. (Under the Clean Coal Technology Demonstration Program the government provides half the total support, the private sector the other half.) In the fall of 1987, however, the pilot and commercial demonstration phases of the project were canceled, because the required private-sector support was not secured.

At the time of this writing other significant efforts are under way to improve the overall economics of gas turbine-based power systems fired with gas derived from coal. Most notably, the Appalachian Project, a U.S. DOE-supported commercial demonstration, involving the use of an air-blown KRW fluidized bed gasifier, hot-gas cleanup, and a 63.5 MW combined cycle power plant, is moving ahead, with operation scheduled to begin in 1993 (66); also various advanced coal gasification/combined cycle concepts are moving ahead outside the U.S., with especially intense activity in West Germany (67). With the collapse of the Lurgi/STIG project, however, there is no ongoing coal gasification/gas turbine project underway to exploit the advantages of aeroderivative gas turbine technology, even though a recent U.S. Department of Energy (DOE) analysis (35) that systematically compared an air-blown, pressurized fluidized bed gasifier/ISTIG unit with various combined cycle-based systems concluded that the ISTIG-based system would be the least costly. The DOE study estimated, for example, that the unit capital cost and fuel requirements per kWh for the ISTIG-based system would be 37% and 15% less, respectively, than for mature Cool Water technology.

6 Environmental Aspects of Advanced Gas Turbine Technologies

Natural gas-fired gas turbines emit negligible amounts of sulfur oxides and particulates. The high combustion temperatures lead to high emissions of nitrogen oxides (NO_x), however. Their uncontrolled NO_x emissions (Table 3) exceed the levels established in the federal New Source Performance Standards (NSPS) for natural gas-fired gas turbines promulgated in the U.S. in 1977[15] and are far in excess of the standards in some states with especially severe air quality problems.[16]

Among available technologies for reducing NO_x emissions, a well-established approach involves the injection of steam or water into the primary combustion zone. NO_x emissions tend to fall exponentially with the ratio of steam or water injected in the primary combustion zone to fuel, as demonstrated empirically for steam in the first field-modified STIG unit derived from the LM-5000 (Figure 11). NO_x is controlled when steam or water is injected into the primary combustion zone because so doing reduces the flame temperature (Table 3).

With water injection, the percentage increase in fuel required to bring the mixture up to the turbine inlet temperature exceeds the percentage increase in power output resulting from the increased turbine mass flow, so that the electrical efficiency is reduced. With steam injection, the electrical efficiency is also reduced for combined cycle systems (where the optimal use of the steam produced in the HRSG is for power generation in the steam turbine) (50), but not for STIG or ISTIG cycles, where NO_x control is an automatic side benefit.

Dramatic reductions in NO_x emissions are also possible with chemically recuperated gas turbine cycles, with and without additional steam injection for NO_x control, as indicated for chemically recuperated STIG and ISTIG (CRSTIG and CRISTIG) cycles in Table 3. As with steam and water injection, the chemically recuperated cycles achieve low NO_x emissions levels because

15 For utility gas turbines consuming more than 100 million BTU per hour (approximately 10 MW) the standard is 75 ppm x (n/25) by volume (@ 15% O_2), where n is the turbine efficiency in percent. The standard for all other continuous-duty gas turbines consuming more than 10 million BTU per hour and producing less than 30 MW is 150 ppm x (n/25).

16 The toughest proposed standards in the U.S. are those proposed in August, 1988, by the South Coast Air Quality Management District (SCAQMD) in California (69): an emission level of 9 ppm x (n/25) for all new stationary gas turbines with capacities greater than 0.3 MW (tentative adoption date: January 1989) and an emission level of 12 ppm x (n/25) for existing turbines, within 30 months of the adoption date. The proposed standards are part of the proposed SCAQMD plan to bring the greater Los Angeles area into compliance with the U.S. ambient air quality regulations established under the Clean Air Act.

Table 3. Estimated NO_x emissions for alternative cycles based on the LM-5000.[a,b]

	Adiabatic Flame Temperature (T°K)		NO_x Emission Rate (ppm @ 15% O_2)[c]	
1. Uncontrolled Simple Cycles[d]	2530		260	
2. Steam-Injected Cycles	*STIG*	*ISTIG*	*STIG*	*ISTIG*
Dry[e]	2548	2471	301	159
S/F[f] = 1.0	2406	2329	89.5	43.6
S/F[f] = 2.0	2274	2199	25.3	11.6
S/F[f] = 3.0	2158	2085	7.4	3.2
3. Chemically Recuperated, Steam-Injected Cycles	*CRSTIG*	*CRISTIG*	*CRSTIG*	*CRISTIG*
Dry[e]	2223	2161	15	7.6
S/F[f] = 0.2	2107	2048	4.1	2.0
S/F[f] = 0.3	2055	1997	2.2	1.0
S/F[f] = 0.5	1959	1905	0.63	0.29

[a] The adiabatic flame temperature (AFT) is calculated for stoichiometric conditions–for methane @ 300 °K in simple and dry STIG/ISTIG cycles; for methane @ 400 °K in wet STIG/ISTIG cycles; for a mixture (@ 950 °K) of 10% methane, 62% steam, 22% hydrogen, 1% carbon monoxide, and 5% carbon dioxide (a typical fuel composition expected from the steam-reforming of natural gas) for CRSTIG/CRISTIG cycles. (A reheat turbine would be needed to achieve the turbine exhaust temperature needed for steam reforming. Because of the reduced oxygen level in the reheat combustion "air", the AFT and hence NO_x formation in the reheat combustor would be significantly lower than in the primary combustor. This effect is neglected here.) The temperature of steam for NO_x control is assumed to be 578 °K. For the calculated AFT, the indicated NO_x emission level is obtained from an empirical relationship between AFT and NO_x emissions (68). For the lowest AFTs considered, empirical data are not yet available, and the indicated NO_x emissions, obtained by extrapolation, may underestimate actual emissions. While thermal NO_x emissions are strongly correlated with AFT, prompt NO_x emissions are not. At high AFT, thermal emissions dominate; at low AFT, prompt NO_x emissions become more important.

[b] For a compression ratio of 27.6:1, 31.3:1, and 33.5:1, and an air temperature at the compressor exit of 787 °K, 816 °K, and 658 °K, for the simple cycle, STIG/CRSTIG cycles, and ISTIG/CRISTIG cycles, respectively (27).

[c] For a typical turbine, 1 ppm is approximately equivalent to 1.7 grams per GJ of fuel.

[d] The uncontrolled NO_x emissions rate for this simple cycle turbine is higher than for the turbines used in combined cycles, because the low pressure ratio of the gas turbines used in combined cycles leads to a lower compressor exit temperature and hence a lower AFT.

[e] "Dry" means that all steam is injected with the dilution air (i.e. no steam is injected into the primary combustion zone).

[f] S/F is the molar ratio of steam/fuel for steam injected into the primary combustion zone for NO_x control.

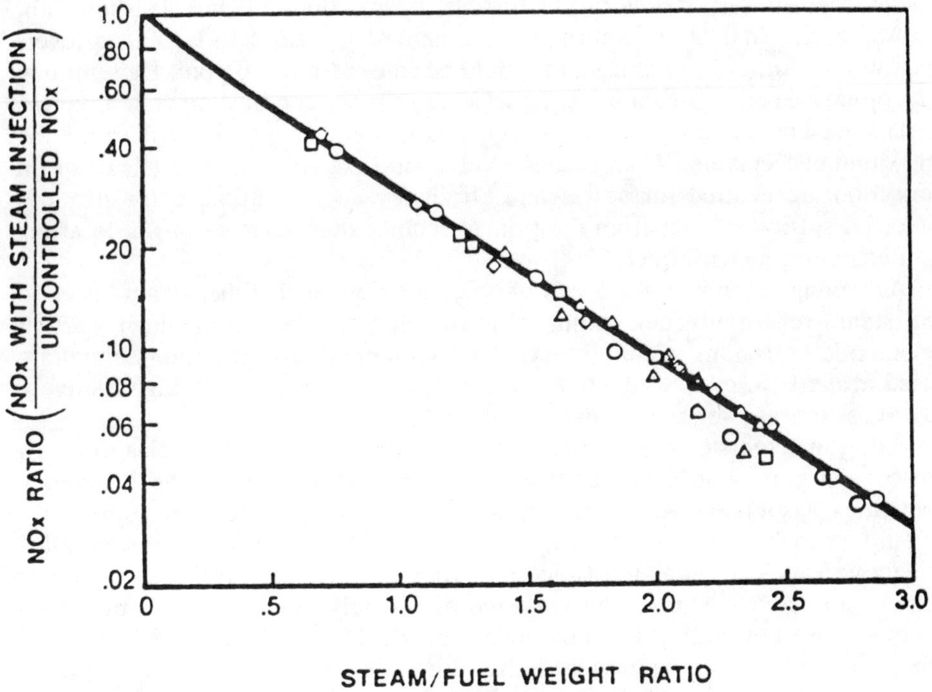

Figure 11. The NO_x reduction ratio as a function of the ratio of the weight of the steam injected into the primary combustion zone to the weight of the fuel, measured at the first field-modified STIG unit based on the LM-5000, at Simpson Paper Company, Anderson, California (59).

with steam-reformed fuel, flame temperatures are lower than with natural gas fuel (since the heating value of the steam-reformed fuel is much lower *per mole* than that of the fuel from which it is derived). CRSTIG and CRISTIG cycles are also more efficient electricity generators than cycles that do not use steam-reformed fuel (64). The combination of low NO_x emissions and high efficiency achievable with these cycles has led the South Coast Air Quality Management District (SCAQMD) in California and the California Energy Commission to examine chemically recuperated gas turbine cycles as a promising technological strategy for reducing NOx emissions in Southern California while simultaneously controlling costs (personal communication from Jack Janes, California Energy Commission, December 15, 1988)

CO emissions increase as the level of steam injection increases, while NO_x emissions are declining. (Similarly, increased CO emissions might arise with chemically recuperated gas turbine cycles, since a lower flame temperature is the mechanism responsible for the higher CO emissions.) Figure 12 shows, for example, that if NO_x is controlled in a natural gas-fired LM-5000 to levels below 25 ppm, the CO emissions would be in excess of 50 ppm. Determining the optimal level of steam injection into the primary combustion zone involves balancing considerations of both NO_x and CO emissions. If, by the time the maximum acceptable CO emissions level is reached, still higher levels of steam injection are desired for power and efficiency augmentation, extra steam is injected sufficiently far from the primary combustion zone so as not to affect pollutant emissions further.

Achieving extremely low levels of NO_x emissions with either steam injection or steam-reformed fuel might require supplemental control of carbon monoxide emissions. Catalytic oxidation of the exhaust gas (the technology used in the U.S. to control CO emissions from cars) is a promising, relatively low-cost strategy for accomplishing this (68).

Using injected steam to reduce NO_x emissions of combined cycles to levels below 10 ppm would probably not be pursued because of the efficiency penalties associated with steam injection. Instead, the favored approaches would probably involve a combination of water or steam injection plus selective catalytic reduction (injecting ammonia into the turbine exhaust gas stream to reduce NO_x in the presence of a catalyst) or staged or premixed combustion (not widely available today) plus selective catalytic reduction (or the equivalent) (68)–approaches that are inherently more costly ways to control NO_x than the approaches available for aeroderivative turbines.

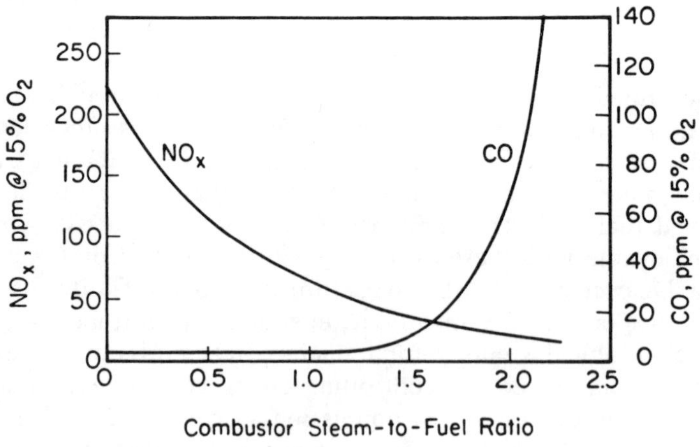

Figure 12. Typical effects of combustor steam injection on NO_x and CO emissions at full load operation (28).

With coal gasification even coal plants can achieve low pollutant emissions. The Cool Water project demonstrated emissions levels for SO_2, NO_x, and particulates far below U.S. New Source Performance Standards (Table 4). With advanced coal gasification technologies that involve hot-gas cleanup, NO_x control will be more difficult, however. Many coals contain 0.5 to 2 percent nitrogen by weight (dry basis), much of which is converted into ammonia in the gasifier. At Cool Water, this ammonia condensed out at the low temperatures realized with the "scrubber" gas cleanup technology used, but with hot gas cleanup the ammonia will be carried over into the combustor, where it can be converted to NO_x. Catalytic decomposition of ammonia or staged combustors will probably be needed to control NO_x emissions from this fuel-bound nitrogen in coal gasification systems using hot-gas cleanup (35).

Still another important environmental benefit of advanced gas turbine power generating technologies is the potential for reduced emissions of carbon dioxide, which is desirable in light of heightened concerns about the global greenhouse warming. In general, emphasis on natural gas fuel for the transition to the post-fossil fuel era would help slow the atmospheric build-up of carbon dioxide. Burning one energy unit of natural gas releases just 0.55 times as much CO_2 as the combustion of one energy unit of coal. Furthermore, generating electricity with natural gas in gas turbines as efficient as ISTIG would release just 0.4 times as much CO_2 per kWh generated as a conventional coal-fired steam plant.

Even efficient coal-gas fired turbines would be better in this regard than conventional coal-fired steam plants. A Lurgi/ISTIG unit fired with coal would emit only 0.8 times as much CO_2 per kWh as a conventional coal-fired steam plant. Applying advanced gasification technology to biomass grown on a renewable basis (71) would lead to zero net emissions of carbon dioxide, since the carbon dioxide released in combustion would just balance the carbon dioxide removed from the atmosphere in growing the biomass.

Table 4. Actual air pollutant emissions from the Cool Water demonstration power plant vs. EPA New Source Performance Standards.[a]

	Measured Emissions[b]	*EPA New Source Performance Standards*
SO_2	95 percent removal (0.014 kg/GJ)	90 percent removal (maximum = 0.52 kg/GJ)
NO_x	0.026 kg/GJ	0.26 kg/GJ
Particulates	0.00043 kg/GJ	0.013 kg/GJ

[a] The Cool Water plant produces 94 MW at an average efficiency of 30.2 percent, using the Texaco gasifier (70).

[b] For Utah (SUFCO) design coal.

7 Potential Applications of Advanced Gas Turbines

7.1 Natural Gas Resources

Natural gas will dominate initial applications of advanced gas turbines, because gas supplies will be abundant in many parts of the world in the decades immediately ahead. According to estimates by the U.S. Geological Survey, there is about as much conventional natural gas left in the world as conventional crude oil, but globally gas is used at just half the rate oil is (Table 5). Remaining gas resources for the U.S. and Canada are about 50% greater than oil resources, while for all industrialized countries remaining gas resources are more than twice as large as remaining oil resources (Table 5).

The outlook for gas is especially promising for developing countries. Natural gas resources exist in about 50 developing countries, including 30 that import oil (75). Also, gas resources are large in relation to gas production in developing countries; although they have about as much gas as industrialized countries, they produce it at just one-fifth the rate it is produced in industrialized countries. Outside the Middle East, developing country gas resources are equivalent to more than a 200-year supply at the current rate of production (Table 5).

7.2 Natural Gas/Coal-Gas Central-Station Power Applications in the U.S.

Even in the U.S., where natural gas resources are more fully developed than in most of the rest of the world, natural gas-fired gas turbines can play important roles in power generation in the decades immediately ahead.

One important initial application for ISTIG technology would involve replacing the 127 GW of existing oil and gas-fired steam-electric plants expected to be operating in the U.S. in 2000 (24) with natural gas-fired ISTIG units. While these are load-following steam plants typically operated at low capacity factor [projected by DOE to average 46% in 2000 (24)], they are so inefficient (32%) that it would be worthwhile replacing them with ISTIG units, even with fuel prices as low as $2 per GJ ($12 per barrel of oil equivalent).[17]

[17] The breakeven price is determined by setting the levelized busbar cost from a natural gas-fired ISTIG unit equal to the operating cost of an existing steam-electric plant, assuming a 46% capacity factor for the ISTIG unit, the same fuel price for both plants, and an O&M cost of 4.0 mills per kWh for the existing steam plants [the average value in the U.S. in 1985 (18)].

Table 5. Natural gas and oil resources and production.

	Natural Gas		Crude Oil	
	Resources[a] (EJ)	Production[b] (EJ/year)	Resources[a] (EJ)	Production[b] (EJ/year)
Industrialized Countries				
U.S. / Canada	1079.0[c]	21.0	720.2	23.3
Western Europe	481.6	7.3	318.8	8.6
Australia / New Zealand	149.3	0.6	36.7	1.3
U.S.S.R. / Eastern Europe	2757.6	27.1	991.9	25.9
SUBTOTALS	4467.5	56.0	2067.6	59.1
Developing Countries				
Central America	228.3	1.3	342.7	6.5
South America	290.4	1.7	523.8	7.7
Asia	779.5	3.9	641.3	11.6
Africa	620.9	2.0	627.2	11.1
Middle East	2316.8	2.6	3175.1	23.0
SUBTOTALS	4235.9	11.5	5310.1	59.9
GLOBAL TOTALS	8703.4	67.5	7377.7	119.0

a Proved reserves plus estimated reserve appreciation in discovered fields plus estimated recoverable undiscovered resources, as of January 1, 1985 (72).

b Production in 1985 (73).

c The resource estimate given here for the U.S. (650 EJ) may underestimate remaining gas resources. A more recent assessment carried out for the U.S. Department of Energy under the auspices of the Argonne National Laboratory (74) estimated that remaining gas resources in the lower 48 states of the U.S. recoverable at wellhead prices less than \$2.75/GJ amount to 633 EJ, with 189 EJ of additional resources recoverable at costs in the range \$2.75/GJ to \$4.60/GJ.

Doing so for all oil and gas-fired steam plants in the U.S. would lead to producing the same amount of electricity as DOE projected for such plants in 2000 (24) while saving the fuel equivalent to 0.82 million barrels per day of oil. Considerations of both the cost savings potential and the large NO_x reduction potential (especially important in Southern California) of ISTIG technology, led the staff of the California Energy Commission to a recommendation along these lines–that Southern California Edison Company retire some 5700 MW of existing gas-fired steam-generating capacity in favor of ISTIG plants (30). Supported by the California South Coast Air Quality Management District, this recommendation was under active consideration by the Southern California Edison Company at the time of this writing.

Another potential application of advanced gas turbines for central station power generation in the U.S. is as an acid rain control strategy. The large quantities of SO_2 and NO_x emitted by existing coal-fired steam-electric power plants are among the most significant pollutant emissions leading to acid rain (76). Growing concern about acid rain will probably lead to U.S. legislation in the not-to-distant future aimed at curbing these emissions (77). While much of the emphasis in the ongoing debates is on requiring the retrofitting of control systems such as stack gas scrubbers on these old power plants, the only benefit that would be gained thereby is a reduction of the pollutant emissions, at considerable cost.

An alternative would be to encourage scrapping existing coal-fired steam plants (regardless of their remaining useful lives) and replacing them at the same sites with new advanced gas turbine power plants, but using where possible the old plants' equipment (78). With all the incremental costs of the new facilities allocated to sulfur removal (giving no credit for reduced NO_x emissions or other benefits), the cost of this scrap-and-build strategy can be expressed as the cost per tonne of SO_2 removed and compared to the cost of SO_2 removal with scrubbers. Figure 13 shows that SO_2 removal with Lurgi/ISTIG would cost about $400 per tonne, compared to $340 to $640 per tonne for scrubbers. The remarkable result that Lurgi/ISTIG would be competitive with scrubbers in most circumstances is due to the fact that, despite the higher capital cost of the Lurgi/ISTIG option, there would be a significant coal savings because of the higher plant efficiency. It is noteworthy that the much less electricity-efficient Lurgi/STIG and Texaco gasifier/advanced combined cycle (Texaco/ACC) options, costing respectively some $690 and $810 per tonne of SO_2 removed, would not be competitive with scrubbers.

Of course, the advanced Lurgi/ISTIG option is constrained by the fact that it would not be commercially available for about a decade, if the development program went ahead. But ISTIG units could be fired initially with natural gas. Doing so for 20 years before switching to coal would lead to a cost of sulfur dioxide removal during this period of $330 per tonne and $640 per tonne, with natural gas prices of $3 per GJ and $4 per GJ, respectively, prices which are 50% and 100% higher than the average 1987 U.S. utility gas price.

7.3 Natural Gas/Coal-Gas Central-Station Power Applications in Western Europe

Outside of the Netherlands and Italy, natural gas has been little used for power generation in Western Europe (accounting for less than 5% of all electricity produced in 1984), largely because of the belief that natural gas should be saved for "more noble purposes," since electricity can be readily produced from abundant coal and nuclear energy sources. This situation could change radically, however, because of: (*a*) the growing availability of natural gas

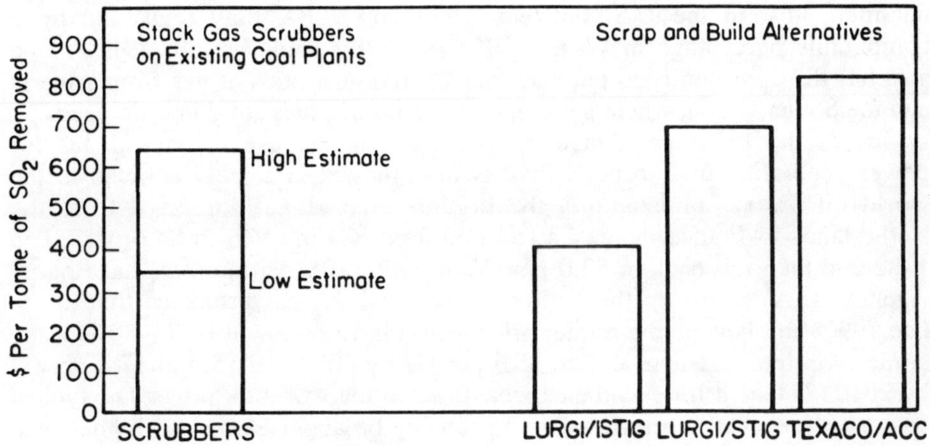

Figure 13. The cost of sulfur removal with alternative acid rain control strategies for coal with an average sulfur content of 3.1% (78). The bar on the left shows the range of estimated costs for putting stack gas scrubbers on existing coal steam-electric plants, for capital costs in the range \$193 to \$344 per kW, extra O&M costs in the range 0.42 to 0.83 cents per kWh, a 60 percent average capacity factor, and 85% sulfur removal. The three bars on the right are for cases in which existing plants are retired and new coal gas-based gas turbine plants are built at these sites, allocating all the incremental cost to sulfur removal, assuming 95% sulfur removal with coal gasification. All the cost and performance characteristics for the new plants are from Table 2, except that capital costs are reduced 10 percent because some of the existing facilities can be used in the new plants. In taking credit for the operating costs of the retired plants, it is assumed that the efficiency of the existing plants is 33.2 percent, the average for coal plants in the U.S. in 1985, and that the O&M cost for the existing coal plants is 0.38 cents per kWh, the average for Midwest coal plants without scrubbers. A 6.1% discount rate and a 30-year lifetime are assumed for all cases.

The major cost uncertainties for the Lurgi/ISTIG option are the costs for the chemical hot-gas cleanup system. The uncertainty in the capital cost of the cleanup system is ± 25%, or ± \$27 per tonne of sulfur dioxide removed. At the targeted catalyst consumption rate of 0.5%, the contribution of the cost of the catalyst to sulfur dioxide removal is \$41 per tonne. But the consumption rate could be as high as 1% or as low as 0.25%, so that the catalyst cost could range from \$21 to \$83 per tonne of sulfur dioxide removed. Thus, the cost of sulfur removal with Lurgi/ISTIG is probably in the range \$352 to \$469 per tonne, with the most likely value being about \$400 per tonne (78).

supplies in Europe, (*b*) the favorable economics of power generation based on advanced gas turbines, and (*c*) growing concerns about coal and nuclear power.

Estimated remaining recoverable natural gas resources are about three-fourths as large in Western Europe as in the U.S. (72), but European gas resources are much less developed. While proved reserves of natural gas declined 30% in the U.S. between 1973 and 1987, they increased by a comparable percentage in Western Europe in this same period (79). Europe also has the opportunity to purchase substantial quantities of gas from Algeria and the Soviet Union, where gas supplies are far in excess of domestic needs.

One of the important considerations limiting the use of natural gas for power generation in Europe is that its price has been closely coupled to the world oil price. For example, the border price of gas imported from the Netherlands to France increased from $2.0 per GJ in 1976 to $4.4 per GJ in 1982 and then fell back to $2.0 per GJ in 1986 (80). An International Energy Agency (IEA) report on the outlook for natural gas has projected that in the late 1990s and beyond the border price of gas in Europe will be 73–80% of the crude oil price, reaching $3.5 to $5.5 per GJ by 2000 and $5.9 to $7.2 per GJ by 2010 (81), as it follows the expected rise in the world oil price. The typical gas price for electric utilities would probably be about $1 per GJ higher than the border price, to cover transmission and distribution costs from the border to the power plants.[18] Such projections discourage the use of natural gas for power generation, because lifecycle gas prices must typically be no higher than $3.7 to $4.4 per GJ for natural gas-fired ISTIG units to be competitive with new coal or nuclear steam-electric plants in Europe (Table 6).

The high European gas prices projected by the IEA are a result of the assumption that competition between gas and oil at the point of end use will determine the market price of gas. However, if marginal costs were low and there were major markets in which natural gas could compete with coal, competition with coal might instead determine the price of gas.

Current and projected gas prices are indeed far in excess of the marginal cost of bringing forth new gas supplies in Europe, according to a 1986 study by the International Gas Trade Project at MIT, which has analyzed the costs of increasing supplies from the major sources of gas for Western European markets on a source-by-source basis (80). Gas supplies at low marginal costs appear to be so large over the coming decades that gas could become a major fuel for power generation.

To illustrate the possibilities, suppose that all incremental power generation in Western Europe after 1995 were based on natural gas-fired ISTIG units. If overall electricity production were to grow at 2.5% per year after 1995 [in the midrange of projections made in 1985 by the IEA (3)], then by 2010 180 GW of electric power capacity would be advanced gas turbines, accounting for 30%

[18] The weighted average cost of transmission and distribution inside borders in Europe was about $1.9 per GJ in 1984 (81). In the U.S., the difference between the retail price and the wellhead price averaged about $2 per GJ between 1981 and 1985; in this same period the difference between the gas price for electric utilities and the wellhead price averaged $1 per GJ (82). Thus, a reasonable estimate for the price of gas to electric utilities in Europe is $1 per GJ more than the border price.

Table 6. Levelized costs (cents/kWh) for electric power generation from alternative sources in Western Europe.[a]

	Levelized Busbar Cost	
Nuclear[b]		
	6-year construction	10-year construction
Capital	2.01	2.25
Fuel	1.00	1.00
Operation & Maintenance	0.54	0.54
TOTAL	3.55	4.09
Coal/Steam w/FGD[c]		
Capital	1.46	
Fuel	1.88	
Operation & Maintenance	0.54	
TOTAL	3.88	
ISTIG w/Gasified Coal[d]		
Capital	1.22	
Fuel	1.55	
Operation & Maintenance	0.60	
TOTAL	3.37	
ISTIG w/Natural Gas[e]		
Capital	0.47	
Fuel	0.766 x P	
Operation & Maintenance	0.29	
TOTAL	0.76 + 0.766 x P	

[a] All costs are for a 6% discount rate, a 30-year plant life, and a 70% capacity factor. All taxes and subsidies are neglected.

[b] Based on an International Energy Agency (IEA) study assessing the outlook for electricity in IEA member countries (3). For a site with two 1100 MW(e) units and start-up in 1990, the unit capital cost is estimated to be \$1700/kW (\$1900/kW) for a 6-year (10-year) lead time.

[c] For a site with two 600 MW(e) units with flue gas desulfurization (FGD) and start-up in 1990, the unit capital cost is estimated to be \$1230/kW, assuming a 4-year lead time (3). The efficiency is 35% (36%, LHV basis). The coal price is assumed to be \$52/tonne (\$1.82/GJ).

[d] For a 110 MW(e) ISTIG unit fired with gas derived from coal in a Lurgi gasifier with hot gas clean-up, the unit capital cost is estimated to be \$1030/kW, assuming a 2-year lead time. The coal price is assumed to be \$52/tonne (\$1.82/GJ). The efficiency is 42.1% (43.4%, LHV basis). See Table 2.

[e] For a 114 MW(e) ISTIG unit fired with natural gas, the unit capital cost is estimated to be \$400/kW, assuming a 2-year lead time. The efficiency is 47% (52%, LHV basis). Here P is the average lifecycle cost of natural gas (\$/GJ). See Table 2.

of all electricity production. Under this scenario, gas turbines would consume 8.5 EJ of natural gas in 2010, or about two fifths of total gas demand then, assuming gas demand in other sectors grew at 1.1% per year [the average of the high and low projections made by the IEA in 1986 (81)]. Even with this very large increase in aggregate demand, the marginal cost of gas would still be only about \$2.5 per GJ in 2010, according to the MIT analysis (Figure 14).

If this natural gas-based power generating technology were "backstopped" with a capability to shift to gas derived from coal, the result could be a capping of natural gas prices at levels far below what is being forecast. The backstop gas price, obtained by equating the cost of converting an ISTIG plant to coal with the cost of continuing to operate the ISTIG unit on natural gas, would be about \$3.6 per GJ,[19] and the corresponding border price would be about \$1

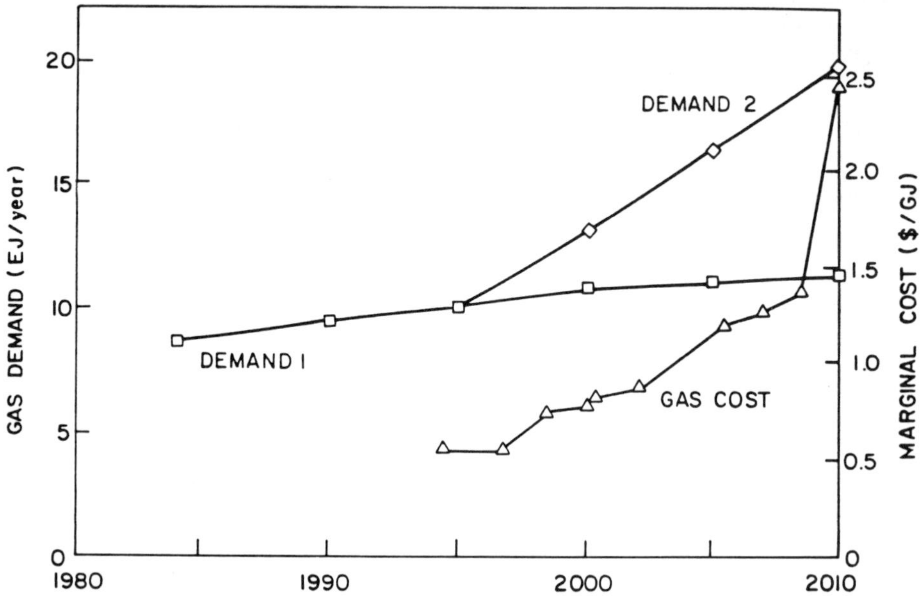

Figure 14. Two alternative projections of natural gas demand in Western Europe. Demand 1 is the average of the high and low projections made in 1986 by the International Energy Agency (81); Demand 2 superimposes on the Demand 1 projection additional demand for gas associated with providing all incremental electricity production with natural gas-fired ISTIG units after 1995. It is assumed that total electricity production grows at 2.5% per year [in the midrange of projections presented in a 1985 International Energy Agency Study (3)], from 1888 TWh in 1984 to 3588 TWh in 2010, of which 1110 TWh is accounted for by ISTIG units. The marginal cost curve shown for natural gas was developed in the 1986 MIT International Gas Trade Project (80) and is for the gas supply levels of the Demand 2 projection.

[19] The coal price is assumed to be \$52 per tonne (\$1.82 per GJ). This is the average price of imported coal and also the price of Australian coal imported into the European Economic Community between 1984 and 1986 (83). In light of the abundance of coal, the multiplicity of sources, and the low cost of transporting coal long distances, it is assumed here that the coal price remains stable in the coming decades. With these assumptions and the other parameters indicated in Table 6, the backstop utility gas price is estimated to be \$3.6 per GJ.

per GJ less, far below the $5.9 to $7.2 per GJ forecast for 2010 by the IEA (81).

The availability of this backstop technology would put a ceiling on the market price of gas because, if gas producers tried to set a higher price, they would lose a large gas market to coal, and the backstop price would still probably be higher than the marginal cost of gas through the first decade in the next century.

If the gas price could be controlled at this backstop price, the result would probably be lower electricity costs through 2010 from natural gas-fired ISTIG units than would be the case for either nuclear power plants (even when nuclear power plants are built quickly, i.e., in only six years) or coal steam plants with flue gas desulfurization (Table 6).

In addition to these utility benefits, the benefits from lower gas prices for gas consumers in Europe other than electric utilities would be worth perhaps $40 billion per year in 2010, for the midrange natural gas demand forecast by the IEA (81).

Of course, similar benefits could be derived from technology that is already commercially ready, e.g. the Texaco/ACC technology as a coal-gas backup for advanced combined cycle units fired with natural gas. However, the backstop gas price with this technology would be about $1 per GJ higher than with Lurgi/ISTIG. Thus, the annual benefit to nonutility consumers in Europe in 2010 would be some $10 billion greater if Lurgi-ISTIG were the backstop instead. It would seem that Western European governments have a strong stake in bringing advanced coal gasification/gas turbine backstop technology to commercial readiness, even if there were little prospect of its being used for two or three decades!

7.4 Natural Gas/Gas Turbine Power for Developing Countries

While the abundance of their natural gas resources (Table 5) suggests that natural gas could play a major role in the energy economies of many developing countries, the use of natural gas is presently inhibited by the lack of gas transmission and distribution infrastructure, which is costly to develop. This problem could often be overcome if power generation were emphasized as an initial market. Some of the large revenues generated in power sales could be used to pay for the construction of the gas delivery system, thereby helping to make gas available to other users at reasonable cost (84).

Its low capital cost makes the gas turbine an especially attractive technology for developing countries in light of the unaffordability of capital investments for electricity based on conventional sources (Table 1).

Not only are the overall capital requirements small for these advanced gas turbine power plants, but also, many industrializing countries could draw on indigenous management and engineering talent for much of the design and construction effort required. The power turbine, the heat recovery steam generator, and the electrical generator, as examples, are system components that can be readily manufactured in many parts of the world. The part of the system for which it may be difficult to avoid expenditures of foreign exchange is the "gas generator," the "high technology" part of the system that is derived from a jet engine. The gas generator actually accounts for only a modest fraction of the total power plant cost; the mass-produced CF6 jet engine, from which LM-5000 STIG and ISTIG units would be derived, costs only about $6 million–thus contributing only $53 per kW to the cost of the ISTIG unit (see footnote 7).

The scale characteristics of aeroderivative turbines are also well suited to developing countries. In most developing countries, the total utility grid capacity is too small to be well matched to much larger hydroelectric or steam-electric power plants. Adding new capacity in small increments with gas turbines makes it possible to avoid the alternating periods of power glut and power shortage associated with utility planning based on large plants and can lead to improved system reliability.

The compact, modular nature of aeroderivative turbines makes it possible to replace failed parts and even whole engines quickly with replacements flown or trucked in from centralized maintenance facilities (see footnote 9). This feature of the aeroderivatives is especially attractive for many developing countries, where sophisticated maintenance capability is typically unavailable at power generating sites. The required maintenance network is already in place in most developing countries that have their own commercial airlines; their planes are typically maintained through centralized lease-pool arrangements. This advantage is reflected, for example, in the fact that of the 210 General Electric LM-2500 aeroderivative turbines in service throughout the world as of 1986, 54, 8, and 26 were being used in developing countries of Latin America, Africa, and Asia, respectively (personal communication from L. Gelfand, Marine and Industrial Turbine Division, General Electric Company, Cincinnati, Ohio, February 1987).

8 Conclusion

For the next two to four decades, advanced gas turbines offer multiple benefits for power generation. The prospects of reducing electric power costs in both industrial and developing countries, of reducing local air pollution and acid rain emissions to low levels, of reducing carbon dioxide emissions to levels considerably below those associated with coal-fired steam-electric plants, and of avoiding the risks of expanded dependence on nuclear power, are benefits not easily matched by alternatives.

While the benefits resulting from the wide use of heavy-duty industrial gas turbines would be large, there are good reasons for also bringing advanced aeroderivative turbines into wide use. It appears that controlling pollutant emissions responsible for acid rain to very low levels would be more readily accomplished with various advanced aeroderivative turbines than with combined cycles. Also, with aeroderivatives the advantages of high efficiency and low unit capital costs can be extended to modest scale, resulting in greater flexibility in capacity planning, improved reliability, and ease of maintenance. And the small size of aeroderivatives makes it possible to reverse the trend in power technology toward costly field construction and bring most construction work back to the factory, where the economies of mass production can be exploited. Moreover, aeroderivative turbines will continue to benefit more directly from improvements in jet engine technology than heavy-duty industrial turbines.

Wide use of advanced gas turbines would not solve the electric power problem for all time. Eventually, the tightening of world gas supplies and concerns about the atmospheric buildup of carbon dioxide will limit the attractiveness of further expanding the use of these engines for power generation with fossil fuels. However, a major shift to fossil fuel-fired gas turbines for power generation in the decades immediately ahead would buy time to develop alternative clean power sources for the long term, e.g., the direct conversion of sunlight into electricity using amorphous silicon solar cells (85).

References

1. Hass, J.E., Mitchell, E.J., Stone, B.K. 1974. *Financing the Energy Industry.* Cambridge: Ballinger.
2. Fisher, J.C. 1974. *Energy Crises in Perspective.* New York: Wiley.
3. Int. Energy Agency. 1985. *Electricity in IEA Countries: Issues and Outlook.* Paris.
4. *Technical Assessment Guide. 1: Electricity Supply – 1986.* 1986. Palo Alto, Calif: Electr. Power Res. Inst.
5. Williams, R.H., Larson, E.D. 1988. Aeroderivative turbines for stationary power. *Annual Review of Energy* 13: 429–89.
6. Schneider, H.K. 1987. *Investment Requirements of the World Energy Industries 1980–2000.* London: World Energy Conference.
7. Loose, V.W., Flaim, T. 1980. Economies of scale and reliability: the economics of large versus small generating Units. *Energy Systems and Policy* 4 (1–2): 37–56.

8. Duffy, T.E., Schneider, P.H., Campbell, A.H. 1986. *Advanced high performance steam systems for industrial cogeneration. SR86-R-5323–06*. Report prepared by Solar Turbines Int. for the U.S. Dept. Energy.
9. Electric Power Development Co., Ltd. 1988. *Wakamatsu 50 MW Demonstration Plant for Fluidized-Bed Combustion Boiler and Ultrasuper Critical Turbine*. Kyushu, Japan: Wakamatsu Coal Utilization Center.
10. Combustion Eng., Inc. 1981. *Combustion: Fossil Power Systems*. Windsor, Conn: Rand McNally.
11. Fraas, A.P. 1982. *Engineering Evaluation of Energy Systems*. New York: McGraw-Hill.
12. Wolfe, W.R. 1976. *Energy conversion alternatives study (ECAS), Westinghouse Phase I report, vol. XI: Advanced Steam Systems. NASA CR- 134941*.
13. El-Sayed, Y.M., Tribus, M. 1986. *A theoretical comparison of the Rankine and Kalina cycles*. Cambridge, Mass: Cent. for Adv. Eng. Study, Mass. Inst. Technol.
14. Mech. Technol., Inc. 1986. *Technology assessment of advanced power generation systems II – Kalina bottoming cycle. EPRI AP-4681*. Palo Alto, Calif: Electr. Power Res. Inst.
15. Rankin, B. 1987. Revolutionary power plant to get its day in the sun. *New Technology Week* Dec. 21: p. 5.
16. Gluckman, M.J. 1988. *The future of gas turbine-based power generation in the United States*. Paper presented at the 2nd Int. Symp. on Turbomachinery, Combined Cycle Technologies and Cogeneration. Montreux, Switzerland: Am. Soc. Mech. Eng.
17. *Electric Power Annual Report*. Annual. Washington, DC: Edison Electr. Instit.
18. Energy Inf. Admin. 1987. *Historical plant cost and annual production expenses for selected electric plants 1985. DOE/EIA-0455(85)*. Washington, DC: U.S. GPO.
19. Off. Electr. Power Regul. 1987. T*he qualifying facilities report: a cumulative list of filings made for small power production and cogeneration facilities through December 31, 1986*. Washington, DC: Fed. Energy Regul. Comm.
20. Off. Ind. Programs, U.S. Dept. Energy. 1984. *Industrial cogeneration potential, 1980–2000, DOE applications of four commercially available prime movers at the plant site*. Washington, DC.
21. Bur. Census, U.S. Dept. Commerce, *Statistical Abstract of the United States 1987*. Washington, DC: U.S. GPO.
22. Directorate for Inf. Oper. and Rep. Annual. *Prime contractor awards by service category and federal supply classification*. Washington, DC: U.S. Dept. Defense.
23. Wilson, D.G. 1984. *The Design of High-Efficiency Turbomachinery and Gas Turbines*. Cambridge, Mass: MIT Press.
24. Energy Inf. Admin. 1988. *Annual energy outlook 1987, with projections to 2000. DOE/EIA-0383(87)*. Washington, DC: U.S. GPO.
25. Brandt, D.E. 1986. Heavy-duty turbopower: the MS7001F. *Mech. Eng*. July: pp. 28–36.
26. Larson, E.D., Williams, R.H. 1987. Steam-injected gas turbines. *ASME J. Eng. for Gas Turbines and Power* 109 (1): 55–63.
27. Eng. Dept., Pacific Gas and Electr. Co. 1984. *Scoping study: LM5000 steam-injected gas turbine*. Based on work performed by the Mar. and Ind. Engine Proj. Dept. (Evendale, Ohio) of the Gen. Electr. Co.
28. Horner, M.W. [Mar. and Ind. Eng. and Serv. Div. (Cincinnati, Ohio) of the Gen. Electr. Co.]. 1988. *Position statement – intercooled steam-injected gas turbine*. Testimony presented at the Committee Hearing for the 1988 Electricity Report of the Calif. Energy Comm., held at the So. Calif. Edison Co., November 21–22, 1988.

29. Soroka, G.E. (Bechtel Eastern Power Corp.). 1987. *Modular remotely operated fully steam-injected plant for utility application.* Paper presented at the 2nd Int. Symp. on Turbomachinery, Combined Cycle Technologies and Cogeneration, Switzerland: Am Soc. Mech. Eng.
30. Calif. Energy Comm. Staff. 1988. *Resource case analysis report.* Testimony presented at the Committee Hearing for the 1988 Electricity Report of the Calif. Energy Comm., held at the So. Calif. Edison Co., November 21–22, 1988.
31. Corman, J.C. 1986. *System analysis of simplified IGCC plants.* Report prepared for the U.S. Dept. Energy by GE Corp. Res. and Dev. Schenectady, NY.
32. *Cool Water coal gasification program: fourth annual progress report.* 1986. *EPRI AP-4832.* Palo Alto, Calif: Electr. Power Res. Inst.
33. Pillai, K.K. 1989. *Pressurized fluidized bed combustion*, this volume.
34. Merrick, D. 1984. *Coal Combustion and Conversion Technology.* New York: Elsevier.
35. Notestein, J.E. (Dept. Energy, Morgantown Energy Technology Cent.). 1988. *Update on Department of Energy Hot Gas Cleanup Programs.* Paper presented at the EPRI Coal Gasification Conf., Palo Alto, Calif.: Electr. Power Res. Inst.
36. Fluor Eng., Inc. 1986. *Planning data book for gasification-combined cycle plants: phased capacity additions. EPRI AP-4395.* Palo Alto, Calif: Electr. Power Res. Inst.
37. Catina, J.L., Fortune, H.J. Jr. (Virginia Power), Soroka, G. E. (Bechtel Eastern Power Corp.). May 31-June 4, 1987. *Repowering Chesterfied 1 and 2 with combined cycle.* Paper *87-GT-12.* Presented at the Int. Gas Turbine Conf. and Exhib. Anaheim, Calif: Am. Soc. Mech. Eng.
38. *Can gas turbines be twice as good?* April 1987. *Global Gas Turbine News.* Atlanta, Ga: Int. Gas Turbine Inst. Am. Soc. Mech. Eng.
39. Ungeheuer, F. 1988. They make good things for flying. *Time.* May 2: p. 55.
40. Jackson, J. 1984. *Aircraft-derivative gas turbine maintenance practices.* Gas Turb. Ref. Lib. (GER-3424). Schenectady, NY: Gen. Electr. Gas Turbine Div.
41. Knorr, R. 1984. *Heavy-duty gas turbine maintenance practices.* Gas Turb. Ref. Lib. (GER-3412A). Schenectady, NY: Gen. Electr. Gas Turbine Div.
42. NJ Energy Conserv. Lab. 1986. *Report on the NJECL workshop on steam- injected gas turbines for central station power generation*, ed. E. D. Larson, R. H. Williams, Princeton, NJ: Cent. Energy and Environ. Stud., Princeton Univ.
43. Larson, E.D., Williams, R.H. 1985. *A primer on the thermodynamics and economics of steam-injected gas-turbine cogeneration. PU/CEES Report No 192.* Princeton, NJ: Cent. Energy and Environ. Stud., Princeton Univ.
44. Leibowitz, H., Tabb, E. 1984. The integrated approach to a gas turbine topping cycle cogeneration system. *ASME J. Eng. Gas Turbines and Power.* 106: 731–36
45. Allen, R., Kovacik, J. 1984. Gas turbine cogeneration – principles and practice. *ASME J. Eng. Gas Turbines and Power* 106: 725–30.
46. Touchton, G. 1985. Influence of gas turbine combustor design and operating parameters on effectiveness of NO_x suppression by injected steam or water. *ASME J. Eng. Gas Turbines and Power* 107: 707–13.
47. Diamant, R.M.E. 1970. *Total Energy*, New York: Pergamon.
48. Hayward, R.W. 1980. *Analysis of Engineering Cycles*, New York: Pergamon.
49. Maslennikov, V., Shterenberg, V.Y. 1974. Power generating steam turbine-gas turbine plant for covering peak loads. *Thermal Eng.* 21 (4): 82–88.
50. Fraize, W.E., Kinney, C. 1979. Effects of steam injection on the performance of gas turbine power cycles. *J. Eng. Power* 101: 217–27.

51. Davis, F., Fraize, W. 1979. *Steam-injected coal-fired gas turbine power cycles. MTR-79W00208*. Mclean, Va: Mitre Corp.
52. Brown, D.H., Cohn, A. 1981. An evaluation of steam-injected combustion turbine systems. *J. Eng. Power* 103: 13–19.
53. Kosla,L., Hamill, J., Strothers, J. 1983. Inject steam in a gas turbine – but not just for NO_x control. *Power*. Feb.
54. Digumarthi, R., C.N. Chang, C. N. 1984. Cheng-cycle implementation on a small gas turbine engine. *J. Eng. Gas Turbines and Power* 106: 699–702.
55. Jones, J.L., Flynn, B.R., Strother, J.R. 1984. *Operating flexibility and economic benefits of a dual-fluid cycle 501-KB gas turbine engine in cogeneration applications*. Paper *82-GT-298*. NY: Am. Soc. Mech. Eng.
56. Nicolin, C. 1951. A gas turbine with steam injection. *Swedish Patent Application No. 8112/51*. Stockholm.
57. Cheng, D.Y. 1978. Regenerative parallel compound dual-fluid heat engine. *U.S. Patent No. 4,128,994*. Washington, DC.
58. Cheng, D.Y. 1981. Control system for Cheng dual-fluid cycle engine system. *U.S. Patent No. 4,297,841*. Washington, DC.
59. Burnham, J.B., Giuliani, M.H., Moeller, D.J. 1986. *Development, installation, and operating results of a steam injection system (STIG) in a General Electric LM 5000 gas generator*. Paper *86-GT-231*. Presented at the Int. Gas Turbine Conf. and Exhib., Dusseldorf, West Germany: Am. Soc. Mech. Eng.
60. Oganowski, G. 1987. *LM 5000 and LM 2500 steam injection gas turbine*. Presented at the Int. Gas. Turbine Cong. Tokyo: Gas Turbine Soc. Japan
61. Janes, C.W. 1979. Increasing gas turbine efficiency through the use of a waste heat methanol reactor. *Proc. 14th Intersoc. Energy Convers. Eng. Conf.* 2: 1968–1972.
62. Tsuruno, S., Fujimoto, S. 1987. *Gas turbine cycle with steam reforming of methanol*. Paper *87-Tokyo-IGTC*-83. Presented at the 1987 Tokyo Int. Gas Turbine Cong. Tokyo, Japan: Gas Turbine Soc. of Japan.
63. Klaeyle, S., Laurent, R., Nandjee, F. 1987. *New cycles for methanol-fueled gas turbines*. Paper *87-GT-12*. Presented at the Int. Gas Turbine Conf. and Exhib. Anneheim, Calif: Am. Soc. Mech. Eng.
64. Mar. and Ind. Eng. and Serv. Div. (Cincinnati, Ohio) of the Gen. Electr. Co. 1988. *Advanced chemically recuperated gas turbine cycle evaluation*. Work sponsored by the Calif. Energy Comm.
65. Cohn, A. 1988. Steam-injected gas turbines versus combined cycles. *EPRI* J. 13(5): 40.
66. M.W. Kellogg Co., Bechtel Dev. Co. 1988. *The Appalachian Project*. Houston, Tex., San Franscisco, Calif. 35 pp.
67. Peters, W. 1989. *Coal gasification technologies for combined cycle power generation*, this volume.
68. Sidebotham, G.S., Williams, R.H. 1989. *Technology of NO_x control for stationary gas turbines*. Princeton, NJ: Cent. Energy and Environ. Stud., Princeton Univ.
69. South Coast Air Quality Management District (El Monte, Calif.). 1988. *Proposed Rule 1134 – Control of Oxides of Nitrogen Emissions from Stationary Sources*.
70. Alpert, S.B., Gluckman, M.J. 1986. *Coal gasification systems for power generation*. Palo Alto, Calif: Electr. Power Res. Inst.
71. Larson, E.D., Svenningsson, P., Bjerle, I. 1989. *Biomass gasification for gas turbine power generation*, this volume.

72. Masters, C.D., Attanasi, E.D., Dietzman, W.D., Meyer, R.F., Mitchell, R.W., Root, D.H. 1987. *World resources of crude oil, natural gas, natural bitumen, and shale oil.* Paper prepared for the 12th World Petrol. Cong. Houston, Tex.
73. Energy Inf. Admin. 1987. *International energy annual 1986. DOE/EIA-0219(86).* Washington, DC: U.S. GPO.
74. Argonne National Laboratory. 1988. *An assessment of the natural gas resource base of the United States.* Washington, DC: Off. Policy, Planning, and Analysis, U.S. Dept. Energy.
75. World Bank. 1983. *The Energy Transition in Developing Countries.* Washington, DC.
76. Off. Technol. Assess. 1984. *Acid Rain and Transported Air Pollutants: Implications for Public Policy.* Washington, DC.
77. Shabecoff, P. 1988. Cuomo and Governor of Ohio join in proposal on acid rain. *The New York Times.* June 6: p. 1.
78. Williams, R.H. 1988. *Using acid rain controls to promote U.S. competitiveness and innovation in the electric utility industry.* Princeton, NJ: Cent. Energy and Environ. Stud., Princeton Univ. (draft).
79. Am. Petrol. Inst. 1987. *Basic Petroleum Data Book: Petroleum Industry Statistics.* VII (2). Washington, DC.
80. Adelman, M.A., Lynch, M.C. 1986. Natural gas supply in western Europe. *Western Europe Natural Gas Trade, Final Report.* Cambridge, Mass: Int. Natural Gas Trade Project, Cent. Energy Policy Res., Energy Lab., Mass. Inst. Technol.
81. Int. Energy Agency. 1986. *Natural Gas Prospects.* Paris.
82. Energy Inf. Admin. 1987. *Monthly energy review. DOE/EIA- 0035(87/04).* Washington, DC: U.S. GPO.
83. Int. Energy Agency. 1987. *Energy prices and taxes, fourth quarter 1986.* No. 2. Paris.
84. Schramm, G. 1984. The changing world of natural gas utilization. *Natural Resour. J.* 24: 405–36.
85. Carlson, D. 1989. *Low-cost power from thin-film photovoltaics,* this volume.

Pressurised Fluidised Bed Combustion

Krishna K. Pillai

Abstract

Pressurised fluidised bed combustion (PFBC) is a coal combustion technology just being commercialised for the generation of electricity and heat. As a pressurised process, it inevitably requires the use of a gas turbine for the expansion of the exhaust gases and this, in combination with a conventional steam turbine, gives a combined cycle plant and its associated high efficiency. As a fluidised bed technology, PFBC provides the opportunity to burn coals – even the so-called "bad" coals – in an environmentally acceptable manner. PFBC plants are now under construction and the performance of these will determine the role PFBC will play as a "clean coal technology" through the nineties and beyond.

1 The Perspective

Some 30% of the world's population today consumes over 70% of the electricity generated, and the overwhelming majority aspire to levels of consumption significantly greater than they enjoy today. The perils of nuclear energy – both actual and perceived – have dimmed hopes we might have had of nuclear providing the "universal solution". The environmental consequences of fossil fuel use are now receiving the attention they deserve. Acid rain precursors, emissions of gases enhancing the greenhouse effect, waste products and their impact on ground-water contamination, and emissions of carcinogenic compounds are setting limits as to how and where and when fossil-fired electricity generation may be used. The "renewable" sources of energy – sun, wind and waves – remain of only marginal significance.

Within this conflicting scenario of constraints and demands, the electricity generation and supply industry faces the challenge of inventing acceptable solutions.

The use of fossil fuels for electricity generation is intimately linked to their availability. Coal, oil and gas are, and will continue to be, readily available throughout the world. Oil and gas however, are premium products whose value as a chemical feedstock or as premium fuels for transport together with their limited reserves mitigate against their profligate use purely for electricity production. Coal remains the dominant fuel which will be used for electricity generation in virtually every part of the world for the forseeable future.

The great challenge will be to carry out the electrification of the world while using coal, but without poisoning our atmosphere or raising the sea level to a catastrophic point. The solution lies not in the unconditional and immediate shut-down of operating power plants, nor indulgence in end-of-the-world scenarios. Rather, what is needed is the judicious use of legislation to stimulate and promote the development of environmentally and economically sound,electricity-generating technologies.

The future production of electricity from coal thus requires:

- emissions of sulphur be kept below stringent levels as set by legislation;
- emissions of nitrogen-dioxide be kept below strictly legislated limits;
- the integration of heat and electricity generation be maximised, if necessary by legislation;
- the waste products – ash, slag, waste-water etc. – be at the very worst, benign and, at best, usable;
- the efficiency be as high as possible – and perhaps regulated by legislation; and
- a global view on emissions and environment be taken.

Pressurised fluidised bed combustion (PFBC) is now beginning its commercial life. Being based on a fluidised bed process, the technology has the potential of utilising all coals – even the so-called "bad" coals, i.e. those with high-ash, high-sulphur, low heating value or low reactivity – in an environmentally acceptable manner. It uses a combined cycle for power generation and this provides the step-change in efficiency when compared to conventional coal combustion techniques. The use of a pressurised fluidised bed process also means that compact plants, capable of being modularised to a high degree can be built. This in turn makes it possible to "clean-up" existing coal-fired electricity generation sites by means of PFBC retrofits which exhibit the added advantages of increasing power output and efficiency simultaneously.

Figure 1 illustrates the routes available for electricity production from coal and the position occupied by PFBC in this hierarchy.

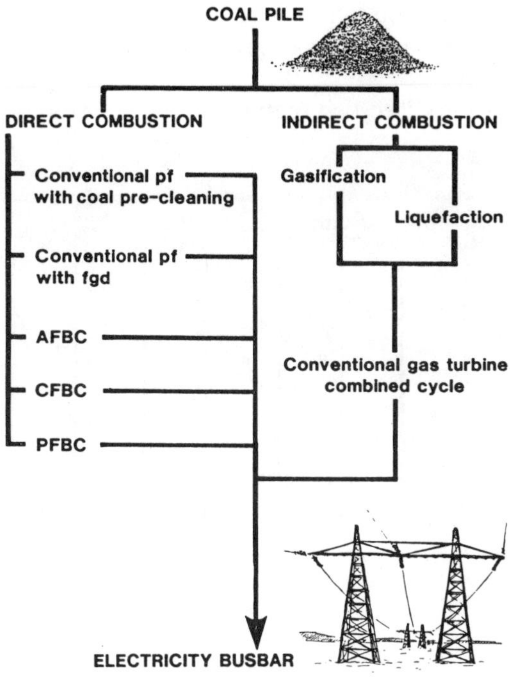

Figure 1. From Coal to Electricity: Possible routes.

2 The Technology

2.1 Atmospheric Fluidised Bed Combustion (AFBC)

Fluidised bed combustion is the only new coal combustion technique to have appeared since pulverised coal combustion was introduced.

A bed of inert particles can be made to exhibit many of the properties of a fluid – fluidised – by passing a gas upwards through the bed at a velocity sufficiently high for frictional drag to support the weight of the particles, but not so high as to initiate pneumatic transport of the particles (1,1a). The fluidised regime thus exists between the packed bed and the pneumatic transport states. A two-phase, gas/solid system exists with one phase consisting of solid particles together with the gas just necessary to support them and the second phase consisting of the excess gas passing through the bed as bubbles. It is these bubbles and their motion which give a fluidised-bed its characteristic hydrodynamic and thermodynamic properties which are analogous to those of a liquid (2).

Figure 2 illustrates the fluidised bed regime.

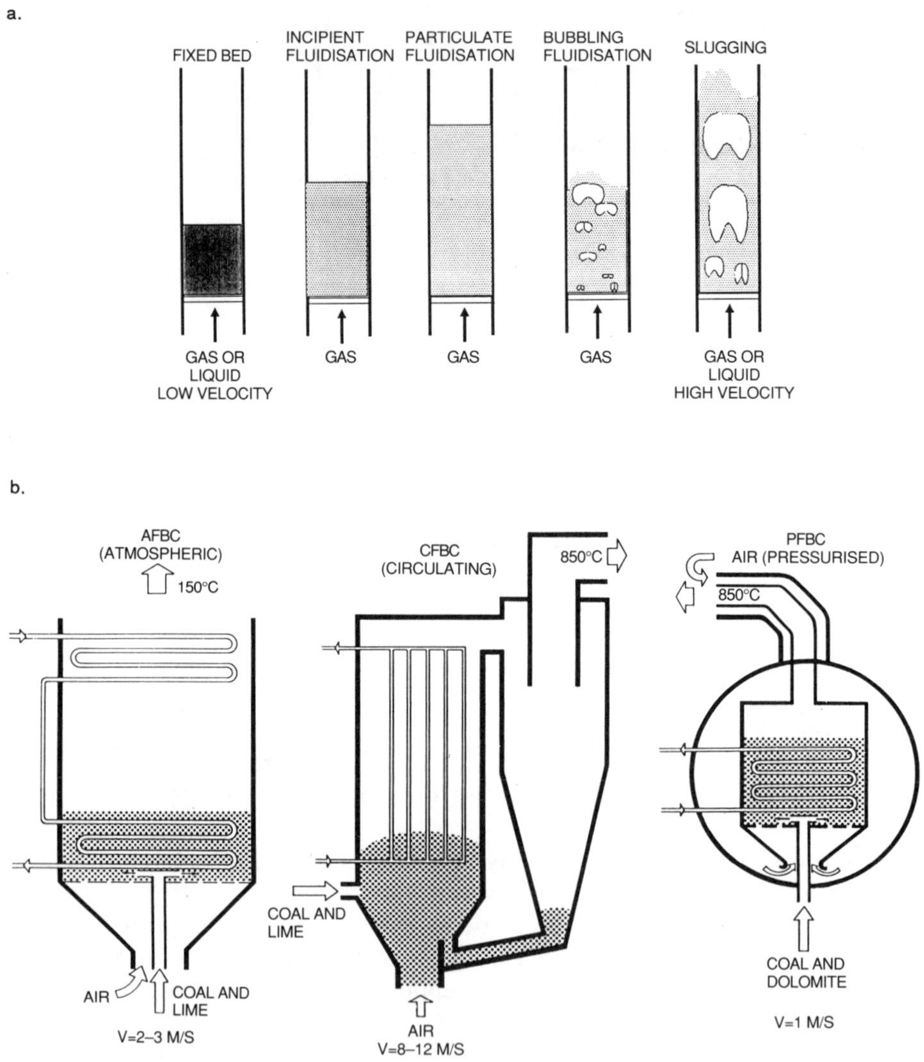

Figure 2. Fluidisation Regimes
a. Types of fluidisation
b. Types of fluidised combustion.

If such an air-fluidised bed of inert particles is heated to a temperature greater than the ignition temperature of some fuel, and the fuel then introduced into the bed, the fuel burns without it being necessary to have the high temperatures usually associated with a flame. Thus, nearly all fuels can be burnt in a fluidised bed with a temperature of less than 1000°C instead of the usual flame temperatures of over 1600°C. This is because the mixing rate of the fuel with the rest of the bed particles is so high that heat is carried away from the reacting fuel so as to prevent the temperature increasing. Of course this implies that some cooling medium be available to carry away all heat generated in excess of that necessary to heat the fuel and associated gas upto the bed temperature. This cooling medium can be excess air in sufficient quantities, or cooling tubes immersed within the bed particles.

In practice, bed temperatures of between 800 and 950°C are sufficient for the effective combustion of virtually all fuels (3).

It is because of this low temperature combustion that the emissions of nitrogen-dioxide from fluidised beds are low. The fixing of atmospheric nitrogen is virtually non-existent at temperatures below 1300°C and thus fluidised beds give rise to no thermal NOx. The conversion of fuel nitrogen, however, can proceed at these temperatures and emissions are dependant upon the excess air level being used, the forms of fuel nitrogen present and the NOx destruction reactions which proceed in the bed on the surfaces of the solid particles present. Typically, NOx emissions from a fluidised bed consist of zero thermal-NOx and about 10% of the fuel-nitrogen as NOx (4).

A further characteristic of fluidised beds is that the combustion rate of the fuel within the bed is sufficiently high that, even with close to stoichiometric combustion, only around 1% of the bed particles are active fuel particles (or droplets in the case of liquid fuels or bubbles in the case of gaseous fuels). This leads to the fuel flexibility associated with fluidised beds. In theory even fuels with extremely low calorific values can be used. In practice, it is hardware limitations on fuel feeding or ash removal which limit the type of fuels which can be burnt.

The third major feature of fluidised beds – sulphur capture – is also a consequence of the operating temperature range. At these temperatures, any sulphur dioxide evolved from fuel-sulphur can be absorbed by calcium bearing solid particles within the bed to produce calcium sulphate. Limestone and dolomite are typical calcium materials which are fed into the bed in sufficient quantities so as to absorb as much of the sulphur as desired. The sulphate produced is a granular, solid material which is relatively easily handled.

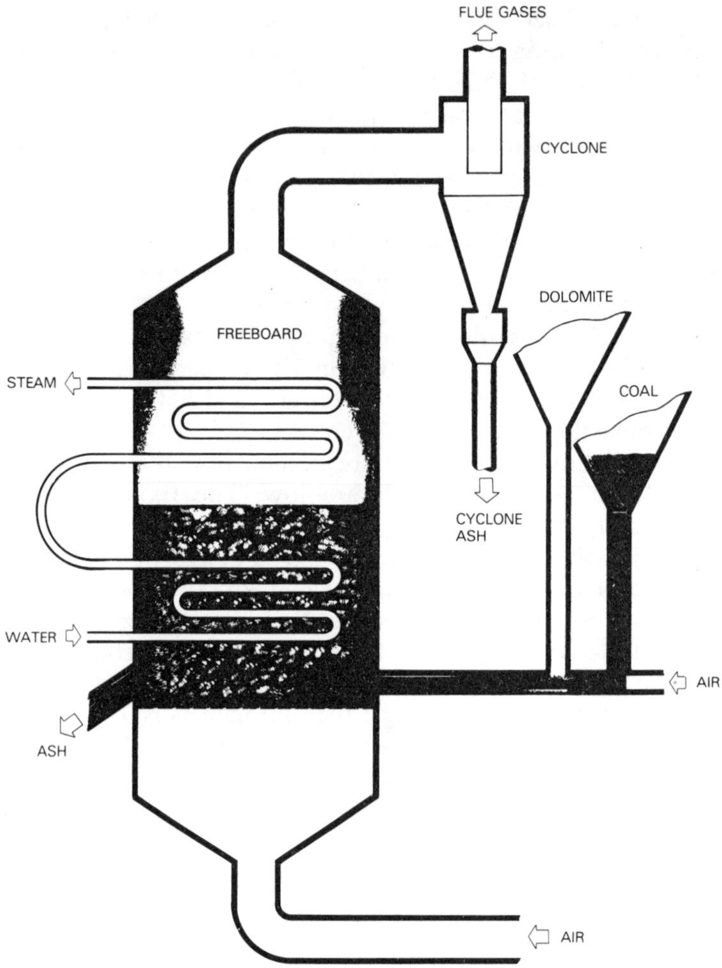

Figure 3. Principles of Fluidised Bed Combustion.

Figure 3 illustrates the main features of a fluidised bed for the combustion of coal. It should be noted that the gas/solids motion within a fluidised bed gives rise to very high heat transfer coefficients to immersed cooling tubes. Even though the low operating temperature gives reduced temperature differences as the heat transfer driving force, the requirements for expensive, high temperature materials is significantly reduced in comparison with conventional boilers (5).

With coal as the fuel, it only needs to be crushed to granular form (typically less than 10mm) rather than having to be finely pulverised. The sulphur sorbent is fed as granular limestone together with the fuel. The bed material therefore consists mainly of inert, coal-ash and partly-calcined, partly-sulphated limestone.Thus, atmospheric fluidised bed combustion (AFBC) is seen not only as an alternative to conventional coal burning techniques, but also as a means of extending the range of fuels for industrial use. AFBC is, in other words, a direct alternative to conventional coal-fired electricity generation since the boiler could be coupled to a conventional steam plant for electricity generation. With AFBC the thermodynamic steam cycle remains unchanged from that of a conventional coal-plant, and the benefits accruing are those due to the above-mentioned characteristics of the fluidised bed. AFBC is now commercially available at the industrial scale and efforts to commercialise the technology at the utility scale are in progress (6).

2.2 The PFBC Process

Advances in the efficiency of conventional power plant have been brought about by innovations in materials technology which have permitted higher operating temperatures (and pressures) to be used. However, steam temperatures have now reached an operating plateau and materials corrosion and stressing limitations make it difficult for further improvements to be achieved. Combined gas/steam cycles offer the most realistic possibility of significantly improving power generation efficiency; the margin of improvement increasing with the temperature at which the gas turbine can be brought to operate (7,8,8a).

Though coal-fired gas turbines have long been sought, it is only the advent of fluidised bed combustion which has made this goal a real possibility.

Fluidised bed combustion as a coal-fired boiler concept attracted considerable attention in the 1960's, primarily from the emissions control point of view. AFBC studies revealed some difficulties with size and coal feeding when considering large utility applications. Furthermore, the absence of any significant efficiency advantage for AFBC led to the development of the PFBC concept, which with its combined cycle efficiency offered an incentive for development.

Pressurised fluidised bed combustion (PFBC) allows a large increase in specific plant output, but still retains all the characteristic features of fluidised beds. Indeed, many of the characteristics are significantly enhanced due to the use of elevated pressures. Combustion is enhanced to the level that fuel concentrations in the bed are significantly reduced (typically less than 0.5%). Pressure drops are relatively low in comparison with AFBC, and this permits the use of deep beds with low fluidising velocities which result in correspondingly long gas residence times. Gas-solids contact is consequently improved and results in better combustion efficiency and more effective sulphur capture. For reasons not fully understood but probably connected with the gas-solids mixing, NOx emissions are substantially reduced at higher pressures (3,9).

Inevitably, PFBC implies the use of a gas turbine where the pressure energy in the exhaust gases from the combustion process can be recovered.

As for AFBC's, the range of bed temperatures is 750 to 900°C. The upper limit is set by the need to prevent ash fusion temperatures being reached and by the need to keep vaporised alkali levels low enough to satisfy corrosion considerations (3,9,10,11,11a). This gives a gas turbine entry temperature of also less than 900°C.

In 1968, the BCURA (British Coal Utilisation Research Association) placed a fluidised bed within an existing pressure vessel to create the first experimental PFBC. Testing began in 1969 and the facilities at Leatherhead, UK remained a centre for PFBC development for the next 15 years. The early work was primarily funded by the United States Department of Energy (and its predecessors) and the US Environmental Protection Agency (12).

Interest in the concept grew rapidly in the US, Sweden and Germany. A number of PFBC experimental rigs and pilot plant facilities were built through the 1970's and extensive testing was carried out. Feasibility studies for utility application were undertaken and provided the motivation for continuing with the development. Table 1 summarises the particulars of the most important of these various facilities.

In 1976 an agreement for the development of PFBC was entered into by the American Electric Power Service Corporation (AEPSC) and ASEA STAL (then Stal Laval and now part of the Asea Brown Boveri Group, ABB STAL with its subsidiary ABB Carbon). This was of particular significance because it represented an agreement between a utility and a supplier (12a). In due course, it led to the construction of a wholly private pilot plant facility – the 15MWth Component Test Facility (CTF) in Sweden – which has provided the major impetus for the commercialisation of the technology (13,14,15). The largest test facility was the experimental PFBC unit built at Grimethorpe, UK, funded as an International Energy Agency (IEA) project by the United Kingdom, the United States and the Federal Republic of Germany (16,16a).

Throughout the development of PFBC, the central issue was the survivability of the gas turbine in the combined cycle process. The use of cascades of turbine (static and rotating) blades exposed to actual PFBC exhaust gases has shown (9,11,11a,17,17a) that with temperatures of less than 900°C in the bed and at the turbine, vapour phase alkali is either not present or already condensed on dust particles. The low alkali levels measured have provided the confidence that corrosion of gas turbine blades is manageable. Erosion was

Table 1. Important PFBC test plants.

Organisation	Site	Operated	Thermal Input MWth	Max. Pressure bar
UK NCB	Leatherhead	1969–1984	4.5	6
UK NCB	Leatherhead	1981–1984	1.2[a]	20
ANL, USA	Argonne, IL.	1973–	0.15	10
EXXON, USA	Linden, NJ	1974–1979	1.7	10
CPC, USA	Menlo Park, CA	1974–1975	8.7	4
Curtiss Wright, USA	Wood Ridge, NJ	1977–1982	2.3[b]	7
Curtiss Wright, USA	Wood Ridge, NJ	–	40[b,c]	7
IEA	Grimethorpe, UK	1981–1984	64–45[d]	12
UK NCB/CEGB	Grimethorpe	1985–	30[d]	12
GE, USA	Malta, NY	1982–1984	0.5	10
ABB Carbon	Malmö, Sweden			
	Finspång, Sweden[e]	1982–	15	17
New York University	Westbury, NY	1983–	5	10

a. Moved to CRE, Cheltenham, UK

b. Air cycle PFBC

c. Never operated due to disposal of site

d. Derated as fluidising velocity was reduced

e. The CTF pilot plant was moved to Finspång, Sweden in 1986 and is still in operation.

generally considered the greater danger. However, PFBC dust was found to be relatively soft (not having been heated up to sufficiently high temperatures to make glassy material), and, provided all large dust particles (greater than 10 microns) were removed, erosion of gas turbine blades was not expected to be a major problem. Two stages of high-efficiency (relatively high pressure-drop) cyclones in series were found to satisfy the particulate specification developed for the gas turbine blades (17b).

The other major theme running through PFBC development has been the emissions performance. Gas residence time was found to be the major parameter affecting combustion efficiency and sulphur capture (3,17c). Similarly, excess air levels were found to be the major factor determining NOx emissions. In all cases, the effects of pressure were found to be beneficial (9) in comparison to the performance of atmospheric fluidised beds.

PFBC cycles can be classified as air-cooled or steam-cooled cycles. There are two principal variants of the air-cooled cycles (17d), but it is widely accepted (7,8,8a) that the air-cooled cycles are not competitive either with conventional pulverised coal-fired plants or with the steam-cooled variants of the PFBC cycle.

Steam-cooled PFBC cycles also come in two varieties.

There is the supercharged boiler cycle where steam is raised by in-bed tubes and the gas turbine entry temperature is virtually the bed temperature. Exhaust gas from the turbine is cooled to stack temperature by means of feed-water heating.

Second there is the so-called "turbocharged" boiler cycle where additional steam cooled heat exchanger surface in the freeboard above the bed reduces the gas turbine entry temperature to such an extent that only sufficient power to drive the compressor is generated at the gas turbine. Since no gas turbine power is available this is not strictly a combined cycle.

Generally, steam-cooled schemes utilise as low an excess air level as is compatible with good combustion.

The supercharged boiler concept has evolved as that providing the best potential and is the one currently being commercialised.

Figure 4 is a schematic diagram of the PFBC combined cycle. The gas-turbine compressor supplies air to a pressure vessel within which is contained a bubbling fluidised bed. Boiler tubes are immersed within the bed for steam generation. Coal and sulphur sorbent are fed to the fluidised bed combustor where combustion takes place at about 850°C. The exhaust gases from the combustor pass through two stages of cyclone cleaning designed as protection for the gas turbine. The exhaust gases are then expanded through the gas

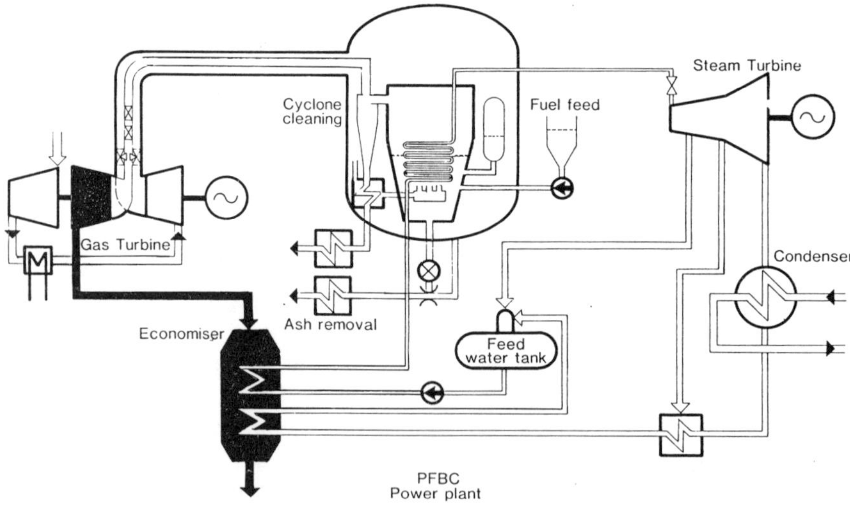

Figure 4. Schematic of a PFBC Combined Cycle Plant.

turbine which drives the compressor and generates some surplus power. The exhaust gas from the turbine passes over a heat exchanger where the gases are cooled to the desired stack temperature while heating up boiler feed water. Steam can be reheated in the PFBC if desired and used in a conventional steam generator. The exhaust gases are passed through a final filter before entering the stack. Solid waste products (coal ash and sulphated limestone) are removed from the bed and from the cyclones.

About 20% of the power output is generated at the gas turbine and about 80% at the steam turbine. This combined cycle is untypical in that the steam raising occurs during the combustion process, i.e., before the gas turbine, rather than in the gas turbine exhaust. This permits steam conditions to be decoupled from the exhaust gas conditions and thus the process places no restrictions on the steam temperature. The system is limited by materials considerations rather than process conditions.

With typical steam conditions of about 180 bar, 540°C 540°C, the PFBC cycle exhibits a net plant efficiency of 42 to 44% based on Lower Heating Value and a cooling water temperature of 10°C. This represents about a 10% improvement over a conventional coal plant having the same steam cycle (but

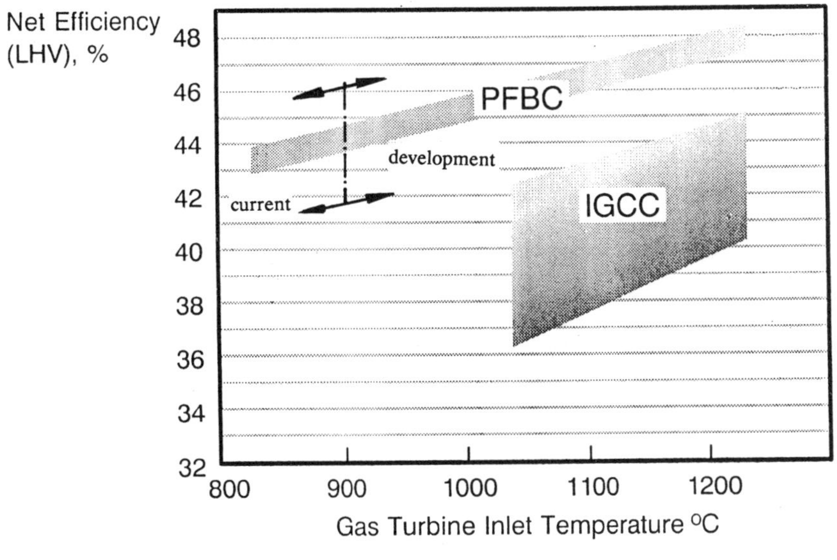

Figure 5. Efficiency Comparison: PFBC–IGCC.

of course no gas turbine) and having no losses attributed to any flue gas scrubbing process.

Figure 5 shows the effect of gas turbine entry temperature on plant efficiency for PFBC and for plants based on coal gasification as a rival "clean coal technology". A gas turbine entry temperature of around 900°C represents the present state of the art for PFBC with the region above this temperature being PFBC's development potential once hot gas cleaning has been further developed.

The pressure level and air flow in the PFBC determine the amount of coal that can be fired and therefore the power output from the plant. They in turn are functions of the gas turbine. Hence, the choice of gas turbine is crucial for the PFBC plant. However, since gas turbine development (or, more accurately, compressor) development costs are so high, gas turbines have not been developed specifically for PFBC's. Existing gas turbines have been modified instead for PFBC applications.

2.3 Design Features of PFBC Power Plant

Today, PFBC represents a coal-fired combined cycle process which incorporates a pressurised bubbling fluidised bed and a gas turbine operating at a conservative turbine entry temperature. The deep (circa 4m) fluidised bed together with the relatively low fluidising velocity (circa 1m/s) lead to a gas residence time of around 4 seconds in the bed and a further 3 to 4 seconds in the hot freeboard above the bed. This in turn provides excellent sulphur capture and combustion efficiency performance without having to resort to recycle arrangements. (Recycling of fines is only called for when dealing with very unreactive fuels – petroleum coke or anthracite, for example – or when particularly low values of minimum load are called for). Unreactive fuels can require an increased gas residence time or recycling of unburnt fines to achieve a satisfactory combustion efficiency level. At very low loads, gas residence time is reduced and recycle provides an artifice for increasing the contact time between gas and solids to achieve the required combustion and sulphur capture performance.

Table 2 details the performance capabilities of PFBC. Figure 6 illustrates the emissions performance expected of PFBC plants.

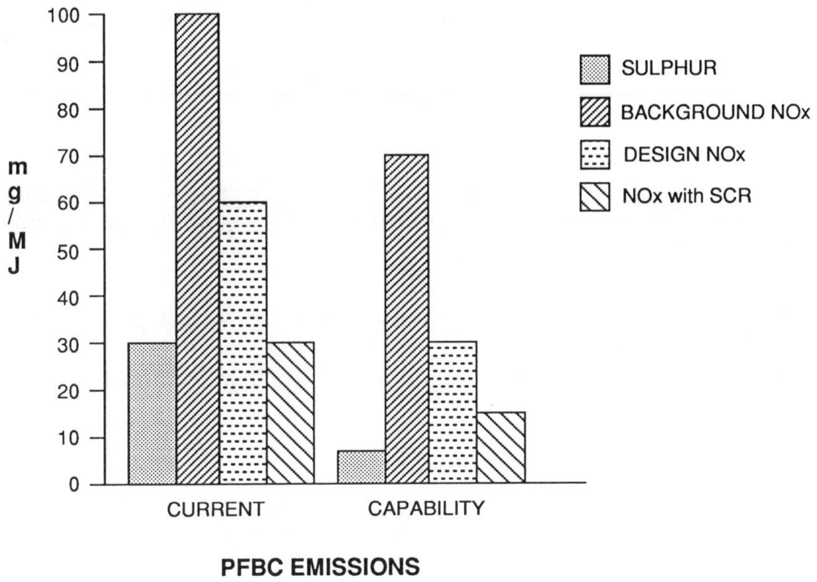

Figure 6. Emissions from PFBC Plants.

Table 2. Performance capabilities of PFBC plant.

Combustion efficiency	Greater than 99% with any coal
Sulphur Capture	90% with a Ca/S ratio of 1.5 with dolomite 90% with a Ca/S ratio of 1.5–2.0 with limestone Upto 98% with a 2% Sulphur coal is practical
(NO + NO2) emission as NO2	Less than 100 mg/MJ (basic design) 60 mg/MJ with NOx reduction techniques (but without catalytic reduction)
Net efficiency (based on LHV, 10°C cooling water temperature and reheat)	42% at 80MWe 44% at 350MWe

PFBC design has evolved over the years and now incorporates the following features:

- The separation of pressure and temperature barriers.
- The enclosure of all hot parts by cooler air at a higher pressure so that all leaks are directed inwards.
- A two-shaft gas turbine with a small intercooler between the low-pressure and the high-pressure compressor to limit the final air temperature and thus the temperature of the pressure vessel.
- Low carbon steels for the pressure vessel.
- The absence of moving parts within the pressure vessel to facilitate maintenance and service.
- The absence of refractories in the hot gas path to facilitate load changing and start-up.
- Hot gas cleaning (cyclones) within the same pressure vessel as the boiler to prevent temperature losses before the gas turbine.
- The bed storage vessel used for load changing also within the same pressure vessel to enable rapid load changes with minimal thermal shock.
- A ruggedised industrial gas turbine with the hot parts strengthened to withstand the PFBC exhaust gas.
- A bed temperature below 880°C to limit NOx emissions, avoid coal ash sintering and avoid the vaporisation of corrosive alkalis.
- A conservative gas turbine entry temperature of around 840°C.
- Integral gas-air ducting and isolation valves between the PFBC and the gas turbine.
- Advanced steam conditions with reheat if desired.
- The possibility of integrating heat demands into the steam cycle.

Figure 7 shows a PFBC combustor and illustrates some of these features.

Two methods of coal feeding are available. Coals with ash contents of less than about 25% and which have a relatively high heating value can be pumped into the combustor in the form of a coal-water-paste. The paste, which has no additives and has a high viscosity compared to a slurry, needs a total water content of only 20 to 25%. At this level of water content, net efficiency is virtually unchanged compared to the feeding of dry coal because:

- part of the evaporative loss is recovered as additional mass flow through the gas turbine
- coal drying is eliminated;
- lock hopper pressurisation and inertisation is eliminated;
- combustion performance is somewhat improved especially at part load conditions (18); and
- emissions performance is unchanged.

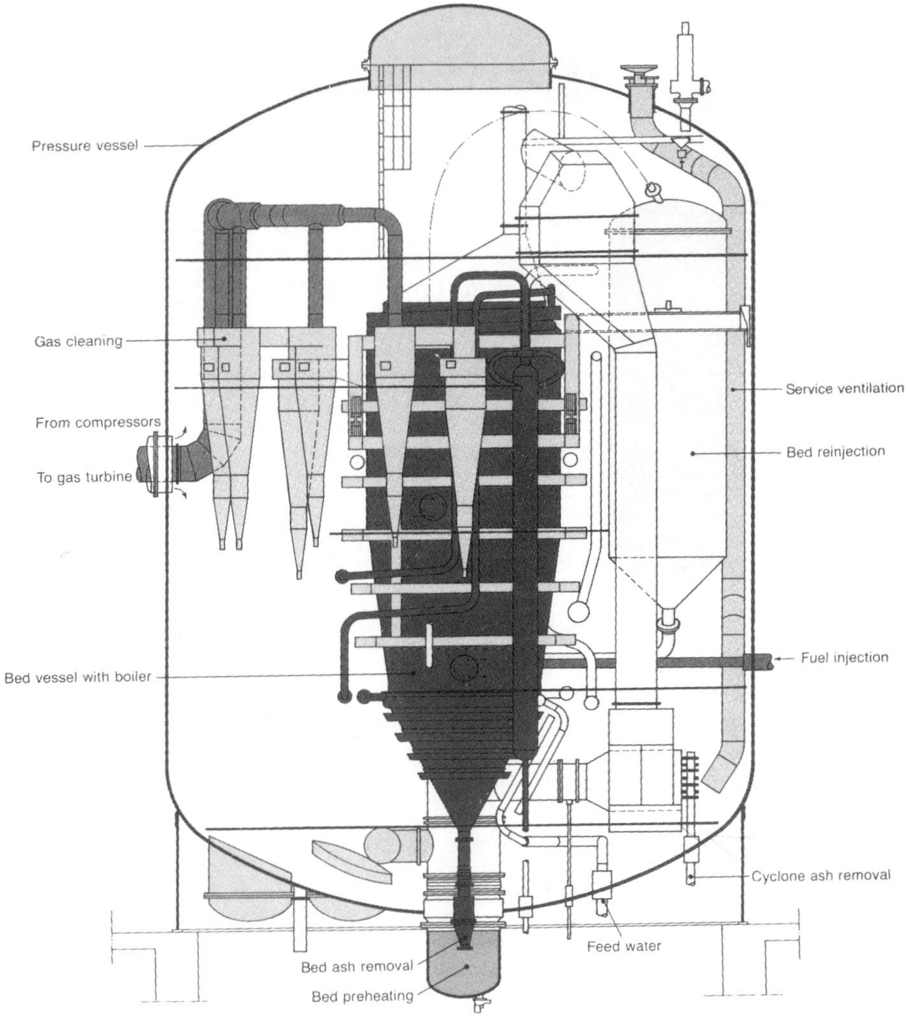

Figure 7. A PFBC combustor.

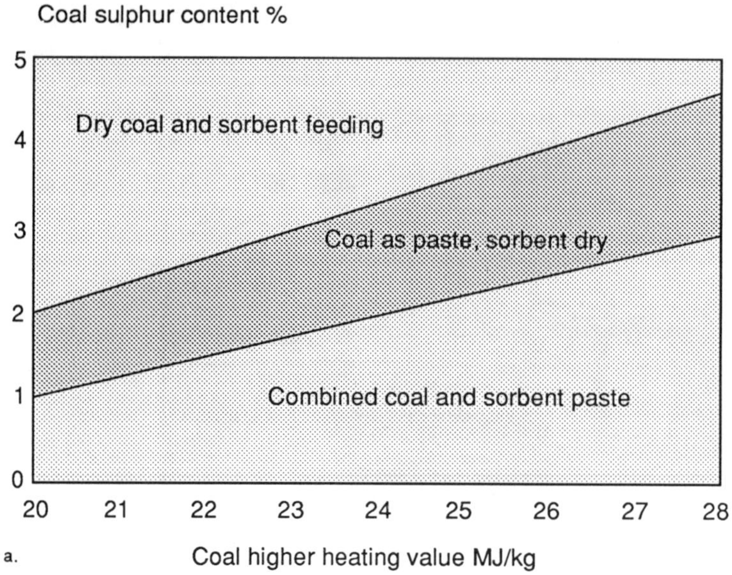

a.

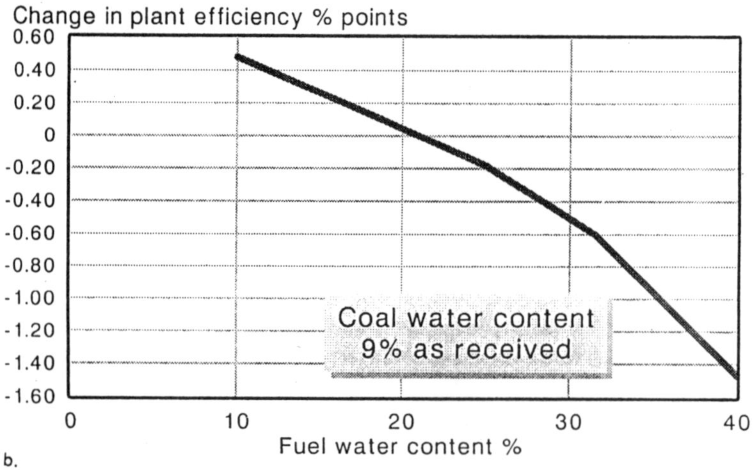

b.

Figure 8. Selection of Coal Feeding Method
a. Coal and sorbent feeding methods
b. Effect of as-fired water content on plant efficiency.

Of course, the stack loss due to evaporation increases. However, the benefits of lower auxiliary power consumption compensate, provided that total water content does not exceed about 25% (19). For coals with high ash contents and low heating values, the efficiency penalty becomes unacceptable when feeding as a paste because the water content needed to feed the inert material absorbs its heat of evaporation in the combustion chamber. Instead, pneumatic feeding via a lockhopper is used.

With low sulphur coals (less than about 1% sulphur) the sorbent material can be mixed directly into the paste. For higher sulphur coals, pneumatic feeding of the sorbent must be used to avoid an excessive efficiency penalty. Figure 8 illustrates the choice of feeding method.

The gas turbine exhaust gas is used for feed-water heating but represents only about 10–15% of the heat transferred to the steam cycle. The remainder, including all evaporation, superheating and reheating (if applicable), comes from the in-bed tube bundle. Intensive development of tube bundle design (20,21) has taken place following the identification of tube erosion as a potential problem in fluidised beds. This development has led to design values

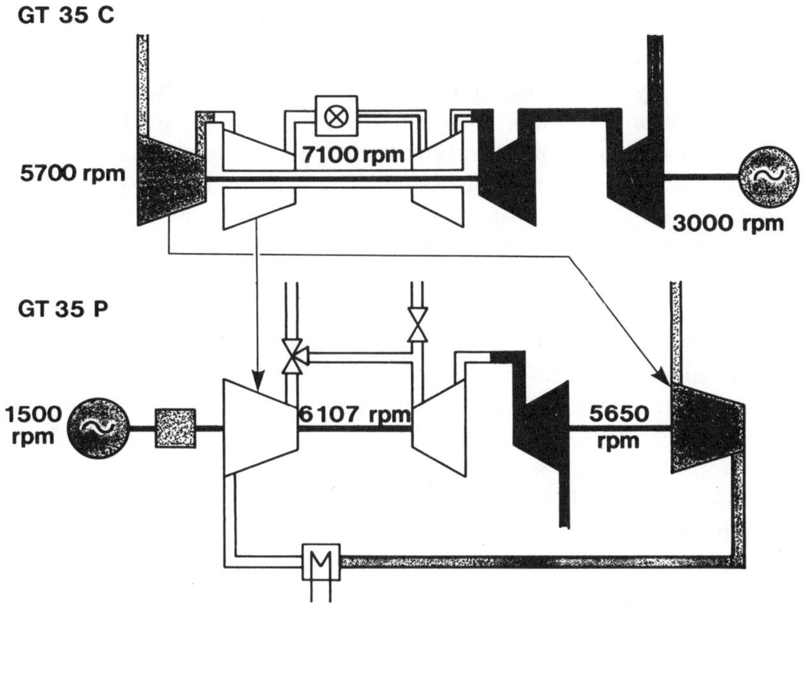

Figure 9. Development of the GT35P Gas Turbine from the GT 35 C machine.

of tube bundle life of around 50,000 hours which is compatible with the usually acceptable values for time between major overhauls for power plant components.The once-through boiler concept is preferred, though, a drum boiler may also be used. At the steam conditions currently being used for modern steam turbines (180 bar and 565/565°C) corrosion does not limit tube bank design. Tube materials currently available are suitable for steam temperatures of up to about 600°C. However, at higher temperatures and consequently, higher steam pressures, the high heat fluxes in a fluidised bed lead to materials limitations from the thermal stress viewpoint. Materials development is now being advanced by powder metallurgy techniques, and new erosion-resistant materials capable of withstanding much higher thermal stresses are currently being tested.

At present, a two-step gas cleaning process is used, with the first step being used for gas turbine protection (cyclones) and the second step at low temperatures and pressures providing protection of the atmosphere (bag filter or electrostatic precipitator). Though development is under-way, a single-step high temperature, high-pressure gas cleaning device is not yet commercially available.

Figure 9 illustrates how an existing gas turbine has been modified for PFBC application. The ruggedisation of these machines together with good design of cyclone cleaning of the PFBC exhaust gas are expected to give a commercial design life for the turbine components. Ruggedisation has consisted of maintaining low velocities in the turbine, reducing the load and turning angle per turbine stage by increasing the number of stages, and thickening sensitive areas of the blade profile. Here, too, modularisation of the turbine components is carried out to a high degree to allow for the rapid replacement of complete modules which can then be refurbished off-line to provide a new spare.

Bed level alteration at constant bed temperature is used for load changing. Bed storage vessels are placed within the pressure vessel and bed material is transported pneumatically between the storage vessel and the fluidised bed. When bed material is removed to storage, in-bed tubes are exposed in the freeboard. This leads to both steam generation and the gas turbine entry temperature being reduced. The two-shaft gas-turbine with its free wheeling low-pressure shaft gives a constant volume flow over much of the load range. Air flow and operating pressure reduce in such a manner that fluidising velocity, cyclone inlet velocities and excess air levels remain constant over much of the load range (17).

Load changing rates are around 4% of full load per minute and minimum load is typically around 30% of full load. A cold start takes around 5 to 6 hours and a warm start can be accomplished in around 2 hours.

2.4 PFBC Plant Sizes and Their Performance

PFBC plant sizes are determined by the capacity of the gas turbines chosen. Figure 10 details the module sizes available based on the gas turbines adapted for PFBC application. One or more PFBC modules can be coupled to a single steam turbine. As shown in Table 3, the current configurations span the range from 80MWe upto 700MWe. Table 3 also details plant efficiencies, and it can be seen that the larger modules exhibit slightly higher efficiencies, primarily due to the higher operating pressure (16 bar instead of 12 bar), the higher component efficiencies, and the economies of scale for auxiliary power needs. Both modules exhibit a small efficiency increase when coupled in pairs to a large steam turbine.

Table 3. Current PFBC plant sizes and efficiency.

Plant Size (net) MWe	No. of PFBC modules	Steam turbine MWe	Gas turbines MWe	Net efficiency % LHV
80	1 x P200	67	1 x 17	42.5
162	2 x P200	135	2 x 17	42.9
350	1 x P800	286	1 x 80	43.8
705	2 x P800	573	2 x 80	44.1

Note: Efficiencies are based on LHV, 10°C cooling water temperature and reheat steam cycles (540/540°C). Gas turbine entry temperatures are 830°C in all cases.

Figure 11 illustrates the difference in boiler size between a conventional pulverised coal boiler and a PFBC boiler.

Currently coals ranging from lignite to low-volatile bituminous can be burned in the PFBC. Anthracite is acceptable but needs some fines recycle if high combustion efficiency (greater than 99%) is to be maintained. Ash levels of 40 to 50% are acceptable (depending on how much sorbent must be fed too), with the limitation being placed by the dimensioning of feed and ash removal systems.

Carbon monoxide levels are negligible (less than 20ppm) as to be expected under pressurised operation where oxidation rates of CO are roughly proportional to pressure (22). The long gas residence times at relatively high temperatures mean the carcinogenic compounds (Polycyclic Organic Materials and Polycyclic Aromatic Hydrocarbons) are at or below detection levels (23,23a). Sulphur capture levels of around 98% (99% with low sulphur coals) can be achieved if sufficient sorbent can be fed. Limestone is the preferred sorbent though dolomite can also be used. Typically, 90% sulphur capture can now be achieved with a Ca/S molar ratio of around 1.5 with limestones of

PFBC modules

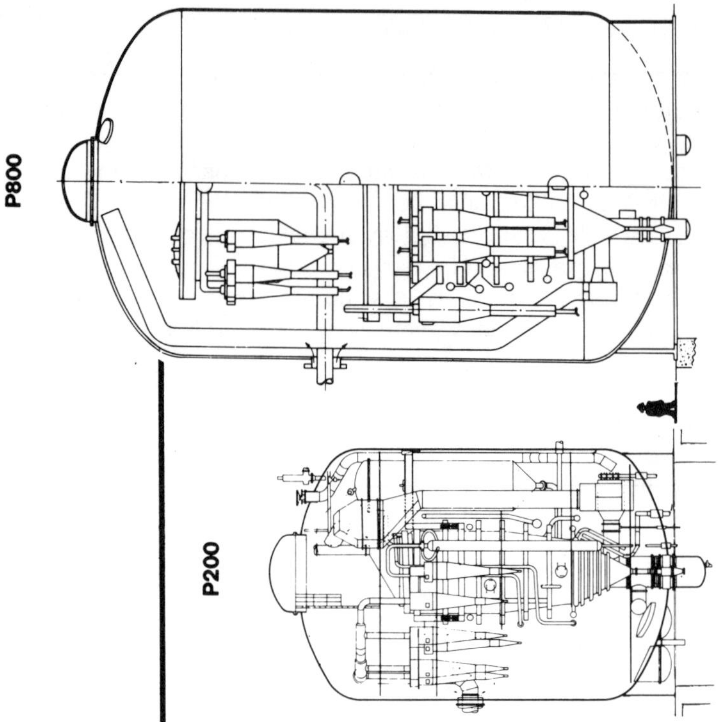

DESIGNATION	P200	P800
Gas Turbine	GT35P	GT140P
Net Power MWe	80	350
Gross Power MWe	84	366
GT Power MWe	17	80
ST Power MWe	67	286
Net Efficiency %	42.5	44

Figure 10. PFBC Module Sizes.

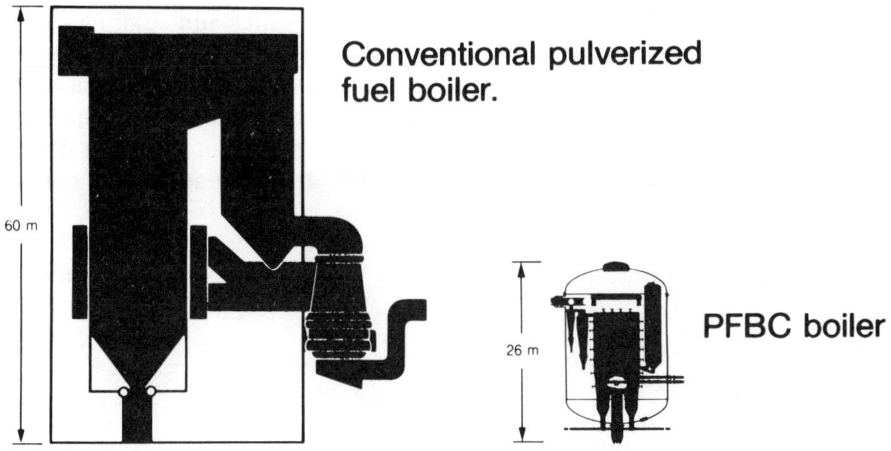

Figure 11. Size Comparison: PFBC-Conventional boiler for lignite firing.

"average" reactivity but with an optimised size distribution. Sulphur levels of around 8% in the coal can be accepted provided that feed systems and ash removal systems can be designed to cope.NOx emissions (Nitric oxide and nitrogen-dioxide measured as the latter) are typically around 100mg/MJ at the operating excess air levels (about 20% or 15% for lignite). These emissions can be reduced to about 60mg/MJ by simply injecting reducing agents in small quantities. NOx formation and destruction mechanisms are not yet fully understood. It appears that excess air controls NOx formation, with pressure being a secondary variable reducing the amount of NOx formation (3,4). NOx destruction, it is supposed, then occurs as a surface reaction on coal-ash particles in the bed. This would explain the lower emissions from pressurised beds where gas-solids contacting rates are a magnitude higher than in atmospheric beds (2,9). One feature of the pressurised process is that NOx, SOx and CO are very weakly linked (23), and thus very low emissions of all three pollutants can be achieved simultaneously.

Sulphur exits the plant as solid sulphate with the ash or as sulphur-oxides. Typically, 5 to 10% of the gaseous emission is as the trioxide. The availability of excess oxygen means that hydrogen-sulphide levels are at or below detectable levels. Similarly, no Carbonyl-sulphide can be detected either (23a).

The ash produced contains a mixture of partially sulphated sorbent and coal ash (23b). Virtually no free lime (CaO) exists and sulphites and sulphides are also negligible. The ash exhibits self-binding properties and after setting with water, the heavy metals in the coal are essentially non-leachable from the ash. The strength of the ash after setting is comparable to concrete, and with the addition of small quantities of binder (cement for example) can approach those of a good quality concrete (24,25,26,27,28). One disadvantage when using dolomite (calcium-magnesium carbonate) as the sorbent is that the slow hydration of the magnesium oxide (MgO) present causes the strength to deteriorate with time. Limestone-based ash is much more stable. Development efforts have concentrated on finding practical uses for the material. The material has been found suitable for:

- road building material;
- the manufacture of synthetic gravel;
- the manufacture of formed bricks or tiles;
- protective roofing or flooring material for the
- dumping of other hazardous materials.

It seems probable, therefore, that as PFBC plants come into operation and markets develop, ash sales could help to reduce electricity production costs.

3 The Commercial Status of PFBC Technology

PFBC plants are under construction in the US, Spain and Sweden. These plants will be taken into operation through 1990. Tables 4 and 5 summarize the key features of these projects.

Table 4. Summary of PFBC Projects

Site	Escatron	Tidd	Värtan
Utility	ENDESA	AEP	SEP
Supplier	ABB Carbon+BWE	ASEA Babcock	ABB Carbon
Purpose	PFBC Demo	PFBC Demo	Cogeneration
Output	79 MWe	73 MWe	135 MWe + 220 MWt
Unit	1 x P200	1 x P200	2 x P200
Steam Turbine	Existing	Existing	New
Start up date	Mid 1990	Mid 1990	#1: Late 1989 #2: Early 1990

Table 5. PFBC Projects: Technical Data

Site	Escatron	Tidd	Värtan
Coal	Lignite	Ohio	Polish
Coal-S	7%	4%	1%
Ash	36%	10%	15%
Coal feed	dry	paste	paste
Sorbent feed	dry	dry	mixed as paste
Sorbent	Limestone	Dolomite	Limestone Dolomite
Feed points	16	6	6
Cyclones	9 x 2	7 x 2	7 x 2
Steam data:			
Pressure	95 bas	90 bar	137 bar
Temperature	510°C	496°C	530°C
Exess air	15%	25%	20%
Fluidising velocity	0.9 m/s	0.9 m/s	0.9 m/s
Bed height	3.5 m	3.5 m	3.5 m
Pressure	12 bar	12 bar	12 bar
Coal feed	18.0 kg/s	7.2 kg/s	2 x 8.0 kg/s
Sorbent feed rate	7.0 kg/s	2.3 kg/s	2 x 0.75 kg/s
Steam flow	60.0 kg/s	56.0 kg/s	114.0 kg/s

3.1 Sweden

Stockholm Energi, the main producer of district heat for the city of Stockholm, with an annual heat production of about 4,000 GWh, has decided to install a PFBC cogeneration plant at Värtan in the centre of the city (29,30).Like most cities in Sweden, Stockholm is to a great extent served by a district heating system and the system is expanding rapidly. It is targetted to supply most of the city's heat demands in the 1990s. Värtan is the largest heating plant with an installed capacity of about 1200MWth and 210MWe.

Like all other Swedish municipal utilities, Stockholm Energi is required by law to plan for the secure supply of heat and electricity, at reasonable cost and with special consideration for the environment.The Swedish Parliamentary decision to shut down the 10GW of nuclear capacity by the year 2010, together with the policy of reduction of dependance on oil imports have had a major impact on Stockholm Energi's planning for generation and supply. Stockholm Energi has been considering the introduction of a coal-fired cogeneration unit at Värtan for a long time. However, the decision making process has been protracted due to the sensitivity of the location and the lively environmental debate.The project has undergone changes of scope and detail in the permitting process, during which public negotiations were held with the Environmental Board.

Figure 12. The Construction Site of the Värtan PFBC plant.

The Government approved the project in December 1986 and the final investment decision was taken by the Stockholm city council in January 1987. The contract was finalised in April 1987 and excavation work started in summer 1987. Hot commissioning will start in autumn 1989, and the plant will go into commercial operation in 1990 (31).

Figure 12 shows the construction site at Värtan where the foundation rings for the two pressurised combustors are clearly visible. The plant consists of two P200 PFBC modules supplying steam to a common, new steam turbine. Each module consists in turn of a fluidised bed combustor, a GT35P gas turbine/generator, and auxiliary equipment. A low-sulphur, imported coal will be used and will be pumped into the plant as a paste. Limestone will be used as the sulphur sorbent and the emissions requirements to be satisfied by this plant are probably the most stringent ever placed on a coal-fired plant. Underground coal storage was one of the conditions included in the permits obtained for this site. Open coal piles were out of the question at this location and large, above ground buildings were also to be avoided.

This PFBC plant will produce about 135MWe and over 220MWth. Steam conditions are 137 bar and 530 C. Reheat is not used for this district heating application.

The compactness of the PFBC system is well suited to this constrained site. The emission requirements of 30mg/MJ of sulphur, 60mg/MJ of NOx and 10mg/MJ of dust mean that with the PFBC plant in operation, capacity at Värtan will be increased with no additional environmental impact from the site.

3.2 United States

American Electric Power (AEP) has decided to install one P200 PFBC module for the repowering of its mothballed Tidd power plant at Brilliant, Ohio. The roughly 70MWe plant is to be a demonstration of PFBC technology for future commercial plants expected to be between 300MWe and 700MWe in size.

AEP is one of the largest electrical utilities and coal consumers in the US. It consumes over 40 million tons of coal annually to produce some 85% of its electricity generation, which is then distributed over seven states. AEP's search for future clean coal technologies, which generate electricity cleanly and economically, has led to its interest in and development of PFBC technology.

The coal to be used at Tidd is a 4% sulphur coal, which will be pumped as a paste. The objective with future commercial plants is to tap the vast reserves of high-sulphur coal in Ohio and Indiana.The Tidd project has a P200 module with its combustor and gas turbine connected to an existing steam turbine.
Site work started in April 1988 and start-up is scheduled for summer 1990 (32).Both the US Department of Energy and the Ohio Coal Development Office have selected this project for support within their respective Clean Coal Technology Programs.

3.3 Spain

Empresa Nacional de Electricidad S A (ENDESA) is the partly state-owned Spanish utility producing about 25% of Spain's electricity, mainly from coal and lignite. The high-sulphur, high-ash black lignite reserves in the Aragon province total more than 1000 million tons, with known recoverable reserves of more than 400 million tons. These lignites contain 4 to 8% sulphur, 25 to 45% ash and about 20% moisture. The annual consumption in 1985 was around 5 million tons, and it is planned to increase this to about 10 million tons annually by 2000. This corresponds to about 2000MWe of installed capacity and presupposes the availability of a clean coal technology capable of burning this coal (33).

After investigating the various developing technologies, ENDESA has chosen to repower an existing 62MWe steam turbine with a P200 PFBC unit at the Escatron plant near Zaragosa. The repowered unit will produce 80MWe and will serve as the demonstration for future units enabling increased use of the Aragon lignite as planned.

Limestone will be used as the sulphur sorbent and at least 90% sulphur capture will be achieved. The potential is that electricity generation using this difficult fuel could be doubled from 1000MWe to about 2000MWe, while simultaneously reducing the sulphur emission to 20% of its current value.Site work for this plant starts in summer 1988 and the plant will be commissioned in spring 1990 (34).

3.4 Repowering with PFBC

Repowering with PFBC means the replacement of an existing boiler (oil or coal) with a PFBC module and retaining the use of an existing steam turbine and auxiliaries. Depending upon the existing steam conditions and the goodness of match to the available PFBC modules, the retrofit may also need topping steam turbines and/or a parallel steam turbine. Figure 13 illustrates a typical repowering scheme and shows how existing equipment could be utilised.

Since the PFBC recovers gas turbine exhaust heat in an economiser, some existing bleed-steam feed-water heaters will become redundant. Another consequence is that the elimination of some bleed-steam flows increases the mass flow through the steam turbine at the low pressure end. Clearly, limitations on turbine swallowing capacity or exhaust flow must be considered. Similarly, condenser capacity may be limiting. In those cases where low steam conditions justify the inclusion of a topping turbine, generator capacity may be limited and a new generator for the topping turbine may be required. In cases where a parallel turbine is justified, it will generally need a new generator. In all cases the PFBC repowering provides at least the extra power represented by the gas turbine.

The heat rate improvement provided by a PFBC repowering is a function of the steam cycle already existing. For the steam conditions typical of a 30 year old boiler, an efficiency improvement of 4 to 6% points is possible. For modern steam cycles the improvement is around 2 to 3% points.

Stringent environmental requirements can, of course, be met after the repowering. Of particular importance for existing sites is that no space is required for the sulphur capture and NOx emission limitation functions (35).
A good example of the improvement brought about by repowering is the Escatron project as detailed in Table 6.

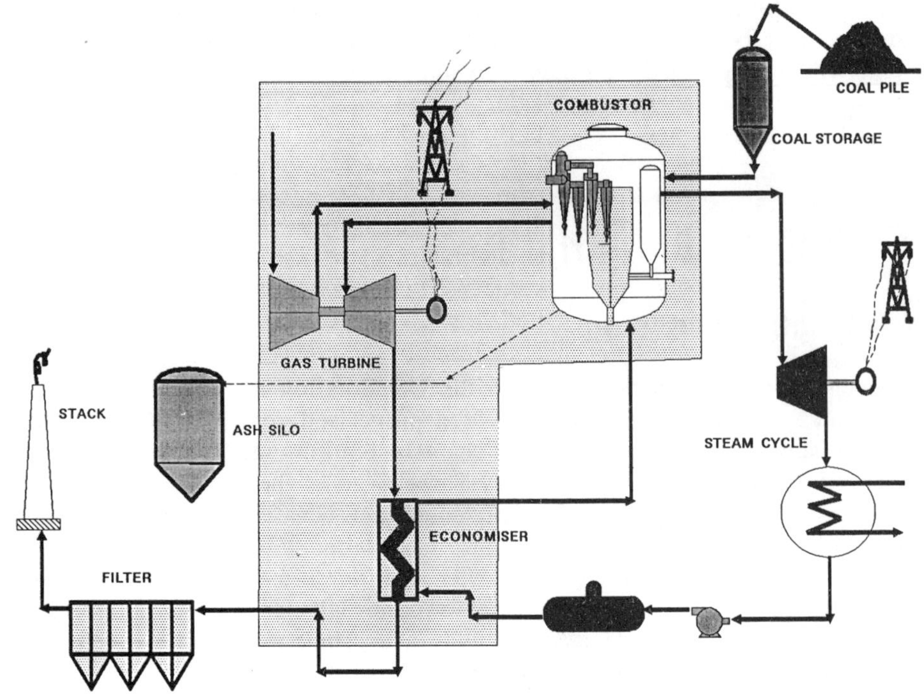

Figure 13. Flow schematic for repowering with PFBC.

Table 6. Escatron Retrofit Project

	Original Plant	With PFBC Retrofit
Coal	Black Lignite	Black Lignite
Coal-S	7%	7%
Sulphur capture	0	90%
Steam data	95 bar/510°C	95 bar/510°C
Plant Output		
–steam turbineMWe	62.5	62.5
–gas turbine MWe	0	17
Net Power MWe	c.55	76 (+38%)
Emissions		
–S mg/MJ	c.5000	500
–(NO + NO2) as NO2 mg/MJ	c.400	100

Coal handling equipment, electrical switchgear and ash handling equipment can generally be reused or modified for reuse in a repowering project. Dust filters which may be existing have also some potential for reuse.

4 PFBC's Potential Role

4.1 Relative Economics

A recent study (36) performed with the US situation in mind compares the economics of a PFBC unit to that of a pulverised-coal plant fitted with a sulphur scrubber but without any de-NOx equipment. Table 7 gives the assumptions for the comparison and Table 8 gives the results. It should be noted that this comparison is internally consistent, but that care should be taken when comparing these results with those from other studies. Actual equipment costs will only be determined in a truly competitive bid situation, and therefore the study gives PFBC no credit for a potentially lower equipment cost. Nevertheless, the compactness of the equipment leads to a shorter construction time and a reduced capital requirement for the PFBC case.

In this conservative scenario, even with no credit being taken for a potentially lower investment cost, the PFBC system is projected to have about 9% lower capital costs, an 11% better efficiency, and a 6% lower power generation cost. No credit is taken for the lower emissions due to the higher efficiency (environment regulations being based on heat input and thus giving no credit for efficiency) or for the substantially lower NOx emissions from the PFBC.

Similar estimates have been seen in other studies (37,38). European experience indicates that including de-NOx would approximately double the power generation cost benefit and the capital cost benefit accruing to PFBC. Such studies, however, are just an indication as to the technology's potential which now remains to be realised.

4.2 Technology Risk Assessment

As a new technology just being commercialised, the performance of PFBC power plants remains to be verified in the field. To assess this risk, it is appropriate to consider hardware performance and process performance separately.

Table 7. Economic format: Comparative economics of clean coal combustion technologies.

(Extracted with permission from Ref. 36)	
Constant Dollars	1987$
Project Life	30 years
Operating factor	330 days/year
Total Capital Investment ($)	
-cost of capital	10$
-working capiyal	60 day inventory 60 day cash supply
-land required	$10000/acre
-Start-up expense and organisation	2% of capital investement
Operating Costs	
-Coal	1$/MMBTU
-Limestone	20$/ton
-Utilities	
--Water	0.6 mills/kWh
--Natural gas	11 mills/kWh
-Operating labor	
--skilled	4000$/year
--unskilled	30000$/year
--supervision	
20% of labor	
-Maintenance	5% of investment
-Administrative labor	20% of all other labor
-Payroll extras	40% of all other labor
-Insurance	1.5% of investment
-General administrative expenses	2% of investment
-Taxes	50% of net profit
-Depreciation	Straight line
Coal analysis	
-Proximate	
--moisture	6%
--Volatile matter	21%
--Fixed carbon	40%
--ash	33%
-Ultimate analysis	
--carbon	46.41%
--hydrogen	3.60%
--sulphur	2.50%
--oxygen	7.50%
--nitrogen	0.90%
--chlorine	0.04%
--moisture	6.00%
--ash	33.05%
-Heating value	8000 BTU/lb

Table 8. Comparative economics of clean coal technologies.

(extracted from Ref. 36)		
Plant Type	PFBC	PC/FGD
Capital cost $/kw	1310	1350
Capital investment (incl. AFUDC) $/kw	1560	1700
Plant life years	30	30
Capacity factor %	65	65
Plant capacity MWe	680	680
Plant heat rate BTU/kWh	8360	9320
Ca/S mol. ratio	1.6	1.1
Capital carrying charge %	17.8	17.8
Annual generation kWh	3871.92×10^6	3871.92×10^6
Plant Investment $	1060.8×10^6	1156×10^6
Land cost $	5×10^6	5×10^6
Start-up cost $	21.216×10^6	23.120×10^6
Total plant investment $	1087.016×10^6	1184.120×10^6
Total capital investment $	1102.116×10^6	1199.262×10^6
Cost of Electricity mills/kWh		
Variable operating cost		
-coal	8.36	9.32
-limestone/dolomite	2.25	1.00
-water	0.6	0.6
Total variable costs	11.21	10.92
Total fixed costs	13.12	13.47
Fixed carrying charge	50.67	55.13
Cost of Electricity	*74.99*	*79.52*

4.2.1 Hardware performance

Pressurised fluidised beds themselves have been used in the oil industry for catalytic cracking units since the 1940's (38a). Such beds operate at similar temperatures to PFBC's but at somewhat lower pressures–4 to 6 bar instead of the 10 to 16 bar in PFBC's. Fluidised beds in the oil industry and atmospheric fluid bed combustors are larger in area than those required for PFBC. There is a high level of confidence, therefore, that the fluidised bed itself, in terms of design and construction, poses no insuperable problems.

The combination of steam generation with coal feeding into such a bed is new. Coal feeding techniques into a pressurised container have been used in the oil and chemical industries and, in themselves, are not new (38b,38c,38d). The requirements on control of coal feed rates to meet utility load following requirements however, are more rigorous than previously encountered. This has been adressed by only permitting scale-up by a factor of about 2 from proven components. Thus, scale-up from pilot plants to the commercial plants being considered involves the use of several, slightly larger components (pumps, valves, nozzles, etc.), operating in parallel but under the same process conditions and with identical control philosophies.

Ash handling is the necessary corrollary of a coal feeding system. Here, too, the same scale-up philosophy is used so that components and process conditions for commercial plant are close to the conditions used in comprehensive pilot plant testing. The availability of pressure drop as a design parameter allows parallel paths to be designed with some confidence as to their behaviour.

As noted previously, the potential erosion of boiler tubes is a risk area but one where the pilot plant development work has eliminated the risk for catastrophic failures. What remains is uncertainty about how long tubes will actually last. This uncertainty is only significant for evaporator tubes; and superheater or reheat tubes exhibit very low erosion levels (20,21). One criterion for judging whether the magnitude of such an uncertainty is acceptable is that design lifetimes must be of the order of one or two major maintenance periods (30,000 to 60,000 hours). The uncertainty itself must be addressed, and this is done by designing for rapid inspection, repair or replacement in order to minimise forced outage rate. The evaporator tubes constitute about 35% of the tube bank area which itself is relatively small (about 4000sq. m for a 350MWe plant) due to the high in-bed heat transfer coefficients (5). Furthermore, only a fraction of the tubes at risk would need to be replaced or repaired at any one time so that maintenance costs are not prohibitive.

Construction of the steam generator tube bank follows the same scale-up philosophy of a two-fold scale-up followed by a multiplication of components. Corrosion of boiler tubes is not seen to be the determining factor for tube-life (39), though material selection is important. Some materials are specially vulnerable (high Ni alloys for example) and must be avoided. The high oxygen partial pressure in a PFBC is an advantage here. In general, a number of materials are suitable for each of the heat transfer duties from the erosion, corrosion and strength viewpoints, so that final selection can be done on the basis of cost. Depending upon steam conditions, austenitic materials may be necessary for the hottest parts of the superheater tubes.

The use of a gas turbine with direct coal combustion is new at the commercial level (39a). However, the use of a gas turbine in similar environments is not unknown. Here, too, catalytic cracking units have used expander turbines in dust laden atmospheres (39b,39c). The concern is primarily for erosion and corrosion of gas turbine blades with secondary concern for the fouling risk. Here the success of cyclones in protecting expander turbines from the very erosive catalyst dust is very relevant. PFBC dust has been shown to be much softer (10,11), and the use of high efficiency cyclones has shown that erosion of static and rotating, unclad, blade cascades can be virtually eliminated by eliminating all particles larger than about 10 microns (40). The cyclones themselves are scaled-up in the same manner as other components to ensure their function. For commercial plants, a further safety factor is introduced by using coated blades (though with conventional gas turbine coating materials) which have an erosion resistance of 3 to 6 times greater than unclad blades. These coatings provide extra corrosion resistance as well (41,42). At the temperatures used in the PFBC cycle, the risk for alkali vaporisation is first minimised and second is condensed on dust particles in preference to gas turbine blades. Pilot plant testing has been at identical gas conditions and corrosion rates are so low as not to be the blade-life determining factor. Here, too, a modularisation approach has been adopted to address the inevitable uncertainties when extrapolating from data from a few thousand hours of exposure to lifetimes of a few tens of thousands of hours. Rapid inspection, repair and exchange of turbine modules has been designed to keep forced outage risk to a minimum.

Secondary erosion of gas turbine blades (in the downstream stages) has been simulated but not tested as thoroughly as primary erosion. There remains a risk that the simulations (both experimental and theoretical) are not truly representative of commercial plant operation. However, the determining factor for the design lifetime of the blades remains primary rather than secondary erosion considerations.

While hardware risks are bound to remain, they are sufficiently low to allow the development of the technology to continue into its commercial stage. Furthermore, the remaining uncertainties are such that they can now only be resolved by long-term operation of commercially-sized plants.

4.2.2 Process performance

Process conditions tested in pilot plants remain unchanged in commercial plants, and the risks are those of fluidization behaviour being affected by scale-up considerations. Here the two-fold scale-up rule is invoked again and pressure drop is the parameter which provides confidence that fluidisation, mixing, and heat and mass transfer behaviour will remain unchanged. Provided fluid-hydrodynamic behaviour remains unchanged, there is confidence that combustion and emissions performance will live up to expectations from pilot plant tests (1,2,38a).

Plant efficiency values as quoted represent values being commercially guaranteed and so have a degree of conservatism. For efficiency calculations, stack temperatures have been maintained at 140°C, ash cooling is assumed to cease at 250°C, and combustion efficiencies are assumed to be 99%, even if predictions are for somewhat higher levels. In addition, a relatively simple rather than the most efficient coupling between the gas turbine exhaust and the steam cycle is used. Fouling factors for gas turbine efficiency and economiser effectiveness are built into the calculations. The component efficiencies of the steam turbine pose little risk. Thus there is good confidence with plant efficiency predictions.

Plant availability cannot, of course, be proven, till commercial plants have been in operation for some time. However, reliability analysis techniques developed for use with the Swedish nuclear program are being used as a design tool, to identify systems and components which contribute to non-availability. The criterion used is that calculated reliabilities are compared to those calculated with the same technique for a conventional coal plant (without a scrubber) and engineering development is concentrated on those systems contributing to significant non-availability. The fault frequency calculated for each system is used to develop the need for redundancy. Similarly, those components contributing to non-availability are included in the spares list and in the maintenance schedules. Costs for implementing the maintenance requirements can then be included by the user in his analysis of investment and operation and maintenance costs. Reliabilities calculated for PFBC plant are therefore equal to or better than for a comparable pulverised coal plant. These apply, of course, to a mature plant and the early experience with commercial PFBC plants will determine the rate at which designs mature.

4.3 The Market for PFBC

Over the 1990's, the worldwide increase in electrical generating capacity is forecast to be around 600GWe as shown in Figure 14. Of this some 360GWe will be new coal-fired plant and around 60 GWe will be for repowering coal plants. It can be expected that these 420GWe of coal capacity will be dominated by "clean coal technology", especially in the latter half of the decade (44).

It is here that the role for PFBC exists. However, as a new technology, market penetration cannot be predicted accurately. Depending upon the success of the first PFBC plants coming on-line in 1990, it can be expected that PFBC has the potential to capture 20 to 25% of the orders placed in the year 2000 and perhaps up to 50% of the orders placed in 2010.

The promise of PFBC is primarily connected with the increase of generating capacity while simultaneously decreasing the emission of acid rain precursors. The high efficiency provides a small step in setting a brake on the development of the "greenhouse effect".

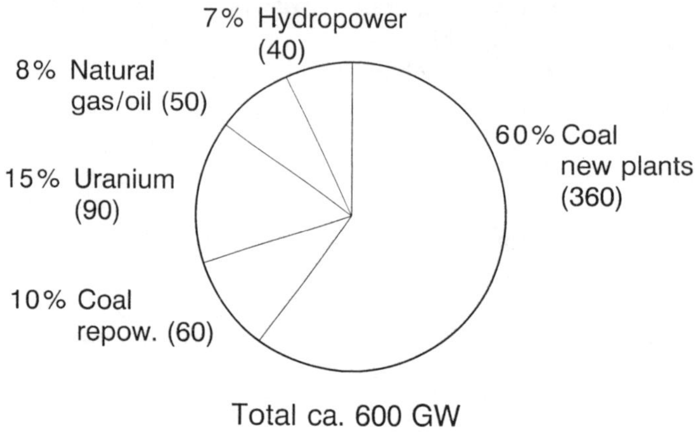

Figure 14. The World Market for Electricity Production:1990–2000 (Ref. 44).

5 Further Development

With no change in the current concept and with much the same gas turbine entry temperature (840–870°C), significant improvements to efficiency and environmental performance can be made. It can be expected that plants built through the 90's will exhibit these improvements to an increasing degree.

Since superheat and reheat surface requirements in a PFBC are only 10 to 20% of that required in a conventional pulverised boiler, the techno-economic optimal steam conditions for a PFBC are different than those of a conventional boiler. In general, the optimum for PFBC lies at higher temperatures and lower pressures.

As materials become available to design the optimal PFBC steam cycles, efficiency gains of over 1% point can be realised by this means alone.

Environmental performance will also improve. The methods being developed to further reduce NOx emissions consist of modifying combustion patterns within the bed (43), and the use of in-bed or gas path additive injection (though still avoiding the use of catalysts).

As shown in Figure 6, the combustion modifications could reduce background emissions to about 60 mg/MJ and gas phase additive injection could then further reduce this to about 30 mg/MJ. Though a primary objective is to avoid the use of catalytic de-NOx reactors, they could be used to reduce emissions still further.

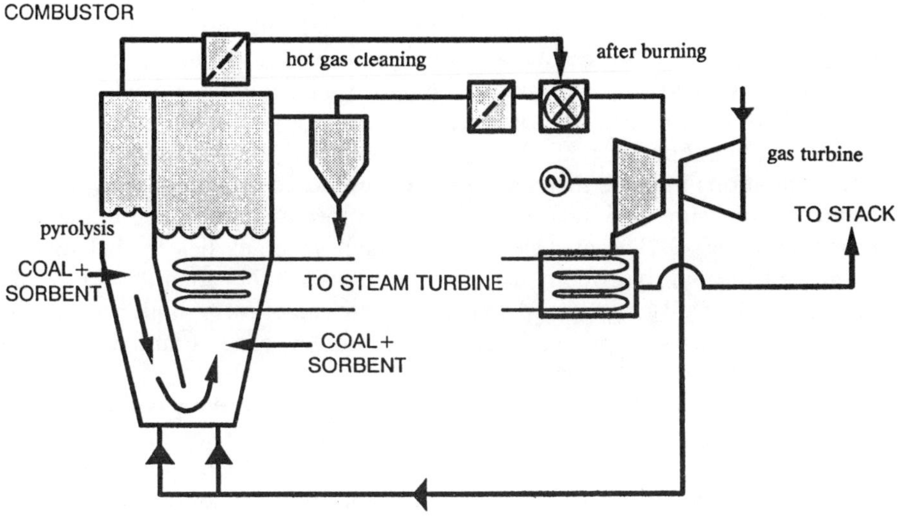

Figure 15. The PFBC/Temperature Topping Cycle.

While sulphur emissions pose no significant limitations, there is a strong incentive to reduce sorbent consumption to a minimum, and development to maximise sorbent utilisation is in progress. Significant progress has been made in recent years so that a Ca/S molar requirement of around 1.5 with limesone suffices for 90% sulphur capture. This has been achieved by modifying limestone preparation techniques to optimise the size distribution of the limestone to suit the sulphur capture processes in the bed, i.e., one dominated by the gas phase residence time and the in-bed process controlled by the solids residence time – the inert material feed rate.

Further development is now expected to concentrate on the maximising of the sulphur capture level, especially with high sulphur coals, so that 99% capture is practical for all coals and not only low-sulphur (< 2%) coals.

Continued development of efficiency requires that some means be devised for increasing gas turbine entry temperature. Since the bed temperature is limited to around 900°C, this implies the use of an after- burning process. Here processes such as pyrolysis or mild gasification could come into play, to produce the gas for turbine temperature topping (Figure 15). The PFBC is retained for the combustion of the residues from the pyrolysis process. These hybrid schemes have the advantage that the PFBC is not only available for steam generation but also as the sink for partially utilised coal and for heat losses from the pyrolysis or mild gasification stage. The modifications needed to be made to a PFBC to include an initial stage where pyrolysis takes place, with the coal and sorbent being fed to the pyrolyser and then proceeding to the fluidised bed combustor, requires engineering development, but has the potential to be a compact and simple arrangement within the pressure vessel. Initial calculations indicate that such a system could be accomodated within the same size of pressure vessel as at present and, moreover, even within the same size of boiler vessel. The attractiveness of such schemes lies in that thermal efficiency of the pyrolyser/combustor is above 98% rather than the 85 to 90% typical of pure gasification, and that development of very high gas turbine entry temperatures can proceed at a much slower pace. Plant efficiencies approaching 50% then become possible (45).

Depending upon coal type, such schemes would permit the gas turbine entry temperature to be raised to between 1100 and 1200°C. Though the fluidised bed and the fuel gas process would both be under 900°C, rigorous cleaning would be needed before the temperature topping process. Thus parallel development of hot gas cleaning (of particulates) is needed. In fact, this development is under way and filter materials capable of withstanding the temperatures and the environments needed appear to be available. Considerable work remains, however, in the engineering of the systems that could be offered commercially.

A further point is that such hybrid schemes allow for the potential reduction of sulphur and NOx emissions even from their presently low levels. Sulphur capture is achieved both during the pyrolysis process and during the fluidised combustion process so that sorbent needs are, at worst, unaltered and at best somewhat improved. Fuel NOx formation mechanisms can potentially be decoupled from temperature and controlled by excess air alone. Furthermore, gas-phase NOx reduction can be at more suitable temperatures than at present. Thermal NOx remains insignificant.

Acknowledgement

This article is published with the permission of ABB Carbon, but the views expressed are those of the author and not necessarily those of ABB Carbon.

References

1. Geldart , D. ed. 1983. *Gas Fluidisation*. New York: John Wiley and Sons.

1a. Kunii, D., Levenspiel, O. 1969. *Fluidisation Engineering*. New York: John Wiley & Sons.

2. Davidson, J.F., Harrison, D. 1971. *Fluidisation*. London: Academic Press.

3. Howard, J.R. ed. 1983. *Fluidised Beds, Combustion and Applications*. London: Applied Science Publishers.

4. Shaw, J.T. 1983. Howard, J.R. ed. *Fluidised Beds, Combustion and Applications*, Chapter 6.

5. Botterill, J.S.M. 1975. *Fluid-Bed Heat Transfer*. London: Academic Press.

6. Skinner, D.G. 1971. *The Fluidised Combustion of Coal*. London: Mills & Boon.

7. Davidson, B.J., Holmes, J., Keeley, K.R., Merrick, D., Pillai, K.K. 1983. *Proceedings of CoalTech*. Amsterdam.

8. Roberts, A.G., Pillai, K.K. 1983. *Proceedings of 15th International Congress on Combustion Engines*. CIMAC, Paris.

8a. Graves, R.L. 1980. ASME Paper 80-GT-102.

9. Podolski, W.F. et al. 1983. *Pressurised Fluidised Bed Combustion Technology*. Noyes Data Corporation, Park Ridge, New Jersey.

10. Pillai, K.K., Wood, P. 1980. *J Inst. Energy*, pp. 159–72.

11. Roberts, A.G. et al. 1980. *Proceedings of Fluidised Combustion: Systems and Applications*. Institute of Energy, London.

11a. Gray, D.A. et al. 1980. *Proceedings 6th International Conference on Fluidised Bed Combustion*, Atlanta.

12. Hoy, H.R. 1988. *Proceedings of Seminar on Fluidised Bed Combustion for Utility Applications*. EPRI, Pala Alto California.

12a. Markowsky, J., Wickström, B. 1979. *Proceedings of 6th Energy Technology Conference*. Washington.

13. Olesen, C., Kraemer, W., Carlsson, S. 1984. *Proceedings of 4th European Coal Utilisation Conference,* Coal Technology Europe, Essen, W. Germany.

14. Pillai, K.K., Kreij, S-E., Paulsson, R. 1985. *Proceedings of 8th International Conference on Fluidised Bed Combustion*, Houston.

15. Jansson, S.A. 1987. *Proceedings of 9th International Conference on Fluidised Bed Combustion,* Boston.

16. Wright, S. et al. 1988. *Proceedings of Seminar on Fluidised Bed Combustion for Utility Applications. EPRI*. Palo Alto, California.

16a. Carls, E.L. et al. 1980. *Proceedings of 6th International Conference on Fluidised Bed Combustion,* Atlanta.

17. Pillai, K.K., Wickström, B., Tjellander, G. 1986. *Proceedings of 2nd Biennial PFBC power Plants Utility Conference*. EPRI, Milwaukee.

17a. Hoy, H.R., Roberts, A.G., Raven P. 1978. *Proceedings of 5th International Conference on Fluidised Bed Combustion*, Washington.

17b. Roberts, A.G. et al. 1980. *1000 Hour Test Program in a Pressurised Fluidised Bed Combustor*. NCB CURL, Vols 1 to 4, US DOE Reports FE-3121–15-a,b,c,d.

17c. Stantan, J.E., Scott, R. 1985. *Proceedings of 8th International Conference on Fluidised Bed Combustion,* Houston.

17d. Moskowitz, S. et al. 1979. *ASME Paper ASME 79-GT-193*.

18. Pillai, K.K., Andersson, M.C. 1987. *Proceedings of 9th International Conference on Fluidised Bed Combustion,* Boston.

19. Pillai, K.K., Nilsson J. 1985. *Proceedings of Fluidised Combustion: Is it achieving its promise?*. Institute of Energy, London.
20. Jansson, S.A. 1985. *Proceedings of 8th International Conference on Fluidised Bed Combustion,* Houston.
21. Jansson, S.A. 1982. *Proceedings of Conference on Corrosion-Erosion-Wear of Materials in Emerging Fossil Energy Systems*. Berkeley, California.
22. Field, M.A., Gill, D.W., Morgan, B.B., Hawksley P.G.W. 1967. *Combustion of Pulverised Coal.* BCURA, Leatherhead.
23. Statens Energiverk. 1985. *Coal and Sorbent Tests in ASEA PFBC's pilot plant CTF.* Report on Project 37 70 17, Stockholm (in Swedish).
23a. Henschel, D.B. 1980. *Proceedings of 6th International Conference on Fluidised Bed Combustion,* Atlanta.
23b. Sun, C.C. et al. 1978. Westinghouse R & D Center. *Report to EPA*. Report No. EPA 600/7–78–049.
24. AEP Service Corp. 1986. *Tidd PFBC Demonstration Plant: PFBC Bed Ash and Flyash Testing Results*, Support Document for Exemption from the Solid Waste Regulations ORC 3734 Appendix B.
25. Nilsson, C., Gustafsson, T. 1984. *Värmeforsk Stiftelsen for Värmeteknisk Förskning.* Report No.172, Stockholm.
26. Nilsson C. 1987. *Värmeforsk- Stiftelsen för Värmeteknisk Förskning*, Report No.276, Stockholm.
27. Sventsam, G. 1988. VIAK AB (SSAB). *Report No. 4910.01.1440–08*, Private Communication.
28. Lundgren, T. 1988. Statens Geotekniska Institut, Terranova. *Report No. 8018*, Private Communication.
29. Almquist, P. 1987. *Proceedings of 9th International Conference on Fluidised Bed Combustion,* Boston.
30. Almquist, P. 1988. *Proceedings of Seminar on Fluidised Bed Combustion Technology for Utility Applications*. EPRI, Palo Alto, California.
31. Pillai, K.K. 1988. *Modern Power Systems*, July.
32. Mudd, M.J., Bauer, D.A. 1987. *Proceedings of 9th International Conference on Fluidised Bed Combustion,* Boston.
33. Rubio, J., Tome J., Pillai, K.K. 1987. *Proceedings of 9th International Conference on Fluidised Bed Combustion*, Boston.
34. Menendez, E., Bencomo, V., Cuenca, A. 1987. *Proceedings of 9th International Conference on Fluidised Bed Combustion,* Boston.
35. Pillai, K.K., Jansson, S.A. 1987. *Proceedings of EPRI Workshop on Fossil Plant Retrofits for Improved Heat Rate and Availability,* San Diego, California.
36. Jensen, A.D. 1987. *Proceedings of Conference on Comparitive Economics of Clean Coal Technologies, Grand Junction*, Colorado.
37. Yeager, K.E. 1986. *Proceedings of 15th IERE General Meeting*, Tokyo.
38. Mudd, M.J., Guha, M.K. 1985. *Proceedings of 8th International Conference on Fluidised Bed Combustion*, Houston.
38a. Zenz, F.A., Othmer D.F. 1960. *Fluidisation and Fluid-Particle Systems*. New York: Reinhold Publishing Corporation.
38b. Reintjes, H. 1977. *Proceedings of Conference on Coal Feeding Systems*, Jet Propulsion Labs. No. JPL-77–55.
38c. Lewis, R. et al. 1977. *Proceedings of Conference on Coal Feeding Systems,* Jet Propulsion Labs. No. JPL-77–55.

38d. Funk, E., Reinart, R. 1977. *Proceedings of Conference on Coal Feeding Systems*, Jet Propulsion Labs. No. JPL-77–55.

39. Stringer, J. 1986. *Proceedings of 2nd Biennial PFBC Power Plants Utility Conference,* EPRI, Milwaukee.

39a. Australian Dept. of Minerals and Energy. 1973. *The Coal Burning Gas Turbine Project.* Australian Govt. Publishing Services, Canberra.

39b. Stettenbenz, L.M. 1972. *Proceedings of 47th Annual Fall Meeting*, Western Gas Processors and Oil Refiners Association. Newport Beach, California.

39c. Mathers, W.G., Schonewald, R. 1980. *Proceedings of 6th International Conference on Fluidised Bed Combustion,* Atlanta.

40. Jansson, S.A., Nilsson, N.G., Malm, B.O.1980. *Proceedings of 6th International Conference on Fluidised Bed Combustion,* Pittsburgh.

41. Goebel, J.A. 1979. *Proceedings of 1st Conference on Advanced Materials for Alternative Fuel Capable Directly Fired Heat Engines*, CONF-790749.

42. Dapkunas, S.J. 1978. *Proceedings of 3rd Annual Conference on Materials for Coal Conversion and Utilisation*, CONF-781018.

43. Taylor, T.E. 1980. *Proceedings of 6th International Conference on Fluidised Bed Combustion*, Atlanta.

44. Gustavsson, O. 1988. ABB Carbon, *Private Communication.*

45. Pillai, K.K. 1987. *Proceedings of Seminar on Pressurised Fluidised Bed and Gasification Power Systems*, Otaniemi, Finland.

Low-Cost Power From Thin-Film Photovoltaics

David E. Carlson

Abstract

Thin-film photovoltaic technology has developed rapidly in the last several years, and in 1987 amorphous silicon solar cells accounted for about 37% of the total world market for photovoltaics. Conversion efficiencies of 12% have been attained for single-junction, amorphous silicon cells, and efficiencies in the range of 13.5% to 15.6% have been observed for stacked-junction cells. With improvements in reliability and decreasing manufacturing costs, amorphous silicon solar cells are now displacing crystalline silicon solar cells for low power, outdoor applications. the development of large-scale, computer-integrated-manufacturing lines should decrease the manufacturing costs of amorphous silicon solar cells to less than $1/$W_p$ by the early 1990's, leading to the development of both residential and central station utility applications in the mid to late 1990's.

1 Introduction

Thin-film photovoltaics (PV) is a rapidly developing, relatively new technology for generating electricity. Photovoltaic power generation is an attractive power source since it involves the direct conversion of sunlight into electricity without any moving parts or pollution. Hence, photovoltaic power plants require no fuel, no operators, and little maintenance. Moreover, photovoltaic power can be brought on line in increments so that capital investment can be used to generate power almost immediately.

This paper will review recent technological developments in the area of thin-film photovoltaics with an emphasis on amorphous silicon (a-Si) solar cells. Recent progress in reducing manufacturing costs and the impact of cost reduction on market development will also be reviewed.

1.1 A Brief History of Photovoltaics

While the first observation of a photovoltaic effect was made by Becquerel in 1839, it was not until 1954 that three scientists at Bell Laboratories fabricated the first relatively efficient solar cells (1). Their devices were single crystal silicon p/n junctions that converted sunlight into electricity with an efficiency of about 6%. Since that time the performance of crystalline silicon solar cells has improved steadily, and recently conversion efficiencies as high as 22% have been reported for 1 sun illumination (~100 mW/cm^2 with sunlight directly overhead) and as high as 28% for concentrated sunlight (at an intensity 150 times normal sunlight)(2). Single crystal gallium arsenide (GaAs) solar cells have attained very similar efficiencies under normal and concentrated sunlight.

The first thin-film solar cells employed a heterojunction of copper sulfide/cadmium sulfide – Cu_2S/CdS – and were fabricated shortly after the Bell Laboratories' work. Commercial Cu_2S/CdS cells appeared on the market briefly in the mid 1970's, but production was discontinued a few years later due to reliability problems.

A-Si solar cells were first made at RCA Laboratories in 1974 (3), and the first commercial product using a-Si appeared in hand-held calculators made by Sanyo in 1980. Today, the major commercial PV markets are dominated by solar cells made from single-crystal silicon, polycrystalline silicon and amorphous silicon.

1.2 Description of the Photovoltaic Effect

Semiconductors absorb sunlight by ejecting electrons from the chemical bonds between atoms. A certain amount of energy (the bandgap or energy gap) is required to create a free electron in a specific semiconductor material. In crystalline silicon, this minimum energy is 1.1 eV corresponding to a wavelength of light of 1.13μm. The ejected electron leaves behind a positively charged hole that can also move by diffusion or under the influence of an electric field.

An electric field can be created in a semiconductor by the formation of a semiconductor junction such as a p-n junction or a Schottky barrier. The photovoltaic effect occurs when free electrons and holes are photogenerated in the vicinity of the semiconductor junction, and the electric field of the junction separates the oppositely charged carriers to create electric power. A schematic of this effect is shown in Figure 1 where E_V represents the energy level of the valence or bonding electrons (where free holes move) and E_C represents the

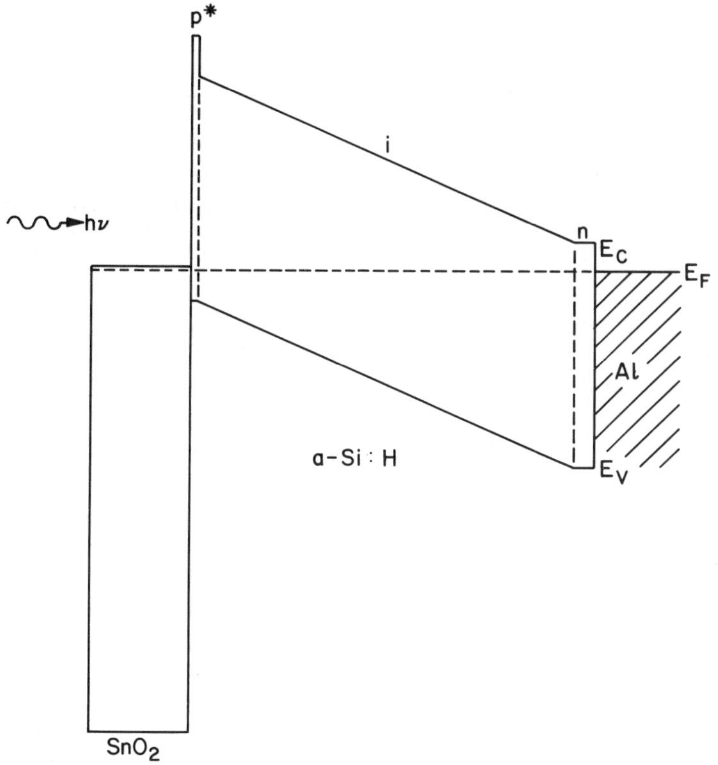

Figure 1. The energy band diagram for an a-Si p-i-n solar cell.

energy level where free electrons move. The difference E_C-E_V is the energy gap (E_g) or the minimum energy required to produce an electron-hole pair. (E_F denotes the Fermi level, which is constant throughout the solar cell structure at equilibrium).

Figure 1 represents the band diagram of an a-Si p-i-n solar cell in which the electric field is relatively uniform throughout most of the cell. The p layer is usually doped with boron and the n layer with phosphorus. These doped layers are charged and create an electric field across the undoped or intrinsic layer (i layer). The power created is I·V where I is the photogenerated current flowing through the potential barrier of the junction.

The maximum power is produced when the load is adjusted to maximize the product of the photocurrent and photovoltage. Figure 2 shows a typical I-V curve for an a-Si solar cell under 1 sun illumination. The conversion efficiency of such a cell is defined by the expression

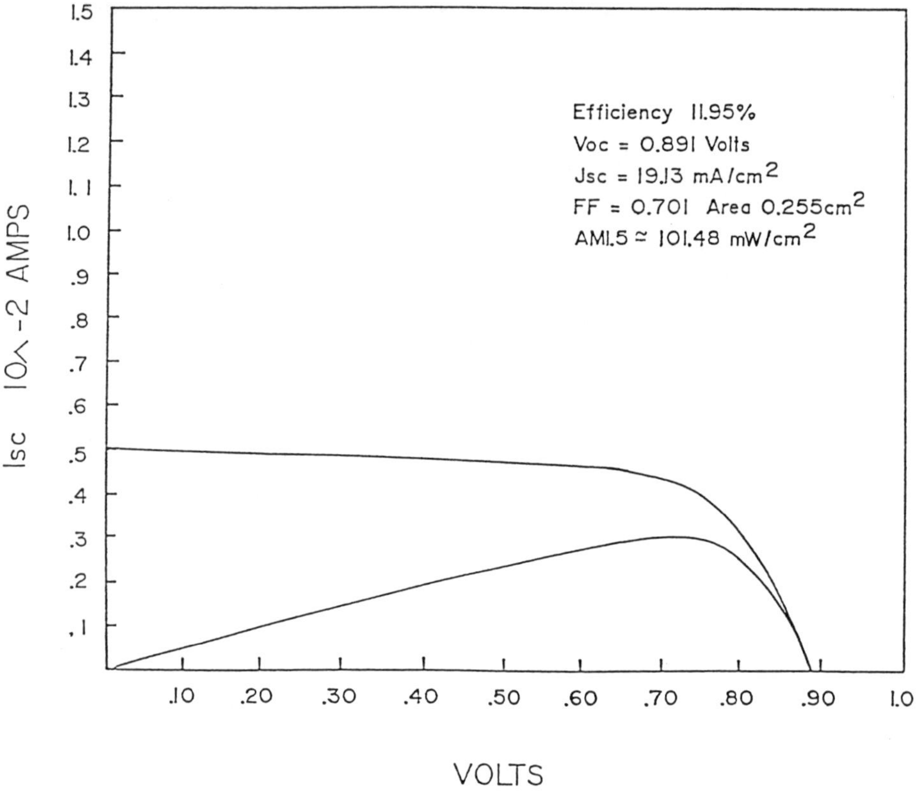

Figure 2. Current-voltage characteristics of an a-Si p-i-n cell under approximately 1 sun illumination (AM 1.5 ≈ 101.48mW/cm²). The lower curve is the power-voltage characteristic.

$$\eta = I \cdot V/P \cdot A = (FF)J_{sc} \cdot V_{oc}/P \qquad (1)$$

where P is the power density of sunlight (~100mW/cm²), A is the area of the solar cell, J_{sc} is the short-circuit current density (I_{sc}/A) and V_{oc} is the open-circuit voltage. The fill factor (FF) is defined by Equation (1) and approaches unity as the I-V curve approaches the power rectangle represented by $J_{sc} \cdot V_{oc}$. For an ideal diode, the fill factor is limited by diffusion effects in the junction and the maximum value at room temperature is about 0.89 for a device with $V_{oc} \simeq 1.1V$(4).

The maximum theoretical value of Voc is determined by the built-in potential of the junction and the recombination of electrons and holes. Tiedje (5) has estimated that the maximum value of V_{oc} in a-Si solar cells is about 1.1V due to recombination in band-tail states (caused by the disorder). This same effects limits the fill factor in a-Si cells to about 0.86. The maximum value of J_{SC} in a thin-film solar cell is determined by the optical properties of the semiconductor and by the geometric path length of weakly absorbed light. For a-Si with a bandgap of ~1.7eV and a solar-cell structure with an effective thickness of 3μm, the maximum possible value of J_{SC} is ~22mA/cm^2 (6).

Thus, from Equation. (1) the theoretical limit for the conversion efficiency of a single-junction a-Si solar cell is ~(0.86)(22)(1.1)/100 = 0.21 or 21%. Since the power in sunlight incident on 1m^2 is ~1kW, an ideal a-Si solar module of that size would generate 210W of electricity.

1.3 Stacked-Junction Solar Cells

Even higher conversion efficiencies can be obtained in principle by stacking cells of varying bandgaps on top of each other. In a 2 junction stacked cell as shown schematically in Figure 3, the top cell will have a wide bandgap and

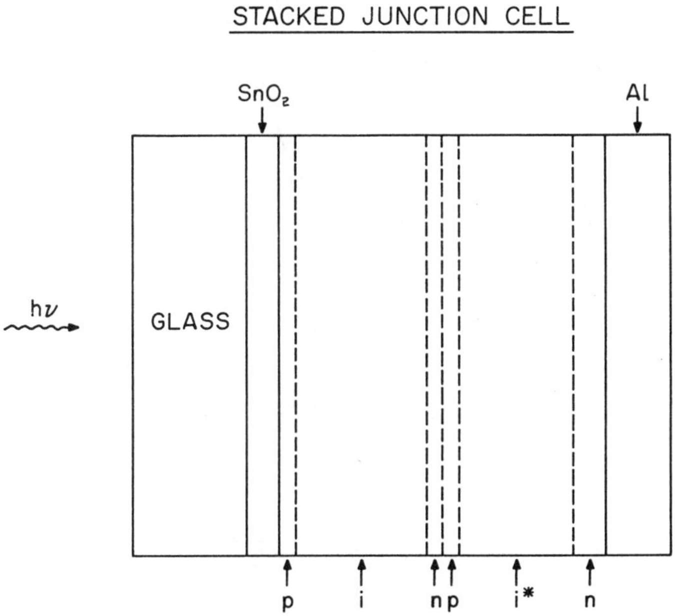

Figure 3. A schematic diagram of an a-Si stacked junction cell where the i* layer of the bottom cell is an a-SiGe alloy (narrow bandgap).

only absorb high energy photons (or short wavelength light, such as blue and green light). Longer wavelength light such as red and infrared light will be transmitted to the bottom cell which will have a smaller bandgap and be able to absorb most of the transmitted light.

Amorphous silicon can be alloyed with germanium to form a narrower bandgap material while alloying with carbon creates a wider bandgap material. The data in Figure 4 shows that the optical properties of these amorphous semiconductor alloys can be readily varied by changing the alloy content. Presently, research efforts are focused on improving the electronic properties of these alloys for inclusion in stacked-junction structures.

In a 4 terminal stacked-junction structure, the cells are usually fabricated separately, and the top cell is made with transparent contacts and optically coupled to the bottom cell. In a simple 2 terminal device, the component cells

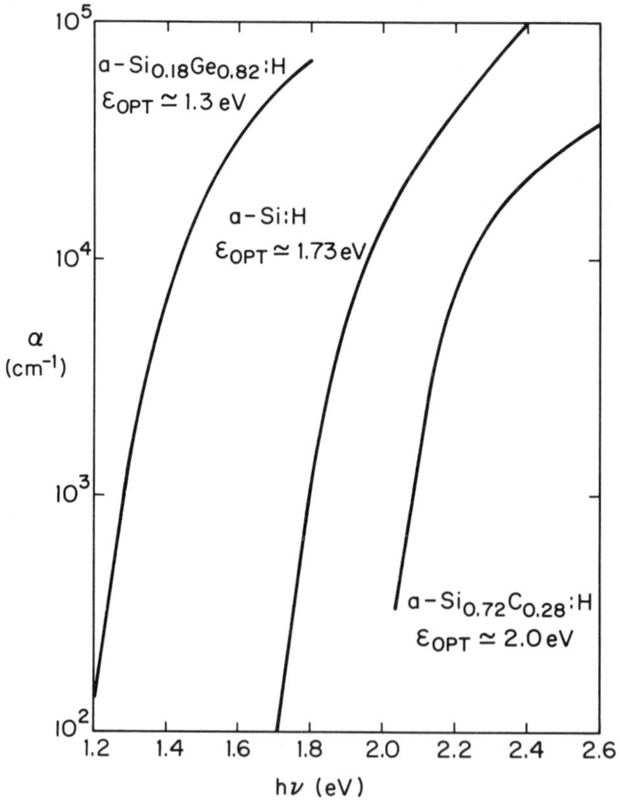

Figure 4. Optical absorption coefficient as a function of photon energy for films of amorphous silicon, amorphous silicon-germanium and amorphous silicon-carbon.

are configured so as to produce equal currents under illumination, and since the cells are in series, the voltages add. Therefore, stacked-junction cells are able to absorb more of the radiation in sunlight and convert it into a photocurrent at a higher net voltage than a single-junction cell.

2 Thin-Film Photovoltaic Technologies

For the purposes of this paper, thin-film solar cells are defined as those containing semiconductor films that are a few microns thick or less. The leading material systems are a-Si, copper-indium diselenide ($CuInSe_2$ or CIS) and cadmium telluride (CdTe). As shown in Table 1, all three of these materials have been used to fabricate laboratory cells with conversion efficiencies well in excess of 10%.

Table 1. Performance of Thin-Film Solar Cells.

Solar-Cell Structure	Conversion Efficiency (%)	Cell Area (cm^2)	Reference
CdS/CdTe (heterojunction)	12.8	1	[7]
CdS/$CuInSe_2$ (heterojunction)	12.5	4	[8]
a-Si (p-i-n)	12.0	0.26	[9]
a-Si/$CuInSe_2$ (stacked junction)	15.6	4	[8]

2.1 Amorphous Silicon

Amorphous silicon solar cells were first introduced as a commercial product in 1980 when Sanyo started selling a-Si solar-powered calculators. Mainly as a result of rapid growth in consumer applications, a-Si solar cells captured about 37% of the total PV market in 1987. These cells are now widely used in battery-charging applications involving cars, boats, recreational vehicles, electric fences, etc.

In addition, a-Si is now being used in a number of commercial applications other than solar cells. Thin layers (few hundred Å thick) of a-Si are deposited on soda-lime-silicate glass by chemical vapor deposition for use as heat-reflecting windows. Thick layers (several tens of microns thick) of photoconductive a-Si are found in many of the office copiers produced in

Japan. Photoconductive a-Si is also used in high-speed laser printers. Linear diode arrays of a-Si are employed in page readers, electronic "whiteboards" and facsimile machines. Recently, a-Si thin-film-transistors have appeared in active-matrix, liquid crystal displays for small-area televisions. The total sales of a-Si components for non-PV applications has been estimated to be about $500 million in 1987 (10), which is about 5 times larger than the sales of a-Si solar cells in 1987.

In addition to the applications mentioned above, researchers are also investigating the possible use of a-Si in integrated circuits, light-emitting diodes and neural networks for advanced computers. In other words, there are a large number of scientists and engineers working on further developing a-Si technology for both PV and non-PV applications. This large research base gives a-Si photovoltaics a significant advantage over PV systems based on other thin-film materials. Table 2 shows the author's estimates of the total investment in various PV technology by private industry through 1987. While the cumulative investment in crystalline silicon solar cells is by far the largest, the total investment in a-Si solar cells is surprisingly large. This is due in part to the significant commitments to a-Si PV technology made by large oil companies such as AMOCO, ARCO, Shell and Standard Oil of Ohio (now part of British Petroleum) and by Japanese companies such as Sanyo, Fuji Electric, Sharp, Kyocera and Kanegafuchi. In addition, total government support for a-Si PV research through 1987 totaled about $200 million, with the U.S. and Japanese governments contributing the most. (Since 1975, the U.S. government has invested more than $800 million in all PV technologies).

For commercial electronic applications, a-Si is grown by the glow-discharge decomposition of silane gas (SiH_4); this process is also called plasma-enhanced chemical vapor deposition (PECVD). This deposition process allows one to coat large-area (>1m^2) substrates with a-Si films that are uniform in thickness to within a few %. Moreover, the glow discharge process promotes conformable coating so that thin (~100Å) pin-hole-free layers can be deposited on textured substrates. In the case of a-Si solar cells, the manufacturing process can be readily automated so that both high quality and low cost can be assured in the finished product (see Section 6).

Table 2. Investment in PV Technologies.

Technology	Est. Investment (through 1987) ($M)
Crystalline Silicon	1070
Amorphous Silicon	600
Copper Indium Diselenide	50
Gallium Arsenide	45
Cadmium Telluride	40

Perhaps the most significant limitation associated with a-Si photovoltaics is the relatively low conversion efficiency of present power modules. All a-Si PV modules exhibit some degree of degradation in the first few months of operation before stabilizing (see Section 3.3). Currently, the best quality a-Si modules will stabilize at an efficiency in the 5–6% range. While this performance is certainly adequate for many applications such as consumer products, values on the order of 15% have been projected as necessary for utility central station power plants (11). A high conversion efficiency is necessary to reduce the total cost of electricity in such applications where "the balance of system cost" is high (see Section 6.4); balance of system refers to ancillary items such as land, power conversion equipment, structural supports, fences. etc.

Thus, further improvements in the performance of a-Si PV modules are required before large-scale, grid-connected, central station power plants become a reality. However as we shall see later in Section 5, the results currently being obtained with a-Si stacked-junction cells in the laboratory indicate that *the utility market may open up for a-Si PV in the mid to late 1990's* (see Section 7).

2.2 Copper-Indium-Diselenide

As shown in Table 1, conversion efficiencies as high as 12.5% have recently been obtained in thin-film solar cells using copper-indium-diselenide (CIS). The highest performance CIS cells have been fabricated with the structure: glass/Mo/CIS/(Cd,Zn)S/ZnO. The CIS layer is typically a few microns thick while the (Cd,Zn)S window layer is about 500Å thick. For 30cmx30cm CIS modules, conversion efficiencies as high as 9.7% have been obtained where the efficiency is based on an aperture area (which excludes the frame) of 835cm^2 (8). While CIS PV modules are not commercially available, they should appear on the market in the next year or two.

Recently, conversion efficiencies as high as 15.6% have been obtained for a-Si/CIS stacked-junction cells (4cm^2 in area) (8). These cells are fabricated by placing a transparent a-Si cell (ZnO/p-i-n /ZnO) on top of a CIS cell so that the cells are optically coupled but electrically independent. Thus, the stacked-junction structure has 4 terminals, 2 from each component cell. As discussed in Section 1.2, the a-Si cell with a bandgap of ~1.7eV in this stacked junction structure will absorb the high energy photons (blue and green light) in the solar spectrum while the CIS cell with a bandgap of 1.04eV absorbs the lower energy photons (red and near infrared light). Conversion efficiencies of 12.3% have been obtained for a-Si/CIS modules with aperture areas of 835cm^2.

CIS cells must be encapsulated so that they are protected from oxygen and water vapor, but they are stable under illumination. Moreover, their preliminary field tests are encouraging – no significant changes in performance have been noted in modules after more than 1 year outdoors (8).

CIS films can be grown by several techniques: coevaporation of copper, indium, and selenium; sputtering of Cu and In in a hydrogen selenide – H_2Se – atmosphere; and electrodeposition of Cu and In followed by a high temperature treatment in H_2Se. Sputtering and electrodeposition can be readily scaled up to coat large-area surfaces, but coevaporation is more difficult to apply to large areas. Since the CIS films are polycrystalline, it is likely that these films will need to be at least a few microns thick in large-area PV modules in order to minimize electrical shorting and shunting (leakage in the semiconductor junction).

Hence, about an order of magnitude more material may be required for CIS PV modules than for a-Si modules. Moreover, indium is a relatively expensive material so that the total materials costs for CIS modules are expected to be greater than for a-Si modules.

Another concern related to CIS solar cells is that of toxicity associated with the manufacturing process or with contamination of the environment through accidents or improper handling. While H_2Se is an extremely toxic gas and is used in some methods of growing CIS, manufacturing processes can usually be designed and controlled so that toxic effluents are negligible. However, contamination of the environment with toxic materials such as cadmium, cadmium sulfide and selenium could occur as the result of a fire in the vicinity of a CIS PV array or as the result of improper disposal of broken or obsolete CIS PV modules.

Nevertheless, the rapid progress made in developing CIS technology may make CIS solar cells competitive with other PV technologies in the next few years.

2.3 Cadmium Telluride

Another promising polycrystalline material for thin-film solar cells is cadmium telluride (CdTe). CdTe films have been grown by evaporation, closed space sublimation, sputtering, screen-printing and electrodeposition. Conversion efficiencies as high as 12.8% have been obtained by screen-printing CdTe and then sintering the CdS/CdTe heterojunction (7). Matsushita has been producing CdTe solar cells (by screen printing) for low light level applications (e.g., calculators) for the last few years.

One problem with CdTe solar cells has been the difficulty in forming good stable ohmic contacts for power applications. Recently, good results have been obtained with a CdS/CdTe/ZnTe structure, and conversion efficiencies as high as 10.7% have been attained (12). Preliminary results indicate that recent CdTe cells should exhibit good long-term reliability. As with CIS cells, CdTe cells do not exhibit light-induced degradation effects as observed for a-Si solar cells.

However, the modest research effort on CdTe indicates that it may be several years before large-scale manufacturing will become a commercial reality. As in the case of CIS, toxicity issues involving the manufacturing, use and eventual disposal of CdTe solar cells are major concerns.

2.4 Other Thin-Film Materials

Researchers have been investigating a number of other crystalline materials such as Zn_3P_2, $CuInS_2$, GaAs, Cu_2S, CuO, HgZnTe, HgCdTe, etc., for possible use in thin-film solar cells. Of these, GaAs is perhaps the most promising with conversion efficiencies as high as 14% obtained in GaAs/Si structures (13). However, this device is not a regular thin film cell since the substrate is single crystal silicon. Conversion efficiencies as high as 18.4% have been observed for GaAs solar cells using thin single crystals grown by the cleft process (14). However, GaAs cells are probably too expensive for widespread terrestrial use.

The research effort involving other polycrystalline materials is relatively small so it is unlikely that commercial PV modules using these materials will appear on the market before the mid-1990's.

3 Properties of Amorphous Silicon

Those a-Si properties that are relevant to its use in solar cells will now be discussed.

3.1 Optical Absorption

An important property of a-Si is its ability to absorb a significant fraction of the energy in sunlight within a thin film about one micron thick (6). As shown in Figure 5, the optical absorption coefficient of a-Si depends on deposition conditions such as the substrate temperature (T_S) but is generally about 10 times larger than that of crystalline silicon in the visible light range. (An absorption coefficient of $10^4 cm^{-1}$ or greater means that a film 10^{-4}cm or 1μm thick will absorb most of the radiation at that wavelength of light).

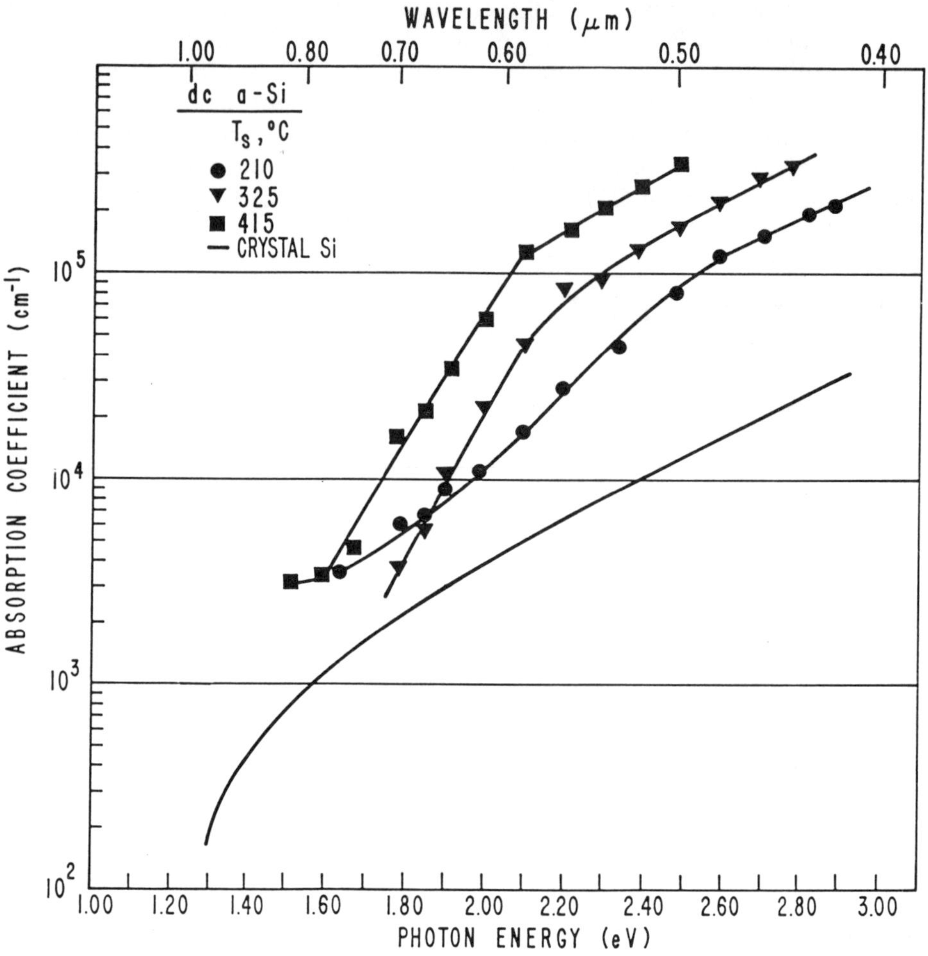

Figure 5. Optical absorption coefficient as a function of the wavelength of light for a-Si films grown at substrate temperatures of 210°C, 325°C and 415°C. The absorption coefficient of single crystal silicon is also shown.

The variation in properties of a-Si with deposition conditions is due to changes in its microstructure and composition. Good quality a-Si films typically contain about 5–15 atomic percent of hydrogen, and the material is actually an amorphous silicon-hydrogen alloy (a-Si:H). The microstructure is largely determined by the H bonding and the size and distribution of microvoids. The H can bond to Si as isolated Si-H bonds, clustered Si-H groups, Si-H_2 groups and (Si-H_2)$_n$ polymer chains. The best quality films are usually dense and consist of mainly isolated Si-H bonds. Good quality films are generally formed by low-power glow discharges in SiH_4 gas (~0.1 – 0.5 millibar of pressure) at substrate temperatures of about 250°C.

3.2 Electronic Properties

Efficient photovoltaic energy conversion requires that both photogenerated electrons and holes be collected by the semiconductor junction. In a-Si p-i-n cells, this condition is met if the drift length of photogenerated carriers is greater than the thickness of the i layer. The drift length is given by the expression

$$l_d = \mu\tau E \tag{2}$$

where μ is the mobility of the electron or hole, τ is its lifetime and E is the average electric field in the i layer. (The product μE is an average drift velocity of the photogenerated carrier moving across the i-layer).

The drift mobility for electrons in a-Si is typically $1.0 cm^2/V$-s while that for holes is $\sim 10^{-3} cm^2/V$-s. The recombination lifetime for photogenerated carriers in good quality a-Si is $\sim 10^{-5}$s (15) so that $l_d \simeq 2\mu m$ for holes if $E = 2x10^4 V/cm$. This estimate of the field applies to a p-i-n cell under short-circuit conditions with a built-in potential of 1V across an i layer 0.5µm thick. Under these short-circuit conditions, most of the photogenerated holes and electrons would be collected. As the cell goes into forward bias (e.g., with a load applied), the average electric field decreases and becomes more nonuniform so that at the maximum power point (see Figure 1), l_d may become comparable to the cell thickness. (Under open-circuit conditions, no net current is collected as the photocurrent is balanced by recombination currents).

Photogenerated carriers can still be collected by diffusion even when the electric field in the i layer is very small. The diffusion length is defined by the expression:

$$L_D = \left(\frac{\mu\tau kT}{q}\right)^{1/2} \tag{3}$$

where q is the electron charge and $kT \simeq 0.025eV$ at room temperature. Thus, for a hole mobility of $10^{-3} cm^2/V$-s and lifetime of 10^{-5}s, the diffusion length is about 0.16µm so that most photogenerated holes would thermally diffuse across a weak field region of this thickness without the help of a drift field.

The diffusion length (and drift length) in a-Si are limited by electron-hole recombination through defect states; i.e., the photogenerated electron and hole recombine mainly at dangling bond defects with the liberation of heat so that they do not contribute to the collected photocurrent. Hydrogen plays a major role in a-Si by forming Si-H bonds and reducing the number of dangling bonds to less than $10^{15}cm^{-3}$ in good electronic grade material. (Without H, the density of dangling bonds or broken chemical bonds is $\sim 10^{19}cm^{-3}$.)

The low density of defect states in hydrogenated a-Si allows one to electronically dope the material by adding dopant gases such as phosphine (PH_3) or diborane (B_2H_6) to the SiH_4 discharge (3, 16). However, the doping efficiency is low in a-Si as many of the dopant atoms do not go into electronically active substitutional sites. Nevertheless, the dark conductivity can be varied from $\sim 10^{-12}$ to $10^{-2}\Omega^{-1}cm^{-1}$ by varying the amounts of boron and phosphorus in the material. Thus, as mentioned in Section 1.2, one can form semiconductor junctions in a-Si (hydrogenated) by doping, and properly fabricated junctions can function as efficient solar cells.

3.3 Light-Induced Effects

As mentioned earlier in Section 2.1, a-Si solar cells generally exhibit some degree of light-induced degradation during the first few months of outdoor operation before stabilizing. This light-induced degradation, called the Staebler-Wronski effect (17), appears to be associated with the creation of metastable defects by recombination or trapping of photogenerated or injected carriers. These metastable defects appear to be mainly dangling bonds, and they can be removed by annealing the a-Si film for several minutes at a temperature of ~200°C.

These light-induced defects act as recombination centers and reduce the diffusion length so that the performance of a-Si solar cells will gradually degrade during the first few months of operation. Some of these metastable defects also trap charge so that the internal electric field in the solar cell may be distorted leading to the creation of a weak field region in the center of the cell. After about 1000 hours of exposure to sunlight, the density of metastable defects is on the order of $10^{17}cm^{-3}$, and the annealing rate at operating temperatures (~50–60°C) becomes comparable to the degradation rate so the effect stabilizes.

The light-induced degradation of several types of a-Si solar cells is shown in Figure 6. As shown in the figure, the degradation can be reduced by using single-junction a-Si solar cells with thinner i layers or by using stacked-junction cells with relatively thin i layers. Cells with thinner i layers possess larger internal electric fields that help to separate and collect photogenerated

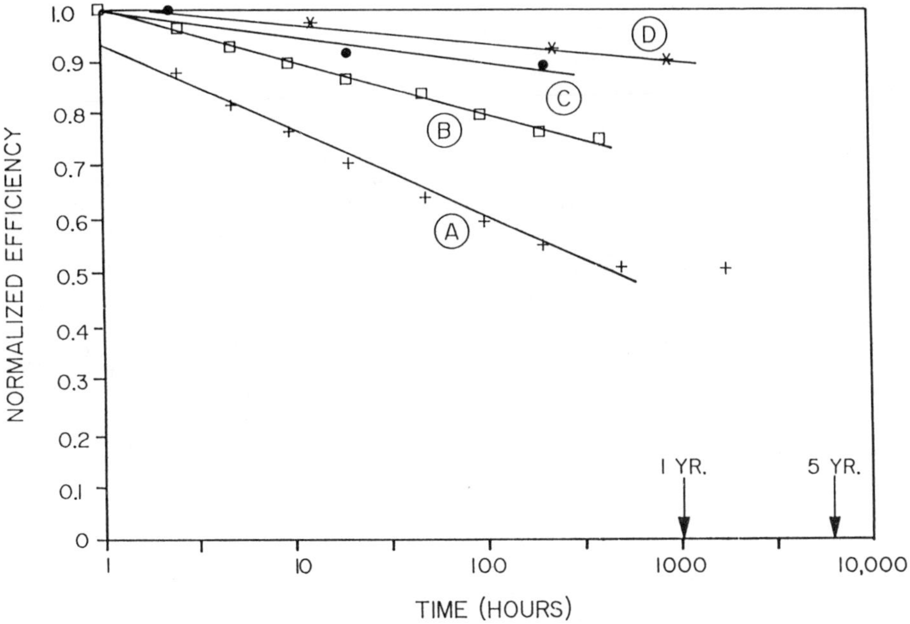

Figure 6. Normalized conversion efficiency as a function of exposure time to 1 sun illumination for p-i-n cells with (A) 0.6µm thick i layer, (B) 0.25µm thick i layer, (C) stacked-junction modules (1000cm^2) and (D) recent stacked-junction cells; all initial efficiencies were in the 6 to 8% range.

carriers before they can recombine to create metastable defects. The absorption of light in thin cells can be increased by utilizing light-trapping geometries such as textured tin oxide (see Figure 7).

While the microscopic nature of the metastable defects is not clearly defined, the density of these centers can be reduced by reducing the concentration of impurities such as oxygen and carbon (18), and also by optimizing the deposition process to minimize SiH_2 and $(SiH_2)n$ groups (19).

4 Manufacturing Amorphous Silicon Power Modules

Amorphous silicon solar cells are fabricated commercially on both steel foil substrates and glass substrates. In the first case, the cell configuration is usually ITO/p-i-n (a-Si)/stainless steel or a stacked structure, ITO/p-i-n/p-i-n (a-

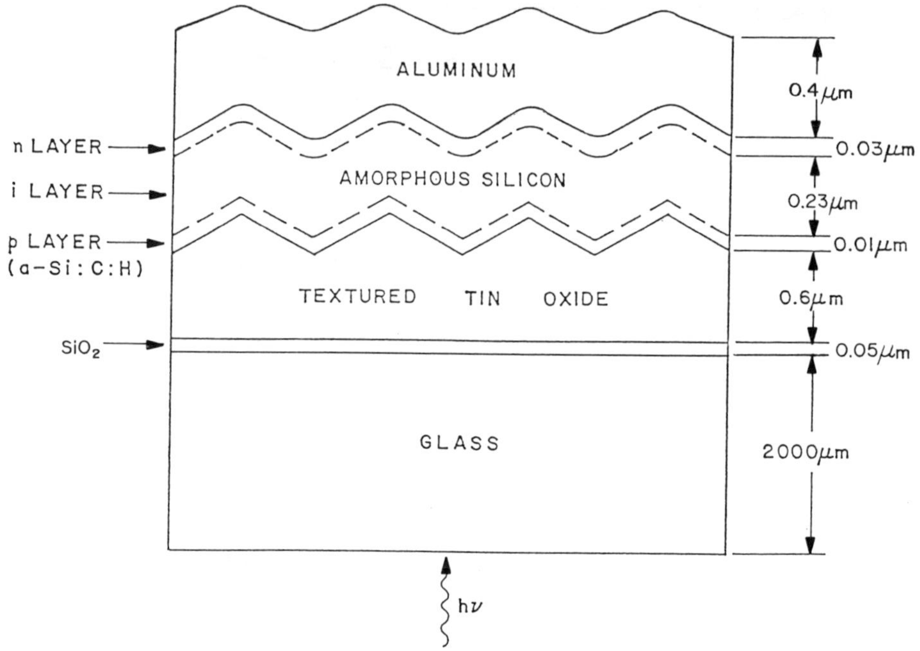

Figure 7. A schematic diagram of a commercial a-Si solar cell fabricated on glass.

Si)/stainless steel where ITO is indium-tin-oxide. However, most commercial a-Si cells are made in the configuration shown in Figure 7 where the structure is glass/SnO_2/p-i-n (a-Si)/Al.

The manufacturing process steps used by Solarex Corporation to fabricate a-Si PV modules (or power modules) are listed in Table 3. Soda-lime-silicate glass (window glass) is first cleaned before coating with a barrier layer of ~500Å of silicon dioxide (SiO_2) and ~0.6μm of tin oxide (SnO_2) by atmospheric chemical vapor deposition. The tin oxide is deposited from a vapor containing tin tetrachloride ($SnCl_4$), and the process is controlled so as to form a textured surface with submicron dimensions. A computer-controlled laser scribing system (Nd-YAG laser) is then used to pattern the tin oxide into about 30 segments (~1cm wide) for a 1000cm^2 module. The substrates are then cleaned again before depositing the p-i-n a-Si structure by glow-discharge decomposition of SiH_4 at a substrate temperature of ~250°C.

The p layer is usually an amorphous silicon-carbon alloy, ~100Å thick, containing ~10–20 at% of carbon and ~1 at% of boron. The i layer is ~0.23μm thick while the n layer is ~300Å thick and contains ~1 at% of phosphorus (see Figure 7).

Table 3. Manufacturing Process Steps for a-Si Power Modules.

1.	Clean glass
2.	Deposit 500Å of SiO_2 and 6000Å of SnO_2:F (textured) by chemical vapor deposition (CVD)
3.	Laser-scribe tin oxide
4.	Clean substrate
5.	Deposit a-Si p-i-n layers by glow discharge (2300Å i-layer)
6.	Laser-scribe a-Si layers
7.	Sputter-deposit aluminum (4000Å)
8.	Laser-scribe aluminum
9.	Test and cure segments
10.	Ultrasonically bond aluminum power leads to the end segments of the modules
11.	Roll-on encapsulant layer
12.	Frame
13.	Final test

The a-Si films are then laser scribed at locations next to the laser scribes in the tin oxide (see Figure 8) before depositing ~0.4μm of aluminum (Al) by magnetron sputtering. The Al is then laser scribed to complete the patterning. All 30 segments in a 1000cm^2 module are now series-connected as the Al back contact of one cell makes a contact to the SnO_2 front contact of the adjacent cell through the laser scribe in the a-Si.

Each cell is then tested under illumination and if necessary subjected to a reverse bias voltage to burn out defects. Al power leads are then ultrasonically bonded to the Al film at the edges of the module, and an encapsulation layer is rolled onto the back of the module. Next, the module is framed and a final test is performed.

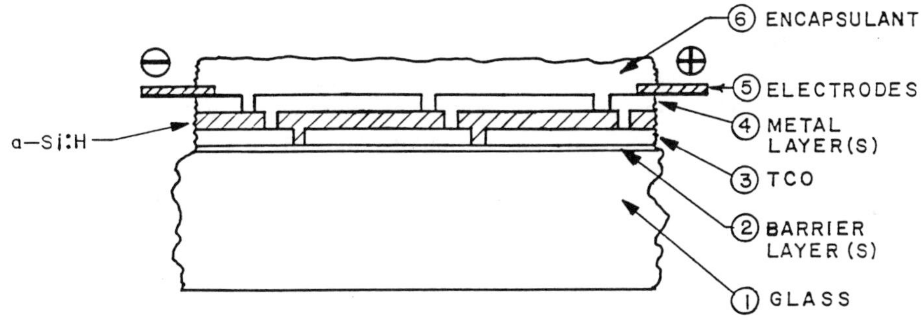

Figure 8. A schematic diagram of a series-connected a-Si PV module. The barrier layer is usually SiO_2; the TCO (transparent conductive oxide) is usually SnO_2.

Recently, Solarex has developed a 2-step laser patterning process that allows one to deposit the a-Si and the Al sequentially before completing the patterning process. This approach is being combined with a machine vision system to create a computer-integrated-manufacturing (CIM) line. The CIM line will be totally integrated and will use an automated loader to feed the line and an automated unloader to accumulate finished product at the end of the line. This approach assures high-throughput production of high quality a-Si PV modules with minimal labor costs.

5 Performance of Amorphous Silicon Power Modules

As shown in Table 1, conversion efficiencies as high as 12% have been obtained in a-Si p-i-n solar cells in the laboratory (9). These cells employed textured tin oxide to trap the weakly absorbed light thus increasing the effective thickness of the cells. In addition, the carbon content was graded over a region of a few hundred angstroms in the vicinity of the p/i interface, and a highly reflective back contact of indium-tin-oxide/silver was used.

Recently, conversion efficiencies as high as 13.5% have been obtained for stacked-junction cells with the structure: ITO/p-i-n (a-Si)/p-i-n (a-Si)/p-i-n (a-SiGe)/metal, where the bottom cell contains an amorphous silicon-germanium alloy (a-SiGe) with a bandgap of 1.5eV (20). Also, as previously mentioned in Section 2.2, conversion efficiencies as high as 15.6% have been obtained in a-Si/CIS stacked-junction cells (8). Figure 9 shows how the efficiency of a-Si laboratory cells has progressed historically; efficiencies of ~18% are projected for stacked-junction cells by 1990.

While the performance of a solar cell is usually characterized by its conversion efficiency, another common measurement involves determining its spectral response. A typical spectral response curve for a single-junction a-Si solar cell is shown in Figure 10 – a quantum efficiency of 0.80 means that 80% of the incident photons of light at a particular wavelength create electron-hole pairs that are collected by the semiconductor junction, while the remainder are lost due to reflection, recombination, etc. The short-wavelength response is limited mainly by absorption losses associated with the tin oxide and the p layer while the long-wavelength response is limited by the absorption coefficient of the a-Si (see Section 3.1) and the optical path of the light. The long-wavelength response can be improved by using a-SiGe alloys with narrower bandgaps than a-Si and by using light-trapping geometries (e.g. textured tin oxide and highly reflective back contacts) that increase the effective thickness of the cell.

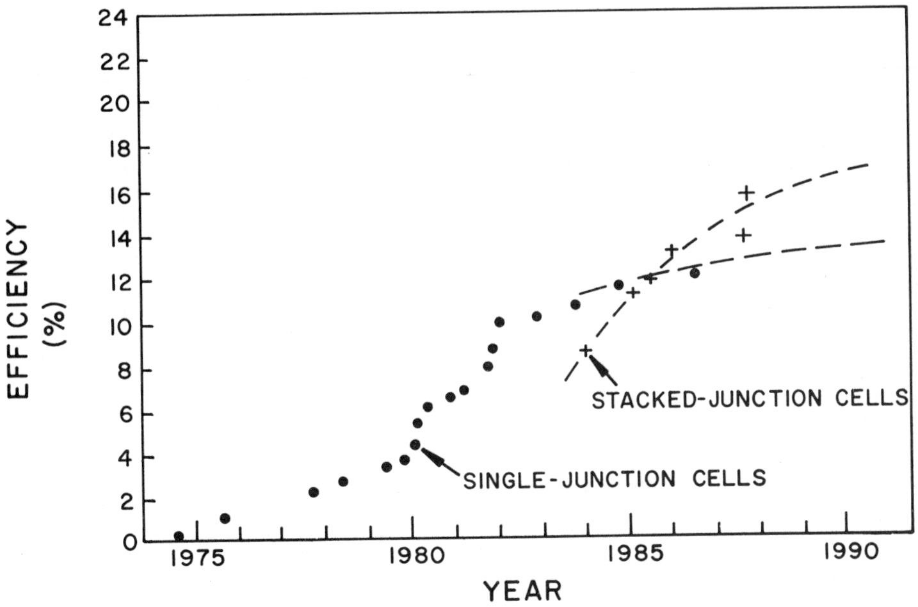

Figure 9. Historical performance of a-Si laboratory cells. The highest performance data point refers to an a-Si/CIS stacked-junction cell (η=15.6%). All data are initial efficiencies.

For large-area a-Si PV modules (>800cm^2), the best reported conversion efficiencies are in the range of 8–9% for 1 sun illumination. (Solarex is presently involved in a 3-year, 50% cost-shared program with the Solar Energy Research Institute (SERI) with the goal of producing 13% efficient modules by 1990.) Figure 11 shows the distribution of conversion efficiencies obtained for a typical manufacturing run of large-area (1000cm^2) modules. Since the cutoff was at a conversion efficiency value of 5.6%, the electrical yield was ~97%. Manufacturing yields at Solarex have averaged greater than 90% since the introduction of a 3-step laser scribing process in December 1986. Rejects are often caused by handling problems (e.g., scratches) so that even higher yields may be obtained with the computer-integrated-manufacturing line.

The current-voltage characteristics of a 30 segment, large-area (1000cm^2) module are shown in Figure 12. Since 30 a-Si cells are connected in series, the open-circuit voltage of the modules is 25.5V while the short-circuit current is 0.42A. Maximum power is delivered to the load when the operating voltage is

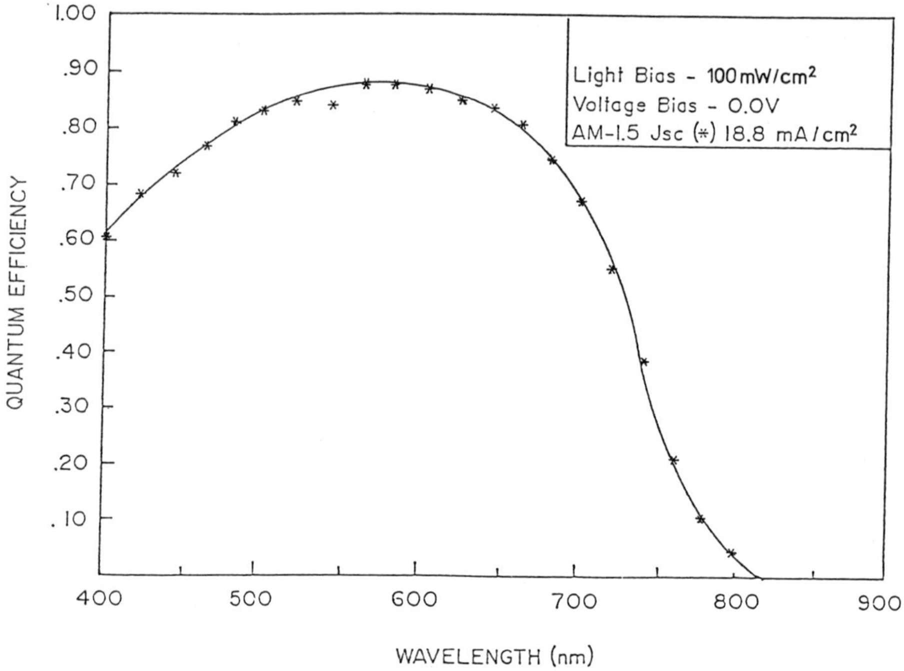

Figure 10. Spectral response curve of an a-Si p-i-n cell.

~20V and the current is ~0.37A (corresponding to a FF of 0.68). For large PV arrays, a-Si modules are connected in series to obtain higher voltages and in parallel to obtain higher currents.

As mentioned in Section 3.3, the conversion efficiency of present a-Si PV modules will decrease by ~10–20% in the first few months of operation before stabilizing. Some of the earliest a-Si modules were constructured with relatively thick i layers (~0.6μm) and therefore degraded about 50% before stabilizing (see Figure 6). Thus, in planning for the construction of a-Si PV arrays, one should use stabilized power ratings, not initial power ratings.

Preliminary tests indicate that properly fabricated a-Si PV modules should exhibit good long-term reliability. Field experience with crystalline silicon modules has shown that encapsulants such as EVA/Tedlar can protect the solar cells for at least 10 years in an outdoor environment. Present commercial modules pass a battery of accelerated environmental tests such as thermal cycling (250 cycles from –40°C to 90°C), humidity soak testing (4 weeks at 85°C, 85% relative humidity), high-voltage testing (<50μmA at 1500V), hail impact testing, etc. (21). Further field testing and improvements in processing should lead to a-Si modules with lifetimes >30 years.

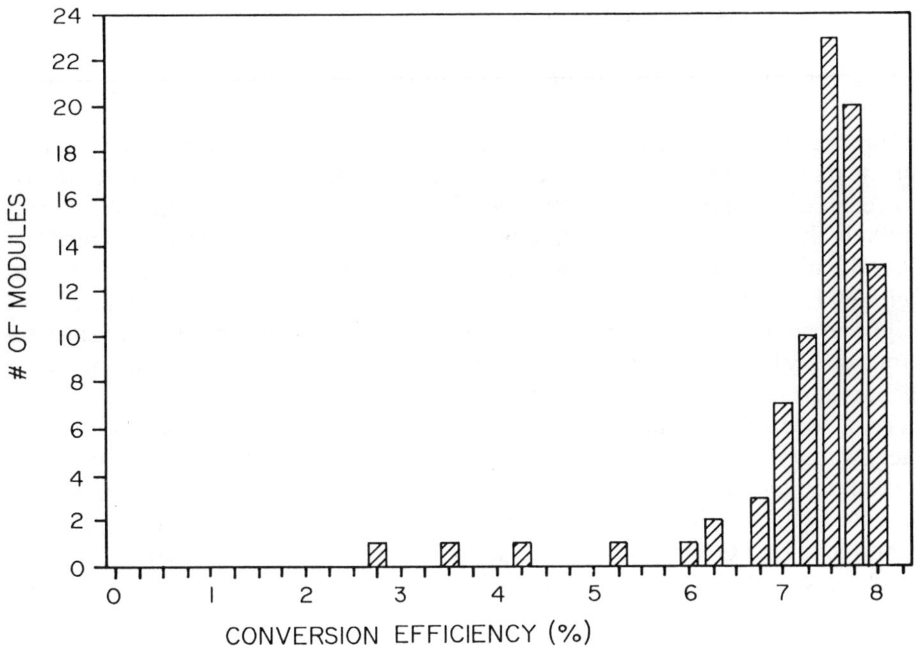

Figure 11. The distribution of conversion efficiencies for a manufacturing run of large-area (1000cm^2) a-Si PV modules.

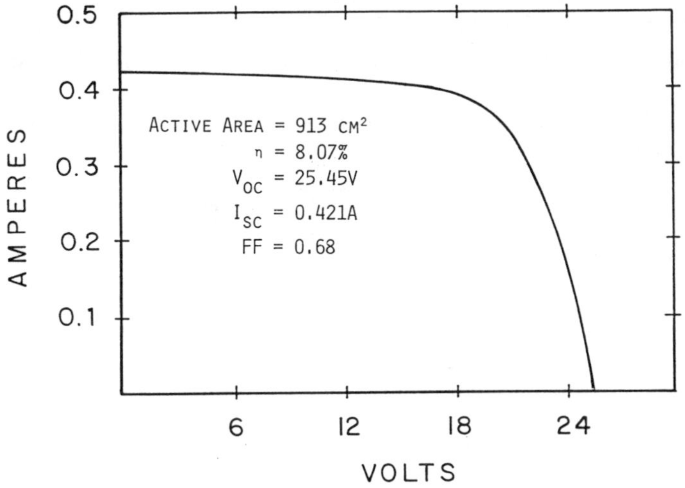

Figure 12. Current-voltage characteristics of a large-area (1000cm^2) a-Si PV module.

6 Manufacturing Costs

While the price of PV modules has fallen from ~\$14.50/W_p in 1980 to ~\$3.80/W_p in 1988, the module selling price must fall to about \$2W_p or less before significant penetration of the electric utility market will occur. (\$2/W_p means that it would cost \$2 to buy a solar cell that produced 1 watt of electric power in direct sunlight of 100mW/cm^2; W_p stands for peak watt.) While it is unlikely that conventional crystalline Si solar cells can be produced at costs less than \$2/W_p, it is now clear that *a-Si solar cells may be manufactured at costs below \$1/W_p by the early 1990's.* (Manufacturing costs of less than \$2/W_p might be attained with other PV technologies such as concentrators or stacked-junction cells using crystalline semiconductors, but a review of these technologies is outside the scope of this paper.)

As a specific case, an a-Si PV manufacturing facility with a capacity of 10 megawatts/year is used here. It is assumed that this plant is producing 1'x4' monolithic modules (glass substrates) with 6% stabilized conversion efficiency and 84% line yield. The plant operates 2.5 shifts/day, 260 days/year and produces 1.67x10^{5}m^2 of PV modules per year. A plant meeting all these conditions could be operating within 2 years.

6.1 Cost of Materials

The materials that are used to fabricate the a-Si PV modules are listed in Table 4 (21). Also listed are the costs of the materials in \$/W_p and the percentage each material contributes to the total cost of materials. (Thus, the cost of chemically strengthened glass in a 1 kW_p array is estimated to be about \$230.)

Table 4 shows that the major cost drivers for materials are glass and silane

Table 4. Cost of Materials for monolithic modules with 6% stabilized conversion efficiency (10MW_p/yr Facility)

Material	Cost (\$/W_p)*	% of Total
Glass (chemically strengthened)	0.23	50
Silane	0.07	15
Encapsulant	0.04	9
Frame	0.03	7
Diborane, Phosphine	0.03	7
Stannic Chloride	0.02	4
Wire, Other Process Gases	0.02	4
Aluminum	0.02	4
Total	0.46	100

* W_p= peak watt

gas. The glass cost could be reduced to about 0.06$/$W_p$ if chemical strengthening (an ion exchange process) is not required. The glass cost could be reduced still further if a float glass manufacturing plant were set up to feed an a-Si PV plant. However, since float glass plants typically produce more than one million m^2 of glass/year, the a-Si facility would need to have a capacity of greater than 60 megawatts/year.

Silane costs have decreased gradually in recent years and may continue to do so as the processing technology improves and the demand increases. Also, the cost estimate for silane in Table 4 was based on a utilization of 5%, i.e., only 5% of the silane in the discharge atmosphere contributes to the growth of the a-Si film on the substrate. The utilization can be increased by an order of magnitude through cryogenic trapping of silane gas in the vacuum pump exhaust and recycling the trapped gas. This approach may become cost effective at the 10 megawatt/year production rate.

Framing costs might be essentially eliminated by designing large-area, frameless modules that would be clamped to new types of support structures designed for large-scale utility plants. Even encapsulation costs might be reduced further by developing lower-cost, spray-on encapsulants. If all these cost reduction steps were implemented, the total cost of materials would fall by more than a factor of 2.

Of course, the cost of materials can be reduced significantly by improving the stabilized conversion efficiency of the modules. Doubling the efficiency from 6% to 12% causes the cost of all materials (in $/$W_p$) to fall by a factor of 2 and doubles the output of the plant (in MW_p/yr). Thus, material costs as low as $0.11/$W_p$ might be achieved for very large plants producing 12% efficient modules.

6.2 Labor Costs

The present model assumes that the manufacturing facility (10MW_p/yr) is partly automated but would still require 82 operators to run the 2.5 shifts/day. In addition, 4 supervisory personnel and 13 technical support personnel would be require to staff the plant. The direct labor costs would be about $1,512,000/yr including fringes or about $0.16/$W_p$, while the indirect labor costs would be about $0.07/$W_p$.

In-line automation or computer-integrated-manufacturing (CIM) of the entire process would reduce the number of operators to about 27 or about $0.06/$W_p$ for direct labor costs. The indirect labor costs would remain at about $0.07/Wp so the total labor costs would be ~$0.13/$W_p$. Increasing the size of the plant to a production rate of 100MW_p/yr would reduce the total labor costs to about $0.07/$W_p$ for a CIM facility.

As before, doubling the efficiency to 12% would cause the labor costs to fall by a factor of 2 (to ~$0.035/W_p$).

6.3 Depreciation Costs and Other Overhead Costs

The capital investment for a 10MW$_p$/yr plant would be about $16,500,000 (21) Assuming that the processing equipment is depreciated over 5 years, the annual depreciation cost is $3,950,000 or $0.41/W$_p$. Increasing the capacity of the plant to 100MW$_p$/yr requires almost a ten-fold increase in processing equipment so there is no significant reduction in depreciation costs. However, as before, doubling the efficiency of the modules causes the depreciation costs to fall by a factor of 2.

As shown in Table 5, direct materials and depreciation are the major cost drivers with the former accounting for 49% and the latter for 43% of the total costs for product produced by a 100MW$_p$/yr CIM plant.

There are other overhead or indirect expenses such as building rent: 2,000m^2 at $100/m^2 for a 10MW$_p$/yr facility or $0.02/W$_p$. Other overhead expenses are indirect materials such as purging gases (e.g., nitrogen) and vacuum pump oils. Still other expenses are electricity, telephones, janitorial services, security, freight, etc. As shown in Table 5, these indirect expenses total about $0.06/W$_p$ for a 10MW$_p$/yr plant.

The net result is that the total manufacturing cost for a-Si PV modules ranges from $1.16/W$_p$ for a 10MW$_p$/yr plant producing 6% efficient modules to $0.47/W$_p$ for a 200MW$_p$/yr plant producing 12% modules. If the other cost reduction steps mentioned in Section 6.1 are implemented, then the total cost

Table 5. Cost Summary ($/W$_p$)

Plant Output	1.67x10^{5}m^2/yr		1.67x10^{6}m^2/yr	
	10MW$_p$/yr	20MW$_p$/yr	100MW$_p$/yr	200MW$_p$/yr
(Stabilized conversion Efficiency)	6%	12%	6%	12%
Variable Costs				
Direct Labor & Fringes	0.16	0.08	0.06	0.03
Direct Materials	0.46	0.23	0.46	0.23
Fixed Costs				
Indirect Labor	0.07	0.035	0.01	0.005
Indirect Expenses	0.06	0.03	0.01	0.005
Equipment Depreciation	0.41	0.205	0.40	0.20
Total Cost	1.16	0.58	0.94	0.47

could fall to about $0.35/W_p$ in case of the $200MW_p$/yr plant. And the total cost could be reduced even further to about $0.25/W_p$ in the larger plant if one assumes that a mature technology would eventually allow one to use a 10 year depreciation schedule. (These estimates of operational costs do not take into account the effects of taxes or interest rates).

Figure 13 shows estimates by various sources of total manufacturing costs as a function of annual production capacity. The overall trend is clear: manufacturing costs are high for small plants, mainly due to high depreciation costs, but as the size of the plant increases and the level of automation increases, the manufacturing costs decrease sharply to less than $2/W_p$ for plants with capacities in excess of a few MW_p/yr.

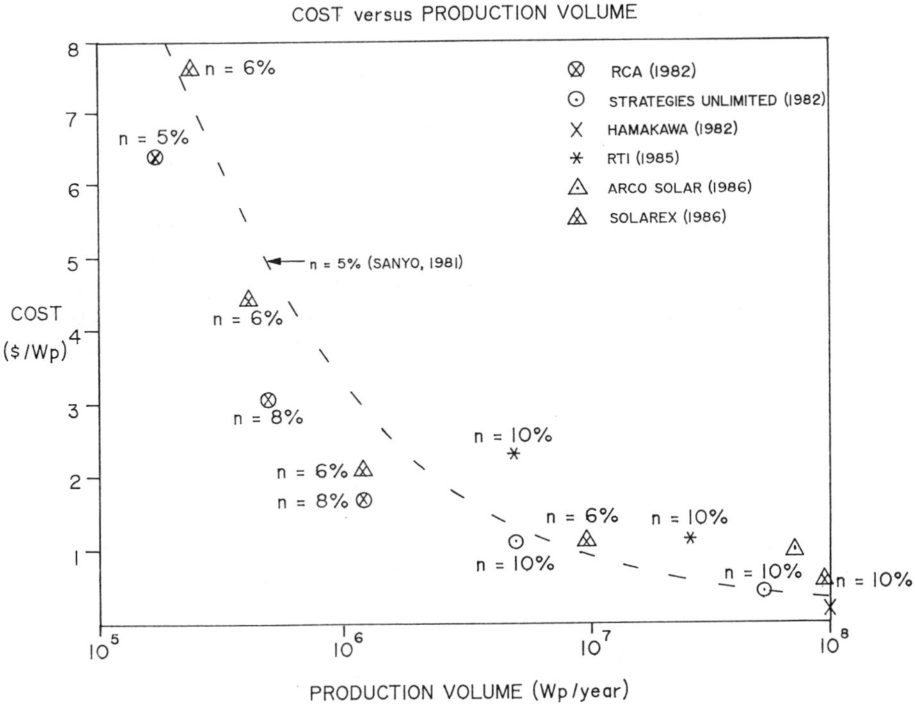

Figure 13. Estimates of manufacturing costs as a function of annual production volume.

6.4 Costs Associated With the Balance of System

The factors that contribute to the total cost of the a-Si PV module have been discussed in previous sections, but for most applications additional expenses are required to complete the PV system. For a central station power plant, the balance of system (BOS) would include the land, preparation of the site, support structures, wiring for the array, power conditioning equipment for converting dc power to stable ac power, fencing, etc. The total BOS costs for a 1MW$_p$ PV array (Sacramento Municipal Utility District in California) were \$2.13/W$_p$ excluding engineering costs of \$2.41/W$_p$ (22). The PV modules cost for this array was \$5.45/Wp so the total system cost was \$9.99/W$_p$ (1984\$).

The U.S. Department of Energy goals for flat-plate PV systems for the year 2000 are listed in Table 6 (23). The area-related BOS consists of the support structure, land, module installation, roads, fences, wiring, site preparation and other civil engineering. Power-related BOS consists of the dc subsystem, power conditioning, ac subsystem, and the switchyard. The BOS overall efficiency is estimated to be about 85% so that an array with 15% efficient modules would exhibit an array efficiency of about 12.75%. For a 10MW$_p$ central station, the cost of the PV modules should be in the range of \$0.30 – \$0.53/W$_p$ assuming area-related BOS costs of \$0.39/W$_p$, power-related BOS costs of \$0.15/W$_p$ and a levelized cost of electricity of 6¢kWh. The levelized cost is based on a weighted average fuel contribution over 30 years of 5¢/kWh and a power generating capacity displacement contribution of 1¢/kWh (23). The annual solar availability is assumed to be 2400kWh/m^2-yr (e.g. southwestern USA). Thus, the total system cost for a fixed, flat-plate PV array would be in the range of \$0.84 – \$1.07/W$_p$ for a 10MW$_p$ central station power plant.

Central station PV plants as well as residential PV roofing will first become cost effective in areas where electricity costs are high. (Residential electricity rates are presently >12¢/kWh in some regions of the U.S. such as Philadelphia and Long Island.) Table 7 shows how higher costs of electricity as well as higher module efficiencies allow more expensive PV modules to become cost

Table 6. U.S. Department of energy goals for fixed, flat-plate PV systems (year 2000).[1,2]

Module Conversion Efficiency (25°C)	15–20%
Module Cost	\$45–\$80/m^2
Area-Related BOS Costs	\$50/m^2
Power-Related BOS Costs	\$150/kW
System Life Expectancy	30 years

1 1986\$

2 Targets based on a levelized cost of electricity of 6¢/kWh in the year 2000.

Table 7. Estimates of PV module selling prices required to meet different levelized costs of electricity for central station PV plants[1, 2].

Levelized Cost of Electricity (¢/kWh)	PV Module Efficiency	Required Selling Price of Modules ($/$W_p$)
6	10	0
6	15	0.30
6	20	0.39
6	10	0.51
9	15	0.69
9	20	0.81
12	10	0.90
12	15	1.09
12	20	1.29

1 Based on a fixed, flat-plate PV array
2 For more details, see Reference 23

effective. If the cost of fossil fuels increases more rapidly than predicted above (2.7%/year escalation rate), then the levelized cost of electricity will exceed 6¢/kWh and thus allow more rapid market penetration by PV. If the levelized cost of electricity is 12¢/kWh in the year 2000, then 10% efficient modules could be used in a central station array if the module cost is about $0.90/$W_p$ or less (23). The BOS cost for such a system would be about $0.74/$W_p$ so the total system cost would be about $1.64/$W_p$ or less.

7 Development of Markets

Amorphous silicon solar cells are now widely used for consumer products such as calculators, clocks, watches, battery chargers and walk lights. Another consumer product that is under development is the PV sunroof for automobiles which powers a small fan that keeps the car cool on hot, sunny days. An a-Si PV module is favored for this application because of a more pleasing appearance.

The factors involved in the market development of stand-alone applications and grid-connected applications for a-Si PV are now considered.

7.1 Stand-Alone Applications

Stand-alone applications refer to PV installations that are not connected to the utility grid, and they include microwave repeater stations, remote lighting systems, cathodic protection of pipelines, remote houses and villages, emergency telephones, irrigation systems, etc. Most stand-alone systems require batteries to store electrical energy so that the system can be used at night or on cloudy days.

Amorphous silicon PV panels are starting to be used in some stand-alone applications such as electric fence chargers, but to-date their penetration into stand-alone markets is relatively small. For many remote applications, a high priority is placed on the long-term reliability of the power source. Since crystalline silicon solar cells have an excellent track record of supplying reliable power over long periods of time, most companies supplying stand-alone systems still use crystalline silicon cells.

The reliability of a-Si PV modules has improved significantly in the last few years, but since the technology is relatively new, the amount of actual outdoor data is relatively short term. Sanyo has reported on the performance of a $2kW_p$ a-Si array that has been operational outdoors for more than $4\frac{1}{2}$ years and continues to perform well (24). As mentioned in Section 5, all a-Si modules exhibit some light-induced degradation, but the performance of the modules usually stabilizes after a few months of outdoor exposure. As the experience base with outdoor applications of a-Si PV grows, the acceptance of these modules for stand-alone application will increase, because they are rapidly becoming cost competitive with crystalline silicon modules.

At present, a-Si solar cells are less expensive than crystalline cells for small power ($<5W_p$), high voltage (>6V) applications, and as larger, more automated a-Si manufacturing plants come on-line, the average selling price will continue to decline dramatically (see Figure 13). On a cost basis alone, a-Si PV should start to penetrate applications such as cathodic protection first, followed by applications such as water pumping and remote villages. Significant penetration of remote applications presently using gasoline or diesel engines should start by 1990. By that time, the average selling price for large-area a-Si modules should be less than $\$3/W_p$.

7.2 Grid-Connected Systems

Grid-connected systems may be either distributed or located in a central station. Distributed systems consist of PV arrays on structures such as houses, high schools, hospitals, commercial buildings, etc., that are connected to the local utility grid. Since grid-connected, distributed PV systems do not become cost effective until module prices fall below about $\$2/W_p$ (25), the existing systems have mainly been demonstrations subsidized by government agencies.

Beacause a-Si solar cells can be fabricated on a number of substrates such as glass, metal foils and certain plastics, there is a high probability that specially designed, a-Si PV roofing will appear on the market in the early 1990's. Sanyo has developed prototypes of a-Si PV roofing tiles on glass substrates. Moreover, the construction of multi-megawatt/year a-Si PV manufacturing plants should lower the average selling price of the modules to less than $\$2/W_p$ by the early 1990's.

The most difficult market for a-Si solar cells to penetrate will be that of the grid-connected central power station. As mentioned in Section 2.1, the module efficiencies may need to be about 15% or greater and the module selling price less than about $\$1/W_p$ for large-scale central station applications. However, penetration may start at about the $\$2/W_p$ level in some parts of the U.S. where electricity rates are high, peaking power (mid-day to early afternoon) is in demand, solar insolation is high, and environmental constraints impede the construction of conventional power plants, e.g., coal-fired plants.

Because of the promise of high efficiencies with a-Si stacked-junction cells and the great potential of significant cost reductions in the near future, a number of utilities in the Unites States such as Alabama Power, Detroit Edison, Florida Power & Light, Pacific Gas & Electric, Philadelphia Electric and Toledo Edison are experimenting with grid-connected a-Si PV arrays. Utilities formulate their strategic plans with a time horizon of about 10 years. This means that the present experiments will allow utilities to become familiar with a-Si PV arrays and be ready to use them in large scale when they become cost effective in the mid- to late 1990's.

8 Future Directions

The present technology trends indicate that a-Si power modules will be available in the early 1990's with stabilized conversion efficiencies of about 10%. The manufacturing plants at that time will probably have annual capacities of about $10MW_p$ so that the manufacturing costs are projected to be on the order of $\$0.70/W_p$ (see Figure 13). These modules should capture PV markets such as irrigation, consumer products, remote villages, and general battery-charging applications in the early 1990's. *Significant penetration of the communications market and the development of the grid-connected PV roof market will probably occur in the mid-1990's.*

As mentioned in Section 2.1, module conversion efficiencies of about 15% are probably necessary to address the grid-connected central station market

(11). The attainment of this goal requires that stabilized efficiencies of ~17–18% be demonstrated first in laboratory cells. This, in turn, requires significant improvements in the electronic properties of materials such as a-SiGe alloys. Thus, significant penetration of the central station, utility market will probably not occur until the late 1990's.

One barrier to the widespread utilization of PV is the relatively high capital cost associated with the purchase of the PV system even though life-cycle costs may be competitive with other alternatives. This issue is partially circumvented in the case of residential PV roofing since the cost of the PV system would be included in the selling price of the home, and mortgage financing allows the payments to be stretched out over periods up to 30 years. Long-term financing would also be the conventional way to handle industrial PV roofing arrays.

Another issue that can impede the commercialization of PV is the added cost of normal distribution channels. For large-scale projects such as central station utility plants, however one might choose to build the PV manufacturing site next to the central station location. The utility might also choose to own and operate the PV manufacturing plant as well as the central power station. This approach would minimize transportation costs and allow the central station to incrementally add new PV arrays with little delay in producing power. The capacity of the PV plant could be tailored to the projected local demand for electricity based on peak power requirements or other criteria.

Improvements in other technologies could have significant impact on the commercialization of PV. For example, the development of low-cost storage of electricity such as new types of batteries or superconducting magnets might adversely affect the central station market for PV while at the same time enhancing remote, stand-alone applications such as village power. Efficient energy storage would allow utilities to level their loads by using base load power, e.g., from coal-fired plants, for energy storage during low demand periods and then supplementing the base load with the stored energy during peak demand periods.

The cost of competing sources of energy can also have a significant impact on the commercialization of PV. Oil prices are likely to rise rapidly in the next decade or so as more and more oil producing nations start to deplete their reserves and pass through their period of peak production. As in the 1970's, increasing oil pries should lead to an increase in the price of other fossil fuels, and this will increase the use of PV.

Finally, the commercialization of PV may well be accelerated by environmental considerations. There is increasing concern about the build-up of gases such as CO_2 in the atmosphere and the possible adverse effects on the global climate (26). The accumulation of these gases in the atmosphere is attributed mainly to the burning of fossil fuels. Photovoltaics is one of the few clean energy options available to alleviate impending detrimental changes to the earth's environment.

Acknowledgements

The author thanks G.F. Wilkins for providing some of the cost data. Also, some of the Solarex research reported in this paper was partially supported by the Solar Energy Research Institute under Subcontracts No. ZB-4–03056–3 and ZB-7–06003–2.

References

1. Chapin, D.M., Fuller, C.S., Pearson, G.L. 1954. J. *Appl. Phys.* 25: 676.
2. Sinton, R.A., Swanson, R.M. 1987. *Proceedings of the 19th IEEE Photovoltaic Specialists Conf.* NY: IEEE, p. 1201.
3. Carlson, D.E. 1977. U.S. Patent No. 4,064,521.
4. Hovel, H.J. 1975. *Semiconductors and Semimetals, Vol. II, Solar Cells.* Academic Press p. 61.
5. Tiedje, T. 1982. *Appl. Phys.* Lett. 40: 627.
6. Carlson, D.E., Wronski, C.R. 1976. *Appl. Phys.* Lett. 28(11): 671.
7. Matsumoto, H., Kuribashi, K., Uda, H., Komatsu, K., Nakano A., Ikegami, S. 1984. *Solar Cells* 11: 367.
8. Morel, D.L., Blaker, K., Bottenberg, W., Fabick, L., Felder, B., Gardenier, M., Kumamoto, D., Rinker, D., Rifai, R. 1988. *8th E.C. Photovoltaic Solar Energy Conf.* Florence, Italy, May 9–13.
9. Catalano, A., Arya, R.R., Fortmann, C., Morris, J., Newton J., O'Dowd, J.G. 1987. *Proceedings. of 19th IEEE Photovoltaic Specialists Conf.* NY: IEEE, p. 1506.
10. LeComber, P. 1988. 8th E.C. *Photovoltiax Solar Energy Conf.* Florence, Italy, May 9–13.
11. DeMeo E.A., Taylor, R.W. 1984. *Science* 224(4646): 245.
12. Meyers P.V., Liu, C.H. 1988. 8th E.C. *Photovoltaic Solar Energy Conf.* Florence, Italy, May 9–13.
13. O'hara, T., Kadota Y., Moriya, K. 1987. *Proceedings of 19th IEEE Photovoltaic Specialists Conf.* NY: IEEE, p. 338.
14. Gale, R.P., King B.D., Fan, J.C.C. ibid. p. 293.
15. Tiedje, T., Cebulka, J.M., Morel, D.L., Abeles, B. 1981. *Phys.* Rev. Lett. 46: 1425.
16. Spear W.E., LeComber, P.G. 1975. Solid State Communications 17: 1193.
17. Staebler D.L., Wronski, C.R. 1977. *Appl. Phys.* Lett. 31: 292.
18. Crandall, R.S., Carlson, D.E., Catalano A., Weakliem, H.A. 1984. *Appl. Phys.* Lett. 44(2): 200.
19. Fortmann, C.M., Lange, S., Farley M., O'Dowd, J.G. 1987. *Proceedings of 19th IEEE Photovoltiac Specialists Conf.* NY: IEEE, p. 296.
20. Sabisky, E. private communication (stacked-junction cell made by Sovonics and measured at SERI).
21. Carlson, D.E. 1988. *8th E.C. Photovoltaic Solar Energy Conf.* Florence, Italy, May 9–13.
22. Kern, E.C. Jr. 1984. *Proceedings of 17th IEEE Photovoltaic Specialists Conf.* NY: IEEE, p. 1196.

23. U.S. Department of Energy, DOE/CH10093–19, DE88001141, January 1988.
24. Tsuda, S., Nakamura, N., Watanabe, K., Nishikuni, M., Ohnishi, M., Nakano, S., Kishi, Y., Shibuya H., Kuwano, Y. 1984. *Proceedings of 1st Photovoltaic Science and Engineering Conf.* Kobe, Japan, p. 213.
25. Maycock, P.D., Stirewalt, E.N. 1981. *Photovoltaics, Sunlight to Electricity in One Step.* Andover, Mass: Brick House Publ. Co., p. 90.
26. Ramanathan, V. 1988. *Science* 240: 293.

Fuel Cells

A review of fuel cell technology and its applications.

Leo J. M. J. Blomen

Abstract

This chapter assesses the status of fuel cell technology for power generation. Fuel cells are expected to have major application in the near future, in both centralised and decentralised electricity generation. However, in order to achieve sufficiently low costs, especially for smaller systems, the mass production of fuel cell stacks will be a necessity; so, too, will be the serial production of entire, integrated fuel cell plants, i.e., fossil fuel processors and fuel cell stacks and power conditioners. Both of these developments seem likely today. It appears that fuel cells will be quite competitive with both new generating technologies and the more conventional ones in terms of the cost of produced electricity. And, fuel cells come with some very important, additional advantages, i.e., virtually no emissions, modular construction, and stable power characteristics that maintain high efficiency even at partial load. Hence, the more emphasis society and electricity producers put on fuel efficiency and reduced emissions, the greater will be the market penetration of fuel cells.

1 Introduction to Fuel Cell Technology

Fuel cells are electrochemical devices that continuously convert the chemical energy of a fuel – typically hydrogen from a hydrogen-rich gas mixture – and an oxidant – typically oxygen such as from air – into electrical energy. It is the reverse of water electrolysis. This principally clean form of combustion is an isothermal process within the fuel cell, implying that no Carnot cycle limitations exist. As a consequence, fuel cells have a high fuel to power conversion efficiency, ranging from 40 to 70% based on the lower heating value (LHV) of the fuel to the fuel cell (1). The energy not utilised for electric power production is available as waste heat, and should be removed from the cell to maintain the isothermal environment. Through integration, it is possible to use a large part of this waste heat in other parts of the fuel cell system, e.g.

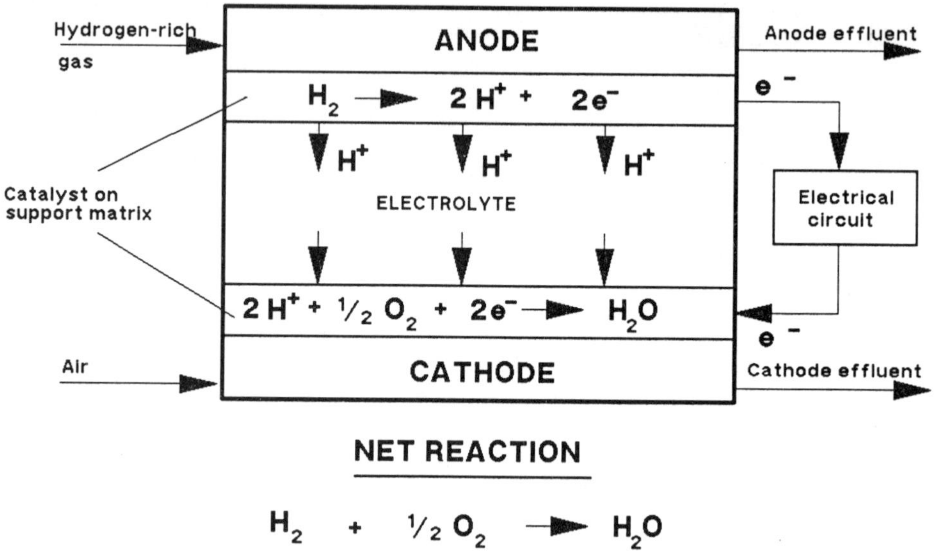

Figure 1a. Schematic showing the principles of fuel cell operation, Phosphoric Acid Fuel Cell (PAFC).

to produce hydrogen from a fossil fuel, increasing overall system efficiency considerably.

The basic fuel cell structure consists of an electrolyte within a support matrix sandwiched between two electrodes (see Figure 1a). Each electrode contains a specifically developed catalyst to promote the appropriate electrochemical reactions.

The provision of a hydrogen-rich gas to the fuel cell stack usually first requires the conversion of a commercially available fuel, such a natural gas, naphtha, biogas or coal. Therefore, most fuel cell power plants will comprise three main subsystems: a fuel processing unit, the fuel cell stack power section, and in many cases a DC to AC power conditioner (see Figure 1b). In addition, an energy recovery system is usually provided to recover heat and power for process heating and gas compression requirements, thereby improving overall system efficiency. In certain instances (large systems), extra electric power may also be economically produced in this energy recovery system by a bottoming cycle.

The principle of the fuel cell technology is 150 years old; few modern alternative electricity producing technologies can boast such a long history. After its initial discovery by Sir William Grove, the principle of fuel cell operation remained essentially unused for almost a century. Fuel cells were seen as a technical curio, and considered only as a theoretical possibility for

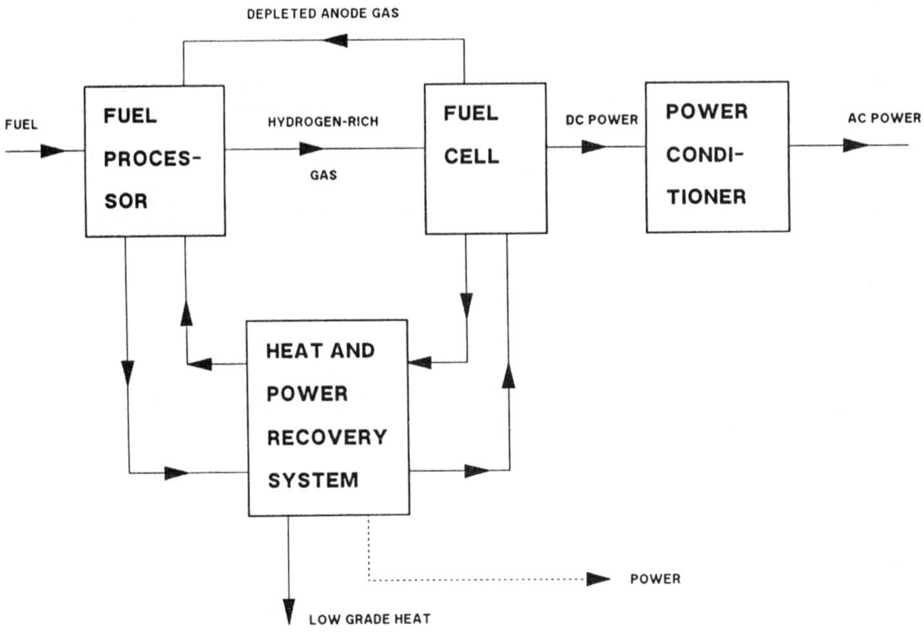

Figure 1b. A Fuel cell power plant.

use in energy storage and re-usage. In the middle of the twentieth century, Europe became very active in research and development of several fuel cell types, and some twenty years ago a commercial interest arose, when researchers realised that the fuel cell technology might become a useful method for stationary electric power production as well as an interesting power source in space, submarine and other transport applications. Early applications were developed for the Gemini and Apollo space programs, and the Space Shuttle is also equipped with fuel cell systems. Small systems have been developed for test vehicles and for military applications, the latter especially with low temperature fuel cells that make infra-red detection difficult.

In the last decade, the interest in fuel cells for stationary power production has grown strongly. Today more than 65 demonstration plants in the power range of 10 kW to 5 MW are or have been operational, almost all of them in the United States and Japan. In the very near future, one or two 10–11 MW systems will become operational. The largest scale demonstration units were the two electric power plants producing 4.5 MW electricity built by United Technologies Corporation (UTC) in New York and Tokyo. The plant in Tokyo has clearly demonstrated that fuel cell technology fulfills all expectations with respect to high efficiency, flexibility during operation, and minimal or even almost zero environmental costs. On the basis of these two large units, as well as an extensive on-site demonstration program with more than 45 units of 40 kW each, UTC has decided to commercialise multi-kW power plants, in a joint venture with the Japanese company Toshiba. They are calling it International Fuel Cells (IFC). This event marked the beginning of large scale commercialisation plans, and several companies have since announced commercialisation of especially small scale fuel cell systems, to be followed in the future by larger systems.

2 Advantages of Fuel Cell Systems for Power Generation

2.1 Requirements for New Power Producing Techniques

Many current power producing techniques have considerable drawbacks. Small units in the range of 20 to 100 kW such as battery powered systems and diesel engines have disadvantages that include very low efficiencies, unacceptably high environmental emissions or leaks, and inferior part-load characteristics. Power producing or cogeneration units in the range of 100 kW to a few MW, such as gas engines, diesel engines and gas turbines, also have these disadvantages, though their efficiencies are somewhat better. Especially noteworthy are the environmental emissions of these technologies. This disadvantage will cause the current cost of produced electricity to rise if legislation requires increased reduction of emissions. In the multi-megawatt range, most important in many areas of the world because of centralised power producing structures, the efficiencies of electricity generating technologies are also low and the environmental costs high. Safety issues and environmental pressure have also made costs for nuclear power plants prohibitive, at least in many countries. Very strict measures for reducing emissions in conventional technologies are under consideration or have already been implemented in several countries.

The various disadvantages of the current technologies have sparked interest in new power generating technologies, especially those which can provide: higher efficiency (at partial loads too), fuel flexibility, as well as flexibility to varying heat and power demands in cogeneration applications, low noise and emission characteristics, fast load response, low maintenance, high reliability and possibility for modular construction. Fuel cell systems are expected to compare very favourably with most competitive technologies in providing these characteristics.

2.2 Efficiency

The electrical efficiencies of fuel cell systems exceed those of conventional power generation technologies. Depending on the type of fuel cell involved, total system efficiencies can range from 40 to 60%, whereas only the most modern combined steam and gas turbine systems can achieve maximum efficiencies between 40 and 50%. In fact, the United States average for present day power plants is about 35% (LHV), (1), and the equivalent figure for power production in the Netherlands is about 39% (2). A unique characteristic of fuel cell systems is the fact that efficiency hardly decreases at partial load, even taking into account the slightly inferior performance of the fuel processing system integrated with the fuel cell stack. Typically, the efficiency of the total system is more or less constant throughout the range of 50 to 100% of capacity, and only drops significantly below 30% load.

2.3 Flexibility

Fuel processing subsystems can be designed (on the basis of current state-of-the-art technology without much further development) to produce either pure hydrogen or synthesis gas of variable composition, from almost any hydrocarbon fuel. While natural gas is most commonly used, other fuels have been studied as well, including pure methane, biogas (3), LPG, naphtha and coal (via coal gasification) (4). This multi-fuel flexibility could be very important in penetrating markets in countries where liquid hydrocarbon transport is expensive, natural gas pipelines are not available, and/or biogas can be made available locally.

In order to achieve a high electrical efficiency, a fuel cell system must be designed so that the maximum electric power is produced while simultaneously recovering the waste heat for cogeneration purposes. This makes fuel cell technology attractive in applications where only electric power is generated, but also where heat and electricity are needed, especially with high flexibility with respect to heat demand variations. Simple design modifications, such as oversizing of the steam reformer convection section, can easily result in a system that satisfies the local demand for electricity and heat simultaneously, without additional investments. Alternatively, one might install cheap additional steam boilers that may even use lower cost fuels. However, the overall electrical efficiency of the system will always have to be maximised during design.

Figure 2 shows the effect of cogeneration on the cost of electricity (COE) produced by different types of power producers at 250 kW capacity. The economic assumptions are consistent with those used throughout this chapter (see Tables 5 and 6), and Figure 2 details the reduction on the COE possible when credit is given for the released heat. With the gas engine, the COE is relatively high if the waste heat is not utilised. The ratio between electricity and electricity+congenerated heat, R, is then 1. For R<1 (i.e. cogeneration), waste heat up to a maximum level of 500°C can be generated but with increasing heat utilisation the temperature level of the available heat drops. In space heating applications, one may utilise all the heat down to a temperature level of about 100°C, which level is reached for this gas engine technology at a value of R=0.35. If more heat is demanded, additional investment is required to install extra heat generating capacity. However, this only marginally increases the COE, because of the relatively low investment of the extra boiler. This is indicated in Figure 2 by an almost horizontal line running to the left from the point on the cogeneration curve where the temperature level of available heat has dropped to 100°C. If, on the other hand, heat is required at a temperature level above 100°C, say at 300°C, the COE curve will be followed only until the R-value at which the gas engine's cogenerated heat is available at 300°C and above. For any further increase in cogenerated heat demand a

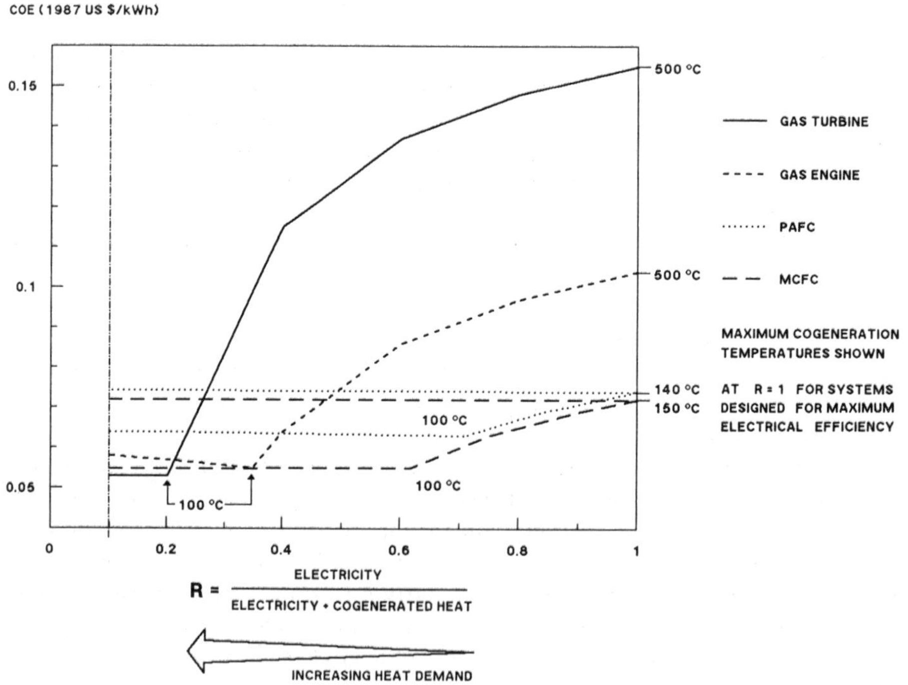

Figure 2. The effect on the cost of electricity (COE) of variation of the temperature of the cogenerated Heat and of the degree of cogeneration for the 250 kW power plants.

boiler has to be installed and the COE curve is then almost horizontal for all lower R-values (parallel to the line drawn for 100°, but starting at a higher level).

Similarly for the gas turbine, if all cogenerated heat can be utilised to 100°C (at R=0.2), the COE even becomes slightly cheaper than that produced by the gas engine, though this difference is negligible. However, at R>0.2 the gas engine is the cheaper of the two alternatives, whatever the temperature level of heat required.

With respect to the flexibility of fuel cell systems, the Phosphoric Acid Fuel Cell System (PAFC) and the Molten Carbonate Fuel Cell System (MCFC) curves show that these systems (in high volume production) compare favourably, and have lower cost of electricity especially when R>0.4. Both systems in this case, after optimisation for maximum power production, have only heat available at low temperatures (below 140 or 150°C, respectively). If heat can be utilised down to 100°C, the MCFC, and to a slightly less degree the PAFC, have very low COE, which stay low, even if extra heat generating capacity has to be installed. It is important to note that the fuel cell systems can also produce heat at 500°C by simply installing a boiler for the purpose, or by increasing the reformer size, resulting in nearly horizontal COE curves right from R=1, throughout the whole R-range, which imply production of competitively priced electricity to about R=0.5 (cf. gas engine) and R=0.3 (cf. gas turbine). Figure 2 demonstrates that fuel cell technology for this capacity can be competitive to gas turbine and gas engine technology in many applications, especially if heat demand is not continuous and excessive. The fuel cell systems provide more flexibility, and even long term changes in heat demand can be very cheaply dealt with. Gas engine and gas turbine systems only become cheaper in cases where heat demand is large and continuous, and even in that range the difference in COE is only marginal for MCFC.

2.4 Environmental Considerations

By combusting essentially hydrogen with oxygen to water, the fuel cell stacks themselves are clean sources of energy, and produce only high purity water as well as electricity and waste heat. Especially in lower temperature fuel cells, no extra NO_x or other noxious compounds are being produced. However, the fuel processor requires a lot of heat which will normally be supplied by furnace burners (if methanol is converted into hydrogen-rich gas, then lower grade heat sources can be used). The burners are a major source of NO_x, carbon monoxide, unburnt hydrocarbons and soot particles. Modern burner design can guarantee acceptable carbon monoxide, soot and unburnt hydrocarbon emissions. NO_x in case of fuels with low nitrogen content is mainly formed from the combustion air, and can be extremely low in a fuel cell system because of the following measures:

- Fuel cell systems will mainly use the anode off-gas in the burners, which is a very lean gas containing the unutilised hydrogen, carbon monoxide and some traces of unreacted hydrocarbons only. The lean gas burns with low flame temperatures and hence low NO_x formation.
- Special burners are in use which, by limiting the actual flame temperature (either via catalytic combustion, or in modern ceramic fiber or metallic fiber matrices), produce very low NO_x levels.
- Many fuel cell systems in the future will employ pressurised combustion, which is known to further suppress NO_x formation (5).

Recent studies (see Table 1), based on computer simulations of NO_x formation in different type of burners, have predicted very low NO_x emissions from PAFC systems (between 1 and 10 ppm at 3% oxygen), and even 5 to 10 times lower in MCFC systems (4). The emissions actually measured in PAFC systems have confirmed these predictions (6,7).

Fuel cell systems are also expected to produce acceptable noise levels, and this has been confirmed by demonstration systems. Noise is mainly generated by rotating equipment, and therefore strongly depends upon system design rather than on the fuel cell stacks themselves.

Table 1: NO_x emissions of fuel cell systems g/GJ_e (LHV).

	Equilibrium[a] Emissions	Standard conv. burners	Modern burners	Best low NO_x burners
PAFC/AFC[b]				
25 kW	207	7.1	3.4	0.6
250 kW	218	7.4	3.7	0.6
MCFC				
25 kW	17 (with oxygen enrichment)	0.6	0.3	0.05
250 kW	32	0.9	0.9	0.3
3.25 MW	55	1.8	0.9	0.09
100 MW	190 (with carbon dioxide removal)	6.6	2.8	0.53

[a] Equilibrium evaluations assume that the NO_x formation reactions proceed until they approach their equilibrium states. The reality is, however, quite different, as kinetic and mixing effects must be accounted for.

[b] AFC = Alkaline Fuel cell.

2.5 Load Response

The fuel cell, being an electrochemical device, reacts to load variations very quickly. The hydrogen fuel processing system integrated with the fuel cell stack, however, has fairly slow response times, requiring special control of the integrated plant, and special compact steam reformer design.

2.6 Reliability and Maintenance

Although fuel cell systems are expected to be extremely reliable and have low maintenance costs, current demonstration programs have not yet proven this. Small cell stacks have achieved lifetimes of about 40,000 hours, with 5–10% performance loss over this period. Fuel processing systems, similar to those normally employed in hydrogen plants, have very long lifetimes, and top performing plants have had trouble-free operation for more than 120,000 hours. As a consequence, the lifetime of fuel cell systems will mainly depend on fuel cell stack life, once reliable system designs are being utilised. However, in demonstration programs up to now, the fuel cell stacks always performed more than adequately, whereas peripheral systems have shown inferior performance. About 98% of the failures in one fuel cell plant were reportedly due to small off-the-shelf components (8), while others have shown major component failures in the fuel processing units, e.g. the reformer, instrumentation, and controls, as well as mechanical systems (9). The main reason for this is that most demonstration plants have been built by electric component manufacturers, which apparently have had only limited experience with hydrogen plant technology. In recent years, most of the world's fuel cell manufacturers have started to involve specialised engineering contractors.

Maintenance costs for a mature fuel cell technology are expected to be low. Stacks will probably have to be replaced every five years (assuming 8000 hours per year continuous operation), and catalysts will have to be replaced every 10 years (reformer catalysts, shift catalysts, and so on). Apart from stack development, other maintenance costs are expected by the author to be similar to hydrogen plants, and can be estimated at 3% of the total capital requirements for the fuel cell plant for the first series of production, gradually going down to less than 2% for producing larger numbers (typically 20), to below 1% for production series of typically 200. For robotised production of smaller systems, these maintenance costs may go down to below 0.5% of total capital requirement. However, these figures will have to be proven in practice, because the current stage of development of the fuel cell technology does not allow confirmation of these estimated percentages.

2.7 Modular Construction

Repetition of many equal fuel cell stacks in larger capacity plants and the fuel cell stack itself being a modularised component, as well as the combination of several reactors (reformer, shift reactors, vessels) and heat exchangers in all system capacities will encourage modular construction of fuel cell units from the very start of commercialisation. The very first commercial systems that are expected on the market – small scale, methanol and natural gas fueled power producers in the 1 to 50 kW range – will be built like containers and modular construction is absolutely necessary to decrease costs, maintain heat losses at an acceptably low level, and insure reliability. In case of serial production, necessary to reduce costs for especially smaller sized units, modular construction is even more essential.

3 Types of Fuel Cells

3.1 Principle of Operation

As in the battery world, fuel cell technology shows an enormous variety in types of fuel cells. However, all types are based on the same electrochemical principle, illustrated in Figure 1a for the acid fuel cell. The principle of each fuel cell is the use of two electrodes, separated by an electrolyte, through which ions can migrate and electrons cannot. A hydrogen-rich gas (or sometimes high purity hydrogen) is supplied to the anode, where the hydrogen molecules dissociate into hydrogen ions and electrons. In the PAFC, hydrogen ions migrate through the electrolyte to the cathode compartment. In this compartment air or oxygen is being fed, allowing a reaction with the migrated hydrogen ions and the electrons that are supplied by the external circuit. Water is formed as a consequence of this reaction. In alkaline fuel cells (AFC), hydroxide ions form at the cathode and migrate to the anode. In the MCFC, a different mechanism is at work: carbon dioxide and oxygen react in the cathode compartment with the supplied electrons, forming carbonate ions that migrate through the molten alkali carbonate electrolyte to the anode. Other types of fuel cells, that will not be discussed in detail here, use other ions for transport between the two electrodes. All fuel cell systems (see Table 2), however, owe their electricity generating capacity to that stream of electrons continuously transferring from the anode to the cathode via the external circuit. In addition, in all fuel cell types, the final overall reaction is the isothermal oxidation of hydrogen to water.

Basically, the cell voltage that can be obtained from this water forming reaction is the difference in electrochemical potential between hydrogen and oxygen, minus all losses as a consequence of diffusion limitations at the electrodes, losses in the electrolyte, etc. As will be explained in Section 4 of this chapter, this cell efficiency is not at all equal to the total system efficiency after integration into a hydrogen producing plant. (Excellent references, such as 1 and 10, are available on cell electrochemistry.)

Table 2. Main characteristics of fuel cells.

Type	Stack Efficiency		Operating temp	State of the art
	Current	Future (expected)	°C	
AFC	60% (pure H_2)	65% (pure H_2)	15–120	50 kW (pure H_2) 20,000 h
PAFC	40% (SG)	47% (SG)	170–210	4500 kW (nat. gas, methanol) 40,000 h
MCFC	55% (SG)	65% (SG)	600–650	25 kW (nat. gas, methanol) 10,000 h R&D only
SOFC	50% (SG)	60% (SG)	1000	5 kW R&D only
DMFC	40% (CH_3OH)	45% (CH_3OH)	60	R&D only
SPFC	55%[a] (pure H_2)	60% (pure H_2)	15–80/300	R&D only

SG = synthesis gas (CO/H_2), with or without $CO–CO_2$ shift.
[a] Lower on SG.

During the electrochemical process, a typical fuel cell will generate between 0.5 and 1.0 V depending upon current density. With a defined surface area per cell, this leads to a specific power for each cell. In a cell stack, a combination of individual cells is made by stacking cells on top of each other, and the electrochemical process takes place in each cell individually. Gas has to be distributed via manifolds to all the cell compartments, and electricity from each cell is combined by putting a certain number of cells in series and several such series in parallel. The waste heat from the electrochemical reaction is removed via separate cooling plates or cooling tubes, specially designed to fit between the cells. To design individual cells for high performance – materials, long-term behaviour, fabrication procedures, equal distribution of reactants over all parts of the cell – is quite difficult, but the design of the fuel cell stacks poses some very special difficulties which have to be overcome. These include the need for hydrogen-tight sealing of large surface areas (especially in pressurised cells), production methods with extremely low tolerance, and special manifold designs; in addition, there are the corrosion problems in cooling systems that have an extremely large surface area per unit of volume. (See references 11, 12 and 13 for some good reviews of fuel cell and stack technology.)

3.2 Alkaline Fuel Cells (AFC)

The alkaline fuel cell has a well proven history in space applications. AFC production capacity of several megawatts per year already exists, and it includes the automated manufacture of electrodes. In Europe production is centered at Siemens, FRG, and at Elenco, Belgium. Stack fabrication is partly by manual production techniques.

Alkaline fuel cell technology, needing pure hydrogen as a fuel, has been demonstrated in vans and forklift trucks, as well as in a geological survey vehicle. A 200 kW prototype system for submarines was demonstrated in Sweden in 1964. Presently, applications are being developed for city bus systems, several other kinds of trucks as well as for stationary applications such as uninterruptable power supply (UPS) systems, chemical industry for DC power generation, etc. In addition, various defense applications are being developed, and very high power densities are now being achieved in some of these applications. Pressurised alkaline fuel cells are under development, but few efforts have been undertaken with respect to operation on commercial fuels other than pure hydrogen (14).

3.3 Phosphoric Acid Fuel Cells (PAFC)

PAFC stacks have been under development for about 25 years, and can be considered as the most mature type of fuel cells. Total investment in developing PAFC systems is about half a billion US dollars. Throughout the years, PAFC performance has increased with temperature from about 60–80°C to 200°C. The present status is to demonstrate commercial viability. Main R&D areas include fuel processor system analysis, electro-catalysts, structures of the electrodes and alternative electrolytes.

3.4 Molten Carbonate Fuel Cells (MCFC)

Renewed development in MCFC area has been going on for more than 10 years now. However, MCFC development is 5 to 10 years behind PAFC, partly due to several problems related to: electrode kinetics and mass transport; the development of conductive ceramics; the existence of corrosion; anode creep and cathode dissolution. In addition, there are electrolyte management problems as well as gas cross-over and matrix cracking, and sulfur tolerance problems to be overcome. Demonstrations are taking place with stacks of 10 to 40 kW, and operating times of about 10,000 hours have been demonstrated. Studies are being done in system development, carbon dioxide transfer devices (since the carbonate ion transfers the electrons from one electrode compartment to the other, a recycle of carbon dioxide is necessary), and it is hoped that simpler overall system designs will further reduce the costs of MCFC.

3.5 Other Types of Fuel Cells

Table 2 also contains some very brief data on three other types of fuel cells – the so-called solid oxide fuel cells (SOFC), direct methanol fuel cells (DMFC) and solid polymer fuel cells (SPFC). These three types of fuel cells are in the research and development phase, and the largest stack for solid oxide systems is a 5 kW demonstration. Extensive studies of conventional fuel based systems have not yet been undertaken, though both the solid oxide and SPFC offer opportunities: the first because of the extremely high temperature of the available heat from the fuel cell, and both because of the possibilities of converting the fuel directly into hydrogen within the catalyst itself (internal reforming), and of using methanol as fuel. SPFC, in particular, enables extremely high current densities (~4A/cm^2). (For more details on these and other fuel cells see references 11–13, 15–20).

3.6 Current Status of R, D and D Programs

The main research, development and demonstration efforts are concentrated in the United States, more recently Japan and very recently Europe (especially in Holland and Italy).

In the United States, from 1958 to 1978 the government has contributed between 1 to 2 million dollars per year, whereas after 1978 this amount more than quadrupled (4 to 9 million US dollar per year). Major efforts, mainly in PAFC and MCFC, have been undertaken by United Technologies Corporation (UTC) – later International Fuel Cells (IFC) in joint venture with Toshiba. Moreover, Engelhard Corporation is working on PAFC, Energy Research Corporation on PAFC and MCFC, and the Institute for Gas Technology (IGT) on MCFC.

Japan has begun a major development effort, and is spending more than 200 million US$ government funds in the period of 1987 to 1995. Major efforts on both PAFC and MCFC have been undertaken by Fuji, Toshiba, Hitachi and Mitsubishi. Several other companies are concentrating mainly on MCFC systems – IHI, Sanyo, Matsushita, NKK, Kawasaki – although most of these also spend R&D on other types of fuel cells.

In Europe, the European Community modestly supports R&D on several types of fuel cells – over 1 million dollars a year – and both Italy and Holland have larger development and demonstration programs, with each probably spending about 40 million US$ between 1987 and 1995 or 1992, respectively. These efforts should enable Europe to develop a strong position, both in the system design, engineering and construction aspects as well as in the MCFC stack technology.

4 System Design

4.1 General Considerations on Design

Although fuel cell stacks constitute the heart of the power plants, most first require a conversion of commercially available fossil fuels into synthesis gas that can be used in the fuel cell stacks. The primary conversion step is a reaction of the hydrocarbons with steam, to form carbon monoxide and hydrogen (steam reforming). Sometimes other techniques can be used, such as partial oxidation or gasification, depending on the fuel being used. In most cases, part of the carbon monoxide is shifted with water to more hydrogen plus carbon dioxide, in one or more so-called "shift-reactors".

With molten carbonate fuel cells, solid oxide fuel cells and some other fuel cells, internal reforming is possible. This means that clean natural gas and steam can be fed directly to the anode. Internal reforming is in a rather early development stage, and imposes some extra problems on the R&D of the fuel cell stack materials and catalysts. Due to economies of scale, internal reforming is not expected to be used in very large scale systems, though in the longer term it seems very promising for smaller capacities (<1–2 MW) (21).

Hence, most systems using fossil fuel feedstocks will need an external steam reformer, and several processing steps to purify the hydrogen-rich synthesis gas mixture to within the allowable specifications imposed by the fuel cell stack. The fuel processor section may consist of between 15 and 30 different pieces of equipment, and it connects to the electrochemical section. (See Figure 1b). During the steam reforming process, steam is needed, and it can be made by recovering the fuel cell stack's waste heat. In other words, the process lends itself to a certain degree of process integration, something engineering companies in this field have become expert in. Also, the DC current produced by the fuel cell system has to be, in many cases, converted to AC power. In very large capacity systems, extra heat generated either in the reformer or in the fuel cell, can sometimes be utilised to produce more electric power by means of a bottoming cycle (22), resulting in a higher total system efficiency (4).

Aside from the fuel cell stack, the steam reformer is the most important reactor in the total plant. In the reforming process, catalytic conversion of hydrocarbons with steam to hydrogen and carbon monoxide occurs. This process, especially in small hydrogen plants, requires a very specific knowledge of materials. The synthesis gas that comes from the reformer contains a lot of carbon monoxide (typically 10%), as well as some impurities. For application in alkaline fuel cells, the carbon oxides must be removed so that at least 99.98% pure hydrogen is obtained. For use in phosphoric acid cells (23) and solid polymer cells (24), the carbon monoxide will have to be decreased in shift reactors to levels of <1 to 2 percent of volume. Higher concentrations of carbon monoxide can be tolerated in molten carbonate fuel cell stacks.

The fuel processor also contains heat exchangers, a control system, equipment for desulfurisation, start-up, further purification steps of the fuel, water treatment and a cooling system for the fuel cell stack. This cooling system should be integrated with the heat exchangers network. In principle, countless variations of flowsheets, each with their own efficiency and operational characteristics, are possible (25).

A systematic design approach is necessary to achieve an acceptably performing unit. Different criteria for optimisation are chosen for different systems. For example, a small 25 kW phosphoric acid fuel cell system needs to be designed with an acceptable efficiency, aiming at a minimum amount of equipment (only 6 heat exchangers are possible). A 100 MW unit, on the contrary, will benefit from achieving slightly higher efficiencies just by installing many more heat exchangers. In addition, such a large capacity unit can benefit economically from use of a bottoming cycle. A systematic optimisation will be based on a trade-off between the number of units, cost of exchanger surface area, and electrical efficiency.

The importance of process integration is indicated in Table 3. Total system efficiency is the product of the efficiencies of the fuel cell stack itself, of the electrical inverter, and of the fuel processor (mostly a reformer) multiplied by a factor for system integration. The latter combination of the integrated fuel processor results in a so-called "design efficiency". As Table 3 shows, PAFC stack efficiency can be expected to vary from 40% to 45%. Current stacks already produce 40% efficiency, so future R&D can only be expected to increase it to 45% – a relative gain of only 12%. The electrical inverter needs only limited development (especially for small inverters due to special requirements for low voltage stacks); its efficiency can be expected to vary from 90 to 99%, mainly dependent upon the capacity of the system – a relative variation of only 10%. Most significantly, however, the total design efficiency can vary from about 70% to more than 110% (LHV) – a relative variation of 57%.

Table 3. Sensitivity of contributing factors to system efficiency.

System Efficiency	=	Fuel Cell Efficiency	*	Inverter Efficiency	*	Fuel Cell Processor Eff.	*	System Integration Efficiency
η	=	η_{fc}	*	$\eta_{inverter}$	*			η_{design}
PAFC	0.25–0.49	0.40–0.45 12% variation		0.90–0.99 10% variation				0.70–1.10 57% variation
MCFC	0.32–0.74	0.50–0.60 20% variation		0.90–0.99 10% variation				0.70–1.25 80% variation
		Dependent on Stack R&D		Dependent on Capacity				Dependent on Fuel C/H Ratio and System Design

In the case of MCFC, the efficiency variations are more pronounced. As Table 3 shows, the MCFC stacks have a relative efficiency variation of 20% and the inverter a relative efficiency variation of 10%, but the relative variation that can occur due to process design and system integration is far more significant – typically 80%. Engineering contractors have an excellent opportunity in fuel cell systems development for this very reason, and many system studies and fuel processor developments have been carried out to this end (4, 26, 27, 28, 29, 30, 31).

Figure 3 shows an example flowsheet of a 250 kW PAFC power plant designed to optimise process integration. Natural gas after compression is desulfurised and sent to the reformer (a so-called regenerative reformer which contains an internal heat exchanger), and then the reformed gas is sent to the shift reactors, after which part of the water is removed and the gas flows into the fuel cell anode compartment. The spent anode gas, still containing some hydrogen, carbon monoxide and traces of methane, is used in the burner(s) of the reformer, after which the exhaust gas is expanded and sent to the stack. Further, air is compressed and separated into streams to the fuel cell cathode and to the burners. The cathode exhaust air is sent through an expander before venting. Although extremely simple, this flowsheet still contains approximately ten heat exchangers to make optimum use of the fuel cell stack cooling for all steam demands of the system.

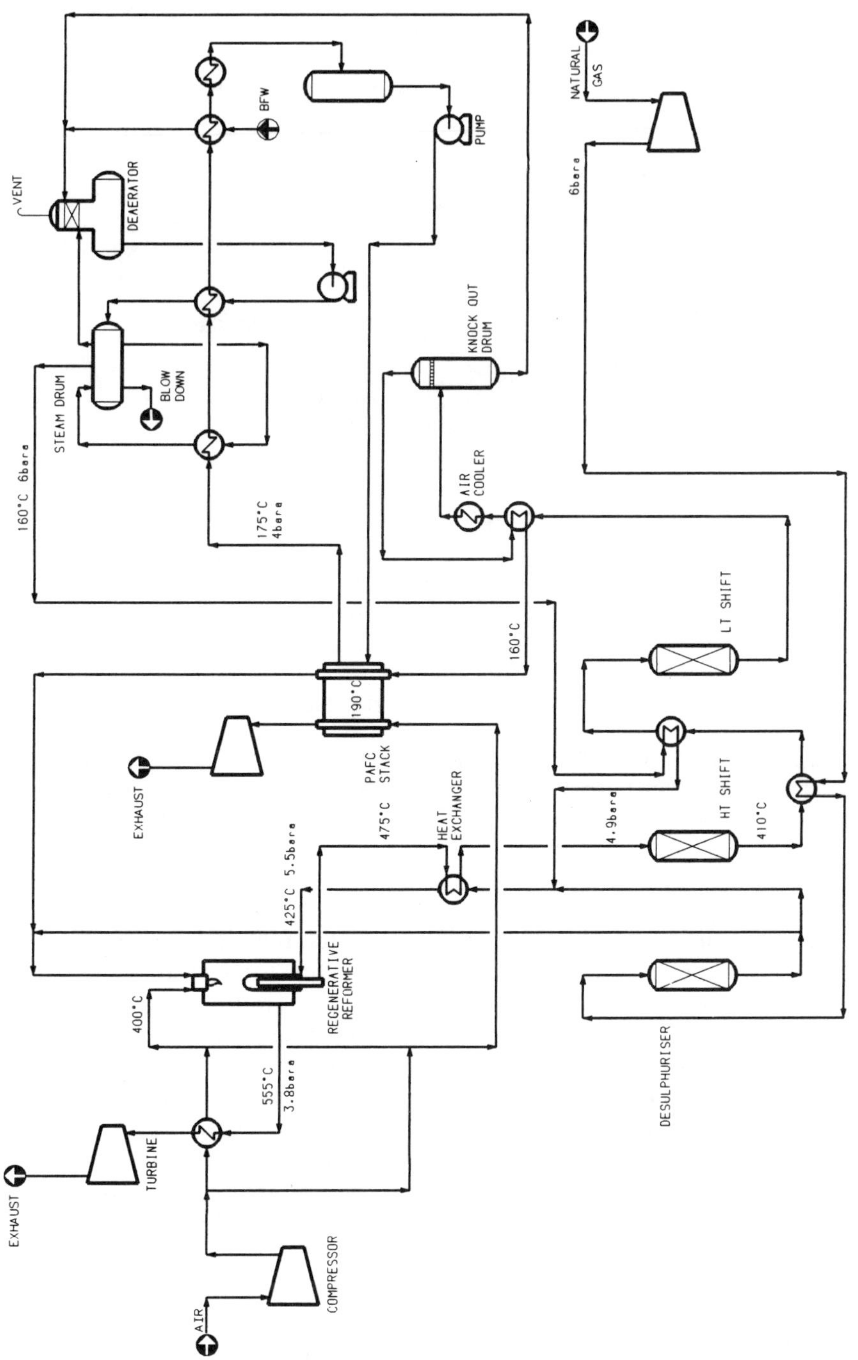

Figure 3. Flowsheet of a 25 kW PAFC power plant.

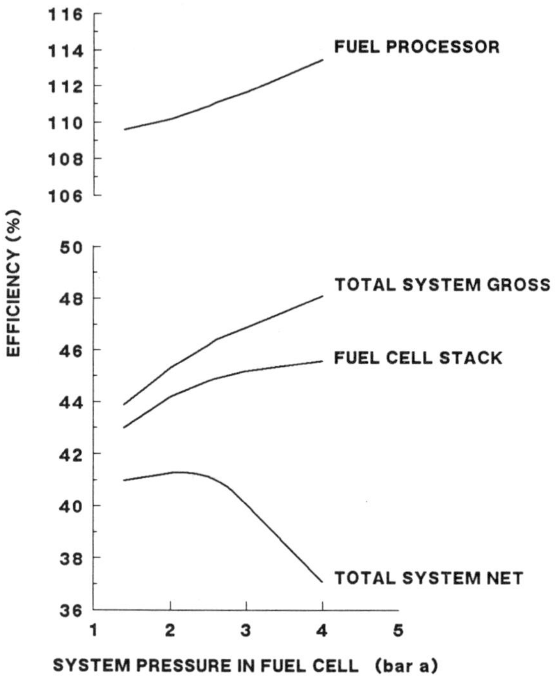

Figure 4. Efficiencies in a fuel cell system (based on lower heating value of total feed to the system). The curves refer to a 25 kW Phosphoric acid fuel cell system with reforming temperature 700°C, radiant efficiency 67% and steam to carbon ratio 6.5.

Many parameters influence the efficiency of a fuel cell. Figure 4 shows how changes in just one parameter – pressure in the fuel cell – affect the efficiency of the fuel processor, fuel cell stack and the total system. Keeping other parameters constant (such as reforming temperatures, reformer geometry, steam-to-carbon ratio), the pressure in the fuel cell itself (and that in the furnace) was varied from atmospheric to about 4 bar. In this case, a 25 kW PAFC system was used (of a slightly different design than in Figure 3), with fewer heat exchangers and only one shift reactor. Current PAFC performance shows a maximum efficiency of about 45.6% at a pressure of 4 bar(a). The fuel processor efficiency, at increased system pressure, increases from 109 to 114% (defined as lower heating value of anode feed divided by total feed and fuel to the fuel processor). The total system efficiency is the product of the fuel processor, fuel cell, and inverter efficiencies (93%), and it increases as a function of pressure from 44 to above 48%. However, after subtracting parasitic power demands, the net system efficiency optimises at about 41.4%, at a pressure of 2.5 bar(a). The reason the net efficiency drops rapidly as fuel

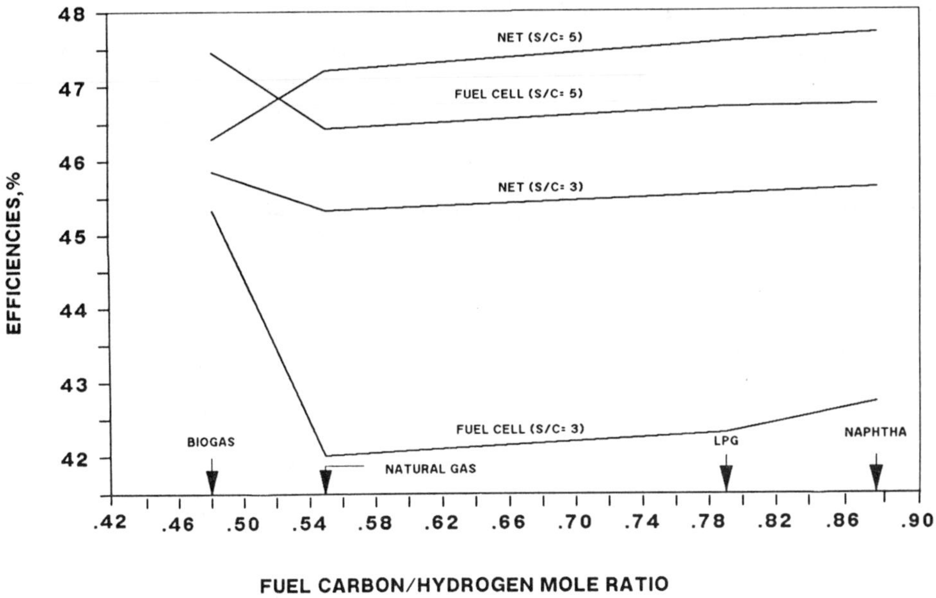

Figure 5. 250 kW Phosphoric acid fuel cell: Effect of varying fuels on efficiencies. The pressure is 4 bar (a), reforming temperature 750°C, and inverter efficiency 96%. The calculations have been done for two steam-to-carbon ratios (s/c) in mol per mol.

cell operating pressure increases beyond 2.5 bar(a) is that to avoid condensation, the expander outlet pressure has to be raised and thus not all the energy in the expanding gases can be recovered. In other words, the overall system operates more efficiently at a different optimum pressure than the fuel cell itself. This underscores the importance of process integration in optimising the system's performance.

4.2 Feedstocks

The effect of changing the type of fuel to a 250 kW PAFC power plant is illustrated in Figure 5. The calculations have been performed for two steam-to-carbon ratios. Although the fuel cell efficiency is strongly dependent (especially at low steam-to-carbon ratio) on the actual feed in use, the net system efficiency after individual optimisation of each particular design is much less dependent on the steam-to-carbon-ratio. As a consequence, one can conclude that the economics of fuel cell systems for different fuels will depend more upon the price of the plant and the price of the particular fuel, rather than of the system performance as such. At current prices, biogas and natural gas have the advantage (and especially biogas in countries where it is cheaply available) over more expensive fuels such as LPG or naphtha.

4.3 Effect of Scale

In general, increased capacity of a fuel cell system allows for more complicated equipment to squeeze out a little extra efficiency at the expense of increased investment. However, the effect of capacity on efficiency is not great. Of more importance are the fixed and variable costs of the plant, which will be illustrated in next section.

5 Commercial Aspects

5.1 System Economics

Based on optimised designs, it is possible to make detailed cost estimates of the fuel cell plants. This can be done by combining actual manufacturers' quotations, cost estimating techniques used in engineering offices, and database information. Table 4 shows a cost price breakdown of 250 kW PAFC units. (see reference 4 for extensive economic data analysis). Costs are based upon 1987 US dollars. Materials and erection costs have been specified for all equipment, and engineering is shown. Costs are included for the fifth unit in a First Series of five (FS), the twentieth unit in a series of hundred (LVP) and the 200th unit in a series of one thousand (HVP).

As can be seen in Table 4, installed costs decrease sharply with volume production. However, several assumptions had to be made, such as the costs for actual production of the phosphoric acid fuel cell stacks, the costs for the power conditioner and the number of engineering hours. Moreover, cost reduction factors had to be determined for larger series, taking into account an 85% learning curve for the reduction of direct labour cost components in the construction and installation of the plants.

Table 4. Cost breakdown for 250 kW PAFC system
Cell type = PAFC Unit size = 250 kW (nominal)

Cost basis: 1987 (US$)	First Series (5th unit) of 5		Low Volume Production (20th unit) of 100		High Volume Production (200th unit) of 1000	
	mat. (a)	erect. (b)	mat. (a)	erect. (b)	mat. (a)	erect. (b)
total fuel proc. equip.	181.310	12.760	137.780	11.480	96.520	10.210
fuel cell stacks	184.000	12.760	85.870	11.480	61.330	10.210
piping	51.520	73.600	41.220	53.180	30.910	30.910
instrumentation	146.960	37.780	88.100	27.280	54.760	15.850
power conditioner	110.400	7.730	79.740	5.580	61.330	4.290
elect./insul./paint	39.250	43.670	35.330	31.550	29.440	18.350
engineering	–	223.260	–	31.890	–	5.100
sub-total	713.440	411.560	468.040	172.440	334.290	94.920
total of a+b	1.125.000		640.480		429.210	
contingency/profit	112.500		32.020		8.580	
Grand total (installed plant cost)	1.237.500		672.500		437.790	
installed cost per kW	4.400		2.390		1.560	

additional information: actual capacity = 281 kW; overall efficiency= 47.1% (LHV)

Table 5 contains a summary of the assumptions that have been used in the economic investment cost calculations for several different sizes of phosphoric acid and molten carbonate systems ranging from 25 kW to 100 MW. The table shows assumptions on the number of units to be produced, directly

Table 5. Assumptions for economic investment cost calculations.

Number of Units					
Capacity kW	First Unit FU	First Series FS	Low Volume Production, LVP	High Volume Production, HVP	Robotised Production, RP
25	–	5	20	200	2000
250	–	5	20	200	–
3,250	–	5	20	100	–
100,000	1	5	20	–	–

PAFC Stack Costs, US$/kW				
Stack Size	FS	LVP	HVP	RP
25 kW	980	490	340	200
250 kW	740	340	250	–

MCFC Stack Costs, US$/kW					
Stack Size	FU	FS	LVP	HVP	RP
25 kW	–	880	390	250	150
250 kW	–	640	250	150	–
1000 kW	340	150	100	–	–

Power Conditioner Costs, US$/kW					
Number	Capacity, kW	25	250	3200	10,000[a]
5		830	440	170	100
50		490	320	120	90
500		390	250	100	–

Engineering Hours (@ 64 US$/h)					
Capacity, kW	FU	FS	LVP	HVP	RP
25	–	2000	400	50	(490 US$/unit)
250	–	3500	500	80	–
3,250	–	7000	1700	450	–
100,000	60000	28000	22000	–	–

[a] for the 100 MW plant, 10 units of 10 MW are to be used.

Table continued on the following page.

Table 5. (cont'd)

Cost Reduction Factors

Capacity, kW[a]	Total Fuel Processor Equipment		Piping		Instrumentation		Electrical/insulation/ painting	
	M	E	M	E	M	E	M	E
25[d]								
LVP	0.5	0.5	0.6	0.5	0.8[b]	0.4	0.75	0.5
HVP	0.16	0.16	0.2	0.125	0.4[c]	0.2	0.5	0.2
RP	0.1	0.05	0.1	0.05	0.1	0.05	0.1	0.05
250[d]								
LVP	0.76	0.9	0.8	0.723	0.9[b]	0.723	0.9	0.723
HVP	0.53	0.8	0.6	0.42	0.8[c]	0.42	0.75	0.42
3,250[d]								
LVP	0.845	0.9	0.8	0.723	0.9	0.723	0.9	0.723
HVP	0.69	0.8	0.6	0.5	0.8	0.5	0.75	0.5
100,000[e]								
FS	0.85	0.9	0.8	0.686	0.9	0.686	0.9	0.686
LVP	0.7	0.8	0.6	0.5	0.8	0.5	0.75	0.5

M = Materials E = Erection

[a] the production rates relate to the first unit for the 100 MW case, and to the first series for the others

[b] remove gas analysers

[c] remove PC as well

[d] with respect to FS

[e] with respect to FU

Source: Reference 4

reflecting reasonable production series with respect to the size of the potentially accessible markets in the short term, as well as PAFC and MCFC stack costs for different stack sizes that will likely be used in the different plants. PAFC stack costs are based on actual current developments in the USA, where the cheapest stacks are currently being produced. MCFC stack costs are extrapolations, based on the current status of PAFC stack costs, and corrected for the cheaper materials in the MCFC (nickel rather than platinum catalysts, though the nickel price is unstable), and assuming similar stack construction problems for both fuel cell types. Table 5 also contains assumptions on the power conditioning costs (32), reflecting reasonable estimations of the cost and cost reductions that can be expected for these power conditioning systems. Table 5 includes cost estimates for the number of engineering hours, as estimated by an engineering company (4). Finally, the table contains the cost reduction factors, in material and erection, for the different production volumes of all system sizes.

Using the standard EPRI "TAG methodology" (33) for calculating the cost of electricity (COE), systematic calculations have been done. Table 6 contains the assumptions for these calculations, that vary of course as a function of size. Based on these assumptions, levelised cost of electricity have been calculated for all different systems, and Figure 6 contains an example for the 250 kW phosphoric acid fuel cell systems. The figure contains the COE as a function of production volume for PAFC power plants, but also includes similar bars for molten carbonate power plants on natural gas and on naphtha feed. It is evident

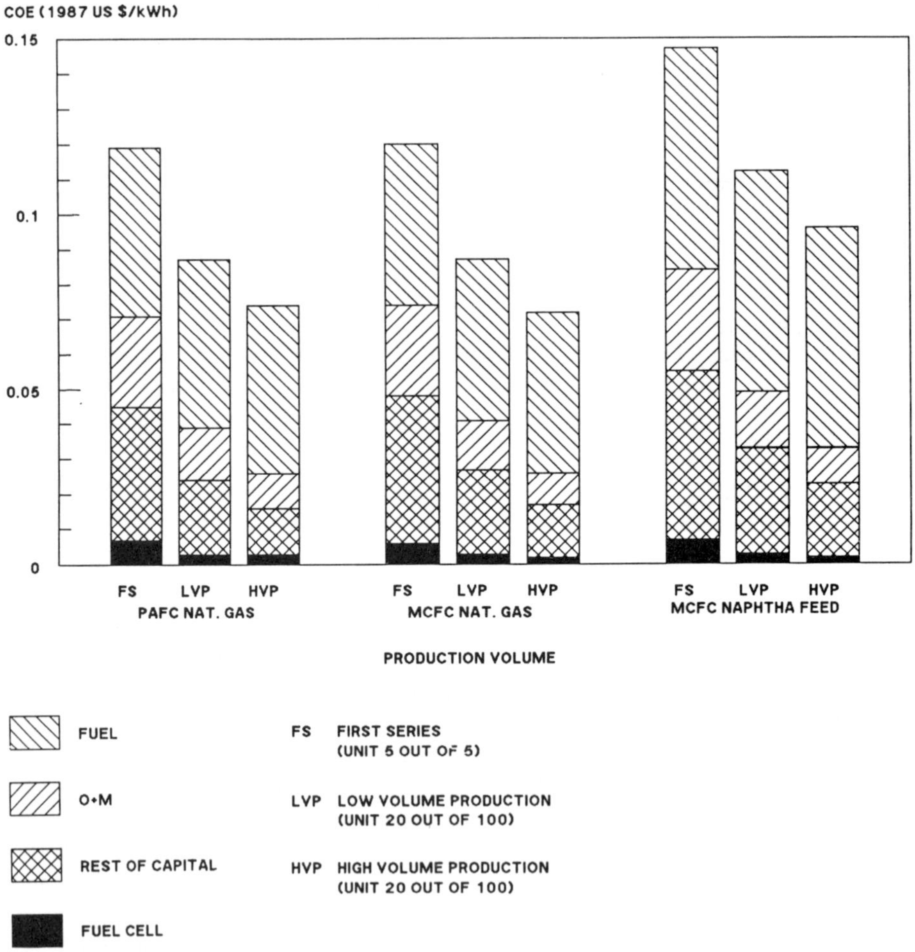

Figure 6. Cost of electricity for 250 kW FC power plants. The assumed plant life is 20 years, gas price 6.2 and naphta price 8.8 US$/GJ and the operating time is 8592 hours per year.

from this figure that the PAFC and MCFC systems' COE at this capacity are very similar. Though MCFC systems have a slightly lower energy cost because of the slightly higher efficiency, this is offset by slightly higher capital requirements. This is an important observation because, although the stacks for MCFC are cheaper (nickel instead of platinum), the MCFC system contains more rotating equipment which means a significantly higher investment requirement. It is also evident from this figure that naphtha-fuel of plants have a significantly higher overall COE because of the higher cost of naphtha feedstock, despite the higher system efficiency with this fuel.

Figure 7 shows the sensitivity of the COE for the 250 kW MCFC system at high volume production to variations in various parameters. This analysis shows that gas price has a very strong influence on the economic calculation. Capital charges as well as carrying charge rate also have a considerable influence, though of a second order effect only. Other important effects come from the system capacity and electrical efficiency. The COE is less sensitive, however, to changes in the operational and maintenance costs.

For the purpose of this assessment, some comparative results, based on the same assumptions as cited in Tables 5 and 6, are provided in Figures 8 through 11.

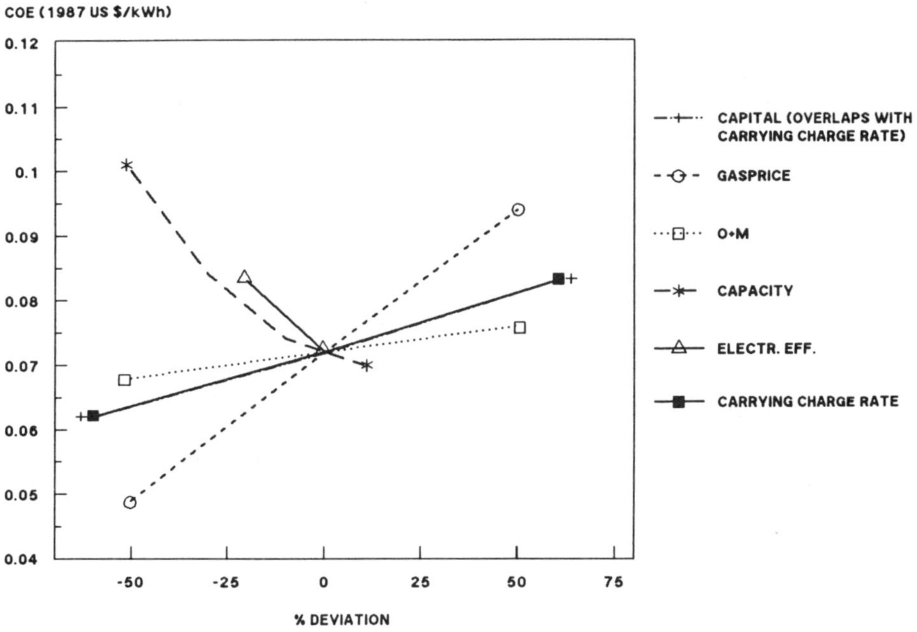

Figure 7. Sensitivity of cost of electricity to variations in various parameters at high volume. Production 250 kW MCFC.

Table 6. Assumptions for cost of electricity calculations.

Inflation rate	0%
Real discount rate	6%
Income tax rate	0%
Properly tax and insurance rate	0%
Investment tax credit	0%
Straight line depreciation assumed	

booklife (yrs)	levelised carrying charge	levelising factor for fuel, O&M costs
15 (25 kW)	14.35% of Investment/yr	1.210
20 (250 kW; 3.25 MW)	12.85% of Investment/yr	1.264
30 (100 MW)	11.55% of Investment/yr	1.344

System	Gas price (US$/GJ LHV)	Service time (hrs p.a.)	Operators
25 kW	7.8 excl VAT	8592	1pp, 1hr/wk
250 kW	6.3	8592	1pp,1hr/wk
3.25 MW	5.7	8000	1pp,3 shifts
100 MW	5.4	8000	3pp, 3 shifts

Labour costs	20 US$/hr
Naphtha costs	9 US$/GJ
Boiler feed water	1.35 US$/ton
Cooling water	2.5 UScts/ton

Table 7. Data used for conventional technologies.

Power output (MW)	type of plant	capital cost per unit power (1000$/kW)	unit fuel cost ($/GJ(LHV))	overall electrical efficiency based on LHV of fuel (%)
3.25[a]	GE	1.23	5.7	33
	GT	0.74–1.03	5.7	25
	STIG[c]	0.90	5.7	37
600[b]	NGPP	0.73	2.0	40
	CCPP-NG	0.49	2.0	50
	CPP	1.36	1.6	38
	CCPP-CG	1.50	1.6	42

[a] reference 4

[b] reference 34

[c] Steam Injected Gas Turbine. For other abbrevations, see Figure 9a.

Source: reference 37.

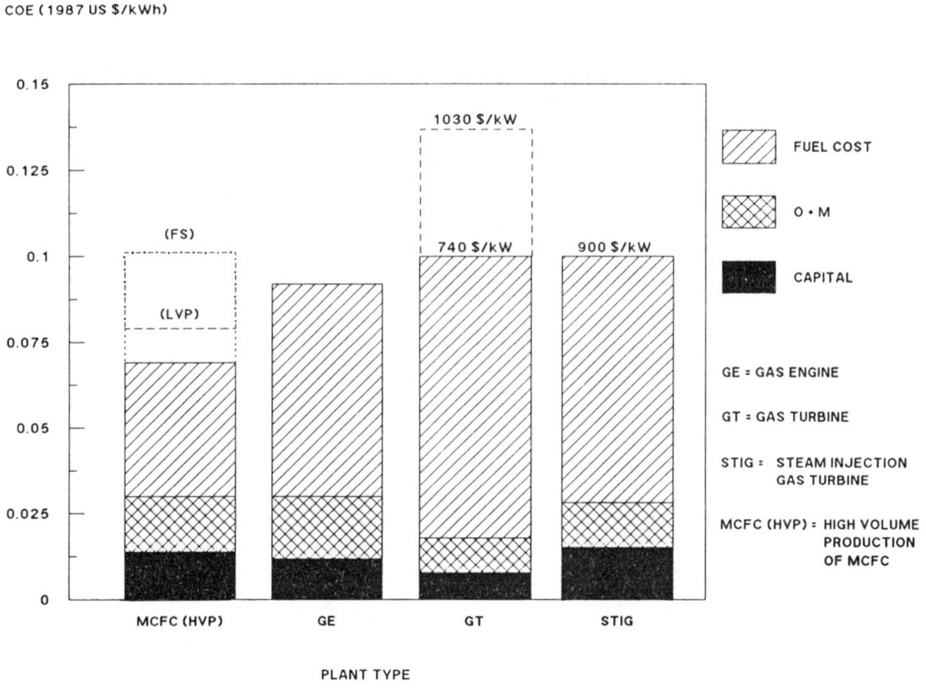

Figure 8. Comparing fuel cells with other technologies: 3.25 MW power plant.

Figure 8 shows the comparison of the COE for a 3.25 MW power plant and its current competitors, mainly gas engines and standard gas turbines. Cost and efficiency assumptions for these competitive technologies are given in Table 7. Although the capital cost of the molten carbonate fuel cell systems, even in high volume production, is higher than the gas engine and the gas turbine and the operation and maintenance are not expected to be much different, the much lower fuel cost because of the higher efficiency make even current performance MCFC systems look very competitive. Even low volume production units are more attractive than the competing technologies, and the first series of units already has a competitive chance. Also, it should be kept in mind that no credit has been given to the better environmental performance of the MCFC unit, which further increases its competitiveness versus the gas engine and gas turbine systems.

Figure 8 does also contains a comparison with a so-called "STIG-unit" (*ST*eam *Injection G*as turbine) (37). Despite the low investment costs, its COE are higher than for larger series of MCFC units.

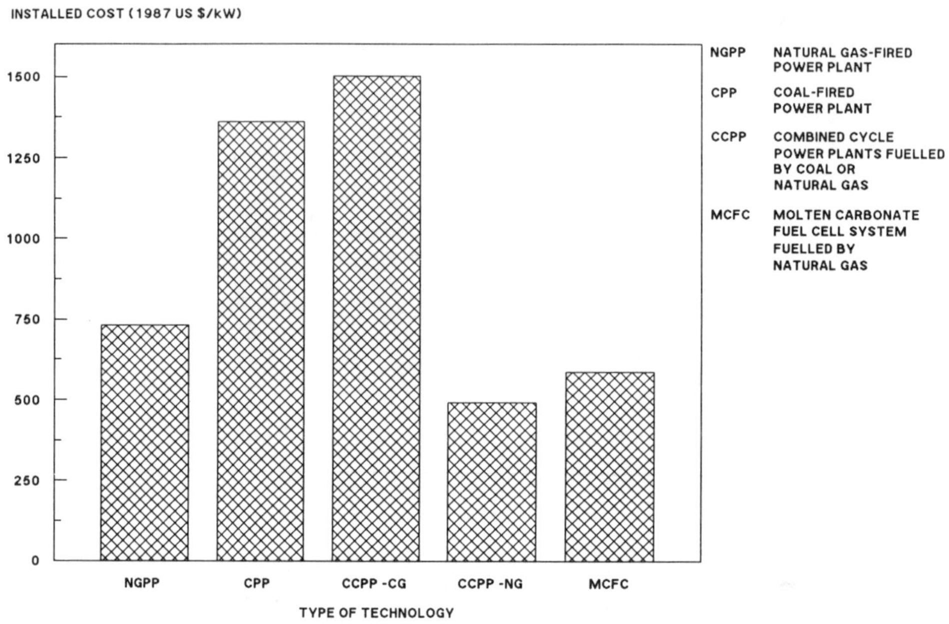

Figure 9a. Unit capital cost for 600 MW power plants.

Figures 9a and 9b show the unit capital costs for 600 MW power plants and the COE in those plants respectively, for different types of technology. These include natural gas power plants, coal-fired power plants, combined cycle power plants based on natural gas and coal gasification, and MCFC in a production series of 20 units of 100 MW each (no further cost reduction for a 600 MW unit has been assumed versus 6 units of 100 MW each). The cost and efficiency assumptions for the other competitive technologies are given in Table 7 and reference 34. These figures clearly show that both investment-wise and in terms of the projected COE, the MCFC power plants are already quite competitive. However, the current technology level does not yet permit large production volumes of stacks, which is clearly the determining step for the economy.

Figure 10 summarises the levelised COE for power production for several sizes, production series of both MCFC and PAFC systems, including values for the competitors. In the 25 kW range, the production series of both cell types have been compared to gas engines and battery systems (uninteruptable power supply), for 250 kW and 3.25 MW systems the gas engine and gas turbines are included as competitive technologies. Conclusions from this figure are clear cut: small systems, high volume production and robotised production

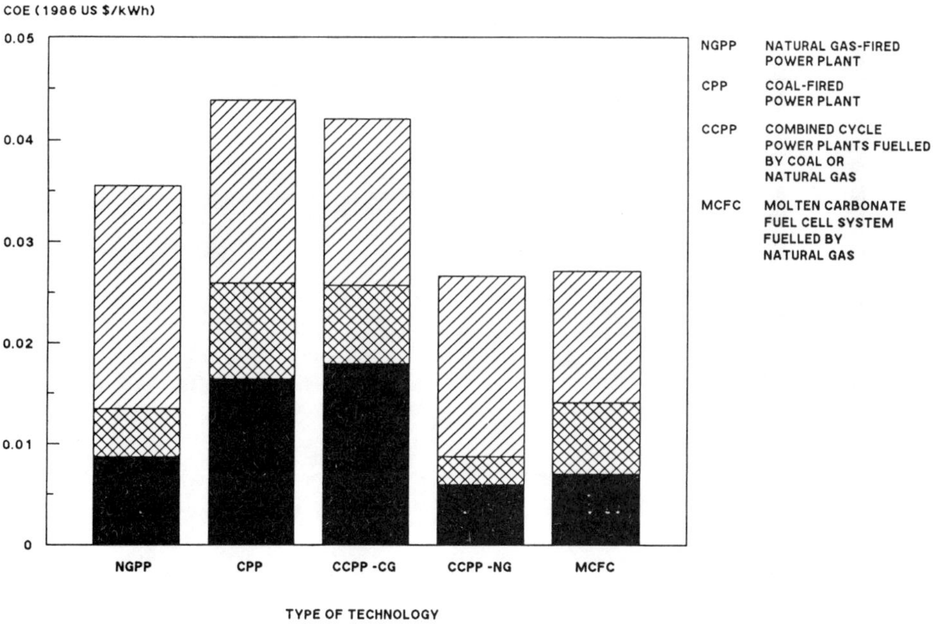

Figure 9b. Cost of electricity for 600 MW power plants.

make fuel cell systems competitive, but at large sizes already the first series and certainly the low volume production series are competitive. Since cost estimates in Figure 10 are based on consistent assumptions (Tables 5 and 6) and were arrived at using standard 1987 cost estimating practices from engineering firms working in the market, these costs are realistic. Even if stack costs would, for instance, be twice as high as estimated, or power conditioner costs would be higher than anticipated, this would have negligible influences on the COE values indicated.

It is clear that fuel processor costs, as compared to fuel cell stack costs, are the primary contributor to the COE (see, for example, Table 4), and fuel costs, which are related to energy price in general, are a secondary contribution. Of course the higher the energy prices, the more favourably fuel cell technology (with higher efficiency) will compare to conventional forms of technology. Also, in low energy price scenarios fuel cell technology will show certain benefits for most applications due to the higher efficiencies of the plant, and due to the additional advantages of fuel cell technology.

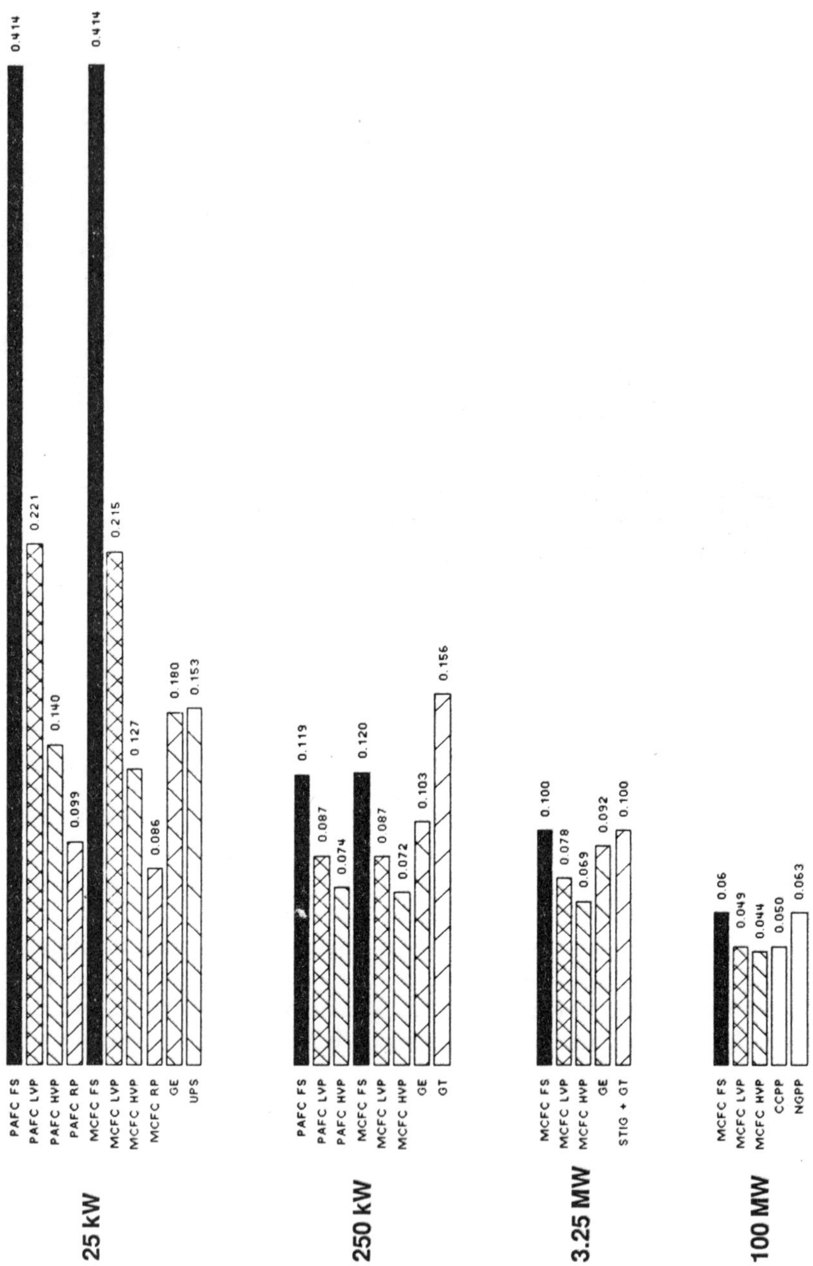

Figure 10. Cost of electricity for power production only. Comparison fuel cell and other technologies, abbrevations and assumptions as outlined in earlier Figure, COE in 1987 US $/kWh.

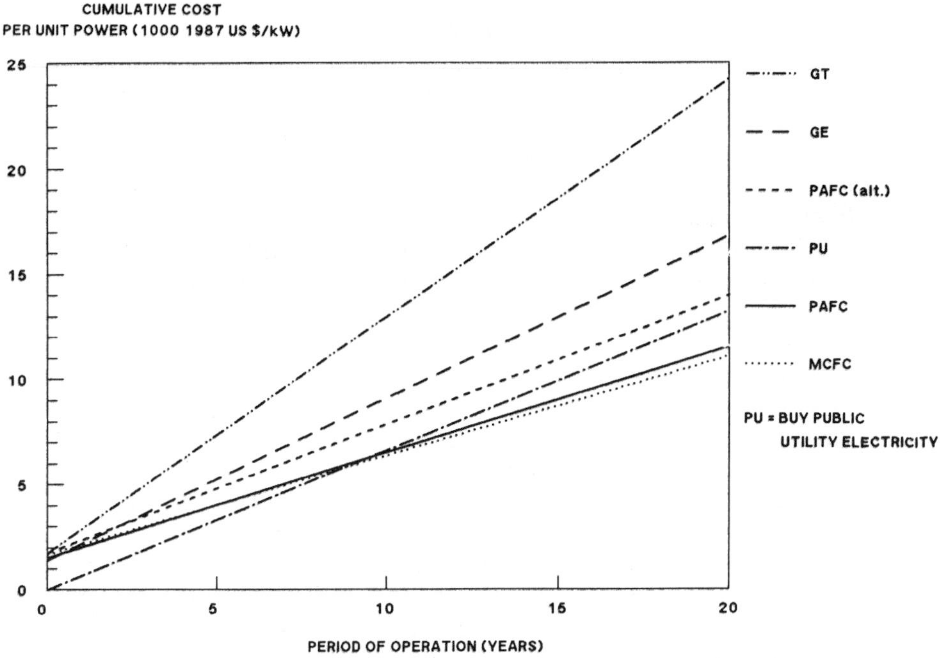

Figure 11. Simple cumulative cost plot for 250 kW power plants over a 20 year period and for buying public utility electricity. Full year operation is assumed.

Figure 11 presents a different way of looking at the cost of fuel cell and competing systems, by showing simple cumulative costs per unit power produced over a longer term period. The figure shows these cost calculated over a period of 20 years, and a comparison is made of the optimised PAFC and MCFC systems, utility electricity, gas engine and the gas turbine. An alternative PAFC design is also included, and the figure clearly demonstrates that both the MCFC and PAFC systems are much cheaper over the period of operation, even when compared to normally purchased public utility electricity. The widely used gas or diesel engine for this capacity range, and therefore the main competitor, is over 30% more expensive throughout the 20 year period than the PAFC and MCFC systems. It should be stressed that both the COE calculations and the life cycle cost calculations are much more realistic means of evaluating new technologies than the often mentioned investment cost per kW of electric power produced.

Fuel cell technology will become even more competitive than shown in Figures 8 through 11 if future improvements in system performance are taken into account. If better catalysts and better materials are used, not only will the stack price go down (which as noted earlier has only a slight effect on the

COE), but the efficiency of the system will improve as long as a proper design is being used in combination with an improved stack performance. However, improvements in competitive technologies as well, which will lead to higher efficiencies for instance in gas turbine generators, may (partly) offset the cost reductions of fuel cell improvements.

Finally, stricter and more stringent environmental regulations will increase the cost of competitive technologies, but not the fuel cell system cost predictions presented here.

In short (cf. Figure 10), fuel cell technology is competitive in most size ranges, but for smaller sizes (below MW scale) only in serial production.

5.2 Operation Scenarios

As is clear from Table 6, the assumptions used in the previous section have been based on almost continuous operation (with the exception of scheduled maintenance) throughout the year for all systems. This would suggest, however, limitation of the use of fuel cell systems to applications where full-time power use at maximum capacity is necessary. Although this is not realistic for many applications, initial segments in the fuel cell market will have to be found that utilise as much as possible the available potential of the new fuel cell unit, since such a unit will give the highest efficiency for electric power production. In designing a system, it is therefore logical to try to use the fuel cell system for as many hours as possible, using existing conventional plants for peak load situations, etc. Of course, one may reverse this scenario, and use a completely different approach, one based on the fact that fuel cells have excellent part-load efficiency. Within the scope of this paper, it is impossible to credit all the different scenarios that can result from such an approach.

However, one should consider other practical implications as well. As long as public utilities do not charge extra high prices for purchased electricity by a private consumer who has his own power generator, it becomes attractive for the consumer to produce as much power as possible by himself. Such a client will preferably use a fuel cell unit for bottom peak demand, and purchase only electricity in cases of failure of the fuel cell unit, or in cases of peak power use that is larger than the installed capacity allows for. This also implies almost free back-up power in, e.g., a computer centre, hospital, hotel, etc.

5.3 Catalyst Availability Issues

The catalyst availability issue arises mainly in the instances of platinum for phosphoric acid fuel cell stacks and nickel as well as nickel compounds for the MCFC fuel cell stacks. Moreover, reformer catalysts containing nickel compounds are also important. Current nickel price fluctuations are of influence, though it is clear from Section 5.1 that these prices will have a negligible effect on the total cost price of the systems, and certainly on the cost of produced electricity. Even in the case of platinum, the effect of increased prices may be negligible, though availability of platinum may certainly be a factor of influence. For example, current phosphoric acid fuel cells have a platinum loading of about 5 grams per kW electricity, and this may decrease by 50% or more in the future. One of the world market leaders in platinum supply (Johnson Matthey) has predicted that increased use of fuel cells might double the demand for platinum in the world (35). However, a lot will depend on recycling opportunities, and studies have already indicated that at least 95% of the platinum within fuel cell systems could be recycled (36).

5.4 Applications and Markets

Up until now, the commercialisation of fuel cell systems has not proceeded as fast as originally anticipated, mainly due to political and financial problems, as well as delays in the technical development of fuel cell stacks. This slower-than-anticipated development is partly a consequence of the fact that the largest private interests in fuel cell system research, development and demonstration are utility companies, and they have been interested mainly in large systems for central power generation, and consequently they go through a very expensive learning curve which in fact has been prohibitive to demonstrate a size larger than 10 MW until now. Only recently some have decided that they also are interested in more de-centralised, small-scale power generating applications. Moreover, due to the large number of problems with peripheral components in the demonstration projects to date, fuel cell systems have not yet inspired confidence in utility policy makers, i.e. their reliability has been suspect. A strong change in this respect is expected in the near future, as demonstrations will concentrate on proving reliability and secondary advantages rather than the already proven basic concepts of clean technology and high efficiency.

IFC has made the most ambitious effort to-date to commercialise large fuel cell systems. Through its "PC-23 program", IFC marketed an 11 MW system that allowed full load variations in less than ten seconds and used natural gas and naphtha almost simultaneously. However, the sales price of these systems turned out to be a stumbling block, which necessitated IFC to choose a new financing arrangement – selling to a consortium of several utilities rather than to a single utility. Since the initial costs involved were too high, even this has led to a very limited committment up to the present date (one demo only). Moreover, many companies in the world have come to the conclusion that the smaller system units can be commercialised earlier, due to the greater cost

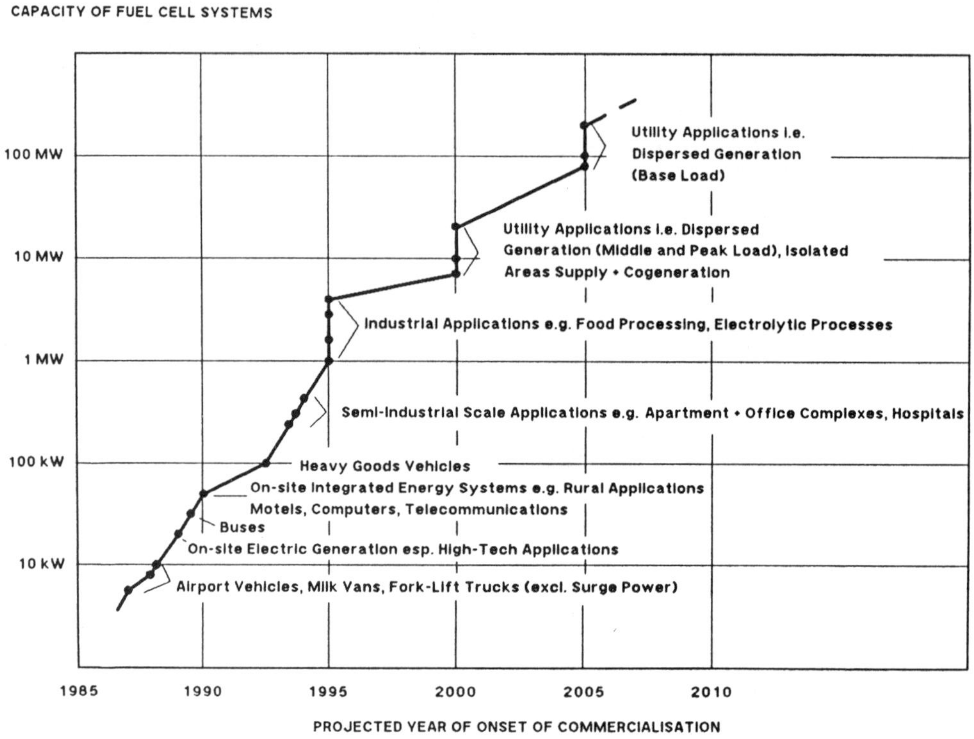

Figure 12. Fuel cell system commercialisation expectations.

reductions possible with mass production. This conclusion is valid mainly for PAFC systems. Alkaline fuel cell systems, which until now have in most cases run on pure hydrogen, mainly have been developed and applied for small scale applications, such as space and transportation. Fuel processing technology, either from methanol, ammonia or from fossil fuels, is commercially available, and this may make AFC systems competitive in larger scale applications as well. The other types of fuel cell are not yet at a stage to allow commercialisation.

Figure 12 shows the expected commercialisation of fuel cell systems in different sizes over the next few decades. Though it is difficult to pinpoint the precise onset of commercialisation and many companies do have different views with respect to their application of fuel cell systems, the figure suggests that every three to five years on the ladder of commercialisation system capacity will increase by about ten times. This type of learning curve makes it plausible that the small scale applications will be the first to be commercially available.

The first commercial system may be the 5.6 kW forklift truck application, as developed by Fuji in cooperation with Engelhard Corporation. Small on-site generators (25 to 100 kW) will soon be commercialised as well; so, too, IFC's 200 kW units. In fact, more than 50 of these 200 kW units are already on order. Markets will mainly be in computer systems, for baseload power delivery in hotels, office buildings, etc., and for various mobile applications, such as airport vehicles, golf cars, etc. There will also be a market for primary electricity generation at locations where power failure is critical. A few years later, industrial applications, especially for cogeneration, will become major markets; followed by larger scale systems for localised power plants and larger industrial applications.

As previously noted, societal concerns about the environmentally-damaging emissions of conventional electricity-generating technologies as well as their relatively inefficient use of finite fossil fuel supplies could greatly enhance the market penetration of fuel cell systems in the coming years. In addition, market introduction of fuel cell systems in areas where the system can be simplified – e.g. where pure hydrogen, pure oxygen or a combination is available, or where DC can be used instead of AC – will give these systems a real boost. Once it can be shown that sufficient reliability of a total system is achievable, then the market penetration of fuel cell will grow significantly, even at current system costs. Availability of stacks in sufficient quantities at realistic prices will be necessary for the commercialisation of larger systems. When serial production of smaller systems starts, this will not provide additional problems.

6 Conclusions on Fuel Cell Technology Prospects

Here are some overall conclusions on the state-of-the-art of fuel cell technology and the future prospects of this technology in different applications.

- Over the long run, the economics of fuel cell systems will depend mainly on the cost for hydrogen producing technologies, that will be integrated with fuel cells in most applications, and, to a much less extent, on the prices of the fuel cell stacks. As a consequence, R + D on fuel cell systems should focus primarily on the reliability and efficiency of system operation, and less exclusively on cost reduction of the stacks.

- For fuel cell systems, an economy of scale is clearly evident: the smaller the capacity, the higher the cost of electricity produced. However, comparable systems have the same disadvantage, and the comparison with competitors shows an advantage for fuel cell systems (at least PAFC and MCFC, but similar studies can also be done for AFC, SOFC, etc.).
- Small systems (25 to 500 kW) are expected to be commercial soonest, and they will most readily benefit from cost reduction due to serial production. Conversely, for larger scale power plants, the fuel costs are more crucial to economics.
- Small fuel cell systems should be built with fewest and simplest items of equipment available. Large systems can benefit from more complicated flowsheet design, in order to achieve a slightly higher efficiency as a bonus.
- An increase in gas prices will affect the cost of electricity produced by fuel cell systems less than that of the conventional technologies. This is because of the higher efficiency and, hence, lower fuel consumption of fuel cell systems. This will even increase competitiveness of fuel cell technology.
- Extremely low NO_x emissions – probably below 10 ppm with modern burners for PAFC and even below 1 ppm for MCFC – are expected. Environmental advantages may provide the impetus for fuel cell technology's big market breakthrough, especially in conjunction with their easily attainable competitive cost of produced electricity.

References

1. Penner, S.S., ed. 1986. Assessment of research needs for advanced fuel cells. *Energy, The International Journal*, II (1/2).
2. Poly-energie zakboekje. 1986. p. 931.
3. Mullick S., Pietrogrande P., Warren D. 1986. *Performance evaluation for a 11 MW fuel cell power plant which utilises a generic landfill gas feedstock.* KTI Corp. report prepared for EPRI.
4. Mugerwa, M.N. et. al. 1988. *Fuel cell systems.* Design optimisation and environmental aspects of fuel cell systems. KTI BV report prepared for PEO, Utrecht.
5. Blomen, L.J.M.J. et. al. 1987. *Pressurised combustion.* Final report. KTI BV report prepared for PEO, Utrecht, App. IV.
6. Gas Research Institute. 1986. *40 kW On-site Fuel Cell Field Test Program (Jan. '82-May '86).* Final report.
7. Tokyo Gas Company. May 1987. private comm.
8. Tokyo Electric Power Company. May 1987. private comm.
9. Gas Research Institute. 1986. *40 kW On-site Fuel Cell Field Test Program (Jan. '82-May '86).* Final report.
10. McDougall, A. 1988. *Brennstoffzellen*, Technik und Zukunft der elektrochemischen Stromerzeugung.

11. Penner, S.S., ed. 1986. Assessment of research needs for advanced fuel cells. *Energy, The International Journal,* II (1/2).
12. NTIS. 1978. *Phosphoric Acid Fuel Cell Stack and System Development.* Technical progress report PC A04/MFAO1
13. United Technologies Corp. July 1985. *Molten Carbonate Fuel Cell System Verification and Scale-up.* EPRI EM-4129.
14. Exxon Enterprises Inc. 1977. *Application of the Alstrom/Exxon Alkaline Fuel Cell System to Utility Power Generation.* EPRI EM-384.
15. Penner, S.S., ed. 1986. Assessment of research needs for advanced fuel cells. *Energy, The International Journal,* II (1/2): 95–135.
16. Penner, S.S., ed. 1986. Assessment of research needs for advanced fuel cells.*Energy, The International Journal,* II (1/2): 13–94.
17. Penner S.S., ed. 1986. Assessment of research needs for advanced fuel cells. *Energy, The International Journal,* II (1/2): 153–208.
18. Penner, S.S., ed. 1986. Assessment of research needs for advanced fuel cells. *Energy, The International Journal,* II (1/2): 209–29.
19. Cameron, D.S. et al. 1987. CEC Direct Methanol Fuel Cells Research. CEC – Italian Fuel Cell workshop, Proceedings, pp. 201–12.
20. Penner, S.S., ed. 1986. Assessment of research needs for advanced fuel cells. *Energy, The International Journal,* II (1/2): 137–52.
21. Haskoning and IME Consult for the Dutch Management Office for Energy Research (PEO). 1985. Market and System Studies for Molten Carbonate Fuel Cell systems. Project No. 19.11–010.10.
22. Fluor Engineers and Constructors. 1980. *An Economic Comparison of Molten Carbonate Fuel Cells and Gas Turbines in Coal Gasification-Based Power Plants.* EPRI AP-1543.
23. Benjamin, T.G. et al. 1980. *Handbook of Fuel Cell Performance.* Institute of Gas Technology.
24. Poster presentation. 1987. *Ballard Solid Polymer Fuel Cell Development.* Fuel Cell Technology and Applications, International Seminar, The Netherlands.
25. Westinghouse Electric Corp. 1987. *Effect of Alternate Fuels on the Performance and Economics of Dispersed Fuel Cells.* EPRI EM-1936.
26. Catalytic Associates. 1977. *Assessment of Fuel Processing Alternatives for Fuel Cell Power Generation.* EPRI EM-570.
27. KTI Corp. 1980. *Assessment of Fuel Processing Systems for Dispersed Fuel Cell Power Plants.* EPRI EM-1437.
28. General Electric Co. 1981. *Fuel Cell Power Plant Integrated Systems Evaluation.* EPRI EM-1670.
29. United Technologies Corp. 1985. *Fuel Processor Developments for 11-MW Fuel Cell Power Plants.* EPRI EM-4123.
30. KTI Corp. 1985. *Assessment of Centralized Medium-sized Fuel Cell Power Plants.* EPRI EM-4332.
31. Haldor Topsoe Inc.Westinghouse Electric Corp. 1987. *Demonstration of a High-Efficiency Steam Reformer for Fuel Cell Power Plant Applications.* EPRI AP-5319.
32. Holec Nederland B.V. 1988. private comm.
33. Electric Power Research Institute (EPRI). 1982. *TAG: Technical Assessment Guide.* P-2410SR.
34. Williams, R.H. Larson, E.D. 1989. *Aeroderivative Turbines for Power.* Chapter in this publications.

35. Sabbioni F., Steel M.C.F. et. al. 1986. *Phosphoric acid fuel cell for electrical power generation: development and penetration in the EEC countries*. Final report. TESI/Johnson Matthey report prepared for Commission for the European Communities, D.Gr. XII.
36. Van Bogaert, G., Van den Broeck, H., Elenco, N.V. 1987. *Platinum and Platinum-Alloys for Alkaline Fuel Cells*. International Precious Metals Institute, 11th Proceedings, Belgium.
37. Larson, E.D., Williams, R.H. 1987. Steam-injected gas turbines. *J.Eng. for Gas Turbines and Power*. 109:55–63.

Coal Gasification Technologies for Combined Cycle Power Generation

Werner Peters

Abstract

The introduction of gas turbines to power generation heralds a major event for the electric industry. The use of coal in such plants requires conversion of coal to a clean, high pressure gaseous fuel. While gasification of coal is a very old art, the application in power plants requires substantial advances which this chapter covers in detail. The prospects for three different processes – fixed bed gasification, fluidized bed combustion and entrained flow gasification – are assessed, with emphasis on recent applications in the field, and the developments in the entrained flow gasification technology are found to be the most impressive of all. Combined cycle coal-gasification power plants are compared to conventional coal-fired plants (with desulphurization and Denox facilities) in terms of efficiency and cost. It is concluded that the commercial realization of advanced combined cycle power plants cannot be expected before the turn of the century.

1 Basic concept

1.1 Limits of Conventional Power Station Technology

The conventional coal-fired power station, operationally proven in units of up to 800 MW, realizes the simple principle of electricity generation from coal. In a single large boiler firing, all combustible coal components are quickly and directly converted into inert flue gases which are vented into the atmosphere after cooling-down and dedusting.

The problem is, of course, that these flue gases contain pollutants such as fine particulates, sulphur dioxide and nitric oxides. Almost every country in the world has moved to reduce these emissions in order to protect public health and the environment. Techniques have been developed in recent years to remove some pollutants from waste gases by filtration, but their application is raising the price of coal-fired electricity generation: the equipment for flue gas purification accounts for up to 25% of the investment costs of a power station and it reduces the plant's efficiency by 2 to 3 percentage points – 39% to 36%(1). Thus, there has been a growing interest in the development of new "clean coal technologies" for electricity generation.

1.2 Principle of the Combined Cycle Power Station

One way of attacking the problem is to reduce the pollutant carriers in the fuel prior to combustion. Measures aimed at removing the pollutants prior to combustion inevitably lead to the conversion of the solid fuel into combustible gases. In the gaseous fuel the sulphur, for example, exists mainly in form of H_2S, the nitrogen of the coal in form of NH_3, and both can be much more easily attached to adsorbents than the oxides in the flue gas. Moreover, the volume of purified gas is four to six times less than the flue gas volume of the same feed coal. In addition, because gasification processes are today operated at a pressure of several 10 bars, H_2S, NH_3, HCl etc. have to be removed from 40–150 times lower volume where the gasfication route is used. Thus, much less effort is required compared to flue gas desulphurization. Also the reduction of NO_x formation from the nitrogen in the air in the gas flame is guaranteed thanks to burner technology, so a Denox plant for the waste gas is no longer required.

Another potential benefit of converting coal into gas prior to combustion is the possibility of using gas turbines for electricity generation, at least in some cases. Gas turbine power stations are now a proven technology. Their efficiencies are up to 50% higher than the ones of conventional coal-fired generating plants, and they produce a practically pollutant-free waste gas.(1,37).

Based on these considerations, the novel concept of the combined cycle coal-fired power station (Figures 1, 2) was developed. It provides for:

- the conversion of the solid fuel, e.g., coal, oil shale, waste, biomass etc., into combustible gases;
- the removal of pollutants from the primary gases;
- the low-NOx combustion of the purified gases in the combustion chamber of a gas turbine; and
- the utilization of the sensible heat of the waste gases in a steam boiler with adjacent steam turbine (35).

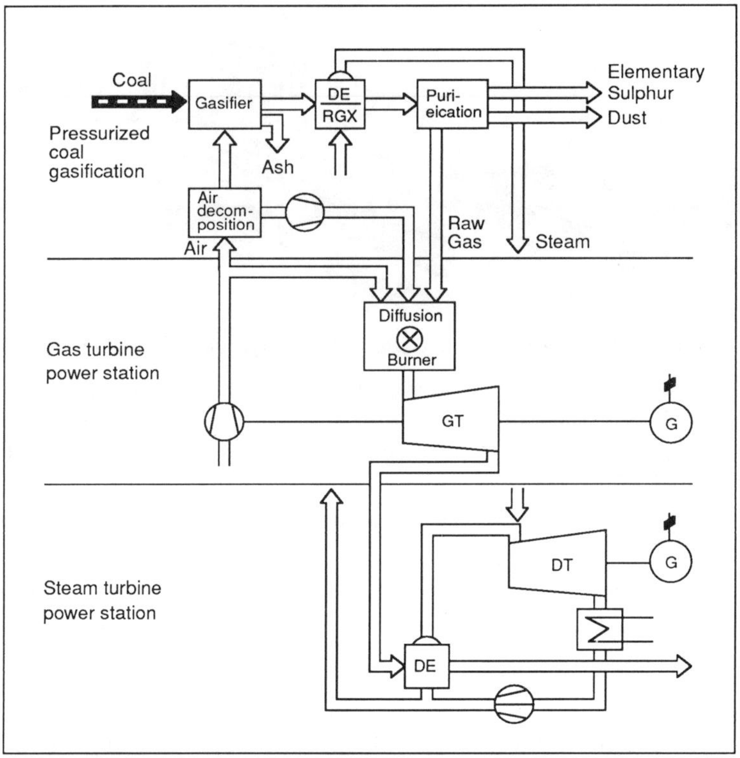

Figure 1. Simplified flow chart of a combined cycle power station with pressurized coal gasification.

2 Theory of the Conversion of Solid Fuels Into Gas

2.1 Degasification, Pyrolysis

Natural gas was produced during the mineral oil and coal genesis when low-molecular gas, in particular methane, which accumulated in geologically favourable formations, split off from vegetable or animal decay products in the course of geological cycles. Some methane adhered adsorptively to the coal itself. During coal mining, when the internal pressure collapses due to mechanic destruction and the diffusion paths are exposed, a certain amount of this methane is released. The emitted methane may account for up to 3% of the calorific value of the coal. To avoid explosions during excavation, the methane

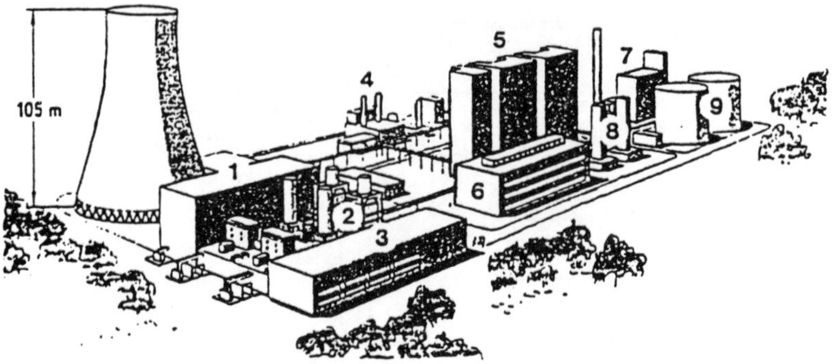

Ground area approx. 400x200 meters (Comparable to a pulverized coal-fired power station)

① Power house steam turbine

② Gas turbo sets with waste heat steam generator

③ Central control room and switch-board station

④ Gas purification

⑤ Coal gasification

⑥ Compressor house

⑦ Coal grinding and drying plant

⑧ Air decomposition plant

⑨ Oxygen and nitrogen tank

Figure 2. Architectural drawing of a 700 MW combined cycle power station with pressurized coal gasification (2, 3).

has to be drained as far as possible prior to coal mining. The drainage efficiency may, however, attain only 50%. The energy content of the drained methane may be used in combustion plants or gas turbines. In fact approximately 600 million cubic meters of methane drained from coal mines are used in the Federal Republic of Germany each year (4).

Methane is also produced on landfill sites, i.e. both by aerobic and anaerobic bacteria. The process has a certain similarity to coal genesis, but here it takes only a few decades in contrast to millions of years in case of coal. The product is almost identical so that the same equipment may be used to recover its energy content.

Coal pyrolysis at temperatures of more than approximately 450°C is significantly more efficient than degasification (33). Depending on the coal type up to 40% and more of the coal substance may be converted into gases, oil and tar vapours. The same holds true for refuse, biomass, oil shale, tar sands etc. It has been tried successfully to pyrolyse the power station coal prior to its utilization in the boiler firing. The gas and the condensed tars were fed to a chemical processing unit whereas the hot coke was supplied to a wet bottom boiler (5,6,7) (Figure 3). Instead of the chemical utilization the pyrolysis gases may also be supplied to a gas turbine (7,32) (Figure 4). Thus, the most simple form of a combined cycle power station is realized through pyrolysis. It

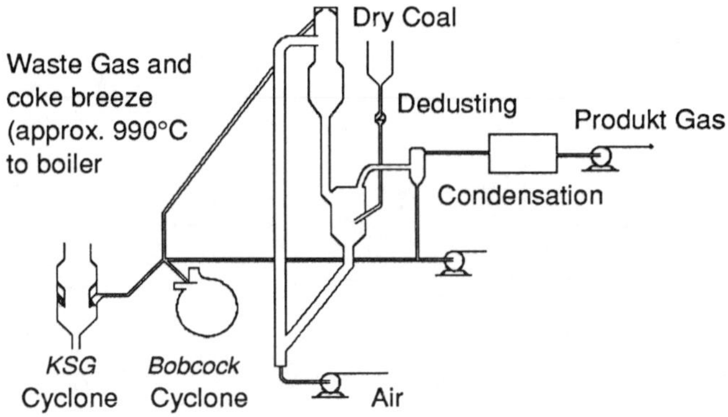

Figure 3. Devolatization according to the Lurgi-Ruhrgas (LR) process (5).

is by far easier than gasification of solid fuels like biomass and oil shale, when a high distillable fraction is available. However, one may want to achieve higher gasification efficiencies.

2.2 Gasification

The total gasification of coal is more consequent. Its first step during heating-up to reaction temperature is always a pyrolysis. The residue, the coke, is completely converted into hydrogen and carbon monoxide, except for the mineral components, by a reaction with a gaseous gasifying agent, mainly oxygen and steam. Essentially, the following primary reactions take place (8):

Water gas reaction:
$C_{fix} + H_2O \text{ --> } CO + H_2 + 118.5$ kJ/mol
Partial combustion:
$C_{fix} + 1/2\ O_2 \text{ --> } CO - 123.1$ kJ/mol

Under technical conditions the equilibrium is on the right hand side of the equations, i.e. the reactions take place until the complete consumption of the carbon or the gasifying agent. In the gasification reactor, the gasification products CO and H_2 mix with the pyrolysis products. Depending on the temperature secondary homogeneous gas reactions occur in this context, the result of which may be precalculated by means of basic thermodynamic rules if

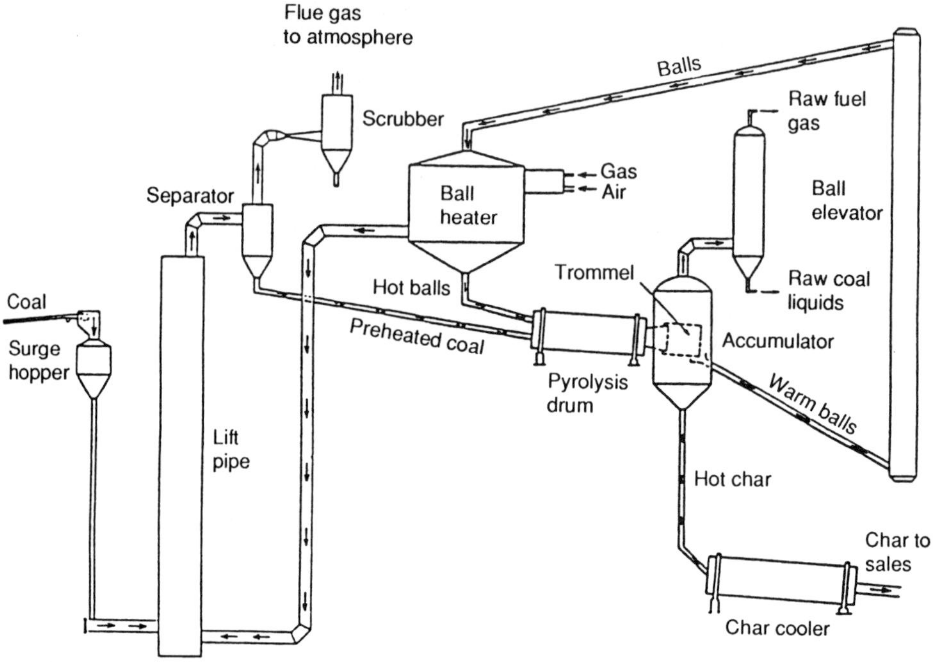

Figure 4. Toscoal process (7).

the residence time is sufficient (8). At relatively low temperatures, the crack reactions of the higher hydrocarbons are rather arrested and a product gas with a relatively high calorific value is produced. At very high temperatures the products ultimately become more and more similar to those of the above-mentioned gasification reaction. In this case, CO and H_2 are produced almost exclusively.

The mixing ratio of oxygen and hydrogen in the gasifying agent is governed by the heat requirement. Oxygen is added to such an extent that the exothermal sensible heat of the partial combustion provides the heat required for the endothermal water-gas reaction for the heating-up of the reactants to the reaction temperature and for the heat losses of the plant. This operating mode is termed autothermal. In contrast, in the allothermal operating mode, the entire heat requirement is supplied via a heat exchanger from an exterior source (10). In this case, it is possible to recover the full energy content of the feed coal in form of chemically bound heat in the product gas.

Whilst the heat chemically bound in the gas is utilized in the gas turbine, the sensible heat which may not be used for the preparation and preheating of the reactants has to be used as far as possible in a steam turbine for steam production and electricity generation. Because the efficiency of a gas turbine at very high temperatures (approximately 1200°C) is significantly higher than the one of a steam boiler, particular attention will have to be paid to this aspect when selecting the gasification process and the combined cycle process control(34).

From the kinetics point of view, the primary gasification reactions are heterogenous reactions between the gaseous gasifying agent and the solid residue. The kinetics is considerably influenced by the transport of the reaction gas to the reactive solid surface. Therefore, small-sized and porous particles generally react with most intensity. This is, however, temperature-related (12).

At lower gasification temperatures (<1000°C), it may be assumed that the gasifying agent is transported into the coke pores quicker than the conversion at the pore walls. In this case, the gas concentration throughout the interior of the pore structure is identical to the exterior gas concentration and the total conversion follows the Arrhenius law :

$$r = r_m \, e^{-A/R\,T} \text{ for } T < 1000°C$$

A the activation energy, and R the universal gas constant, r_m denotes the specific reactivity of the coke, a material constant measured under standardized conditions, and r is the reactivity of the coke (11). Generally, cokes produced from younger coals like lignite have the highest reactivity due to their distinct pore system. In contrast, cokes from bituminous coal are chemically inert. The activation energy A of 200 to 350 kJ/mol is very high, thus indicating the very rapid rise in the gasification rate at a temperature increase in the range of below about 1000°C. An approximate 50°C higher temperature causes the gasification rate to double, in this so-called kinetic temperature range.

At higher temperatures (1000 - 1300°C), the gasification agent may not diffuse into the core interior in such amounts that it will make up for the amount consumed there. In this case, the gasification rate is related to the diffusion rate of the gasifying agent in the voids and pores of the coke. The mathematic model derived for this case shows that the gasification rate also follows the Arrhenius Law, but with half of the activation energy (Figure 5). This means, however, that the gasification rate increases to a much lesser extent with increasing temperature. Also in this case young coals have an advantage due to the larger number and width of the pores. Here, also the sizes are of importance: coarser particles have longer diffusion paths and thus react more slowly.

Eventually, at very high temperatures (>1500°C) an entirely different mechanism takes over: the molecules of the gasifying agent react immediately at the first impact on the exterior coke particle surface, and the amount of the gasifying agent reaching the surface per time unit depends only on the diffusion rate through the adhering gas envelope around the individual coke grains. Since the diffusion rate increases only slightly with the temperature

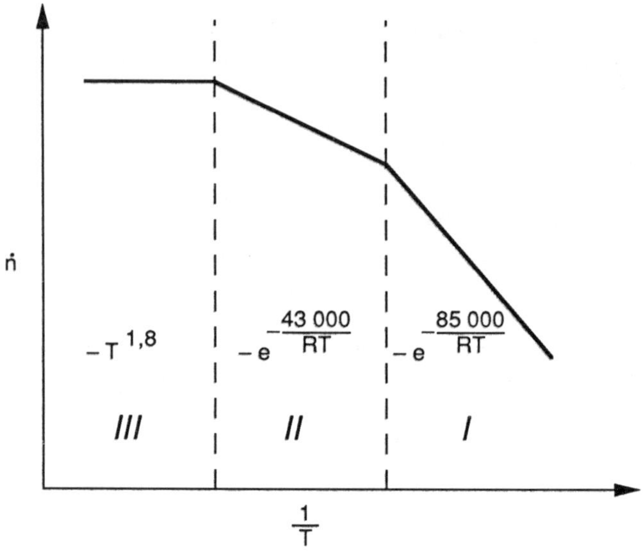

Figure 5. Arrhenius Diagramme: Gasification Rate as a function of the temperature.

($T^{1.8}$), a further temperature increase is not very effective. Aerodynamic measures such as reversing, shock waves, etc., which may reduce or even remove the adhering gas envelope, are much more efficient. In this way, the gasification rate may be increased by a factor of ten or more. In this reaction range, the coal type is not important, but the grain size is important due to the exclusively controlling specific exterior surface.

There is a conclusive theory which is described in a mathematic model (12) of this complicated gasification kinetics. For all technically relevant ranges the effect of the main parameter, the temperature, may be represented in an Arrhenius diagramme (Figure 5), which is very valuable also for the practician.

3 Gasification Technology

3.1 Reactor Types

The total gasification of coal has been known since the last century. Initially, only the cyclically charged generator was known, but it was replaced by continuous processes after the turn of the century. Advances in high pressure process and air decomposition technologies contributed greatly to coal

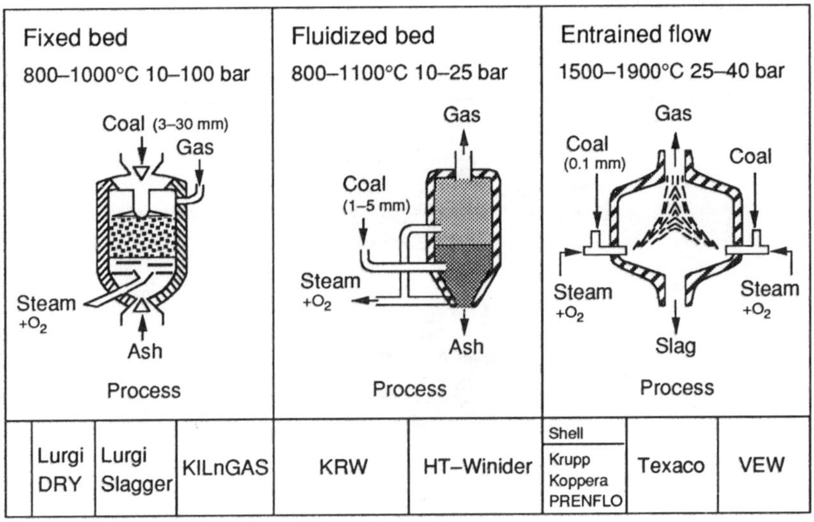

Figure 6. Basic types of coal gasification reactors (1,2,3).

gasification development. Milestones were set in the twenties when Lurgi developed the pressurized continuous fixed-bed gasification. Almost at the same time, Winkler succeeded in developing fluidized bed gasification, although under normal pressure. After World War II, the entrained flow operating mode, especially using the Koppers-Totzek process, gained importance (9).

These three fundamentally different process approaches (Figure 6), often described in papers on coal gasification, have been copied in various ways. Today, approximately 40 gasification processes (8) have been tried in test plants. Only a few of them, however, have reached the operational stage. With a few exceptions, all of them may be traced back to the three basic types shown in Figure 6.

The fixed-bed reactor according to Lurgi (Figure 7) (13) is a tube-shaped pressure vessel in which a uniform coal charge is maintained by means of the upper pressure lock. An agitator avoids possible agglomerations of caking coals. The bed rests on a grate at the reactor bottom through which the gasifying agents oxygen and steam are supplied. The reaction gas flows countercurrently through the charged column and is drawn off at the top. The oxygen reacts in this is lower layer with the coke. This is the hotest zone. Above, the water gas reaction starts consuming heat and cooling down the charge increasingly towards the top. Below approximately 700 to 800°C, when the gasification reaction has almost ceased, the feed coal is degassed, preheated and dried. Under ideal circumstances the waste gas has a temperature of

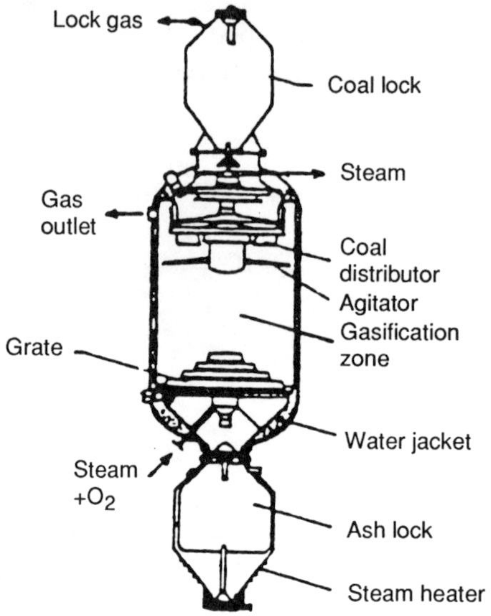

Figure 7. Pressurized Gasification according to Lurgi GmbH Frankfurt (8).

several 100°C. A lower limit is only established by the condensating tar which might produce blockages. The Lurgi pressurized gasification is by far the most common process applied on technical scale. So far, maximum unit sizes amount to approximately 20 tounes per hour (t/h) hardcoal feed giving 60,000 m^3 (STP)/h product gas.

In fluidized bed gasification processes (Figure 8), also called Winkler gasification after their inventor, the gasifying agent is also fed through the reactor bottom. It is distributed over the cross-section as uniformly as possible so that the flow rate almost corresponds to the suspension rate of the average grain size. The coal with a grain size of 0 to 5 mm fed from the side at the bottom will then float to the side and upwards and downwards in a relatively dense phase within the reactor chamber under strong exchange movements. In this context, the pyrolysis, combustion and gasification take place under practically isothermal conditions. The process has worked well in numerous cases, mainly for lignite. Maximum unit size so far: 35 t/h lignite giving 60,000 m^3 /h synthesis gas.

In the entrained flow gasification process, very fine-grained feed coal <0.5 mm is injected into a very hot reaction chamber together with the gasifying agent so that combustion and gasification start spontaneously and are ter

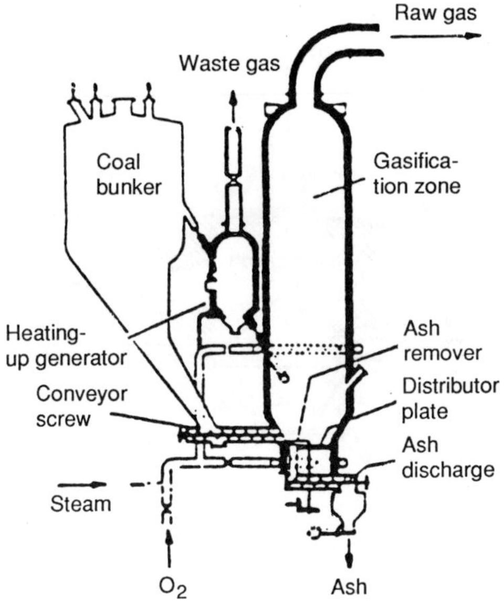

Figure 8. Fluidized bed gasification according to Winkler, as built by Davy Powergas Ltd., Cologne (8).

minated in seconds or even fractions of a second. Sharp deflections or impact like coal feed accelerate the process even further. The operating units built by Koppers (Figure 9) so far reached unit sizes of 21 t/h lignite respectively 32,000 m^3 /h synthesis gas. Units of twice the capacity are planned.

One of the very few new developments that deviates from the three processes above is the molten iron bath generator of Klöckner and Sumitomo. It is probably only applicable for combined cycle processes in steel works and is by far not fully developed (14) (Figure 10).

All processes mentioned so far are autothermal, i.e. the required reaction heat is supplied by feeding oxygen into the reactor. In contrast the process of Bergbau-Forschung is allothermal. Here, the heat, originally produced by a high-temperature reactor, is supplied to the gasification bed which is operated in form of a fluidized bed by means of immersed tubes. The Bergbau-Forschung test plant for approximately 100 kg/h was operated with great success. Before its commercial application, though, years of development work on larger scale are still required (10, 15). To apply this process in a combined cycle power station the combustion chamber of the production gas may be used as heat source to heat the allothermal gasification generator. A demonstration plant of this type is in the planning stage.

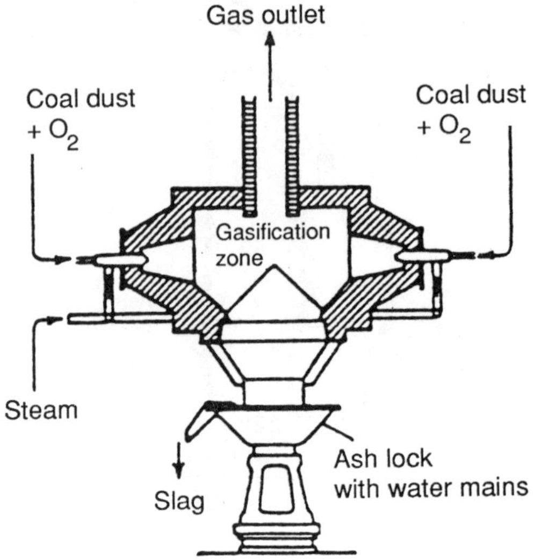

Figure 9. Entrained flow gasification according to Koppers-Totzek, as built by Krupp-Koppers GmbH, Essen (8).

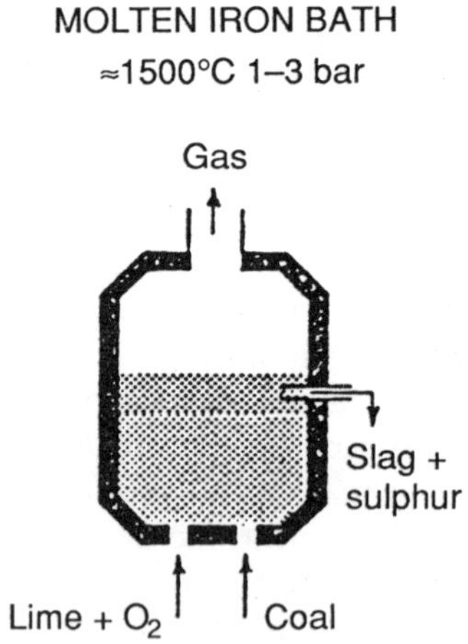

Figure 10. Molten iron bath generator according to Klöckner and Sumitomo (14). The generator operates at a pressure of 1–3 bar and a temparatur of about 1500°C.

3.2 Pressure and Temperature Setting

The choice of the gasification process for a planned combined cycle process depends on the given operating conditions. If a gas turbine is going to be used, then a higher gasification pressure, at least above approximately 10 bar, preferably between 20 and 30 bar, is required from the beginning. A higher pressure is also favourable from the gasification point of view: up to approximately 10 bar the gasification rate is strongly increased and additionally the higher gas density provides for a smaller reactor volume. Therefore, practically all processes have undergone further development for higher pressures. In the Lurgi process peaks of 100 bar were demonstrated, but this seems to be much higher than the optimal pressure (16).

In contrast to pressure, the operating temperature may not be set voluntarily from the beginning. For the fixed bed gasifier, which has to be operated at temperatures of above approximately 1000°C due to the conversion rate, normally the melting point of the coal ash imposes a limit which ranges between 900°C and 1300°C. In the largest Lurgi plant at Sasol in South Africa, the coal used does not begin to melt, i.e. soften until 1300°C – an almost ideal situation.

To reach higher temperatures – and thus higher plant capacities – and to be able to gasify coals with low ash softening points, British Coal Company developed the Slagging Gasifier in cooperation with Lurgi. It overcomes initial slagging in the lower section of the reactor by using a moving grate which enables the slag to drip off in liquid form (17).

The fluidized bed reactor is even more sensitive to coal ash softening than the fixed bed reactor. Hence, only very reactive fuels such as lignite, peat and biomass may be gasified normally. For them, a low temperature of approximately 900°C is sufficient. However, experience at higher pressure is still insufficient.

In contrast, the entrained flow process prefers a lower coal softening point because the ash is drawn off in liquid form. At very high operating temperatures of 1500°C and more, an ash problem does generally not exist. If specific operating modes such as the utilization of air instead of oxygen or a coal feed with aqueous suspension requiring lower operating temperatures, solidified slag zones may occur, they represent a problem.

3.3 Coal Properties

Besides the softening point of the slag, other coal properties are important as well. The caking propensity of the coal may, for example, lead to lump formation if the coal passes through the temperature range of 400–500°C in the bed of the fixed bed gasifier. As a remedy, Lurgi developed an agitator construction.

In the Winkler process, even coal with a low caking propensity agglomerates. The lumps form on the distributor plate and interrupt the process. Here, too, thought a remedy is available: the jet feeder developed by Ruhrgas, which was further developed by Bergbau-Forschung (18).

The entrained flow process is not sensitive to caking coals.

The fixed bed process depends on a larger coal grain size of more than 5 mm with a permitted 30% percentage of material <5 mm, so that an efficient amount of gas will flow through the bulk material column. Fines may be briquetted or pelletized prior to being fed. This will, of course, increase the costs of the process.

Fluidized beds require grain sizes of up to 5 mm and should only contain a minimum amount of fines <0.1 mm.

Feed coals for entrained flow reactors have to be very finegrained because the gasification rate depends solely on the exterior surface. However, it also turns out that excessively small sizes may be unfavourable because there is practically no relative velocity between the grain and the gas and the relatively wide gas envelope around the individual grains results in longer diffusion paths and low gasification rates. The residence time in the reactor is also shorter in case of small grain sizes. The fineness of grinding of power station coal, however turned out to be favourable. Entrained flow processes are thus practically universely suitable for all coals.

The desired qualities for coal also apply to all other feed substances such as waste and biomass if the above-mentioned processes are applied. With the development of much smaller gasification plants, however, other development lines are being persued (19, 28), but they will not be detailed here.

3.4 Gas Composition

The composition of the product gas is very important if the gas is to be used in the chemical industry. Combustion chambers of gas turbines, in contrast, may be adjusted to all gas types. However, a particular gas type has to be adhered to after a specific design is chosen. The different gasification process produce gases of different compositions. (Figure 11) (8).

In the fixed bed process, higher hydrocarbons and high-boiling tars are discharged due to the countercurrent flow principle. They have to be washed out in a particular way to be used as liquid fuels. If appropriate, however, they may be recycled to the gasifier until their complete thermal cracking. Fixed bed reactors achieve the highest gas calorific values of up to 15,000 kJ/m^3.

NAME	LURGI	WINKLER	KOPPERS-TOTZEK
REACTOR TYPE	FIXED BED	FLUIDIZED BED	ENTRAINED DUST CLOUD
TEMPERATURE °C	1000/700 (PROFILE)	APPROX. 1000	MORE THAN 1300
PRESSURE bar	35	1	1
FEED COAL	LIGNITE TO MODERATELY CAKING HARDCOAL NO SMALLS	LIGNITE TO MODERATELY CAKING HARDCOAL SMALLS	ALL SOLID FUELS SMALLS
RAW GAS % BY VOL. (IF OPERAND WITH PURE O_2) H_2	37/39	35/46	31
CO	20/23	30/40	58
CO_2	27/30	13/25	10
CH_4	10/12	1/2	0.1
GAS GENERATOR COAL t/h	20	40	12
RAW GAS m^3/h	35000/60000	~≈60000	20000

Figure 11. Examples for the gas composition when gasifying different coals in the three fundamentally different reactor types.

In contrast, entrained flow reactors always produce pure CO/H_2 mixtures, with only slight admixtures of CH_4, etc., due to their high temperature peaks. In most cases, CO predominates. At temperatures of 2000°C or higher, almost pure CO with only a little CO_2 and H_2 is produced. (25).

The gas properties of the products of fluidized bed installations range between those of the other two processes. The resulting pyrolysis products are mainly cracked. Moreover, the primary gases comprise an enormous variety of gas compounds in low concentrations, often only as traces. Due to their toxic character and to the possible damage they may cause to downstream installations, many of them have to be removed.

4 Gas purification

4.1 Liquid Purification

It turns out that the removal of dust and pollutants from the product gases represent a critical challenge in the combined cycle process. At first glance, this is surprising because similar tasks are routinely solved in the chemical industry. However, the problem there is simplified by the fact that practically in all cases the gas has to be purified after cooling it below the dew point.

First, the dust-containing, high temperature raw gas passes through cyclones in which the coarse dust is separated, which is generally recycled with the raw coal to the reactor. Then, the fine dust precipitates in the first wash cooler, retaining residual contents of <5 mg/m^3. Prior to H_2S separation, a pre-scrubbing for hydrogen cyanide, rhodanide, phenols, ammonia and fatty acids is recommended, depending on the existance and compatibility with the subsequent scrubbing processes. For the removal of H_2S, both physical and chemical processes are available (20).

In the physical scrubbing processes, H_2S is dissolved in the scrubbing liquid, which may be polyethylene at 20°C (Allied Chemical), methanol at 20°C (Rectisol of Lurgi) or propylene carbonate at 40°C (Fluor-Solvent of Fluor). The scrubbing liquid is removed and expanded so that H_2S and frequently also CO_2 become volatile. Then, the H_2S is fed to the Claus oven for reduction, and the final product is highly pure elementary sulphur.

In the chemical gas scrubbers, a binding of the H_2S amine takes place at 40°C (Sulfinol scrubber of Shell), of amine-methyle-pyrolidon at 40°C (Pyrosol process of Lurgi) or of potash at 110°C (Benfield scrubber of the U.S. Bureau of Mines).

4.2 Dry Scrubbing Processes at Higher Temperatures

The hot potassium scrubber is an example. It is operated at temperatures above 100°C. All other scrubbers operate below this temperature, and this is the drawback of all of these proven scrubbing processes when being used in combined cycle power stations, because if the raw gas is cooled below the dew point the steam contained in it condenses. The steam content of the raw gas is relatively high due to: the utilization in the gasifying agent; the coal moisture; the formation water; and finally the use of steam as cooling and driving agent. If it remains in the raw gas its thrust force has a significant impact during the expansion in the gas turbine. If it is precipitated, the overall efficiency of the process is reduced. Therefore, it is highly desirable to design the system so that the hot dedusting and hot desulphurization units can be avoided. Both of them are not yet state of the art.

To remove dust without scrubbing, work is being carried out relating to the development of solid filters, gap-type candle filters or metal filters. The dust removal from the filters themselves is critical. For desulphurization above the dew point, dry processes using molecular sieves (Union Carbide and Davison Chemical Company), activated carbons (Bergbau-Forschung) and iron oxides (Ruhrgas AG) are being examined. Their operating temperatures range only slightly above the dew point at the most. Real hot gas desulphurization, in contrast, might be succeful – applying the hot iron-oxide process (U.S. Department of Energy, Morgantown) at 815°C or the Dolomite process (Conoco, Airproducts and Chemical) at 915°C. In the interest of energy recovery, it would be very attractive if both dedusting and desulphurization might be accomplished at these high temperatures. Then, the purified gas could be led directly to the turbines without losses and without waste heat boiler.

4.3 The NO_x and Alkali Problem

Combustion plants produce nitric oxides with different valencies such as N_2O, NO, NO_2, etc., together termed "NO_x". They are noxious and their concentrations in waste gases of combustion plants have therefore to be reduced below the limits prescribed by different countries.

Two mechanisms have to be distinguished regarding NO_x formation. Either nitrogen and oxygen react in the very hot flame zones or primarily ammonia forms from the nitrogen of the feed coal, the ammonia is then burnt to NO_x.

NO_x formation from the combustion air can be suppressed by setting the burner so that overheated zones are avoided. In gas firings, this is possible; in coal firings, however, only to a limited extent. Above all for coal which is difficult to burn, like anthracite, this is rarely sufficient. To this the nitrogen in the coal has to be added.

Coal, and by the way heavy fuel oil, contains approximately 1% nitrogen. In a combined cycle power station ammonia is primarily formed during gasification and is mostly washed out during wet scrubbing along with the traces of higher nitrogen compounds such as hydrogen cyanide (Section 4.1). For the purified gas, the same applies as for natural gas: the NO_x emissions may be suppressed by a suitable burner setting. The hot gas scrubber, in contrast, which should be used for energy-related reasons, cannot remove NH_3 at present as far as it is known. Given the state of the art, a Denox plant is required in this case.

A similar situation exists in terms of avoiding alkali traces in the waste gas because of the danger to the highly sensitive turbine blades. Apart from many other elements, coal contains sodium and potassium, which are part of the gas produced during pyrolysis. They can be removed by wet scrubbers, but most probably not by dry gas purification. The problem is exasperated by the fact that available analysis equipment can not determine alkali traces in the waste gas. A retention technology is not yet in sight.

To close this gap, British Coal started to prove alkali in the waste gas and to develop a retention technology (38). So, unsolved problems still exist in gas purification and work in this context is most important.

4.4 Air or Pure Oxygen

For coal gasification purposes in the chemical industry, oxygen is always used as gasifying agent to obtain a working gas as pure as possible for the synthesis to produce fertilizers, fuels, plastics etc. Oxygen recovery from air is expensive. As shown in Figure 16, separation plants i.e., plants for air decomposition, account for approximately 25 % of the total investment of a combined cycle power station. Moreover, their compressors consume a certain amount of the produced energy. Therefore, the question arises whether it is better to do without an air decomposing unit and use air directly with steam addition as gasifying agent. For all three basic processes, it has been proven that this is feasible.

The use of air instead of oxygen has considerable consequences. A five-fold compression energy is required which is, however, partly recovered when expanding the turbine. As the volume of the gasification reactor has to se about trippled, the unit performance is lower. The same holds true for gas purification plants.

In the combustion chamber of the gas turbine, there are no disadvantages because an addition of the nitrogen produced in the air decomposition plant or of steam is provided for anyway to limit the temperature. Finally, the flue gas composition corresponds to the flue gas produced during operations with oxygen. A general conclusion on the relative advantages or disadvantages of the operation with air versus oxygen cannot be made at this time. More experimental work with reactors using air-steam as a gasifying agent needs to be done.

5 Development Projects

5.1 Fixed Bed Gasification

The Lurgi gasification process in the fixed bed at high pressure is the most widely applied process worldwide to-date. It is mainly applied to produce synthesis gas from coal. Due to the Lurgi process application of high pressures and its more advanced development state, it was the obvious choice for combined cycle power stations.

The first operational trials in full scale were carried out between 1972 and 1977 at Steag's Kellermann/Lünen power station (Federal Republic of Germany) (21). The total capacity of the demonstration power station amounted to 170 MW_{el}, 74 MW_{el} of which were produced by the gas turbine and 96 MW_{el} by the steam turbine. Air and steam were used as gasifying agents. The trials furnished the proof that a combination of coal gasification and gas and steam turbine is feasible. Unfortunately, the designed efficiency of 36.7%, which was still relatively low, could never be achieved. One reason for this was that the plant only reached 70% of its rated capacity due to a higher than anticipated gasification air consumption. Moreover, the feed coal grain size was too small for the Lurgi gasifier and did not meet the expectations. Due to the uneconomic results, the demonstration plant had to be given up in 1977.

As mentioned in section 3.2, the application of the Lurgi gasification process for a combined cycle power station has received a significant boost by the development work done at Westfield/Scotland by the British Gas Cooperation (BGC) together with Lurgi (17). By raising the gasification temperature, the reactor performance increased significantly. Recently, BGC tested its slagging gasifier in combined cycle power stations. A gasification reactor with a process capacity of 25 t/h hardcoal was connected to a gas turbine type SK 30 - Rolls Royce Olympus. The short total running time was sufficient to furnish the proof that this approach can work.

Moreover, BGC has studied the further development of the Lurgi slagging gasifier, i.e., a composite slagging gasifier with high operating temperature which is characterized by pulverized coal combustion below the gasifier grate. Based on the Lurgi process with dry or liquid slag discharge using air or oxygen, several system studies were carried out (22). The results unanimously confirm the expectation that efficencies of 39% – in advanced gas turbines with high inlet temperatures of 42% – may be achieved. The specific investment costs (as per 1984) are quoted as 2,261 DM/kW for a 200 MW combined cycle power station and 1,711 DM/kW for an 800 MW combined cycle power station, including the Lurgi gasifier (1, 2, 3). These costs are considerably lower than the ones of conventional coal-fired power stations with flue gas desulphurization and Denox plants.

5.2 Fluidized Bed Gasification

A 30 tonnes per day (t/d) test with a fluidized bed gasifier plant was commissioned in Waltz Mill, Pa, by the Kellog, Rust and Westinghouse (KRW) group. After a gasification project for 1,200 t/d coal feed at Sasol in South Africa could not be realized in 1980, the emphasis was shifted to the utilization for combined cycle power stations. In 1986 the Electric Power Research Institute (EPRI) in the U.S. produced a study for a 540 MW power station, operating five gasifiers at 30 bar and using Illinios coal. A 35-38% efficiency at a gas turbine inlet temperature of 1200°C is expected. At present, a 60 MW demonstration plant is being planned (Figure 12).

In the Federal Republic of Germany a further development of fluidized bed gasification was made by the "Rheinische Braunkohlenwerke AG" (Cologne) around 1979 (23). Of major importance was above all the application of higher pressures, initially 5, later on 10 bar. Due to the temperature limit imposed by the melting point of the ash, mainly lignite was used as fuel. Initially, only a synthesis gas production was planned. To this end, a plant was commissioned in 1986 in Ville/Berrenrath for 30 t/h raw lignite to operate a methanol synthesis unit in the near-by chemical factory of UK-Wesseling with the gas.

A new development is now scheduled in order to utilize this successfully operated process for combined cycle power station. The pressure shall amount to 25 bar and it is intended to use an air/steam mixture as gasifying agent. To use also less reactive fuels such as hardcoal, a coke recycling along the lines of the circulating fluidized bed combustion is under discussion. A study produced in cooperation with KWU (Erlangen) and Uhde GmbH (Dortmund) provided the input for a 600 MW combined cycle power station. It is thought that an efficiency of 45% may be reached in the long run (Figure 12).

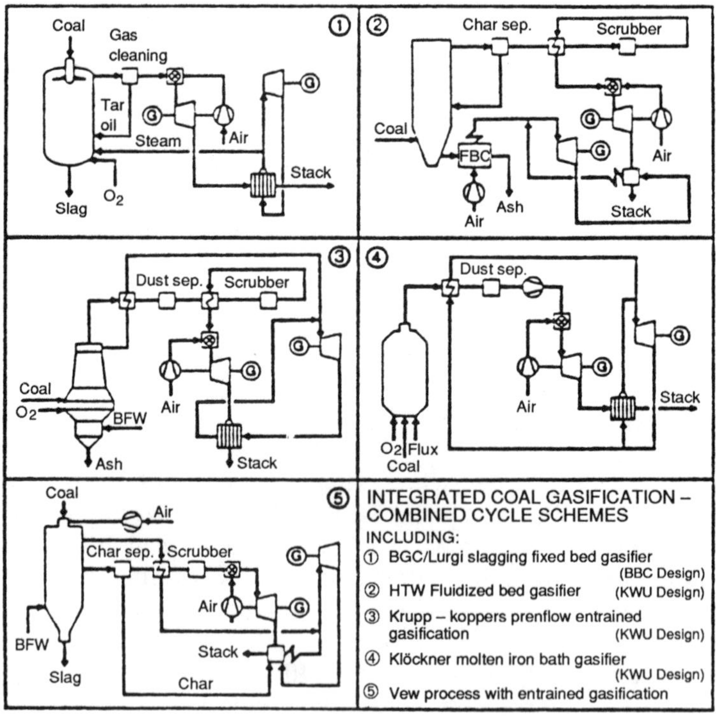

Figure 12. Integrated coal gasification – combined cycle schemes (34)

5.3 Entrained Flow Gasification

From a process-related point of view, entrained flow gasifiers are closest to conventional pulverized dust combustion systems. Basically, they are understoichiometrically operated pulverized dust combustion systems operating according to the basic principle described above, but with very different designs. R & D activities relating to the entrained flow processes are the most intensive of all the coal gasification R&D efforts.

At the end of the seventies, the Koppers-Totzek process described above was further developed by a group including Saarbergwerke AG, Shell AG and Krupp-Koppers GmbH in the Saarland. The initial objective was to create an efficient process for synthesis gas production by applying higher pressures of up to 25 bar. In 1978, a first plant for 150 t/d was constructed in a very short time at the Shell refinery in Hamburg, and it was operated successfully for a short time (25). The group separated in 1981. The Saarberg/Krupp-Koppers group constructed a 48 t/d test plant in Fürstenhausen/Saarland, which is still being operated (26). Today, the process is termed "Prenflo". Based on the results a study was produced in cooperation with KWU. This presented in references 1–3.

In contrast Shell shifted its activities to the US (27). There, a demonstration plant for 250 t/d hardcoal was constructed by Lummus Combustion Engineering and Shell in Deer Park/Houston, Texas, which was commissioned in 1987. Both this and the Fürstenhausen/Saarland plant operate with almost pure oxygen and a small amount of steam so that a product gas with a very high CO content is produced at a very high flame temperature. Due to the high gasification efficiency, the integration into a combined cycle process seems to be favourable according to an EPRI study (25). The Lummus–Shell group plans to build a large power station with a capacity of 1100 MW_{el}, perhaps in 1993/94. An efficiency of 45% is expected.

Moreover, in the fifties the development of the Texaco process was started, initially in a small test plant in Morgantown (West Virginia/US). Later, this effort was transferred to Montebello (California/US). The operating mode is characterized by the coal feed using a coal-water-suspension at the top of a pressurized reactor (Figure 13). The disadvantage of the heat loss caused by water evaporation is practically balanced out by the simpler operating mode. The technical impetus for this development came from the "Arbeitsgemeinschaft Ruhrkohle/ Ruhrchemie" (Essen-Oberhausen), which initially constructed a test plant in Oberhausen and commissioned a production plant for the production of synthesis gas from hardcoal in 1986. Based on this experience, plants with the same objective were built in Kingsport, Tennessee/US (820 t/d); Ube, Japan (3 x 500 t/d); and Lu Nan, People's Republic of China (380 t/d) (29).

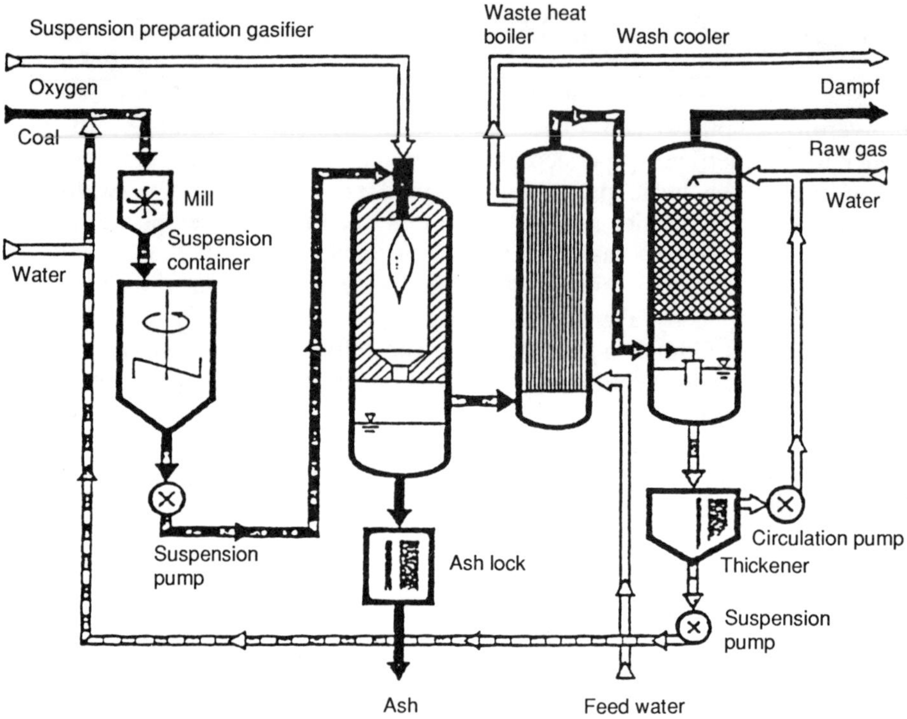

Figure 13. Texaco process (29).

At the beginning of the eighties, a group consisting of EPRI, Bechtel, General Electric (GE), and Japanese companies built a "Cool-Water" plant in Daggett, Cal., which processes 900 t/d coal. It is equipped with a steam turbine and has an electric output of 94 MW (37). Since it was primarily intended to demonstrate the process with available technology, all components were designed conventionally and features such as hot gas scrubber, high turbine inlet temperature etc. were disregarded. Hence the efficiency is correspondingly moderate. It is the aim of the Daggett group to improve it.

At the same time, but independent from this, the development of a similar entrained flow process was started at Dow-Chemical, Houston, Texas from 1979 onwards. Two opposite nozzles (as in the Koppers-Totzek process) supply a coal-water suspension (as in the Texaco process). A test plant of 30 t/h in Plaquemine/Baton Rouge in Louisiana, US, was the first major step in this development work. A prototype plant with a capacity of 450 t/d raw lignite was operated between 1983 and 1985. In 1987, a demonstration plant (2000 t/d) was commissioned. The present plans call for the construction of units with 100 MW gas turbines and 160 MW steam turbines and the combination of

10 of these units for lignite-based electricity generation is planned for a large power station. In the future it shall replace the present natural gas-fired power station on gas turbine basis at the Baton Rouge site (30) (Figure 14).

A process development of Ruhrgas AG in Essen (31) at the end of the forties and the beginning of the fifties fell into oblivion. Generally, it is very similar to the objective of combined cycle power station. At that time, it was the task to produce a clean lean gas at normal pressure by gasifying fines with air. The gas should be used in the underfiring of coke ovens to make the coke oven gas, which would otherwise have to be burnt, available to the energy industry. The process has proven well with two plants of approximately 100 t/d feed coal each in Suderwich (operator Ruhrgas AG, Essen) and Wath upon Werne, Yorkshire (operator National Coal Board, United Kingdom). During cyclone gasification, coal suspended in air is supplied tangentially to the reactor through several nozzles. In the reactor, the slag, as in all entrained flow processes, is drawn off in liquid form into a seal pot. The reactivity of hard

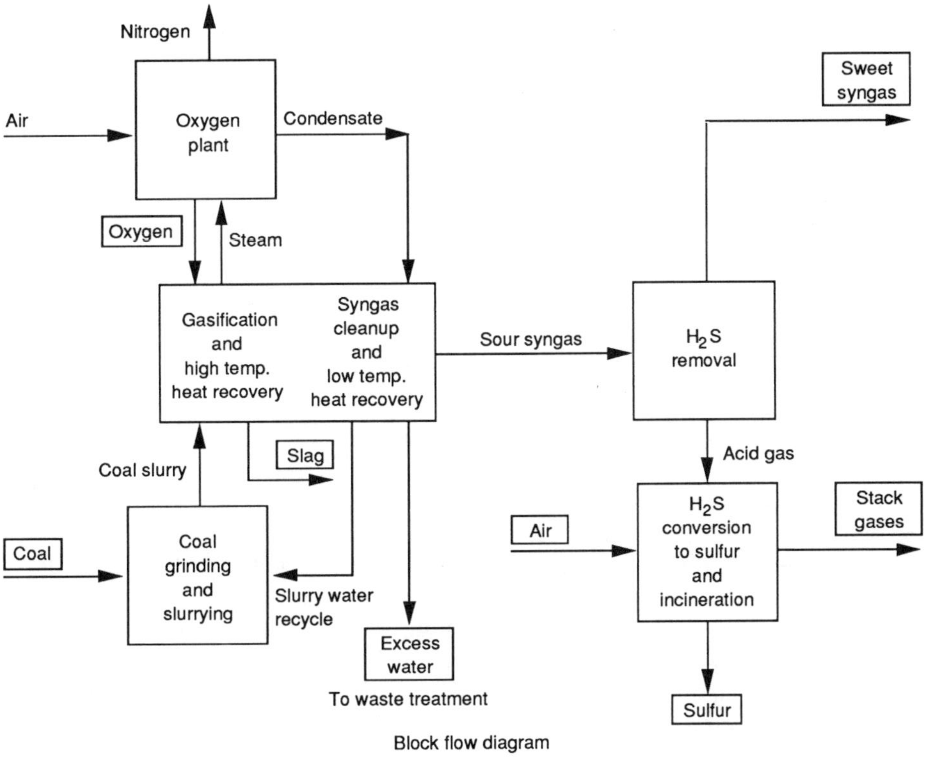

Figure 14. The Dow Chemical gasification process (30).

coal, which is too low, resulted in a high discharge of non-combusted fine coke in the gas, which could, however, be recycled successfully to the reactor so that the coal was gasified totally in the end.

The development project of VEW (Dortmund) (32) is based on the development of a pyrolysis process for steam coal of Steinmüller in Gummersbach and was started in the early fifties. The reactor consists of a long vertical tube to which preheated air with steam addition and coal dust is supplied from the bottom. During the pneumatic upward lifting, the pyrolysis and incomplete gasification, with a remaining residue, takes place. At the top, the reactor opens into a cyclone system, in which the residue of fine-grained coke is collected.

After extensive trials at VEW in the fifties with a test plant for 25 t/d hardcoal, a larger test plant for 240 t/d was commissioned in the VEW Gerstein power station in 1984. The residual coke is supplied to a boiler combustion system which is equipped with a flue gas desulphurization unit. The gas is used to develop a hot gas scrubber, later on it is to be supplied to a gas turbine.

The devolatilization tests carried out at VEW are part of a comprehensive development programme for a combined cycle power station. To study the connections, a 750 MW block of the Werne power station is equipped both with a conventional hardcoal-fired combustion system and a gas turbine for natural gas. It is hoped that the first large power station integrating coal gasification and a gas turbine with efficiencies of 45% may be built in the nineties (Figure 12).

6 Concluding Evaluation

6.1 Efficiencies and Coal Consumption

High efficiencies in a coal gasification, combined cycle power station can only be achieved by the subsequent use the sensible heat in the gas after gasification. This will require efficient heat exchangers. The efficiency of the gas turbine, in particular the permissible gas inlet temperature, is also of decisive importance. The gas turbine is dealt with in the chapter by Williams and Larson (37) and will not be discussed further here.

In terms of gasification efficiency, the percentage of the energy content of the feed coal chemically bound in the production gas is of decisive importance. As R. Müller (KWU, Erlangen) reports (Figure 15) the BGC-Lurgi slagging gasifier achieved the highest gasification efficiency of 89%. The reason for this is, as noted in section 3, the countercurrent flow principle of the fixed bed

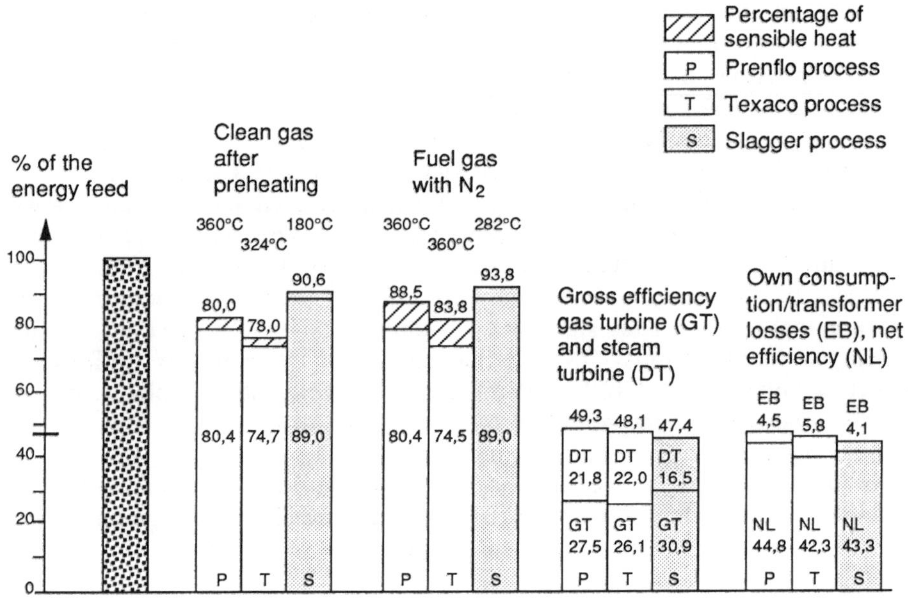

Figure 15. Energy balance of power station with different gasifier types (gas turbine inlet temperature 1220°C) (3).

reactor. In the two entrained flow processes compared to it in Figure 15, the gasification efficiencies are much lower due to the co-current flow principle: 80.4% in the Prenflo and 74.7% in the Texaco process. The last figure is the lowest because the coal is supplied in water suspension and the consequent water evaporation consumes a lot of heat. Figure 15 also shows that the electric efficiencies of the gas turbine with a relatively high assumed inlet temperature of 1220°C are correspondingly lower. With 30.9%, the slagging gasifier reaches the highest values. The fact that the efficiency of the Texaco process is lower than the Prenflo process – 26.1% versus 27.5% – can probably be attributed to the thrust force of the evaporated slurry water in the one of the Prenflo process which makes up for a part of the heat losses caused by evaporation.

High gasification efficiencies are contrasted by lower sensible heat yields so that their subsequent use in the steam turbine may create a balance. This is confirmed by the net efficiencies shown in Figure 15: slagging gasifier 43.3%, Prenflo 44.8%, Texaco 42.3%. Note that the Prenflo process is slightly more efficient than the slagging gasifier. The small deviation of the three values from one another should, however, not be overrated. They are calculated results of a study rather than the measured results of operational plants.

If it is possible to realize these net efficiencies in practical operations,these plants will be 15-20% more efficient than conventional power stations with flue gas desulphurization and Denox plants. In other words, they will be able to produce the same amount of electricity with 15-20% less feed coal.

Based on an EPRI study, however, Williams and Larson calculated lower efficiencies (37) for a conventional power station – 34.6% versus the 36% used here – and they estimate that a combined cycle power station using the existing turbines will achieve an efficiency of only 37.9% (11% more than the conventional plants). They state that it is only by using advanced gas turbines developed from jet units (ISTIG) that combined cycle power stations can achieve an efficiency of 42.1% – 22% more than conventioned plants. In such systems, the useful heat would be totally utilized in the gas turbine, thereby eliminating the need for steam turbines.

As noted earlier, however, the actual advantage of combined cycle technology, i.e., higher efficiencies, is still speculative because actual data from full-scale operating plants is not yet available.

6.2 Plant Size

In gasifiers using oxygen, unit performances of approximately 30 t/h hardcoal were achieved. Further development work carried out on all three basic types may well double this performance. At a 40% efficiency, 30 t/h coal produce approximately 100 MW electricity, i.e., 200 MW if the performance is doubled. In a large power station, several gasifiers will be available. According to the present state of the art, 8 units with 2 back-up units are likely for a 800 MW power station. Existing experience indicates that by switching individual gasifiers on and off, with adequate back-up available, a continuous production is possible.

Generally, it is possible to switch from part to full load and vice versa in minutes. Due to the lower fuel inventory in the reactor, the entrained flow process should be the most flexible process. A combined cycle power station does not have to be inferior to a conventional one in terms flexibility.

Due to the relatively small unit size of the gasifier and the gas turbines available today, combined cycle power stations have a flatter cost-vs.-capacity curve than conventional power stations. This may be advantageous as Eric H. Reichl states (39):

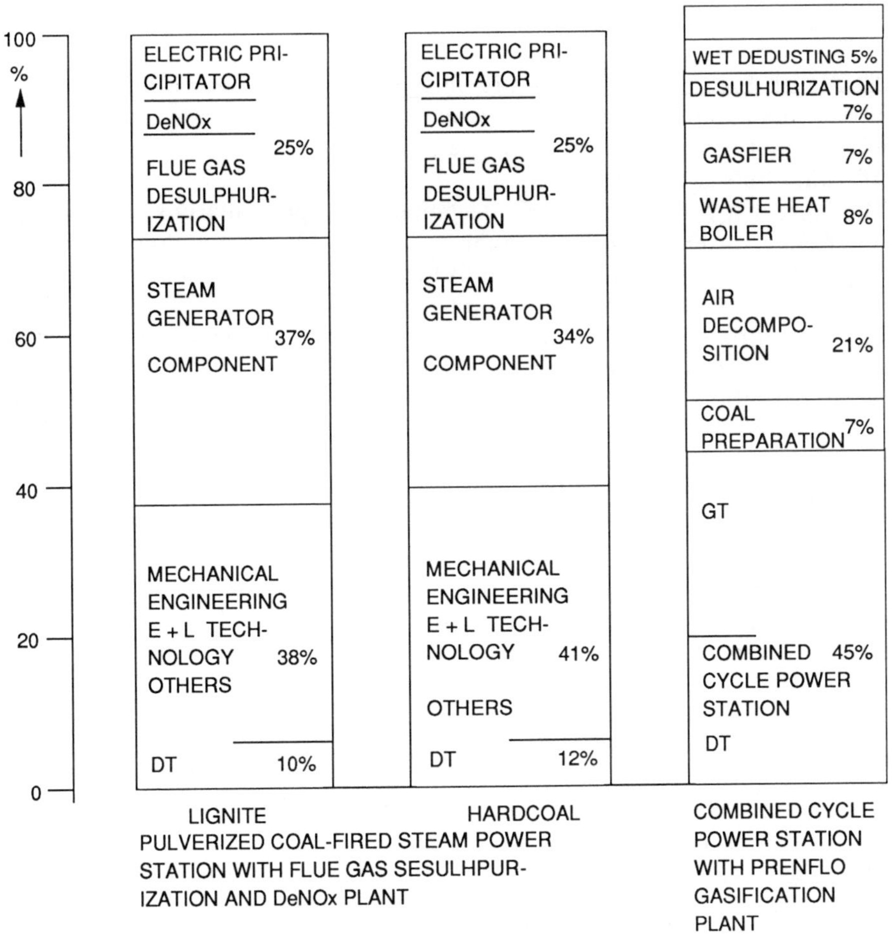

Figur 16. Comparison of the cost structure of a conventional power station with flue gas desulphurization and $DeNO_x$ plant versus that of a combined cycle power station with integrated coal gasification. (3, 1).

> "A particularly attractive feature of the gasifier – combined cycle power plant is the rather flat cost-vs.-capacity curve; thus the economics of a small, say 100 MW, unit or module is almost as attractive as that of a 500-600 MW block. This gives the power company a most welcome flexibility to tailor capacity additions to the rate of growth in demand. This feature alone represents a substantial saving to the consumer when compared to the much larger steps required for additons of conventional steam-electric plants. This advantage will become increasingly significant as we enter a period of reduced growth in demand of energy."

6.3 Plant Costs

As shown in Figure 16, the cost structure of a combined cycle power station differs considerably from that of a conventional power station (1, 3). It confirms that fundamentally new paths are tread with the use of coal gasification for electricity generation.

The high cost percentage for air decomposers is particularly striting. But if they are dispensed with, additional expenses due to the lower specific throughputs will bi incurred. Because of the lack of data it cannot yet be stated which approach will be the most economical.

The very different cost structure of a combined cycle power station means that considerable inaccuracies are possible in calculating the costs for a large plant. A variation range of ±20% has to be expected in the estimates of combined cycle power stationcosts, if they were not produced carefully. This should be kept in mind when considering the following figures.

Several system studies (22) are based on the Lurgi process using air or oxygen and dry or liquid slag removal. For advanced gas turbines with higher inlet temperatures, specific investment costs (basis 1984) of 2.261 DM/kW for a 200 MW combined cycle power station and of 1.711 DM/kW for a 800 MW combined cycle power station are estimated (1, 2) 3)[1]. This is slightly less costly than combined power stations with flue gas desulphurization and Denox plants.

Based on the Texaco gasification process, EPRI extimate plant costs of $1,630 per kW for a 500 MW combined cycle power station, while putting the cost for a conventional power station of the same capacity at $1,360 per kW. According to a study of General Motors, a price of less than $1,000 per kW might only be realized if ISTIG gas turbines are used.

The considerable differences among the cost statements in different studies are not only due to the variation range of ±20% and the exchange rates, but also to different fundamental assumptions. Logically, the same holds true for the calculated operation costs which were influenced by the different, yet-to-be-verified efficiencies estimates. In addition, the coal costs, which may differ by the factor 7 worldwide, are of major importance.

In conclusion, therefore, only the following can be stated with any degree of certainty: Advanced combined cycle power stations designed to available plans are likely to provide for a reduction of both plant and operation costs, but commercial realization of this prospect cannot be expected before the turn of the century.

[1] Due to the presently unrealistic exchange rates the costs in DM were not converted into Dollar

7 Summary

The gasification of coal has been well proven for decades in the production of synthesis gas in the chemical industry – in particular in the fertilizer industry. The fixed bed gasification according to Lurgi (Frankfurt), the fluidized bed combustion according to Davy Power Gas (Cologne) and the entrained flow gasification according to Koppers-Totzek (Essen) were competing. The three very different operating modes form the basis of approximately 40 process applications, some of which were derived from them. Exceptions are the molten iron bath gasification of Klöckner (Duisburg) and Sumitomo (Tokio) and the allothermal gasification of Bergbau-Forschung (Essen) – two new approaches which are still in the initial phase (8).

During the oil crises in 1973/74 and 1979/80, a number of major institutes intensified their work relating on gasification technology in an unprecedented way (36). In cooperation with Sasol (South Africa), Ruhrgas (Essen), British Gas Company (Scottland) and others, Lurgi worked on the extension of the reactor unit and the pressure increase. In cooperation with Uhde (Dortmund), the "Rheinische Braunkohlenwerke AG" (Cologne) tried to apply higher pressures in the fluidized bed gasification of lignite. Westinghouse (US) developed a similar multi-step, fluidized bed concept. The activities in the field of entrained flow technology are the most impressive ones. Krupp-Koppers was the starting point for considerable process improvements at Saarberg (Saarbrücken) and Shell at Hamburg, Amsterdam and Houston. Texaco (US) is the most successful company constructing commercial synthesis gas plants in cooperation with Ruhrkohle/Ruhrchemie (Essen/Oberhausen). Its process is characterized by a coal supply in form of a coal-water suspension. In parallel to this, Dow Chemical tried to developed a process that represents a synthesis of the Krupp-Koppers-Saarberg-Shell process and the Texaco-Ruhrkohle-Ruhrchemie process.

Generally, all processes resulting from this development work, which often advanced the state of the art, are also suitable for the conversion of coal into a fuel gas for gas turbines as components of modern combined cycle power stations. The first experimental approaches in this direction were advanced most in the KDV plant of Steag (Essen) on the basis of the Lurgi process and in the Cool-Water plant (Daggett, California) on the basis of the Texaco process. In both cases, encouraging concepts have to be acknowledged, but technical problems remain to be overcome because the integration of a gasification system into a power station is governed by totally different prerequisites compared to a synthesis gas plant. The use of air instead of oxygen as gasifying agent may, for example, be reasonable. A purification of the hot gas should be aimed at to avoid heat losses. The calorific value and the composition of the gas is not of decisive importance. Hot gas purification is one particular field in which much more process-related work will be done.

Numerous studies which were carried out and published by interested companies with outside financial support (DOE, EPRI, NCB, BMFT) deal with the consideration of these and other questions. Unanimously, all of them confirm the expectation that the efficiency may be improved significantly by the use of high-temperature turbines. The waste gas is by far less pollutant-loaded than the gases emitted by conventional plants. Moreover, investment costs are slightly lower than for conventional power stations with flue gas desulphurization and Denox plants.

References

1. Frewer, H. 1986. *VGB-Kraftwerkstechnik* 66: 4.
2. Müller, R. 1984. *VGB-Kraftwerkstechnik* 64: 12.
3. Müller, R. 1987. *Energiewirtschaftliche Tagesfragen* 37: 3.
4. Handbuch Grubengasabsaugung, Verlag Glückauf Essen, 1980.
5. Sommers, H. 1974. *VGB-Kraftwerkstechnik* 54: 5.
6. Peters, W. 1960. *Chemie-Ing.-Technik* 96:16.
7. Cortez D.H. (Tosko Los Angeles), Answer '82 Conference Brüssel.
8. Schilling, H.-D., Bonn, B., Kraus, U. 1982. Peters, W. Editor Essen: Glückauf Verlag.
9. von Gratkowski, H.W. 1958, 1970. Ullmann Encyclopädie 10 u.
10. Kirchhoff, R., van Heek, K.H., Jüntgen, H., Peters, W. 1984. *Nuclear and Design.* North Holland Publishing Amsterdam 78.
11. Peters, W., Echterhoff, H. 1961. *81st Furnace Conference,* Philadelphia. K.H. van Heek. 1982. VDI Forschungsheft No.612.
12. Peters, W., Jüntgen, H. 1965. Brennstoffchemie Essen, Heft 2.
13. LURGI Handbuch Frankfurt 1972.
14. Henrich, G., Pook, H., Waldhecker, H.D. 1986. *Energie* 38: 8.
15. van Heek, K.H. 1978. *Technische Mitteilungen* 71: 3.
16. Preyer, H., Röbke, G. 1980. *Chem.Ing.* 9: 401.
17. Vierrath, H., Kriebel, T. 1984. *Coal Technology Europa Conf.*, Essen.
18. Döring, H., Sommers, H., Rammler, R. 1975. *Erdöl und Kohle* 28: 5.
19. Jones, J.L., Radding, S.B. 1980. ACS Symposium Series 130.
20. Hydrocarbon Processing, April 1984.
21. Dorstewitz, U., Wien, H. 1987. *VGB Konferenz* TB 201.
22. Corman, J.C. 1986. DOE-Study DE-AC-80 ET 14928 and *Cool Water Conf.*, Veen Ceelen, Kreutzkamp, BMFT Forschungsbericht T 86-047.
23. Lambertz, J., Schrader, L., Teggers, H. 1986. *Chem.Ing.Techn.*58: 8.
24. Osthaus, K.H. 1974. *VGB-Kraftwerkstechnik* 54: 221.
25. Runge, H.C., von den Berg, G.J., Vogt, E., Wetzel, R. 1980. *11. Weltenergiekonferenz* München. Fluor Engineers Inc. 1983. EPRI AP-3129.
26. Pohl, H.Ch., Röhm, H.J. 1985. *Chem.Ind.* 37 t.
27. Lüke, H.W., van der Burgt, M.J. 1987. *Proceedings Coal Power Conf.* Düsseldorf.

28. Larson, E.D., Ogden, J.M., Williams, R.H. 1987. Princeton University PU/CEES, No.217.
29. Ruprecht, P., Cornils, B., Langhoff, J., Dürrfeld, R. 1987. Stahl und Eisen 105, No.217.
30. Bornemann, G.A., Sundstrom, D.G. 1986. *American Chem.Soc.* Preprints New York, 31: 2.
31. Nistler, F. 1957. *Coke and Gas*, 19: 54.
32. Knizia, K. 1985. VGB-Kraftwerkstechnik 65: 6. Weinzierl, K. 1982. *VGB-Kraftwerkstechnik* 62: 5.
33. van Heek, K.H. 1988. *Glückauf*, 124: 4/5.
34. Schilling, H.D. 1986. *Ind.Chem.Coal Techn.* Congress St. Louis.
35. van Heek, K.H. 1978. *Techn. Mitteilungen* 71: 3.
36. Peters, W. 1980. *Brennstoff Wärme Kraft* 32: 9.
37. Williams, R.H., Larson, E.D. This volume.
38. British Coal, Progress Report No.1, 1987. ECSC Project No.7220-EC/835.
39. Reichl, Eric H. 1988. Princeton – USA, personal communication.

Biomass Gasification for Gas Turbine Power Generation

Eric D. Larson, Per Svenningsson, Ingemar Bjerle

Abstract

This chapter evaluates the technical requirements and status of biomass gasification and hot gas cleanup for gas turbine power generation. The economics of gasifier-gas turbine systems are also discussed. Gas turbine technologies, including combined and steam-injected cycles, are of interest because they have relatively high efficiencies and low unit capital costs in the small sizes appropriate for biomass installations. Gasifier-gas turbine systems are not available commercially, but they could probably be developed in a few years. Extensive past work related to coal gasification and pressurized combustion of solid fuels for gas turbines would be relevant in this effort, as would work on pressurized biomass gasification for methanol synthesis. Biomass-gasifier gas turbine systems, if they are developed, could play important roles in the future supply of electricity in both industrialized and developing countries. In Sweden, electricity could be cogenerated at gas turbine district-heating plants at costs estimated to be competitive with new fossil-fuel central station plants. One quarter or more of the current national electricity production could be cogenerated from biofuels in the long term. The potential electricity generation is also very large in many developing countries, where the economics would be even more attractive due to lower biomass production costs.

1 Introduction

Biomass has attractive characteristics as an energy resource. It can be recovered as a byproduct or produced specifically for energy. The climate and soil in tropical countries are especially well suited for its production, but it can also be grown in temperate climates such as in Sweden. The production and use of biomass can be structured to cause only minor environmental impacts compared to the exploitation of fossil resources. Most biomass contains much less sulfur than coal or oil. Its use results in no net addition of heavy metals and other toxic elements to the biosphere, as is the case with most fossil fuels. And, if produced and used renewably, biomass would make no net contribution to atmospheric carbon dioxide.

Biomass is already an important energy source in many developing countries, but it is used very inefficiently. And, in industrialized countries such as Sweden and Denmark, it accounts for a significant portion of primary energy supply, but is predominantly converted to heat or steam in boilers. The low efficiency of photosynthesis will ultimately lead to land use constraints on the level of bioenergy use. However, this constraint can be relaxed if emphasis is given to energy efficiency in bioenergy conversion processes and end uses.

Thermochemical gasification is one of the simplest and most energy-efficient means for upgrading biomass. While gasification is conceptually simple, practical systems must be tailored to the end-use of the gas. Biomass gasification has been used with varying degrees of success with industrial furnaces and boilers, internal combustion engine-generators, and methanol synthesis plants. This paper considers biomass gasification for electric power generation using gas turbines. This application has received little attention, although there is considerable interest and ongoing development work in coal gasification for gas turbines (1,2,3).

Gas turbines appear well-suited for biomass applications because their natural scale is small (5–100 MW_e), as required for biomass facilities by cost constraints on fuel transport. Efficient gas turbine technologies, including combined cycles and steam-injected cycles, are discussed briefly in this chapter and in greater detail elsewhere (3).

In Section 2, we review the technical requirements and status of biomass gasification and hot gas cleanup technology for gas turbine applications. In Section 3, we present some preliminary performance and cost estimates for biomass-gasifier gas-turbine systems. In Section 4, some energy and economic implications of the widespread use of biomass-gasifier gas turbine systems are considered for Sweden and other regions of the world where biomass can be grown.

2 Biomass Gasification Systems for Gas Turbines

2.1 Background

In a biomass-gasifier gas turbine system, the feedstock would be gasified in a pressurized air-blown reactor and the products cleaned of particulates at elevated temperature before being burned in the gas turbine combustor. Hot gas cleanup is required to avoid cost and efficiency penalties that would reduce economic attractiveness. Pressurized gasification is required to avoid losses associated with compressing the fuel gas after gasification. Air gasification is dictated by the high cost of oxygen plants at the small scale required for biomass installations.

A representative biomass-gasifier gas turbine cycle, based on a steam-injected gas turbine (STIG), is shown in Figure 1. The basic components of the gasification system would be unchanged even if an alternative gas turbine technology were used in place of the STIG, e.g., a simple cycle, a combined cycle or an intercooled steam-injected gas turbine (ISTIG).

Past applications of gasification have related to industrial heating, internal combustion (IC) engines, and methanol synthesis. Gasifiers feeding industrial boilers, which have been successful in North America (4), Brazil (5), and elsewhere (6), typically operate with air at atmospheric pressure in "close-coupled" arrangements where gas cleanup is generally not a major concern.

Air-blown-gasifier/IC engines, which were widely used for vehicles during World War II in Europe (7) and are receiving renewed attention today (8,9), include extensive gas cleaning (to remove troublesome particulates and tars) and cooling for proper engine operation.

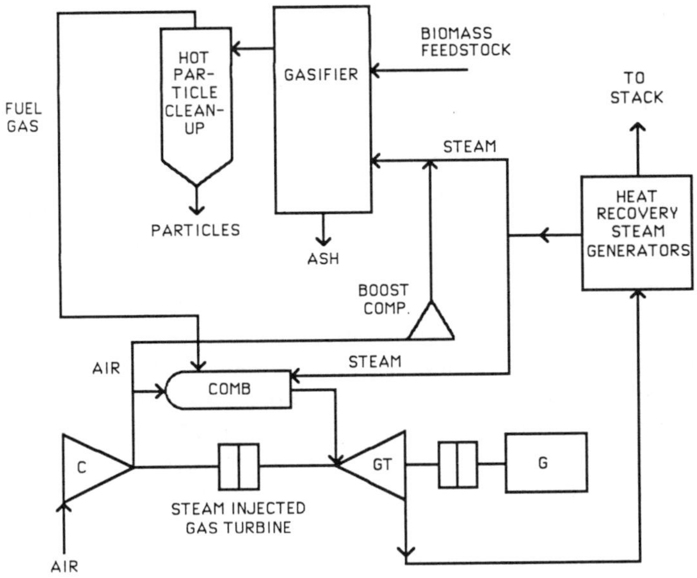

Figure 1. A biomass-gasifer gas turbine cycle. For illustration a steam-injected gas turbine is shown. An alternative gas turbine technology, e.g., a simple-cycle, a combined cycle, or an intercooled steam-injected unit could be considered as well.

A methanol synthesis plant typically includes: a pressurized oxygen-blown gasifier, necessitating large-scale operation to exploit scale economies; cleaning to eliminate all particulates and condensibles; a shift reactor to obtain the desired gas composition for methanol synthesis (2:1, H_2:CO); and a still higher-pressure catalytic reactor for the final conversion to methanol.

The technical requirements of gasification for gas turbines are quite different from those for previous applications. Particulate cleanup requirements are much stricter than for heating applications and comparable to those for IC engines and methanol synthesis. However, with hot cleanup, neither gas cooling nor tar removal are required. Furthermore, there are no strict requirements on gas composition (as with methanol synthesis), providing fuel flammability limits are met and the heating value of the fuel is above 4 MJ/Nm^3.[1] Overall, the requirements for gasification for gas turbines are less stringent than for other non-heating applications, which suggests that supplying fuel gas to gas turbines may in many ways be simpler and, perhaps, less costly.

2.2 Gasification Fundamentals

Gasification consists of two basic sets of reactions. Pyrolysis refers to a complex set of reactions during which the volatile components of the feedstock vaporize at temperatures below about 600°C and leave behind fixed carbon (char) and ash (10,11). Biomass consists of 75–85% volatile matter compared to half or less this level with coal (Table 1), so pyrolysis plays a larger role in biomass gasification. In addition, biomass generally pyrolyzes at lower temperatures than coal (Figure 2a).

Products of pyrolysis include water vapor and heavy hydrocarbon compounds (tars and oils) that condense at relatively high temperatures. In some gasifiers, the tars and oils constitute an important energy component of the raw gas. In other cases, gas-phase reactions occur in higher temperature regions of the gasifier, converting most of the primary pyrolysis products into lower-molecular-weight permanent gases.

Char conversion refers to the gasification and/or combustion of the fixed carbon that remains after pyrolysis. With biomass, a relatively small amount of char remains after pyrolysis due to the low fixed carbon fraction (Table 1). Much of the char burns to provide heat for pyrolysis and to gasify any remaining char. Biomass chars gasify much more easily than coal chars, because they are 10–30 times more reactive (12). Thus, biomass gasifiers can typically operate at lower temperatures than coal gasifiers to achieve the same char conversion (Figure 2b).

[1] One Nm^3 (normal cubic meter) is one cubic meter at standard temperature and pressure.

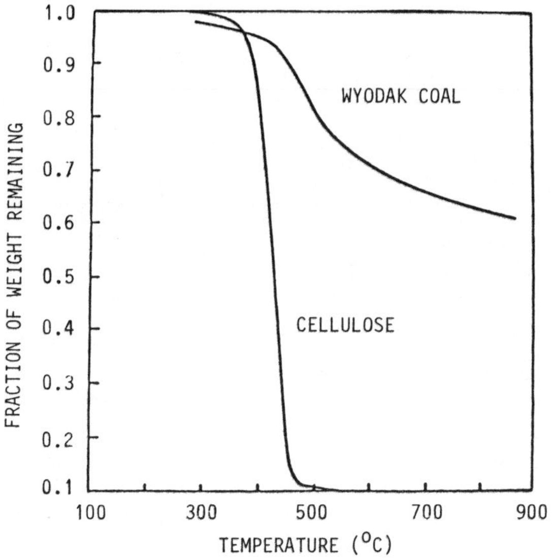

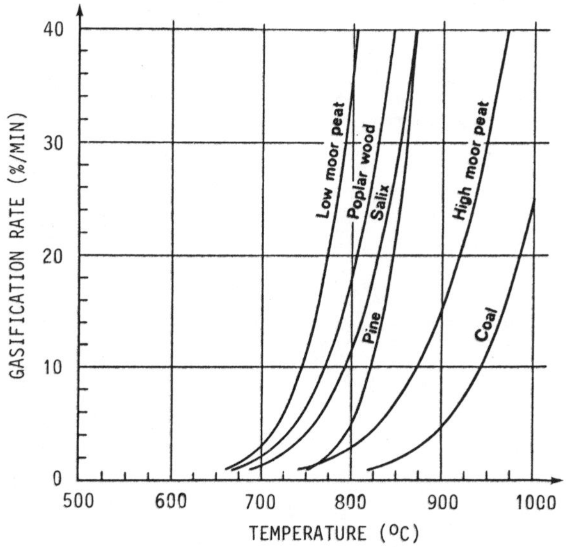

Figure 2. *(a)* Weight loss as a function of pyrolysis temperature for coal and cellulose (a major component of biomass) (83) and *(b)* Gasification rates in steam of chars from different feedstocks (84).

Table 1. Typical proximate and ultimate analyses of gasification feedstocks (weight percent, dry basis).

Proximate analysis	Volatiles	Fixed Carbon	Ash	*Ultimate analysis* C	H	N	S	O	Ash	Heating Value (MJ/kg)[d]
Biomass[a]										
Douglas Fir										
Wood	86.2	13.7	0.1	52.3	6.3	0.1[c]	0.0	40.5	0.8	21.0
Bark	70.6	27.2	2.2	56.2	5.9	0.0	0.0	36.7	1.2	22.0
Sawmill waste										
Redwood shavings	76.2	23.5	0.3	53.4	6.0	0.1	0.1	39.9	0.6	21.3
Oak chips	74.7	21.9	3.3	49.5	5.7	0.2	0.0	41.3	3.3	19.2
Other										
Sugar cane bagasse	74.0	23.0	3.0	45.0	6.0	0.2	0.0	46.0	3.0	19.1
Municipal solid waste (USA, 1970)	65.9	9.1	25.0	47.6	6.0	1.2	0.3	32.9	12.0	19.8
Coals[b]										
Texas Lignite	38.9	44.5	16.6	65.1	4.8	1.1	1.2	16.9	10.8	24.3
Illinois No. 6	35.7	54.3	10.0	68.0	5.2	1.2	3.8	9.8	12.0	29.6

[a] From (49a), except for sugar cane bagasse (49).
[b] From (49b).
[c] Nitrogen content of biomass can vary substantially and could be as high as 0.5% in some cases.
[d] Higher heating value, moisture free basis.

The chemistry and practical operational aspects of gasification are influenced by four main parameters: type of oxidant, temperature, pressure, and presence of moisture.

Oxygen or air is the oxidant in most gasifiers, and steam is sometimes used as an additional reactant. Using oxygen produces a gas with a heating value of typically about 11 MJ/Nm^3 (compared to 38 MJ/Nm^3 for natural gas). Air gasification produces gas with about half this energy content, due to the diluting effect of the nitrogen, but this is still sufficient for gas turbine combustion.

Temperature affects gasification rates and, more importantly, reactor design. Operating temperatures influence the design of ash removal systems. Both slagging and dry-ash gasifiers have been developed for coal. Biomass and coal ash have comparable melting temperatures (typically 1100–1200°C), but since biomass can be gasified at lower temperatures (Figure 2b), most biomass gasifiers utilize dry ash removal systems. Ash removal from biomass gasifiers is further simplified by a much lower ash content than for coal (Table 1).

Pressure has a modest effect on gasification chemistry, but an important effect on system design and cost. Higher pressure permits higher processing rates for a given reactor size.[2] The ability to use smaller reactors leads to savings in material and in field construction costs. Drawbacks of higher pressure operation are the increased complications of feeding and the increased risk of clinker formation in hot spots within the oxidation zone of the gasifier.

Moisture plays a role in char gasification. In some gasifiers, steam is injected to provide some of the required moisture,[3] since water in the feedstock is driven off during pyrolysis and is thus unavailable for reaction. In some gasifiers, additional steam is used to maintain the reacting bed below a specified temperature, typically that of ash melting.

Excess moisture acts to dilute the heating value of the raw gas. The lower limit on the heating value of gas for stable gas turbine combustion appears to be about 3.9 MJ/Nm^3 (13). This sets an upper limit on the acceptable moisture content of the feedstock of 25–30%.[4]

2 In a fluidized-bed gasifier, throughput (tonnes per hour) increases from m_o to m as pressure increases from p_o to p, approximately according to $m/m_o = (p/p_o)^{0.6}$. Thus, for example, fuel throughput would be seven to eight times as large when operating a reactor at 30 bars instead of atmospheric pressure.

3 Because of the inherently higher moisture content and reactivity of most biomass feedstocks compared to coal, external steam requirements are generally lower for biomass gasification than for coal.

4 Unless explicitly indicated otherwise, feedstock moisture contents are expressed on a wet weight basis (percentage of the total wet feedstock weight).

2.3 Gasifier Designs

Three gasifier designs have traditionally been considered for coal: entrained-bed, fixed-bed and fluidized-bed. Several entrained-bed gasification systems have been brought to advanced stages of development for coal-gasifier combined-cycle applications, but appear ill-suited for biomass-gas turbine applications.[5] Air-blown fixed-bed and fluidized-bed gasifiers appear to be more promising candidates in terms of performance and near-term commercial viability. Table 2 gives some comparative performance characteristics based primarily on pilot-scale measurements.

2.3.1 Fixed-bed gasifiers

Performance

The most interesting of several fixed-bed (also called moving-bed) gasifiers for gas turbine applications is the updraft or counter-current design, in which the feedstock successively undergoes drying, pyrolysis, char gasification and char combustion (Figure 3a).[6] The blast air is largely consumed at the grate during combustion, leading to peak temperatures at this point in the bed. Blast air requirements are relatively low (Table 2). Steam is used to maintain the bed below the ash melting temperature and to a lesser extent to facilitate char conversion. The combustion products and steam drive char gasification reactions just above the grate. The gas cools as it travels up through the bed, pyrolyzing the incoming feedstock.

The product gas exits near the top with relatively low temperature and laden with tars and oils (in vapor form). The gas exits with relatively low particulate concentrations because of low exit velocities and the filtering effect of the fuel bed. The raw gas heating value is high due to the tars and oils. Good carbon conversion (99%) is achievable because of the high peak temperatures and the efficient (counter-current) heat transfer from the gas to the fuel bed.[7] Hot-gas efficiency can also be high (90–95%), but cold-gas efficiency is typically low (65–75%) because tars and oils are excluded from this measure.

[5] One drawback is the relatively low heating value of raw gas from an air-blown entrained-bed unit.

[6] Other types of fixed-bed gasifires include downdraft (co-current) and rotary kiln gasifiers.

[7] See Table 2, note (g) for definition of efficiencies.

Table 2. Operating characteristics of fixed and fluidized bed gasifiers with biomass feedstocks.[a]

	Fixed-bed (updraft)[b]	Fluidized-Bed[c]
Solids residence time (s)	300–900	5–120
Throughput (dry tonne/m^2-hr)	2.1	7.0
Typical blast (kg/kg$_{dry\ feed}$)		
Air	0.8	1.3
Steam	0.3	0[d]
Peak temperature	1100	827
Raw-gas		
Exit temperature (°C)	550	800
Yield (kg/kg$_{dry\ feed}$)	2.7	2.9
Typical Composition (Vol.%)		
(w/15% feedstock moisture content (mc))		
CO	13.0	8.1
H_2	17.1	6.9
CH_4	5.1	4.8
C_2H_4	–	0.2
C_2H_6	–	0.3
C_6H_6	–	0.5
CO_2	13.1	14.1
N_2	29.5	36.3
H_2O	28.0	30.0
Tars/oils (wt %)	3.3	0.8
Particulates (g/Nm^3)[e]	1	100
Higher heating value (MJ/Nm^3)		
Wet gas	5.0	4.6
Dry-gas	6.9	6.5
Efficiencies[f,g]		
Carbon conversion (%)	99	98
Hot efficiency (%)	93	95
Cold efficiency (%)	67	81

a The performance described here is based primarily on pilot-scale testing. Gasifiers can be operated over a wide range of parameters, and performance can vary accordingly.

b Unless otherwise noted, these are preliminary results for gasification of 14.9% mc wood pellets at 21.4 bar in a gasifier designed for gasification of 24 t/day of coal (14). The carbon conversion efficiency is that estimated for oxygen gasification of coal at the same pressure (44).

c Unless otherwise noted, these are for gasification of 14.5% mc whole-tree hardwood chips at 21.8 bar, 827°C in air and steam in the IGT 12 tonnes/day bubbling fluidized-bed gasifier without char recycle (17).

d Based on (50).

e Estimates range over an order of magnitude or more. The upper-end value is given here (30).

f Efficiencies for the fluidized-bed unit are for 9.1% mc whole-tree chip gasification in enriched air and steam at 911°C, 21.6 bar in the IGT gasifier without char recycle (17).

g Carbon conversion is the fraction of input carbon converted to gas. Hot efficiency is the chemical and sensible energy in the raw gas divided by the total energy input (chemical plus sensible). Cold efficiency is the chemical energy in the gas excluding condensibles (tars/oils), divided by the total input.

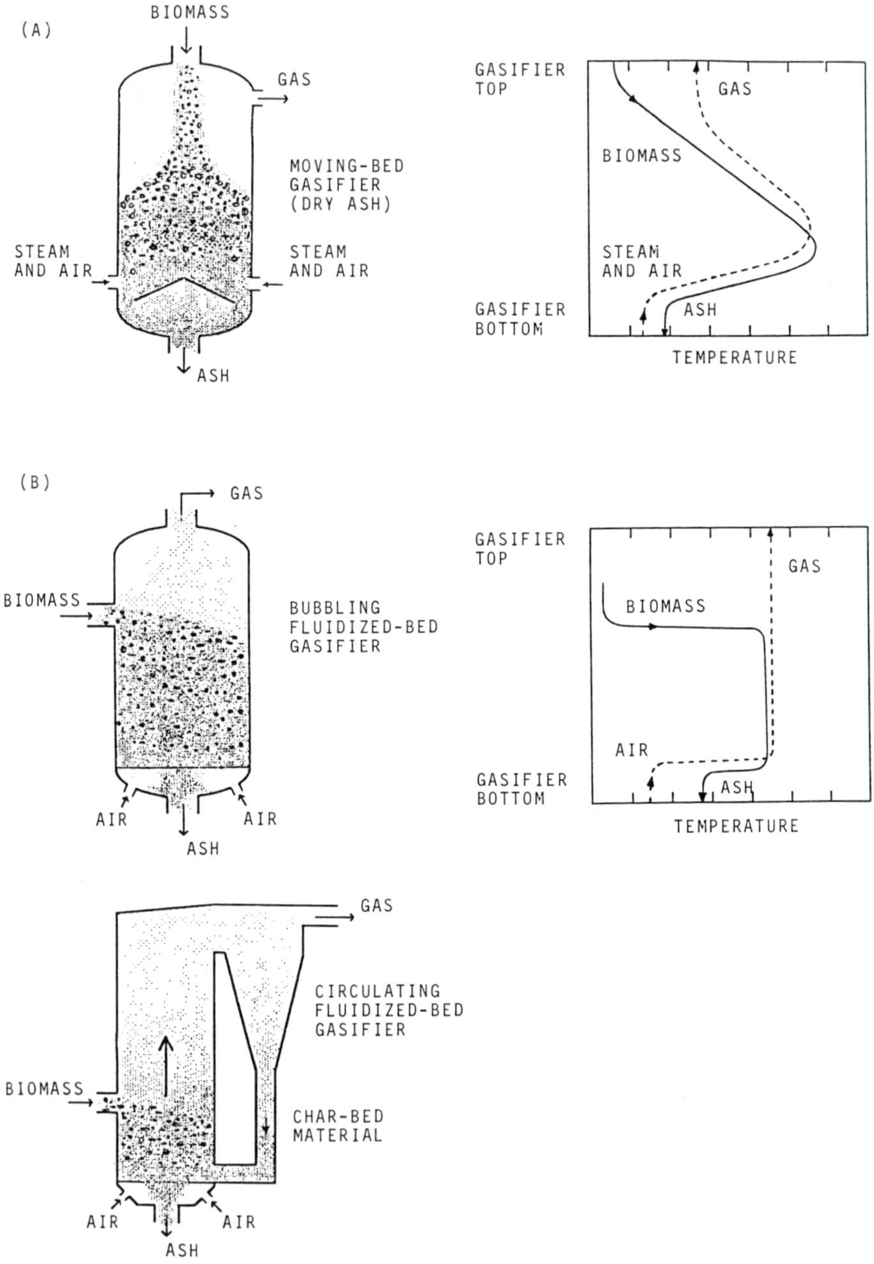

Figure 3. Operating principles and temperature profiles for (*a*) fixed-bed and (*b*) bubbling fluidized-bed gasifiers. Also shown is the operating principle of a circulating fluidized-bed gasifier.

Feedstock requirements

Relatively large, dense, and uniformly-sized feedstocks are preferable for fixed-bed gasifiers to help prevent bridging, reduce feed losses, and reduce carryover (Table 3). Some designs have rotating agitators to assist the fuel flow. Feedstocks with as high as 50% mc (moisture content) can be gasified, but the moisture will dilute the heating value of the raw gas as discussed earlier. Suitable fuels appear to be wood chips and mill residues (Table 3), though some size screening of the residues would probably be required. Most agricultural residues would require densification for use in fixed-bed reactors, which would increase fuel costs.

Status

Because of their simplicity and high efficiency, updraft gasifiers are widely used commercially with both coal and biomass. The pressurized Lurgi dry-ash gasifier, which has the most commercial operating experience of any gasifier with coal and which has been considered for coal-gasifier STIG, ISTIG, and combined-cycle systems (13), appears to also be a good candidate for biomass applications. Successful, but limited, pilot-scale testing of a Lurgi-type unit has been carried out by the General Electric Company (GE) using biomass pellets (14) and RDF/coal briquettes (15) (Table 4). More extensive testing would be required to demonstrate the feasibility of operating on various feedstocks, e.g., wood chips and densified agricultural residues, and of removing particulates at elevated temperatures (see below). Such testing might be undertaken at low incremental cost using the GE pilot gasifier or a comparable existing facility.

2.3.2 Fluidized-bed gasifiers

Performance

In a fluidized-bed gasifier, the feedstock is typically fed continuously and is "fluidized", together with an inert heat-distributing material, e.g., sand, by oxidant and/or steam injected from below. The bubbling bed (Figure 3b) was the first fluidized-bed design developed. A variation of this, the circulating fluidized bed (CFB) (Figure 3b, lower), has been developed more recently. In any fluidized bed, pyrolysis and char conversion occur throughout the bed. The excellent heat and mass transfer leads to relatively uniform temperatures throughout the bed, better fuel-moisture utilization, and faster reactions than in fixed beds. The latter leads to much higher throughput capabilities per unit volume for fluidized beds (Table 2).

Table 3. Required biomass feedstock characteristics for alternative gasifier designs and properties of actual biomass feedstocks.

Gasifier feedstock requirements:	Fixed-Bed Updraft	Fluidized-Bed Bubbling	Fluidized-Bed Circulating[a]
Length (mm)[b]	13–75	0–50	0–100
Max. thickness (mm)[b]	6–13	0–3	0–50
Size variability	small	large	larger
Moisture (wt%)[c]	< 30%	< 30%	<30%
Bulk density (kg/m^3)	> 240	(d)	(d)

Actual properties:	Length (mm)	Thickness (mm)	Bulk Density[e] (kg/m^3)
Wood chips			
Hardwood	16–22	1.5–3	280–480
Softwood	16–22	3–6	200–340
Mill Residues	maximum size < 50 mm		290–400
Biomass Pellets	6–19	6–19 (diam.)	610
Non-Woody Biomass			
Pits (cherry, olive, etc.)			360
Straw (baled)			160–300
Bagasse			160
Rice hulls			130
Shells (walnut, coconut, etc.)			64

a From (50).

b For fixed and bubbling fluidized-bed, see (51).

c Fuel moisture level to produce raw gas of suitable quality for gas turbines. The moisture level at which gasification can be sustained is about 50%.

d The minimum acceptable bulk density for stable operation of fluid beds is much lower than for fixed beds. Fuel handling and feeding considerations are likely to set the lower limit.

e Bulk densities vary with moisture content. For wood chips, the range given here is for 15 to 50% mc. Bulk densities for mill residues (hog fuels) vary widely. The indicated range is for 50% mc samples. Pellets and non-woody biomass have 10–15% mc, execpt the figure for bagasse, which is for 50% mc. The baled straw range is for standard and "double compressed" bales.

Reaction temperatures are kept relatively low (to prevent ash melting) by control of blast air rates and good mixing of the bed. As a result, little or no steam is needed for cooling, unlike with a fixed bed. While peak temperatures are lower than in a fixed-bed, average temperatures are higher (Figure 3b), leading to conversion of a significant amount of tars and oils into permanent gases.

Carbon conversion when gasifying coal appears to have a practical maximum of about 96% (for lignite when char carried out of the reactor with the gas is recycled to the bed (16)). With the more reactive biomass, however, conversion in excess of 98%–comparable to that in fixed-beds–has been achieved in pilot-scale tests without char recycle (17). Hot-gas efficiencies are also comparable to those for the fixed-bed, but cold efficiencies are higher (80–85%) because of the low tar/oil production (Table 2).

The particulate loading in the raw gas is one to two orders of magnitude greater than with fixed beds (Table 2).

Feedstock requirements

Fluidized-bed gasifiers can accept much smaller, less dense, and less uniform feedstocks than fixed-beds (Table 3). The minimum acceptable bulk density for fluid beds is much lower than for fixed-beds. Requirements for fuel handling and feeding, which is more difficult the less dense the fuel, will most likely set the acceptable lower limit. Fuel moisture considerations are the same as those for fixed-beds. The range of acceptable feedstock characteristics is greater with circulating fluidized beds than with bubbling beds (Table 3). Most of the feedstocks listed in Table 3, including minimally processed agricultural residues, appear to be acceptable fuels.

Status

The major advantages of fluidized-bed gasifiers over fixed-bed units are their greater fuel flexibility and higher throughput capability. The higher throughput capability leads to lower unit capital costs than fixed-beds in most of the size range of interest for gas turbines–greater than 15 MW_{fuel} (18).[8] The main drawback of fluidized beds appears to be higher levels of particulates in the raw gas, which makes gas cleanup more challenging (see below).

8 Fixed-beds have a capital cost advantage below about 10 MW_{fuel} because of the relatively high cost at this small scale of the blowers, continuous feed systems, control systems and other instrumentation associated with fluidized beds.

Table 4. Gasifiers that have operated with biomass feedstocks.

	Status[a]	Capacity (t/h)	Pressure (bars)	Blast[a]	Tested Feedstocks
Fixed-bed					
Lurgi Updraft (Dry Ash)[b]	P	2	21	A/S	Wood pellets/RDF briquettes
Syn-Gas Downdraft[c]	P	2	10	A,O	Wood chips/Wood pellets/Corn cobs/ Peat pellets
Wright-Malta Kiln[d]	P	0.2	21	S	Wood chips/MSW
Fluidized-bed					
Rheinbraun- Uhde HTW[e]	D	3	10	A/S	Wood chips/Peat
	C	27	13.5	O/S	Peat
ASCAB[f]	D	5	25	O/S	Wood chips
Biosyn[g]	C	6	8	A	Wood chips
	D	10	14	A,O	Bark/Sawdust/ Wood Pellets
IGT Renugas[h]	P	1	34	A/S,O/S	Wood chips/Corn stover/Pulp mill wastes
MINO[i]	P	0.5	28	O/S	Wood chips/Peat/ Lignite
TRC Finland[j]	L	0.09	10	A	Peat pellets
Lurgi CFB[k]	C	16	1	A	Bark/Wood wastes
Götaverken CFB[l]	C	8	1	A	Bark/Sawdust/Wood wastes
Ahlström CFB[m]	C	8	1	A	Wood waste/Peat

[a] C=commercial, D=demonstration, P=pilot, L=laboratory, A=air, O=oxygen, S=steam.

[b] The pilot facility is at the General Electric Corporate Research Center, Schenectady, New York.

[c] From (52).

[d] In the Wright-Malta rotary kiln gasifier, external heating provides most of the energy for gasification. High moisture feedstocks are used (40–50%), with a residence time of 1 hour. The gasifier is being developed by Wright-Malta with Zurn Engineering (Erie, Pennsylvania) to drive a gas turbine, the exhaust heat from which would drive the gasifier. Commercial availability is anticipated in 1991 (53).

[e] The pilot plant, which is located at the Rheinisch-Westfalische Technische Hochschule in Aachen, FRG, has been operated primarily with lignite, though some tests have been made with biomass as well (54). The commercial unit has recently started operation at an ammonia production plant in Oulu, Finland (55).

——————>

Fluidized-beds appear to be good candidates for gas turbine applications. A major coal-gasifier gas turbine demonstration project in the USA will utilize a fluidized-bed gasifier (19). Development of fluid-beds for biomass-gas turbines could build on such efforts and on the extensive previous development efforts on large-scale pressurized gasification for methanol-from-biomass (Table 4). Of the units described in Table 4, the most extensive development and testing has been done with the Rheinbraun/Uhde HTW, Biosyn, IGT, and MINO units. Of these, only the Rheinbraun/Uhde unit is currently being actively developed (for coal applications), with a 25 bar pilot plant (for combined-cycle applications) scheduled for start up in late 1989 (20). As with the fixed-bed, additional development work to tailor biomass-fueled fluidized-beds to gas turbines could be undertaken at low incremental cost in one of the existing units.

Also noteworthy is the commercial availability of three atmospheric-pressure CFB gasifiers (Table 4). One manufacturer of CFB gasifiers, Ahlström, has recently announced plans to startup a 10 MW_{fuel} pressurized CFB combustor in the spring of 1989 (21). The Ahlström project is intended to demonstrate the feasibility of CFB combustion for direct firing of gas turbines with solid fuels, including the use of ceramic hot gas filters (22). Since CFB gasifiers are similar to CFB combustors in many respects, successful demonstration of a pressurized CFB combustor/hot filter would be a major step toward successful demonstration of a pressurized CFB gasification system for gas turbines.

f The unit, located in Clamecy, France, has undergone only limited testing to date (50). It was originally designed for methanol production.

g The demonstration facility is located in St. Juste de la Breteniere, in southern Quebec Province, Canada (56). The commercial unit, which supplies gas to a 6.7 MW diesel engine in Cayenne, French Guyana, was scheduled for startup in early 1987 (57).

h The facility is located at the IGT research center in Chicago, Illinois (17,58).

i Testing, including some with hot gas filtering, was undertaken between 1983 and 1986 at the pilot plant facility at Studsvik AB, Nyköping, Sweden (23,50).

j The unit is located at the Technical Research Center of Finland, Espoo, Finland. Related work is ongoing on gasifier feeding systems, gas cleaning, and gas turbines fired with gasified peat (59).

k This circulating fluidized-bed (CFB) supplies fuel gas to a lime kiln at a pulp mill in Pols, Austria (60).

l The Götaverken unit supplies fuel gas for operations at a pulp mill in Värö, Sweden (18).

m Ahlström units are operating in Finland, Sweden, and Portugal (22).

2.4 Gas Cleanup

Cleaning raw gas is required to avoid damage to turbine blades and to meet emissions regulations. The main gas contaminants of concern from biomass gasifiers are tars/oils, alkalai compounds, nitrogen (as a component of various chemical compounds), and particulates. The most effective cleaning systems involve cold wet scrubbing, which complicates a plant and reduces efficiency. Thus, only processes for cleaning gas at elevated temperature are considered here. A limited amount of work has been done on pressurized hot cleanup of biomass gas, for methanol production (23) and more recently for direct firing of gas turbines using pressurized combustion (22,24). Extensive work has been done on hot cleanup for coal-gasifier gas-turbine systems and for directly-fired systems using pressurized-fluidized-bed combustors (PFBC) (25–29). The coal-based work is relevant for biomass, but cleaning biomass gases is simpler, because most biomass gas does not require sulfur removal (Table 1).

2.4.1 Tars and oils

Tars and oils are a problem only if the temperature of the gas falls below their dew point, which ranges from 150°C to 500°C for biomass tars (30). In condensed form, tars and oils foul surfaces, leading to increased maintenance costs and performance degradation. Tar and oil condensation is of greatest concern with updraft gasifiers, which produce the largest quantities and at the lowest gasifier exit temperatures (Table 2). Tar condensation can be avoided, however, by keeping the gas temperature above the tar/oil dew point until it reaches the gas turbine combustor. Such close-coupled arrangements have been incorporated into designs for advanced coal-gasifier gas turbine systems (13,31).

2.4.2 Alkalai compounds

Alkalai compounds in fuel gas, which form from potassium and sodium in the feedstock, are a concern because they can accelerate hot corrosion and cementing of particulate deposits on turbine blades (32). The limit for alkalai metal in coal-derived gaseous fuel for a gas turbine has been given as 0.1 to 0.2 ppm by weight (32,33). Limits for biomass-derived gas are probably similar to those for coal-derived gas. Thermodynamic calculations suggest that for pressurized gasification at 800–900°C, raw-gas concentrations of alkalai would be well in excess of the suggested limits and largely in vapor form, making their removal more difficult (32,34).[9]

Alkalai compounds in coal gas appear to condense on entrained particulates at temperatures of 500–650°C and, therefore, can be reduced to acceptable levels by particulate cleanup at these temperatures (13,29,31). Measurements are required to determine whether this would be the case for alkalais in biomass-derived gas as well. If alkalai removal by this mechanism is insufficient, then the use of an inorganic sorbent (e.g. aluminosilicates) after the particulate removal system might be feasible to collect ("getter") much of the remaining alkalai (32). Additional work is needed to fully understand alkalai production and removal.

2.4.3 Nitrogen oxides

Nitrogen oxides (NO_x) can be produced in the gas turbine combustor from nitrogen in the fuel (fuel-bound NO_x) or in the combustion air (thermal NO_x). The peak combustion temperatures of coal or biomass-gas are relatively low because of their low heating value, so that thermal NO_x levels (produced by dissociation of N_2) would be relatively low (33). In any case, thermal NO_x can be controlled to a large extent by the use of low-NO_x combustion technologies, of which steam-injection is the most widely used today. As NO_x regulations tighten in the future, the development of dry low-NO_x combustors is anticipated, using staged or catalytic combustion (35).

Fuel-bound NO_x ($FBNO_x$) is produced from nitrogen compounds, primarily ammonia, which form during gasification. In the gas-turbine combustion of gas derived from bituminous coal, $FBNO_x$ has been estimated to be produced at a rate of about 0.3 grams per MJ of fuel burned (33). $FBNO_x$ from the combustion of biomass gas has been measured to be significantly less than this (36), which would appear consistent with the lower fraction of nitrogen in most biomass compared to coal (Table 1). With either coal or biomass, however, the $FBNO_x$ levels would probably exceed most emissions regulations, and would therefore need to be reduced. Staged or catalytic combustion technologies are under development to limit $FBNO_x$ in coal-gasifier gas-turbine systems (29,37). Such technologies for biomass might be simpler because of the lower fuel-bound nitrogen levels.

9 This is in contrast to the calculations assuming oxidation conditions, which would be representative of pressurized fluidized-bed combustion (PFBC). At temperatures below about 950°C, the alkalai are present as condensed sulphates, which can probably be removed with the particulates (32). At higher temperatures, the alkalais would be present in vapor form.

2.4.4 Particulates

Cleanup of particulates is required to prevent erosion of gas turbine blades, although the extent of required cleanup appears uncertain. Updraft biomass gasifiers produce two orders of magnitude less particulate matter than fluidized-beds, but also much smaller particles (Table 5). Estimated permissible particulate loadings for gas turbines vary widely (Table 5), as do particle size distributions. Table 5 gives one size distribution specification developed by General Electric (GE). More recently, the Electric Power Research Institute (EPRI) in the USA has proposed a particulate specification which takes account of the greater damage done by the larger particles (38):[10]

Particle size (micron)	> 20	10–20	4–10
Maximum ppm (weight)	0.1	1	10

Two technologies are usually considered for hot particulate removal. The most common and least costly is the cyclone. A high efficiency cyclone will remove most particles larger than 5–10 microns, but is ineffective for smaller particles. Because of relatively large particle sizes in gas from an updraft *coal* gasifier (Table 5), cyclones have been shown in pilot scale tests to provide adequate cleanup to meet the particle size specification developed at GE (Figure 4). For biomass applications, cyclones would appear not to be sufficient to meet the GE specification because of the large fraction of submicron particles (Table 5). They may be sufficient, though, to meet the criteria proposed by EPRI. Particles from fluidized-bed biomass gasifiers are larger but much greater in number than from fixed-bed gasifiers, and therefore may require the use of more efficient particle removal technology.

Barrier filters provide more complete filtration than cyclones, but have not been commercially proven for gas turbine applications. They are typically constructed of either sintered metal alloys, e.g. inconel or hastalloy, which can operate at 700–900°C, or of ceramic materials, e.g. silicon carbide, which can operate at still higher temperatures. A variety of filter designs have been developed, most of which operate by initially trapping large particles within a fine mesh (0.5–100 micron pore sizes), but then quickly begin trapping much smaller particles as a layer of dust builds up. When the pressure drop across the filter reaches a specified level, the filters are cleaned by a reverse jet blast.

For hot cleanup of biomass gas, sintered metal filters appear promising, based on previous pilot-scale work in Sweden (23). For coal-gas cleanup, the

[10] Smaller particles will tend to follow streamlines around the turbine blade. They will do little damage even if they hit the blade, since the angle of impact will be shallow and their velocity will be slowed in passing through the fluid boundary layer around the turbine.

Table 5. Particle loadings and size distributions from updraft and fluidized-bed biomass and coal gasifiers and specifications for gas turbine operation.[a]

	Particulates in raw fuel gas			*Suggested turbine inlet requirements*
	Biomass[c] Updraft	Biomass[c] Fluid-Bed	Coal[d] Updraft	
Load (g/Nm^3)[b]	0.1–1	10–100	0.5–0.94	0.002–0.02[e] 0.1–0.2[f]
Distribution (wt% less than)				
500 microns			98	
250			82	
100			50	
50		99	13	
30		97	10	
20	99+	87	9	98[g]
10	97	38	8	83
5	96	17	3	45
2	95	2	2+	7
1	89		2	0.3
0.5	81			

a The particulates in raw fuel gas are at the gasifier exit, while the gas turbine specifications are for the gas entering the turbine expander, i.e., after combustion and dilution with cooling air. Cooling air would dilute the fuel gas particulate concentration by about a factor of six.

b 1 g/Nm^3 is approximately 1000 ppm by weight.

c From (30) for atmospheric-pressure gasification.

d The updraft results are based on operation of a pilot-scale pressurized Lurgi-type gasifier (33). Data for fluidized-bed coal gasification were not available.

e From (30).

f From (61).

g The size distribution is from (25) and was developed by General Electric in the late 1970s as part of a US Department of Energy supported program in pressurized fluidized-bed coal combustion for locomotive gas turbine applications.

emphasis has been on silicon-carbide candle filters (39). The collection efficiencies with either metal or ceramic filters would be high. In pilot-scale tests with coal-gas from the KRW fluidized-bed gasifier, collection efficiencies well in excess of 99% have been reported, using commercially available silicon-carbide candle filters (26). Similar efficiencies have been reported for removal of simulated gasifier char (4.5–5.5 micron size), using a silicon-carbide/Nicalon filter developed at the Oak Ridge National Laboratory (40). Since collection efficiency appears well established, how rapidly ceramic barrier filters reach commercial readiness will probably depend more on the demonstration of their ability to withstand thermal and mechanical shock and

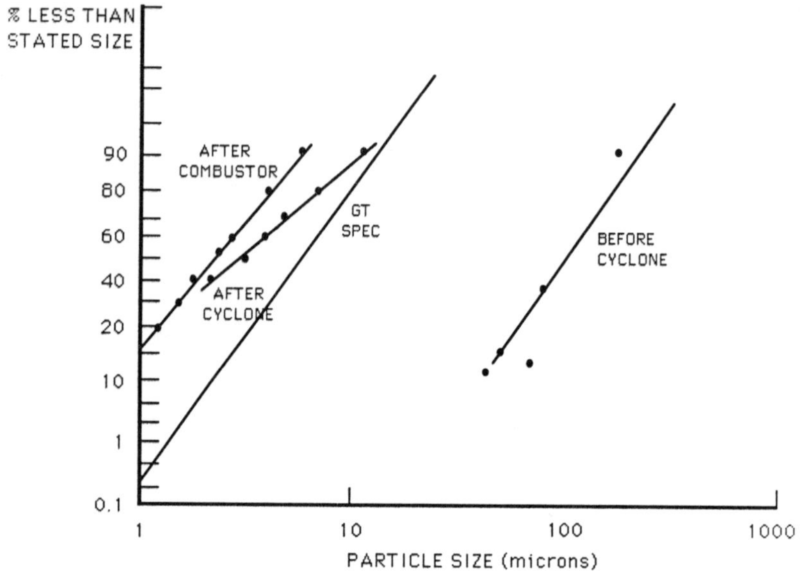

Figure 4. Measured particulate size distribution in tests with a pilot-scale fixed-bed dry-ash coal gasifier (25). The data are for the raw gas at the gasifier exit, after the cyclone, and after a simulated gas turbine combustor. Also shown are recommended gas turbine specifications (see Table 5, note g).

to maintain effective seals in gas turbine systems. Pilot projects in the USA (19) and Finland (22) are intended in part to demonstrate the thermal and mechanical integrity of ceramic candle filters for such applications.

The cost of barrier filters for gas turbines would be high compared to cyclones, but would still represent a small part of the total plant capital cost.[11] Operating costs, for pulse cleaning and periodic replacement, may be of greater concern.

A number of other cleanup technologies for hot applications are at various early stages of development (41), including ceramic cross-flow filters, granular bed filters, and various electrostatic precipitators.

11 In one design study for a 100-MW coal-gasifier steam-injected gas turbine, cyclones were estimated to cost $10/kW out of a total installed plant cost of $1290/kW (13).

2.5 Summary

Five main conclusions can be drawn from the above discussion:

1. The gasification of biomass for gas turbines should be simpler and probably less costly on a unit basis than for most other applications; restrictions on gas quality are less severe, and pressurization leading to higher specific throughputs would help reduce unit capital costs.
2. Biomass gasification is easier in many respects than coal gasification: lower temperatures yield comparable gasification rates, ash levels are much lower, and sulfur removal would not be required.
3. Fixed and fluidized-beds are both interesting candidate gasification technologies for biomass-gas turbines. Fixed-beds offer greater simplicity and less challenging particulate cleanup, while fluidized-beds offer greater fuel flexibility, higher throughput rates, and lower unit capital costs in the size range of interest with gas turbines.
4. Particulate and alkalai cleanup have not been adequately demonstrated for biomass-gasifier gas turbine systems. Pilot-scale demonstration would probably resolve these uncertainties.
5. The development of biomass gasification systems for gas turbines would be facilitated by "piggy-backing" on: existing coal-gasifier gas-turbine work, past work on pressurized biomass gasification/hot gas cleanup for methanol synthesis, and ongoing work on solid fuel PFBC gas turbine/gas cleanup systems.

3 Applications for Power Production

Gas turbines can operate with significantly higher peak temperatures than steam turbines and, thus, have the potential for producing power more efficiently from solid fuels. Gasification would permit the full efficiency benefit of higher temperatures to be realized both in today's gas turbines and in future units with even higher firing temperatures.[12] Both central station and cogeneration applications are of potential interest.

12 Gas turbines could also be fired directly with biomass, e.g. using pressurized fluidized-bed combustors (PFBC). The limit for turbine inlet temperatures used in PFBC systems today is about 900°C, which is well below the rating for the best commercial gas turbines – about 1200°C. At combustion temperatures higher than about 900°C, cleanup problems associated with ash melting and alkali release would become more serious. Unless these problems can be overcome, therefore, the full efficiency potential of gas turbines cannot be realized with direct combustion.

3.1 Central Station Power

For central station power generation, the most interesting gas turbine technologies are combined cycles and steam-injected cycles (3). Efficiencies for advanced gasifier-fueled versions of these cycles would probably be comparable. However, unit capital costs are likely to be higher for combined cycles in plant sizes less than about 200-MW_e, because of the high cost of smaller-scale steam turbines. Since most biomass applications will probably be smaller than this, we limit the discussion here to steam-injected cycles.

Biomass-gasifier steam-injected gas turbines (STIG) and intercooled steam-injected gas turbines (ISTIG) would be similar in many respects to the coal-gasifier STIG and ISTIG systems which have been proposed for baseload electricity generation in the USA.[13] In a detailed design study, capital costs for 100-MW coal-fired STIG and ISTIG plants that would use Lurgi fixed-bed gasifiers and General Electric LM-5000 gas turbines have been estimated to be $1290/kW and $1040/kW,[14] respectively, with efficiencies of 35.6% and 42.1% (higher heating value basis)[15] (13).

Detailed capital cost estimates have not been made for biomass-gasifier STIG or ISTIG plants, but costs are likely to be lower than those estimated for the above coal STIG and ISTIG systems for several reasons. Hot sulfur removal equipment, which accounts for 15–20% of the estimated capital cost of the coal-fired systems, would not be required. In addition, the less harsh reactor conditions required for biomass gasification would contribute to lower costs. Furthermore, substituting a fluidized-bed gasifier for the fixed-bed would lead to some cost reduction.

Thus, a reasonable capital cost estimate for a biomass-gasifier STIG based on the LM-5000 gas turbine would be in the range of $1100/kW. Such a system would produce about 53 MW of electricity, and an estimate of its efficiency is 32.5%. This efficiency estimate was made by research engineers at the General Electric Corporate Research Center (Schenectady, New York) based on limited experimental data from pilot-scale fixed-bed biomass gasification, which indicated lower gasification efficiency than with coal.

[13] STIG units are commercially available with natural gas firing in sizes from 1.8 to 51 MW. ISTIG units have yet to be commercialized. See (3).

[14] Costs in this paper are expressed in constant 1987 US$. The US gross national product deflator has been used to convert costs originally given in other-than-1987 dollars. Costs in Swedish kronor have been converted at 6.5 SEK/$.

[15] Higher heating values (HHV) are used for fuels in this paper. For natural gas, the HHV is approximately 10% greater than the lower heating value (LHV). For coal the difference is about 3%. The HHV of biomass on a dry matter basis is independent of its moisture content (mc), unlike the LHV. For biomass with 0%, 15%, 30%, and 50% mc, the HHV is greater than the LHV by approximately 5%, 11%, 18%, and 25%, respectively.

However, there is no obvious reason biomass gasification efficiency, and hence overall cycle efficiency, would not be at least as high as for coal. Thus, the efficiency estimate here can be considered conservative.[16]

A similarly conservative estimate for the efficiency of a 110-MW biomass-gasifier ISTIG is 38.4%. Its estimated capital cost is $880/kW.[17]

If these cost and conservative performance estimates are realized in practice, 50–100 MW central station biomass gas turbine plants would be competitive on a lifecycle cost basis with much larger coal-fired plants, including conventional steam-electric plants and advanced combined-cycles based on gasification (IGCC) or pressurized-fluidized bed combustion (PFBC). Figure 5 shows calculated total lifecycle electricity production costs for fuel costs ranging from $1.8/GJ to $3.6/GJ, which covers a range of possible future coal and biomass costs. (Costs of biomass are discussed further in Section 4.) In many cases, the biomass option would be less costly even if biomass fuel cost more than coal.

3.2 Cogeneration

In a biomass-gasifier gas-turbine cogeneration system, the high-temperature exhaust from the turbine would be used to raise steam, which could be (*a*) used directly for heat (simple cycle), (*b*) used to drive an absorption heat pump, while low-temperature heat in the gas turbine exhaust after the waste heat boiler provides the pump with a heat source, or (*c*) passed through a steam turbine first (combined cycle). Because of scale economy considerations, combined cycles are likely to be of interest only in relatively large sizes, unlike the other options. Option (*b*) might be of particular interest where only low-grade heat is required, e.g., as for hot-water district heating. With this option, latent heat in the exhaust (arising from the relatively high moisture content in fuel gas from biomass) could provide a significant heat source for the heat pump.

All three gas turbine systems could be designed with a flexibility to boost electricity production and electrical conversion efficiency when heat demand falls. In cases (*a*) and (*b*), this would require a steam-injected gas turbine

16 The efficiency estimate assumes an input biomass moisture content of 15%. For additional discussion, see (42).

17 Consistent estimates for biomass-gasifier combined cycle systems have not been made. One design study indicates a capital cost of $1270/kW for a 205-MW unit based on a system using a fluidized-bed gasifier and ASEA-STAL GT-200 gas turbine (43). Based on the scale-sensitivity of coal-gasifier combined cycle systems discussed in (44), a 100-MW biomass-gasifier combined cycle would cost about $1530/kW.

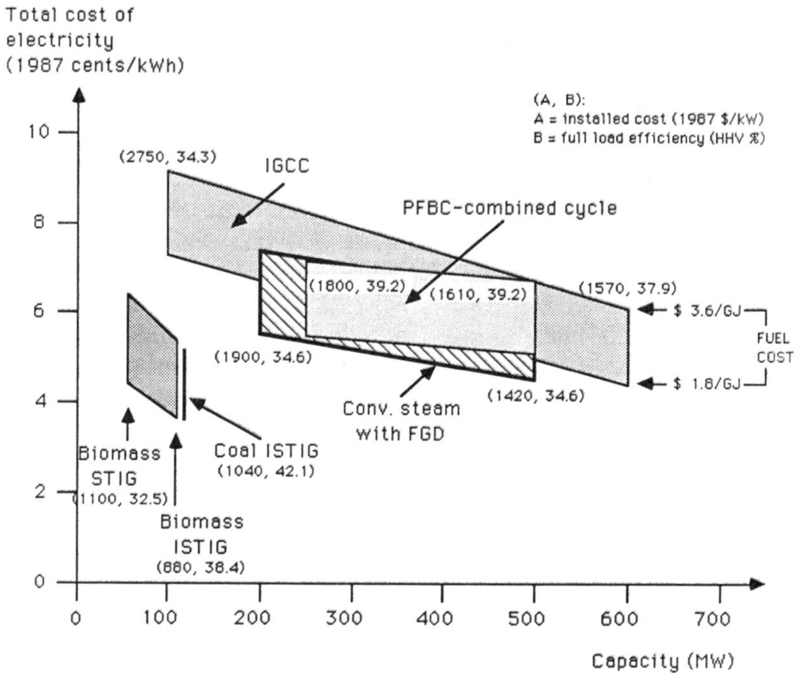

Figure 5. Calculated levelized lifecycle cost (including capital, operating and maintenance, and fuel) of electricity generation with coal and biomass as a function of plant size. A 6% discount rate, 30-year life and 70% capacity factor are used. No taxes or tax incentives are included. For simplification, linear relationships between the costs for the largest and smallest units are assumed. The cost range for each technology at a fixed size assumes a fuel cost of $1.8/GJ to $3.6/GJ. The lower fuel cost is the average utility coal price projected for the USA in 1995 by the US Department of Energy (85).

Performance and cost estimates for the conventional steam plant, the PFBC-combined cycle, and the IGCC are from the Electric Power Research Institute (86). The conventional steam plant uses wet flue gas desulfurization. The PFBC-CC has a gas-turbine inlet temperature of 843°C and steam conditions of 163 bar, 538°C. The IGCC utilizes oxygen-blown Texaco gasification with cold gas cleanup and gas-turbine inlet temperatures of 1093°C and 1204°C for the 100-MW and 600-MW units, respectively.

Performance and cost estimates for the coal-ISTIG, utilizing an air-blown dry-ash Lurgi gasifier, hot sulfur and particulate cleanup, and an intercooled steam-injected LM-5000 gas turbine, are from (13).

Performance and cost estimates for the biomass-STIG and ISTIG are discussed in the text.

The PFBC and IGCC have undergone successful technology demonstrations. The STIG-based systems have not.

(STIG) to be substituted for the simple-cycle gas turbine. With a STIG, steam not needed for heating is instead injected into the gas turbine combustor and/or expander to raise electrical output and efficiency. For example, a gasifier-gas turbine system based on the General Electric LM-5000 would produce an estimated 53 MW_e at 32.5% efficiency with full steam injection (42), compared to 39 MW_e at 28.6% efficiency with no injection (Table 6). The added flexibility with steam injection may be advantageous in the many potential biomass applications where heat demands vary widely, e.g., crop-processing facilities or district heating plants that operate seasonally, and where excess electricity can be marketed, e.g., sold to a utility.

Table 6 shows performance estimates for two sizes of simple cycles and one simple/heat pump cycle compared with those for two different sized conventional back-pressure steam turbines.[18] The gas turbine based systems would have higher electrical efficiencies than steam turbines. Their electricity-to-heat production ratio (E/H) would also be higher, which means they would produce more electricity while supplying a given amount of heat. The simple-cycle option without the heat pump would have a relatively low total efficiency compared to the other options.

The economic viability of alternative cogeneration options would depend on relative biomass, heat, and electricity prices. Gas turbines would tend to be favored on a capital cost basis, particularly in smaller sizes. For example, estimated installed unit costs for biomass-gasifier systems based on steam-injected gas turbines are lower and less sensitive to scale than those for steam turbines (Figure 6). The lower sensitivity to scale is particularly important since there are potentially many small biomass applications.

4 Implications for Regions with Biomass Resources

4.1 Sweden

In the decades ahead, Sweden will be seeking alternative fuels and electricity-generating technologies to replace existing nuclear capacity. Sweden has no indigeneous fossil fuel resources and limited unexploited hydroelectric

[18] Consistent estimates for combined cycles are not available. However, see (43) for some discussion of biomass-gasifier combined cycle cogeneration.

Table 6. Estimated full-load output and efficiencies of biomass-fired cogeneration systems (higher heating value basis).

	Electricity		Heat		Electricity-	Total
	MW	Eff. (%)	MW	Eff. (%)	to-heat ratio	Efficiency (%)
Simple-cycle[a,b]						
LM-5000	39	28.6	37	27.3	1.05	56
LM-1600	15	27.1	17	30.7	0.88	58
SC + Heat pump[c]						
LM-5000	39	28.6	55	40.0	0.71	69
Steam turbine[d]						
Back-pressure	18	25.9	36	51.8	0.50	78
Back-pressure	4.3	20.4	12	57.1	0.36	78

[a] From (42) for systems using Lurgi-type fixed-bed gasifiers with 15% moisture content biomass fuel. The heat would be generated as steam at 20 bar, 316°C in a heat recovery boiler. The full steam production would be about 20% greater than indicated, but some of the steam would be used for cooling in the gasifier. With a fluidized-bed gasifier more steam would be available for heating since the gasifier would not require any.

[b] These are aeroderivative turbines made by General Electric for natural gas applications. The LM-5000 is available with steam injection. The turbine inlet temperature and compression ratio are approximately 1200°C and 25:1, respectively, in both machines.

[c] This system would be a modification of the simple cycle LM-5000. Some steam produced in the heat recovery boiler would be used to drive absorption heat pumps with an assumed coefficient of performance of 1.7. The heat source for the heat pump would be water at about 55°C generated by recovery of the sensible and latent heat in the gas turbine exhaust between 80°C and 35°C (using a direct-contact condensing heat exchanger.) (A small amount of steam would also be used to reheat the cool, dry exhaust before it leaves the stack.) The total heat production of 55 MW is the sum of the output of the heat pump (31 MW), the sensible energy recovered from the turbine exhaust (after the waste heat boiler) between 140°C and 80°C (8 MW), and the portion of the steam not used to drive the heat pump or heat the stack gas (16 MW). In a district heating system, the 55 MW of heat would raise water from typical return temperatures of 55–60°C to delivery temperatures of 85–90°C.

[d] Estimated for new plants in Sweden using 15% moisture content wood chips, based on (62).

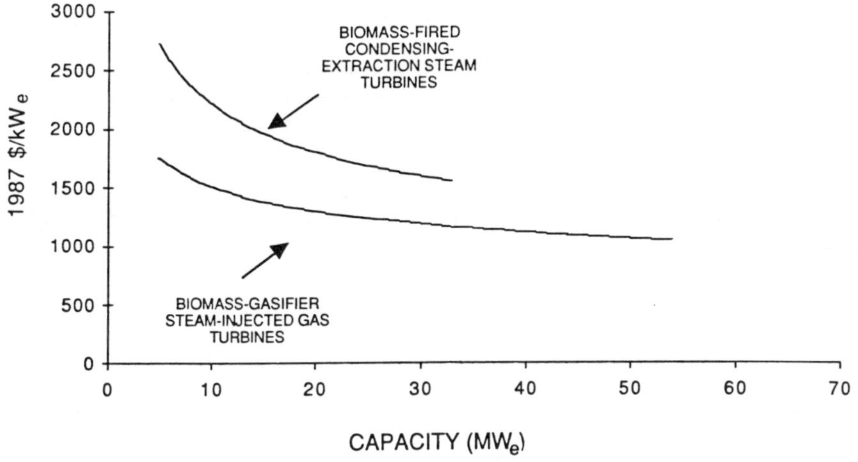

Figure 6. Estimated installed capital costs (1987 $/kW$_e$) for biomass-gasifier steam-injected gas turbine and condensing-extraction steam turbine cogeneration systems (42).

potential. And increasingly stringent environmental regulations will decrease the attractiveness of using coal.[19] Imported natural gas will likely be available, but costs are uncertain. Wind might supply some electricity, though the economics are also uncertain.

Sweden has significant biomass resources which might be used with gas turbines to generate some of its future electricity needs. Biomass fuels are already used to a significant extent today for heat supply, but generate less than 2% of Sweden's electricity. The potential production of electricity from biomass gas turbine central station or cogeneration units would depend on the commercial readiness of the technology and on the availability and cost of biomass.

4.1.1 Biomass supply and cost

Sweden's potential sustainable production of biomass energy is estimated to be some 730 PJ per year, about three times its current use (Table 7). Some 360

[19] In the spring of 1988, the Swedish Parliament passed a recommendation that in the future net carbon dioxide emissions should not exceed the present level. In addition, most new electricity and/or heat producing plants must comply with very stringent sulfur dioxide and nitrogen oxide emission regulations: allowed emissions of sulfur and nitrogen are half (or less) of those allowed by the New Source Performance Standards in the USA.

Table 7. Current use of biomass fuels in Sweden and estimated long-term potential supplies and costs.

	Used in 1987[a]		Long-Term Potential Supplies		Costs
	(PJ)	(TWh)	(PJ)	(TWh)	($/GJ)[b]
Forest-industry residues	*212*	*59*	*356*	*99*	
Forest residues[c]	54	15	198	55	2.0–2.6[d]
Pulping liquors[e]	104	29	104	29	0
Other byproducts[f]	54	15	54	15	0–1.7
Plantation fuelwood[g]	*0*	*0*	*302*	*84*	2.4–3.4
Other	*14*	*4*	*68*	*19*	
Refuse-Derived Fuel[h]	14	4	14	4	0
Straw[i]	0	0	54	15	?
Total	*234*	*65*	*726*	*202*	

[a] From (63).

[b] Higher heating value basis for 50% mc chips, including 30-km transport for plantation fuelwood and 50-km transport for forest residues. [For comparison, $3.4/GJ (HHV basis) is SEK 0.1/kWh (LHV basis).] For drying to 15% mc, the extra cost would be $0.5 to $0.6/GJ (64).

[c] Forest residues include mainly tops and branches from Norwegian Spruce, Scots Pine, and birch, and whole trees from thinning. The long-term potential reflects limits imposed by environmental considerations: *no important irreversible effects and no deterioration of the long-term soil productivity* (65). The potential accounts for ecological reductions following recommendations in (66). Thus, the actual volume of tops and branches produced is about 70% greater than the estimated recoverable volume. In addition, no removal of stumps is assumed.

[d] Production costs assume integrated recovery of energy and industrial feedstocks using the tree-section method (67). Total costs are fairly sensitive to transport costs: the cost is increased by 18% for 100-km vs. 50-km transport.

[e] Byproduct of the pulp and paper industry, currently used for steam and electricity production. The long-term potential is assumed to be today's use. Zero cost is assumed since it now has no important alternative uses.

[f] Currently approximately 9 TWh are used in the pulp and paper industry (mostly bark and other moist products) and 6 TWh in sawmills (bark and relatively dry sawdust) (63). The long-term potential is assumed to be the same as today's use. The upper cost limit is the current market price for off-site use (68). Zero cost is assumed for on-site use.

[g] Intensively cultivated Salix or Populus on agricultural land with an assumed average yield of 17 tonnes dry matter per hectare per year (45). The indicated potential assumes the use of some 1 million hectares of surplus agricultural land. Currently, some 300,000 hectares could be considered surplus. Some 500,000 hectares could be available by 1990 (69). Continued agricultural productivity growth would result in 1 million hectares by 2000 (70). Costs are based on (45,71).

[h] Current use is in district heating (63). The total long-term potential is assumed to be today's use.

[i] Potential is what would be available after reduction for soil humus preservation, cattle stable use, etc (70).

PJ/year are generated today as forest residues or industrial byproducts, 60% of which are currently used. Short-rotation energy plantations of willow and poplar trees, which are currently under development (45), could provide an additional 300 PJ/year. The development of this potential must be seen in the perspective of decades, however, because it requires a major restructuring of the use of agricultural land.

The costs of producing biomass in Scandinavia are generally higher than in many regions of the world (see Section 4.2). The present market price for wood chips from forest residues in Sweden, about \$3.4/GJ,[20] reflects the current costs of recovering the residues separately from other forest-industry feedstocks. Integrating the recovery processes would lower the cost for such chips to \$2.0–2.6/GJ (Table 7). The cost of industrial byproducts (bark and sawdust) would be for handling and transport, implying essentially zero cost for onsite use. Chips from short-rotation fuelwood plantations are estimated to cost \$2.4–3.4/GJ.

4.1.2 Biomass-based electricity production

Biomass-based gas turbine electricity generation could be considered for Sweden. A commitment to large-scale use of biomass would probably be accompanied by a shift in forestry practices toward more economically efficient integrated residue recovery and a concerted effort to develop energy plantations. A reasonable long-run cost of biomass would therefore be in the neighborhood of \$2.5/GJ–at the upper end of estimated costs for integrated recovery of residues and the lower end of estimated costs for plantation fuelwood (Table 7). Drying the fuel, as required for gasification, would raise this cost to about \$3/GJ. (See Table 7, note (b).)

Central-station power

With biomass costing \$3/GJ, and based on the cost and performance estimates for 50–100 MW steam-injected gas turbine systems given in Section 3.1, central station plants would produce power at a cost in the range of 4.8 ¢/kWh to 5.7 ¢/kWh (Figure 5). With lower-cost forest residues (\$2.5/GJ), the busbar cost range would be 4.3–5.2 ¢/kWh.[21] With these costs, biomass power plants

20 Note that higher heating values are used for fuels in this paper. (See footnote 15.) The \$3.4/GJ is for wood chips with 50% moisture content.

21 Larger combined cycles (> 200-MW_e) may produce electricity at comparable, but not lower costs.

would be competitive with much larger condensing coal-steam plants and advanced coal-fired PFBC and IGCC systems (Figure 5). The biomass option would clearly have the advantage if a cost were assigned to the net emissions of CO_2 from coal-fired plants.

Cogeneration

Gas turbine cogeneration might also be considered as part of a biomass-based electricity supply strategy in Sweden. The higher electricity-to-heat production ratios compared to steam turbines would permit larger quantities of electricity to be cogenerated while meeting a given heat demand. Potentially important applications include those in the pulp and paper industry, a well-established user of bioenergy, and in district heating (DH). DH applications are considered here. Some DH plants use biomass for fuel today, but more widespread biomass-based DH could be considered since most communities are surrounded by actively managed forests or agricultural land that could be used to produce biomass.

District heating (DH) systems today produce about 40 TWh/yr of heat,[22] primarily from plants providing heat only. A typical load duration curve for heat production in a DH system includes a temperature-dependent component superimposed on a relatively constant component (domestic hot water supply) (Figure 7). A mix of individual plants is operated to provide the required heat while minimizing operating costs. Heat that is produced with the lowest operating cost typically comes from refuse-derived-fuel incinerators, industrial waste heat, or sewage-water heat pumps. The next increment is typically supplied by baseload plants (including cogeneration units in some cases) sized to operate for about 5000 equivalent full-load heating hours per year.[23] Such plants would annually provide about 3/4 of the heat required in the system (shaded area in Figure 7). Oil or, in some cases, natural gas-fired boilers with minimal capital costs are typically used to supply peak demand.

If a biomass-gasifier combined-cycle or STIG-heat pump cycle were designed as a baseload DH plant, it would follow heat demand and could increase electricity production when heat demand falls. To estimate a cost

22 Total demand for district heat is not expected to rise significantly in the future. Increases in district heating demand due to growth in the number of connected buildings are likely to be offset by increases in building energy efficiencies. Efficiency improvements could even lead to a reduction in total heat demand.

23 The output of an individual DH plant varies over the year, depending on heat demand. A unit operating for 5000 equivalent full-load heating hours produces cumulatively as much heat in 8760 hours (one year) as the same plant would produce if operated at full heat output for 5000 hours.

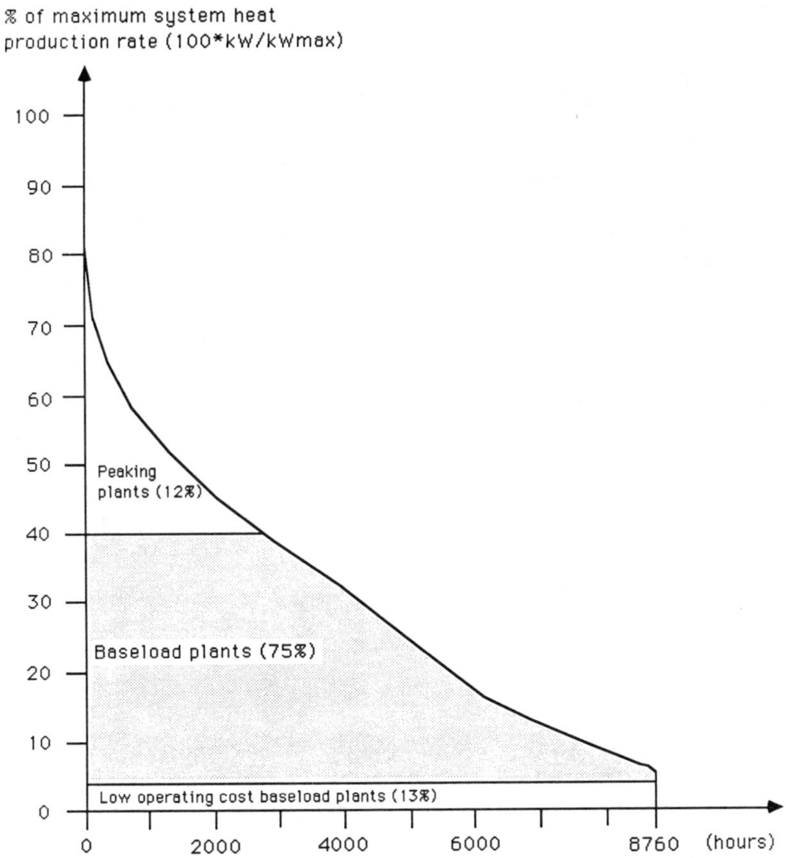

Figure 7. Typical load duration curve for heat production in a Swedish district heating system, representing the combined output of a number of individual plants. The fraction of the system's maximum heat production rate is shown versus the number of hours per year that production reaches this level or higher. The lowest operating-cost plants, e.g. refuse-derived-fuel incinerators, industrial waste-heat recovery, and sewage-water heat pumps, typically supply the first increment of baseload heat, in this case 13% of total annual heat production. Other baseload plants provide most of the heat (75% in this case), as indicated by the shaded area, which corresponds to operation of the baseload plants for 5000 equivalent full-load heating hours (see footnote 23). Higher operating-cost units, e.g., oil or natural gas-fired boilers, produce heat during peak demand periods.

range for the electricity that would be cogenerated, we consider a STIG-heat pump DH plant with the characteristics described in Table 6, the heat production of which is assumed to match the shaded area in Figure 7. The plant would produce heat at its maximum rate of 55 MW for approximately 2700 hours per year. Excess heat would be available as steam during the rest of the year. The excess steam could be injected to raise electricity output from 39 MW_e (Table 6) up to a maximum of 53 MW_e (if no steam at all were used for heating). If all steam not needed during the year for heating were used to boost electricity output, the resulting annual electricity-to-heat production ratio (E/H) for the system would be 1.4.[24] If only one-third the potential electricity output were generated during the three summer months, when electricity demand is lowest in Sweden, the annual E/H would be 1.1.

For this latter operating scenario, the cost of cogenerating electricity with the gas-turbine systems would range from 4.5–5.6 ¢/kWh (3.9–5.0 ¢/kWh), with biomass costing \$3/GJ (\$2.5/GJ) and assuming a credit for heat based on an avoided fuel cost of \$2.0–5.0/GJ (Figure 8). (Higher avoided fuel costs lead to larger credits for heat and, thus, lower costs of power.) The lower displaced fuel cost would be representative of heavy fuel-oil (at \$20/bbl for crude), while the higher prices might be representative of future natural gas prices. These electricity costs would be competitive with efficient central station power based on natural gas for a gas price higher than about \$4.5/GJ (with biomass at \$3/GJ) or higher than \$4/GJ (with biomass at \$2.5/GJ) (Figure 8). (For comparison, Sweden's National Energy Administration indicates that gas prices for perhaps the next decade may range from \$3.5/GJ to \$5.0/GJ (46).) Even assuming low gas prices, the cogenerated power would be less costly than that from new, large coal-fired central station power plants (Figure 5).

4.1.3 Overall potential

On a per-capita basis Sweden's biomass resources are large relative to those of most other industrialized countries. Sweden is therefore in an enviable position of having the option to use some of these resources to meet future energy needs. This option may become increasingly attractive as the use of fossil fuels grows increasingly troublesome due to rising costs and environmental problems such as the global greenhouse warming. The biomass resource is limited, however, so using it efficiently would be extremely important.

[24] If the plant were to operate with full heat output year-round, the E/H would be 0.71 (Table 6). The E/H of 1.4 assumes the heat output varies as shown in Figure 7 and that electricity output increases linearly as the heat output is decreased. (Zero heat output would correspond to 53 MW_e.)

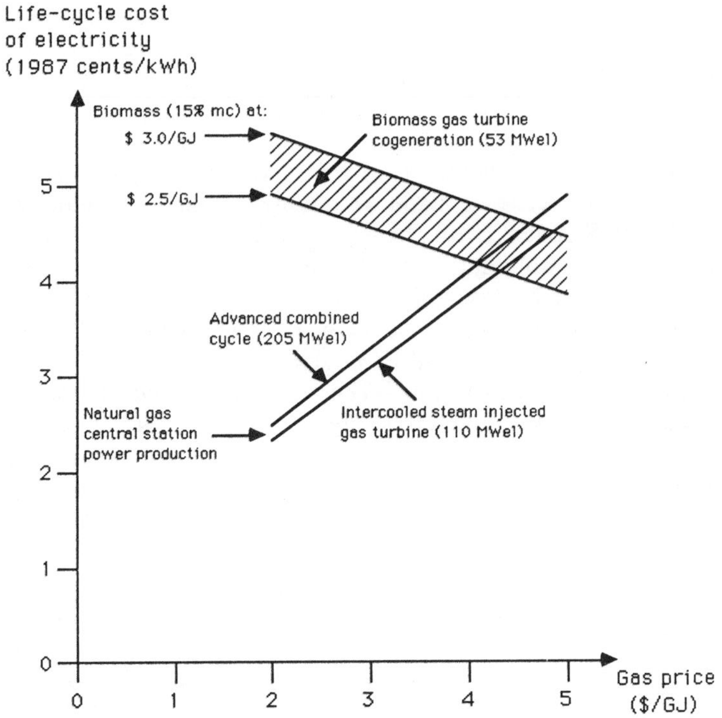

Figure 8. Calculated lifecycle electricity production costs with biomass-gasifier gas turbine district-heating cogeneration compared with the costs of efficient natural-gas fired central station power as a function of the assumed gas price.

The cogeneration system is based on a STIG-heat pump system (Table 6, note c), the heat production of which follows Figure 7, with a total of 5000 equivalent full-load heating hours (see footnote 23). With full heat production, the system performance would be as in Table 6. When producing no heat, the system would produce 53 MW at 32.5% efficiency by steam injection. Electricity production is assumed to increase linearly with decreasing heat production. As discussed in Section 4.1.2, the assumed system operating strategy yields an annual electricity-to-heat production ratio of 1.1 (and an electrical capacity factor of 62%).

The total estimated installed capital cost for the cogeneration unit is $67 million ($1100/peak kW_e for the STIG (Figure 6) plus $280/$kW_{th}$ for 31 MW_{th} of absorption heat pump.) Based on (43), fixed annual maintenance costs are assumed to be 3% of capital costs, variable costs are 2 mills/kWh_e, and labor costs are $0.65 million per year. The calculations assume a 6% discount rate, 30-year life, and 90% equipment availability.

The credit for heat is the cost to produce the same amount of heat in a stand-alone natural gas boiler with a capital cost of $77/$kW_{th}$, non-fuel operating costs of $3.1/$kW_{th}$-yr, and a higher heating value efficiency of 84% (92% LHV).

The assumed installed capital costs (efficiency) for the natural gas central station power plants are $420/kW (47%) and $510/kW (45%) for the ISTIG and advanced combined cycle, respectively, and operating and maintenance costs are 0.3 ¢/kWh (3). A 6% discount rate, 30 year life, and 70% capacity factor are used.

To illustrate the potential contribution of biomass-gasifier gas turbine systems in the long term, we consider the use for electricity production of the potentially available biomass resources in Sweden that are not currently utilized, i.e., a biomass fuel use of up to 500 PJ (Table 7).

A biomass cogeneration strategy might be considered, e.g., based on gas turbine district heating systems. An estimate of the potential electricity cogeneration in Sweden can be made by assuming Figure 7 to represent the DH profile for all of Sweden, with the shaded portion (75%) supplied by gas turbines with the operating characteristics of STIG-heat pump systems. In this case, the annual E/H of 1.1 discussed in Section 4.1.2 would apply to Sweden as a whole. For an assumed total district heat demand of 40 TWh, therefore, some 33 TWh of electricity would be cogenerated (0.75*40*1.1 = 33), which would require about 400 PJ of fuel.[25] The approximate monthly distribution of electricity production with such a strategy would be as shown in Figure 9 superimposed on the 1986 Swedish electricity production. Figure 9 reflects the cogeneration operating scheme discussed in Section 4.1.2, but a variety of other strategies are conceivable, since the overall electricity-to-heat ratio for gas turbine systems could be varied over a wide range.

Alternatively, with a central station strategy, e.g., based on biomass-gasifier ISTIG technology with an efficiency of 38%, the 500 PJ of fuel could produce some 53 TWh of electricity.

The potential 33–53 TWh of electricity from biomass would represent 25–40% of current total generation, or 50–80% of nuclear production. Thus, *in the long term, if electricity demand were reduced through end-use efficiency improvements (47), hydro and biomass sources combined could provide all of Sweden's electricity needs.*

How rapidly biomass-gas turbine electricity production could be introduced would depend firstly on the successful commercial development of the technology. It would also depend on the availability of feedstocks and the extent to which they are committed to power generation. (If a cogeneration strategy were pursued, the rate of capital stock turnover would also be important.) Forest industry residues are an attractive initial fuel source. Some 160 PJ are currently recoverable but unused (Table 7), which alone would

[25] An alternative cogeneration strategy based on back-pressure steam turbines could also be used to meet the district heating demand and would produce electricity at lower cost than the gas turbine strategy considered here. However, because of the much lower electricity-to-heat ratio of steam turbines (0.3–0.5), these would need to be augmented by other electricity sources to produce the same amount of power as with the gas turbines. To maintain overall biomass-based electricity costs and resource use comparable to those with the gas turbine cogeneration strategy, efficient central station power would be required, e.g. based on gasifier-ISTIG or combined cycle technology.

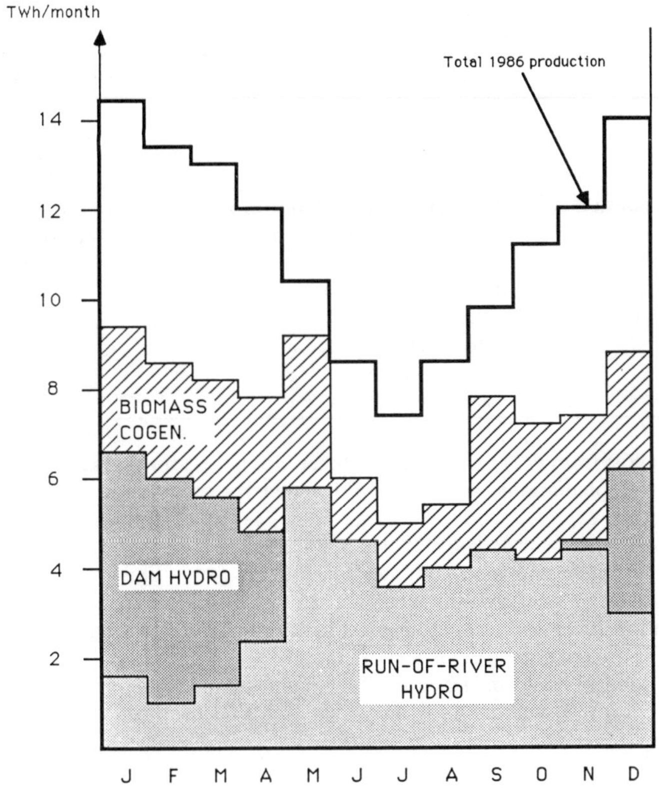

Figure 9. The potential electricity supply in Sweden from biomass-fueled gas turbine district heating cogeneration plants is shown superimposed on the total 1986 Swedish electricity production. Nearly all non-hydroelectricity was supplied by nuclear plants in 1986. The distribution of the hydroelectric supply can vary to some extent from year to year depending on rainfall variations and the operating strategy chosen.

support the production of some 14 TWh of cogenerated electricity. This initial use of residues would provide time to fully develop fuelwood plantations.

4.2 Worldwide

Biomass-based gas turbine power generation would be even more economically attractive in the many regions of the world where better climate and lower labor rates lead to lower biomass production costs, e.g., in many developing countries (Table 8).

Table 8. Estimated cost of producing biomass chips from fuelwood plantations.[a]

Region	Species	Cost (1987$/GJ)
Sweden	Salix, Populus	2.4–3.4[b]
Finland	Birch	3.1[c]
USA, Hawaii	Eucalyptus	1.4–2.0[d]
India, Uttar Pradesh	Eucalyptus, Acacia, others	1.9
Thailand	Pine, Eucaplytus, Casuarina	1.7–1.9
India, West Bengal	Ecalyptus	1.7
Brazil, Minas Gerais	Eucalyptus saligna	1.5
Brazil, Minas Gerais	Eucalyptus	1.4[e]
Haiti	Leucaena, Albizia, Cassia, others	1.4
India, Gujarat	Albizia, Acacia	1.4
Philippines	Leucaena, Albizia	1.0

[a] Production cost from (72) (except where otherwise noted) for fuelwood from dedicated or multi-purpose plantations, with $0.30/GJ added for chipping and $0.30/GJ added for transport.

[b] For fuelwood from dedicated plantations. See Table 7.

[c] From (73) for birch pulpwood (the lowest cost pulpwood in Finland) delivered to the mill. Delivered forest residues are estimated to cost $1.7/GJ to $2.6/GJ.

[d] For plantation fuelwood, including establishment, harvesting, and chipping. Based on ongoing 300-hectare trial on the island of Hawaii (74).

[e] Based on detailed yield data and establishment, maintenance and harvesting costs for dedicated energy plantations (75), to which $0.60/GJ has been added for chipping and transport.

With biomass costs of less than $2/GJ, central station steam-injected gas turbine plants would be economically competitive power producers (Figure 5). In addition, in many developing countries, the small scale of the steam-injected units would be well-matched to small utility grids which cannot accommodate large (500–1000 MW) new power plants (48). Maintenance characteristics of the technology are also attractive (3). Its low capital intensity is an added attribute, as is its requirement for local rather than imported fuels.

Gas turbines might also be operated as cogenerators fueled by biomass made available as an industrial byproduct. Of course, some of these byproducts would be used for other purposes, as they are today. A detailed analysis of gas turbine cogeneration in the cane sugar industry (49) shows, however, that although bagasse, a byproduct from the crushing of cane, is often fully utilized at present in meeting the on-site energy needs of the sugar factory, it is typically used very inefficiently. Thus, it would be economical for the typical sugar producer to improve process energy efficiency and install a steam-injected gas turbine fired with gasified cane residues. For one Jamaican sugar

Table 9. Estimated potential gas turbine generating capacity (in 1000s of MW) supportable with current biomass residue production.[a]

Residue Source	Asia[b]	Africa	Latin America	Industrial Market Economies	World
Saw mills[c]	32.0	1.2	4.8	44.1	82.1
Sugar cane[d]	13.9	4.9	27.9	4.8	51.5
Pulp mills[e]	1.8	0.1	1.0	16.6	19.5
Corn Stover[f]	10.9	2.7	4.5	30.7	48.8
Rice husks[g]	18.0	0.3	0.5	0.3	19.1
Total	76.6	9.2	38.7	96.5	221.0

a Assuming 33% conversion of residues to electricity and a 75% capacity factor.

b Includes the Soviet Union.

c For the 1983 level of sawnwood production (76), assuming 0.7 bone-dry tonnes of residue per m^3 of sawnwood (77) and 20 GJ/tonne.

d For 1985 level of cane production. Residues include bagasse and approximately half the cane trash (tops and leaves) (49).

e For the 1983 level of chemical pulp production (76), assuming from each tonne of pulp 3.1 GJ of bark and 13.1 GJ of black liquor are derived (78).

f For the 1985 level of corn production (79), assuming each tonne of corn produces 0.5 tonnes of stover (15 GJ/tonne). About 0.75 tonnes of stover are actually produced, but only 2/3 of this can be removed without damaging soil productivity (80).

g For the 1983 level of rice production (81), assuming 0.2 tonnes of rice husks (15 GJ/tonne) are produced from each tonne of rice (82).

factory, it has been estimated that electricity output could thereby be increased more than 20-fold while still meeting onsite energy needs. For the local electric utility, buying the excess electricity would be more economical than building a new central station coal-fired plant. The total amount of cane-based excess electricity that could be produced in Jamaica is estimated to be about two-thirds of that nation's present utility production.

Such cogeneration possibilities appear to exist in many countries. The potential gas turbine electricity generating capacity that could be supported worldwide by the residues of sugar cane, corn and rice, sawmills, and the pulp and paper industry is about 220 GW at current rates of residue production (Table 9).[26] This is equivalent to about 10% of the installed electric generating capacity in the world in 1984 and would represent a total investment of about $300 billion.

26 This estimate does not include a large fraction of the forestry residues that are currently unutilized, which may be a significant resource in some countries, such as in Sweden.

5 Conclusions

The biomass-gasifier gas turbine would be an attractive technology for efficiently utilizing biomass resources for energy, but the technology is not commercially available today. The assessment here of the state of gasification technology and required developments for gas turbine applications indicates, however, that it could probably be developed in a few years, given an appropriate level of effort. Such a short time is conceivable because previous and ongoing work on coal-gasifier gas-turbine systems, PFBC systems for coal and biomass, and pressurized gasification of biomass for methanol production provide a headstart in the development process. In addition, biomass gas turbine systems would not require the commercial demonstration of any fundamentally new technology. In particular, hot sulfur removal, the commercial-scale viability of which has not been proven for coal-gasifier gas-turbine systems, would not be required with biomass. The overall development time would be further reduced since the scale of a demonstration plant would be comparable to that of a commercial plant, eliminating the need for significant scale-up work.

The unique work required for the development of the biomass gas turbine technology would be the demonstration of gasification and hot gas cleanup systems with biomass feedstocks and gas turbines. We have identified some of the key technical considerations that would need to be addressed in such a demonstration project, as well as some potential candidate gasification systems. Sweden might be a promising setting for such a demonstration because of its extensive experience in the development of biomass gasification systems, including pressurized units, and the design, installation and operation of gas turbines.

Acknowledgements

The authors thank Mattias Lundberg for assistance in calculating energy balances for gas turbine/heat pump cogeneration systems.

Partial financial support in the preparation of this paper is gratefully acknowledged by EDL from the Office of Energy of the US Agency for International Development, Washington, DC and by PS from the Swedish Energy Research Commission (Efn).

References

1. Peters, W. 1989. Coal gasification technologies for combined cycle power generation. This volume.
2. Penner, S. S., et al. 1987. Coal gasification: direct applications and synthesis of chemicals and fuels. *Energy*. 12(8/9).
3. Williams, R.H., Larson, E.D. 1989. Expanding roles for gas turbines in power generation. This volume.
4. Hodam, R., Williams, R., Lesser, M. 1982. *Engineering and economic characteristics of commercial wood gasifiers in North America*. Golden, Colorado: Solar Energy Res. Inst.
5. Furtado, P., Antunes, R.C. 1985. *Commercial applications of biomass gasifiers in Brazil*. Presented at 2nd Intl. Producer Gas Conf., Bandung, Indonesia.
6. Baker, E.G., Mitchell, D.H., Williford, J.F. 1980. *Assessment of large scale biomass gasification systems for less developed countries*. Richland, Washington: Battelle Pacific NW Lab.
7. National Res. Council. 1983. *Producer Gas: Another Fuel for Motor Transport*. Wash., DC: Natl. Academy Press.
8. Rensfelt, E. 1987. *Cogeneration using gasification and diesel engines*. Nyköping, Sweden: Studsvik Energy AB.
9. Reed T.B., Das, A. 1988. *Handbook of Biomass Downdraft Gasifier Engine Systems (SERI/SP-271–3022)*. Golden, Colorado: Solar Energy Res. Inst.
10. Antal, Jr., M.J. 1983. Biomass pyrolysis: a review of the literature, part 2: lignocellulose pyrolysis. *Advances in Solar Energy*. 2:175–255.
11. Milne, T. 1979. Pyrolysis: the thermal behavior of biomass below 600°C. In *Survey of Biomass Gasification,* II:95–131. Golden, Colorado: Solar Energy Res. Inst.
12. Graboski, M.S. 1982. Comparison of coal and wood as feedstocks for methanol manufacture. *Proc. Biomass-to-Methanol Specialists' Workshop (SERI/CP-234–1590),* ed. T.B. Reed, M. Graboski, 41–59. Golden, Colorado: Solar Energy Res. Inst.
13. Corman, J.C. 1986. *System analysis of simplified IGCC plants*. Schenectady, New York: General Elec. Corp. Res. Center (for US Dept. of Energy).
14. Furman, A.H. 1987. personal communication. Schenectady, New York: General Elec. Corp. Res. Center.
15. Smith, D.P. 1982. *Feasibility investigation: gasification of simplex briquettes*. Schenectady, New York: Mechanical Technology, Inc. and General Elec. Corp. Res. Center (for New York State Energy Res. and Dev. Authority).
16. Teggers H., Nitschke, E. 1986. *The Rheinbraun high-temperature Winkler (HTW) process and its application for the production of synthesis gas and town gas*. Presented at 4th Adv. Coal Gasification Symp., Suzhou, China.
17. Evans, R.J., Knight, R.A., Onischak, M., Babu, S.P. 1988. *Development of biomass gasification to produce substitute fuels (PNL-6518)*. Richland, Washington: Battelle Pacific NW Lab.
18. Lindman, N. October 1988. personal communication. Göteborg, Sweden: Götaverken Energy AB.
19. Buckman, J.W., Cover, A.E. 1988. The Appalachian project. *Proc. 8th Annual Gasification and Gas Stream Cleanup Systems Contractors Review Meeting, 1988 (DOE/METC-88/6092)*, 1–10. Morgantown, West Virginia: USDOE Morgantown Energy Technology Center.
20. Keller, J. July 1988. personal communication. Dortmund, FRG: Udhe GmbH.
21. A. Ahlström, Inc. Nov. 16, 1988. press release. Karhula, Finland: R&D Center.

22. Viel Lamare, C. Nov. 1988. personal communication. Stockholm: Ahlström Boilers.
23. Sinclair A., Waldheim, L. 1985. *MINO feasibility study, concluding report (Studsvik Arbetsrapport-Technical Note)*. Nyköping, Sweden: Studsvik Energiteknik AB.
24. Webb, P.R. 1987. *Wood-fired turbine emissions measurements*. Columbus, Ohio: Battelle Columbus Lab.
25. Corman, J.C., Horner, M.W. 1986. Hot gas clean-up for a moving bed gasifier. *Proc. 5th Annual Contractors' Conf. on Coal Gasification (AP-4680)*, ch. 14. Palo Alto, Calif: Elec. Power Res. Inst.
26. Cherish, P., Holmgren, J.D. 1988. *Hot gas cleanup process demonstration of ceramic filters for IGCC power generation*. Presented at Elec. Power Res. Inst. Seminar on Fluidized-Bed Combustion Technology for Utility Applications, Palo Alto, Calif.
27. Wright, I. G., Stringer, J. 1986. *Erosion and corrosion considerations for PFBC gas turbine expanders (86-GT-217)*. Presented at Am. Soc. Mech. Engineers' Intl. Gas Turb. Conf., Dusseldorf, FRG.
28. Westinghouse Elec. Co. 1986. *Alternative combustion turbine designs and cleanup systems for pressurized fluidized-bed combustion power plants (CS-4860)*. Palo Alto, Calif: Elec. Power Res. Inst.
29. Notestein, J. 1988. *Update on Department of Energy hot gas cleanup programs*. Presented at 8th Elec. Power Res. Inst. Coal Gasification Conf., Palo Alto, Calif.
30. Brown, M.G., Baker, E.G., Mudge, L.K. 1987. Evaluation of processes for removal of particulates, tars, and oils from biomass gasifier product gases. In *Energy from biomass and wastes X*, ed. D.L. Klass, 655–77. London and Chicago: Elsevier Appl. Sci. and Inst. of Gas Technology.
31. Holmgren, J.D. 1986. *The KRW hot gas cleanup coal gasification process*. Presented at 4th Adv. Coal Gasification Symp., Suzhou, China.
32. Scandrett, L.A., Clift, R., 1984. The thermodynamics of alkalai removal from coal-derived gases. *J. of the Inst. of Energy*. Dec:391–97.
33. Horner, M.W. 1985. *Simplified IGCC with hot fuel gas combustion (85-JPGC-GT-13)*. Presented at Jt. ASME/IEEE Power Generation Conf., Milwaukee, Wisconsin.
34. Mojtahedi, W., Backman, R. 1988. *Alkali metals volatilization in pressurized fluidized-bed combustion and gasification of peat*. Presented at 4th Intl. Fluidized Combustion Conf., London.
35. Sidebotham, G.S., Williams, R.H. 1989. *Technology of NO_x control for stationary gas turbines*. Princeton, New Jersey: Center for Energy and Environ. Studies, Princeton Univ.
36. Leppalahti, J., Stahlberg, P., Simell, P. 1988. Nitrogen compounds in peat and wood gasification and combustion. *Proc. Res. in Thermochemical Biomass Conversion*. London: Elsevier Appl. Sci.
37. Unnasch, S., Chang, R. 1988. Ammonia removal in coal gasification processes using integrated systems approach. *Proc. 8th Annual Gasification and Gas Stream Cleanup Systems Contractors Review Meeting, 1988 (DOE/METC-88/6092)*, 1–10. Morgantown, West Virginia: USDOE Morgantown Energy Technology Center.
38. Meadowcroft, D.B., Stringer, J. 1987. Corrosion in coal-fired gas turbines. *Materials Sci. and Tech.* 3:562-70.
39. Reed, G.P. 1985. Filtration of high temperature fuel gases from a fluidised bed gasifier. *Filtration and Separation*. March/April:106–09.
40. Stinton, D.P., Lowden, R.A., Chang, R. 1987. Fiber-reinforced composite hot gas filters (ORNL/FMP-87/4). *Proc. Fossil Energy Materials Program*, 156–67. Oak Ridge, Tennessee: Oak Ridge National Lab.

41. Clift, R. 1988. Hot gas cleanup: the state-of-the-art and prospects for the future. *Proc. VTT Symp. 83: Pressurized Fluidized Bed Combustion and Gasification Power Systems,* 213–35. Espoo, Finland: Technical Res. Center of Finland.
42. Larson E.D., Williams, R.H. 1988. Biomass-fired steam-injected gas turbine cogeneration. *Proc. 2nd Intl. Symp. on Turbomachinery, Combined-Cycle Technologies and Cogeneration,* 57–66. New York: Am. Soc. Mech. Engineers.
43. Rensfelt E., Waldheim, L. 1986. *Vedbaserad elproduktion med förgasning och kombinerad cykel (Wood-based electricity production with gasification and combined cycle) (BF-86/8).* Stockholm: National Energy Admin. (In Swedish)
44. Synthetic Fuels Assoc. 1983. *Coal gasification systems: A guide to status, applications, and economics (AP-3109).* Palo Alto, Calif: Elec. Power Res. Inst.
45. *Energiskog (Energy plantations) (1985:9).* 1985. Stockholm: National Energy Admin. (In Swedish)
46. *El- och värmeproduktion med naturgas (Electricity and heat production with natural gas) (1987:R5).* 1987. Stockholm: National Energy Admin. (In Swedish)
47. Bodlund, B., Mills, E., Karlsson, T., Johansson, T.B. 1989. The challenge of choices: technology options for the Swedish electricity sector. This volume.
48. *The energy transition in developing countries.* 1983. Wash., DC: World Bank.
49. Larson, E.D., Ogden, J.M., Williams, R.H. 1987. *Steam-injected gas turbine cogeneration for the cane sugar industry: optimization through improvements in sugar-processing efficiencies (PU/CEES 217).* Princeton, New Jersey: Center for Energy and Environ. Studies, Princeton Univ.
49a. Graboski, M., Bain, R. 1979. Properties of biomass relevant to gasification. In *Survey of Biomass Gasification,* II:21–65. Golden, Colorado: Solar Energy Res. Inst.
49b. Fluor Engineers, Inc. 1985. *Cost and perfomance of Kellog Rust Westinghouse-based gasification-combined-cycle plants (AP-4018).* Palo Alto, Calif: Elec. Power Res. Inst.
50. Rensfelt, E. Nov. 1988. personal communication. Nyköping, Sweden: Studsvik Energy AB.
51. Miles T.R., Miles, Jr., T.R. 1988. Reliable feed systems for thermochemical conversion. *Proc. Res. in Thermochemical Biomass Conversion.* London: Elsevier Appl. Sci.
52. Reed, T.B., Levie, B., Graboski, M.S. 1987. *Fundamentals, development and scaleup of the air-oxygen stratified downdraft gasifier.* Golden, Colorado: Solar Energy Res. Inst.
53. Coffman, J. A. April 1988. personal communication. Ballston Spa, New York: Wright-Malta Corp.
54. Teggers, H., Scharf, H.J., Schrader, L. 1985. Gasification of biomasses by HTW-gasification process. *Proc. 3rd Intl. Conf. on Energy from Biomass,* ed. W. Palz, J. Coombs, D.O. Hall. London: Elsevier Appl. Sci.
55. Sipila, K. May 1988. personal communication. Espoo, Finland: Technical Res. Center of Finland.
56. Gravel, G., Chamberland, A. 1987. Testing and evaluation of a pressurized 10 t/h fluid-bed wood gasifier. In *Energy from biomass and wastes X,* ed. D.L. Klass. London and Chicago: Elsevier Appl. Sci. and Inst. of Gas Technology.
57. Drouin, G., Bechard, L., Du Chouchet, P. 1987. Electricity generated by a diesel engine running on lean gas produced from a biomass pressurized fluid-bed gasifier. In *Energy from biomass and wastes X,* ed. D.L. Klass. London and Chicago: Elsevier Appl. Sci. and Inst. of Gas Technology.

58. Onischak, M., Lynch, P.A., Babu, S. 1988. *Process and economic considerations for the gasification of pulp mill wastes using the Renugas process*. Presented at Conf. on Energy from Biomass and Wastes XII, New Orleans, Louisiana.
59. *The Otaniemi pressurized fluidized bed combustion and gasification research project*. 1988. Espoo, Finland: Technical Res. Center of Finland.
60. Herbert, P.K., Baguss, I., Loeffler, J.C. 1988. *Low-btu gas from bark and waste via CFB gasification*. Presented at Conf. on Energy from Biomass and Wastes XII, New Orleans, Louisiana.
61. Haldipur, G.B., Cherish, P., Lewandowski, D.A., Pavel, J. 1986. Hot gas cleanup using the KRW gasifier. *Proc. 5th Annual EPRI Contractors' Conf. on Coal Gasification (AP-4680)*, ch. 11. Palo Alto, Calif: Elec. Power Res. Inst.
62. *Småskalig kraftvärme (Small-scale cogeneration) (1986:1)*. 1986. Stockholm: National Energy Admin. (In Swedish)
63. *Energiläget i Siffror (Energy in Sweden: facts and figures)*. 1988. Stockholm: National Energy Admin. (In Swedish)
64. Munter, C. Sept. 1988. personal communication. Göteborg, Sweden: Svensk Exergiteknik AB.
65. Parikka, M. 1987. Hur mycket trädbränsle finns det? (How much wood fuel is there?). *Skogsfakta Konferens 10: Proc. Conf. on Forestry in Energy Supply*. Uppsala: Swedish Univ. of Agric. Sci. (In Swedish)
66. *General recommendations on limitations of recovery of tree sections in excess of stemwood on forest land (memorandum)*. Dec. 2, 1985. Jönköping, Sweden: National Forestry Board (In Swedish)
67. Hildeman, A., Rutegård, G., Tibblin, G. 1986. *Träddelsmetoden-integrerad produktion av skogsbränsle och massaråvara (The tree-section method–integrated production of wood fuel and pulp feedstock) (Forest-Industry-Market Study No. 19)*. Uppsala: Swedish Univ. of Agric. Sci. (In Swedish)
68. Lönner, G. Sept. 1988. personal communication. Uppsala: Forest-Industry-Market Studies, Swedish Univ. of Agric. Sci.
69. 1983 års livsmedelskommitte (1983 government committee on food production). 1984. *Slutbetänkande (final report)*. Stockholm: Ministry of Agric. (In Swedish)
70. Stjernquist, I. et al. 1986. *Miljö och energi – framtida möjligheter och begränsningar (Environment and energy – future possibilities and limitations) (Rpt. 3238)*. Stockholm: Swedish National Environ. Protection Board.
71. Parikka, M. 1988. *Ekonomisk analys av energiskogsodling (Economical analysis of short rotation forestry)*. Uppsala: Forest-Industry-Market Studies, Swedish Univ. of Agric. Sci. (In Swedish)
72. Spears, J. (World Bank Forestry Div.). 1985. *Summary of World Bank supported projects with fuelwood components*. Presented at World Resources Inst. Conf. on Biomass Energy: Building Blocks for a Sustainable Agric., Airlie House, Virginia.
73. Marjokorpi T., Leppa, K. 1983. Northern European approach to wood energy. In *Industrial Wood Energy Forum '82, Vol. II*, 363–67. Madison, Wisconsin: Forest Products Res. Society.
74. Crabb T. (Bioenergy Dev. Corp.). 1987. Presentation at 2nd Pacific Basin Biofuel Workshop, Kauai, Hawaii.
75. Rodrigues de Almeida, M. 1987. personal communication. Belo Horizonte, Brazil: Charcoal, Technical Subproducts Department, Acesita Energetica.
76. Food and Agric. Org. 1985. *Yearbook of Forest Products*. Rome: United Nations.

77. Bergvall, J.A., Bullington, D.C., Gee, L., Koss, W. 1978. *Wood waste for energy study*. Olympia, Washington: Wash. State Dept. of Natural Resources.
78. Fors, J., Nord, B. 1980. Energianvändingen inom massa- och pappersindustrin, SCA-Nordliner i Munksund (Energy use in the pulp and paper industry, SCA-Nordliner factory in Munksund (STU 174–1980). Stockholm: Swedsh Board for Technical Development. (In Swedish)
79. US Dept. of Agric. 1986. *Agricultural Statistics 1986*. Wash., DC: US Govt. Print. Office.
80. Kresovich, S., McClure, T., Scantland, D. 1979. Field crop feedstock as renewable resources for energy production. *Proc. EPRI/GRI Workshop on Biomass Resources and Conversion,* 6–12. Palo Alto, Calif: Elec. Power Res. Inst.
81. US Dept. of Commerce. 1986. *Statistical Abstract of the United States*. Wash., DC: US Govt. Print. Office.
82. Beagle, E.C. 1978. *Rice-husk conversion to energy (Agric. Services Bull. No. 31)*. Rome: Food and Agric. Org., United Nations.
83. Antal, M.J. 1978. *Biomass energy enhancement*. Princeton, New Jersey: Dept. Mech. and Aerospace Engineering, Princeton Univ.
84. Waldheim, L., Rensfelt, E. 1982. Methanol from wood and peat. *Proc. Biomass-to-Methanol Specialists' Workshop (SERI/CP-234–1590)*. ed. T.B. Reed, M. Graboski, 239–59. Golden, Colorado: Solar Energy Res. Inst.
85. Energy Inform. Admin. 1987. *Annual energy outlook 1986, with projections to 2000 [DOE/EIA-0383(86)]*. Wash., DC: US Govt. Print. Office.
86. Planning and Evaluation Div. 1986. *Technical assessment guide, vol. 1: electricity supply–1986 (P-4463-SR)*. Palo Alto, Calif: Elec. Power Res. Inst.

Implementing Electricity Conservation Programs: Progress Towards Least-Cost Energy Services Among U.S. Utilities

Howard S. Geller

Abstract

A growing number of utilities in the U.S. are investing large sums in end-use efficiency, thereby cutting load growth by up to 1.4% per year and reducing the need for new power supplies. These utilities generally find that efficiency is much more cost effective than supply alternatives, even without taking into account environmental and social costs. This paper reviews the U.S. utilities' experience with efficiency incentive programs, including conservation purchase through rebates, bidding, loans, and direct installation. Also covered are the nonfinancial aspects of utility programs and the planning and regulatory options for stimulating investments in efficiency. Least-cost utility planning can lead to more extensive conservation efforts, but regulatory reform is needed to provide utilities with a financial incentive to move towards least-cost energy services.

1 Marketplace Trends and the Role of Utilities

The overall energy efficiency of the U.S. economy has dramatically improved during the past 15 years. Overall energy consumption per unit of gross national product (GNP) fell nearly 27% between 1973 and 1987 (1). Regarding electricity, total consumption has continued to rise in recent years, although at much lower rates than prior to 1975. During 1973–87, total electricity use grew at an average rate of 2.6% per year, while electricity use per unit of gross domestic product (GDP) rose at an average rate of 0.2% per year. But since 1982, electricity use per unit of GDP fell 0.5% per year.

For certain electrical end-uses, there is a clear shift towards greater efficiency. For example, the average refrigerator sold in 1987 consumed 42% less electricity than the average model sold in 1972 (2). During the same period, average refrigerator size increased by 14%. Likewise, the efficiency of new air conditioners increased 25–35% on average between 1972 and 1987 (3).

Regarding commercial buildings, electricity use per square foot of floor area in 1983 was 18% lower in buildings constructed during 1980–83, compared to buildings constructed during 1974–79 (4). Total energy use per unit of floor area dropped even more when comparing post-1980 buildings with late-1970s vintage.

Lighting technology is another area where overall efficiency is improving. The market share for energy-efficient fluorescent lamps as of 1987 was estimated to be about 40%, while the market share for energy-efficient ballasts was around 33% (5). Also, newer conservation technologies such as electronic ballasts and optical reflectors are beginning to be adopted on a large scale, although their overall impact is still relatively low.

In other areas, however, adoption of electricity-conserving end-use technologies has been limited. For example, there has been minimal improvement in the average efficiency of new residential electric water heaters during the past 15 years (6). Sales of heat pump water heaters are still very low, even though this technology has been available for over five years. In the industrial sector, it is estimated that less than 5% of motors in use in 1985 were of the energy-efficient type (7). Likewise, the adoption of electronic adjustable speed drives for industrial motors is proceeding slowly and is far from the cost-effective potential (7, 8).

The market penetration of energy-efficient technologies is determined by a complicated mix of factors, including energy prices, manufacturing decisions, consumer behavior, institutional considerations, regulations, and conservation programs. Numerous studies and reviews have shown that there is widespread underinvestment in end-use efficiency among consumers in all sectors in the U.S. (9, 10, 11, 12). Inhibiting the purchase and use of energy-efficient equipment are:

- lack of information or awareness among consumers, especially about recently commercialized or rapidly evolving technologies,
- uncertainty regarding savings and payback,
- lack of capital and resistance to buying equipment with a greater first cost,
- rapid payback and high rate of return requirements of building owners, occupants, and businesses,

- separation of responsibilities for making capital investments and paying operating costs.

Utilities can play an important role in overcoming the barriers to conservation by educating consumers, offering financial incentives or financing efficiency investments, or paying for the installation of conservation measures. All of these activities are occuring to a significant extent in the U.S. A survey of 123 utilities completed in 1987 by the Investor Responsibility Research Center (IRRC) found that 80% of utilities have implemented formal conservation programs (13). Some of these programs are over 10 years old and have already cut peak demand by hundreds and in some cases thousands of megawatts (MWs).

In the IRRC survey, 76 utilities which reported funding levels spent about $582 million on conservation programs and $135 million on load management efforts in 1985 (13). Seven utilities accounted for the majority of overall conservation program expenditures in 1985, but large-scale investment in end-use efficiency and load management has expanded to a broader group of utilities since then.

Utilities are actively promoting and investing in electricity conservation for a number of reasons (14). First, constructing new generating capacity has become costly (i.e., marginal costs now exceed average costs). Over the long run, it is generally less costly to save a kWh or kW through increasing end-use efficiency rather than supplying an additional kWh or kW by capacity expansion. Increasingly, regulators are requiring that utilities attempt to provide their consumers with "least-cost energy services" (see section III). Second, construction of new central-station power plants is now subject to careful regulatory scrutiny. Utilities fear that if they invest billions of dollars in a new power plant, regulators will question the prudency of that investment and will not allow the utility to recover the full amount from ratepayers. Third, promotion of conservation helps utilities to reduce the rate of load growth and uncertainty about future demand, thereby providing a more stable operating environment. Fourth, end-use efficiency is a flexible resource that can be added in small increments as needed.

Section II of the paper reviews the U.S. utilities' experience with different conservation incentive programs. In addition, U.S., utilities have extensively performed energy audits and engaged in other types of educational programs to promote conservation. These activities are beyond the scope of this paper, but are reviewed elsewhere (15, 15a, 15b, 15c). Section III of the paper reviews planning and regulatory options for stimulating investments in efficiency and proceeding towards least-cost energy services.

2 Utility Incentive Programs

Utilities have adopted a variety of strategies involving financial incentives for promoting greater end-use efficiency.

2.1 Conservation Purchase Through Rebates

Rebate programs generally involve a utility offering to make cash payments to consumers who purchase specified energy-efficient equipment. In some cases, a rebate is paid to the equipment seller instead of (or along with) the buyer. A survey of utility rebate programs conducted by the American Council for an Energy-Efficient Economy and the Consumer Energy Council of America in 1986–87 (hereafter referred to as the ACEEE survey) found 59 utilities offering rebates to purchasers or sellers of energy-efficient equipment (16). Most of these programs are offered by electric utilities and/or apply to electrical equipment. Rebate programs are underway throughout the country, although the ACEEE survey shows the largest concentration of programs in regions of the country with relatively high economic and population growth rates and/or below-average electricity supply reserve margins.

Reducing peak demand, which in turn reduces the need for new generating capacity, is the primary reason given for offering rebates. Because of the emphasis on reducing peak demand, efficient air conditioners and heat pumps are the most common products for which incentives are offered. Nearly 60% of the utilities with residential programs include heat pumps and 39% include central air conditioners. In addition, some programs for residential customers cover efficient refrigerators, freezers, water heaters, or lighting products.

While rebates for residential equipment are more common than rebates in other sectors, commercial and industrial (C&I) programs are gaining in popularity due to the large amount of electricity use and the savings opportunities in these sectors. The ACEEE survey found 24 utilities with C&I rebate programs. About two-thirds of these programs include lighting products such as energy-efficient fluorescent lamps, efficient lamp ballasts, and optical reflectors for fluorescent fixtures. Nearly half of the C&I programs include efficient space conditioning equipment, one-third include efficient electric motors, and a few include industrial process modifications.

The ACEEE survey found over 90% of the utilities providing rebates to the purchaser and about 24% offering rebates to the seller. Qualification levels and rebate amounts vary from program to program. In some cases, fixed rebates are given; in other cases, the rebate amount increases with efficiency. This sliding scale approach encourages the production and purchase of highly efficient models, and is preferred by equipment manufacturers because it tends to promote a broader range of models rather than those just above the minimum qualification level.

Regarding program scale, rebate programs vary from pilot programs with funding of under $100,000 per year to full-scale, multimillion dollar programs. The ACEEE survey found five residential and two commercial programs with budgets over $5 million per year. Approximately half the utilities include their rebate program costs in the rate base (i.e., they capitalize the investment and earn their allowed rate of return on it) and about half treat them as an operating expense (i.e., they quickly recover program costs without earning a profit).

Table 1 presents the budgets and other characteristics for six of the largest and most effective utility rebate programs underway as of 1986. All of these utilities are offering rebates to residential, commercial, and industrial

Table 1. Characteristics of selected major utility rebate programs, 1986.[1]

Utility	Products[2]	Annual budget (million $)	Peak demand savings (MW/yr)	Fraction of peak saved (%/yr)	Annual energy savings (GWh/yr)	Fraction of energy saved (%/yr)
Austin, Texas Elect. Dept.	Res: AC, HP C&I: LT, HVAC, MO	5.9	19.3	1.38	NA	NA
Florida Power and Light Co.	Res: CAC, HP, WH C&I: LT	23.7	59.3	0.57	188.6	0.35
Northern States Power Co.	Res: RF, FR, AC, HP, WH C&I: LT, HVAC, MO	3.8	18.8	0.35	NA	NA
Pacific Gas and Elec. Co.	Res: RF C&I: HVAC, LT, MO, EMS	24.9	56.2	0.40	352.5	0.58
Southern Cal. Edison Co.	Res: RF, AC, HP, WH C&I: HVAC, LT, MO	22.0	81.1	0.55	450.0	0.85
Texas Utilities Co.	Res: AC, HP, WH C&I: HVAC, LT	17.0	140.0	0.88	NA	NA

1. The annual budget and peak demand savings figures apply to both the residential and commercial-industrial programs in the most recent year for which data are available. The fraction of peak demand saved by the rebate program is presented in terms of the percentage of total summer peak demand.

2. Product codes: RF – refrigerator, FR – freezer, AC – air conditioners, HP – heat pumps, WH – water heaters, HVAC – commercial heating, ventilating, and air conditioning systems, LT – lighting, MO – motors, EMS – energy management systems.

Source: *A Compendium of Utility-Sponsored Energy Efficiency Rebate Programs*, EPRI EM-5579, Electric Power Research Institute, Palo Alto, CA, Dec. 1987.

customers, and their average peak demand savings is 62 MW per year. Comparing the savings estimates to actual peak demand for each of the utilities, these six programs are cutting peak demand by 0.35–1.38% per year, with an average reduction of 0.69% per year. For the three utilities where electricity savings data are available, the programs are on average cutting total electricity demand by 0.59% per year.

The program started by the Austin, TX municipal utility in 1982 is having an especially large impact for the following reasons: large incentives are offered, the program is heavily promoted, and both dealers and purchasers are eligible for rebates and have responded to the program. Surveys of air conditioning equipment distributors in the Austin area showed that by 1986, local distributors only stocked air conditioners or heat pumps with efficiencies in the top 40% relative to all models produced in the U.S. (17). The minimum efficiency requirements in both the local building code and the utility rebate program were increased accordingly.

More recently, the Wisconsin Electric Power Company (WEPCO) began a large conservation incentive program for all classes of customers. For commercial and industrial customers, specific rebates or interest-free loans are available on a wide range of lighting, air conditioning, heating, refrigeration, and control improvements (Table 2). In addition, WEPCO offers incentives for other "custom" conservation projects proposed by its commercial or industrial customers. For residential customers, WEPCO offers rebates on a wide range of efficient appliance, lighting, and space conditioning technologies.

WEPCO expects to achieve a peak load reduction of about 140 MW and a total electricity savings of about 600 GWh per year in 1987–88. Their two-year program budget is $84 million ($62 million for rebates, $10 million for loans, $12 million for administration). With these savings and budget levels, WEPCO is cutting its electricity demand and peak load by about 1.4% per year at a levelized cost of around $0.02 per kWh of savings (18).

The ACEEE survey showed that utilities are typically paying $200–300 per kW of avoided peak demand, roughly one-tenth to one-half the cost of new generating capacity (16). In general, rebate programs for C&I customers are somewhat more cost effective than programs for residential customers. A large majority of utilities are satisfied with their rebate programs and very few programs are being stopped.

The issue of "free-riders" in conservation incentive programs has received much attention in the U.S. A "free-rider" is someone who receives a financial incentive for purchasing an efficiency measure, which they would have purchased anyway. Although the utility is still investing in energy efficiency, a high free-rider fraction means that the utility may not be getting much "bang for the buck" and may be unnecessarily burdening non-participants.

Table 2. Selected products and rebate levels in Wisconsin Electric Power Company's financial incentive program for commercial and industrial customers.

Category and Product [1]	Maximum Rebate[2]
Lighting	
Energy-efficient fluorescent lamps	$1–2/lamp
Compact fluorescent lamps	$5/lamp
Energy-efficient core-coil ballasts	$10/ballast
Electronic ballasts	$20/ballast
Optical reflectors	$10/fixture
Heating, Ventilation, Air Conditioning	
Solar shades/films	$0.50/sq.ft.
High efficiency air conditioning	$50–100/ton
High efficiency chillers	$250/kW saved
Heat Recovery	$0.10/annual kWh saved
Economizers	$0.07/annual kWh saved
Variable air volume conversion	$0.10/annual kWh saved
Water heating	
Heat pump water heater	$250–650 each
Controls	
Daylighting	$100/kW controlled
Occupancy sensors	$50/kW controlled
Timeclocks	$0.07/annual kWh saved
Refrigeration	
Strip curtains	$5/linear ft.
High efficiency refrigerators	$50–75 each
Refrigeration heat recovery	$100/kW and $0.10/annual kWh saved
Proces improvements	
Energy-efficient motors	$6–10/HP
Variable speed drives	$100/HP

1. Customers can receive "custom rebates" on other conservation measures.

2. The rebate cannot exceed 50% of the installed cost. Payments related to kW reduction are based on savings between 10 AM and 4 PM, weekdays in July and August. Also, customers are given the option of a zero-interest loan equal to 3.75 times the rebate amount.

Source: "Wisconsin's Smart Money Energy Program", Wisconsin Electric Power Company, Milwaukee, WI, 1988.

Some studies of appliance efficiency incentive programs point to a free-rider fraction as high as 70% (i.e., 70% of those receiving an incentive would have bought an efficient model anyway), but these studies are based on self-reported intentions rather than measured data and therefore may not be accurate (19). Researchers reviewing these studies concluded that it is not possible to draw strong conclusions about levels of free-riders in incentive programs as a whole. Also, incentive programs can be designed in ways that minimize the free-rider problem, particularly by restricting incentives to those products with very high efficiency and/or minimal market penetration.

A few utilities have conducted experiments with efficiency incentive programs. The experiments involve different treatment groups and a control group, with sales data collected to assess the effect of different treatments. These experiments have clearly demonstrated that financial incentives can increase investment in highly efficient appliances and other conservation measures (20).

Table 3 shows the results of one experiment conducted by the New York State Electric and Gas Company (NYSEG) involving different approaches for promoting the purchase of efficient refrigerators. Efficient refrigerators were considered to be the 25% most efficient models offered by the industry at the time. The experiment tested the effects of information and advertising (i.e., bill inserts, store displays, radio and newspaper ads) alone and with a rebate

Table 3. Results from New York State Electric and Gas Corporation's Pilot Refrigerator Rebate Program.

Treatment group	Market share of efficient refrigerators[1] (%) Pre-program	Program
Control area	4.7	14.7
Information and advertising area	2.1	34.8
Info./adver. plus $35 rebate area	5.0	48.6
Info./adver. plus $50 rebate area	8.8	59.7

1. Fraction of all refrigerators sold meeting efficiency criteria.

Source: V.L. Kreitler and T.D. Davis, "High Efficiency Refrigerator Pilot Rebate Program – Final Analysis Report," Load and Market Research Dept., New York State Electric and Gas Corporation, Binghampton, NY, Nov. 1987 incentive. During the program period, the market share for efficient refrigerators was nearly 60% in the area that received information, advertising, and a $50 rebate offer, compared to a 15% market share in the control area that did not receive any information or rebate offer (21). This well-conducted experiment shows that utilities can have a significant impact on end-use efficiency by offering financial incentives together with information and education. In this instance, the proportion of "free riders" was relatively low, about 25% for the high rebate area.

2.2 Conservation Purchase Through Bidding

Some utilities have started using bidding procedures to acquire electricity savings. So-called "demand-side bidding" involves a utility requesting that owners of buildings or industries or independent energy service companies propose undertaking electricity conservation projects. Rather than specifying the level of payment, the utility allows the project sponsor to propose payment rates in a competitive process. Proposals are evaluated by the utility on the basis of cost, technical soundness, reliability, and other factors. The utility then enters into contracts with winning bidders. Efficiency measures are installed and the sponsor is paid according to the contracted rate of electricity and/or peak demand reductions.

A number of utilities in the Northeast have solicited bids for electricity savings. In one program, Central Maine Power Company offered to pay up to 50 percent of the initial cost for conservation projects in commercial or industrial facilities. The objective was to lower the payback period on conservation investments to two years, with actual payments dependent on measured savings. Under this scheme, the utility has been paying 1 to 2 cents per kWh saved (22).

The New England Electric System (NEES) has conducted two auctions and contracted for nearly 14 MW of savings from energy service companies (23). These companies are installing a wide range of efficiency measures in large commercial and industrial facilities. Initially, NEES selected energy service companies for individual facilities and paid on the basis of measured savings. But due to high transaction costs, complex payment analyses, and other problems, energy service companies were selected for geographic areas rather than individual facilities in the second solicitation. The companies are paid on the basis of estimated savings from conservation projects they implement.

The state of Maine has begun to allow "all-source bidding", whereby proposals for electricity supply from independent power producers or cogenerators compete with proposed conservation projects. Central Maine Power Company issued its first solicitation for either conservation or electricity supply projects in early 1988. This resulted in 13 conservation project proposals totaling 36 MW at an average requested price of 3.5 cents per kWh of savings (24). When both conservation and supply projects were ranked by the utility on the basis of price, reliability, environmental impacts, and other factors, conservation projects received 12 of the 15 top rankings.

Demand-side bidding offers the potential for utilities to acquire large amounts of electricity savings as the need arises in a cost-effective manner. But experience with conservation bidding is still rather limited. Many issues remain unresolved, including whether to conduct joint or separate demand and supply bidding, the criteria to use in evaluating bids, the basis for making payments, how to minimize transaction costs, how to maximize savings per dollar expended, and whether to require performance guarantees (25). Additional experiments being planned or implemented as of late-1988 should help to resolve these questions.

2.3 Loan Programs

Some utilities are providing low or zero-interest loans to encourage customers to purchase energy-efficient equipment or to weatherize their buildings. The objective is to reduce the first-cost barrier to investing in conservation and, if possible, to offer a loan repayment plan under which energy bill savings exceed monthly loan payments.

Offering low or zero-interest loans can improve both the degree of participation and the resulting energy savings in residential conservation programs. Examination of utility programs throughout the country shows that retrofit activity can be substantially increased by offering a combination of an audit and a loan in contrast to just an audit. Utilities such as the Bonneville Power Administration and Northern States Power Co. have found that consumers who receive a loan as well as an audit save about three times more energy than consumers who only receive an audit (9, 26).

The Tennessee Valley Authority (TVA) has offered zero-interest and low-interest loans for home weatherization and heat pumps since 1977. By the end of 1984, over 500,000 of these loans had been made with an estimated reduction in winter peak demand of more than 600 MW, about 3% of peak demand (16). Although TVA emphasized the residential sector in its electricity conservation programs during the late 1970s and early 1980, the utility plans to greatly expand its commercial and industrial sector audit and low-interest loan programs.

Experience shows that loan programs can result in high administrative costs and debt-service expenses for utilities (9). Also, when customers are given a choice of a rebate or low-interest loan, they tend to prefer the rebate. Southern

California Edison Co., for example, found that only 2% of the participants in their residential conservation incentive program preferred a loan over a rebate (16). Puget Sound Power and Light had similar results when it offered commercial customers either a cash grant or a zero-interest loan (28), as did Wisconsin Electric Power Company (18). For this and other reasons, some utilities have phased out their loan programs.

2.4 Direct Installation

Direct installation involves a utility or an energy service company hired by a utility installing conservation measures in eligible homes or commercial-industrial facilities. Typically, there is no cost or minimal cost for the building owner or industry. Hence, a high level of participation and adoption can be obtained. The utility benefits by avoiding the high capital investment and operating costs of new energy supplies. Direct installation programs are often tailored for "hard to reach" markets that do not respond well to information or incentive programs, such as low-income and rental housing, or commercial-industrial facilities with quick payback constraints.

Free installation is employed in some utility-sponsored low-income housing weatherization programs (29). Pacific Gas and Electric Co., for example, greatly increased the participation of low-income households when it switched from a zero-interest loan to free installation (30). Moreover, the cost per household turned out to be about the same for the utility because of the high debt-service, outreach, and overhead costs with the loan program.

Utilities in California sponsored an innovative direct installation and audit program in Santa Monica, a city with 78% rental housing. The Santa Monica Energy Fitness program involved door-to-door canvassing of every household, streamlined audits, and free installation of up to three low-cost conservation measures oriented primarily towards natural gas and water savings. During the one-year program, about 35% of the households in the city participated. The program cost only $87 per participant and resulted in 5.4% natural gas savings in participating households, yielding a 3.5 year payback based on the value of the actual energy and water savings (31). Expanding the list of conservation measures might have increased energy savings and program effectiveness.

A utility in New Jersey and Pennsylvania contracted with private companies to provide conservation retrofits in selected electrically-heated households at no cost to the consumer. The utility paid the contractors based on actual electricity savings. This program, known as RECAP, was tested in four communities in 1983–85.

Table 4 lists the results in the four communities. The participation rates are much greater than is customary in financial incentive programs (32). However, the RECAP program resulted in only moderate electricity savings (6.6% on average), in part because some contractors limited their work to the most cost-effective conservation measures, i.e., measures with the highest electricity savings per dollar installed, to maximize their profit. With an average cost of $670 and average savings of 1270 KWh per year in retrofit households, the RECAP pilot program was reasonably cost effective with an average simple payback period of 5.8 years. Also, the program was profitable for three of the four contractors.

A pilot program similar to the RECAP program but targeted to apartment buildings is being conducted in the Chicago area. The local utility is paying a contractor on the basis of verified electricity savings; energy improvements are free to both tenants and building owners. The payments for electricity savings are about 15% less than the cost to produce electricity (33).

Another example of direct installation is the Hood River Conservation Project (HRCP). The demonstration project sought to install as many cost-effective conservation measures in electrically-heated homes in Hood River,

Table 4. Results from General Public Utilities Corporation's Residential Energy Conservation Action Program (RECAP)

Community	Number of participants	Participation rate[1] (%)	Total electricity savings[2] (kWh/yr)	(%/yr)	Average retrofit cost[3] ($)	Simple payback period[4] (yrs)
Covered Bridge, NJ	320	43	683	5.8	270	4.4
Leisure Village, NJ	3000	>90	128	1.7	480	41.7
York, PA	368	16 (5)	2531	10.9	1340	5.9
Reading, PA	1500	25	1748	8.1	595	3.8
TOTAL	5188	36	1273	6.6	670	5.8

1. Fraction of eligible households in which retrofits were performed.

2. Fraction of pre-retrofit electricity saved due to participation in RECAP, based on actual performance and statistical analysis including correction for savings by a control group.

3. Total retrofit cost per household including marketing, materials, and labor.

4. Based on an average residential electricity price of $0.09/kWh, appropriate for this utility as of 1987.

5. Low market penetration was due primarily to early withdrawal of the contractor.

Source: M.A. Brown and D.L. White, "Impact Analysis of a Residential Energy Conservation Shared Savings Program: The General Public Utilities Experience", ORNL/CON-217, Oak Ridge National Laboratory, Oak Ridge, TN, Feb. 1987.

Oregon as possible during 1983–86. The Bonneville Power Administration paid for both audits and the installation of conservation measures.

Of the 3500 eligible households in Hood River, about 91% received an energy audit and 85% had one or more major retrofit measures installed (34). The enormous success of the HRCP in achieving participation was due to utility funding, community-based marketing approaches, extensive word-of-mouth communication among residents, and determination on the part of HRCP staff to reach every household.

Eighty-three percent of the conservation measures recommended in the energy audits were actually installed (35). The measures include relatively inexpensive items such as low-flow showerheads and water heater wraps as well as major measures such as ceiling, wall, and floor insulation. Conservation cost effectiveness was viewed from the societal perspective, thereby permitting homes to be retrofitted to a high degree of thermal integrity.

Analysis of the Hood River pre- and post-retrofit electricity consumption data found a typical savings of 2600 kWh per year (14%) in homes retrofitted (36). Savings in single family homes that used electricity as their primary heating fuel were 4000 kWh per year (16%). Actual energy savings were substantially less than predicted based on initial energy audits, a result that frequently occurs in home retrofit programs. Much of the discrepancy between predicted and actual savings in Hood River is explained by conservation actions taken by residents on their own prior to utility-sponsored retrofits as well as changes in energy-related behavior such as higher thermostat settings and less use of wood fuel after the retrofits (36). (See also Fels and Dutt in this volume.)

2.5 Nonfinancial Inducement

Availability of a financial incentive is just one of many factors influencing whether or not consumers invest in energy efficiency. Maximizing the success of conservation incentive programs also depends on the ability to get the attention of the consumer, communicate to the consumer in an understandable and credible manner, and minimize the effort and risk associated with investing in energy efficiency (37). Therefore, the ways in which conservation programs are organized, marketed, and implemented are extremely important.

The importance of nonfinancial inducement is evident when considering the responses to different programs that contain identical financial incentives. For example, nine utilities in New York offered low-interest loans for home weatherization during 1978–84. There was a 50-fold variation in program participation rates between the most successful and least successful utilities

(38). Likewise, the Bonneville Power Administration offered financial incentives for home weatherization through local utilities in the Pacific Northwest. There the most successful utility weatherized eligible homes at 13 times the rate of the least successful utility (38).

Involving local community groups has proven to be one program element that can greatly increase the effectiveness of residential conservation programs. Experience with residential audits in Minnesota showed that community groups provided four times the response rate and delivered better quality audits then utility-performed audits (38). Involving community groups in promotion, audits, and retrofit work is also used by many utilities to increase participation and adoption rates among low-income households (29, 38).

Marketing and outreach are important factors in implementing conservation programs. Studies of the behavior of owners and managers of commercial buildings suggest the following strategies for maximizing adoption and program effectiveness in the commercial sector (39, 40).

1) Aim programs at those responsible for investment decisions, keeping in mind the decision criteria and process involved as well as the barriers that inhibit investment.

2) Tailor program design and marketing strategy to different market segments, based on building type, ownership and organizational type, company size, degree of technical sophistication, etc.

3) Contact owners and decision makers directly to solicit program participation.

4) Explain the non-energy advantages of energy-efficient equipment or services such as increased comfort, productivity, flexibility, or reliability.

5) Reduce the uncertainty and risk associated with conservation investments through demonstrations, guarantees, etc.

Working with "trade allies" such as equipment wholesalers, retailers, and contractors can also enhance utility conservation programs. Widespread adoption is unlikely to occur unless dealers and contractors stock and encourage consumers to purchase efficient products. Hence, many utilities involve vendors in the design and implementation of their incentive programs. Some even pay dealers or contractors when they sell or install qualifying products (16).

3 Planning and Regulatory Initiatives

3.1 Least-Cost Utility Planning

Least-cost utility planning (LCUP) has emerged in recent years as a planning strategy that can lead to much greater implementation of end-use efficiency and load management technologies. The objective of LCUP is to provide services such as heat, light, refrigeration, and motive power – not energy *per se* – at the lowest possible cost (41). Under LCUP, all feasible "demand-side" resources are evaluated along with new energy supply options and are ranked according to cost effectiveness. Utility regulators encourage – or require – utilities to invest in end-use efficiency and load management measures which save energy at *less* cost than options for generating or purchasing energy. Risk, reliability, safety, environmental impacts, and social concerns can be considered in the choice of options.

Various tests of program cost effectiveness are employed in evaluating conservation and other utility programs, including evaluation of cost effectiveness from the perspectives of the program participant, non-participant, utility, and society (42). Under LCUP, either the utility or the societal perspective is used as the primary test of program cost effectiveness (43). From the utility perspective, conservation programs and other resource options are pursued if they lower overall costs for the utility. While this test does not reflect all costs to society, it is familar to utilities and regulators and is relatively easy to evaluate. The societal perspective combines consideration of utility and non-utility costs in delivering energy services. It is the broadest measure of economic merit, however it introduces factors such as customer costs and social costs that can be difficult to evaluate. The non-participants perspective is often considered when evaluating utility conservation programs, but is not used as a primary criterion for judging program cost effectiveness under LCUP because this test is arbitrary and highly restrictive (44).

LCUP dates back at least to 1980 when it was adopted for the Pacific Northwest region of the U.S. The Northwest Power Planning Council was established in part to develop least-cost plans for the region. In assessing resource options, planners in the Pacific Northwest are required to grant conservation options a 10% economic bonus to reflect the favorable environmental and social impacts associated with conservation (45). This is one of the few cases where these externalities are in some way factored into electricity planning in the U.S.

Utilities in the Northwest spent over $800 million on conservation programs during 1980–86 and have acquired the capability to implement efficiency improvements on a large scale if necessary (46). The most recent Northwest Power Plan calls for cutting electricity use 14% (3700 MW) by 2006, at an average cost of 2.4 cents per kWh if population and economic activity in the region grow rapidly (47). Less conservation is called for if economic and demographic growth remain low.

Since 1980, at least 16 states have adopted least-cost planning policies and requirements (48). But striving for least-cost energy services and high levels of investment in conservation are still far from routine throughout the utility industry. In fact, many utilities are resisting LCUP and are continuing to promote load growth (13). The financial incentives created by traditional utility ratemaking explain why this is the case.

3.2 Financial Incentives for Utilities

Utility regulators can encourage utilities to pursue least-cost energy services by linking profitability to a utility's progress in acquiring cost-effective sources of generating capacity and end-use efficiency. Conventional ratemaking procedures in the U.S., however, financially reward utilities for increasing electricity or gas sales regardless of the impact on energy service costs. This occurs because utility rates are set periodically. In the interim, utility revenues and profits rise as the utility sells more electricity or gas, because short-run marginal revenues almost always exceed short-run production costs (excluding fuel costs, which are directly recovered through an automatic rate adjustment).

In pursuit of least-cost energy services, it is a relatively straightforward matter to periodically change rates to adjust for changes in energy sales so that profits do not exceed those expected from the last ratemaking case. This removes the windfall that utilities would otherwise receive through increased sales. Likewise, rates can be adjusted to prevent utilities from being penalized if conservation programs are very successful and energy savings exceed expectations (i.e., if actual energy sales are below previous projections). The California utility commission has established an automatic revenue and rate adjustment mechanism that does this (49).

Utility regulators could go even further and increase or decrease a utility's revenues and profits in proportion to the degree it is able to lower sales and/or energy service costs (50). Also, regulators can raise or lower the rate of return on either the entire rate base or certain types of utility expenditures, or provide a utility with a financial bonus that is proportional to the net economic benefit resulting from conservation, load management, or alternative generation investments (51). Recently, the Energy Conservation Committee of the National Association of Regulatory Utility Commissioners adopted a position urging regulators to make the least-cost plan a utility's most profitable resource plan (52).

A few utility commissions in the U.S. have used rate of return incentives and penalties to steer utilities towards greater energy efficiency and least-cost energy services. In 1979, the California PUC penalized Pacific Gas & Electric Co. (PG&E) for failing to make reasonable efforts towards developing cogeneration potential in its service territory (53). PG&E subsequently contracted for large amounts of cogeneration and independent power.

In 1986, the Massachusetts DPU reduced the allowed rate of return for Boston Edison Co. (BECO) because of its poor record in developing and implementing conservation/load management programs and its failure to develop an integrated resource plan (54). Since then, BECO has greatly expanded its conservation and load management programs, adopted integrated supply and demand planning, and instituted a competitive bidding process for generating resources.

In 1987, the Wisconsin utility commission adopted performance-based incentives in its effort to move towards least-cost energy services, in part through aggressive utility conservation programs. Having determined that a large amount of electricity savings was more economical than a power supply project proposed by Wisconsin Electric Power Company, the utility commission directed the utility to invest about $84 million in conservation projects. The commission offered the utility a performance incentive of a 1 percent additional return on capitalized conservation expenditures for each 125 MW of load that is saved (55). Also, the commission gave the utility cost effectiveness guidelines for its conservation programs.

Establishing performance-based incentives will safeguard consumers from paying for ineffective or overly expensive conservation, load management, and alternative generation programs. However a rate of return incentive like the one established in Wisconsin may not be adequate to offset the revenue loss disincentive to utility investment in energy efficiency. In other words, the value of lost electricity sales may be greater than the bonus the utility receives on investments in conservation (50, 56).

4 Conclusion

A number of overall conclusions can be drawn from the experience with utility conservation incentive programs and least-cost electricity planning in the U.S. (13, 16, 57).

1. *Efficiency incentives can make a difference.* Some utilities have demonstrated the ability to reduce electricity use and peak load by 0.4–1.4% per year. This level of savings was achieved by spending a small percentage of

total utility revenues on financial incentive programs. Also, utilities that have achieved the most savings have offered incentives on a wide range of conservation measures for all customer classes.

2. *Program design should be flexible.* It is difficult for utilities to anticipate the diverse requirements of their customers when they initially design their conservation programs. Recognizing this limitation has led some utilities to build flexibility into their programs. For example, rather than restricting consumers to a prescribed list of conservation measures, some utilities allow consumers to propose their own "customized" conservation projects. Likewise, utility conservation programs usually need to be modified over time depending on factors such as consumer acceptance and changes in the availability, performance, and penetration of different efficiency measures.

3. *Nonfinancial factors are important.* Offering financial incentives does not automatically lead to a successful utility conservation program. The way in which programs are promoted and marketed can greatly influence how consumers react. Getting the attention of decision makers, addressing their concerns, and making it easy for them to participate are critical for achieving a high level of response. Also, utilities need to work closely with equipment suppliers since they provide the product and are a key source of information for consumers.

4. *Utility conservation programs have not reached maturity.* Most utility conservation incentive programs were started during the 1980s and many issues concerning program design have not been resolved. For example, it is impossible to state how large an incentive is necessary to obtain a given level of response or how best to divide up program funds between promotion and incentives. Also, new program strategies such as purchasing savings through competitive bidding are being tried out. Further research, experimentation, and program evaluations are needed before firm conclusions can be drawn regarding the most effective program designs.

5. *Least-cost planning can boost utility conservation efforts.* By forcing utilities to compare a wide range of demand-side and supply-side resource options and base their investment priorities on lowest cost and risk, least-cost planning is a valuable technique for increasing utility investment in energy efficiency. Utility conservation efforts tend to be more advanced and intensive in regions of the U.S. where least-cost planning is well-established (e.g., the Pacific Northwest).

6. *Many utilities are resisting the least-cost approach and large-scale conservation incentive programs, but this is likely to change.* Even though

some utilities are spending significant sums of money (say, at least two percent of total revenues) on end-use efficiency programs, this level of commitment is still far from the norm among utilities in the U.S. But utilities' resistance to least-cost planning and large-scale utility conservation programs is likely to subside as reserve margins fall and the need for new resources becomes more widespread.

7. *Utility ratemaking procedures are a problem.* At present, most utilities lose revenues and profits when they conduct effective conservation programs that reduce electricity sales. Authorities need to modify ratemaking procedures in order to reward utilities for investing in cost-effective efficiency improvements. In this manner, regulators can give utilities an incentive to move towards least-cost energy services. A few state regulatory commissions have provided their utilities with specific financial incentives for investing in electricity conservation measures, and broader regulatory reform has been proposed.

8. *Much of the potential for electricity conservation remains untapped.* Although there are a few examples of utilities significantly increasing the adoption of specific conservation measures in their service area, newer conservation technologies such as variable speed motor drives, compact fluorescent lamps, electronic lamp ballasts, and optical reflectors are just starting to gain a foothold in the market. In theory, utilities could play a large role in the diffusion of technologies such as these. However, more research is needed on the effect that utilities can have on market acceptance.

Finally, it must be remembered that end-use efficiency has been treated as a legitimate utility resource for a relatively brief period. During the past decade, utilities in the U.S. have made considerable progress in developing and refining strategies for obtaining economical "conservation resources". Today, a growing number of electric utilities are investing large sums in end-use efficiency, thereby cutting load growth and reducing the need for new power supplies. These utilities frequently find that efficiency is a bargain compared to supply alternatives, even without taking into account environmental and social costs. Given these developments, the outlook for utility conservation programs is extremely promising.

References

1. Energy Information Administration. 1988. Monthly Energy Review. DOE/EIA-0035(87/12). Washington, DC: U.S. Dept. of Energy.
2. The Association of Home Appliance Manufacturers, Chicago, IL, Personal communication 1988.
3. The Air-Conditioning and Refrigeration Institute, Arlington, VA, Personal communication 1988. Also, Ref. 2.
4. Energy Information Administration. 1986. *Commercial Buildings Consumption and Expenditures 1983*. DOE/EIA-0318(83). Washington, DC: U.S. Dept. of Energy.
5. Geller, H.S. 1988. Commercial Building Equipment Efficiency: A State-of- the-Art Review. Washington, DC: American Council for an Energy-Efficient Economy.
6. Geller, H.S. 1988. Residential Equipment Efficiency: A State-of-the-Art Review. Washington, DC: American Council for an Energy-Efficient Economy.
7. Moe, R.J. Harrer, B.J., Kellogg, M.A., Lyke, A.J., Imhoff, K.L., Fisher, Z.J. 1986. The Electric Energy Savings from New Technologies. PNL-5665 Rev. 1. Richland, WA: Pacific Northwest Laboratory.
8. Resource Dynamics Corp. 1986. *Electrotechnology Reference Guide*. EPRI EM-4527. Palo Alto, CA: Electric Power Research Institute.
9. Hirst, E., Clinton, J., Geller, H., Kroner, W. 1986. *Energy Efficiency in Buildings: Progress and Promise*. Washington, DC: American Council for an Energy-Efficient Economy.
10. Ruderman, H., Levine, M., McMahon, J. 1987. Energy-Efficiency Choice in the Purchase of Residential Appliances. in Kempton, W., Neiman, M., eds., *Energy Efficiency: Perspectives on Individual Behavior*. Washington, DC: American Council for an Energy-Efficient Economy.
11. Goldemberg, J., Johansson, T.B., Reddy, A.K.N., Williams, R.H. 1987. *Energy for a Sustainable World*. Washington, DC: World Resources Institute.
12. Alliance to Save Energy. 1983. *Industrial Investment in Energy Efficiency: Opportunities, Management Practices, and Tax Incentives*. Washington, DC.
13. Cogan, D., Williams, S. 1987. *Generating Energy Alternatives*. Washington, DC: Investor Responsibility Research Center, Washington, DC.
14. Cavanagh, R.C. 1987. Least-Cost Planning Imperatives for Electric Utilities and Their Regulators. *The Harvard Environmental Law Review* 10(2):299–334.
15. Clinton, J., Geller, H., Hirst, E. 1986. Review of Government and Utility Energy Conservation Programs. *Ann. Rev. Energy* 11:95–142.

15a. Walker, J.A., Rauh, T.N., Griffin, K. 1985. A Review of the Residential Conservation Service Program. *Ann. Rev. Energy* 10:285–316.

15b. Synergics Resources Corp. 1984. *Utility Conservation Programs: Assessment of Implementation Experience*. EPRI EA-3585. Palo Alto, CA: Electric Power Research Institute.

15c. Duberg, J., Hanson, P. 1988. *1987 General Report to Congress on the Residential Conservation Service Program* Residential and Commercial Conservation Branch, U.S. Dept. of Energy, Washington, DC.

16. American Council for an Energy-Efficient Economy and the Consumer Energy Council of America. 1987. *A Compendium of Utility-Sponsored Energy Efficiency Rebate Programs*. EPRI EM-5579. Palo Alto, CA: Electric Power Research Institute.
17. Garrett, Doug Resource Management Department, Austin, Texas, Personal communication October 1987.

18. Hawley, T.H. 1988. Wisconsin Electric and the Smart Money Energy Program. Proceedings of the 1988 ACEEE Summer Study on Energy Efficiency in Buildings, Vol. 6. Washington, DC: American Council for an Energy-Efficient Economy. Also, personal communication with Mr. Tom Hawley, Wisconsin Electric Power Company, Milwaukee, WI.
19. George, S.S., Koved, M.D., McRae, M.R. 1988. Appliance Efficiency Incentive Programs: Literature Review and Recommendations for Pilot Program Design and Demand Forecasting. California Energy Commission, Sacramento, CA.
20. Geller, H.S. 1988. Lessons from Utility Experimentation with Appliance Efficiency Incentive Programs. Proceedings of the 1988 ACEEE Summer Study on Energy Efficiency in Buildings,Vol. 6. Washington, DC: American Council for an Energy-Efficient Economy.
21. Kreitler, V.L., Davis, T.D. 1987. High Efficiency Refrigerator Pilot Rebate Program – Final Analysis Report. Load and Market Research Dept., New York State Electric and Gas Corporation, Binghampton, NY.
22. Personal communication with Mr. Douglas Herling, Central Maine Power Co., Augusta, ME, June 1988.
23. Michaels, H.G. 1988. Bulk Purchase of Conservation by Competitive Bid. Proceedings of the 1988 ACEEE Summer Study on Energy Efficiency in Buildings, Vol. 6. Washington, DC: American Council for an Energy-Efficient Economy.
24. Central Maine Power Co. 1988. Project Rankings, All Source Bidding 1988. Exhibit No. 26, Hydro-Quebec Hearings. Maine PUC Docket No. 87–268. Augusta, ME.
25. Cole, J.W., Wolcott, D.R., Weedall, M. 1988. Competitive Bidding of Demand- Side Management. Proceedings of the 1988 ACEEE Summer Study on Energy Efficiency in Buildings, Vol. 6. Washington, DC: American Council for an Energy-Efficient Economy.
26. Hirst, E. 1984. Household Energy Conservation: A Review of the Federal Residential Conservation Service Program. *Public Administration Review* 44(5):421–30.
27. Tennessee Valley Authority. 1984. Energy Conservation Program. Division of Conservation and Energy Management, Knoxville, TN.
28. Lehenbauer, G.W. 1985. Marketing Energy Management to Commercial Facilities: Utility Program Field Experience. *Meeting Energy Challenges: Proceedings of the Second Great PG&E Energy Expo – 1985*. New York, NY: Pergamon Press.
29. Berry, L., Hubbard, M., White, D. 1986. A Review of Financial Incentive, Low-Income, Elderly and Multifamily Residential Conservation Programs. ORNL/CON-206. Oak Ridge, TN: Oak Ridge National Laboratory.
30. Moulton, D.H. 1984. The Impact of Utility-Sponsored Conservation Loan Programs on Low-Income Households. Proceedings of the ACEEE 1984 Summer Study on Energy Efficiency in Buildings, Vol H. Washington, DC: American Council for an Energy-Efficient Economy.
31. Egel, K. 1986. Final Evaluation of an RCS Program Alternative: The Santa Monica Energy Fitness Program. Proceedings of the ACEEE 1986 Summer Study on Energy Efficiency in Buildings, Vol 10. Washington, DC: American Council for an Energy-Efficient Economy.
32. Brown, M.A., White, D.L. 1987. Impact Analysis of a Residential Energy Conservation Shared Savings Program: The General Public Utilities Experience. ORNL/CON-217. Oak Ridge, TN: Oak Ridge National Laboratory.
33. Personal communication with Mr. Richard Esteves, Sentinel Energy Savings Corp., West Springfield, MA, April 1988.

34. Hirst, E., Goeltz, R. 1986. Dynamics of Participation and Supply of Services in the Hood River Conservation Project. ORNL/CON-210. Oak Ridge, TN: Oak Ridge National Laboratory.
35. Goeltz, R., Hirst, E. 1986. Residential Retrofit Measures in the Hood River Conservation Project: Recommendations, Installations, and Barriers. ORNL/CON-208. Oak Ridge, TN: Oak Ridge National Laboratory.
36. Hirst, E., Goeltz, R., Trumble, D. 1987. Electricity Use and Savings in the Hood River Conservation Project. ORNL/CON-231. Oak Ridge, TN: Oak Ridge National Laboratory.
37. Stern, P.C., Aronson, E., eds. 1984. *Energy Use – The Human Dimension*. New York, NY: W.H. Freeman and Co.
38. Stern, P.C., Aronson, E., Darley, J.M., Hill, D.H., Hirst, E., et al. 1986. The Effectiveness of Incentives for Residential Energy Conservation. *Evaluation Review* 10(2):147–76.
39. George, S.S. 1986. Commercial Customer Acceptance of Demand-Side Management: Conservation Programs. *Productivity Through Energy Innovation: Proceedings of the Third Great PG&E Energy Expo – 1986*. San Francisco, CA: Pacific Gas and Electric Co.
40. Dinan, T.M. 1986. Understanding the Market for Conservation Technologies in the Commercial Sector. Proceedings of the ACEEE 1986 Summer Study on Energy Efficiency in Buildings, Vol. 5. Washington, DC: American Council for an Energy-Efficient Economy.
41. Energy Conservation Coalition. 1987. Least-Cost Electrical Strategies: An Information Packet. Washington, DC.
42. California Energy and Public Utility Commissions. 1983. Standard Practice for Cost-Benefit Analysis of Conservation and Load Management Programs. Sacramento and San Francisco, CA.
43. Hirst, E. 1988. Regulatory Responsibility for Utility Integrated Resource Planning. ORNL/CON-249. Oak Ridge, TN: Oak Ridge National Laboratory.
44. Plunkett, J. 1987. Saying No to the No-Losers Test: Correctly Assessing Demand Side Resources to Achieve Least-Cost Utility Strategies. paper presented at the annual meeting of the National Association of State Utility Consumer Advocates, Washington, DC. Also. Williams, R.H. Designing Electricity Conservation Programs (in this edition).
45. U.S. Congress. 1980. The Pacific Northwest Electric Power Planning and Conservation Act. P.L. No. 96–501. Washington, DC.
46. Northwest Power Planning Council. 1987. A Review of Conservation Costs and Benefits – Five Years of Experience und the Northwest Power Act. Portland, OR.
47. Northwest Power Planning Council. 1986. Northwest Conservation and Electric Power Plan. Portland, OR.
48. Markowitz, P., Kriesberg, J. 1985. Least-Cost Electrical Planning: Is There Really a State Movement?. Critical Mass Energy Project, Washington, DC. Also, Shapiro, L., Markowitz, P., Hirsh, N. 1987. A Brighter Future: State Actions in Least-Cost Electrical Planning. Washington, DC: Energy Conservation Coalition.
49. Cavanagh, R. 1988. Responsible Power Marketing in an Increasingly Competitive Era. *Yale Journal on Regulation* 6, (forthcoming).
50. Moskovitz, D. 1988. Will Least Cost Planning Work Without Significant Regulatory Reform?" paper presented at the NARUC Least Cost Planning Conference, Aspen, CO.
51. Wellinghoff, J.B. 1988. The Forgotten Factor in Least-Cost Utility Planning: Cost Recovery. *Public Utilities Fortnightly* 121(7):9–16.

52. National Association of Regulatory Utility Commissioners. 1988. Statement of Position of the NARUC Energy Conservation Committee on Least-Cost-Planning Profitability. Washington, DC. July 26.
53. California Public Utility Commission. 1979. Decision No. 91107. San Francisco, CA.
54. Massachusetts Department of Public Utilities. 1986. Order No. 85–266-A and 85–271-A. Boston, MA.
55. Wisconsin Public Service Commission. 1986. Opinion and Order 6630-UR-100. Madison, WI.
56. Geller, H.S. 1988. Use of Financial Incentives to Encourage Least-Cost Utility Planning and Energy Efficiency. paper presented at the NARUC Least Cost Planning Conference, Aspen, CO.
57. Gandhi, S., Krause, F. 1988. Program Design and Success: A Review of Two Utility Lighting Programs. Proceedings of the 1988 ACEEE Summer Study on Energy Efficiency in Buildings, Vol. 6. Washington, DC: American Council for an Energy-Efficient Economy.

The Coming Reformation of the Electric Utility Industry

Måns Lönnroth

Abstract

Steam turbines of ever larger scale have formed the lynch-pin of the electricity industry for hundred years. Environmental concerns about CO_2 and nuclear power, however, will increasing force the electricity industry off the steam turbine path. Technological possibilities in natural gas using technologies (in the entire range between 100 MW and 0.1 MW) as well as more efficient end use (10's to 100's percent efficiency improvements) exist.

The electric utility industry will have to evolve from being a supply-only oriented entity which builds and maintains huge electricity production facilities to (1) a supply coordinator and then (2) a systems-oriented, electricity services organization in which the costs of expanding or replacing generating capacity is balanced against the cost of increasing end-use efficiency. The domination of the 1000 MW steam turbine plant is over. The age of small advanced gas turbine, decentralized production (cogeneration), and much more efficient electricity-using lights, motors, appliances, etc., has dawned.

The West European electric utilities are evolving towards the first stage, although at different rates in different countries with Denmark appearing to lead the way towards decentralized production. The second stage is only now emerging as a possibility. Edison's vision of an electricity service industry as opposed to an electricity supply industry waits to be resuscitated.

1 Introduction

The electric utility industry is one of the success-stories of the 20th century.

Electricity was the first truly scientific energy form. While coal was tinkered with far back in the middle ages, electricity emerged out of the early 19th century laboratories as a symbol of the power of enlightenment.

Electricity played a fundamental role in the transformation from agricultural to industrial societies. Electricity freed production from the spatial constraints of other forms of energy, making new forms of industrial organization and thus large gains in productivity possible. Electricity facilitated reading and education and made modern media, such as radio, possible. Thus electricity facilitated the transformation from elite to mass politics and, in the end, democratization. Electricity carried the future.

It still does, but now the symbol is the chip rather than the steam-turbine generator. While the future of electricity remains assured, the organization of electricity does not.

The industrial and societal organization of electricity has been formed by three pressures: the pressure for electrification, the pressure for an ever more efficient supply, and the pressure for an equitable access to electricity.

Now, a fourth formative pressure has been added: the environmental pressure.

Simultaneously, the post World War II consensus on the societal organization of electricity is called into question. Technological developments that reconcile electrification, efficiency, equality and environment will have considerable impact on both the industrial and the societal organization of electricity.

The aim of this essay is to explore the implications of the now four pressures.

2 The Organization of Electricity

At the beginning there was lighting. Edison's vision was an integrated organization that supplied lighting to the user from a (small) central station. Power losses during transmission made duplication necessary; other services – such as traction – were to be organized in parallel. Thus Edison's vision implied expansion through duplication of non-connected, essentially identical providers of (energy) services.

The transformer changed all that. The transformer was probably the single most important invention in determining the organization of electricity and, indeed, the emergence of national utilities as such (1). The transformer made it possible to link distant water falls to urban centers, thereby also linking rural and urban areas. The transformer also opened up unexplored but, as it turned out, extremely fertile technological territories. Harnessing technological fields such as materials, manufacturing, control systems as well as more basic developments in chemistry and physics made spectacular achievements possible: over the decades steam turbine sizes increased a thousand-fold as did the voltage in transmission systems. The efficiency of steam turbines increased by a factor of seven. Previously local electric systems were tied together, and the various economies of scale exploited drove down the real price of electricity over the course of the century.

The transformer made Edison's vision of integrated energy service organizations obsolete.

The present organization of electricity should be analyzed on two (interwoven) levels: the level of the *electricity industry* and the level of the *societal organization of the electric utility.*

The electricity industry – which in Edison's view was one integrated industry – thus consists of three sectors: the *generating utilities*, the important ones being counted in the low hundreds thoughout the world; the *electrotechnical industry*, which supplies the utilities with generating and transmission technology, and consisting of a handful of very large multinational corporations – ASEA-BBC, General Electric, Westinghouse, Siemens, Hitachi, Mitsubishi, Toshiba, NEI, GEC, Alsthom and a few others; the *appliance industry*, which supplies the very large number of small users with electric appliances such as lighting, refrigeration, electronic products, electric engines, electrochemical devices, again consisting of a handful of very large multinational corporations.

This necessarily diverse industry has a very distinct core: those companies between which new technologies are developed, inevitably in a symbiotic relationship between utilities and electrotechnical suppliers. In Sweden ASEA, the State Power board and Sydkraft play this role; in West Germany Siemens and RWE; in France Framatome and EdF, etc. This core, within which there is considerable and growing co-operation, is located in a handful of industrialized countries: the US, Japan, West Germany, the UK, France and Sweden-Switzerland. Around the core one finds peripheral companies, tied to the core through various licensing agreements (2). Technologies that aspire to make substantial impacts on the world electricity industry will have to make a dent in the core.

The appliance industry is segmented into different technological areas, each dominated by a handful of extremely large multinational companies such as Philips in lighting, household appliances, electronics; Electrolux in household appliances; Siemens etc. Historically there were close links between supplying technology to utilities and to users – after all, the name of Edison's company General Electric has German and Swedish equivalents – but gradually these activities grew apart.

The two important relations in the core are the relations between the utilities and the electro-technical giants and the relations between the utilities and the appliance industries. There is considerable activity in the former link – this is where supply technologies are developed – but much less so in the latter. While the electric utilities were very active in the early days of electrification in the development of new appliances this link has become much more passive of late.

Given the new environmental pressure, it will be argued that this link will have to become reactived.

The other level of analysis concerns the societal organization of the electric utility. The electric utilities grew up around the middle classes in the big cities and the large industries in the late 1800's. The economic benefits of higher load factors, to no small degree released by the transformer, made utilities reach out for further users. Gradually, markets grew among poorer urban people, smaller towns, smaller businesses and eventually rural areas. The Great War accelerated integration of the different utilities. By then electricity was as much a social right as a means of production and, moreover, a symbol of progress, industrialization and even education. The Scandinavian slogan "light to the people" was an intended pun.

The growing social importance of electricity triggered legislation and regulation. An early Swedish example concerned the struggle over water rights, where a pre-industrial concept of the rights of individual land ownership had to give way for the majority interest of non-landowners, thus making exploitation possible. Similar struggles took place over right-of-ways for transmission systems. Gradually, a complex societal organization grew up, differing between countries but with essentially the same goal: to ensure that the benefits of electricity should be accessible to everyone – all citizens as well as all of industry. Initial technical fragmentation was replaced by common standards. Rural electrification programmes were initiated. The war experiences underlined the need for coordination and systems perspective over the rights of the individual utility. In some countries, electric utilities became publicly owned, sometimes by local or regional authorities and sometimes by the state. In some countries, rates were regulated in detail. In some countries generating and distribution was separated, in others not. In some countries ownership links within the electricity industry – e.g. Germany and Sweden – were close, in others the activities were separate.

The electric utility industry is now by and large one hundred years old. The various national entities have obviously changed considerably during this period. Given present pressures, one important aspect is the balance between self-reorganization and reorganization through outside, i.e., governmental, intervention. At the risk of gross oversimplification, it appears that among the countries within the core of the electricity industry West Germany, Sweden and Switzerland have tilted to the former while France and in particular the UK and arguably the US have tilted toward the latter. Note that this distinction is separate from the issue of ownership.

The electricity industry and its corresponding societal organization has coped well with the pressures for electrification, efficiency and equality. The added pressure of environment has changed the rules of the games, however, to the point where both the structure of the electricity industry and the societal organization of the electric utility industry is increasingly questioned.

3 Formative Pressures

So far, the electricity industry has reconciled the pressures towards electrification, efficiency and equality through systems integration and economies of scale in steam turbines, high voltage transmission and, where possible, in hydro installations. Environmental pressures will exert powerful changes on the existing balance of forces.

While the list of environmental concerns is long, airborne pollutants such as SOx, NOx, and now also CO_2 are high on it. In most countries, the electric utility industry is the major source of SOx and a major source of NOx – and therefore acid rain – as well as of CO_2.

Two issues are likely to dominate the future debate over electricity and the environment: the role of nuclear power and the climatic effects of CO_2.

A nuclear power program large enough to have a substantial impact on global CO_2 emissions would have to be in the order of a few thousands of GW's based full scale fuel reprocessing with all its subsequent risks of accidental contamination and, possibly, weapons proliferation. It is a system which, if pursued, would totally transform the electricity industry and the electric utility industry into a vast, complex, capital-intensive activity centered around essentially the present nuclear weapons states plus Japan and West Germany. The latter two would probably dominate in the longer run (2). No other countries are likely to be capable of mastering a large scale plutonium fuel cycle. The core would be virtually impenetrable.

International agreement on the reduction of CO_2 emissions will have major impacts on the electricity industry. Today, some 20 Gtonnes are emitted annually from fossil fuels, of which a substantial part emanates from the electric utilities.

Taking into account both population growth and the uneven distribution of energy use – reflecting uneven levels of industrialization – the only way to avoid either a very large scale nuclear power program or major climatic changes is through increasing the efficiency with which energy is used in industrialized countries. Halving the present per capita level is a reasonable goal for the next few decades. Such a strategy also opens up for various solar technologies.

Energy efficiency is the only strategy through which the environmental pressures can be relieved. This holds for electricity production as well as industrial production in general. Environmental concerns in industrial production are to a large extent associated with energy-intensive industries such as materials processing. Reducing environmental damage as well as increasing quality of production calls for "precision use of energy": more finely controlled energy forms; sharper temperature limits in energy intensive processes; using information technology and electro-chemistry, etc. (3). This will be all the more easily done through high quality energy forms such as electricity and (natural) gas, which in turn will likely to lead to stagnating or even declining final use of industrial energy use.

This will have two important effects on the electricity industry. First, the interplay between electricity industry and the natural gas industry will become important. Second, industrial production itself will be increasingly benign environmentally, making the electricity industry (and transport) all the more visible from an environmental point of view.

Environmental pressures will therefore have profound effects on the electricity industry. It will lead to an increasing emphasis on end use energy efficiency, and thus probably to electricity and natural gas taking a larger share of the end use market. In addition, it will lead to a redefinition of economic efficiency of electricity supply. Efficient production and inefficient use will increasingly become an anomaly.

Finally, one of the original formative pressures is changing. The societal organization of the electric utility industry is presently under attack in some countries, in parallel with similar attacks on other regulated industries: air traffic, trucking, telecommunications, even health and education. This is partly driven by technological change – most obvious in the telecommunications industry – and partly by the notion that regulation and efficiency are incompatible. Socially, however, the attack on regulated industries is also an attack on the post WWII consensus that the benefits of industrial activities should be distributed justly.

Thus, the changing nature of the pressures for electrification, efficiency, equality and environment can only be reconciled through a rapid turnover of the existing stock of electricity end-use and production equipment. This would make for better use of today's more advanced technologies as well as opening up for substantial technological exploration and development. To accomplish this, though, will require reorganizing the electric utility industry.

4 Technological Exploration and Industrial Cores

The combination of continuing electrification and increasing pressures for low emission electricity production requires considerable technological development both in use and generation of electricity.

The electricity industry is a future-oriented industry. Searching for new technologies is not primarily a question of tenths of percents of the electricity price; it is much more about opening up new technological territories.

Nuclear power was the main such new territory opened up during this century, with, in the 1950's, deep moral overtones. Again materials sciences as well as manufacturing technology made the new territory habitable – or so it seemed – and pushed steam turbines to still larger sizes.

New territories have recently opened up, for example, in the use of dirty fossil fuels such as coal, heavy or residual oil, as well as gas turbine systems.

The following Table summarizes where successful exploration of new territories might lead.

Table 1. Hypothetical costs of electricity generation. Assumptions: see footnotes and text.

fuel	size MW's	capital cost $/kW	fuel efficiency %	lead time years	note
nuclear	1000	2000–5000		10	(1)
coal	1000	1000–2000	40	5–7	(2)
gas	100	400–600	50	2–3	(3)
gas	10	300–500	40	1–2	(4)
gas	1	300–500	40	1	(5)
gas	.1	200	40	.1	(6)

(1) The lower figure reflects Swedish costs, the higher some US experiences.
(2) higher figures estimates of 98%+ SOx reduction.
(3) various estimates for combined cycle plants.
(4) *guesstimate*, based on large series and extrapolating, from, e.g., jet-engines.
(5) *speculation*, again based on extrapolations from large scale series of marine diesel engines.
(6) *pure speculation*. A car engine costs perhaps $30/kW, a car perhaps $150/kW. High volume production for diesel engines for electricity production could perhaps cost $200/kW.

This table is hypothetical and based on two and a half assumptions: abundant natural gas, volume production rather than single plants and that gas turbines can be improved.

The core of the electricity industry consists, in Western Europe, of perhaps ten companies altogether with a long tradition of co-operation and long established highways of technological exchanges. Westinghouse has dealt with Siemens and Framatome, General Electric with AEG, ASEA and others.

The steam turbine, which is the lynch-pin of this core, appears to have been pushed to its technological limits. The territory is well explored. Efficiency increases have leveled off. Sizes will, if anything, decline rather than increase, as systems reliability and efficiency is taken into account.

The light water reactor technology has essentially been developed within this industrial core. It, too, appears to have reached its technological limits (not to mention its social limits, aptly described in Alvin Weinberg's phrase "the Faustian bargain").

The century old structure of the electro-technical industry is under immense pressures, due to overcapacity in large scale steam turbines and the increasing prospects of gas turbines. Previous competitors are merging or making long-term collaborative deals. The process was triggered by the merger between ASEA and Brown Boveri and has gradually spilled over into North America and possibly also Japan. ABB has lumped power business with Westinghouse; Rolls Royce – a jet engine manufacturer – is buying into NEI; Framatome is approaching Mitsubishi, as is ABB.

Three thrusts into only partially explored fossil fuel technology territories are now taking place.

The first thrust is into combustion technology for dirty (fossil) fuel such as fluidized bed technologies (whether pressurized or atmospheric), gasification and possibly steel-processing derived technologies such as the molten iron process. Each has its own industrial characteristic. Fluidized bed combustion is essentially a proven technology and possible to incorporate into the core, through the present boiler makers. Gasification opens up towards processing technology, including refinery technology, and would create a technological bridge between the electricity and the oil industry. In general, the core will open up towards complex, bulk chemistry.

Partial environmental pressures – SOx's, NOx's, upgrading of sites, etc. – will be resolved but not CO_2.

The second thrust into an only partially explored technological territory is into gas turbines. The thrust would have three simultaneous dimensions: efficiency, size and manufacturing cost. Gas turbines are inherently simpler than steam turbines but until recently also less efficient. Exploring high efficiency gas turbine technology has partly been done by the jet engine industry, an outsider to the electricity core, although GE, for example, has been involved in both, and by and large financed by military needs.

100 MW gas-turbines living up to promises shown in Table 1 (3, 4) offer the prospect of a transformed electricity industry. A large number of gas turbine suppliers are likely to emerge (even if only a handful will hold the key technologies). Also new electricity producers could emerge, transforming the utility part of the core as well. The gas industry, in particular, is a likely candidate. Moreover, the electricity industry and the gas industry will develop closer links. Environmental pressures could be substantially reduced – CO_2 emissions per generated kWh could be cut in half compared to advanced dirty fuel combustion technologies.

Gas turbine technology also holds the promise of down-sizing. A 10 MW gas turbine with 40% electricity efficiency offers the potential of cheap cogeneration, provided capital costs can be held in check. Moreover, if advanced manufacturing technology lives up to its promises, turbines could be tailored to individual users without too much loss in unit cost.

One could well imagine an entirely new industry emerging between the electricity producers and users, based on leasing electricity production units to users. The commercial jet engine industry, already used to sophisticated service and maintenance, is a potential god-father.

Advancements in materials and precision manufacturing may well make 1 MW or even smaller sizes possible, further fragmenting the electricity industry. Again, the transportation and aircraft industry are likely god-fathers.

The third thrust of technological exploration is into the territory of the natural gas-fired small scale – frequently a diesel – engine of 1 to 0.1 MW size.

Considerable anecdotal evidence exists today about micro-cogeneration plants at a price of perhaps $1000–1500 per kW for units of 10–100kW. Volume production should release lower prices. For instance the Volvo engine plant at Skövde produces perhaps some hundred thousands of units of 100kW each per year of engines, or some tens of GW at a unit cost of perhaps $30/kW. The peak capacity of ASEA ATOM was 1 GW per year.

Total use of small scale diesel units is today in Western Europe in the order of several hundreds of MWs and expanding, it appears, rapidly. Buyers would be electricity users – hospitals, shopping malls, hotels, apartment buildings. Maintenance and service would require a totally different industrial concept. The trucking industry comes to mind, because it is use to servicing a large number of units.

Natural gas technology also opens up towards another territory: gasification of biological materials, where a combination of engineering, biotechnology and agrobusiness can provide further non-fossile fuel supplies. In the longer run, gas technology also opens up towards hydrogen as a fuel.

Non-fossil fuel technological territories are also developing rapidly. Photovoltaics and fuel cells require totally different skills, compared to the steam-turbine-dominated core. Semiconductor skills, electrochemistry and advanced manufacturing are required on the industrial side. Skills at integrating dispersed sources of electricity will have to be developed on the utility side. Optimism appears great in both fuel cells and photovoltaics, and they both offer the prospect of transforming the electricity core. Obviously, photovoltaics will first enter sunnier areas, such as the US southwest and perhaps Spain and Japan. Fuel cells will enter where there is abundant natural gas, and they will compete with other technologies for small scale cogeneration.

Wind energy is a rather barren territory with the odd, fertile pocket. Some countries do have substantial potential; Denmark appears to have gone the furthest, with a total of a couple of hundred MWs installed capacity within a few years. Wind energy requires advanced manufacturing of both materials, control systems as well as mechanical components. The Danish village tinkering will have to give way for a professional industry. The experience also underlines that social organization makes a difference: wind mills built by landowners have had much less opposition than wind mills owned by utilities.

The fossil fuel technologies discussed above will be able to resolve some but not all of the four formative pressures of electrification, efficiency, equality and environnment. Turnover of capital stock on the supply side capital stock will be able to substantially reduce SOx and NOx emissions and possibly halt the increase in CO_2 emissions. But, as electrification continues around the world, more efficient use is necessary.

More efficient use of electricity follows from the turnover of old electricity using capital. In general, presently available motors, appliances, lighting, etc., are more efficient than the ones in use. There is, moreover, considerable scope for further improvement. Three areas of electricity use will demonstrate this.

Lighting is the first area (5). Between 10 and 15 % of all electricity is used for lighting. The total production of light is increasing rapidly in industrialized countries, but since efficiency in terms of lumen per watt is increasing as well electricity use for lighting has expanded less rapidly. A survey of the technologial possibilities open to designers demonstrates that it should be possible to double or even treble the efficiency of lighting if and when existing lighting is replaced with systems presently available or under development. Advancements in materials and electronics offer room for improvements along several technologial frontiers, such as fluorescent lamps, high intensity discharge lamps, etc., but also through sophisticated control of lighting systems.

Costs of more efficient lighting systems are generally lower on a life time basis but higher on a unit cost basis. The rate of diffusion will differ between users: commercial buildings, offices, etc., are likely to be early users and households late. It is reported, however, that Southern California Edison at one step found it cheaper to give away high efficiency lamps for free to households rather than expand generating capacity. This illustrates both a problem and a solution: turning over the capital stock – physical as well as mental – of electricity users will require rather innovative marketing skills. The electric utilities will have to rediscover Edison.

The second area is electric motors. Some two-thirds of US electricity use occurs in electric motors: running refrigerators, washing machines, office ventilation, pumps, fans, compressors, etc. (6). As a technological territory, it has been rather quiescent over the past several decades, but as other technological advances now reach maturity – such as materials and electronics – the implications for electric motors are likely to be huge.

The individual motor generally has a high efficiency. Nevertheless, it is always part of a system, perhaps with pumps or compressors, and when the system as a whole is taken into account the efficiency losses are considerable. This opens up opportunities for advances in control technology: matching motor speed to load substantially improves overall efficiency. Also, new magnetic materials are important, both because they could reduce losses in conventional motors and because they offer the possibility for entirely new designs.

Various estimates of costs suggest that for a wide range of improvements the costs of reducing the kW requirements for a given task is considerably less than the cost of adding the same amount of kWs to electric generating capacity.

It is also possible to combine electronic control of motors with utility load management programs; Southern California Edison is reportedly doing just this with 100,000 air conditioners via a VHF-FM radio system, at a cost far below the cost of supplying peak capacity.

The third area of electricity use is household appliances (7). To some extent, it overlaps with electric motors – refrigeration, washing and dishing machines, for instance – but appliances also have their own design problems. The total amount of electricity used for appliances amount to tens of percent in several countries, and the potential for higher efficiency is large. It depends on a variety of factors, some of which may be easier to accomodate than others. Again, it is the technological advance along several frontiers that bear fruition: materials are important in insulation in refrigeration; biochemistry in detergents used for washing in turn decide water temperature, etc. Looking across a wide range of appliances, the average appliance sold uses 80% of the electricity compared to the ones in use, while the best ones available use roughly 50%. Advanced designs can cut use to some 25%. Here again, it appears that costs of improvements are less than costs of expanding generating capacity.

There is thus considerable potential for improvements in lighting, motors and appliances and, in general, in the way electricity is used. These improvements follow from general technological advances as well as more sophisticated use of existing technologies. Nothing says that efficiency improvements will cease. The main concern should be over penetration.

When looking at the electricity system as a whole, from steam turbine to refrigerators and lamps, over the decades one is struck by the interplay between electricity and general industrial advance. Over several decades, electricity stimulated technological and industrial development in materials, manufacturing, control technology, etc. These advances were then increasingly fed back into the electricity system in the form of more advanced components. It appears now that the time is ripe for a massive penetration of technological advances that in turn will have substantial impact not only on the technology of electricity but also on its organization. Moreover, when looking across this, as it were, advancing technological front one is struck by the relative backwardness of the meeting point between supply and use: the electric meter. Compared to the level of electronic sophistication in today's households, the electric meter stands out as a relic of early 20th century technology. As a component in electricity control systems – and not only in financial flows – the potential of intelligent metering has barely been tapped. Perhaps the state of metering technology reflects the electric utilities' perception of customer sophistication.

In order to take full advantage of the advancing technological frontier, the electricity industry will have to be transformed in two senses – and probably in two stages. The first sense – and stage – is from a *production* oriented industry to an industry oriented towards the *coordination* of electricity supply, whether small or large scale, whether in the present form or integrated with the electricity user. This involves both the electricity industry as such and the interplay with the natural gas industry. Thus the old electro-technical structure, created a century ago when the AC and DC systems fought it out, is changing. The large, six or so, manufacturers are rapidly making themselves capable of meeting some of the supply opportunities offered by technological advances as well as increasing environmental pressures.

The second sense – and stage – is from a supply-oriented to a electricity *systems-oriented* view. Here, costs for expanding – or replacing – capacity is balanced against costs for increasing the efficiency with which electricity is used. It involves actively accelerating up the rate of diffusion of more efficient electricity-using goods.

Only through returning to Edison's vision of electricity as a system where supply, distribution and use are seen as a whole can the electricity industry cope with the four pressures of electricication, efficiency, equality and environment.

5 Harnessing the "Dynamic Conservatism"

Power, it was once said, is as important for a social scientist striving to understand human affairs as energy is for the natural scientist striving to understand nature.

There is a considerable literature on the interplay between technological change and organizational structure. The expression "dynamic conservatism" is sometimes used to describe how organizations – or rather people within organizations – respond to new technologies. Organizations strive for stability, stability requires power – and technical change destabilizes.

"Dynamic conservatism" does not mean that all technical change is resisted, however. Organizations based on one particular technology that are threatened by competitors may well introduce improved technologies in order to reduce that threat. Ironically, history is full of accounts of how technologies had their best moments when already doomed by initially inferior competitors who have the technology with more inherent potential. This was true for sailing-ships when steam arrived, for gas-lighting when electricity lighting arrived, for DC systems when AC arrived and for mechanical watches when the quartz technology arrived. It appears to be true also for nuclear power.

Technological change affects the internal structure of an organization, relations between different parts of an organization and indeed between different organizations. What Hughes has called "the battle of the systems" between AC and DC aptly demonstrates this, including the somewhat grisly methods used in advocating the dangers of AC by the DC proponents (1).

A more recent illustration can be found in Sweden where, in the 1960's, nuclear power was the preferred alternative open to the large utilities (not allowed to proceed with hydropower). City-owned utilities, which also supplied district heat, had a different first choice: combined heat and power. Correctly or not, this technology was seen by the large utilities as a threat to the electricity market needed for nuclear power, both because district heat would displace electric space heating and because of the electricity produced. Moreover, the cities represented, through history, a complementary force in Swedish politics. As suggested by the law of "dynamic conservatism", the large utilities fought a drawn out and by and large successful battle to reduce the role of combined heat and power (8). Similar struggles were fought in Denmark, but there the core of the electricity industry was considerably weaker and the cities stronger. The outcome was therefore different.

The history of nuclear power offers several intriguing illustrations of "dynamic conservatism". The struggle over thermal reactor technology was to no small degree a struggle over whether the electrotechnical corporations would remain dominant within the electricity industry core or not (2). Lately, when nuclear reactor accidents have become a major concern, alternatives to light water reactors with inherently lower accident risks are still paper alternatives, implying that perhaps the original choice was incorrect. The most spectacular example of "dynamic conservatism", however, is the amalgamation of large scale nuclear laboratories, the plutonium fuel cycle and the liquid metal fast breeder reactor.

According to "dynamic conservatism", the present electricity industry core would prefer variations on the large-scale steam turbine, such as nuclear power or advanced coal combustion, over other, potentially more disruptive, alternatives. Large scale (100 MW) combined cycle gas turbines represent an intermediate case, linking steam and gas turbine technology but also providing new entries into the core on the electricity production side and on the electrotechnical side. The key relations within this industry, as well as within the electric utilities, would probably remain unchanged. In particular, production of electricity would keep its dominant role vis-a-vi distribition.

The potentially disruptive effects of gas turbines comes from the possibility of miniaturization. Natural gas-driven diesel engines (or other engines) would have a similarly disruptive effect. New major suppliers of equipment would enter into the core. But the most dynamic effect would come from the changed relationship between production and distribution of electricity.

With steam turbines as the core lynch-pin, the role of electricity distribution is by and large to ensure the market for the large scale power plants. More diverse production with smaller scale generating units, particularly when electricity production can be combined with heat production, requires a quite different relation between electricity user and the electricity industry.

This has two important ramifications: for overall grid stability and for distribution.

Grid stability is now by and large organized by the large scale producers through the high-voltage grid. Producers have kept firm control over the grid and developed various means for coordination. With more producers, both small and large, grid stability will require new organizational solutions.

Distribution will have to develop a strategic concept of balancing, on the one hand, small scale production combined with heat production and efforts to stimulate more efficient electricity use with, on the other hand, large scale central production.

Two classes of users should be singled out: the energy intensive materials processing industry, which also is electricity intensive; and the electricity-intensive, advanced commercial "smart" building such as office blocks, hospitals, etc. Within both, electricity use is increasingly controlled through advanced information systems together with heat requirements and other forms of utility services such as water, climate control, etc. Technical skills are considerable. Small, or even larger scale, natural gas based cogeneration is more and more feasible. The control systems will increasingly become "mini-utilities" – resembling Edison's original vision – and increasingly will require sophisticated interfaces with the electric utilities.

For electricity users in manufacturing industry and households, where electricity is but a minor share of total expenditure and attention, self-generation will be further off, unless almost service-free systems become feasible. Nevertheless, the need for efficient use of electricity may well, or should, force electric utilities to assist in the rational management.

As the end-use "mini-utilities" become smarter, requirements will grow on the hitherto crudest element in the total system: the electric meter. The advanced electric meter could well be the next century equivalent of the 19th century transformer – the technology that releases a total transformation of the industry. If computer trading is possible on the global stock-market, why should it not be possible to arrange direct on-line, real time computer-controlled trading between "mini-utilities"? Perhaps we will, in the future, see the dissolution of the centrally controlled, electric utility system as networks of "mini-utilities" based on small-scale gas generating technologies, fuel cells and in some countries also photovoltaics start to trade directly with one another. The central system would then increasingly become back-up rather than sole supplier.

Balancing between supply and use will occur on several levels of the system

– end-use, low tension and high tension – with correspondingly more complex financial flows between different actors.

A widened role for electricity distribution would change the organizational dynamics within the electricity industry and thus the directions of technological exploration.

6 Societal Organization and the "Dynamic Conservatism"

How utilities will adjust to the four formative pressures discussed above will to no small degree depend on national circumstances such as, resources, institutional structure, and political traditions. For example, the US, the UK, the Netherlands, Denmark and Sweden are right now embarking on different paths of restructuring of their electric utility industries. France, on the other hand, persists doggedly on the path laid out some twenty years ago. West Germany is in somewhat of a limbo, torn between environmental opposition and the will to preserve what is seen as national endowments in nuclear technology and coal.

The urge for reform stems from the view that the present electricity industry constrains new entrants into the grid. These constraints can be seen as elements of the "dynamic conservatism", preserving the domination of the present industrial core. The reformers now consist of a somewhat unholy coalition of "anti-egalitarians" longing to do away with that symbol of post WW II consensus, the national utility; the financial industry, looking for new markets; environmentalists, critical of the frequently all-to-monolithic behaviour of utilities with respect to nuclear power and other huge generating plants; and those with industrial or financial interest in cogeneration, e.g., the paper-pulp and wood industries, or in the new generating technology. The relative strength of these groups as well as the resulting main thrust will differ from country to country.

The earliest attack on entry conditions was the Public Utilities Regulatory Policies Act (PURPA) legislation in the US. Considerable expansion of medium scale generating capacity has occurred since the utilities have been forced to buy power generated by independent cogenerating producers at prices equivalent to the estimated avoided cost of new coal fired base load power. The impact appears to have been the greatest in California and Texas, partly for political reasons transformed into regulatory practices and partly because of the availability of both cheap natural gas and large scale cogeneration potential in the oil industry. It appears that other small to medium scale electricity and heat users are becoming increasingly interested in cogeneration, e.g. shopping markets, hospitals, restaurants etc. Given the uncertainty of electricity demand over the next ten or so years, utilities are moreover extremely wary of committing themselves to new large scale coal base load plants (not to mention nuclear energy). The combination of short lead times, low capital costs and a natural gas spot market appears to favor independent power producers based on (100 MW+) combined cycle gas turbines. Also, experiences are growing – particularly on the West Coast – in how to organize programs for increasing the efficiency of electricity use. The spread of such programs will depend on the needs for new capacity by different utilities. It is not unlikely that special companies will evolve, which take on this task (demand side bidding). An extremely wide-ranging debate on the role of conservation programs rages, with almost no equivalent anywhere else (9, 10).

In the absence of strong regulatory support cogeneration is almost wholly dependent on the good will of the electric utility industry. The latter depends in turn on the need for new capacity and possible obstacles in front of new large scale power plants. The experiences in the US are so far inconclusive with respect to the future shape of the electric utility industry. Is is possible that additional generation will be dispersed. It is also possible that additional generation will be in the form of nominally independent power production companies, in reality partially owned subsidiaries of the electric utilities. Two things are reasonably clear: first, the utility industry will "unbundle" into more separate activities such as distribution, transmission and generation; second, even dispersed electricity generation requires active support of the electric utility industry. Thus, although technology may lead it into new directions, the US electric utility industry is likely to remain in control.

The most dramatic upheaval in the world electricity industry is taking place in the UK. This country does have a tradition – not unique to electricity – of major industrial upheavals every now and then. In electricity, World War I resulted in such an upheaval: the grid was established and local systems with widely different frequencies and voltages were integrated. The World War II experiences demonstrated that the pre-war electric utility industry was too fragmented to be able to cope with the pressures of post-war electrification, equality and efficiency. Nationalized out of an organizational blind alley, a surge of creativity was released that made expansion for electrification and equitable access possible. The present upheaval represents the next escape, this time out of the alley that the Central Electricity Generating Board has driven itself into through its dependence on 1000 MW steam turbine plants. The coalition forcing the upheaval consists primarily of the free market ideologues the so-called "Scargill-factor" ("the enemy within"), environmentalists, the City, the oil – and thus natural gas – companies. This "mega muddling through" of an upheaval is solely focussed on the supply side.

The divestiture and privatization of the CEGB will lead to production, distribution and the national grid becoming separate entities, with the area distribution companies owning the grid company. The structure chosen for production – with only two production companies, sometimes called little and big G – reflects the need to privatize nuclear power. The Prime Minister's unequivocal commitment to nuclear power will create great problems in the City. The privatized production companies will shun nuclear power unless the government underwrites the financial risks. At present, combined cycles gas turbines appear to be the preferred option, by independent producers, by possible cogeneration schemes in industry (where estimates by little G indicate that well over 10 GW is feasible) and also by the area boards. Micro-cogeneration on the 100 kW to 1 MW scale is not infeasible, given the estimated 1.5 GW's of standby capacity within the National Health Services alone. Electricity and gas tariffs will determine the economics, however, and it is too early to predict the outcome.

The future of the UK electric utility industry will depend on the availability of natural gas. The organization and regulation of British Gas is a worthy topic of discussion, but suffice it to say that a cross channel link looks increasingly likely.

The Netherlands has also embarked on a reorganization of its electric utility industry, but with more restraint than the UK. The Dutch system has its own peculiarities: two thirds of electricity is produced from natural gas. The electric utility industry is essentially inward-looking and peripheral to the great West European electricity industry. The natural gas industry, however, is very much part of the core – according to some it is the core of the core – of the West European natural gas industry. The electric utility industry has recently been reorganized by the government, with the aim to separate production, distribution and the national grid. Electricity distribution will be integrated with the natural gas distribution into locally and provincially owned companies, the production companies will be owned by the distribution companies, and the grid (in contrast to the UK) by the production companies. Steps have been taken to lower barriers of entry for small scale producers of combined heat and power. Some 250 MW are in place. Future expansion will depend greatly on the electricity and gas tariffs, however, particularly in the block rate system. Since Gasunie supplies both the gas for central electricity production and the gas for local production, its pricing policies will determine the outcome. New large scale entrants of electricity producers are not allowed.

The electric utility industry has, through its grid company, negotiated with Statoil for gas imports. Whether this will ease Gasunie's grip over the market remains to be seen.

The Danish electricity industry is another peripheral member of the West European electricity industry and with some decidedly idiosyncratic characteristics. Historically, electricity grew up around cities and rural cooperatives (the latter much like dairies), and distributors have, compared to other countries, a strong position in relation to production. Moreover, the Danish system is split into two, each one with links to the Scandinavian grid, underlining a basic Danish political dichotomy between farmers and burghers along the Belts. Thus, the Danish electric utility industry is relatively weak and open to outside political pressure; the major parliamentary parties have forced the utilities to accept an extremely generous wind energy scheme and also a PURPA-like scheme for buying electricity from decentralized combined heat and power at a coal condensing power based price. The Danish system, which already has the highest share of combined heat and power (CHP) generation in the OECD, will therefore have by far the highest share of wind energy and small scale CHP in the mid-1990's. According to some estimates, the total capacity could be in the order of 800–900 MW's, divided roughly in half between wind and small scale CHP, or 10 % of total capacity.

The regional gas companies are also operators. Rather desperate to increase the natural gas market, some appear to stimulate micro-scale diesel engine-based units below 150 kW (due to exemption of electricity taxes below this level) in swim-baths, hospitals, schools, etc. A further couple of hundred MW's could come about through this.

The national gas company, DONG, is also trying to create conditions for combined cycle gas turbines in the Copenhagen region.

The future expansion and thus shape of the Danish electric utility industry will be heavily dependent on the gas industry. The latter's rather trying relations with Dansk Underground Consortium will have further importance, as will the desire to create stronger links with the nascent Swedish natural gas network.

Sweden is another very special case. Unlike the Netherlands and Denmark but like the UK, it regards itself as being part of the core of the electricity industry. Unlike the other three countries, it has no indigenous natural gas. Like the UK, it has a large nuclear power component. And unlike all other nuclear power countries, it has decided to phase out nuclear power; the first reactor in 1995, the second in 1996 and the twelfth and last in 2010. Also, unlike the other three countries, there is no strong tradition of direct government interference in the organization of the electric utility industry – which, accordingly, has evolved gradually and organically over the century. On the other hand, half of the electricity is produced by the State Power Board.

Thus, the Swedish electric utility system has its own peculiarities. It is certainly one culture, but at the same time not monolithic. Different voices are heard, and at the same time dissent is limited. Cooperation and competition coexist.

The challenge presented by the government is large: to phase out nuclear power without resorting to hydropower or an increase in CO_2 release from the energy sector as a whole. So far, the consensual style appears to hold, and attention to natural gas is increasing. The State Power Board has initiated its own PURPAstyle scheme – on its own initiative and based on the avoided cost of coal base load power – and hopes to stimulate small to medium scale cogeneration. The State Power Board is also looking into different demand side options in order to stimulate higher efficiency of use. The city of Stockholm, an important utility in its own right, has great problems in expanding large scale production facilities within the city and is also increasing its attention to demand side options. The government is committed to fees on SO_x and CO_2 from fossil fuels, which is likely to further stimulate natural gas based smaller scale production.

The future shape of electric utilities in Sweden will depend on the role of natural gas. In contrast with the Netherlands, for example, the nascent natural gas industry is under the firm control of the electricity industry and is likely to remain so. Thus, natural gas expansion will be part of the broadening of the role of electric utilities rather than an independent competitor.

Two other West European nations are worth mentioning: France and West Germany. Both regard themselves as being very much in the core of the global electricity industry, although the French claim is considerably weaker and resting solely on a domestically strong nuclear industry.

France remains doggedly on the path decided upon some twenty years ago: a Colbertian/Napoleonic/Gaullist expansion of nuclear power way beyond the French market. Thus, France is the ardent supporter of an EC-wide trade in electricity, and Electricité de France is likely to export considerable amounts of electricity to neighbouring countries. This will have an influence on Spanish, Italian, British and possibly West German electric utilities. As long as Gaz de France remains the "little sister" of Electricité de France, France appears certain to remain the Vatican of nuclear power in Western Europe.

West Germany is another matter. Its electricity industry is at present in something of a limbo, torn between the RWE-Siemens commitment to coal and nuclear power on the one hand and the environmental pressures for alternatives on the other. Also, given a considerable over-capacity in electricity, the pressures for new generation is weak. Micro-scale cogeneration – based on diesel engines – appears to be expanding rapidly, with some 30–40 MW per year and totalling 340 MW by the end of 1988. This occurs solely on the basis of present gas and electricity tariffs. Much will depend on the relative strength of Ruhrgas and RWE, and so far the former does not appear to have the will to openly challenge the latter for an increased role of small and medium scale generation. Wind energy is expanding slowly in the northern parts, primarily Schleswig-Holstein and Niedersachsen. The Danish program is contagious.

The historical (West) German tradition of government is not to interfere in the industry, and the industry has accordingly evolved gradually and organically. Similar to Sweden, there is considerable variability in ownership and structure, but the West German electric utility industry appears to be considerably more monolithic than the Swedish. Consensus yes, but from above.

It is worth mentioning a few words about the West European natural gas industry. Some 50 GW's of combined cycle gas turbine central power stations would require roughly 25 % increase in West European natural gas supply, or 50 Bcm. Paradoxically, the more small scale-oriented the utilities become, the larger will the West European natural gas industry become. This industry will have to expand its supply considerably and therefore also its imports from Norway, the Soviet Union and perhaps Algeria and beyond. A third Soviet pipeline input into Western Europe is not unthinkable; bypassing Ruhrgas and reaching directly to the North Sea via Scandinavia may be interesting. Zeebrugge emerges as a critical juncture of West European natural gas trade, with links to Algerian LNG, Norwegian North Sea gas and just possibly across the Channel, if gas indeed is to play a part in electricity production. The break-up of CEGB will speed the eventual connection between the British, somewhat insular, gas system and the continental system.

Viewing the six West European countries discussed here altogether, one can state the following conclusions:

- there is a general move towards the Swedish-German system of separation between production and distribution;
- there is growing interest in separating the task of overall systems stability and the grid from the large scale producers;
- there is considerable variability in ownership structure;
- there is considerable variability in government/electric utility relations, with the UK being the most interfering and Sweden and West Germany the least;
- France stands out as the odd, monolithic man out;
- natural gas will be crucial in shaping the future electric utility system;
- Denmark has gone the furthest in stimulating small-scale production, while Sweden will have the largest need;
- nearly all efforts are supply-oriented and there is a general lack of awareness of the need to launch systematic efficiency enhancing programs.

The societal organization of electricity is changing, although very differently in different countries. Efforts are being made to reduce SOx and NOx pollution as well as opening up the electric utilities to new producers. Thus, the West European utility industry is moving towards a less monolithic and more diverse structure. This will in all likelihood reduce barrier entries for new production technologies, particularly those based on natural gas. The industry, therefore, is gradually moving away from a supply-only oriented vision towards a supply coordination oriented vision. The first stage discussed in section 4 is, in other words, being approached. The second stage – an electricity systems orientation, where steps to add supply are balanced against steps to increase effieciency – is still a long way off. Some sketchy experiences exist in the US. Almost everything remains to be done. Nothing – except dynamic conservatism – says that it is impossible to reach out to the electricity users. Such a strategy is most urgently needed in Sweden and there are signs that it is indeed emerging.

7 Full Circle

Electricity gives people power over their everyday life. Power over electricity is, on the other hand, firmly vested within the core of the present electricity industry.

The electricity industry has come full circle in the hundred years after Edison. Then, the fall in distribution costs triggered by the AC system made energy services a less attractive idea. Centralization and economies of scale promised to resolve all supply problems. Now, environmental pressures make the electricity industry face a juncture: the industry will evolve either towards a large scale expansion of nuclear energy or towards a large degree of diversification and, possibly, decentralization.

The nuclear expansion would reinforce the tendency of centralization, nationally as well as internationally, with West Germany and Japan being the likely core members. Diversification, on the other hand, would reinforce the emerging tendencies of a much wider range of technologies, first based on natural gas and then later based on various solar technologies. Also, diversification opens up towards the service concept, where active measures to increase the efficiency with which electricity is used are taken. This is the only strategy that credibly can cope with the fourth pressure in its totality.

From a social point of view, diversification is the preferable path. It will require opening up the presently somewhat monolithic utility industry and turning it into a more service-oriented industry.

Somewhat blasphemously, the electricity industry of the late 20th century can be compared with the Catholic Church of the mid-15th Century. Then, the great cathedral building was over. Most cathedrals had been completed. One or two had collapsed. Some were still under construction – the Kölner dome, started in the 13th century, was not completed until the 19th century.

The 1000 MW steam turbine power plant is the electricity industry equivalent of the 15th century cathedral. It is the material manifestation of an ideology and a world view which promises cheap electricity in return for dependency.

Like mid 15th century (West Roman) Christianity, which split into several directions, it is now possible to envisage several different types of utility systems, based on different technologies and differing relations between electricity users and the electricity industry.

There will be high churches and low churches, new congregations and village chapels, laymen and ordained. Perhaps we will also see the equivalent of the quakers.

All in a truly ecumenical spirit, it is hoped.

Acknowledgement

The author has benefitted from discussions with T.B. Johansson and P. Steen. This chapter reflects personal views only and not necessarily official views.

References

1. Hughes, T. P. 1983. *Networks of Power*. Baltimore: Johns Hopkins Press.
2. Walker, W., Lönnroth, M. 1983. *Nuclear Power Struggles*. London: George Allen&Unwin.
3. MITI 1987. *Energy in the 21st Century*. Tokyo: Ministry of International Trade and Industry.
4. Williams, R. H., Larson, E.D. 1989. In this volume
5. McGowan, T. K. 1989. "*Energy Efficient Lighting*". In this volume.
6. Baldwin, S. F. 1989. "*Energy Efficient Electric Motor Drive*". In this volume.
7. Nörgård, J. 1989. "*Low Electricity Appliances – Options for the Future*". In this volume.
8. Lönnroth, M., Johansson, T. B., Steen, P. 1979. *Solar versus Nuclear*. London: Pergamon Press.
9. Williams, R. H. 1989. In this volume.
10. Cicchetti, C., Hogan, W. 1988. "*Including Unbundled Demand Side Options in Electric Utility Bidding Programs*". Report to be published. Energy and Environmental Policy Center, Kennedy School of government, Harvard.

Planning Implications of Electricity Conservation: The Case of Denmark

Niels I. Meyer and Jørgen S. Nørgård

Abstract

The chapter uses Denmark as a case study to describe problems and possibilities in the planning and implementation of electricity conservation. The Danish experiences have shown that the largest potential for improvements in electricity efficiency is found in the end-use part of the energy chain, but these savings will not be realized unless electricity efficiency incentives are integrated into national energy plans. For any given level of electricity services in the future, the need for electricity and the primary energy to generate it can vary by a factor of five or more, depending on the technology used in the energy system. The chapter also describes Danish demonstration projects concerned with electricity conservation and cogeneration combined with renewable energy, e.g., windmills, biogas, etc., and district heating.

1 Introduction

In this paper we will use Denmark as a case study to examine the problems and possibilities of electricity conservation.

In most industrial countries, energy conservation efforts have concentrated on space heating and to some extent on transportation and process heat, while electricity conservation in general has been given much less attention. In Denmark, for example, encouraging conservation results have been obtained for house heating by improving the building envelopes of new as well as existing houses and by better control of the heating systems. For instance, the heat consumption for the buildings administered by one of the Danish governmental departments was reduced by 23 per cent from 1975 to 1984. In the same period, however, the electricity consumption in these buildings increased by 75 per cent (1). Such trends are, to a large extent the result of a general ignorance in relation to the potential for more efficient and selective use of electricity.

There are several possible explanations for the neglect of electricity conservation. Firstly, the potential for electricity conservation may not be well explored and documented. Secondly, there may be vested interests that are opposed to a decrease in electricity consumption. Thirdly, the conservation efforts are made difficult by the fact that a major part of electricity consumption takes place in a huge number of different small-scale apparatus and equipment. Each of these units consumes an insignificant amount of electric energy, but together they typically account for about one third of the primary energy consumption in an industrial country. All of these barriers should be considered in devising an electricity conservation plan.

The overall goal of energy planning is to reduce the consumption of fossil fuels, rather than just the consumption of electricity. This goal is primarily based on environmental considerations (2) as well as concerns about the availability and cost of these resources, especially oil. Another reason for conserving scarce fossil resources is to secure that enough will be left to meet the energy needs of the developing countries and future generations in a world with a rapidly growing population. Hence we will describe not only ways to save electricity but also some fuel saving systems for generating the electricity.

This chapter describes the Danish experiences up to the present. Electricity consumption is disaggregated into economic sectors and end-uses. The methodology used in our analyses of electricity conservation potentials is outlined and the results are reported.

Some basic principles are described for electricity planning where conservation potentials are integrated in the process. The electricity service demand and the end-use efficiency determine the demand for electricity. The importance of the electricity service scenario is illustrated by an example.

The potential for introducing decentralized cogeneration units based on local energy resources is discussed. Some examples are given of systems in operation based on renewable energy sources.

The present Danish energy policy is outlined and compared to a more effective policy for implementing the potential savings and renewable energy sources. Finally, we use an alternative Danish energy plan as an example for a sustainable energy future.

2 Facts about Denmark

Some basic data about Denmark, relevant to energy analyses and planning, are shown in Table 1. Figure 1 gives a comparison between the energy balance for Denmark in 1972 and 1985. It shows that the desired shift away from oil has been rather successfull, mainly due to increased use of coal in power plants.

Table 1. Main data for Denmark in 1986. An exchange rate of 1 $ = 6.50 D.kr. has been used.

Population	5.1 million
Area	43.000 km^2
Inhabitants per km^2	119
Mean temperature	7°C
Gross Domestic Product	103 billion $
GDP per capita	20,100 $
Exports+Investment income receipts	33 billion $
Imports+Investment income expenditure	37 billion $
Industrial Production*	18 billion $
Agricultural Production*	5 billion $
Service Production*	59 billion $
Primary Energy Consumption	800 PJ per year
Energy Consumpt. per Capita	156 GJ per year
	= 5.1 kW
Electricity consumption	27 TWh per year
Electr.consumpt. per capita	5.2 MWh per year

* at factor cost

Source: Statistical Yearbook, 1987, Danmarks Statistik, Copenhagen, 1987.

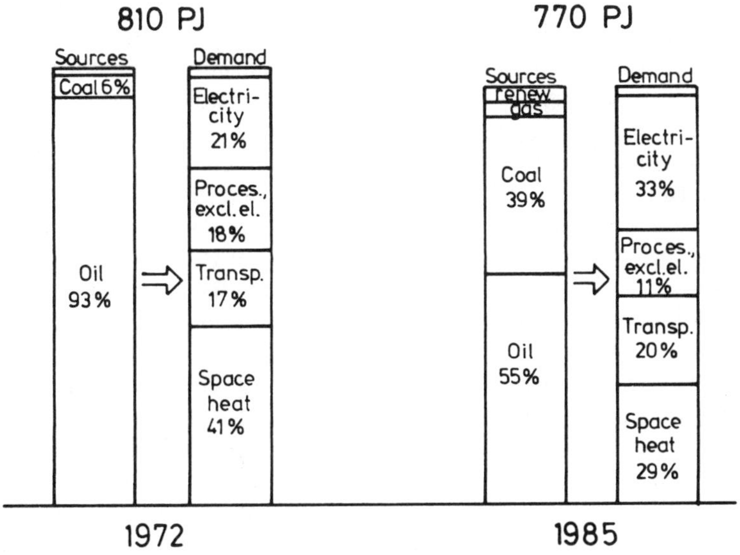

Figure 1. Danish energy balance for 1972 and 1985.

Danish electricity production is presently based on about 95% coal and 5% oil. Electricity production from wind mills is increasing, but is still no more than 1%. Cogeneration with district heating provides about 20% of the Danish heat consumption. It is planned to increase this contribution to 40% before year 2000.

In the period 1972–1985, the energy consumption for room heating was reduced by nearly 30%, while the area heated increased by 30%. This corresponds to about a 45% reduction in average energy consumption per heated m^2.(16)

Total electricity consumption, however, has increased over the period by 50%, partly due to a trend in industry to use electricity for processes in which other energy forms had previously been used.

The Danish GDP has increased by 2.1% per year (real values) over the same period.

3 Methodology for Electricity Conservation Analysis

In considering the potentials for saving primary energy for electricity production, it is essential to look at the energy system as a whole, see Figure 2. Through a chain of technological links the primary energy is converted into electricity services, and the technologies can be grouped into two categories: electricity supplying technology and electricity end-use technology.

We will discuss the options for fuel saving generation technology later. Here we will focus on the potentials for improving efficiency in end-use technology.

The metodology used in analyzing the potential for electricity conservation is also described in the chapter by Nørgård in this volume. Here we will only summarize the main principles used by the energy group at Physics Laboratory III, Technical University of Denmark. It should be noted that in general these principles have not been adopted by the official planning authorities in Denmark.

The methodology includes a complete analysis of the energy chain from electricity generation through different transformation processes to the desired electricity service. In general we have found that the efficiency is lowest close to the final end use. It is important not only to consider the efficiency of the single elements in the chain but to include the interactions of the elements in a complete systems analysis. When energy service demand varies, controlling influences may have to be applied further down the chain in order to achieve the highest efficiency. Examples of this are found in systems for heating, ventilation, lighting, etc.

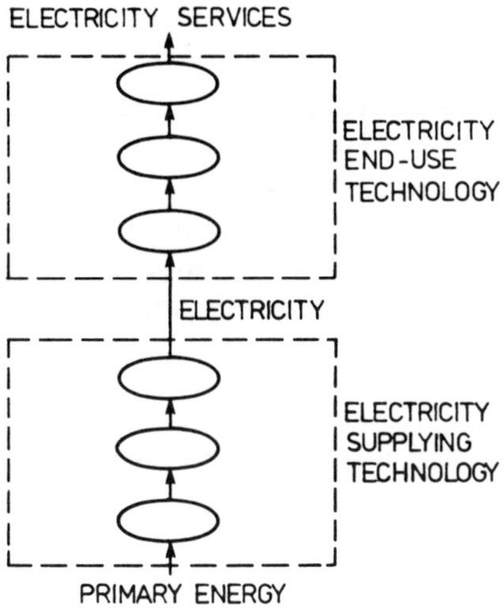

Figure 2. Energy system with links of technology which converts primary energy into electricity services.

Another essential element in analysing the potentials for more efficient use of electricity is the distribution of present use of electricity. Table 2 shows Danish electricity use by economic sectors and Table 3 by various end-use technologies.

The starting point for efficiency analysis of each end-use is a desired, specific energy service, e.g., lighting, space heating, etc. The analysis is then concerned with finding the technologies and systems which can provide this service with the least consumption of electricity. In the final choice between possible solutions, economic considerations are included. In this connection, it is important to distinguish between the long-term economic cost effectiveness for society as a whole in contrast to the short-term individual private economy. National energy planning should include all costs and benefits, including those in the environmental and health sectors. The evaluation of such externalities is rather uncertain, but they seem to be underestimated in the official Danish energy planning.

4 Danish Electricity Consumption Pattern

The data on electricity consumption in Denmark by economic sector, shown in Table 2, are easily available in the official statistics. Data for the technological end-use distribution are not so readily available. They have to be deduced from many different sources. The results presented in Table 3 are rather uncertain (by 15%), but the best available. The end-use categories are chosen with a view to the structure of the Danish economy, but should be generally applicable to most industrial countries.

Table 2. Danish electricity consumption distributed according to economic sector (1986).

SECTOR	GWh	%	kWh/cap
Industry	7,800	29	1,500
Households	9,100	34	1,800
Service	7,300	28	1,400
Agriculture	1,500	6	300
Others	900	3	200
TOTAL	26.600	100	5,200

Source: Ref.5.

Table 3. Danish electricity consumption distributed according to end-use category (1986).

END-USE CATEGORY	GWh	%	kWh/cap
Lighting	5,000	19	950
Pumping	3,700	14	750
Ventilation	3,700	14	750
Cooling	3,500	13	700
Other motors	2,700	10	500
Electronics	1,100	4	200
Comfort heat	3,700	14	750
High temp. process heat	1,600	6	300
Others	1,600	6	300
TOTAL	26,600	100	5,200

Source: The authors.

5 Potentials for Electricity Conservation

It is often postulated that electricity is used relatively efficiently, especially in industry. This turns out not to be the case when the systems and the desired energy services are considered in more detail. For example, electric motors may have efficiencies above 90% when they work at optimum load. In practice they rarely do, and the average efficiency may be 50% or lower. The potential electricity conservation in major end-uses is summarized in Figure 3. In general electricity consumption in Denmark can be *reduced to about one-third* of its present value and still provide the same electricity service. These analyses are based on technology which is known today and which could be brought on the market within 10–15 years. The technology is estimated to be cost-effective when introduced in the course of natural replacement.

Although the results shown in Figure 3 were published in 1983, they were not generally accepted by the official Danish planning authorities or the electric utility companies. On the contrary these institutions have tried to discard the results as "unrealistic". This attitude is slowly changing, though, as more evidence supporting the potential for electricity conservation piles up.

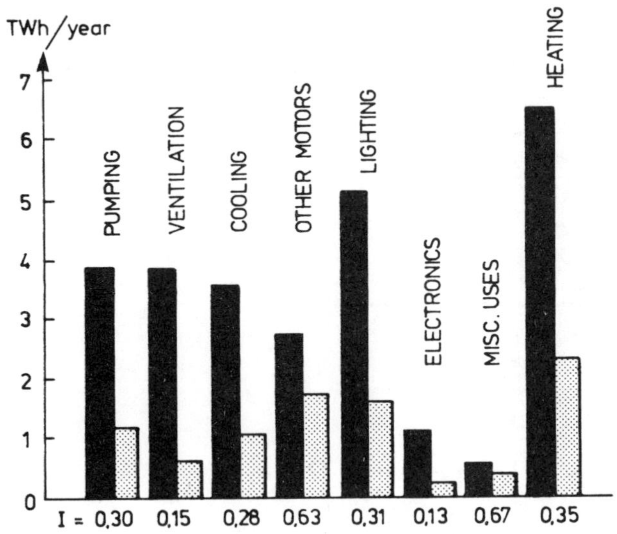

Figure 3. Potential for electricity conservation in Denmark in an earlier study. Dark columns indicate present electricity consumption. Grey columns indicate possible reduced consumption with known technology and same energy service (3, 4) The ratio between the reduced consumption and the present consumption is denoted electricity intensity index, I.

Thus, a new analysis was carried out in 1987 under the auspices of the Danish Ministry of Energy, and it involved energy experts from the utility companies, the Technical University of Denmark and institutions concerned with specific areas of electricity end-use. The potential contributions to electricity savings from changes both in technology and behaviour were evaluated for the best technology on the market and for known technologies expected to enter the market within 10 years. Typical results for the latter are shown in Table 4. They more or less confirm the previous results shown in Figure 3, although they are somewhat more cautious in some areas. The difference in the estimates can be attributed largely to differing views regarding the way in which end- use electricity technologies will be developed and installed. Some of the participants in this recent study took the passive view of what "can be expected to enter the market", while the results in Figure 3 reflect a more active view of what technologies "can be developed and brought into the market". The passive view on end-use technology development and market penetration, we think, underestimates what can be accomplished if electricity efficiency incentives are integrated into energy planning. At present, the technological potentials for electricity conservation shown in Tabel 3 are more or less accepted by Danish decision-makers, but they emphasize the political difficulties involved in implementing these improvements. We will return to this question in a later section.

Table 4. Potential for electricity conservation based on technologies expected to enter the market within the next 10 years (5).

End-use	Conservation potential
Lighting	58%
Ventilation	62%
Pumping	55%
Cooling	66%

Source: Ref.5.

6 End-Use Electricity Planning

Recognition of the significant potential for higher electricity end- use efficiencies logically leads to a new view on electricity planning: For any given level of electricity services in the future, the need for electricity and primary energy to generate it can vary by a factor of five or more, depending on the technology used in the energy system.

The purpose of electricity planning should not just be the safe supply of electricity at the lowest economic cost, as typically claimed by Danish utility companies. The goal of the planning should be to provide electricity services at the least overall cost, including externalities such as environmental and health costs, security of supply etc. This implies that not only the efficiency of generation technology but also that of end-use technology should be subject to active planning. The implementation of higher efficiencies in the two groups of technology, however, require quite different methods.

One consequence of adopting end-use electricity planning is that the demand for electricity services becomes the basic exogenous variable, rather than the demand for electricity itself. An electricity service scenario can be constructed to reflect economic, demographic, lifestyle and institutional developments within a country.

Consequently, electricity planners should be very careful when establishing a scenario for electricity services, and preferably they should apply at least two different scenarios in long-term planning, in order to illustrate sensitivity and flexibility of the plan.

An end-use electricity planning process consists of four main elements:

1 Establishing scenarios for electricity services.

2 Analysing potentials for electricity efficiencies.

3 Analysing options for fuel saving in the electricity production process.

4 Establishing measures to guide the electricity systems in the desired direction.

We have already summarized element 2. In the following we will discuss the other elements of an end-use electricity planning.

7 Electricity Service Scenarios

The main uncertainty in electricity planning is the development of energy services. Several surveys in Denmark indicate a public attitude towards future energy services which conflicts with official expectations (6). A growing majority of Danes seems to be quite satisfied with the material standard of living they already enjoy, and would for instance prefer less work over more income. Similar trends are found in most highly industrialized nations.

If such preferences are gradually reflected in the official planning and economic structure, a saturation in private consumption will eventually spread to production sectors.

Population has stabilized in Denmark and might actually be declining. The family split-up into smaller units has so far lead to an average size household of 2.2 persons and resulted in a growth in number of households. The Danish Statistical Bureau anticipates, however, that this trend will come to an end by the turn of the century. As a consequence, not only will the population decline but so, too, will the number of households.

A stabilization (or even decline) in demand for electricity services is therefore a realistic development within a couple of decades. A scenario on this basis should be included in any long-term planning together with more traditional growth scenarios.

If possible, electricity service scenarios should be disaggregated into demand by the various end-uses in different sectors of society. For instance, one of the consequences of the above mentioned stabilization (and even decline) in the number of households would probably be that house construction and the associated energy intensive sectors producing building materials would continue their downward trend experienced in the last decade.

8 Implications for Supply Planning

A policy based on end-use planning could result in large reductions in electricity consumption. These electricity savings will have significant implications for the choice of primary energy sources and the electricity generation technology.

Efforts to save electricity will lead to less uncertainty about the future need for power. Prognoses can be more precise if, for instance, minimum standards are established for the efficiency of important electric equipment. This higher certainty might save hundreds of millions of dollars in power plant investments in the Danish case.

Furthermore, the introduction of technology to save electricity is a much more flexible solution than the construction of large power plant units of 300 MW or more. Conservation technology comes in units down to a few watts and their penetration rate can be influenced through various incentives, like subsidies and information campaigns. By these and other measures, demand for electricity can be influenced to follow a desired development, without affecting the electricity service. In a first phase, for instance, electricity demand could be coordinated with the depreciation of old power plants and a gradual introduction of more efficient electricity generation systems described in the following sections.

Extensive use of the fuel-saving cogeneration of heat and electricity will be possible if the ratio between demand for electricity and heat is kept within certain limits. Since large savings of heat have been observed and more could be achieved, this calls for a similar reduction in electricity demand in order to keep the right balance. If not, more and more electricity will be produced without cogeneration.

Finally, it should be noted that a relatively low demand for electricity would facilitate the transition to an energy system based on renewable sources in Denmark. The same is true in most countries. Electricity supplied from wind power or photovoltaic cells fluctuates over time. These fluctuations do not follow the demand, and since electricity is difficult to store in large quantities, it should be applied efficiently.

In the following two sections, we will describe the options for conservation of fossil fuels through production of electricity by: cogeneration of electricity and heat and, use of renewable energy sources.

9 Fuel Conservation by Cogeneration

Cogeneration of heat and electricity has a long tradition in Denmark, starting in the 1920s. It is a much more efficient process than production of heat and electricity separately. In a typical example the fuel consumption is about 60% larger in separate production as illustrated in Figure 4.

It follows from the above that fuel-based production systems should include as much cogeneration as possible. Unfortunately, the traditional planning structure in many industrial countries, including Denmark, is based on separate institutions for heat planning and electricity planning. The electricity supply planning is usually done by the utility companies.

Cogeneration has some special technical problems related to the different demand variations for heat and electricity over the day and over the year. This calls for supply installations with great flexibility so that the ratio between electricity and heat output can vary over a wide range. Large electricity-generating plants are usually not very flexible in this respect, which limits the market for cogeneration in the traditional systems. More flexible installations have been developed in recent years in the form of small (0.2 MW to 10 MW) cogeneration units that include gas engines or gas turbines, heat pumps and heat storages (9).

The concept of small, flexible cogeneration units also have important bearings on the potential use of renewable energy. The inclusion of renewable energy sources such as wind and solar in the production of electricity presents new technical problems to the systems. This is mainly due to the fluctuating nature of these energy sources. Hence, the production system has to be able to adapt to the varying input of electricity from renewable sources in a flexible

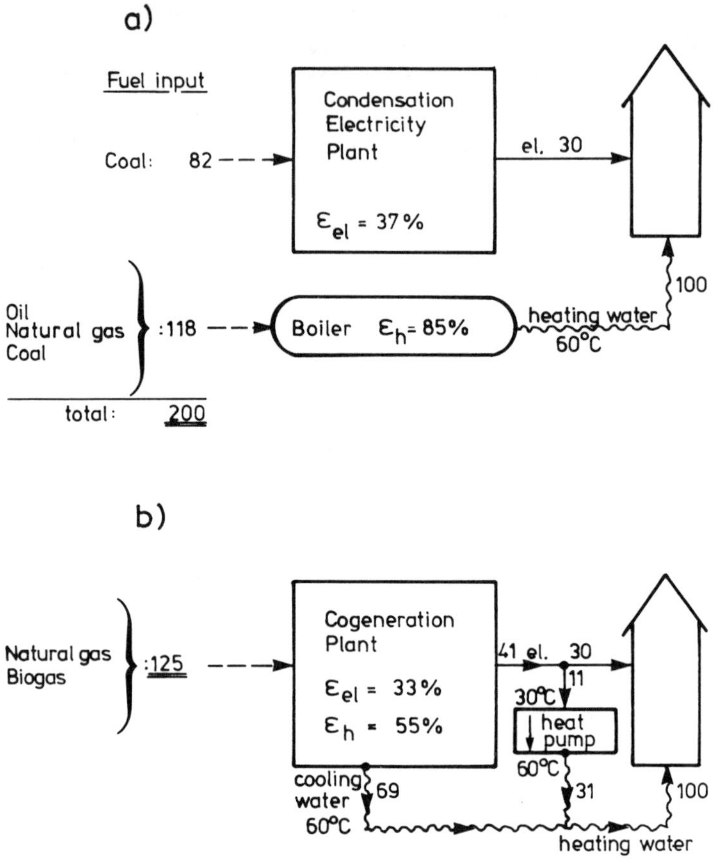

Figure 4. Comparison of fuel consumption for two different production systems with the same energy output.
a) Separate production of heat and electricity.
b) Cogeneration of heat and electricity.

way. This requirement is of the same nature as that posed by the varying demand.

Consequently, the small, flexible cogeneration units described above are expected to work well together with wind power, for example. This turns out to be the case. Utility companies typically claim that traditional electricity production systems can only handle 5–10% wind power. In contrast the flexible cogeneration units can, without difficulty, use 25–30% wind power (8).

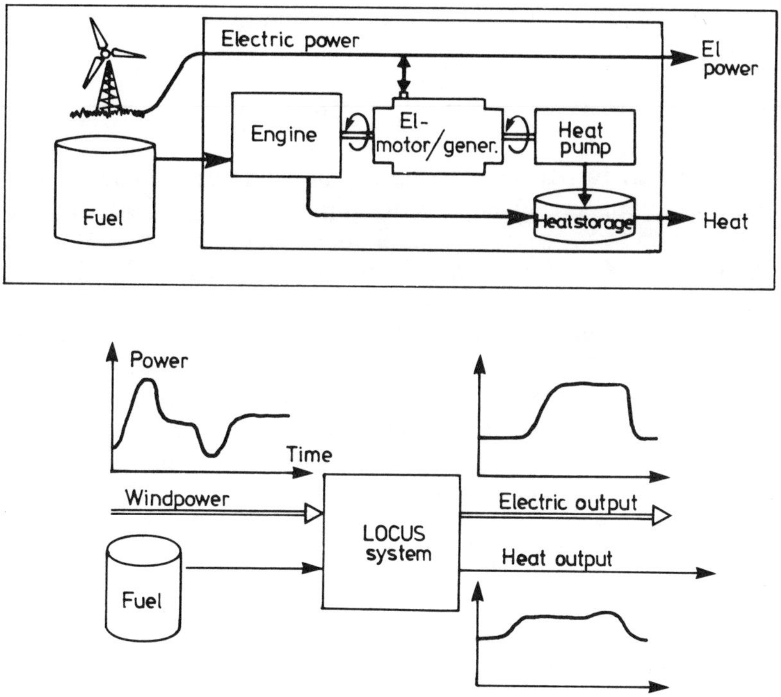

Figure 5. LOCUS (LOcal Cogeneration Utility System) installation including heat pump, wind power and heat storages (9).

An example of such an installation, called a LOCUS-system (9), is illustrated in Figure 5. Different version of this system are presently being tried out in some Danish demonstration projects.

Turn-key units of 0.2 to 10 MW cogeneration units are now available at investment prices around 800 $ per kW. This is competitive with large coal fired-power stations.

The importance of such local cogeneration units can be summarized in the following points:

1 Efficient use of fuels.

2 Large potential market due to flexibility in relative output between heat and electricity.

3 Well adapted to the use of local, renewable energy sources.

4 Smaller transmission losses compared to large electricity and heat production units.

5 Flexible adaptation to changes in electricity demand due to short planning and building times.

In Denmark the official planning authorities have estimated the potential market for small cogeneration units to be about 450 MW (10) This is based, however, on unnecessarily restricted criteria. An alternative analysis (11) indicates an immediate market which is 2 to 3 times larger, i.e., 900 to 1300 MW. This is in addition to the existing central power stations with an installed capacity of about 9000 MW. An extended transition to smaller cogeneration units may gradually occur as the large power stations are phased out.

Under a governmental decision, the Danish utility companies have initiated a construction program of 450 MW small cogeneration units to be completed before 1995. So far, though, the potential for electricity-production by small cogeneration units outside of the utilities has not been realized because the utilities pay too low a price for electricity produced by these units. This is a barrier which can only be removed by political means.

10 Electricity Production Based on Renewable Energy

The main renewable sources for electricity production in Denmark are wind and biomass (manure, straw, waste wood and household waste).

Wind power has been a success story in Denmark, but it has a somewhat uncertain future. At its peak in 1986, Denmark's exports of wind mills totaled about 300 million dollars mainly to California. Now exports are down to about one-third this value. In Denmark there are about 2000 wind mills installed, with a total capacity of about 100 MW. The average size is thus around 50 kW, while most wind mills being installed to-day are in the range from 100–200 kW. As noted the contribution from wind mills to the Danish electricity consumption is still no more than one per cent, but it is steadily increasing. The official goal is that wind power should contribute 10% of the installed power in the year 2000. This will require a much larger support from the Danish government than the present effort in terms of research and development, financing and payment for wind electricity. The most visible action is a programme which obliges the utility companies to build 100 MW of wind power before 1991.

In connection with this programme, a wind mill park at Lem in West Jutland was made operational in the spring of 1988 with a capacity of about 3500 kW. It is planned to increase this park by 35 mills, each with a capacity of 200 kW. The total capacity is thus expected to reach 10,500 kW in 1989. Another example is a single 2 MW mill close to Esbjerg in West Jutland which went into operation in 1988. If successful, it is planned to add more 2 MW mills to this installation.

A barrier for wind power is related to environmental considerations. Areas with large wind potential are in Denmark often also of high nature value. This constitutes a conflict of interest from an environmental protection point of view.

In Denmark the rules concerning possible placement of wind mills in the landscape are quite strict. The official planning authorities have therefore initiated evaluations of possible off-shore placements. The first off-shore wind mill farms are planned for implementation in 1989. If the technical, economic and ecological experiences of these wind mill farms prove satisfactory, then the space constraint on wind power development in Denmark should diminish substantially.

As already mentioned, local cogeneration units should be an essential element in the future Danish supply system. In the transition phase, natural gas from the North Sea could have an important role as a fuel in these installations. In order to reach a sustainable solution, however, natural gas should be replaced by renewable energy sources. The most obvious solution in Denmark is biogas from manure, straw and household waste. The estimated fuel potential from these sources are indicated in Table 5.

This should be compared to a total consumption of primary energy in Denmark of about 800 PJ in 1986.

The data in Table 5 do not take into account the potential for growing energy plants in Denmark. If this were done on a large scale, the total energy in Table 5 could be more than doubled (12).

In 1986 the Danish Ministry of Energy initiated a research and demonstration programme for local cogeneration units based on biogas from manure. These units are supposed to supply electricity and heat to local communities (villages) with district heating systems. At present 4 installations

Table 5. Estimated fuel potential from different types of bio-waste in Denmark in the form of biogas.

Sources	Energy content PJ per year
Biogas from manure	20
Gasified straw	30
Biogas from household and industrial waste	6
Gasified wood waste	3
TOTAL	59

Source: Ref.12.

are operating and 4 more are in the planning stage. They cover an output range from 100 kW to 500 kW (electric). Most of them operate in the mesophilic temperature range. A few installations in the thermophilic range are included in the programme for the sake of comparison.

Besides the production of energy, these systems can yield some environmental advantages through a more suitable redistribution of degassed manure on the fields.

The experiences so far have indicated that the technology is not fully developed. There has been a number of technical failures and shortcomings which have reduced the gas production. It has also been necessary to develop a more efficient transportation system for the manure from the farms to the biogas unit.

It is too early to draw any final conclusions about the economic viability of these systems. It appears, however, that there is no principal obstacle for the success of the systems provided the technical glitches can be remedied.

Gasification of straw has turned out to be more complicated than gasification of wood. Public funding has been given to some development programmes which are well under way. Again no final conclusions can be drawn, but preliminary results are positive (13).

A special demonstration project was initiated in 1984 on the island of Bornholm with a population of 50,000 and a land area of 600 km^2. The goal is to make Bornholm nearly independent of imported fuels over a time horizon of 25 years (8, 19, 20). The project is sponsored by the Danish government and the County of Bornholm. The unemployment is large at Bornholm and the energy project is anticipated to create new local jobs. A striking example is the establishment of a new production of wind mills through cooperation of ten small Bornholm companies. In less than one year they produced their first 100 kW mill. They have now constructed a local wind farm with 10 units from this "Blacksmith mill" (21). Other demonstrations on Bornholm include a biogas system at a pig farm where the biogas is used for heat, electricity and driving fuel in a tractor.

In connection with the energy project, the first district heating system at Bornholm was made operational in 1985. It supplies the heat for 140 dwellings and some commercial enterprises. The system is using waste straw as a fuel.

In September 1988 the city of Aakirkeby on Bornholm has decided to invest in a relatively large LOCUS-system based on biogas. The total investment is around 10 million dollars. This project has been delayed due to uncertainties related to technical problems in the biogas system and uncertainties in relation to the price offered by the utility companies for the electricity produced. The economic uncertainties have recently been increased due to proposals by the European Commission for harmonizing energy taxes in the Inner European Market. A strong reduction of the Danish energy taxes will be quite damaging to the present energy policy in Denmark.

Bornholm county is quite similar in structure to most counties in Denmark. Thus, the experiences from Bornholm are generally applicable to the Danish national energy planning. Projects related to the use of renewable energy in cities have recently been initiated in Denmark under the heading of "green cities" (22).

11 Official Danish Electricity Conservation Measures

In 1986 the conservative Danish minority government made an agreement with the Social Democratic Party about Danish energy policy. This agreement includes the following points related to conservation of electricity and alternative means for production of electricity:

- stronger efforts for electricity conservation;
- building of 100 MW wind power by utility companies before 1991;
- building of 450 MW local cogeneration units by utility companies before 1995.

Unfortunately, the first point concerning electricity conservation was not very well defined in the agreement. It has been interpreted by the Ministry of Energy as implying further analyses of the potential for electricity conservation in the main economic sectors. A study on the influence on consumption from an information campaign and an altered tariff structure in some local regions is also being conducted. No coherent programme for demonstrating and implementing the technological potential for electricity conservation is planned at this time.

If the development is left to the market mechanism, it will probably move in the right direction – but too slowly and not far enough. The decision-making process in the market has a short time horizon (1–5 years) and seldom takes into account environmental costs. Serious problems such as the "greenhouse-effect", the damage of the ozone- layer, acidification of the environment, etc., fall outside the purview of traditional economics. Therefore, energy planning can not be left to traditional economics.

The most significant measure taken by Danish government to save electricity is the special tax on energy. For electricity this tax plus the VAT amounts to around 7 cents/kWh or more than half of the sales price for electricity to the consumers involved. However, only private households and certain private trades are subjected to this special tax, which can be said to represent part of the environmental cost. About half of the Danish consumption of electricity, including that sold to industry, is exempted from paying this tax.

Energy taxes in Denmark have since 1986 been linked to world market prices of fossil fuels. When world market prices decline, this is compensated by increased taxes and vice versa. The purpose is to stabilize fuel conservation efforts,independent of fluctuating fuel prices.

To summarize, the Danish government has not yet applied an end-use approach to electricity planning and policies. In the next section, we will suggest some measures which could be pursued if electricity end- use efficiency became part of national energy policy.

12 Means for Implementation of Electricity Savings

We have concluded in previous sections that present Danish energy policies are insufficient for a rapid implementation of the potential electricity savings.

There is, however, no lack of possible means according to Danish political traditions. A number of these may be summarized under the following headings:

- Information campaigns,
- Labelling,
- Education,
- Demonstrations,
- Subsidies,
- Cheap loans,
- New tariff (rate) structures,
- Selective taxes, and
- Regulations.

All these measures are well known in Danish policy and are being applied today in other areas of Danish economy for other purposes than electricity conservation. See also chapters by Geller and Williams in this volume.

Demonstration of new electricity-efficient systems are needed in order to convince people that the technology is actually working. Thus, demonstration areas including private households, shops and service buildings could be established in a number of cities. Based on public funding, these buildings could be equipped with the most electricity efficient apparatus (refrigerators, freezers, cooking equipment, pumps, ventilation, lighting, etc.) and controls. Such a field test, with well managed demonstration of results, could have far reaching effects.

In Denmark an interesting project of this type was initiated in the 1970s. Analyses had shown and laboratory results in the early 1980s confirmed that it was possible to construct a refrigerator with an electricity consumption of one third of the best refrigerator on the market in 1983 (23). In 1987 a pilot series of 150 units of a low energy refrigerator LER200 was produced by a Danish refrigerator company. In the Spring of 1988 120 units were installed in homes for a detailed technical field test. Preliminary results indicate an average electricity consumption for these 200 litre refrigerators of about the expected 90 kWh per year. The project has been funded by the Danish Ministry of Energy and a number of Danish utility companies with the energy group at Physics Laboratory III, Technical University of Denmark as consultants. Based on the promising results, the Danish refrigerator company has decided to initiate a mass production of these LER200 refrigerators. The first 1000 units were in the shops in September 1988.

In 1988 the Danish utility companies have launched a general information campaign promoting electricity conservation. This campaign includes energy consultancy directed at industrial companies.

Subsidies have been used in Denmark in connection with energy savings (insulation, weather proofing, process heat etc.) for some years. The programmes, however, have been rather unsystematic, fluctuating and with a short time range. In addition, they have not been aimed specifically at electricity conservation. Systematic, long-range programmes should be initiated with direct economic incentives for people to buy equipment and systems with a high electric efficiency.

Tariffs (rates) for electricity can be changed to encourage electricity conservation. In Denmark the tariff structure in most areas include a relatively large fixed annual charge. With a low consumption, the total cost for each kWh used is thus higher than with a high consumption. This should be changed by eliminating the fixed charge or even using a negative fixed "charge". One could then introduce higher kWh prices without affecting the average electricity bill. Also, progressive tariffs could be be applied to households dependent on the size of the family.

As an example of regulation, one could completely ban direct electric heating systems in all new houses. The use of electricity to produce low temperature heat by means of resistance heating is so wasteful that it should be completely avoided. Owners of existing houses with electric heating should be paid to change to a more efficient system, e.g. district heating from cogeneration units. The money might come from utility companies who forego expensive investments in new production capacity as a result of the reduced electricity demand. So far, the Danish government has decided to prohibit direct electric heating in new houses with the exception of super-insulated dwellings. Excepted are also new houses being built outside the reach of district heating or natural gas networks.

13 Planning for a Sustainable Energy Development

The Brundtland Commission (2) stated that a requirement for global sustainable energy development is a reduction by 50% of the energy consumption per capita in the industrial countries over 30 to 40 years. At the same time, the supply system should switch to renewable energy sources.

This development will not occur automatically. It will require strong commitments and systematic planning by the industrial countries. The initial response to the Brundtland report by the Danish Ministry of Energy was to note Denmark's top international position in regard to energy conservation. This response overlooks the fact that it is not enough to be among the least bad when the effort required is an order of magnitude higher. The development in energy consumption anticipated by the Ministry has been quite contrary to the Brundtland recommendation. Electricity consumption, for instance, is expected to increase by 40% in 20 years.

The main lesson to be learned from the Danish experiences is that the technologies for a sustainable energy future are available and economically realistic. The implementation requires, however, a much broader planning effort, including institutional changes, demonstration projects, and changes of taxes and tariffs in order to create economic incentives for electricity conservation and renewable energy.

The Brundtland recommendations are based on global considerations but they are relevant in a national context. In the case of Denmark, the fulfilment of the Brundtland recommendations would result in the following benefits:

- less environmental damage,
- less drain on foreign exchange,
- more Danish employment,
- greater supply security, and
- more price stability in the energy sector.

We do not know of any other policy with so many clear advantages and so few disadvantages.

Already in the 1970s Danish energy researchers recognized the potential of renewable energy resources (17, 18). And in 1983 a group of Danish energy researchers from different universities produced an energy plan which anticipated the Brundtland recommendations (14).

We shall conclude this Danish case study with a summary of the main results from this alternative energy plan.

The methodology is based on the demand side as the driving force. The supply side is then adapted to the demand development based on a system which uses as little fossil fuel as possible.

The analysis of the demand development is based on two separate factors:

1 The level of energy service (m^2 of heated floor space, m^3 of cooled volume, passenger-kilometers etc.) and

2 The energy intensity of a specified energy service (Joules per year per m^2 of heated floor space etc.)

Concerning the level of energy service, there is a clear tendency towards saturation in the basic materials sectors (15). There is also a limit to the number of different activities a human being can manage at one time. These

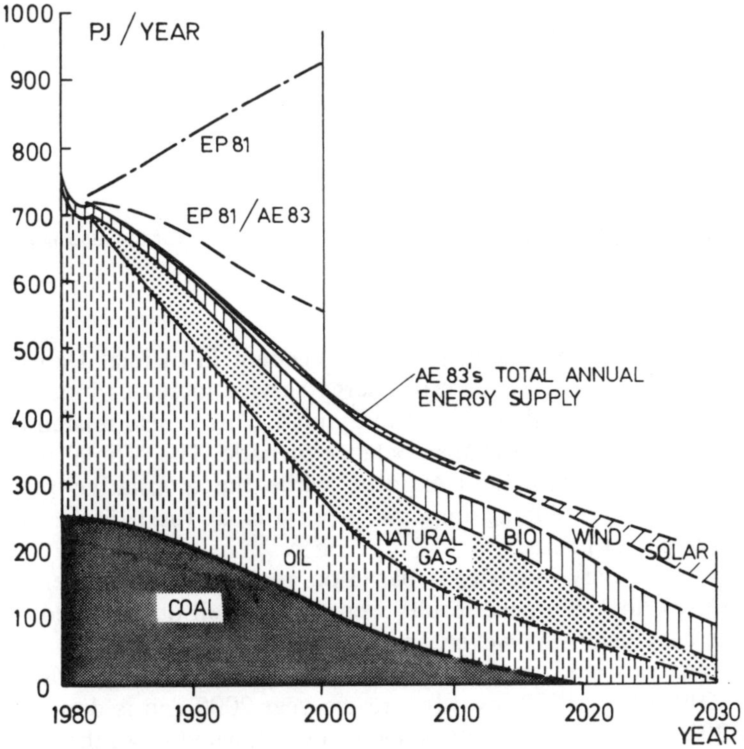

Figure 6. Development of primary energy consumption in Denmark according to alternative energy plan. Detailed calculations up to year 2010, estimated development after that year. For comparison the official Danish energy prognosis EP81 is shown. Furthermore, the results for the same development in energy services as in EP81, but with implementation of efficient technologies is shown as EP81/AE83 (14).

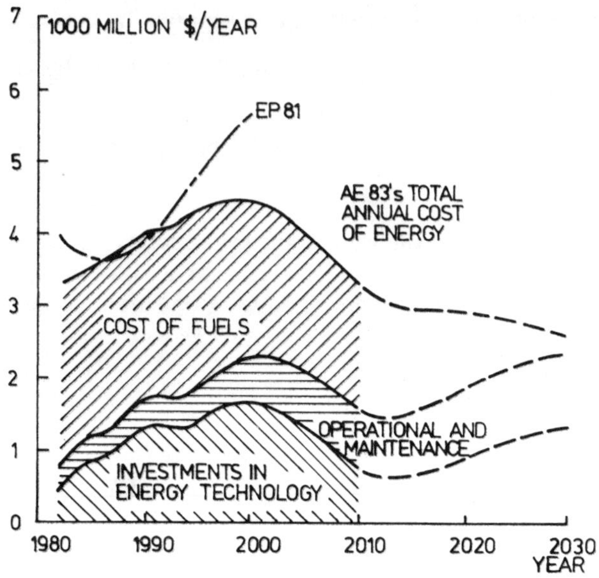

Figure 7. Total energy costs, including fuel, investment, and maintenance for the alternative energy plan AE83 as compared to that of the official plan EP81.

saturation effects have not been taken fully into account in official Danish energy planning. Consequently, the level of energy service increases more in the official planning than in the alternative energy plan. This is also related to the official strategy to improve the Danish balance of trade. The official plan emphasizes increased production and exports, while the alternative plan achieved the same goal by reducing imports, especially of fossil fuels and heavy technology for power plants.

The development of energy intensity is based on analyses of the different technological end uses. In 1983 these results were not as well documented as they are today. The conclusions, however, are more or less unchanged.

The alternative analysis leads to a consumption of primary energy which declines to about 60% of the 1982-level before the year 2000 and is down to 30% by the year 2030. The necessary investments are estimated up to the year 2010. The reduction in electricity consumption is assumed to be only about 30%. With the knowledge we have today this is certainly on the low side.

The supply side according to the alternative plan includes the following main features:

- Transition to local cogeneration units with low temperature district heating. Natural gas is used as transitional fuel with biogas from manure, straw, wood, waste and energy plants as the final fuel.
- Wind power is supplying an increasing part of the electricity consumption up to about 50%.
- Solar energy is supposed to contribute about 50% or 40 PJ per year to the heat consumption in year 2030. This includes both individual solar collectors and collectors connected to district heating systems. The total collector area is estimated to be about 50 million m^2. For comparison, there is an area of about 100 million m^2 on roofs well oriented for solar collectors.

The results in terms of primary energy consumption are summarized in Figure 6.

The alternative plan's total energy costs are illustrated in Figure 7. For comparison, the official plan's total energy costs are also indicated. Except for a brief period of a few years, the alternative energy plan is less costly. The cost differential becomes increasingly larger after 1995.

Finally, it is noteworthy that the pollutant emissions such as SO_2, NO_x and CO_2 are strikingly lower in the alternative plan, and are compatible with the requirements of the Brundtland Commission. The Danish government has announced in October 1988 that they are preparing a comprehensive "Brundtland plan". It remains to be seen how far the government will go concerning real committments to fulfilling the Brundtland recommendations. It is encouraging, however, that the government now includes the Brundtland report in its official program.

References

1. *Proceedings from Seminar on Electricity Conservation.* Danish Agency for Technology Assessment, 1986.
2. The World Commission on Environment and Development: *Our Common Future*, Oxford University Press, 1987.
3. Norgard, J.S., Holck, J., Mehlsen, K. 1983. *Technological Potential for Electricity Conservation.* Report from Physics Laboratory III, Technical University of Denmark (in Danish).
4. Norgard, J.S., Guldbrandsen, T. 1987. *Potentials for Technical Electricity Saving Using Known Technology*, Proc. NATO Advanced Study Institute, Portugal.
5. Research Institute of Danish Counties (AKF), *Technological Electricity Conservation Potential*, 1988 (in Danish).
6. Norgard, J.S., Christensen, B.L. 1984. *Individual Attitudes in Scandinavia point towards a Low Energy Saturated Society.* ACEEE conference, Santa Cruz.
7. Pedersen, H. Ø. 1987. *Danish Energy Planning*, Danish Energy Agency.

8. Jorgensen, K., Meyer, N.I., Pilgaard, H. 1986. *Energy Plan for a Green Island – Project for Renewable Energy on Bornholm.* Borgen Press, Copenhagen (in Danish).
9. Illum, K. 1988. *LOCUS-Systems and Plants*, District Heating Catalogue of Advanced Heat Production Technologies. Aalborg University Center.
10. Danish Energy Agency: *Potential for Decentralized Cogeneration in Denmark*, 1985.
11. Pedersen, S.L. 1987. *Decentralized Cogeneration Installations and Wind Power in Eastern Denmark.* Ph.D. Thesis, Physics Laboratory III, Technical University of Denmark (in Danish with an English summary).
12. Gundermann, J. 1988. *Biomass – a Case Study of Danish Experiences.* Danish Energy Agency.
13. Danish Technology Council. 1988. *Report on Experiments with Gasification of Straw.* (in Danish).
14. Hvelplund, F., Illum, K., Jensen, J., Meyer, N.I., Norgard, J., Sorensen, B. 1983. *Energy for the Future – Alternative Energy Plan for Denmark.* Borgen Press, Copenhagen (in Danish).
15. Goldemberg, J. et al. 1988. *Energy for a Sustainable World.* Wiley Eastern, New Delhi.
16. Danish Ministry of Energy, *Energy in Denmark*, 1987.
17. Sorensen, B. 1976. *The case for using renewable energy sources.* Symposium on Appropriate Technology for the UK, Newcastle.
18. Blegaa, S., Hvelplund, F., Jensen, J., Josephsen, L., Linderoth, H., Meyer, N.I., Balling, N.P., Sorensen, B. 1976. *Alternative Energy Plan for Denmark.* Copenhagen, (in Danish).
19. Jorgensen, K. 1985. *Renewable Energy at Bornholm -selfsufficiency for an Island Community in Denmark*, in *Energy for Rural and Island Community, IV*, J. Twidell et al. (ed), Pergamon Press.
20. County Council of Bornholm. 1985. *Project for Regional Energy Planning – Bornholm Energy System.*
21. Rasmussen, P. 1988. *The Blacksmith-windmill Concept (13–270kW).* Proceedings of the First International Conference on Renewable Energy and Local Production – a Vehicle for Development, Danish Center for Renewable Energy.
22. Copenhagen City Council. 1988. Project City Energy.
23. Heebøll J., Nørgård, J.S. 1986. *Deveopment of energy efficient electrical housekold appliances, past one: Refrigerators*, Report EUR 10449 EN, Commission of the European Communities, Batiment Jean Monnet, Luxemburg.

Energy Service Companies: The French Experience

Olivier de la Morinière

Abstract

A specialized industry comprised of "societes d'exploitation de chauffage" or "energy service companies" has developed in France over the past 50 years. There are currently 140 energy service companies in France. They engage in a wide spectrum of activities aimed at operating and managing heating systems in multi-family and commercial buildings, and in other large building facilities. They offer their services through complex performance-based contracts. The evolution of the industry since the energy crisis, the nature of its contracts, and current industry trends are described in this chapter.

1 Background

On a cold winter night in 1931, in the hospital Villiers-Saint-Denis in the north of France, a sleeping drunk service man let the pressure in the building's boiler get too high. The boiler exploded. Thankfully, the skills and knowledge of the craftsman that installed the boiler enabled him to rapidly repair it, so that it was not necessary to evacuate the hospital's patients. After the incident, the hospital administration decided to sign a contract with the craftsman, Leon Dewailly: Mr. Dewailly became responsible for providing the heating fuel and the operation, maintenance, and major repairs of the hospital's heating system. This was the first energy service contract in France, and Dewailly went on to develop a multinational company, Compagnie Generale de Chauffe. Mr. Dewailly is well known as the founder of the French energy service industry.

The progressive reconstruction of post-war France, mostly through the construction of large buildings (housing, hospitals, schools, etc.), provided an opportunity for the growth of energy service companies. Building managers lacked the skills, the staff, and the motivation for efficiently operating large boilers and district-heating systems. The majority of building owners, therefore, chose to contract for these services through professional energy service companies.

The rapid growth of energy service companies that resulted from the post-war boom in urban construction accelerated the centralization of the industry: through the 1970s, small firms that had been dedicated to local markets were acquired by larger companies, which were in turn acquired by, or became affiliated with, large engineering firms, service groups, and fuel distributors (Compagnie Generale des Eaux, Lyonnaise des Eaux, B.P., etc.). Access to a broader range of technical skills and a strong financial base allowed the industry to bid for contracts involving larger and more sophisticated building heating systems, district-heating systems, and industrial facilities. In the 1970s and 1980s, the industry was able to diversify the building services it offered.

In the past few years, great interest in energy-management contracting and third-party financing has been expressed in government and the industrial and commercial sectors. There are several reasons for this:

- The partial disappearance of opportunities for achieving energy savings through the adoption of measures that were simple and inexpensive and could be carried out independently by building owners.
- Reductions in large scale government funding of energy conservation projects that occurred as a result of the general perception that the energy crisis is over.
- The goal of French policy-makers to increase the leverage of scarce public and private funds devoted to energy conservation has resulted in negative economic evaluations of several energy conservation subsidy programs.
- The trend toward policies in industrialized countries that favor private-sector financing over government subsidy programs.
- The perception in Europe that third-party financing is highly profitable, which resulted from publication of American success stories.

Energy service companies in France have played a unique role in supporting the development of national interest in third-party financing of energy conserving investments in buildings.

2 The Industry Today

2.1 Industry Philosophy and Fundamental Operating Assumptions

Most third-party energy financing companies are in principle *financial institutions*. Their primary goal is to generate revenue. They use energy management as one of many tools for achieving that goal. In contrast, French third-party energy financing companies are in principle *energy management companies*. Their basic skills are building-system design, operation and maintenance. Their primary goal is to promote facilities management. They use third-party financing as a tool for achieving that goal. This is the fundamental difference between third-party energy financing companies in France and those elsewhere.

The French energy service industry's own original philosophy for the use of third-party financing is based on the assumption that the energy efficiency of building mechanical systems can only be maximized with the help of technical skill and continuous care for improving system operation. French energy service companies concentrate intially, therefore, on improving the maintenance, tuning, and operation of existing building systems, and gradually progress toward encouraging energy-conserving system investments. The management services provided by the companies have increased over time from minor and periodic investments in system maintenance repair and tuning, to stimulating large investments in new energy efficient equipment – boilers, distribution-system insulation, geothermal wells, large-scale waste-burning facilities, etc.

French energy service companies are comprised of engineers. They are interested in continuous hands-on work with building mechanical systems. Their primary interest is not in financing the installation of energy-efficient systems and waiting for the equipment to provide a payback. Furthermore, they believe that the durability of energy savings and of building equipment, and therefore the ability to minimize the life-cycle-cost of investments, depend on continuous skilled system operation and maintenance. They see the bankruptcy of some third-party financing companies in the past as a direct consequence of lack of control over the operation and maintenance of the equipment that was installed. The fact that the companies have focused on ongoing operation and maintenance of heating distribution and control systems, i.e., efficient measures for meeting a given heating load rather than building shell improvements or other measures aimed at reducing the heating load is a direct consequence of these interests and beliefs within the industry.

The industry leadership believes that companies should develop toward offering energy efficiency on a large scale, i.e., beyond unrisky investments that do not require their engineering skills and have short payback periods. They are interested in large-scale investments in new technologies which necessitate frequent maintenance and tuning of equipment and constant monitoring and data analysis of system performance. The industry supports, therefore, financing large investments and guaranteeing energy savings and contracted operating costs, but exclusively under contracts where they are given full responsibility for the operation and maintenance of the equipment.

The industry represents:

- Approximately 140 companies, the largest of which are Compagnie Generale de Chauffe (C.G.C.), Cofreth, Montenay, and Esys. Together these four companies comprise more than 50 percent of the industry, and C.G.C. alone represents more than 20 percent. Some of the companies are affiliated with oil companies (Stechenberger with B.P., Sochan with Shell, etc.).
- 19,000 skilled professionals, exclusively dedicated to the operation and maintenance of building heating systems and heat production facilities.
- Specialized research teams in other fields related to energy: remote building system control and operation, waste burning techniques, electric induction boilers, dynamic energy pricing.
- Waste burning capacities of approximately 500 tons of waste per hour.
- A total installed operating capacity of almost 100,000 MW, or roughly seven million dwelling equivalents (one dwelling equivalent equals 70 square meters of non-residential heated floor area).
- Systems with a total annual energy consumption (oil, gas, and electricity) of nearly 82,000 MWh (Table 1). More than 95 percent of the fuel consumed in buildings in France in the 1960s was oil. Because of competition among energy service companies and their commitment to minimizing their customers operational costs, approximately 30 percent of the oil burning capacity in buildings was converted to gas during a 20 year period.

Table 1. Type of energy used.

	MWh	% of total
– Heating		
• Solid – coal and miscellaneaous	3121	4%
• Liquid – oil and miscellaneaous	42834	52%
• Gas	25741	31%
• Other	7615	9%
– Refrigeration	1962	2%
– Space conditioning (Both Heating and Cooling)	666	1%
Total (31 december 1986)	81939	100%

• The breakdown of building types is given in Table 2. Multi-family residential buildings represent about half of the industry contracts. The current trend, however, has been to increase the number of commercial buildings under contract and to enter the market for industrial facilities, offering a larger spectrum of technical services (steam processing, electricity management, etc.).

Table 2. Type of building under energy service contract.

	% of total
– Residential	54.8
– Health Facility	7.1
– Educational Facility	11.4
– Military Establishment	3.1
– Commercial	9.1
– Other	14.5
– Total	100

• The breakdown of plant sizes is given in Table 3. More and more large facilities are being operated and maintained, such as space-conditioning systems in hospitals. Also waste treatment plants for large cities like Barcelona (Spain) and Liege (Belgium) are being operated by French energy service companies.

Table 3. Breakdown of plant sizes.

Heating power (in MW)	% of buildings
• Less than 0.4	85.7
• 0.4 to 1	8.3
• 1 to 3	4.6
• 3 to 10	1.2
• More than 10	0.2
• Total	100

The importance and diverse scope of the energy service industry in France is perhaps best demonstrated through a description of the largest company, Compagnie Generale de Chauffe. This company has evolved to its current state over the past 40 years. The large-scale demand for its services, and the services of companies like it, resulted from (1) the post-war reconstruction of France between 1950 and 1970, and (2) high energy prices after 1972. The French energy service company experience is not directly replicable in other countries today, but there is a need in many counties for highly skilled people to efficiently operate and maintain space-conditioning and heat-production equipment in large buildings. The experience and technical expertise within French energy service companies could be very useful in helping other countries develop energy service companies to fulfill this need.

2.2 Profile of the Compagnie Generale de Chauffe

- **Service Activities:**
  - Management of heat and air-conditioning installations
  - Remote monitoring and control of large building systems
  - Operation and maintenance services for industrial equipment
  - Processing of urban and industrial wastes
  - Treatment of water in closed-loop hydronic heating systems
  - Installation and operation of video-communication networks

- **Installation Activities:**
  - Study, design and implementation of environmental engineering related to fuel flexibility and new energy forms

- **District-heating Networks:**
  - Management of 72 district-heating plants
  - 900 km of network installed
  - 2800 MW of installed power

- **Contracts:**
  - 710,000 multi-family dwellings
  - 3100 teaching establishments
  - 1800 industrial and commercial establishments
  - 1800 public buildings
  - 860 hospitals
  - 365 military buildings
  - 234 swimming pools
  - 25 processing plants for urban and industrial wastes
  - Responsible for 17 million kW of power

- **Staff:**
  - 8252 employees, of which 3143 are outside of France

- **Resources:**
  - 2600 vehicles
  - 9 central remote monitoring and control stations
  - 30 storage centers and workshops
  - a laboratory for water processing
  - an in-house training school

- **Decentralized Organization:**
  - 8 regional offices
  - 40 regional centers
  - 110 local offices

- **International Activities:**

  Operations in nine European countries (West Germany, Belgium, Spain, Great Britain, Luxembourg, Holland, Portugal, Monaco and Switzerland) as well as in Africa (Ivory Coast and Camaroon) and the USA.

3 Energy Service Company Contracts

Energy service companies, following the requirements of government regulation, have developed a complex method of contracting their services. This includes four different classifications of contracts categorized by the type of service desired. Because most of the contracts entered into by the companies offer performance guarantees, several methods have been developed for defining the responsibilities assumed by the parties entering into the contract, and, in turn, the division of the profits produced through the efficiency improvements produced from the company's services.

The categories of contract services are as follows:

3.1 Category P1

Under contracts in this category, the energy service company usually provides the heating fuel required by the customer (oil or coal). It is becoming more common, however, for the energy service company to also assume responsibility for the customer's electric and/or gas utility bill. The energy service company assumes responsibility for the customer's energy consumption. The services (energy savings) produced by the company are determined on the basis of three different calculations:

1 A theoretical estimation of the building's heating load, based on: evaluation of characteristics of both the building and its heating equipment, the average number of degree-days and the length of the heating season for the region, and the indoor-temperature guaranteed by the contractor.

2 A second calculation of the theoretical heating load for the building done after the billing period, correcting the original calculation with the actual number of degree days experienced, length of the heating season, and indoor temperature maintained in the building.

3 The third calculation accounts for the actual amount of fuel consumed in the building over the billing period and is based on the fuel delivered to the building and the change in the stock of fuel on the premises.

The fees paid by the customer for P1 services, and the incentives for both parties entering into a P1 contract, are negotiated in a variety of ways:

3.1.1 P1.MF ("Marché Forfaitaire," fixed price contracts)

Under this type of contract, the service company guarantees to provide a given level of indoor comfort (for example, an indoor temperature of 19°C) for a fixed price irrespective of the weather experienced during the billing period. These contracts are signed for a minimum period of eight years, in order to reduce the impact of the unknown weather factor.

The energy service company assumes all of the risks in the contract, and is *motivated to maximize the energy efficiency of the system because it keeps all of the profits from energy savings that exceed those assumed when computing the customer's fixed price.* If the contract period is long enough, the energy service company has the incentive to continue to make energy efficiency investments in the building – changing the burners, insulating pipes, etc. In addition, the decision-making process with regard to determining what investments to make is improved because the company has the skills available and the motivation for properly carrying out investment measures. The occupants of the building, on the other hand, have no incentive for cultivating energy efficient habits when fixed heating charges are used.

3.1.2 P1.MT ("Marché à la Temperature," temperature adjusted contracts)

Temperature adjusted contracts are similar to fixed price contracts, except that the fee paid by the customer is adjusted based on the actual number of heating-degree-days in the billing period. Because the weather related risks assumed by the contractor are eliminated in this type of a contract, shorter contractual agreements are offered. The advantages and disadvantages of the fixed term contract discussed above also apply to temperature adjusted contracts. These contracts, however, are more complicated to bill and budget for, and the actual number of heating-degree-days in a billing period is difficult to measure and validate.

3.1.3 P1.MC ("Marché à la Chaleur," metered heat contracts)

When metered heat contracts are used, the customer pays a price based on the number of units of heat that are consumed in the billing period. *In this type of a contract, the contractor and the customer share the responsibilities and incentives for reducing the energy consumed for heating the building.* The contractor is reposonsible for providing heat as efficiently as possible (checking fuel quantity and quality, tuning and operating the boiler, and investing in energy efficient equipment) because the less fuel used per unit of heat delivered, the more profits the energy service company makes. The building owner/manager is responsible for using the heat delivered to the building more efficiently, because the less heat that is used, the less the customer pays. Thus responsibility for energy efficiency lies on both sides of the meter: upstream of the meter (where the most skill and experience is required) the energy service company profits from energy efficiency, and downstream of the meter the building owner/manager profits.

3.1.4 P1.CP ("Combustible et Prestations," fuel and services contracts)

The price paid by the customer in this kind of a contract is based on the amount of fuel that is delivered to the building. The contractor acts as a utility and provides as much fuel as is demanded by the client. All of the incentive for reducing the amount of energy used in the building lies with the building owner/manager.

In all of the contracts mentioned above, special provisions are made to shield the contractor from increased costs caused by inflation and/or changing fuel prices (adjustment factors are determined annually by the French government). The contractor assumes the risks associated with the technical measures that it invests in, i.e., those directly related to the contractor's skills. The building owner retains responsibility for the risk of changes in the fuel market.

During the past ten years, however, French energy service companies have, upon occasion accepted responsibility for the risks associated with fluctuations in energy fuel prices. In geothermal applications, for example, most energy service companies guaranteed that the cost of geothermal energy would never rise above that of conventional fuels. That price relationship held true until the price of oil on the world market crashed in 1985. The resulting change in relative fuel prices, and problems caused by acidity in wells, have caused energy service companies with contracts for geothermal power to suffer significant financial losses in recent years. As a result, most energy service companies no longer agree to assume responsibility for changing fuel prices in their contracts. Nevertheless, because energy service companies advise their customers to make investments, and the cost-effectiveness of them is dependent on fuel prices, companies that want to keep customers and stay in business remain commercially, although not legally, responsible for assuming the risks from price changes in the energy market.

3.1.5 Shared incentive contracts – "I" ("Interessment," incentives)

For each of the contract types described above, an additional clause can be added that specifies how the profits from energy savings, or losses from overconsumption, can be shared by the energy service company and the customer. A theoretical heating load is first estimated and later adjusted for actual weather conditions as described previously. In most cases, the building owner receives two-thirds of the profits when actual consumption is less than the theoretical heating load, and the service company gets one-third. When actual consumption exceeds the estimated heating load the customer pays for one-third of the losses and the contractor pays for the rest.

Shared incentive clauses are ideal in theory because both the contractor and the customer have a strong incentive to conserve energy. *In practice, however, they are problematic.* It is difficult to accurately predict the heating load of a building. Building owners frequently contest the results of these calculations after actual consumption data becomes available. Energy service companies claim that building occupants absorb their skillfully produced energy savings in the form of increased comfort. Because of disputes such as these, *most energy service companies have abandoned the use of special incentive clauses in their contracts.*

3.2 Category P2

Under category P2, customers pay energy service companies *a fixed sum for the operation and maintenance* (including minor repairs) of their mechanical systems. Operation and maintenance generally include:

- checking the quantity and quality of the fuel flowing into the system;
- providing for fuel and fluids delivery;
- periodic tuning of combusting burners and boilers;

- basic maintenance;
- organizing the system operation so that quick and efficient intervention can be done in the case of breakdowns.

Category P2 is rapidly broadening, as energy service companies offer more and more services (computer-based remote control of boilers, building security personnel, daily cleaning of offices, etc.). These services are the most essential part of energy service company contracts. *The trend in recent years has been to market efficient and skilled services and to increase the weight given to P2 contracts in relation to that given to the other contract categories.*

3.3 Category P3

Category P3 covers major repair and renovation work on building heating systems. P3 is only offered in conjunction with P2 contracts, and is available in the two forms described below:

3.3.1 Total guarantee coverage

Total guarantee coverage is similar to an insurance policy. It is the most common form of P3 contracts. The building owner pays an annual fee, depending on the condition of the building facilities (type, age, size, etc.) covered. The energy service company provides equipment repair and maintenance when required, regardless of the reason for the system failure. At the end of the contract period, the contractor is required to leave the heating system in good condition and to guarantee its function for at least one heating season. By giving total responsibility (both technical and financial) for the heating system to the contractor, total guarantee contracts make life easier for the building owner. When contractors are financially responsible for repairing damaged equipment the building owner can rest assured that they will operate them with care. In addition, future investments for repairs and replacement of equipment can be factored into the annual fee, thereby making them easier for the building owner to budget for. Although the contractor has the freedom to decide how to repair and/or replace equipment, the building owner retains the right to have the contractor justify expenses at the end of the contract period. When equipment replacement is required, the owner can opt for more efficient equipment and pay for the additional expense or modify the service contract accordingly.

3.3.2 Broad covering maintenance

Under broad covering maintenance contracts, the energy service company is responsible for programable preventive maintenance and urgent repair of building heating systems. The building owner/manager is left with the difficult task of planning and taking responsibility for major repairs and replacement of equipment.

3.4 Category P4

Category P4 is offered as a financial service to building owners. Under P4 contracts, energy service companies finance investments for replacements of, and major additions to, heating systems, and the principle plus interest are repaid by the customer during the contract period. The energy service company releases ownership of the equipment to the building owner at the end of the contract period. The "loan" payments in P4 contracts are made separately from fees collected from other services offered by the contractor.

4 Example Contracts

In the 1950s and 1960s, low energy prices allowed energy service companies enough of a profit margin to offer P2 services, and sometimes P3 services, as part of their basic fuel delivery contracts (P1). Today, each of the four categories of contracts offered by energy service companies is sold separately to building owners: the four categories are not offered as packages (with the exception that P3 services are not contracted without P2 contracts). Investments made with P4 financing often reduce the costs of P1 (fuel delivery services), P2 (operation, maintenance, and other services), and P3 (system failure coverage) because the systems use less fuel, and require less operation, maintenance and repairs. When all four categories, including energy conservation investments through P4 financing, are chosen, the total costs to the building owner are usually *lower* than when only P1, P2, and P3 services are contracted for.

Heating service contracts are most common in public housing.[1] The distribution of heating service contracts by type in public housing is shown in Table 4.

4.1 Example 1: A Typical Unspectacular Example of How the Contractor Operates

The following is an example of an actual contract that was drawn for a large building that was divided into multi-family housing and commercial office space. The entire building was heated by a single boiler plant, operated by an energy service company. Third-party financing was proposed by the contractor in order to save energy through more efficient system operation. This example is typical in that most of the measures considered involved changes to the heating, distribution, and control systems, and no significant changes to the building shell were suggested.

4.1.1 Technical assessment

The heat loss coefficient (G factor) for the multi-family zone of the building (45,505 m^3) was calculated to be 1.19 W/°C/m^3. For the commercial zone of the building (2703 m^3) it was calculated to be 2.00 W/°C/m^3. The total current consumption was:

* 83.2 m^3 of oil for 5,665 m^3 of water heated to 55°C
* 358 m^3 of oil for space heating for 212 days (2,568 heating degree-days).

Table 4. Distribution of contract types.

	1981	1982	Difference
• Fixed Price (P1.MF, P1.MT)	31%	26%	–5%
• Shared Incentive	27%	37%	+10%
• Operation only (no fuel)	23%	25%	+2%
• Owner-operated (no contract)	18%	12%	–6%

Source: Mr. Meunier, CNET-HLM.

1 A large portion of the multi-family dwellings in France are owned and managed by non-profit groups that are unified, both locally and nationally, by organizations such as the National Association of Public Housing (UNOHLM). These non-profit groups offer low- to moderate-rent housing, commonly referred to as "public" or "social" housing.

4.1.2 Current efficiency

The theoretical consumption for the building was calculated as 2.6 GWh (2.4 GWh for the multi-family zone and 0.2 GWh for the commercial zone).
The space-heating efficiency for the building was calculated as:

$$\frac{2.6 \text{ GWh}}{358 \text{ m}^3 \times 1010 \text{ th/ton} \times 1.16 \text{ kWh/th} \times 8.36 \text{ ton/m}^3} = 0.74$$

The actual space-heating efficiency was monitored with a computer and calculated as being composed of:

- Boiler efficiency 0.79
- Basement network efficiency 0.97
- Building network efficiency assumed 1
- Building efficiency assumed 1
- Balance efficiency 0.97

Building space-heating efficiency: 0.79 x 0.97 x 0.97 = 0.74

The water heating efficiency was calculated as being:

- $\Delta T = 45°C$: 45°C x 1th/m^3°C x 1.16264 kWh/th = 52 kWh/m^3

- Current consumption for water heating:

$$\frac{83.2 \text{ m}^3 \times (1010 \text{ th/ton} \times 1.16264 \text{ kWh/th} \times 8.36 \text{ ton/m}^3)}{5665 \text{ m}^3} = 144 \text{ kWh/m}^3$$

The total hot water efficiency was therefore 52/144 = 36%, comprised of:

- Boiler 76% x control 95% x loop 36%

Suggested Improvements

- Changing the burners – efficiency improvement of 5%;
- Sequential control – efficiency improvement of 3%;
- Optimizer to improve comfort (but not save energy);
- Lowering hot water temperature and stopping loop circulation between 1AM and 5AM – efficiency improvement of 13% [2];
- Skilled system operation – efficiency improvement of 2%;
- Changing to a cheaper type of fuel oil: FO2 TBTS; and
- Remote control of the boiler plant for better guaranteed maintenance.

[2] The 13% efficiency calculation took into account the fact that a large part of the heat losses in the circulation loop contributed to meeting the space-heating load.

The total investment made in the building was 350,000 FF which was financed by the energy service company at an annual interest rate of 13 percent to be paid over a 10 year period with an annual annuity of 64,550 FF.

The fuel costs before the energy service company was engaged were 924,470 FF/year; in contrast the future fuel costs were estimated to be 834,796 FF/year – 9.7% saving/year.

The 89,674 FF saved/year (924,470 – 834,296) covered the expenses of the loan annuity (64,500 FF), and a broader contract with the energy service company (10,000 FF/year). The building owner was left with a guaranteed net saving of 15,000 FF/year and a better maintained, guaranteed, building mechanical system.

On the basis of these calculations, a 10 year contract was drawn with the energy service company guaranteeing to produce the estimated savings. The risks associated with the savings prediction were assumed in total by the energy service company.

This example is interesting because it is typical, and it shows that guaranteed savings can be achieved through skilled system operation. Achieving these savings required:

- Frequent operation is the key to finding efficiency improvement opportunities.

- Technical skill to propose and make improvements on each element of the system (boilers, network, control) and to evaluate savings accordingly.

- Skilled and frequent system operation and maintenance, including periodic checks of the building's mechanical system.

Operating systems such as the one in Example 1, and tracking their efficiencies in search of opportunities for improvements, is progressively becoming the greater part of the everyday work of French energy service companies. Competition within the industry has required the companies to hone their skills in this area in order to keep the customers they have and attract new contracts. Although the operation and maintenance of heating systems in large buildings has been the most common type of energy service company contract, many companies have expanded their areas of expertise over the years to include specialised skills in heat and electricity generation using technologies such as geothermal wells and waste recycling and burning. The following example of a geothermal installation is used to illustrate energy service company contracting in these new areas of activity.

4.2 Example 2: Vaux-le-Penil Geothermal System

In 1981, the Compagnie Generale de Chauffe (C.G.C.) negotiated a contract with the Vaux-le-Penil township to create a geothermal facility, by drilling down to the Dogger Limestone Aquifer of the "Bassin Parisien" to a depth of 1700 meters. Water from the aquifer was supplied to a heat exchanger at 72°C, run through a fiberglass district-heating network, and returned to the aquifer (see Figure 1). The Vaux-le-Penil geothermal system provides a partial heat supply to apartments, public buildings, industrial facilities, and the Melun Hospital, producing an annual savings of 3000 TOE (tons of oil equivalent). The drilling work was planned and supervised by Geotherma, and C.G.C. planned and installed the district-heating network. C.G.C. was responsible for operating and maintaining the well and the district-heating network. Together C.G.C. and the township of Vaux-le-Penil created a new company, Geopenil, to operate the well and sell the heat to C.G.C. at a contracted price – 90 FF/MWh. C.G.C. in turn sells the heat to customers on the district-heating network, supplementing the geothermal heat with a decentralized system of gas and oil-burning boilers.

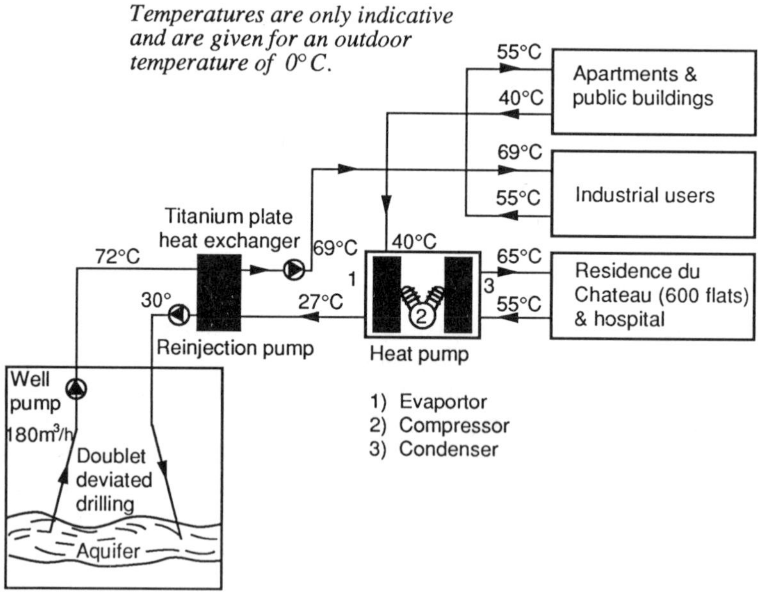

Figure 1.

The estimated price of the heat from the geothermal system was five percent lower in 1981 than the price of heat generated with oil, and experts expected the price gap to increase. The cost-effectiveness calculations for the project were based on the price of oil. Decreases in the price of oil in 1985 made the project uneconomical. Because the contracted price of the geothermal heat rose above that of heat generated with oil-burning boilers, Geopenil was not able to sell as much heat to the grid. Instead of selling the projected 33,000 MWh of geothermal heat per year to an expected 3500 equivalent dwellings, only 22,000 MWh/year are sold to 1200 equivalent dwellings. Because the price of the geothermal heat was guaranteed in the energy service company's contract, Geopenil has had to absorb the financial losses from the project. Although the well and the district-heating network were technical successes (they are considered one of the best French geothermal facilities), the drop in the price of oil destroyed the project.

5 Industry and Customer Points of View

5.1 Industry Views

The following represent some of the most important points of view and trends within the energy service industry in France today:

- Energy management is a skill that is best performed by professionals.
- Building mechanical systems operated by skilled professionals from energy service companies perform better than those run by in-house staff. In some cases, where system operation has been nonexistent or carried out by unskilled operators, energy service companies have been able to guarantee energy savings as great as 15 percent, with only superficial energy audits, and still make a profit from the resulting contracts. Some of the best examples can be found in public buildings where building managers are given no specific incentives for operating systems efficiently.
- The contractor should have overall responsibility for fuel supply and system operation and maintenance, and be paid accordingly. Energy savings and efficiency come from skilled contractors, and can only be achieved by giving contractors a complete opportunity to apply their skills to building mechanical systems.
- Contracts should remain performance oriented, i.e., the contractor guarantees fuel efficiency and maintenance reliability) as they have been in the French energy service industry. "Measure" oriented contracts, therefore, should be rejected.

- Predictable and comprehensive fixed-price contracts (P1.MF and P1.MT) covering both fuel costs and operation can be successfully applied in many buildings and should not be rejected.
- Long-term service contracts should be pursued, in order to give contractors complete responsibility for system operation and maintenance. With long-term P1, P2, and P3 contracts, energy service companies's natural goals are to both maximize energy efficiency and keep equipment in good condition.
- Contracts should be based on confidence: contractors should aim to produce heat under good maintenance conditions and at a competetive price. The customer should not have to worry about the components of the contractor's price (one does not bother asking for the amount of steel or labor that has gone into the product when shopping for a car).

5.2 Customer Views

Customer views are not easy to determine or to describe. The fact that 88 percent of all public housing managers in France maintain contracts with energy service companies can be seen, however, as an indication that customers are satisfied with the services they receive from these companies. The following are diverse examples of customer views:

- Hard technical work should be done by specialists. An energy service company will hire the best specialists and even educate them. Their people will be better than my own staff, so it's legitimate to contract with highly skilled specialists.
- You simply turn over all the boilers to a skilled company and then, since you were careful to draw up a good contract, you never have to worry about anything: they are responsible.
- It simplifies administration, and with fixed and predictable costs, it removes one issue from budget debates.
- It transfers investment decisions to specialists who are more qualified to undertake necessary improvements.
- Once a problem occurs, the contractor will have to deal with it as fast as possible.
- If tenants are unhappy, I will be protected because the contractor is responsible for meeting the tenants' heating needs.

Some customers, however, end up unhappy for a variety of reasons including poorly carried out contracts, overpayments, and lack of cooperation from building managers who prefer to maintain control over the building mechanical equipment. Unsatisfied customers argue:

- Once they got the contract, they never talked about energy efficiency and never proposed any improvements.
- Their interest is to use as little fuel as possible, since the P1 fuel cost is based on a fixed price. They can achieve energy savings by cutting back on services, and our control over the situation will be limited.
- It will be difficult for us to set up the contracting arrangements.
- If we are directly concerned, we will respond more quickly to heating problems.

In fact, as with any service oriented activities, satisfaction depends on both parties' (the contractor and the customer) involvement in the contracting process.

6 Conclusions

It is important to note that contextual factors enabled the French energy service industry to develop quickly and strongly. Because of this, the French experience in developing an energy service industry is not directly transferrable to other countries today. French energy service companies, however, have developed the know-how and skills to efficiently operate and maintain building heating plants and large heat production facilities, and are beginning to export their specialised knowledge and technologies (like waste-burning facilities) world-wide. These are private, business-oriented companies. But in the early years of their development (1950 to 1970), most of their contracts were fixed term P1 agreements so their profits were directly related to achieving energy efficiency. This experience established the goal of achieving energy efficiency as part of the foundation of the industry. The trend toward metered heat based contracts strengthened the industry committment to energy efficiency by connecting profits to efficiency upstream of the meter. These incentives have been key to promoting new companies and encouraging them to find innovative methods for achieving energy efficiency.

These companies have been deeply involved in the development of new technologies – remote system controls, condensing furnaces, heat recovery, etc. New technologies are essential to the continued ability of energy service companies to offer new services to their customers and stay ahead of increasing customer energy demands.

Today, trends are different in the following ways:

- Most of the traditional market – multi-family and public buildings – for energy service company contracts is saturated.

- Seventy percent of the new buildings constructed in France today are electrically heated. Because electric space-heating systems do not need extensive maintenance to remain efficient, these new buildings do not provide traditional energy efficiency business opportunities for energy service companies.

Consequently, energy service companies today are moving in the following new directions:

- Toward expanded skilled services: Because customers are getting more sophisticated and are asking for more advanced technologies, energy service companies will profit from fitting heating plants with automatic controls and connecting large facilities to remote computer control systems.
- Toward a broader range of services: Monetary savings are being achieved by combining technical efficiency with price efficiency. The adoption of Time-of-Use rates by utilities has opened the market for energy service companies to provide their customers with efficient fuel flexibility. In addition, companies representing large blocks of customers are able to bargain with utilities for better fuel rates. Energy service companies are also able to offer individual metering and billing services to customers in large buildings with collective heating systems.
- Toward new markets: Large commercial buildings and industrial facilities provide natural opportunities for energy service companies to diversify their business. In addition, there are international markets for energy service company skills and know-how.

Acknowledgements

The author would like to acknowledge his employer, COMPAGNIE GENERALE DE CHAUFFE, (the largest French energy service company) and the industry professional organisation SNEC (National Association of Heating and Cooling Equipment Management Enterprises) for providing access to internal documents. Jeff Harris (Lawrence Berkeley Laboratory) also provided data and analysis from his extensive research on French building energy performance contracts. Several analyses were provided by CNET – HLM (French National Center – Union of Public Housing Authorities) and the Agence Française pour la Maitrise de l'Energie (French Agency for Energy Management, or AFME). These reports were of immense help in completing this chapter. The author also acknowledges the constructive criticism and editing of Ms. Deborah Wilson.

Innovative Approaches to Marketing Electric Efficiency

Robert H. Williams

Abstract

While the technical and economic potential for electricity demand reduction through energy efficiency imporovements is large, consumer resistance to the adoption of more energy-efficient products is high, so that conservation programs administered by utilities and other appropriate agencies are needed to realize conservation goals.

Although more than a decade of experience with conservation programs has been accumulated, some of the most promising strategies for implementing conservation involve relatively new ideas for making the pursuit of energy efficiency profitable for producers as well as consumers. Especially promising approaches involve the marketing of energy services by utilties, the competitive bidding for rights to sell "saved electricity" to utilities, and persuading manufacturers and builders to routinely incorporate energy efficiency improvements into their new electricity-using products.

1 Introduction

This paper explores the prospects for formulating effective strategies for exploiting the cost-effective potential for improving the efficiency of electricity use. The argument presented here can be summarized as follows:

- The potential for cost-effective electricity conservation is large.
- There are significant barriers that limit the efficacy of the market as an instrument in capturing this potential.
- Electric utilities are well-suited for playing important roles in implementing conservation programs aimed at overcoming these barriers.

- While utility experience with conservation programs is in such an early stage of development that the best approaches cannot yet be specified with confidence, there is evidence that financial incentives are needed to make conservation programs successful.

- The offering of incentives raises questions as to the implications of such incentives for issues of economic efficiency and equity.

- Such questions could be resolved if effective mechanisms were found for marketing energy services, either by utilities or their affiliates or by independent energy service firms. One promising new approach to marketing energy services involves suppliers bidding in auctions to sell "saved electricity" to utilities. The utilities in turn would sell the energy services produced by winning bidders to the conserving customers involved.

- While promoting the development of an energy service industry would be an effective way of marketing commercially available conservation technologies, broad public policies are also needed to promote a continuing flow of new energy-efficient technologies into the marketplace. If these public policies are framed in ways that persuade manufacturers that energy efficiency will be an important long-term societal goal, the result is likely to be new energy-efficent products that tend to "sell themselves," because producers will routinely emphasize energy efficiency in new products.

- Policies that help make it profitable for utilities or independent energy service companies to market electricity services and for manufacturers to routinely emphasize energy efficiency in new products are especially promising approaches for exploiting the large potential for the more efficient use of electricity, because they would tend to increase the effectiveness of market mechanisms in promoting the efficient use of electricity.

2 The Potential for Electricity Conservation

For the purposes of the present analysis the cost-effective potential for improving the efficiency of electricity use is defined as the result of achieving a mix of electricity supply and investments in more energy-efficient end-use technologies that minimizes the lifecycle cost of the energy services provided by electricity (lighting services, comfortable space conditions, domestic hot water, motive power, etc.).[1] In this "least-cost" approach to electricity planning, external social costs (environmental costs, security risks, etc.) as well as private costs can be taken into account.

The opportunities for making cost-effective improvements in the efficiency of electricity use are large, as indicated by the various analyses focussed on specific end-uses in this volume, as well as by the synthesis analysis showing the implications for the Swedish electric sector of systematically pursuing such opportunities (1). Finding the most effective ways of realizing this conservation potential is an important task for electricity planners.

3 Barriers to Electricity Conservation

While the rational market choice by a consumer among alternative options for providing a given energy service would seem to be the one with the least lifecycle cost, where costs and benefits are discounted at a market rate of interest, this choice is typically not made. Many analyses have been done in recent years which show that the discount rates implicit in actual decisions by consumers (individuals or firms) involving choices among different energy efficiency options are far in excess of market interest rates–often in the range of several tens to a hundred percent or more (3–8).

Perhaps the most unambiguous measurement of the individual consumer's reluctance to invest in energy efficiency was made in a study of purchases of new refrigerator/freezers, carried out for a pair of units offered by a major American retailer (3). These units were identical in every way except first cost and electricity requirements. The more efficient unit used 410 kWh per year less electricity, but its purchase price was $60 more. Sales persons were

[1] There is no theoretically "correct" discount rate that should be used in calculating lifecycle costs. One might choose the social discount rate [e.g. the one used by government bodies to evaluate alternative investment opportunities–a measure typically in the range 6% real (e.g. for Sweden) to 10% (e.g. for the US)], the interest rate the consumer would have to pay to borrow money (highly variable but 10% real is typical), or the opportunity cost for capital the consumer might have available to invest–i.e., the rate of return the consumer might receive on his capital if he invested it elsewhere instead of in energy-saving opportunities. While returns on investment opportunities also vary widely, the average total returns on US common stocks (the Standard and Poor's 500 stock index), corporate bonds, and US Treasury bills averaged 14.3%, 9.8%, and 9.0%, 1975–1985 (2), corresponding (for an investor in a 25% marginal income tax rate bracket) to after-tax returns of 10.7%, 7.4%, and 6.8% respectively, and to real (inflation-corrected) after-tax returns of 3.9%, 0.8%, and 0.3%, respectively. Of course, in comparing alternative investment opportunities consideration has to be given to risk, liquidity, hassle, and other factors.

reportedly instructed to advise customers of the savings advantages of the more efficient unit, and a prominent consumer magazine recommended the more efficient unit and listed the dollar value of monthly electricity savings for two alternative electricity rates. Nevertheless, roughly 2/5 of the consumers surveyed behaved as though they had discount rates above 60%, 1/5 between 35% and 60%, and only 2/5 less than 35%.

There are various, well-known reasons for this reluctance to invest in energy efficiency (8–12): uncertainties about future energy prices; inadequate or unreliable information about opportunities for and the potential dollar savings from investments in energy efficiency; the "hassle" of making such investments; and the inadequate availability of capital resources for financing these investments. The problem of limited capital availability often applies to businesses as well as individuals; in case studies of energy-efficiency investments at 15 firms in five energy-intensive industries in the U.S. it was found that two-thirds of the firms surveyed did not undertake energy efficiency investments with attractive rates of return because capital resources were rationed (8). Also, when the buyer of an energy-using device is not the one who will use it, that person is often reluctant to make an extra investment in energy efficiency. For example, landlords and builders who purchase appliances and furnaces and air conditioners often have no interest in the operating costs which tenants or homebuyers will be shouldering. And finally, for some important energy-using activities (e.g. for commercial buildings in the U.S.) expenditures on energy are still such a small fraction of the total cost of the activity that the consumer's concern about energy efficiency is minor.

It is widely recognized that such problems give rise to electricity demand levels in excess of what they would be if lifecycle costs were minimized and to the econonically inefficient diversion of economic resources to energy supply expansion. Hence utility and government conservation programs aimed at promoting cost-effective conservation measures are worth pursuing.

4 The Institutional Challenges of Promoting Energy Efficiency

Programs organized to promote more efficient use of energy are inherently more difficult to implement than those organized to sell electricity or alternative energy carriers like natural gas or oil. Marketing energy carriers is relatively straightforward. Well-established systems exist for making such transactions. The quantitites exchanged are easy to measure, and both producer and consumer understand the values of the commodities involved.

This is not the case for investments in energy savings. The marketing of energy use efficiency is more complicated. A wide variety of relatively novel marketing problems arise. The individual consumer's energy needs must be diagnosed and the most cost-effective means of serving these needs must be identified. The consumer must be educated as to the technical changes needed–a task made difficult because the expected savings are often ambiguous. Often financing will be required for any new equipment or contractor work that may be required–a problem that arises because energy efficiency improvement usually requires increased first costs. In some instances after-purchase servicing of energy-saving will be required. And it will often be necessary to monitor performance of energy-saving equipment in the field to ascertain actual savings, with feedback that can be used to modify energy-saving strategies as needed in light of this experience.

To make matters worse, the markets for energy services are exceedingly complex and ever-changing. The use of energy throughout the economy is determined by a complex set of factors–including user preferences, incomes, and the resourcefulness and technical know-how of those who produce and sell energy-using technologies. Moreover, the technological options for providing particular energy services are varied, and they change as new technologies are developed. In short, the *patterns of energy use* depend on decisions by large numbers of consumers, each of whom is continually confronted with having to make scores of decisions relating to major energy-using activities. This situation is far more complex than that for the *production of energy,* which generally involves only a few different energy forms and a relatively small number of producers.

Such complexities relating to energy services suggest the importance of framing policies so as to help make the market be more effective in implementing energy efficiency improvements, as purely administrative approaches are likely to have limited efficacy.

5 Electric Utilities as Promoters of Energy Efficiency

Utilities that produce and sell electricity to final consumers offer several advantages as institutions for promoting the more efficient use of electricity. Having long-standing relationships with essentially all users of electricity, utilities offer an infrastructure in which to foster the creation of new business ventures that provide efficient electricity-using equipment and related services. Through these relationships utilities have acquired an intimate understanding of consumers' electrical service requirements–an understanding that could be helpful in formulating effective conservation programs. If they were to market electricity efficiency as well as electricity, utilities could build on these relationships to exploit the economies of scale and scope for the transaction costs associated with the identification, marketing, installation, and servicing of appropriate energy-efficient technology. Also, accustomed to accumulating large quantities of capital and familiar with the energy implications of alternative end-use technologies, utilities are in some ways better positioned to help direct capital resources to conservation investments than institutions that have traditionally had the responsibility for allocating capital resources but have little understanding of the energy-saving benefits of efficient end-use technologies (e.g. banks). And finally, the objective of least-cost planning in the electric sector is perhaps best achieved in an institutional framework where planners understand in detail the complexities of both supply and demand issues relating to electricity.

While electric utilities have some inherent advantages relating to the promotion of energy-efficient end-use technologies, their primary expertise relates to electricity supply expansion–not electricity demand reduction. A shift to least-cost planning, where supply expansion and demand reduction strategies are considered on an equal footing, would represent a major break with established practice for most utilities.

The needed changes will involve more than just a reeducation of utility planners. In many instances electric utilities are not well motivated to promote the more efficient use of electricity, simply because the financial reward system in place for utilities will not automatically lead uitlities to adopt least-cost strategies. As pointed out by David Moskovitz of the Maine Public Utilities Commission (13):

> "...Without significant reforms to the ratemaking system we use, least cost planning [won't go] very far... Least-cost planning won't work in most states because it is inconsistent with the type of economic regulation we now use in the electric utility industry...Our current regulatory regime focuses on the price of electricity. Sound least-cost planning requires that we focus on the customer's total bill rather than on price. This seemingly small difference in focus has enormous effects on a utility's investment and planning decisions..."

Motivating utilities to embrace least-cost planning is essential to the success of conservation efforts. While considerable attention is being given to this problem (13–17), the topic is beyond the scope of the present analysis, in light of the fact that the issues involved vary markedly from one utility context to another. In what follows it is assumed that the problem of motivating utilities can be solved, and emphasis is given instead to the issues associated with the formulation of least-cost planning strategies.

6 What Kinds of Conservation Programs?

Growing numbers of electric utilities, especially in the United States, have initiated programs to promote the more efficient use of electricity through various conservation programs that involve: (i) providing information and advice on investments in energy efficiency; (ii) offering to arrange for contractors to carry out such work; (iii) financing such investments with low or zero interest loans; (iv) providing rebates to consumers for the purchase of energy efficient appliances and/or to appliance dealers for promoting their sales; etc. (17–21).

While to date the level of effort committed to promoting the more efficient use of electricity has been modest in relation to supply-side activities, even for those utilities that have pursued conservation the most aggressively, this situation could well change as a result of the increased attention being given to providing a formal "least-cost planning" framework for utility conservation programs (22–24). The significance of this approach to promoting the more efficient use of electricity is indicated in a least-cost planning order issued in 1986 by the Wisconsin Public Utility Commission for electric utilities in that state (25):

> "This order explicitly adopts a new approach to electric utility planning which it calls "Least-Cost Integrated Planning." This term describes a process in which all reasonable options for both supply and demand are assessed against an array of cost and benefit considerations which are defined as broadly as possible. The uniqueness of this approach is that it does not segregate supply side options, such as generating plant or transmission lines, from the demand-side options, such as energy conservation. Instead it seeks to evaluate all options on an integrated equivalent basis. The approach is also unique in the breadth and comprehensiveness of the costs and benefits to be used as measures. In addition to quantifiable economic costs and benefits, the options are evaluated against as many aspects of social and environmental values as possible. The breadth of scope of this analysis requires considerable creative thought and new ways of looking at utility systems..."

As of the end of 1987 at least ten states in the U.S. had adopted a least-cost planning process, six others had major portions of least-cost planning procedures in place, and eight other jurisdictions were actively considering it (15).

While least-cost planning initiatives represent major public policy advances, such initiatives are just the first steps toward the implementation of least-cost planning. There is not yet enough experience with conservation programs to spell out the optimal combination of programs needed to achieve least-cost planning goals. This was in fact recognized by the Wisconsin Public Utility Commission in promulgating its least-cost planning order (25):

> "...Unlike "classical" utility supply side planning, least-cost integrated planning is still in its developmental stages, and must be carefully nurtured if it is to thrive..."

A priori one might think that what is needed is simply to provide consumers with better information about available energy-saving opportunities, because by investing in energy efficiency improvements the consumer would typically be able to save money. But this does not seem to be the case, as indicated by the growing body of field experience with conservation programs. Some examples:

- The Bonneville Power Administration (BPA) has conducted experiments involving the promotion of solar and heat pump water heaters as alternatives to electric resistance water heaters, to understand better the relative importance of promotional efforts involving improving the flow of information (high and low promotional levels) and those involving incentives (high and low levels) (26). The BPA found for this program that, while making better information available can help increase sales, stronger incentives are even more important. The BPA found that sales increased by an order of magnitude as incentives were increased from low to high, for both low and high promotional efforts (Table 1).

Table 1. The impacts of incentive and promotional efforts in the BPA Solar & Heat Pump Water Heater Program.

Market Test Cell		Number of Eligible Households (Thousands)	Sales per 10,000 Households
Incentive Level	Promotional Level		
Low	Low	20.5	1.9
Low	High	174.8	3.3
High	Low	34.7	11.0
High	High	77.5	41.2

Source: Reference 26.

- A survey of appliance dealers' sales data showed that in the New York State Electricity and Gas Corporation's program aimed at increasing sales of more efficient refrigerators, sales of efficient models increased by 40.9% in the district where $50 rebates were offered, by 35.5% where $35 rebates were offered, and by only 22.7% where the utility's effort was limited to advertising only (27).

These experiences are not atypical. In fact, *there is considerable empirical evidence suggesting that simply improving the flow of information to consumers will not be adequate to achieve energy conservation goals* and that financial incentives are important for the success of utility conservation programs. Utilities in the U.S. have experience with a variety of incentive measures, including low-interest loans, rebates for the purchase of energy-efficient products, and installation by the utility of energy-saving measures on the customer's premises at zero or low cost to the customer. The relative merits of these traditional approaches to incentives have been reviewed by Geller (17). Here the focus is instead on the policy issues raised by utility-based incentives programs.

7 Incentives, Equity, and Economic Efficiency

What levels of incentives should be offered to promote the adoption of cost-effective conservation measures? Lovins has argued that:

> "... a kilowatt-hour saved is just like a kilowatt-hour generated...so that they should be treated alike...,"

and, accordingly, has suggested that utilities be willing to "buy" efficiency improvements, beginning with the least costly ones and continuing until the incentive offered for the last purchase equals the marginal cost of supply (28). But this approach to incentives is problematic, as indicated by the example shown in Table 2, Case A. Here a hypothetical investment in energy efficiency would cost society or the utility 4.0 cents per kWh. This would be socially desirable because it would cost less than the marginal cost of supply, some 4.8 cents per kWh. An incentive would be necessary to realize the societal benefit involved, because at the consumer's high discount rate the investment would be unattractive–costing 5.6 cents per kWh, which is more than the price paid by the consumer for electricity, some 5.0 cents per kWh. But if the utility were to pay for the entire cost of this energy efficiency investment, it would suffer a substantial net loss: while it would avoid the cost of having to provide extra supply (4.8 cents per kWh), this gain would be more than offset by the cost of the conservation measure (4.0 cents per kWh) plus the loss of revenue (5.0 cents per kWh). This utility loss would have to be offset either by increased electrical rates or, in the case of privately owned utilities, by reduced profits for stockholders, or both. This approach to incentives raises questions concerning both equity and economic efficiency.

The impact of utility incentives programs on nonparticipating ratepayers has been a source of debate since the earliest days of conservation programs in the U.S. One approach to this equity issue was adopted in the late 1970s by Oregon utilities that financed conservation investments for residential electric heating customers. They were allowed, by the state's Public Utility Commission, to put these investments in their rate bases. The criterion specified for judging how much the utility should invest was an amount less than the difference between the cost of new electricity supply and the average cost or price of electricity (30). This criterion has come to be called the "no-losers test," because when the incentive paid by the utility satisfies this criterion, electricity rates would go up no more than if the equivalent amount of supply were added to the system, so that non-participants in the conservation programs would not be any worse off than if the programs were not implemented.

Recently, Joskow (31) and Ruff (32) have argued that an incentive satisying the no-losers test would also promote the optimal amount of investment in energy efficiency, because it would tell the consumer that the societal value of the conservation is the marginal cost of new supply. They point out that for each kWh saved by investing in conservation the consumer would gain a benefit equal to:

(MC – P) (incentive payment) + P (reduction in the electricity bill) = MC,

where MC is the marginal cost of electricity and P is its price. Thus the consumer would get a windfall for making conservation investments that cost less than MC but would be deterred from making investments that cost more than MC. If instead the utility incentive were as high as MC, the consumer benefit from conservation would be as high as MC + P, which could lead the consumer to invest in energy efficiency measures more costly than MC and thus to levels of energy efficiency that are not optimal.

But there are problems with the no-losers test. In many areas today the difference MC – P is small or even negative and may not lead to incentive levels adequate to promote conservation investments that are cost-effective from society's perspective. For example, adherence to the no-losers test would make it impossible to provide a utility incentive for situations like that indicated in Table 2. What Joskow and Ruff have overlooked in their analysis is the high consumer discount rate problem, which inhibits socially desirable investments in energy efficiency, as illustrated by this example.

Table 2. An example of conservation that is in society's interest but is not in the interest of any single party who has to bear the entire cost.[a,b]

Case A: Benefits (cents per kWh saved) when the utility sells electricity

	Benefit to Utility (Utility Pays for Conservation)	Benefit to Conserver (Conserver Pays for Conservation)	Benefit to Society (Regardless of Who Pays)
Savings[c]	4.8	5.0	4.8
Conservation Cost	–4.0	–5.6	–4.0
Revenue Loss[d]	–5.0	0.0	0.0
Net Benefit	–4.2	-0.6	0.8

Case B: Benefits when the utility pays for conservaion, sells electricity services, and shares savings with the conserver

	Benefit to Utility (Utility Pays for Conservation)	Benefit to Conserver (Conserver Buys Electricity Services)	Benefit to Society
Savings[e]	4.4	0.4	4.8
Conservation Cost	–4.0	0.0	–4.0
Revenue Loss[f]	0.0	0.0	0.0
Net Benefit	0.4	0.4	0.8

a Adapted from Reference 29.

b Here it is assumed that:
Price of electricity = 5 cents per kWh
Marginal cost of electricity = 4.8 cents per kWh
Marginal cost of conservation
For utility and society = 4.0 cents per kWh
For conserver with high discount rate = 5.6 cents per kWh

c The savings to the utility (society) is the marginal cost of not having to produce a kWh of electricity; that to conservers the price of a kWh saved.

d The revenue loss its the revenue that a utility would not receive when a kWh of electricity is saved.

e Here the utility shares the net benefit of not having to produce extra electricity (4.8 – 4.0 cents per kWh) equally with the conserver as an incentive to get the conserver to participate in the conservation program.

f There is no utility revenue loss because the conserver pays the utility for electricity services what he would otherwise pay for electricity.

Recognizing that the no-losers test can dilute conservation incentives to levels considerably below the value to society of the potential energy savings, the Northwest Power Planning Council, in formulating its Regional Conservation and Electric Power Plan in 1983[2] explicitly rejected the no-losers test as a criterion for determining incentives for conservation programs (33):

> "It is the Council's intention that incentives be self-inplementing. Those who respond should benefit. Incentives should not be diluted simply to protect against rate impacts on those who do not respond."

The Public Service Commisssion of Wisconsin took a similar position in promulgating its least-cost planning order in 1986 and went further, arguing that the equity issues arising in incentive programs should be dealt with by making incentive programs broadly available (25):

> "Demand-side options will have an impact on electric rates because they reduce the number of billing units over which some fixed costs are spread. Higher near-term rates will increase bills for non-participating customers. In the long run, demand-side programs reduce the need for new expensive facilities and thus will reduce the the long-run combined average bills for participating and non-participating customers. Some individual non-participating customers bills may increase in the long run. The commission finds that the negative impacts of short-run higher rates (commonly referred to as the "non-participant" or "no-losers" test) should not be used to restrict development of demand-side programs. Rather the negative impacts should be minimized to the extent practicable by offering a broad range of programs to each customer sector. All customers should have the opportunity to receive lower utility bills even though rates might increase in the near term..."

Of course, in practice there will always be some utility customers that will not benefit from utility conservation programs and will have higher utility bills if the incentives offered violate the no-losers test–new firms with state-of-the-art equipment and buildings that do not need to make any energy-saving retrofits, those who made needed retrofit conservation investments before the programs were initiated, those that do not qualify for such programs for technical

[2] The Northwest Power Planning Council (an interstate compact agency made up of two members from each of the states of Idaho, Montana, Oregon, and Washington) was created in 1981 in accordance with the Pacific Northwest Electric Power Planning and Conservation Act of 1980 (Public Law 96–501), to prepare, adopt, and transmit to the Bonneville Power Administration a regional conservation and electric power plan by April 1983. The plan is being implemented by Bonneville, the region's utilities, state and local government, and the private sector.

reasons, and those whose needs are not otherwise well-matched to the incentives.

While the equity issue might be a cause for concern in principle, it would probably not become a serious practical problem, in terms of the adverse impact on rates, until the aggregate level of incentives becomes a significant percentage of total utility revenues. Even utilities that have aggressively pursued conservation programs in the U.S. have spent no more than a few percent of gross revenues on conservation programs[3] And as shown below there are interesting possibilities for motivating conservation investments that do not require large net utility expenditures.

8 Utilities as Energy Services Companies?

Consumers fail to adopt conservation measures to the socially desirable level in large part because there is no well-established industry for marketing energy services like lighting, heating, cooling, and motive power. To the extent that utilities could act to help nurture the development of such an industry, these barriers would be reduced.

One possibility would be for the utility itself to become an energy service company. As such, the utility would make the needed investments in conservation at the customers' premises and charge for the energy services provided an amount per kWh saved that is equal to what the customers would otherwise pay for electricity for the same services. Routinely adding such charges per kWh saved to the customers' utility bills (over a billing period comparable to the expected life of the investment) would convert the extra costs of the conservation measures into operating costs for the customers, thereby overcoming the typical customers' reluctance to incur the extra first cost required of many conservation measures. Doing this alone, however, would probably not motivate many customers to participate in these programs, because they would be no better off than if the conservation measures were not installed. And they could be worse off, if estimated savings was used as the basis for the contractual agreement and actual savings turned out to be less. However, as argued by Cicchetti and Curkendall (29), if the utility's cost of saved energy were less than the marginal cost of supply, the utility could offer to share the savings with the customers. So doing would often make the cost-

[3] In the early 1980s the Pacific Gas and Electric Company's energy conservation programs were among the most aggressive utility conservation programs in the U.S., totalling $115 million, $233 million, and $272 million in 1981, 1982, and 1983, respectively. But as a percentage of gross revenues these expenditures were modest–averaging 2%, 3%, and 3.5% in 1981, 1982, and 1983, respectively (18).

effective conservation investments attractive to both participating customers and the utility, without hurting non-participants with increased rates, even in cases where the marginal cost is less than the price of electricity (Table 2, Case B).

If adopted, the Cicchetti-Curkendall proposal would transform utilities from being simply purveyors of electricity supplies into being purveyors of a broad range of energy services, and they might eventually come to play a role similar to that originally envisioned for them by Thomas Edison when he invented the incandescent bulb. Edison proposed that utilities sell *illumination*, thereby giving them a financial interest to provide this illumination in the most cost-effective way. The basic idea is generating some interest. As of 1986 some nine utilities in the U.S. had formed energy services subsidiaries, and one utility holding company had formed a joint venture with an energy service company (34).

But would utility-based energy service companies be able to provide energy services in the most cost-effective ways available? Would they be the best equipped to take advantage of the continuing flow of new energy-saving technologies onto the market? Or, put another way, would they be able to compete in energy service markets?

To answer such questions it is necessary to characterize the competition. So doing is not easy, because despite the facts that this is potentially a rapid growth industry for many countries, and France has fifty years of experience with non-utility heating services companies (35), experience with energy service companies is still quite limited. Nevertheless, it is possible to make some preliminary observations about the nonutility competition.

In the U.S., high energy prices of the late 1970s and early 1980s led to the formation of numerous energy service companies catering to the energy needs of industrial and large commercial firms. Until recently the major advantages offered by energy service companies in the U.S. have often been financial; many of these companies have emphasized the financing of conservation investments through leasing, joint ventures, shared savings plans, and energy services contracts–by making capital more readily available than would be possible from traditional capital sources through so-called "third-party" financing schemes (36). Energy service companies that are utility subidiaries should theoretically be able to compete effectively with such independent companies, in light their ready access to capital resources. However, it has been argued, for the U.S. context at least, that this advantege might not be significant in practice, because federal anti-trust laws prohibit cross-subsidization by the utility parent, so that the subsidiaries have to stand on their own as viable businesses (34).

Perhaps of greater significance is the emergence of a new generation of technology-oriented energy service companies. Specifically, one manifestation of the growing importance of microprocessor-based control technology for a wide range of business activities is an ongoing boom for energy service companies that originate in the controls industry (37). It will be difficult for utilities to match the energy services expertise offered by such companies.

One of the major reasons for the emergence of the controls industry in energy service markets is that the technology for cost-effectively meeting the energy service requirements of large buildings and factories is becoming increasingly sophisticated, and it is changing rapidly. In the present environment, where businesses must give continual attention to their competitive positions and where the public sector operates with considerable pressure to bring costs under control, some building and factory operators are beginning to recognize that they cannot ignore the new technological opportunities for making their energy systems more energy-efficient, as well as reliable. But the new technologies for these building/factory energy systems–for diagnostics, maintenance, and repair, as well as routine energy system operation–are becoming more and more sophisticated. Either building/factory energy system operators must build up more heavily-equipped, inhouse expertise, requiring frequently updated training, or they must turn instead to outside expertise. Many are beginning to choose the latter course. And the large controls companies themselves (e.g., Johnson Controls, Honeywell) are proving to be strong competitors in these new technology-oriented energy service companies, largely because they are systems-oriented, in the business of delivering comfort and convenience, not just energy carriers like chilled water or steam. With a systems orientation, these controls companies are creating roles for themselves as providers of a comprehensive set of energy services. Typically these new energy service companies offer a wide range of expertise, involving the design, manufacture, and installation of controls, and an array of complementary services aimed at providing efficient, effective operation of the integrated system. The services offered tend to include professional audits of energy service requirements, full-time facility operation, scheduled preventive maintenance, financing equipment retrofits, etc. (37).

9 Competitive Bidding for Conservation?

While it may be difficult for utility-based energy service companies to compete in some sectors of the emerging energy services industry, utilities can still play important roles in hastening the growth of and in shaping this important emerging industry, while promoting least-cost planning.

Least-cost utility planning might be implemented by engaging both independent power producers and energy service companies (or individual energy-using companies) in utility-sponsored auctions, where they would compete to provide the electricity service needs of utility customers (Figure 1). In one version of demand-side bidding, proposed by Lovins (28), winning bidders would install energy-efficient equipment and offer to sell the electricity saved to the utility for a bid price.

At the time of this writing, the thinking about auctions involving independent power producers is much further advanced than that for auctions involving energy service companies. Several auctions involving independent power producers have been held in different parts of the U.S. Moreover, in the spring of 1988 the Federal Energy Regulatory Commission proposed federal rules that, if adopted, would create the option of running auctions as a means of bringing new electricity supplies to market in the U.S.[4]

But there have also been some small auctions leading to utility purchases of saved electricity (38, 39). For example, as of the spring of 1988 the New England Electric System (NEES) [through its operating company in Massachusetts, the Massachusetts Electric Company (MECO)] had conducted two auctions and purchased 13.6 MW in demand side bids from among its large commercial and industrial customer base. The process involved bidding to provide energy services to a specified group of customers by 120 energy service companies (ESCs) that had registered with the utility . The program was limited to customers with connected loads greater than 500 kW, organized into groups of 25 to 50 customers by geographic area. The only criterion limiting bids was the requirement that the bid price + the utility's adminstrative cost be less than the utility's avoided cost. Francises were awarded to five winning bidders. One-time cash payments were made for the winning bids as soon as the measures were installed and inspected by the utility. NEES choose to base bid contracts on engineering estimates of the

[4] The proposed rules are Regulations Governing Bidding Programs (Docket No. RM88–5, March 16, 1988), which would authorize state regulatory authorities and nonregulated electric utilities to implement bidding procedures as a means of establishing rates for power purchases by utilities from cogenerators and small power producers that qualify for such sales under the Public Utility Regulatory Policies Act of 1978 (PURPA), and Regulations Governing Independent Power Producers (Docket No. RM88–4, March 16, 1988), which would relax regulations of those independent power producers that are not qualifying facilities under PURPA. These proposed rules are a consequence of the success of the provisions of PURPA that promote cogeneration (the combined production of heat and power) and small power producers (small-scale power producers who use renewable energy sources). PURPA requires an electric utility to purchase electricity from qualifying cogenerators and small power producers at a price equal to the cost the utility could avoid by not having to provide that electricity. Because the avoided cost provision created windfalls for many cogenerators who could produce electricty at much less cost than utilities, the law has been enormously successful, leading to the creation of a powerful cogeneration industry (40). Now that the PURPA goal has been realized federal policy-makers are seeking ways to make cogenerators and other independent power producers more economically efficient by promoting competition in power generation.

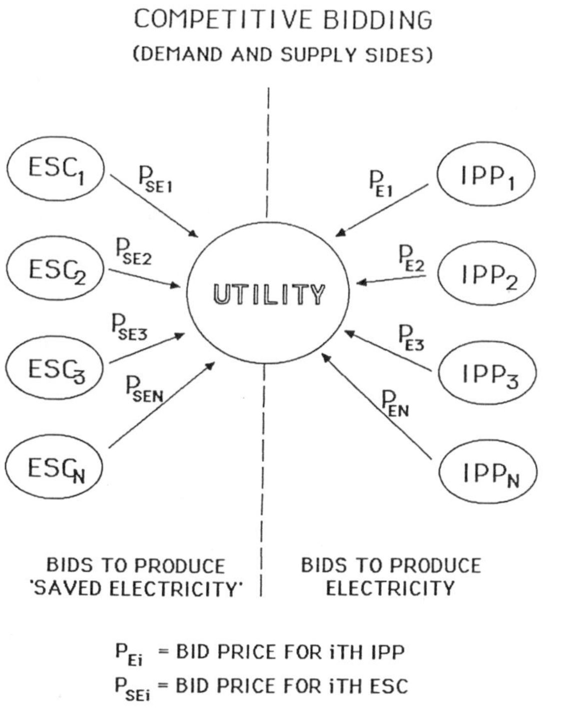

Figure 1. Least-cost utility planning through utility-sponsored auctions in which independent power producers (IPPs) bid to sell the utility electricity and energy service companies (ESCs) bid to sell the utility saved electricity.

savings. For some measures standardized engineering estimates of savings were made. For others (e.g. variable speed motor drives), site-specific estimates were required. To reduce the uncertainty of the savings estimates, the NEES program emphasized a selected a specific set of measures for which they had extensive load databases. NEES also required bidders to provide some kind of security as a guarantee that the benefits estimated in bids would be achieved by successful bidders within a specified period.

Bidding for conservation, as proposed by Lovins (28) and exemplified by the NEES experiment (38,39), has been sharply criticized by Joskow (31) and Ruff (32), on the grounds that bid payments which violate the no-losers test could lead to economically inefficient outcomes and to adverse effects on nonparticipating utility customers through the rate increases involved. They correctly point out that the winning bidders would be paid twice for the services rendered: once through the bid payment and again through the reduced utility bills. They recommend as an alternative to competitive bidding that utilities offer subsidies for conservation investments in the amount per kWh saved equal to the difference between the marginal cost and the price of electricity, if marginal costs are higher than prices.

The problems raised by Joskow and Ruff have been addressed in an analysis by Cicchetti and Hogan (41), who argue that the solution to these problems should not be to reject competitive bidding for conservation but rather to modify the competitive bidding format, as originally proposed. Cicchetti and Hogan propose to supplement the Lovins proposal with the specification that, after purchasing saved electricity, the utility sell the energy services provided by the conservation investment to the customer getting the energy- saving equipment. The price of this energy service per kWh saved would be equal to what the customer would otherwise pay per kWh for the extra electricity needed to provide the same service without the benefit of the conservation investment. In effect the Cicchetti-Hogan proposal combines the Lovins competitive bidding idea (28) with the Cicchetti-Curkendall idea that utilities sell energy services (29). The energy service company involved would probably offer some of its windfall profits (the difference between the winning bid price and the cost of saved energy) as an incentive to get the customer to participate in this arrangement (Figure 2), just as the utility would share savings with the participating customer under the Cicchetti-Curkendall proposal.

In Table 3 both the total costs to society and impacts on utility rates are indicated for alternative ways of meeting the incremental demand for electricity requirements and for four different orderings of the electricity price P, the marginal cost new electricity supplies MC, and the conservation cost CC:

Case A: CC < P < MC
Case B: P < CC < MC
Case C: P < MC < CC
Case D: CC < MC < P

This table shows that *in all cases the Cicchetti-Hogan proposal would lead to both the least costly solution for society and the lowest utility rates*. Under the Joskow-Ruff scheme, costs to society would also be minimized, as long as the marginal electricity cost is higher than the electricity price, but this scheme provides no incentives for cost-effective conservation when the marginal supply cost is less than the electricity price. Also, under the Joskow-Ruff scheme electricity rates would never be lower than if energy services were provided instead through supply expansion, while rates would sometimes be lower with the Cicchetti-Hogan scheme. The Lovins scheme would be the least costly solution, as long as the conservation cost is lower than the marginal cost of supplies. But because the bidder would be paid twice

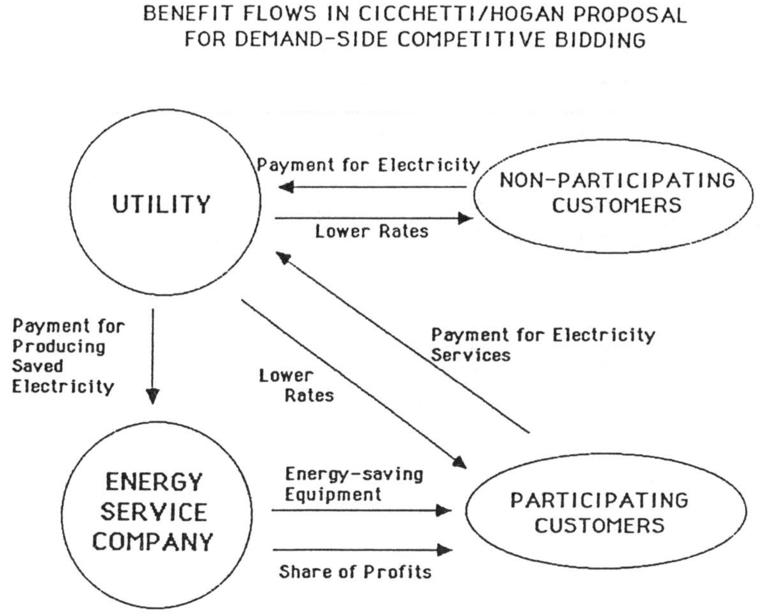

Figure 2. Benefit flows in the Cicchetti-Hogan proposal for demand-side competitive bidding.

under the Lovins scheme, it would still be worthwhile for the bidder to produce some conservation that is more costly than new supplies. Also, rates would always be higher with the Lovins scheme than with the alternative schemes.

The Cicchetti-Hogan competitive bidding scheme would also address the concern raised by the Cicchetti-Curkendall proposal as to whether utility-based energy service companies would be able to provide energy-saving services in the least costly manner. With competitive bidding, the market decides this question. Thus, if practical difficulties could be resolved, the Cicchetti-Hogan competitive bidding proposal would be an elegant market-oriented approach to implementing least-cost energy planning.

One such practical problem is the challenge of devising a satisfactory scheme for determining energy savings, through some combination of energy estimation and measurement, to the satisfaction of all parties involved. While having an effective method of determining energy savings is important to the success of any conservation program, it is crucial for programs that involve monetary payments for savings. Fortunately, remarkable progress is being made in the development of low-cost reliable "scorekeeping" techniques for measuring energy saved in conservation programs (42).

Table 3. Alternative schemes for meeting incremental demand for electricity services with alternative costs for conservation and new supply.[a]

	Electricity Generated (kWh)	Sales of Electricity Services (kWh)	Conservation Payment (cents/kWh)	Revenue Requirement (dollars)	Average Rate (cents/kWh)	Total Cost to Society (dollars)
Case A: cost of saved electricity = 4 cents/kWh; marginal cost = 6 cents/kWh						
Initial State	1000	1000	0	50	5.00	50
New Supply	1100	1100	0	56	5.09	56
Joskow-Ruff[b]	1000	1000	0.9	50.9	5.09	54
Lovins High[c]	1000	1000	6	56	5.60	54
Lovins Low[c]	1000	1000	4	54	5.40	54
Cicchetti Hogan High[d]	1000	1100	6	56	5.09	54
Cicchetti Hogan Low[d]	1000	1100	4	54	4.91	54
Case B: cost of saved electricity = 5.5 cents/kWh; marginal cost = 6 cents/kWh						
Initial State	1000	1000	0	50	5.00	50
New Supply	1100	1100	0	56	5.09	56
Joskow-Ruff[b]	1000	1000	0.9	50.9	5.09	55.5
Lovins High[c]	1000	1000	6	56	5.60	55.5
Lovins Lowc	1000	1000	5.5	55.5	5.50	55.5
Cicchetti Hogan High[d]	1000	1100	6	56	5.09	55.5
Cicchetti Hogan Low[d]	1000	1100	5.5	55.5	5.05	55.5
Case C: cost of saved electricity = 6.5 cents/kWh; marginal cost = 6 cents/kWh						
Initial State	1000	1000	0	50	5.00	50
New Supply	1100	1100	0	56	5.09	56
Joskow-Ruff[b]	1000	1100	0	56	5.09	56
Lovins High[c]	1000	1000	6	56	5.60	56.5
Lovins Low[c]	1000	1000	6	56	5.60	56.5
Cicchetti Hogan High[d]	1000	1100	0	56	5.09	56
Cicchetti Hogan Low[d]	1000	1100	0	56	5.09	56

	Electricity Generated (kWh)	Sales of Electricity Services (kWh)	Conservation Payment (cents/kWh)	Revenue Requirement (dollars)	Average Rate (cents/kWh)	Total Cost to Society (dollars)
Case D: cost of saved electricity = 4 cents/kWh; marginal cost = 4.5 cents/kWh						
Initial State	1000	1000	0	50	5.00	50
New Supply	1100	1100	0	54.5	4.95	54.5
Joskow-Ruff[b]	1000	1100	0	54.5	4.95	54.5
Lovins High[c]	1000	1000	4.5	54.5	5.45	54
Lovins Low[c]	1000	1000	4	54	5.40	54
Cicchetti Hogan High[d]	1000	1100	4.5	54.5	4.95	54
Cicchetti Hogan Low[d]	1000	1100	4	54	4.91	54

a The initial electricity load is assumed to be 1000 kWh, for which the cost is assumed to be 5 cents per kWh. An incremental load of 100 kWh is then met either with new electricity supply or with an investment in conservation.

b Under the Joskow-Ruff scheme the utility would pay the conserver a price for saved energy equal to the difference between the marginal cost and the average cost of electricity to the utility. The utility would charge the conserver only for the electricity purchased.

c Under the Lovins competitive bidding scheme the utility would pay the winning bidder selling saved electricity a price that is greater than or equal to the cost of saved electricity and less than or equal to the cost of new supply. The utility would charge the conserver only for the electricity purchased.

d Under the Cicchetti-Hogan competitive bidding scheme the utility would pay the winning bidder selling saved electricity a bid price that is greater than or equal to the cost of saved electricity and less than or equal to the cost of new supply. The utility would charge the conserver for electricity services.

A related question concerns the extent to which this approach can be relied on to meet conservation goals. In principle, demand-side competitive bidding can be applied to all electricity-using activities. So far, however, auctions have been limited to relatively large-scale electricity users and to a limited set of end-use activities for which the savings can be determined without great difficulty through some combination of engineering estimates and field measurements (38,39). Until energy service companies becomes well enough established to be able to exploit the scale economies in information and other transaction costs, and until reliable, low-cost approaches for determining saved electricity under a wide range of circumstances can be developed, it is likely that competitive bidding will not be a practical approach for promoting electric efficiency for all important electricity-using markets.

Until such practical problems of demand-side bidding can be overcome alternative utility conservation programs will be needed for promoting energy efficiency in the residential and small commercial sectors and for end-use activities for which the determination of savings is not relatively straightforward. The important markets that are not covered by competitive bidding might be covered with more conventional programs (e.g. rebates). Or utility-based energy service companies (as proposed by Cicchetti and Curkendall) might be required to serve them.

10 Toward Energy-Efficient Products that Sell Themselves

The energy service company and demand-side competitive bidding concepts hold forth the promise of being effective mechanisms for marketing energy-efficient technologies that are already commercially available. Consideration should also be given to strategies for promoting a continuing flow of new energy-efficient products to the marketplace.

New energy-efficient technologies warrant focussed attention because otherwise the concerns about the externalities posed by various supplies that partially motivate conservation efforts are likely to increase. Concerns about the global security risks associated with the growing dependence on the politically unstable Middle East will force oil-importing countries to make increasing efforts to reduce oil imports. Concerns about the greenhouse warming will dictate substantial cutbacks in fossil fuel dependence in the decades ahead. And concerns about the nuclear power/nuclear weapons proliferation link, as well as nuclear accidents, will probably limit the global role of nuclear power.

The policies needed to encourage the development and commercialization of new energy-efficient technologies extend well beyond the responsibilies of most utilities. In light of the inherent weakness of market forces in motivating the private sector to conduct research and development (R&D) in areas where the payoff is long-term, where only generic information is created, and where the payoff involves only broad societal benefits, government support for energy efficiency R&D is needed and can often lead to societal benefits that dwarf the costs involved (43). In addition oil taxes and carbon taxes levied by governments to reduce oil and fossil fuel dependence would encourage producers to emphasize energy efficiency in their new products, as would energy performance regulations or "energy-guzzler taxes" on energy-using equipment.

There is already considerable experience with the regulation of energy performance. The power of this policy instrument is illustrated by the impact of automotive fuel economy standards mandated for new cars by the U.S.

Congress in 1975. These standards contributed to the 66% increase in the fuel economy of new cars in the U.S. between 1975 and 1985. If instead U.S. cars in 1985 had 1975 efficiencies, total 1985 oil consumption by all OECD countries would have been 1.6% less than in 1975, instead of the actual level, which was 8.2% lower. The U.S. auto fuel economy standards thus made a significant contribution in the early 1980s to the softening of the world oil price and the easing of world tensions arising from reduced dependency on Middle East oil. Similarly, Swedish building standards have had a powerful impact in forcing the introduction of more energy-efficient building technologies in Sweden (44), and the national appliance efficiency standards that became law in the U.S. in 1987 are expected to have a major impact on the introduction of new energy-efficient appliances in the U.S. in the years ahead (45).

How seriously producers regard the overall policy promoting continuing improvements in energy efficiency in new products will be at least as important as the detailed features of particular measures. If the overall policy convinces manufacturers and builders that energy efficiency will be an important long-term societal objective, they will tend to respond by routinely incorporating energy-efficiency improvements into new products in ways that will automatically attract consumer interest, without much prodding.

To better understand this possibility, it is worthwhile reexamining the high implicit consumer discount rate problem relating to purchases of energy-efficient technologies. While the various reasons for the high implicit discount rate phenomenon mentioned earlier are valid, at a deeper level, the reluctance to invest in energy efficiency may be rooted in the fact that *energy cost is just one of many factors involved in consumer decisions relating to purchases of energy-using buildings or equipment* and often not a very important consideration.

Some insights about how to give considerations of energy efficiency more prominence in consumer decisions can be gleaned from a review of experience in the basic materials processing industries, where improvements in energy efficiency have been routinely incorporated into new product and processes throughout the history of the industrial era. Table 4 shows that for a wide range of basic materials in the United States, the long-term trend has been toward a declining energy intensity of production–a phenomenon that has persisted through periods of stable and declining as well as rising energy prices. An understanding of the process of technological innovation helps to explain this trend.

Economist Robert Solow showed in a 1957 study (46) that innovations in manufacturing in the US economy in the period 1909–1949 tended to be "neutral" with respect to impacts on labor and capital productivity (i.e. innovations tended to improve capital and labor productivities by comparable amounts). Solow's findings were later stated more broadly by Charles Berg

(47), whose case studies of the process of technological innovation led him to conclude that, out of the large number of new inventions that are continually offered, those new processes/products which "make it" in the marketplace tend to be ones that lead to improvements of several attributes over what is offered by existing processes/products. Considering that technological innovation is such a disruptive process–creating new uncertainties regarding costs and benefits for the potential purchaser and threatening many interests vested in perpetuating the status quo–it is quite reasonable that those technologies "making it" tend to be ones that are so superior to existing technologies that potential purchasers cannot afford *not* to adopt them.

In the basic materials processing industries (primary metals, basic chemicals, cement, glass, paper, etc.) successful innovations have tended to be ones that improve energy performance along with other attributes, even in times of constant or declining energy prices, because for these industries energy costs

Table 4. Historical energy requirements per unit of output for selected materials produced in the U.S.

Soda Ash[a] (Solvay Process)		Ammonia[a] (Haber-Bosch Process)		Chlorine[a] (Diaphragm Cells)		Raw Steel[b,c] (Changing Process Mix)	
Date	Energy (GJ/tonne)	Date	Energy (GJ/tonne)	Date	Electricity (kWh/tonne)	Date	Energy[d] (GJ/tonne)
1868	60	1917	93.0	1916	4400	1947	37.5
1894	31	1923–50	81.0	1947–73	3300	1954	32.4
1911	28	1965	52.0	1980	2400	1962	30.0
1925	17	1972	46.5			1971	27.8
1942	15	1978	41.2			1980	27.0
1970	14						

Hydraulic Cement[b,e] (Wet and Dry Processes)		Ethylene Dichloride[a]		Ethylene Oxide[a]		Polyethelene[a]	
Date	Energy[d] (GJ/tonne)	Date	Index	Date	Index	Date	Index
1947	10.3	1967	100	1970	100	1956	100
1955	9.0	1973	15	1973	85	1973	40
1960	8.5			1974	79	1974	18
1965	8.2						
1971	7.5						
1978	6.8						

a Source: Reference 48.

b Data for all all but the last entry are from Reference 49.

c The 1980 data for steel are from Reference 50.

d Electricity requirements are counted @ 3.6 MJ per kWh.

e The 1978 data for cement are Reference 51.

have always accounted for a large fraction of total production costs. An example of this phenomenon is the introduction of the float-glass process, which has captured most of the market for flat glass-making. Previously, the manufacture of high-quality flat glass involved extensive grinding and polishing, which consumed 10 to 20 percent of the glass. The elimination of these steps not only reduced costs but saved both direct energy needs for power and indirect energy needs for making the glass that was ground into waste.

In those parts of the economy outside the basic materials processing industries, where historically the cost of energy has not been a large part of overall costs, reduced energy intensity has not tended to be an important characteristic of many new products. Up until the early 1970s, in the era of declining real electricity prices in most parts of the world, there was often little incentive for producers to make or consumers to buy less energy-intensive equipment. The 2 1/2-fold increase in the electricity requirements for new refrigerators in the United States in the period between the end of World War II and the energy crises of the early 1970s (Figure 3) is illustrative of

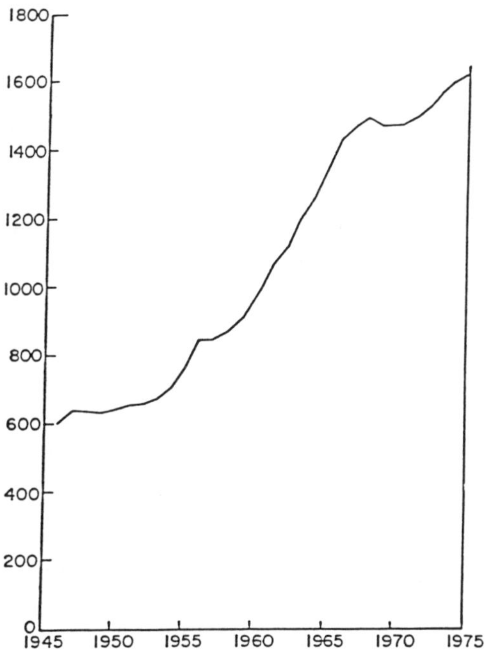

Figure 3. The historical trend in the shipment-weighted average electricity consumption (in kWh per year) for new refrigerators and refrigerator/freezers in the U.S.
Source: Reference 52.

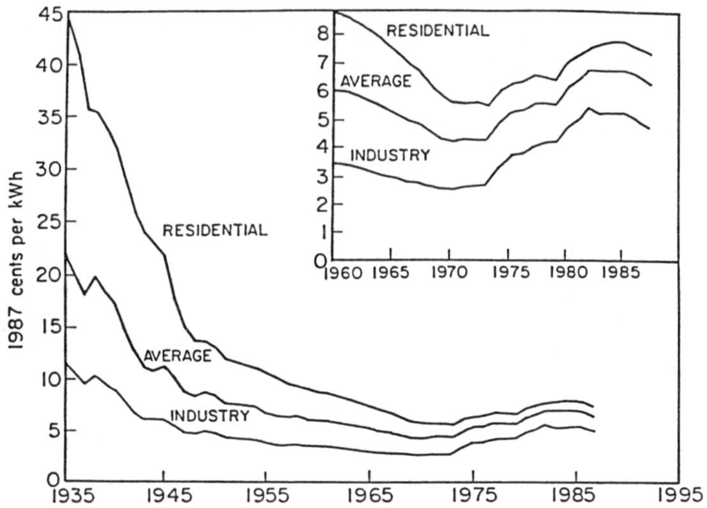

Figure 4. Long-term and recent (insert) electricity price trends in the U.S. In constructing these graphs nominal prices were converted into constant 1987 cents per kWh using the gross national product deflator.

trends for various such technologies; in this period the capacity of refrigerators grew considerably, as did the number of new features, many of which included the addition of electric heaters for purposes such as butter warming and condensation inhibition.

The halting, and in some parts of the world the reversal, of the long-term downward trend in electricity price (Figure 4) has led to greater emphasis on energy efficiency in new products. And while the high consumer discount rate problem shows that in general consumers are not yet readily adopting new more energy-efficient products that are cost-effective, there is some evidence of the beginning of a trend toward new energy-efficient products that offer multiple benefits.

The technological evolution of the compact fluorescent bulb (Figure 5), a version of the fluorescent bulb that can be screwed into incandescent sockets (53), illustrates this trend. The initial "circline" compact fluorescent introduced a decade ago was energy efficient but had little overall consumer appeal: it was bulky (weighing nearly half a kilogram and measuring some 20 centimeters across) so that it would not fit into many incandescent lamp holders; its harsh white light was considered inappropriate for many household applications; and finally many consumers accustomed to the instant-on feature of incandescents found annoying the lag between switching on and lighting up. Each of the three bulbs shown to the right of the circline bulb (Figure 5)

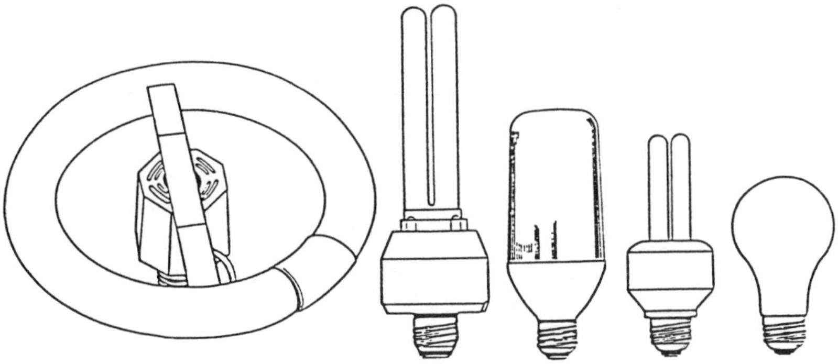

Figure 5. The rapidly changing technology of compact fluorescent bulbs. See text for discussion.
Source: Reference 54.

incorporates improvements that deal with one or more of these problems. The bulb shown next to the incandescent bulb on the right is a compact fluorescent with a high frequency electronic ballast that: provides as much light as a 60 Watt incandescent, while drawing only 11 Watts; lasts six times as long as an incandescent; weighs about 1/4 as much as the circline bulb; fits into most incandescent sockets; has a light color similar to that of incandescents; and, is instant-on. While current high prices for compact fluorescents have limited their penetration of residential markets, these bulbs are selling well in some commercial buildings markets (where a major non-energy-saving appeal has been the labor cost savings of less-frequent bulb changes), and the continuing improvements in the bulb show that manufacturers are trying hard to provide an energy-efficient product offering multiple benefits.

Another example of the trend toward energy-efficient products with potentially broad appeal is the variable speed drive (VSD) for the AC induction motor. In addition to the wide-ranging possibilities it offers for reducing the electricity requirements of the ubiquitous motor (55): the VSD makes possible slow, controlled starts, resulting in less electrical stress on motors, transformers, and switchgear, and less mechanical stress on motors, gears, and driven equipment (pumps, fans, etc.), leading to longer lifetimes for motors, and motor-driven equipment and reduced maintenance costs; its isolation from the power line with a VSD gives the motor a ride-through capability for brief (few-cycle) power failures; with a VSD a motor can operate at higher speeds than the 60 hertz line frequency allows; and once electronic VSD controls have been introduced for an industrial application, it becomes feasible, at very low additional cost, to introduce intelligent controls for fine tuning the motor output to better match processing needs and thereby to improve the quality of the product being manufactured (55).

While many such products are too new to be able to see whether their availability will lead to generally lower implicit consumer discount rates, in the case of at least one class of new energy-efficient products–windows with heat-reflecting coatings–this has happened, since the new windows are actually driving conventional windows off the market! Double-glazed windows with such coatings loose heat at less than 3/5 the rate for conventional double-glazed windows (56). They are even better in this regard than triple-glazed windows–and less costly as well. The use of such windows improves human comfort near the windows in winter by significantly reducing the down-drafts induced by cold window surfaces. Even though such heat-reflecting coatings are relatively new–the R&D on them was initiated only about a decade ago (43)–this coated glass already accounts for a major market share in the U.S., where one major manufacturer has completely phased out conventional double-glazed windows in favor of those with heat-reflecting coatings.

Why are we beginning to see more and more energy-efficient products that have potentially broad appeal? Good product designers understand that the commercial success of any new product requires that the product have broad consumer appeal. The choice of the improved attributes manufacturers design into their new products depends on their perceptions as to what are important attributes from the perspectives of the individual consumer and society at large. Some manufacturers, recognizing that improved energy efficiency is becoming an important attribute for energy-intensive products, are starting to make improved energy efficiency a "design criterion" for the process of product innovation.

It is important to try to nurture this incipient trend, through the formulation of a public policy that convinces producers of society's long-term interest in improved energy efficiency. The result would be not only a more hopeful outlook for coping with the various external societal costs of energy supplies, but also a reduction in the effort required to market energy efficiency, because the new energy-saving products would tend to sell themselves.

11 Conclusion

The problem of high discount rates implicit in consumer decisions relating to purchases of energy-efficient electricity-using equipment dictates that major market interventions are needed to realize the cost-effective potential for energy efficiency improvement. Moreover, empirical evidence indicates that incentives are needed to motivate consumers to adopt energy efficiency technologies. Yet some conventional programs involving significant incentives give rise to concerns about their economic efficiency and equity impacts. But new approaches to implementing conservation goals provide promising ways for coping with these problems, while promoting least-cost utility planning.

Allowing utilities both to buy saved electricity in competitive auctions and sell energy services to the customers who would benefit from the energy-efficiency investments that create this saved electricity would satisfactorily address all economic efficiency and equity issues that have been raised about traditional incentives programs.

It is too early to predict how successful and how widely applicable the competitive bidding concept will be in promoting the efficient use of electricity, but the basic idea is sound. Initially, at least, competitive bidding schemes might be complemented by more conventional conservation programs. Or utility-based energy service companies might be required to serve important markets that are not likely to be targets for bidding.

In light of various environmental and security concerns relating to energy supplies, efforts to improve the marketing of commercially available conservation technologies should be complemented by efforts to promote a continuing flow of new energy-efficient technologies to the marketplace. A promising way of accomplishing this would be the formulation of a public policy which convinces manufacturers and builders that energy efficiency will be a major societal concern for the long term future. If this were done, the likely result would be that producers would tend to routinely incorporate energy efficiency features into their new products. While this approach requires public policies that extend beyond the responsibilities of most electric utilities and/or their regulators, utilities would nevertheless be profoundly affected, because consumers would tend to routinely choose high efficiency products in their purchase decisions, without major prodding from energy service firms, utilities, or government bodies.

The market interventions emphasized in the present analysis are aimed at making the pursuit of improved energy efficiency more profitable for producers as well as consumers. It is difficult to imagine how conservation goals can otherwise be fully achieved, in light of the diverse and ever-changing nature of markets for electricity-using equipment.

References

1. Bodlund, B., Mills, E., Karlsson, T., Johansson, T.B. 1989. *The challenge of choices: technology options for the Swedish electricity sector*, this volume.
2. Sherrid, P. 1986. Growth stocks move to center stage. *U.S. News and World Report*, pp. 54–55.
3. Meier, A.K., Whittier, J. 1983. Consumer discount rates implied by consumer purchases of energy-efficient refrigerators. *Energy, The International Journal*. 8 (12): 957–62.

4. McMahon, J.E., Levine, M.D. 1983. Cost/efficiency tradeoffs in the residential appliance marketplace. *Proceedings of the American Council for an Energy Efficient Economy's 1982 Summer Study of Energy Efficient Buildings, Santa Cruz, California*. New York: Energy Information Center.
5. Hausman, J. 1976. Individual discount rates and the purchase and utilization of energy-using durables. *The Bell Journal of Economics*. 10: 33–54.
6. Gately, D. 1980. Individual discount rates and the purchase and utilization of energy-using durables: comment. *The Bell Journal of Economics*. 11:373–374.
7. Chernoff, H. 1983. Individual purchase criteria for energy-related durables: the misuse of lifecycle cost. *The Energy Journal*. 4 (4): 81–86.
8. Alliance to Save Energy. 1983. *Industrial Investment in Energy Efficiency: Opportunities, Management Practices, and Tax Incentives*. Washington, DC: Alliance to Save Energy. 267 pp.
9. Goldemberg, J., Johansson, T.B., Reddy, A.K.N., Williams, R.H. 1988. *Energy for a Sustainable World*. Delhi: Wiley-Eastern. 517 pp.
10. Williams, R.H., Dutt, G.S., Geller, H.S. 1983. Future energy savings in U.S. housing. *Annual Review of Energy*. 8: 269–332.
11. Komor, P.S., Wiggins, L.L. 1988. Predicting conservation choice: beyond the cost-minimization assumption. *Energy, The International Journal*. 13(8): 633–45.
12. Hirst, E., Clinton, J., Geller, H., Kroner, W. 1986. *Energy Efficiency in Buildings, Progress and Promise*. Washington, DC: American Council for an Energy Efficient Economy. 328 pp.
13. Moskovitz, D. April 12, 1988. *Will least cost planning work without significant regulatory reform*? Presented at the Least Cost Planning Seminar of the National Association of Regulatory Utility Commissioners, Aspen, CO: NARUC.
14. Moskowitz D., Parker, R.B. 1988. *How to change the focus of regulation so as to reconcile the private interest with the public goals of least-cost electric planning*. Presented at the Sixth NARUC Biennial Regulatory Information Conference. Columbus, OH: NARUC.
15. Wellinghoff, J.B. 1988. The forgotten factor in least-cost planning. *Public Utilities Fortnightly*, pp. 9–16.
16. Geller, H.S. 1988. *Use of financial incentives to encourage least-cost utility planning and energy efficiency*. Presented at the NARUC Least-Cost Utility Planning Conference. Aspen, CO: NARUC.
17. Geller, H.S. 1989. Implementing electricity conservation programs: progress towards least-cost energy services among utilities in the U. S., this volume.
18. Rosenfeld, A.H. 1987. Success of conservation. Presented at the Vattenfall Seminar on North American Experiences of the Power Industry's Efforts to Use Electricity Efficiently. Stockholm: Vattenfall.
19. Bales, G. R. 1987. Energy management: experience from a larger utility. Presented at the Vattenfall Seminar on North American Experiences of the Power Industry's Efforts to Use Electricity Efficiently. Stockholm: Vattenfall.
20. Jansson, P.M. 1987. Energy management: experience from a smaller utility. Presented at the Vattenfall Seminar on North American Experiences of the Power Industry's Efforts to Use Electricity Efficiently. Stockholm: Vattenfall.
21. Limaye, D.L. 1987. Demand-side management, the U.S. experience. Presented at the Vattenfall Seminar on North American Experiences of the Power Industry's Efforts to Use Electricity Efficiently. Stockholm: Vattenfall.

22. Sant, R.W., Bakke, D.W., Naill, R.F. 1984. *Creating Abundandce: America's Least-Cost Energy Strategy*. New York: McGraw Hill. 177 pp.
23. Cavanaugh, R.C. 1986. Least-cost planning imperatives for electric utilities and their regulators. *Harvard Environmental Law Review*. 10 (2): 299–344.
24. Schneider, C. 1986. Least-cost utility planning: providing a competitive edge. *Public Utilities Fortnightly*, pp. 15–21.
25. Public Service Commision of Wisconsin. 1986. *Findings of Fact Relating to Order in Docket No. 05-EP-4, Pursuant to the Power Plant Siting Law, Chapter 68, Laws of 1975*.
26. Bonneville Power Administration. September 1986. *Solar & Heat Pump Water Heater Program, Second Interim Report*.
27. Davis, T.D., Kreitler, V. 1987. Integrating quasi- experimentation with utility appliance programs: the NYSEG experience. Presnted at the First National Conference on Appliance Efficiency. Arlington, VA: American Council for an Energy Efficient Economy.
28. Lovins, A.B., 1985. Saving gigabucks with negawatts. *Public Utilities Fortnightly*. pp. 19–26.
29. Cicchetti, C.J., Curkendall, S.M. 1987. Are energy-efficiency programs worth it? *Papers and Proceedings of the Eighth Annual North American Conference of the Internation Association of Energy Econoimists*, D. O. Wood, ed. Cambridge, MA: Massachusetts Institute of Technology.
30. "Application for Approval of Proposed Tariff Rider and Associated Accounting and Rate-Making Treatment," submitted by C. P. Davenport, Vice President, Pacific Power and Light, to the Public Utility Commission of Oregon, in the matter of the *Proposed Residential Energy Efficiency Rider of Pacific Power and Light*, April 14, 1978.
31. Joskow, P.L. 1988. Testimony before the Subcommittee on Energy and Power, House Committee on Energy and Commerce.
32. Ruff, L.E. 1988. Least-cost planning and demand-side management: six common fallacies and one simple truth. *Public Utilities Fortnightly*. pp. 19–26.
33. Northwest Power Planning Council. 1983. *Regional Conservation and Electric Power Plan*. Portland, OR.
34. Weedall, M. 1986. Delivering energy efficiency through energy services diversification. *Proceedings of the Thirteenth Energy Technology Conference*, Washington, DC: US Government Institutes, Inc., pp. 575–583.
35. de la Moriniere, O. 1989. Energy service companies: the French experience, this volume.
36. Alliance to Save Energy. 1982. *Third-Party Financing: Increasing Investment in Energy-Efficient Industrial Processes*. Washington, DC: Alliance to Save Energy. 118 pp.
37. Bevington, R. 1988. The service connection. *Heating* / Piping / Air Conditioning, pp. 103–107.
38. Cole, W.J., Wolcott, D.R. 1988. Competitive bidding of demand side management. *Proceedings from the Panel on Utility and Private Sector Conservation Programs of the 1988 ACEEE Summer Study on Energy Efficiency in Buildings*, pp. 6.11–6.26.
39. Michaels, H. 1988. Bulk purchase of conservation by competitive bid. *Proceedings from the Panel on Utility and Private Sector Conservation Programs of the 1988 ACEEE Summer Study on Energy Efficiency in Buildings*, pp. 6.141–6.152.
40. Williams, R.H., Larson, E.D. 1989. Expanding roles for gas turbines in power generation, this volume.
41. Cicchetti, C. Hogan, W. 1988. *Including unbundled demand side options in electric utility bidding programs*. Energy and Environmental Policy Center Report No. E-88-07. Cambridge, MA: J.F. Kennedy School of Government, Harvard Univ.

42. Dutt, G.S., Fels, M.F. 1989. Keeping score in electricity conservation programs, this volume.
43. Harris, J., Mills, E., Mowris, R., Rosenfeld, A., Geller, H., Ledbetter, M. 1989. The importance of Goverment-supported research and development in advancing energy efficiency in the U.S.buildings sector, this volume.
44. Elmroth, A. 1989. Building design and electricity use in single-family house, this volume.
45. Geller, H.S. 1987. *National appliance efficiency standards: utility and consumer impacts*. Presented at the Third National Conference on Utility DSM Programss. Houston, TX.
46. Solow, R.M. 1957. Technical change and the aggregate production function. *The Review of Economics and Statistics*. XXXIX: 312–20.
47. Berg, C.A. 1979. *Energy Conservation in Industry: the Present Approach, the Future Opportunities*. Washington, DC: President's Council on Environmental Quality.
48. Ayres, R.V. 1983. Final report on future energy consumption by the industrial chemicals industry (SIC 28). Appendix to *The Chemicals Industry*, vol. 5, in *Industrial Energy Productivity Project Final Report*, prepared by Energy and Environmental Analysis, Inc., Arlington, Va., for the Asst. Secretary for Conservation and Renewable Energy, US Dept. Energy. DOE/CS/40151–1. Washington, DC: U.S. Dept. Energy.
49. The Conference Board. 1974. *Energy Consumption in Manufacturing*, a report to the Energy Policy Project of the Ford Foundation. Cambridge, MA: Ballinger.
50. Energy and Environmental Analysis. 1983. *The Iron and Steel Industry*, vol. 4 in *Industrial Productivity Project Final Report*, prepared for the Asst. Secretary for Conservation and Renewable Energy, U.S. Dept. Energy. DOE/CS/40151–1. Washington, DC: U.S. Dept. Energy.
51. DiKeou, J.T. 1980. Cement, in *1978–79 Minerals Yearbook, Volume I: Metals and Minerals*. Washington, DC: U.S. Bureau of Mines.
52. Ross, M.H., Williams, R.H. 1981. *Our Energy: Regaining Control*. New York: McGraw-Hill. 354 pp.
53. McGowan, T.K. 1989. Energy-efficient lighting, this volume.
54. Goldemberg, J., Johansson, T.B., Reddy, A.K.N., Williams, R.H. September 1987. *Energy for a Sustainable World*. Washington, DC: World Resources Institute. 119 pp.
55. Baldwin, S.F. 1989. Energy-efficient electric motor drive systems, this volume.
56. Granquist, C.G. 1989. Energy-efficient windows: options with present and forthcoming technology, this volume.

The Importance of Government-Supported Research and Development in Advancing Energy Efficiency in the United States Buildings Sector*

Howard S. Geller, Jeffrey Harris, Marc R. Ledbetter, Mark D. Levine, Evan Mills, Robert Mowris, Arthur H. Rosenfeld

Abstract

The United States government has played a major role in advancing energy efficiency through research and development (R&D) programs sponsored by the Department of Energy Office of Buildings and Community Systems (BCS). This chapter discusses the importance of government-supported R&D in the U.S. buildings sector, and gives an overview of successful R&D projects. The chapter also examines case histories of seven successful R&D projects sponsored by BCS in lighting, windows, and appliances. These seven energy-efficient technologies promise to save U.S. consumers almost $15 billion per year by the year 2010.

1 Introduction

The United States government has played a major role in both the direction and advancement of technology. Government-supported research and development (R&D) has produced "generic technology", as well as breakthroughs in basic science that have provided a foundation for technological development by private industry.

* This chapter is based on Geller *et al* (1), and has been updated where possible.

But why does government get involved with R&D? In most cases the government gets involved with R&D in response to specific needs which are not being met by the private interests or state governments. Often these needs involve high risks or societal benefits that would be impractical investments for private firms or individuals. In some cases the societal need for government-supported R&D is only recognized when a crisis erupts. This happened following the oil crisis of 1973, and the United States government responded by establishing the Department of Energy (DOE) to manage its energy R&D efforts.

During is 12 year history, the DOE has spent about 90 percent of its R&D budget on expensive energy "supply-side" R&D and about 10 percent on "demand-side" energy-efficiency R&D. Despite low funding levels, energy-efficiency R&D has been very successful. In the buldings sector alone, innovative and cost-effective technologies resulting from government-supported R&D will save consumers billions of dollars annually when fully implemented in the U.S. economy.

In this chapter we will discuss the important role played by the government in advancing energy efficiency in the U.S. buildings sector. We will give an overview of successful R&D projects sponsored by the DOE Office of Buildings and Community Systems (BCS), and finally we will examine case histories of seven energy-efficient developments in lighting, windows, and appliances.

2 The Government Plays an Essential Role in End-Use Energy-Efficiency R&D

Both private and public sector R&D are investments in the future. Industries and nations that invest too little in R&D, or fail to pursue potentially valuable projects to their logical conclusions, will lose opportunities for competitiveness, innovation, and productivity. Such losses are usually not apparent for some years, as industry lives off past research and dissipates historically developed goodwill, but a lack of high-quality R&D inevitably cripples industries and weakens whole economies.

There are widely differing views on the appropriate federal role in energy conservation R&D. The following policy statement in the DOE *FY 1988 Energy Conservation Multi-year Plan* reflects DOE's approach (2):

> "The federal role in energy research and development is concentrated in areas where the incentives for and availability of private investment are severely limited or nonexistent.

The success of federal research and development programs in providing options for private investment and commercial development depends on effective technology transfer."

In addition, however, the process of developing usable technologies should involve prospective users in all stages of the R&D process, from formulation to demonstration.

DOE emphasizes federal research and development in those areas where the private sector has too little incentive or too few resources. This policy must be interpreted in light of recent findings on the limited extent of private R&D investments in many sectors of U.S. industry. A recent study of appliance manufacturers and markets worldwide (3) showed that mature U.S. industry spent only 1%–2% of their 1983 sales revenue on all R&D[1] (not just that aimed at energy efficiency), in contrast to an estimated 3%–4% spent by Japanese appliance manufacturers. Rapidly advancing high-technology industries in the United States by contrast, typically allocate 4%–7% of revenues to R&D (5). Traditional U.S. manufacturers have historically been insulated from world competition and sponsor only limited research.

Other reviews of manufacturing R&D have found that private firms often have limited incentive or resources to undertake their own research because of near-term pressures on market share and profitability, high up-front research costs, risks of failure, and uncertain competitive advantage from a new advance, which might be rapidly adapted or copied by other firms. Nor is it apparent that U.S. federal tax incentives provided for industrial R&D have had any significant impact (6).

With regard to the building industry, the recent Energy Research Adivsory Board Demand Panel report (7) states that:

> Many small design firms, building companies, and component manufacturers comprise the building industry. Most private firms will undertake basic R&D activities if there are substantial and immediate monetary gains for the firm, and this is seldom possible in a fragmented industry of small firms. As a result, too little R&D is performed and [it] must be supplemented by the government if it is to be accomplished on any substantial scale.

However, the U.S. government support for Buildings and Community Systems R&D has dwindled from a high in 1980 of $140 million down to about $35 million in 1987 (1987$). The total money spent in the buildings sector of the U.S. economy is over $1,000 billion per year (includes construction, energy, maintenance, and finance) (8), so an R&D budget of $35 million represents only 0.004%. This is 250 to 500 times less than the 1%–2% for mature industry (9).

1 This is small compared to the national average of 2.5% (4).

In FY 1987 DOE sought to enlarge its joint programs with private industry both to sustain the scope of its research effort and to enhance the probability that industry will use the results. Collaborative, cost-shared research is a sound idea, and in many cases is already a feature of successful DOE-sponsored energy-efficiency R&D. But unfortunately, the private sector tends to reduce its own participation in step with any withdrawal of federal R&D support.

Discussions with industry R&D leaders, including the two largest industry organizations doing research on energy efficiency in buildings (the Electric Power Research Institute and the Gas Research Institute), reveal no prospect of private industry funds that might substantially replace federal funding for efficient use of energy. Even if such private funding could be generated, it would take many years to change industry priorities and to organize new collaborations between the federal government and individual firms or industry research establishments.

We must underscore one other essential element of DOE-funded R&D: the work on fundamental, "technology base" activities, as opposed to specific products or devices. The benefits of this research are more difficult to quantify than the product-oriented R&D discussed above, but they are no less crucial. For example, basic research on phosphors such as those used in fluorescent lamps could lead to a doubling in the efficacy of those phosphors. Industry depends on such government-sponsored research to provide a basis from which it can build innovative, marketable new products. The relatively long-term nature of such research, the highly specialized expertise required, and the lack of directly marketable products resulting from such research severly limit the ability of most industries to pursue it in the absence of a leading federal role.

The DOE *FY 1988 Energy Conservation Multi-year Plan* points to the benefits of a federal technology transfer program. Successful R&D needs to get into the market to have an impact on more efficient use of energy, and technology transfer must be intrinsic to the research effort and be supported adequately and explicitly at a high level. Successful research is meaningless unless it is used.

How much government money is spent by other countries on improving energy efficiency?

Table 1 shows the 1986 energy budgets of selected member countries of the Organization for Economic Co-operation and Development (OECD) as reported to the International Energy Agency (IEA). The countries spent a total of $519 million on energy-efficiency R&D–only 7% of the total $7,345 million spent on energy research. Sweden is currently spending 37% of its total energy research R&D budget on efficiency, while Denmark, the Netherlands, and Norway are spending about 25%. The U.S. is spending 9% of its total energy research budget on efficiency R&D and has decreased its energy-efficiency R&D budget by over 50% since 1980, when it spent almost $470 million (1987$). Japan is spending only 2% of its total research budget on energy efficiency R&D, but has been consistent in funding since 1983. Japan has focused almost exclusively on industrial programs contracted out to private companies and guided by the Ministry of International Trade and Industry (MITI).

There is a very real need for international collaboration in energy efficiency R&D. The European Economic Community (EEC) has the largest joint effort with funding of almost $30 million annually (11). Collaboration with Soviet and eastern block nations will help ease world demand for energy supplies and reduce the problems of global warming and pollution. Additionally, the particular problems of efficient end-use of energy in the developing world must be considered and collaborative R&D efforts with developing countries initiated.

Table 1. 1986 Government energy research and development budgets in selected OECD countries. Source: IEA (10).

Country	Total Energy R&D (million 1987$)	Efficiency[a] R&D (million 1987$)	Efficiency as Fraction of Total (%)
Japan	2,380	53	2
United States	2,328	219	9
Italy	784	43	6
West Germany	583	14	2
United Kingdom	389	44	11
Canada	346	33	10
Netherlands	166	45	27
Sweden	81	30	37
Norway	30	7	24
Denmark	14	4	28
Total OECD	7,345	519	7

[a] Includes electric power conservation

3 Energy-Efficiency R&D has Achieved Dramatic Successes and Holds the Promise of Continued Success

Table 2 shows the dramatic returns on investment generated by DOE-sponsored R&D in the buildings sector. Each of the technologies shown in Table 2 is yielding large payoffs as it achieves market penetration. Three have already demonstrated significant commercial success by 1985 – solid-state ballasts for fluorescent lights, low-emissivity window films, and high-efficiency refrigerator compressors.

1 The total government cost of these three projects is about $6 million, thanks to DOE cost-sharing with private industry.
2 The three projects have already produced annual net savings[2] far greater than the total federal R&D investment over several years. In other words, a one-time federal investment of $6 million has already generated an annual return of $25 million.
3 The near-term benefits are small, however, compared with the eventual impact of full market saturation by these technologies. Taken together, the projects in Table 2 will ultimately save U.S. consumers more than $10 billion per year.
4 Even if we assume that these technologies would have eventually been developed without public R&D support, the government can take credit for accelerating their commercialization by 2–5 years, contributing to net savings of about $40 billion. This still represents a 3000-to-1 return to the taxpayer on an initial investment of only $16 million.

Table 2 illustrates another important point: the long lead-times involved in research, development, and commercial adoption of a new technology. DOE research on solid-state electronic ballasts, for example, began in 1976 and lasted until 1980, but the ballasts probably will not capture 50% of new ballast sales until the mid-1990s. Since installed ballasts last for 10–15 years, it will take at least another decade before the older core-coil ballasts disappear.

Additional examples of high-impact R&D projects outside the buildings sector are provided by DOE in its *FY 1988 Energy Conservation Multi-year Plan*. Although the number differ slightly form those presented in Table 2, DOE estimates for 13 successful buildings sector R&D projects to date show total annual savings of $15 billion by the year 2010.

[2] Net savings = annual energy savings minus the increase in (annualized) capital cost to buy the more efficient products.

Table 2. Lead-times and net savings for sucessful U.S. DOE Office of Buildings and Community Systems R&D projects.

Project	Solid-state ballasts	Low-E[b] window films	Residential absorption heat pump	Advanced electric heat pump	High-efficiency refrigerator compressor	High-efficiency refrigerator/ freezer	Heat-pump water heater
1 DOE project duration	1976–1980	1976–1990	1978–1988	1977–1986	1977–1981	1978–1983	1977–1982
2. Estimated date of 50% penetration for sales	1995	2000	2001	1998	1990	1996	1995
3. Number of years that DOE advanced commercialization	5	5	5	2	2	2	2
4. Cost of conserved energy (CCE)[c]	2¢/kWh	$1.75/GJ	$2.20/GJ	$2.45/GJ	1¢/kWh	3¢/kWh	5¢/kWh
5. Cost of DOE project	$3M[e]	$2M	$6.8M	$2M	$1M	$0.8M	$0.7M
6. Net annual savings in 1985[a]	$11M	$14M	–	–	$0.4M	$0.2M	$6.5M
7. Net annual savings at saturation (i.e. 10–15 years after 50% penetration)[d]	$5,000M	$3,000M	$2,400M	$2,500M	$1,100M	$850M	$1,800M
8. Cumulative net savings (Line 7 x line 3)	$25,000M	$1,300M	$1,200M	$5,000M	$2,200M	$1,700M	$3,600M
9. DOE project return on Investment (ROI = line 8 ÷ line 5)	8000:1	7000:1	1500:1	2500:1	2200:1	2000:1	5100:1

a All dollar savings in energy costs are net of increased costs for purchase of the improved equipment (ballast, window, etc.).

b Low emissivity.

c Line 4. Cost of conserved energy, CCE. We use a real discount rate d= 6%, and the useful life of the product is noted in Table 4, line 3 of Geller et al (1) (e.g. 10 years for ballasts, 20 years for windows). The CCE of low-E films, now still a novelty, is currently $3.50/GJ, but as the market matures, we estimate that the CCE will drop to $1.75/GJ, and in Massachusetts it has already dropped to zero in 1988. Cowell, S. 1988. Personal communication. Massachusetts Fair Share Development Corporation. Boston, MA.

d Line 7. Net annual savings at saturation are in 1985 dollars, uncorrected for growth in the building stock, changes in real energy costs, or dicounted future values. We decided not to account explicity for these factors, since all three are uncertain and their combined impact probably small, as shown by the following calculation. To translate savings in 2000 or 2010 to 1985 terms, one multiplies by exp [(g+e–d)T], with the following annual rates: g=growth rate of building sector (3%), e= real price escalation for electricity or fuel (1–2%), d = real discount rate (4–7%), so that g+e–d=(–3% to +1%), and T=years between 1985 and the estimated saturation date, 20–25 years.

e M=million

Energy efficiency R&D, as these examples make clear, is far from an academic exercise. The total federal investment in energy conservation R&D, has already paid off dramatically in bottom-line economics and has planted the seeds for future energy savings.

3.1 What About Failures?

Not all government energy conservation research projects, however, have been successful. Within the Office of Buildings and Community Systems some projects have failed to work as hoped and others have been slow in achieving commercial acceptance, but many of the projects have succeeded, as documented in Table 2. This pattern is to be expected in research programs. In fact, without risky projects that industry is unlikely to pursue by itself, there would be no justification for federal involvement. Nonetheless, federal researchers must weed out once-promising research concepts that are not meeting technical or economic expectations or are being surpassed by newer, more-promising innovations. They also have to make efforts to ensure links between their research and its prospective implementation by industry through effective technology transfer programs.

Perhaps paradoxically, the failures in energy conservation R&D illustrate an important advantage of research on end-use energy efficiency: these R&D efforts are diversified so the failures have been offset by successes. In contrast, when government or private resources are concentrated on a few large-scale projects – as with some energy supply technologies (e.g., the breeder reactor) – any setback is very costly, in both lost dollars and lost time. The simple payback on DOE Buildings and Community Systems R&D illustrates this point. Cumulative government spending on BCS R&D from 1976 through 1987 was about $620 million (12). By the year 2010, the expected savings from the 13 projects will come to more than $15 billion per year. If we conservatively assume that BCS R&D advanced these technologies by 3 years, then the net savings to U.S. consumers is about $45 billion, yielding a simple payback of 75-to-1 on the U.S. government's *total* R&D investment, including major successes, partial successes, and failures.

Although this is a convincing argument for past spending, future success may not be as dramatic. Much of the "cream" has been skimmed over the past decade, and future R&D is going to be much more challenging. Work has already begun on the next generation of lighting systems, windows, appliances, etc., but recent budget cuts have hampered these efforts. The success or failure of future R&D work on improving end-use energy efficiency will largely depend on continued funding, and the ability to attract young scientists, engineers, and researchers who will meet the challenge.

4 The Case Studies

We now present seven case studies of DOE Buildings and Community Systems R&D success stories in lighting, windows, and appliances. Many other equally impressive success stories could be recited. By "success," we mean that the improved technology is already in or near commercial production, and will achieve 50% market penetration in 5-15 years. We can then evaluate the annual savings at market saturation, which occurs roughly 30 years after project initiation.

The payback calculations for these developments do not attempt to estimate any secondary benefits such as the improved competitive position of U.S. industry or enhanced national energy security.

The seven BCS cases are summarized in Table 2. A typical feature of these seven cases is their low cost of conserved energy, i.e. the cost of providing the desired service by improving efficiency instead of purchasing electricity or gas. The cost of conserved electricity ranges from 1 to 3 ¢/kWh vs about 10 ¢/kWh for new power today; for natural gas the cost of conserved energy ranges from $2–3 per GJ versus $7.50/GJ for retail gas today. Thus, for these seven BCS cases improved energy efficiency is 2–10 times cheaper than new energy supplies.

An important common element of these seven cases is the early involvement of private firms, generally through subcontracting. In most instances, private firms did most of the development. Often small (sometimes very small) businesses played key roles. They did the early development, and when the innovation demonstrated commercial viability, larger firms became involved through buy-outs or partnerships. Government R&D efforts were the catalyst in this public/private sector partnership, supplying the financial resources, technical support, and independent evaluation through the National Laboratories. This support was sustained long enough for the product to successfully enter the market.

4.1 Lighting

4.1.1 Energy-efficient solid-state ballasts for fluorescent lighting

In 1976, DOE established a lighting research program at the Lawrence Berkeley Laboratory (LBL) to accelerate the commercialization of energy-efficient lighting technologies. LBL researchers soon focused on the potential of solid-state ballasts, which power fluorescent lamps at high frequency, to reduce the energy needed for lighting by 15–30%. Calculations showed the potential of high-efficiency, solid-state ballasts to reduce the energy needed for fluorescent lighting by $5 billion per year as a result of reducing energy demand by an amount equal to the output of 10 typical large (1000 MW) power plants.

None of the major ballast manufacturers showed any interest in the program or replied to a 1976 request for proposal (RFP) that LBL issued for cost-sharing subcontractors to develop circuitry for the ballasts. However, about 14 small entrepreneurial firms did apply. This pattern appears in many of the R&D case studies discussed here. Large established firms have shown little interest in the early stages of product development, while small companies jumped at the opportunity to conduct joint R&D with the government. It appears, therefore, that modest federal funding helped speed valuable research that large firms would not have pursued on their own initiatives and that small firms working alone could not afford. The results of this public-private partnership are impressive, both for the potential energy savings to society and for the business successes they have generated.

Throughout the period when the ballasts were being developed, LBL published its findings and made numerous presentations to the lighting community and general public. Major U.S. ballast manufacturers continued to show little interest, although two foreign companies, O. Y. Helver (Finland) and Toshiba (Japan), began producing solid-state ballasts in 1980. From 1980 to 1984 a dozen small domestic manufacturers struggled to establish a market for solid-state ballasts and to improve the reliability of their products. The relatively conservative lighting industry was slow to adopt these innovative designs, even though foreign firms were already entering the domestic market with energy-efficient lighting products. LBL's program support was reduced during this period, but it continued to publicize the benefits of solid-state ballasts and to test the new products for reliability. LBL also worked with the manufacturers and the Federal Communications Commission to develop voluntary standards to ensure that electromagnetic radiation emitted by the high-frequency ballasts would not interfere with computers, communications devices, and other electronic equipment.

By the end of 1984, one million solid-state ballasts for the F40 (40-watt) fluorescent lamp had been sold in the United States. Though this was a small number in comparison to the total annual demand for 70 million units, the proven credibility of the products, in concert with LBL's R&D support, information dissemination and technology transfer, had finally brought the solid-state ballast to the brink of commercial success.

In 1984, General Electric and GTE announced lamps designed to be operated with a solid-state ballast. This signified that the major lamp manufacturers had recognized an important new market. In 1985, General Electric announced a solid-state ballast, and Universal Manufacturing announced a low-cost dedicated solid-state ballast. In 1986, Advanced

Transformer, the largest U.S. ballast manufacturer, announced plans to market a solid-state ballast for two 40-watt F40 fluorescent lamps. Small companies are still entering the field expecting to obtain a market share for special applications. The market for solid-state ballasts is expected to grow rapidly, and should achieve 50% penetration by 1995 and saturation by the year 2000.

The successful strategy and perseverance of the LBL/DOE program, with the coordinated and cost-shared investment of small companies, has created a new, efficient lighting technology. DOE funding for this program during the past nine years was only $2.7 million, approximately half of which was subcontracted, with subcontractors providing an additional $1 million.

Today, two million solid-state ballasts are in operation in the United States, each saving on the average 25 watts for 4000 hours annually, a total savings of 200 million kWh per year. At a typical energy cost of $0.075/kWh, the annual savings are worth more than $15 million. Within 30 years, the cumulative net savings from the government's participation will total approximately $25 billion. This achievement reflects how an effective strategy to promote technological innovation can succeed even in a restricted environment dominated by a few large corporations with a seemingly captive market.

4.1.2 Future possibilities

The high-frequency solid-state fluorescent ballast is the key technical element allowing more widespread use of dimming controls. Standard core-coil ballasts can dim fluorescent lamps by conditioning their input power (voltage reduction or phase control). Solid-state ballasts dim lamps by low-voltage signals, reducing the cost and increasing the flexibility of the control system. Solid-state ballast control systems independently dim single lamps as well as large banks of lamps. Conventional core-type ballast systems, in contrast, can dim only banks of lamps at greater expense.

Thus, in addition to catering to individual demand preferences, such as scheduling, adjusting the light to match the task, and compensating for lost brightness of old lamps and dirty fixtures (strategies that bring the energy savings to 30–40%), solid-state ballast dimming systems can effectively take advantage of daylight and load-shedding strategies. The latter strategies can reduce energy consumption by additional 30–40%. The total potential energy savings, considering the intrinsic 25% gain and all the other demand strategies, are 60–70%, compared to the best core type ballast dimming system.

4.2 Windows

4.2.1 Advances in high-performance windows: The successful development of low-emissivity coatings for windows

The annual resource energy used to heat and cool U.S. buildings is about 13 EJ of resource energy[3]. One quarter of this is required to compensate for unwanted heat flow through windows. Windows are perceived as weak thermal links relative to insulated walls and roofs, but this narrow perspective ignores potential window benefits.

In 1976, DOE initiated a program at LBL to develop guidelines for more efficient use of windows and to develop more energy-efficient window components and window systems. LBL researchers set out to develop a window that would behave thermally like an insulated wall. They sought to develop a selective window coating that was transparent to visible light and solar radiation but blocked heat flow, so that energy loss through windows could be dramatically reduced.

In 1976 the principle behind such low-emissivity (low-E) coatings was understood, but no commercial low-E products were available. Some manufacturers knew about the potential of such coatings, but they doubted that durable products with adequate optical properties could be produced at low cost and high volume, so the window industry essentially ignored low-E coatings. It was not until the federally funded research program at LBL demonstrated the promise of this technology that the window industry began to aggressively develop and ultimately market low-E products.

The $2 million federal investment stimulated more than $100 million in private investment in low-E film production technology. In 1985, low-E window coatings were commercially available and saving consumers $14 million per year on their energy bills. By the year 2000, cumulative savings from low-E coatings used in the residential sector alone should be worth billions of dollars.

This major advance in window technology began with a series of research grants that LBL issued, with DOE funding, to several small research firms in 1976. Their task was to investigate potential coating systems and deposition processes for low-E films. LBL undertook studies to determine the optimal use of coatings in multi-element window systems and to determine the potential benefits of these hypothetical windows in residences.

Coating technology consultants to LBL reviewed various deposition processes to determine if any had the potential to deposit the types of materials

[3] Table 3-1 in reference 2 gives matrices for the residential and commercial buildings sectors, of energy end use by type of fuel, or electricity. Total heating and cooling adds up to 12.7 EJ of resource energy.

required for low-E coatings at high deposition rates and with good uniformity. A process known as "magnetron sputtering" appeared to meet the requirements and was gaining popularity during the late 1970s for small area coating in the electronics industry. The major U.S. producer of large magnetron sputtering systems, Airco Solar Products, was trying to sell these new large deposition systems to the U.S. glass industry for producing solar control coatings for architectural applications. Based in part on the promising initial results of the DOE-supported studies, Airco developed suitable low-E coating systems that could be licensed to the purchasers of its deposition systems. Airco had to do this work because many U.S. glass companies did not have the resources to undertake R&D on low-E coatings, and none had the sputtering systems capable of producing the coatings.

By the early 1980s, several large sputtering systems had been sold to major U.S. manufacturers, but none yet marketed a commercial low-E product. In Europe, with its higher energy prices, more restrictive building standards, and a tradition of greater initial investment in buildings, several versions of low-E coatings had been successfully marketed.

At this time, LBL staff gave presentations at industry association meetings and trade shows, and met privately with research and marketing staff from a number of major window manufacturers to build confidence in low-E technology. LBL also developed new tables of heat transfer data for low-E windows, which were published in the 1985 ASHRAE Handbook of Fundamentals.

Interest and confidence in low-E coatings advanced when they were used in a small test building at MIT. Since low-E windows were one of several innovative building technologies being demonstrated in the facility, it was not possible to isolate the effects of the low-E glazing. However, many interested parties first looked through windows with low-E coatings at the test building.

4.2.2 Market breakthrough

A major market breakthrough occurred in 1983, when Airco installed a large sputtering plant for low-E coatings for Cardinal IG, the firm that supplies the sealed insulating glass units for the largest window manufacturer in the United States, Andersen Corporation. This was the first time a large glass coater had been operated in the window industry by anyone other than a glass producer; Cardinal IG's move stimulated additional purchases of this type. Once Andersen announced the availability of a low-E window, the generic product gained new credibility in the eyes of consumers, builders, and specifiers, which placed pressure on other window manufacturers to supply low-E windows. In 1984 the first significant sales of low-E glass occurred; by 1985 industry estimated that 7 million m^2 per year, or more than 5% of the market, had been captured.

By the mid-1980s, industry had invested millions in facilities that could produce the new generation of low-E coatings, and virtually every major glass and window company offered a low-E product. The marketing directors for several major glass firms estimate that 25–50% of the residential market could be low-E by 1990, a very high penetration rate for a new technology.

Although the product was originally developed with the northern residential market in mind, low-E coatings will also penetrate the residential sunbelt market and some fraction of the nonresidential building market. Traditional, high transmission low-E coatings can be modified to selectively transmit daylight but reject solar near-infrared energy to reduce cooling loads. In the sunbelt, this modification offers control of winter heat loss and summer heat gain without excessive loss of view. This application represents a major new market opportunity.

In nonresidential buildings, these modified low-E coatings will also make it easier for architects to use daylighting to save electric lighting energy and to reduce peak electrical demand charges. LBL simulation studies were the first to explore extensively the complex trade-offs between daylighting benefits and heating and cooling loads and to make recommendations on the best coating types for different climates.

The proliferation of low-E coatings on the market is a sign of their commercial success, but they create problems for designers and specifiers, who may be overwhelmed by confusing and conflicting data. LBL helped co-sponsor a Low-E Industry Roundtable in August 1985, at which more than 100 industry representatives expressed an interest in developing standardized calculation procedures so that low-E claims could be compared on a consistent basis. A new DOE/LBL microcomputer model, was released in response to that need. "WINDOW" allows users to quickly and accurately calculate the thermal properties of low-E windows of almost any design. Performance data from LBL field tests have also been used to help validate simulation models of low-E performance, to develop reliable design guidelines, and to gain the confidence of developers and specifiers of this new technology.

4.2.3 Sales, costs, and savings

An estimated 5 million m^2 of low-E windows were sold in the United States in 1985 at a retail cost of \$0.20–0.30 per m^2 more than normal double-glazed windows. These were primarily sputter coated on glass and plastic. Several U.S. glass companies have now introduced pyrolytic low-E "hard coats." These can be used on single glazing and in nonsealed multiple glazing units, thereby extending sales to the large retrofit market. The emittance properties of the pyrolytic coatings (with one exception) are not currently as good as those of the sputtered "soft coats." Additional coating research and development in the United States and overseas in the years ahead will continue to improve optical and thermal properties and cost factors. It should be possible eventually to reduce coating manufacturing costs to \$0.01–0.02 per m^2, which should reduce the additional retail cost to less than \$0.10 per m^2

The potential energy savings from full market penetration of this technology are enormous. In the residential sector, windows with low-E coatings are now specified for new construction, additions, and renovations, a market with total annual window area sales of about 65 million m^2. The recent introduction of the pyrolytic hard coats provides additional opportunities in the retrofit market, such as storm windows and add-on glazings. The multitude of new low-E products and their varied potential applications makes it difficult to predict precisely their total energy-saving potential. It seems likely, however, that annual heating energy savings by 1995 will be worth more than \$400 million[4] and that annual energy savings by 2010 may total more than \$3 billion.

In summary the U.S. Department of Energy invested approximately \$2 million in research efforts to accelerate development and commercial introduction of low-E window technology. This helped leverage a much larger private-sector investment in new deposition systems and in development of new window product lines. Builders, architects, engineers, and homeowners are, in

4 We estimate total savings based on the following assumptions: a) existing stock is a mixture of single and double glazing; b) single can be replaced by single with low-E or by double with low-E; and c) double can be replaced by double with low-E or by other multi-glazed and gas-filled configurations with low-E. For the residential sector, we calculate the improvement in heat transfer rate for each and derive an overall weighted average improvement of about 1.42 W/m^2-K. Since our interest is in heating energy, we select a typical climate that is 20% colder than average. In a climate with 3333 centigrade heating degree days, the improved windows will reduce heat loss by about 409 MJ/m^2-yr, which will save about 584 MJ/m^2-yr after heating system efficiency (about 70%) is accounted for. At an average 1987 natural gas price of \$5 per GJ, the savings is about \$2.90/m^2 per year. If industry projections of capturing more than 50% of the residential window market by 1995 are realized and if total annual residential window sales climb slowly to 800 million square feet, this would result in 1995 low-E sales of 37 million square meters. The annual heating energy savings are derived by multiplying the cumulative amount of low-E sold by 1995, about 140 x 10^6 m^2, by the savings/m^2 or 140 million m^2 x \$2.90/m^2 = \$400 million. Actual savings will be larger since we neglect storm window retrofits and the higher value of savings in electrically heated homes and ignore potential cooling-load savings in the sunbelt. Low-E glazing use in the commercial sector will further increase savings. It seems likely, that annual heating energy savings by 1995 will be worth more than \$400 million and that annual energy savings in 2010 may total more than \$3 billion.

turn, investing to purchase and install these new systems. The ultimate benefits accrue to homeowners, in the form of lower utility bills and improved comfort and amenity; to the fenestration industry, in the form of new production technology, increased employment, and new value-added products; and to the United States, in the form of decreased dependence on imported fuel, an improved balance of payments, and maintenance of a strong buildings sector that can compete effectively in the international market.

4.3 Building Equipment

4.3.1 Overview

The objective of the Building Equipment Research (BER) program at Oak Ridge National Laboratory (ORNL), sponsored by DOE's Office of Buildings and Community Systems, is to develop a technology base for improving the energy efficiency and load characteristics of equipment for space heating and cooling, water heating, and other appliances used in residential and commercial buildings. This program has been sponsored by DOE since 1976. During the late 1970s and early 1980s, the philosophy of DOE and its predecessor, the Energy Research and Development Administration (ERDA), had been to work on projects that showed the potential for near-term commercialization of energy-efficient appliances. As a result of the research sponsored by DOE, a number of new products were developed and successfully marketed by manufacturers.

The change in administrations brought a change in philosophy within DOE, and the charter of DOE-sponsored research shifted to long-term, high-risk research with the potential for high payoffs. Under this philosophy, activities with near-term commercialization potential were phased out, and research shifted to equipment that might represent products to be introduced in the 1990s. Emphasis has been on high-risk activities not likely to be carried out by the private sector.

Below are five case studies. Three represent projects that were completed in the early 1980s, and resulted in more efficient products on the market. Two represent research that is currently under way and that, if successful, will result in new products in the 1990s. They represent clear cases in which DOE-sponsored research has been instrumental in accelerating the development of technologies. The DOE-sponsored research either produced the first product or prompted other manufacturers to enter the market, leading to further product improvements.

4.3.2 Refrigerators

In 1977, ORNL funded Arthur D. Little, Inc., to work with Amana Refrigeration, Inc., to design a high-efficiency refrigerator. From a list of 18 energy-saving options, six were chosen for the prototype model. These changes, including separate evaporators for the freezer and fresh food compartments, thicker cabinet insulation, improved door gaskets, relocation of the fan motor outside the freezer, and better defrost controls, resulted in a 60% reduction in energy use. Twenty-five prototypes were assembled and marketed to test consumer acceptance and field performance. Response was so positive that Amana began producing the new design commercially in 1981, selling more than 16,000 units by 1983 and producing net savings to consumers of more than $200,000 per year. Other manufacturers soon followed with their own high-efficiency refrigerators employing some of the options listed above.

While the Amana project was under way, ORNL also funded the Kelvinator Compressor Company to develop and field test 20 units of a prototype, high-efficiency refrigerator compressor. Since the compressor accounts for 70–85% of the energy demand of a refrigerator, improvements in this component were seen as critical to the goal of markedly improving overall appliance efficiency. By implementing several design changes in the motor and suction muffler, Kelvinator was able to increase compressor efficiency by 44%. The company manufactured 30,000 of these efficient compressors and then made a business decision to buy rather than manufacture its own compressors. Those 30,000 compressors yielded annual savings of $12 each, repaying DOE's cost for the project every three years. Many other manufacturers, including a production facility in China, have since adopted similar high-efficiency designs.

4.3.3 Heat pump water heater

The concept of a heat pump water heater (HPWH) is decades old, but until the 1970s energy prices were too low to justify such a heater's relatively high cost. In 1976, Energy Utilization Systems, Inc. (EUS) began to develop a residential HPWH, with assistance from DOE and ORNL beginning in 1977. By the end of 1978, EUS had completed design studies and had put 12 test units through a one-year laboratory performance evaluation. A concurrent marketing study by EUS indicated that an attractively short payback period could be achieved.

A 100-unit, year-long field test was then conducted with the cooperation of 20 electric utilities. This test, completed in 1980, produced useful information on the design and operating characteristics of these units. It also showed an installed average coefficient of performance (COP) of 1.93, compared to about 0.85 for conventional electric water heaters. EUS began producing HPWHs for sale in 1980, and was the first to market this product. Within a year, several other manufacturers offered HPWH units, and by 1984, 15 firms had entered the market. We believe the EUS project accelerated development of this market by two to five years. HPWH sales are still very modest, with 16,000 units sold in 1985. However, the gains have been substantial, with increases of 50% in 1984 and 33% in 1985.

4.3.4 Heat pumps

ORNL has devoted about $9 million to cooperative research with manufacturers and industry R&D groups such as the Gas Research Institute and the Electric Power Research Institute to advance the efficiency of gas and electric heat pumps. These devices have the potential to heat and cool interior spaces more efficiently than conventional furnaces and air-conditioners. However, they rely on sophisticated thermodynamic cycles which have been difficult to engineer at low cost. ORNL has played a key role in identifying the most promising advanced designs and in supporting R&D with private firms to put these designs on the market.

Because the commercial heat pump market has been highly competitive in the past, with low profit margins, major domestic manufacturers have not devoted extensive resources to improving their products. General Electric and Westinghouse have both sold their HVAC (heating, ventilating, and air-conditioning) manufacturing businesses, presumably because of low profitability, and Borg-Warner is in the process of spinning off its HVAC division for the same reason. Consequently, many heat pump manufacturers are turning to foreign suppliers for their components.

With ORNL support, several U.S. manufacturers, including Trane Corp., Carrier Corp., and Phillips Engineering, are on the verge of marketing gas-fired absorption heat pumps with coefficients of performance 40% greater than those of the best models available to date. These devices will trim $250–$300 per year from the average home's heating and cooling costs.

Electric heat pump efficiencies have recently been climbing by about 2.5% annually, but have still reached only 25% of their theoretical maximum efficiency. ORNL is now working with several firms to develop variable-speed compressors, and to reduce dynamic losses in frosting, defrosting, and cycling that have thus far kept advanced designs from reaching their full potential. It is too early to tell how much electric heat pumps can be improved, but the payoff from efficiency improvements already achieved is substantial. Manufacturers acknowledge that federal R&D support has advanced heat pump development by two to five years, yielding the prospect of $17 billion in savings when high efficiency heat pumps saturate the space-conditioning market.

5 Conclusions

The U.S. DOE Office of Buildings and Community Systems (BCS) has played an important role in advancing energy efficiency in the U.S. buildings sector. Although not all of BCS R&D has been successful, the 13 projects mentioned in this chapter will eventually save consumers $15 billion per year when fully incorporated into the U.S. enonomy. Some of the success of these projects is due to cost-sharing with private industry as well as to effective technology transfer efforts. But in highly fragmented industries like the U.S. building industry it is very unlikely that substantial, continuing R&D energy efficiency would be carried out without government support.

In recent years, U.S. government support for energy-efficiency R&D has dwindled. Private industry has also relaxed its interest in energy-efficiency R&D, duel to low oil prices. This has penalized both the U.S. enconomy and, indirectly, other countries. Abandoning or deferring R&D investments needed now will hurt long-term U.S. economic competitiveness because of the long lead-times, 20 years or more, for development and commercialization of advanced energy-saving technologies. Less productivity in the U.S. economy can in turn have adverse impacts on other countries – not to mention the serious consequences of continued high levels of fossil fuel consumption, with their implications for global warming.

In the U.S. alone, the scale of the problem is impressive, given the sheer numbers of energy-using facilities, equipment, and components that are built and sold each year, many with expected lifetime of 10 to 50 years. Just in the buildings sector, annual additions to the stock include 8 million refrigerators, 2 million single-family and multi-family dwellings, and 100 m^2 of commercial building space. Improving energy efficiency will require a strong and sustained government commitment to R&D, technology transfer, and the development of effective, market-enhancing energy policies. The successes mentioned in this paper are only a beginning, compared to what could be done if the U.S. applied its full technical prowess to creating an energy-efficient, sustainable future.

References

1. Geller, H., Harris, J., Levine, M., Rosenfeld, A. 1987. The role of federal research and development in advancing energy efficiency: A $50 Billion Contribution to the US Economy. *Annual Review of Energy*. 12:357–95.
2. U.S. Dept. of Energy, Office of Conservation. 1986. *FY 1988 Energy Conservation Multi-Year Plan*. Washington, DC.

3. Sterling Hobe Corporation. 1984. *Comparative Analysis of U.S. and Selected Foreign Household Appliance Industries*. Prepared for the US Department of Energy. Building Equipment Division.
4. Sterling Hobe Corporation. 1984. See Ref 3.
5. National Science Foundation 1987. *R&D Indicators*. Washington, DC.
6. Baily, M. N., Chakrabarti, A. K. 1985. Innovation and U.S. competitiveness, *The Brookings Review,* 4(1), Fall.
7. Energy Research and Advisory Board (ERAB). 1985. *Guidelines for DOE Long Term Civilian Research and Development,* Vol. 5. Dec. DOE/S-0042.
8. U.S. Dept. of Commerce, Bureau of the Census. 1987. *Satistical Abstract of the United States 1988*. 108th Ed. Dec.
9. Rosenfeld, A. 1988. *National Institutes for the Built Environment*. Draft LBL report. Lawrence Berkeley Laboratory. Berkeley, CA 94720. February.
10. International Energy Agency (IEA). 1987. *Energy Policies and Programmes of IEA Countries,* 1986 Review. OECD. Paris, France.
11. International Energy Agency. 1987. See Ref 10.
12. Sissine, F. 1987. Science Policy Division, Congressional Research Office, U.S. Congress.

The Challenge of Choices: Technology Options for the Swedish Electricity Sector

Birgit Bodlund, Evan Mills, Tomas Karlsson, and Thomas B. Johansson

Abstract

In this chapter we illustrate alternative future developments of the Swedish electricity sector, based on the new electricity supply and end-use technologies discussed in the other chapters in this volume. We develop integrated electricity supply/demand scenarios in which all electricity sevices associated with a 1.9% annual growth in gross domestic product are delivered, and in which all nuclear power plants are phased out by 2010, as decided by the Swedish Parliament. The scenarios vary in the degree of end-use efficiency assumed and in the reliance on fossil-fuel supply technologies. See the conclusions section for a comparison of the resulting costs and carbon dioxide emissions.

1 Introduction

The power production industry in Sweden is facing a major change. Following a public referendum in 1980 and several subsequent Parliamentary decisions, nuclear power, which now provides about one-half of the country's electric energy and one-third of it's generating capacity, will be phased out by the year 2010. Other related national objectives include continued economic growth, improved environmental quality, reduced oil consumption, and increased use of sustainable, preferably renewable and domestic energy sources.

Environmental goals place strict demands on the electricity system in Sweden, some of which are potentially conflicting in nature. Hydroelectric power now provides the other half of Sweden's electricity. After some minor expansion, the construction of additional hydropower capacity will be forbidden by law. Fossil fuels are not an easy alternative in light of a Spring 1988 decision by Parliament to set general guidelines restricting carbon dioxide emissions to their present level, and encouraging their overall reduction. Meanwhile, official forecasts of electricity demand indicate continued growth from today's value of 130 TWh/year to 145 TWh/year in 2010, including losses (1), with a broad range of uncertainty.[1] This forecast implies a 0.5%/year demand growth rate, much reduced from the 3.9% annual rate in the 1970-82 period and the 5.2%/year rate between 1982 and 1987.

All the above-mentioned factors will have great implications for the strategy of utilities in Sweden as they plan for the future. The global importance of the CO_2 problem could lead to a much larger-scale planning challenge involving many countries. If nations follow the recommendations of the 1988 Toronto conference – to reduce CO_2 emissions by 20% by the year 2005 – key strategies will address the end-use efficiency of energy as well as the choice of primary energy sources and conversion technologies (2).

Around the world, development of technologies for electricity production and end-use efficiency is proceeding, and consumer-oriented efficiency improvement programs are emerging. Most of the existing experience with demand-side management lies with the North American utilities. However, in Sweden, Vattenfall (the Swedish State Power Board) is conducting the "Uppdrag 2000" project to help demonstrate efficient end-use technologies and identify the country's efficiency potential by working directly with power and heat customers, suppliers, and equipment manufacturers.

1.1 Conceptual Approach

One way to gain perspective on the current situation is to construct scenarios that shed some light on the various possibilities and their implications. This chapter presents scenarios that focus especially on the technological options. Our objective is to examine the technologies and strategies that could be chosen to meet the multiple objectives i.e., with respect to the nuclear phase-out, keeping CO_2 emissions constant, and maintaining economic growth. Our analysis makes use of the assessments provided in the other chapters of this

[1] This is the Vattenfall extension of the current forecast interval for 1997 to the year 2010. The 145-TWh demand level represents the center of a 120-170 TWh/year range projected for the year 2010. The corresponding range of average growth rates is -0.4%/year to 1.2%/year.

book and draws as well on other state-of-the-art studies of electricity-efficient end-use and generation technologies.

We begin with the simple observation that electricity *per se* is not of interest, but rather that the demand for electricity is a reflection of the demand for the *services* it can provide: hot showers, cold herring, clean clothes, illumination, motive power, maintenance of a comfortable indoor climate, data storage/retrieval and so on. This perspective has led us to adopt end-use oriented methods in the preparation of our scenarios. Framing the analysis in terms of energy services helps to identify options and constraints that affect the electricity demand-supply balance, while delivering all of the energy services demanded by electricity users in a growing economy. This was in fact the perspective of Thomas Edison in starting the first electric utility. His ambition was to sell *illumination*, rather than electricity, realizing that as the efficiency of lamps inevitably increased that his profits would also increase.

The boundary conditions of our study include all present (1987) electricity services from the direct uses of electricity in Sweden, and, because of the linkages in the supply sector, the energy services from district heating and from the heat produced in industrial cogeneration. Our analysis, then, does not include the energy services derived from fuel presently used in transportation, space heating, and industry, except that consumed in cogeneration. Today's interruptible power (about 10 TWh/year, consumed in large electric boilers used in the district heating systems and industry), is also excluded from the analysis.

The scenarios developed below are all based on the same growth in these services, resulting from projections of 1.9% real growth in gross national product over the 1987-2010 period. This growth rate reflects average growth, which is the same as that used in the long-term economic forecasts made by the Swedish Ministry of Finance (3).

Our approach employs electricity demand models and techniques that are used in the planning work at Vattenfall, with the exceptions noted below. We make use of the same future activity levels in all sectors and changes in lifestyle as those used in a recent Vattenfall demand scenario study.

The demand-side scenarios are constructed in the following way:

- *Reference Scenario.* This is one of several scenarios developed by Vattenfall for use as input into the planning of future power production needs (4). We chose a scenario close to the middle of the current official forecast interval for 2010 (5). This scenario is based on past Swedish experience of the market's behavior, and assumes no policy measures (e.g., incentives, standards, information) beyond those already existing and an anticipated 50% real electricity price increase. Thus, adoption of efficient end-use technologies is based on the assumption that the decision makers – e.g., electricity consumers, retailers, consultants, and utilities – will act on their own accord and invest when motivated to do so by their individual perceptions of payback time, discount rates, and non-economic factors.

To prepare for developing the remaining scenarios, we employ a computational aid that we call the "Frozen-Efficiency baseline". The demand level corresponding to frozen efficiency is a projection of demand as it would develop given the increase in long-term economic activity, and associated structural change within the anticipated GDP growth, but no improvement in efficiencies compared to the base-year levels. The Frozen-Efficiency baseline thus represents the "electric energy service" level in the target year, based on efficiences in 1987. With the following three scenarios, we then explore potential impacts of existing and emerging technological opportunities discussed in this book. The efficiency levels of electricity-efficient technologies are assumed to become average efficiency levels in the stock at the rate of capital turnover and expansion. Fuel switching, from electricity to oil and natural gas, is also incorporated. In contrast to the Reference Scenario, these scenarios utilize efficient end-use technologies that are cost-effective from a societal prespective (6% real discount rate), rather than from a private perspective.

- *Efficiency Scenario.* This scenario is based on high penetration of the most efficient end-use technologies commercially available[2] and cost-effective in comparison with new electricity supply.
- *High-Efficiency Scenario.* This scenario is based on the previous scenario plus adoption of selected end-use technologies for lighting, motors, and several household appliances, in advanced stages of development or already developed but not yet commercialized.
- *Advanced-Technology Scenario.* This scenario is intended to show the effect of adopting additional technology measures and strategies, still at the research and development stage, that can not at this time be judged to be cost-effective.

It must be stressed that scenarios are not forecasts. The last three scenarios represent in no way what *will* happen, but rather what *could* happen if information, education and training, regulations, and incentives are orchestrated to steer the decisions of the many electricity users as well as investors and designers in this direction (see Section 6).

On the supply side, we determine the amount of new generation necessary to meet the difference between the demand levels resulting from each scenario and the existing non-nuclear capacity. We adhere to Sweden's standards of electrical system reliability, insuring that it remains at or above today's criteria for reliable operation. The power generation technologies are assumed to be

[2] By "commercially available" we mean technologies that are available in today's marketplace.

among the most efficient available and meet Sweden's environmental guidelines planned for 1995. Various options to provide the required electricity are then organized into three generation scenarios:

- *Economic Dispatch.* The first scenario is based on the traditional notion of economic dispatch – new power plants are constructed and utilized in order of increasing costs per kilowatt-hour supplied, including all costs for complying with present environmental regulations.
- *Natural Gas/Biomass.* The second scenario includes currently unutilized forest residue resources and some wind, and then proceeds with economic dispatch, excluding coal-based technologies.
- *Environmental Dispatch.* The final generation scenario specifies new power plants in order of increasing net carbon emissions per unit of electricity supplied. Biomass fuels from energy plantations are utilized in this scenario.

To facilitate comparisons between investments in conservation and in new generating capacity, all economics evaluations of the scenarios are made using a 6% real discount rate and do not incorporate the effects of direct energy taxes. For efficiency improvements, we use the annualized investment cost divided by annual savings as our cost-effectiveness indicator. For traditional power plants and wind, we adopt cost estimates used by the Swedish power industry. For new technologies, cost estimates (which by definition have higher uncertainty due to lack of experience) are taken from other chapters in this book. We use the 1987 world prices for oil and coal. The price of natural gas is selected to equalize the busbar costs from gas- and coal-fired condensing power stations based on the assumption that future gas prices will be negotiated this way. The cost estimate for domestic biomass is also based on a 6% real discount rate, and is therefore below the present market price. Real energy prices are assumed to be constant during the 1987-2010 planning period. The basic assumptions are presented in Box 1.

Finally, by integrating our analyses of the demand- and supply-side options, we describe a range of choices and their key economic and environmental implications.

2 Electricity Demand Scenarios

2.1 Methodology

To develop scenarios of future electricity demand, we employ four overview electricity demand models used at Vattenfall: one each for residential

Box 1. Key economic, demographic, and structural assumptions in the scenarios. (All growth statistics are for the 1987-2010 period)

General

- Population growth (million): from 8.41 to 8.60 (0.1%/y).
- Gross domestic product: 1.9% real annual growth (index: 100 → 154).
- Private consumption: 1.5% real annual growth (index: 100 → 141).
- Public consumption: 0.9% real annual growth (index: 100 → 123).
- Nuclear power phase-out period: 1995-2010.
- Energy prices: real world market oil prices, $20/barrel crude (equals $26 heating oil) [the Reference Scenario uses $30/barrel crude in 2010]. Fuel prices for power generation (fuel basis, lower heating value): Coal, 0.60 cents/kWh_{th}; natural gas, 1.10 cents/kWh_{th}; biomass, 1.25 cents/kWh_{th}. All energy prices reflect costs to Sweden, net of direct taxes. 1 US$= 6.3865 Swedish kronor.

Household Appliances

- The number of households (million): single-family – grows from 1.77 to 1.88; multi-family – grows from 1.95 to 2.24.
- Persons/household: decline from 2.82 to 2.67 (single-family) and from 1.75 to 1.60 (multi-family).
- Degree of appliance saturations: Trends in saturation reflect lifestyle changes that include the increased consumption of white goods. (See footnote 3).

Service Sector

- Floor area (million m^2): 148 existing in 1987, 23 additions, 5 demolitions.
- 1987 end-use shares and annual increase in electricity service to 2010: lighting (31%, 2.3%); motors, pumps, fans, etc, (37%, 3.1%); food preparation and refrigeration (17%, 1.6%); office electronics (14%, 4.5%); other (1.5%, 4.5%).

Industry

- Structure: overall output increases by 2.6%/year (value added). The output of mining and textiles declines, respectively, by 1.5% and 1.0% annually. The greatest growth is in the engineering sector – 3.6% annually; chemicals 2.2%/year. Steel industry output 0.8%/year; The largest electricity use is in paper and pulp where output grows at 1.6% annually.

Electric Space Heating and Domestic Hot Water

- Electric heating equipment, GWh and million m^2 in 1987:
  [Electric resistance; electric boilers; heat pumps; multi-fuel furnaces]
  - Singly-family homes – 8,700 (69.0); 4,300 (31.0); 390 (5.9); 5,800 (76.6).
  - Multi-family homes – 900 (5.1); 500 (2.5); 500 (3.6); 2,600 (19.5).
  - Service sector – 1,600 (8.0); 720 (4.4); 100 (1.7); 860 (15.7).
- Multi-fuel heating equipment in single-family homes: net conversions 13.5 million m^2, new homes 20.5 million m^2, demolitions 9 million m^2 (Reference scenario only.)

appliances, services, industry, and space plus water heating. These models are end-use oriented and are essentially based on the notion that total electricity demand is the product of specific use (efficiency) and activity level (structure and utilization), summed over all activities in society. The models incorporate many sources of data on the structure and intensity of electricity demand in Sweden. Total electricity demand is obtained by adding the results from the four models. When forecasting, Vattenfall uses these models plus more detailed material and supplementary models. Where possible, the results of interviews with energy users, large buyers, and firms supplying energy-using equipment are also incorporated. The size of Sweden makes it possible to incorporate customer-specific data, especially within the industrial sector.

The models vary, however, in their suitability for estimating the impacts of changes in end-use efficiency. The residential appliance model is the most well-suited for this purpose because it explicitly accounts for the efficiency of many different appliance end-uses and degree of saturation, lifestyle factors affecting them, and for their lifetimes in the stock. Because of lack of data, the service sector model is the least developed; it provides only minimal end-use breakdowns without explicitly separating efficiency and structural-change factors.

This section will briefly describe the concept and structure of each model. Section 2.2 will then describe the scenarios that we have developed using the models.

2.1.1 Residential sector appliance model

In this model, appliance saturation, stock turnover, and efficiency are specified for 35 residential end-uses, including "cottage industry", well pumps on farms, and even "weekend apartments" rarely occupied. Trends in private consumption, home area, persons per household, mix of single- and multi-family homes, utilization, saturation, and technical improvements are all specified in the model. In addition, small adjustments are made to account for a slight deterioration of efficiency as individual appliances age and for product-change factors such as the anticipated increased use of microwave ovens.[3]

[3] Structural change plays a large role in the development of household appliance electricity use between 1987 and 2010. For example, in single-family homes, the saturation of combination refrigerator/freezers increases from 9% to 40%; the saturation of single electric stoves decreases from 84% to 30% with a shift to microwave/electric stove combinations – increasing from 14% to 70%; the use of electric clothes dryers grows from 18% to 85%; and 45% of the homes are expected to have *two* combination refrigerator/freezers in the year 2010 versus only 6% of the homes in 1987. The model also accounts for growth in electricity services such as increasing TV size and viewing time, refrigerator volumes, etc.

For some of the major appliances (refrigerators, freezers, and combination units), a separate stock/vintage model is used to trace each year's cohort of new appliances through their lifetime in the stock. A logit function reflects that some appliances remain in use for much longer than the average lifetime.

Surveys conducted by National Bureau of Statistics, and interviews with large buyers (e.g., housing management companies) are used to incorporate consumer preferences into the estimates future appliance type, saturation, and efficiency. The most attention is given to refrigeration, especially in multi-family buildings where they represent 40% of non-heating electricity use. Price effects are not accounted for explicitly and are assumed to play a much smaller role in determining efficiency than does technological change.

2.1.2 Service sector model

In this model, electricity demand in 17 sectors is separated into lighting, motors, food preparation and refrigeration, general office electronics, and "other" end-uses.[4] Future demand is estimated based on official projections of changes in employment, floor area per employee, number of firms, and so on (6). Sub-calculations are made for floor-area-independent demand (e.g., trains). The existing end-use and sub-sector data are so limited that no clearly separable efficiency effects are included in the forecasts made with this model. Instead, trended projections based on historic electricity-per-employee statistics are the determining factors. The growth rates prior to 1982 – when the promotion of electric heating was escalated in Sweden – are used by Vattenfall to estimate non-heating electricity trends for the future. Price effects are not formally included in the model.

2.1.3 Industrial sector model

Industrial electricity demand is modeled using a series of linked matrices, each containing parameters relevant to electricity demand. The matrices include fuel share, efficiency improvement, and production. For each matrix, different rate-of-change assumptions are made for four time periods between 1987 and 2010. The model distinguishes among ten sectors plus "small industry", defined as firms with five or fewer employees. End-uses include transport, lighting, processes (four sub categories), and a space heating and "other" category for offices. Demand for each end-use is assumed to grow in direct proportion to

[4] This sector includes a wide variety of customers and they are grouped in the model into two categories: floor-area dependent (e.g., retail stores, wholesale stores, banks, athletic facilities, insurance, and defense) and floor-area independent (e.g., street lighting, trains, agriculture, and construction companies).

production and is then modified by an efficiency parameter. Production is measured in monetary terms, but for the paper and pulp and steel industries production is also measured in physical units. Price effects are not explicitly accounted for in the model. (When used for forecasting, high future prices are assumed to result in the elimination of some electricity-intensive processes, e.g., certain electric arc furnaces, and sometimes to precipitate the closure of entire firms.)

2.1.4 Space heating and domestic hot water model

Space and water heating in all sectors, except industry, are incorporated into one model. Single-family homes, multi-family homes, vacation homes, the service sector, and large heat pumps are treated separately. Virtually no detail is included on water-heating for domestic purposes, an end-use that represents roughly 5 TWh_e per year in single-family homes. For space heating, an attempt is made to separate the effects of projected stock additions, demolitions, efficiency improvements, and price-induced conversions to and from electricity. In addition, about 50% of all efficiency improvements in the appliance model are counted as increased electric heating in homes that have electric heating, because the reduced waste heat from these appliances will increase the need for electric space heating during part of the year.

There is a large quantity of "hidden heating" in Sweden, i.e. electric heating used by customers that are counted as fuel-heating customers in the national statistics but in fact use electric heaters as well as (or instead of) fuel. Vattenfall estimates that single-family homes had at least 2.1 TWh of hidden heating in 1986[5] and multi-family homes 1 TWh.[6] This special type of heating is included in the heating model. Electric space heating in industry is included as an end-use in the industrial model, rather than in the space heating model.

2.2 The Demand Scenarios

The scenarios we develop in the following pages are intended to illustrate a range of choices. The demand scenarios differ in the assumptions made about end-use efficiency and the choice of energy carrier for heating, and in the

5 The PREDECO model and data bases (Lars-Göran Carlsson, personal communication. October 10, 1988) gives a somewhat higher figure for single-family homes, 2.4 TWh for 1986. Note also that electricity consumption in vacation homes has risen from an average of 2,718 kWh/year in 1972 to 4,281 kWh/year in 1986. These homes are not subject to the national thermal construction standards.

6 Personal communication, Gunnar Larsson, Vattenfall, October 31, 1988.

conservation investment criteria used, as explained in Section 1.1. All other economic growth, demographic, and structural factors are held at values based on long-term studies from the Ministry of Finance (see Box 1). The choice of the year 2010 as an "end-point" is made because considerable attention is focused on this date in other energy studies and discussions in Sweden. Of course, the technology changes described here are ones that would occur gradually and the exact year of full implementation would be a function of many factors.

In the following scenarios, we allow ample time for capital turnover. Net additions to the buildings stock are taken directly from those assumed in the Vattenfall model. The 23-year time horizon of each scenario (1987-2010) is longer than the lifetime of household appliances and is well within the time necessary to formulate and deploy programs, incentives, etc., designed to stimulate adoption of more efficient technologies at the time when replacement decisions are being made. This time period is also longer than the lifetime for many technologies in the industrial sector, especially given that equipment is often replaced because there is an economic gain from doing so, rather than because it is "worn out". Aside from these considerations of capital turnover, the exact shape of the path from 1987 to 2010 is not analyzed.

All calculations and assumptions are shown in (7), a detailed companion document to this chapter. Brief supporting information and sources are given in the footnotes. The assumptions are based on chapters in this book and other published sources.

2.2.1 Frozen-Efficiency electricity demand

The level of electricity *services* required in 2010 can be thought of as today's activity level plus all growth in activity that would occur if the economy expanded as shown in Box 1 and end-use electricity efficiency remained at it's 1987 level – e.g., a 10% increase in steel production and no change in electricity use per ton of steel yields a 10% increase in the electricity services provided and a 10% increase in the electricity required. New technologies are thus assumed to have the same electricity efficiency as the present average.[7]

[7] Determining the Frozen-Efficiency baselines for space heating and for the service sector is not as straight-forward as for other sectors. In developing the heating baseline, we assume that there are no efficiency improvements beyond 1987, conversions *to* electricity occur as shown in the Reference Scenario, new homes are constructed at today's average use per home, and internal heat gains increase according to the Frozen-Efficiency baseline for household appliances. The baseline for the service sector is an extrapolation of the 1970–1987 growth trend. Although an attempt has been made to remove electric heating from the historic data

This "Frozen-Efficiency" baseline is included as an analytical tool only, in order to be able to isolate the impact of efficiency in the scenarios from that of other factors embedded in the models.

The resulting level of electricity services (measured for the remainder of the chapter in units of TWh equivalent, TWh_{eq}) forms a basis from which to measure prospective efficiency improvements, without double-counting those embedded in any previous forecast or scenario. Box 2 contains the efficiency assumptions used in each of the following scenarios.

2.2.2 Reference Scenario

The Reference Scenario estimates electricity demand in 2010, assuming market-driven conservation behavior in an environment with real electricity prices 50% higher than today's.[8] Conservation effects are included, based on individuals' investment perspectives. Fuel-switching in homes is based on an economic comparison of electric resistance heating only, multi-fuel systems, or oil-only systems. Given time-of-use rate structures, customers are assumed to switch heating systems if there is at least a $85/year net benefit from doing so.[9] This is intended to reflect historic consumer behavior. The scenario assumes the implementation of 90% of those changes fulfilling this criterion. An average tariff (*mellantariff*) is used for those with only electric resistance systems. The scenario results in a net 3 TWh of conversions to electricity, mostly as a result of rate structures that favor the use of electricity for summertime water heating. An additional 0.6 TWh increase in space heating demand results from the net demolition/construction of buildings.

New buildings in the service sector are assumed to use 140 kWh/m^2-year (excluding electric space heating), due to the increased number of electric end-uses. As a result of retrofits in existing buildings, the stock average remains at the present level of ~100 kWh/m^2-y in 2010. Developments in the industrial sector are quite varied. Efficiency as well as production volume decline in some sectors and increase in others. The average efficiency of household appliances in 2010 is expected to be near today's best, due to technological change that is only partly driven by design changes intended to reduce electricity use.

series upon which this estimate is based, the absence of high-quality statistics leaves open the possibility that heating growth in the 1980s has biased the extrapolation.

8 This is one of six scenarios (Scenario "F") developed at Vattenfall in the Spring of 1988 (22).

9 Defined as the difference between the value of annual energy cost savings and the levelized capital cost of the conversion. Typical wintertime furnace efficiencies of approximately 80% are assumed.

Box 2. Efficiency assumptions in the Scenarios.* (Except where noted otherwise, the numbers shown refer to the percent of base-year intensity)

Appliances[a]
("Reference"), "Efficiency", "High-Efficiency", and "Advanced-Technology" Scenarios
Food storage (69), 49, 24, 24
Food preparation (86), 85, 69, 69
Lighting (61), 60, 59, 59[b]
Dish and clothes washing/drying (56), 55, 19, 19
Miscellaneous appliances (89), 89, 76, 76

Services
Lighting (55), 36, 30, 30[b]
Motors, pumps, fans, compressors (70), 65, 60, 60[c]
Refrigeration and food preparation: 60 in all cases[d]
Office electronics (65), 32, 20, 20[e]
New buildings 140 kWh/m^2, 50 kWh/m^2 in all other cases (excludes space heating)[f]

Industry[g]
Lighting: (71), 45, 40, 40[b]
Motors, pumps, fans, compressors (85), 70, 60, 50[c]
Mechanical pulping: (included in total motors), 90, 80, 70[c]
Process heating: (60)**, 80, 80, 80[h]
Electrolysis: (75)**, 85, 85, 85[i]
Steel scrap-melting: (included in process heat), 80, 25, 25[j]
Conversions from electricity to natural gas (TWh_e): (0), 2.1, 3.6, 3.6[h]

Space and water heating
Multi-fuel furnaces – conversion to oil (CO) or earth-source heat pumps (HP) [COP 2.7 Efficiency Scenario, 3.3 High-Efficiency and Advanced-Technology]: (0% (CO) and 0% (HP)), 72% and 18% penetration in all other cases[k]
Existing heat pumps estimated to have COPs ~ 2.3 and are replaced (over 23 years) with the efficient models described in the previous note.
Electric boilers – earth-source heat pumps in homes with electric boilers: (0%), 90% penetration in all other cases[l]
Large heat pumps: (60% conversion), 80%, 80%, 100%[m]
Envelope retrofits in homes with direct electric resistance heating (penetration): (not indicated), 50% roof retrofits, 20% wall retrofits in all other cases[n]
New homes: (8,000 kWh/year for space and water heating), 2,500 kWh/year space heat and 2,500 kWh/year water heat in all other cases[o]

*Notes to the Box are located before the references section, page 940.
**Includes price-induced plant closures (Reference Scenario only).
The Reference Scenario values are enclosed in parentheses to distinguish them from those for the other scenarios. The Reference Scenario numbers are *implied* by the Frozen Efficiency results, i.e., they were not used to derive the scenario.

2.2.3 Efficiency Scenario

In this scenario, we retain all of the structural factors from the Reference Scenario and explore the effect of high penetration of the best commercially available and cost-effective efficiency technologies described below and in greater detail in Box 2.

Household appliances

We assume that the efficiencies of presently best-available appliances become average in the stock.

Service sector

In the service sector, we estimate the potential efficiency improvements from new lighting technologies; adjustable-speed drives for motors, pumps, fans, compressors (and more efficient motors and systems design); and retrofit of food refrigeration and preparation equipment in existing buildings, and from high-efficiency office electronics. Measured results from 206 U.S. buildings show median electricity savings of 19% from retrofits, roughly 90% of which had payback times of five years or less (23).[10] New buildings are assumed to be constructed with electricity intensities comparable to the values described by Abel in this volume (24) and from the measured performance for several hundred of the best buildings known (25, 26).

Industry

In industry, we estimate the increased energy productivity opportunities from improved lighting, adjustable-speed drives and more efficient motors, used in well-designed systems moving gases or liquids, process-heat conservation measures and fuel switching (2.1 TWh_e), and efficiency improvements in "office" equipment.

[10] Including fuel and electricity, this study contains 311 buildings (schools, small offices, hospitals, large offices, other services). The retrofits include operations and maintenance, lighting, ventilation, envelope measures, water heating, load management, and others. The economic data are available for a subset of 134 buildings with electricity-saving retrofits. See the reference for general information and results; additional electricity-specific data are from Kathy Greely and Mary Ann Piette, Lawrence Berkeley Laboratory, memorandum of November 9, 1987.

Space heating and domestic hot water

Savings in this scenario are derived from a combination of building envelope and heating system retrofits, plus fuel switching. For this purpose, we segment the stock into four parts, based on annual surveys by the national bureau of statistics in Sweden. The first segment has electric resistance heating and no ability to use an alternative fuel. The second segment heats with hydronic electric-only systems. Other buildings use electric heat pumps. The fourth segment currently has the ability to heat with two or more fuels, including electricity.[11]

To estimate penetration in households, we assume, as in the Reference Scenario, that 10% of the homes do nothing whatsoever. For those remaining: electric-resistance homes install insulation and upgrade their glazing if other renovation work is performed (27).[12] The very-high-use homes (>30,000 kWh/year) install ground-source heat pumps (COP 2.7); homes with electric-only boilers install heat pumps; homes that can switch fuels are assumed to do so or to install heat pumps.

To be conservative, we assume that all new homes use electric heating. These homes conform to the state-of-the-art performance described in the chapter by Elmroth.[13] Triple-pane windows with coatings or gas fill, described in (28), are included in new energy-efficient homes in the scenario. The current Swedish construction standard requires triple-glazing, but not coatings or gas fill.

We do not analyze envelope efficiency improvements in buildings that switch fuels. There is a potential quick increase in electricity demand due to re-conversion if the electricity-to-oil price ratio for any reason decreases (and policies do not respond).

[11] All base-year values for electric heating are 1987 consumption, normalized to average-year weather conditions (Personal communication, Anders Sjögren, Vattenfall, January 15, 1989).

[12] According to Elmroth's chapter, of all homes heated with electric-resistance systems and constructed before 1982, only 25% of the window area (6.57 million square meters) is triple-glazed. Roughly 1% of the area is single-glazed windows and 1% is quadruple-glazed. Even more remarkable, only 60% of the window area installed between 1975 and 1981 was triple-glazed.

[13] According to the Reference Scenario, the increased consumption for single-family homes is 1.7 TWh/year by 2010. Were all of these homes heated at the efficiency described in Elmroth's chapter (2,500 kWh/y for space heating and 2,500 kWh/y for water heating) the resultant demand growth would be 1.1 TWh.

2.2.4 High-Efficiency Scenario

In this scenario, we assume (in addition to the previous scenario) slightly more effective lighting system retrofits in services and industry, adjustable-speed drive applications, and several advanced household appliances that are not commercialized at this time but can be designed using existing components. Heat pump COPs increase from 2.7 to 3.3 based on (29). We make a special scenario for steel, based the Plasmascrap process described in (30, 31). Plasma beam heating is discussed more generally in (32). Conversions to natural gas in industrial process heating increase from 2.1 to 2.8 TWh_e in 2010.

2.2.5 Advanced-Technology Scenario

The final scenario makes use of measures that are technically possible but can not necessarily be judged cost-effective at present. We convert to oil and conserve a total of 0.5 TWh of electric heating in vacation homes; (out of a total of 2.4 TWh), 0.5 TWh of electricity used in large heat pumps, and an additional 1 TWh of electric heating in single-family homes. In the steel industry, we assume the introduction of direct casting (33) which saves 1.3 TWh by eliminating 75% of the estimated need for motor power in the rolling process.

2.3 Electricity Demand Results

2.3.1 Overview

The level of electricity services corresponding to the Frozen-Efficiency baseline is equivalent to 194 TWh/year by 2010. The Reference Scenario estimates generation at 140 TWh_e – which means that 54 TWh_{eq} equivalent of the required services will be provided with increased end-use efficiency and alternate fuels. In the Efficiency, High-Efficiency, and Advanced-Technology Scenarios, electricity generation falls to 111 TWh, 96 TWh, and 88 TWh, respectively.

The historic and projected sectoral components of total supply and demand are shown in Figures 1a and 1b. In light of the policy to phase Sweden's nuclear plants out of operation by 2010, the question immediately arises as to when new capacity will be required. The stepped curves in Figure 1a reflect

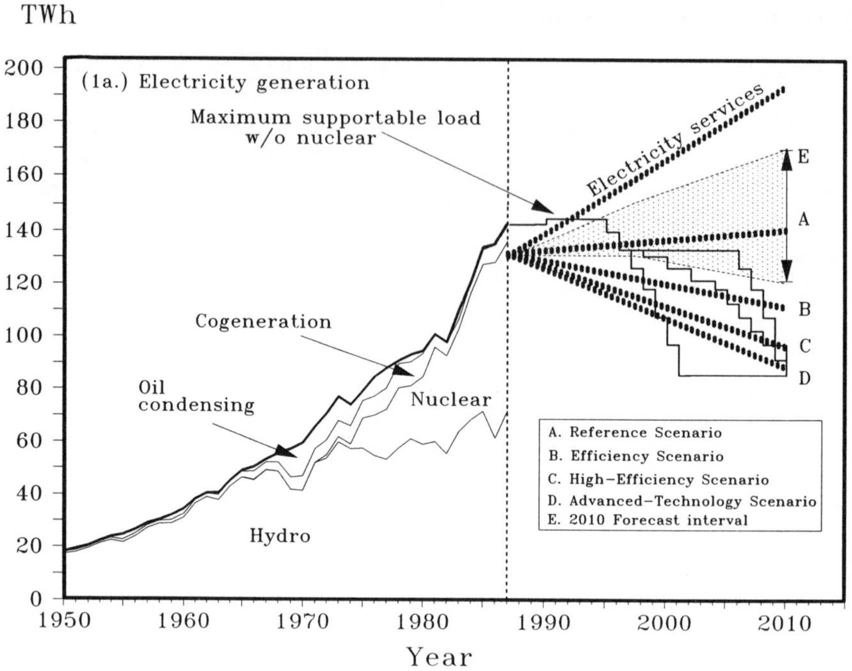

Figures 1a and 1b. Historic electricity generation and demand scenarios for the future. Figure 1a shows the evolution of Sweden's generation mix up to 1987. Peaking turbines have provided only about 0.1 TWh/year since the late 1960s. The three stepped curves represent the range of proposed schedules for phasing out Sweden's 12 nuclear reactors. The curves show the level of electricity demand acceptable with the current system reliability levels and no new capacity additions. The small vertical gap between 1987 and the scenario starting point represents interruptible power. As shown in Figure 1b, since the beginning of the nuclear building period, electric space heating has risen from almost zero to over 30 TWh/year.

three possible nuclear phase-out programs.[14] The middle case reflects a linear phase-out plan, and faster and slower cases are included for comparison. These three curves have clearly distinct impacts on the schedules for the construction of new generating capacity, which is required when demand falls below the stepped curve.

[14] The curves represent today's hydro and fuel-based generation plus that from the nuclear plants remaining in each year of the period up to 2010, at which time nuclear generation will be zero. The height of the curves includes today's accepted reliability margin, using the rule of thumb that the maximum tolerable demand = 0.80 x hydro production potential + 0.90 x thermal production potential. (Environmental restrictions will require additional investments in the existing fuel-based power plants if they are to be used much more heavily than is the case today.)

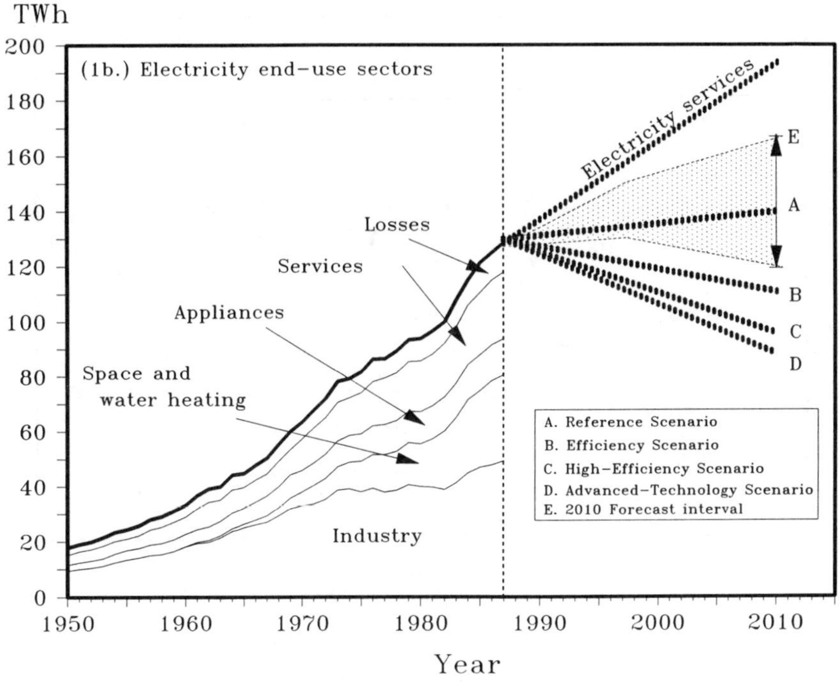

The sectoral breakdown for each of the four scenarios is shown in Figure 2. Additional detail on the sector-specific supply of electricity is shown in Table 1 and in Figures 3 through 6. Following are some key observations for each sector, and the overall reduction in specific needs (Efficiency, High-Efficiency, and Advanced-Technology Scenarios):

- Substantial increases in efficiency occur in the household appliances sector. Yet structural changes (increased numbers of appliances) offset much of the savings. Food storage appliances offer perhaps the greatest efficiency potential. Overall specific needs decline to 66%, 47% and 47% of their 1987 levels.
- In the service sector, demand falls, due mostly to improvements in existing buildings. New buildings represent 15% of the total floor area in 2010. The greatest efficiency gains are in motors, etc., and in lighting. Overall specific needs decline to 50%, 44% and 44% of their 1987 levels.
- In the industrial sector, electricity demand in the Efficiency Scenario remains about the same as in 1987. Motors, fans, pumps, etc., are by far the most important end-use (64% of all industrial electricity in 1987) and paper and pulp is by far the most important sub-sector. Overall specific needs decline to 70%, 60% and 53% of their 1987 levels.

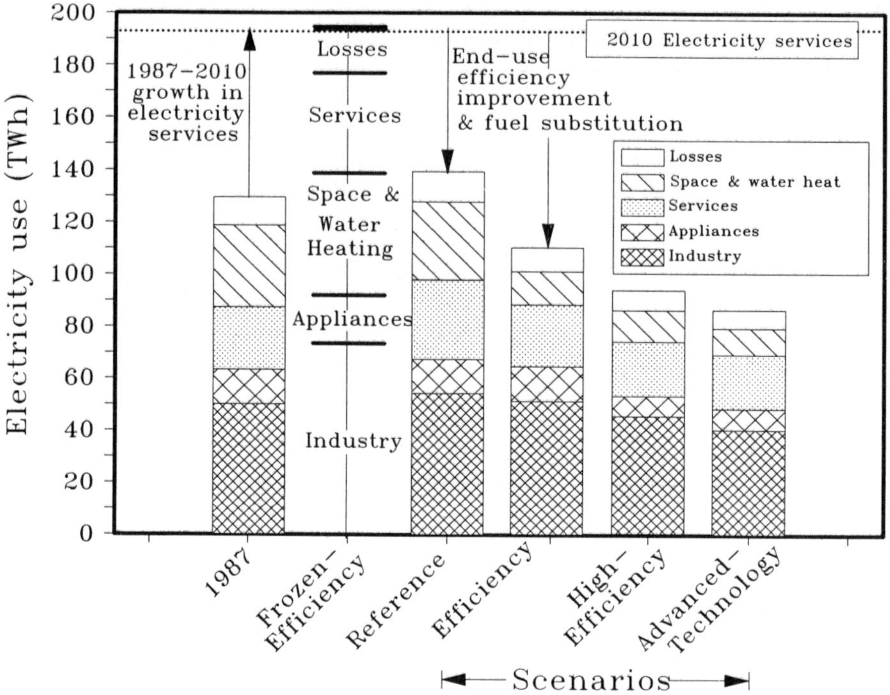

Figure 2. Total electricity demand plus losses for 1987 and for the four scenarios in 2010. The level of electricity services supplied (from the Frozen-Efficiency calculations) is shown as the horizontal dotted line. The height of the bars represents electricity use and the gap between the top of each bar and the horizontal line represents the electricity services provided with efficiency improvements and fuel substitution.

- Existing fuel and equipment flexibility in the space heating sector contributes to substantial electricity reductions via fuel substitution and the use of heat pumps in existing hydronic systems. The Reference Scenario also includes conversions to electricity of 3 TWh_e (because a crude oil price of $30/barrel is assumed). The Efficiency Scenario incorporates 6.9 TWh_e of fuel switching in buildings, which would increase Sweden's 1987 total oil imports by 4%.[15] Overall specific needs decline to 39%, 37% and 32% of their 1987 levels.
- Transmission and distribution losses are 9.2 TWh in the Efficiency Scenario, versus 10.6 TWh in 1987. Both numbers are based on losses equal to 9% of demand.

[15] Based on an oil boiler efficiency of 80%.

Table 1. Summary statistics for the scenarios.

TWh_e	1987	2010	Average %/y growth rates (23 yrs)
Frozen-Efficiency baseline			
Industry	49.3	73.9	1.8%
Household appliances	13.2	18.1	1.4%
Services	24.2	46.9	2.9%
Space and water heating[a]	31.3	38.7	0.6%
Losses	10.6	16.0	1.8%
Total	**128.6**	**193.6**	1.8%
Reference Scenario			
Industry		54.2	0.4%
Household appliances		13.2	0.0%
Services		30.5	1.0%
Space and water heating		30.0	– 0.2%
Losses		11.5	0.4%
Total		**139.5**	0.4%
Efficiency Scenario			
Industry		51.4	0.2%
Household appliances		11.9	– 0.4%
Services		23.5	– 0.1%
Space and water heating		14.9	– 3.2%
Losses		9.2	– 0.6%
Total		**110.9**	– 0.6%
High-Efficiency Scenario			
Industry		44.2	– 0.5%
Household appliances		8.5	– 1.9%
Services		20.7	– 0.7%
Space and water heating		14.5	– 3.3%
Losses		7.9	– 1.3%
Total		**95.9**	– 1.3%
Advanced-Technology Scenario			
Industry		39.0	– 1.0%
Household appliances		8.5	– 1.9%
Services		20.7	– 0.7%
Space and water heating		12.5	– 3.9%
Losses		7.3	– 1.6%
Total		**88.0**	– 1.6%

[a] In 1987, the end use shares were: single-family homes (61.0%), milti-family homes (14.3%), vacation homes (6.3%), services (10.4%), municipal heat pumps for district heating (8.0%). Values may not add perfectly due to rounding errors.

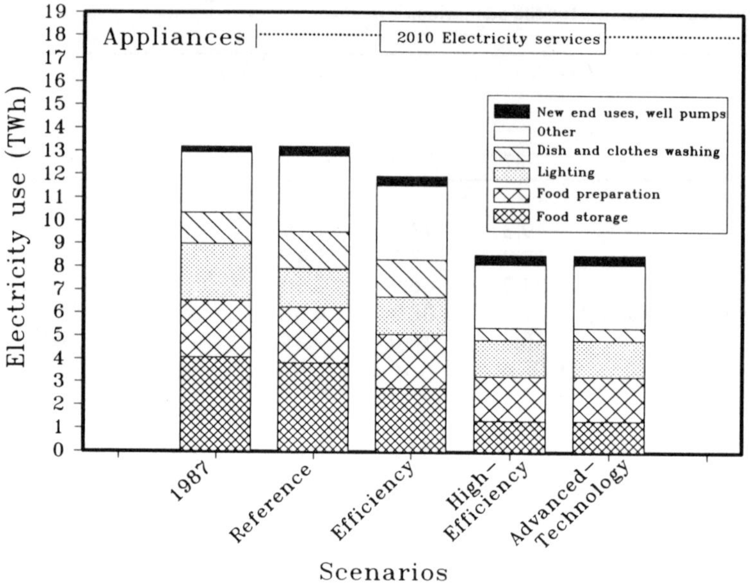

Figure 3. Residential appliance electricity demand results for 1987 and for the four scenarios in 2010. Six end-uses are shown.

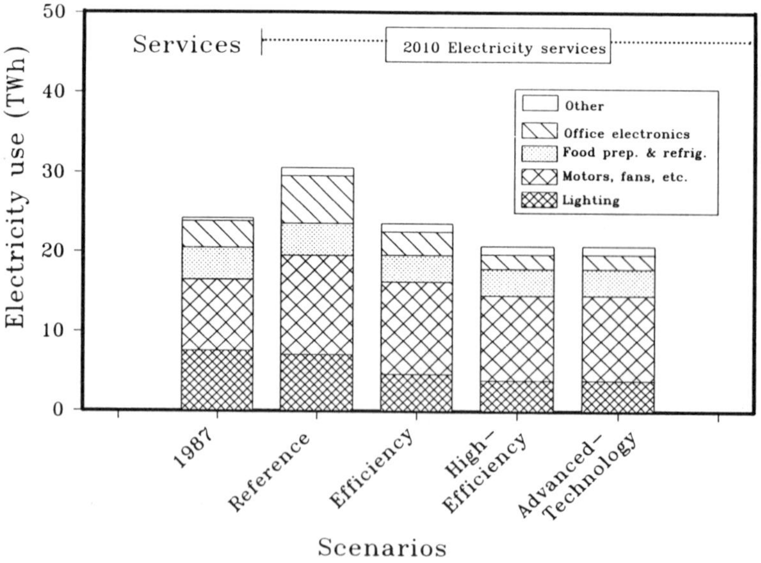

Figure 4. Service sector electricity demand results for 1987 and for the four scenarios in 2010. Five end-uses are shown.

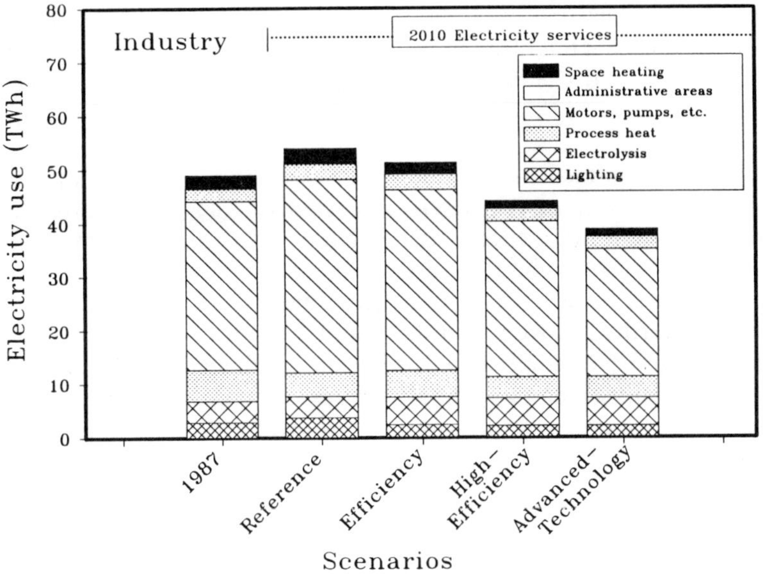

Figure 5. Industrial sector electricity demand results for 1987 and for the four scenarios in 2010. Six end-uses are shown.

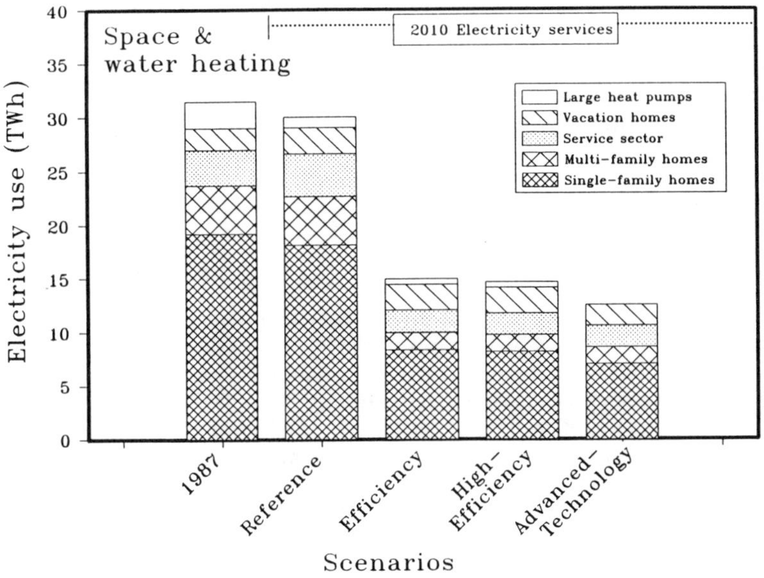

Figure 6. Space and water heating electricity demand results for 1987 and the four scenarios. Five sub-sectors are shown. The main source of the difference between the Reference and the two Efficiency Scenarios is fuel switching and use of heat pumps.

Our results are consistent with those from a number of other similar studies in Nordic countries and elsewhere. Recent scenarios made for Denmark resulted in overall end-use electricity intensities at 66% of the 1986 level for presently commercialized technology and 51% for advanced technologies that could be made available within 10 years (34). A similar, and more detailed, study has been completed for Finland (35). The authors identify a potential to reduce demand to 60% of the level forecasted for the year 2020. There are many examples from the United States, most of which focus on the residential sector and include both energy and peak-demand potentials (36, 37). At the State level, studies have been done for Michigan (38), Texas (39), the four-state Pacific Northwest Region (40)[16] and California (41, 42). Studies of electricity demand potentials have also been conducted in Canada (43, 44). Many of these analyses employ supply curves of conserved electricity to rank the various conservation measures in order of levelized cost per kilowatt-hour.

2.3.2 Efficiency economics

For most demand-side technologies in the Efficiency Scenario, we used the cost of conserved electricity (CCE) statistic to calculate cost effectiveness.[17] The CCE is the annualized investment in increased efficiency (and incremental operations and maintenance, if any) divided by the annual reduction in electricity use required to provide the same service. Thus, the result has the same units as the price of electricity (cents/kWh) and can be easily compared

[16] Perhaps most relevant to the Nordic situation, this study evaluated the electric demand-side potentials in a large hydropower-based region of the Pacific Northwestern United States where a large generating company sells power to many small distributors. This study determined that the base case forecast of residential sector demand by the year 2000 (7,728 average megawatts) could be reduced to 58% of their forecasted levels if all technically feasible conservation measures costing less than 15 cents/kWh were implemented. Of this total, 87% could be realized at a cost of 4 cents/kWh.

[17] The CCE is effectively a "busbar" cost indicator for conservation technologies and can thus be directly compared with levelized cost (fixed + variable) of new electricity supply. To ensure compatibility, we account for the reinvestment costs if efficiency technologies must be replaced during the 30-year economic lifetime of a reference power plant. (Note: the economic calculations in this chapter are based on 1987 US dollars. We use a 6% real discount rate to determine the choice of end-use strategies, except in the Reference Scenario where consumer investment behavior corresponds to a significantly higher discount rates.)

with retail prices or with the busbar cost of new supply.[18] Table 2 summarizes the results. Note that several measures are shown at zero cost because they involve fuel switching. The cost of the new fuel is counted instead in the final economic assessment in Section 4.

From the table, we can see that substantial cost-effective efficiency resources are available in Sweden, i.e., are available at a fraction of the cost of new electricity supply. In Section 4.1 the demand-side resources we have identified will be ranked along with electricity supply technologies in a grand supply curve for the year 2010.

Although capital investment is commonly required to achieve savings, some efficiency-improving technologies can increase electricity efficiency at zero or even negative cost. This can result from:

1 gains derived from no- or low-cost changes in operations of existing technology,

2 more careful selection among appliances available in the marketplace,

3 technological change that has a "side-effect" of improved efficiency,

4 reductions in labor and/or replacement costs that exceed the investment, or

5 complex engineering interactions that create "bonuses" by reducing the energy demand in an end-use other than that targeted with the efficiency technology.

Today's marketplace for refrigerators and freezers in Sweden provides a clear example of the sometimes imperceptible incremental cost for selecting the best-available appliances (point 2 in the list above). As shown in Figures 7a to 7c, there is no correlation between the price and the energy efficiency of refrigerators in Sweden.[19] This also holds for freezers and combination refrigerator/freezers.

[18] For a given pool of electricity consumers that make investments to increase end-use efficiency, there will be a range of costs and savings. Thus, one can envision a corresponding distribution of costs of conserved energy. From a societal perspective, over-investment in efficiency may occur if some participants in the right-hand tail of this distribution have CCEs higher than the cost of new energy supply. This risk is low when the average CCE is itself well below this "break-even" cost and can be further minimized by careful definition of the "target group". Some analysts allow the more economic projects to "carry" the less economic ones, but our methodology is to judge each investment on its own merits.

[19] Mills, E. 1988. *Energy and the marketplace for residential appliances in Sweden.* Department of Environment and Energy Systems Studies. Lund University. Manuscript in preparation. Scatter diagrams of retail price versus total annual electricity use also show little or no indication of an extra cost for higher efficiency. This is true even within narrow size ranges (e.g., among 200-250 liter models). We have observed the same result for residential clothes washers and for stand-alone domestic water heaters.

Table 2. Selected cost-effectiveness calculations for the Efficiency Scenario.

End-use technology or strategy	Annual savings (TWh)	Cost of conserved or switched electricity (cents/kWh)[a]
Measures requiring incremental investment		
Lighting (all sectors)[b]	12.0	
compact fluorescents (3000 h/y)[b]		1.0
fluorescent systems (2000 h/y)[b]		1.6
Motors & pumps, compressors, etc[c]	18.5	1.4
Commercial food refrigeration[d]	2.4	2.5
Residential heat pumps[e]	5.7	
of which electric resistance home[e]	(2.0)	6.8
of which electric boiler home[e]	(3.7)	4.5
Appliances (excl. lighting)[f]	5.1	0.6
Subtotal	43.7	1.9
Measures requiring little or no investment		
New home construction[g]	0.6	very low
Miscellaneous office equipment[h]	6.4	very low
Large heat pumps[i]	2.0	0.0
Fuel switching for space heating[j]	6.9	0.0
Transmission & distribution losses[k]	6.8	0.0
Subtotal	22.6	
Total	66.3	1.2
Residual non-monetized savings[l]	16.3	?

[a] Electricity savings are measured here as the difference between the Frozen-Efficiency Baseline and the Efficiency Scenario results. Fuel substitution is shown at zero cost because there is no capital cost and the costs of the alternate fuels themselves are counted instead in Section 4.2. For lighting, the simple average of the two values shown is used. All CCEs are comparable to the levelized cost of a power plant with a 30-year economic life and 6% real discount rate. The weighted average excludes the TWh in the last row.

[b] For compact fluorescents, the lifetime of an improved lamp is 10,000 hours (14-Watt), versus 1,000 hours (60-Watt) for the replaced lamp. Thus, 10 incandescents need not be purchased. We assume a retail price of $10 for the high-efficiency Satco lamp (45) and an avoided cost of $0.80 for the incandescents. These costs represent recent prices in the U.S., which have fallen by more than a factor of two in just a few years. A similar development should occur in Sweden. For ballasts, we assume 260 8-hour days/year operation, plus costs and savings according to the case study in (46).

[c] Capital costs from ABB Corporate Research (personal communication, Lars Gertmar, December 15, 1988) and represent costs relevant to the Swedish setting: $296/kW (18.5 kW motor) and $104/kW (750 kW motor). Installation costs assumed at 23% of capital costs for the

small drive and 8% for the large drive. A 15-year lifetime is also assumed in the calculations (47). The corresponding CCEs for 3,000 and 4,000 hours/year are 2.3 cents/kW and 1.8 cents/kWh, respectively. In his chapter, Baldwin suggests substantially lower installed costs: on the order of $100/kW.

[d] Savings result from a combination of measures described in (48). The costs of conserved energy are the average from the following list: retrofit glass doors, plastic curtains, unequal compressor systems, variable-speed drives, and floating head pressure controls.

[e] Earth-source heat pumps with closed-loop working fluid. Applied to high-use electric-resistance homes (parts/labor $13,500) and to average multi-fuel homes with pre-existing storage/distribution systems ($9,400) (49). COP assumed = 2.7; lifetime 15 years.

[f] The CCE is the average for all appliances from the chapter by Nørgård in this volume (50).

[g] The average electrically-heated home in the stock uses approximately 17,000 kWh/year for space and water heating (and this is the frozen-efficiency assumption), and the current building Norm results in 8,000 kWh/year (Elmroth's chapter in this volume). Most of these savings beyond the Norm (3,000 kWh) come at little or no incremental cost.

[h] The savings estimate is based on technological changes that are happening naturally in the marketplace. The chapter by Norford *et al* refers to several examples where incremental costs in today's marketplace are zero.

[i] Assumes that some large heat pumps are not replaced at the end of their useful lifetime.

j Fuel switching assumed to occur at zero capital cost because we consider only homes with multi-fuel systems. The total cost for heating energy declines as a result of switching. In some homes an investment might be required, for example, to revitalize oil burners. Electricity prices from the Reference Scenario provide a strong incentive to switch fuels during on-peak times: 11.1 cents/kWh on-peak electricity, 5.4 cents oil.

[k] Transmission and distribution losses (9% of generation) are reduced because total demand is reduced. There is no additional cost for these savings.

[l] The residual includes industrial process heat (3.4 TWh); industrial space heating (2.3 TWh); industrial electrolysis (0.9 TWh), triple-pane or low-e windows (0.3 TWh), residential building envelope retrofits (1.4 TWh), commercial buildings construction (2.5 TWh) [Commercial buildings are constructed based on an estimate from the chapter by Abel in this volume. The performance target is arrived at on the basis that it is cost-effective from the perspective of the building owner]. There is no reason to assume that the CCEs for these measures are higher than those for the other measures in the table.

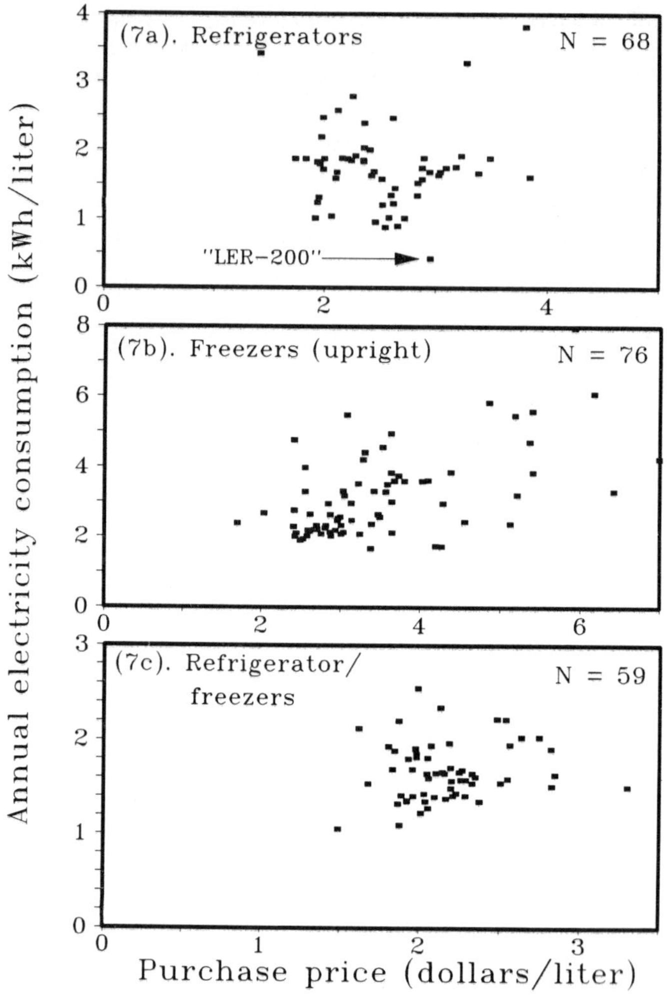

Figure 7a-7c. The non-correlation between appliance retail price and energy-efficiency. Each data point is a refrigerator, freezer, or combination refrigerator-freezer model sold in Sweden as of August 1987. Note that for a single price level, efficiency, measured as kWh/liter-year, can vary by a factor of two or three.

Each model is government-tested according to a standard procedure (25°C ambient, no door openings, no auto-defrost cycle). To compensate for varying freezer volumes in refrigerators with small freezer compartments and in combination units, freezer volumes are adjusted to their refrigerated-volume equivalents according to the U.S. Department of Energy testing methodology. No such adjustment is made for freezers. A number of specialty models are excluded. The 200-liter Danish LER-200 (now marketed in Denmark by Brdr. Gram A/S) is shown for reference. A model of nearly identical size and price, but with twice the electricity use, is made by the same manufacturer.

2.3.3 Electricity and economic well-being

Assuming that Sweden's GDP continues to grow at 1.9%/year, electricity demand would decline at 0.6%/year, according to the Efficiency Scenario. Thus, the ratio of electricity use to GDP would decline sharply while economic growth continues.[20] Sweden, like Canada and Norway, has enjoyed relatively low electricity prices and, partly as a result of this situation, many electricity-intensive industries and widespread use of electric space heating in their cold climates have contributed to a high level of electricity use. However, assuming the development shown in the Efficiency Scenario, Sweden's 2010 electricity intensity will be similar to today's values for many other industrialized countries, as shown in Figure 8.[21]

Comparisons of electricity use per capita are also interesting. In 1987, the intensity was about 17,400 kWh/capita. This may be compared with that of Japan (5,600 kWh/capita), West Germany (6,700 kWh/capita), or the U.S. (11,000 kWh/capita). The comparison is shown in the left-hand set of bars in Figure 8 for the OECD as a whole and for 13 member countries.

Figure 9 puts each of the four scenarios in perspective with historic electricity demand and GDP. An interesting characteristic of the Frozen-Efficiency baseline is that the aggregate result for all sectors show electricity use per unit of GDP as almost the same in the year 2010 as in 1987. This is in line with our assumptions insofar as electricity use per GDP can be considered a measure of national electricity use. This implies a break in the historic trend of increasing electricity use per unit of economic activity. This comes about because of structural changes in the economy towards less electricity-intensive activities, and in much-reduced growth in electric space heating. In the 2010 Reference Scenario, electricity/GDP has fallen by 30%; in the Efficiency Scenario it falls by about 50%.

2.3.4 Factors that could lead to other scenarios

No forecast or scenario can incorporate every determinant of energy use, nor can it account for all of the possible permutations of it's assumptions. The scenarios in this chapter are constructed within a given set assumptions about the economy and about technologies. Following is a list of factors that could contribute to scenarios of either higher or lower electricity use. Some factors

[20] Gross domestic product statistics from (51) through 1980. Data for 1981 to 1987 are from (52). For the 1987 to 2010 estimate, we use the standard 1.9% annual growth estimate.
[21] Decoupling of electricity and GDP is occurring in some other industrialized countries. See Rosenfeld & Koomey (53) for an illustration of trends in the United States.

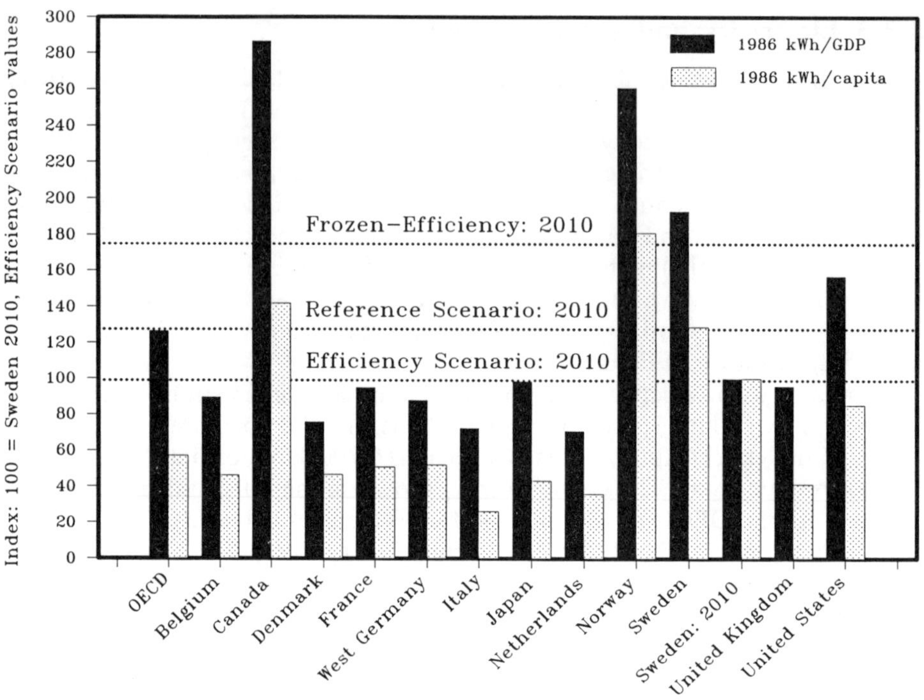

Figure 8. Comparison of electricity intensity (measured as kWh/capita and kWh/GDP) in the Efficiency Scenario with that of other industrialized countries. The results for Sweden in 2010 are transformed into an index = 100 (third pair of bars from the right). Sweden's intensities in 1986 as well as those in the other countries are thus compared with the Swedish 2010 values. Horizontal lines for two other 2010 demand levels – relative to the Efficiency Scenario – are also shown.

reflect differences in energy service levels that would be associated with different developments of society and the economy:

- Alternate levels of energy services demanded by consumers
- Alternate GDP growth trends
- Changes in lifestyle other than those assumed in the structural and utilization factors in the appliance model and implied in the growth of electricity services in the Frozen-Efficiency baseline for the service sector
- Alternate levels of immigration and domestic population growth
- Different patterns of structural change in industry, e.g., a larger shift to more electro-mechanical pulping in the paper and pulp industry, than the 44% increase in production assumed in the scenarios

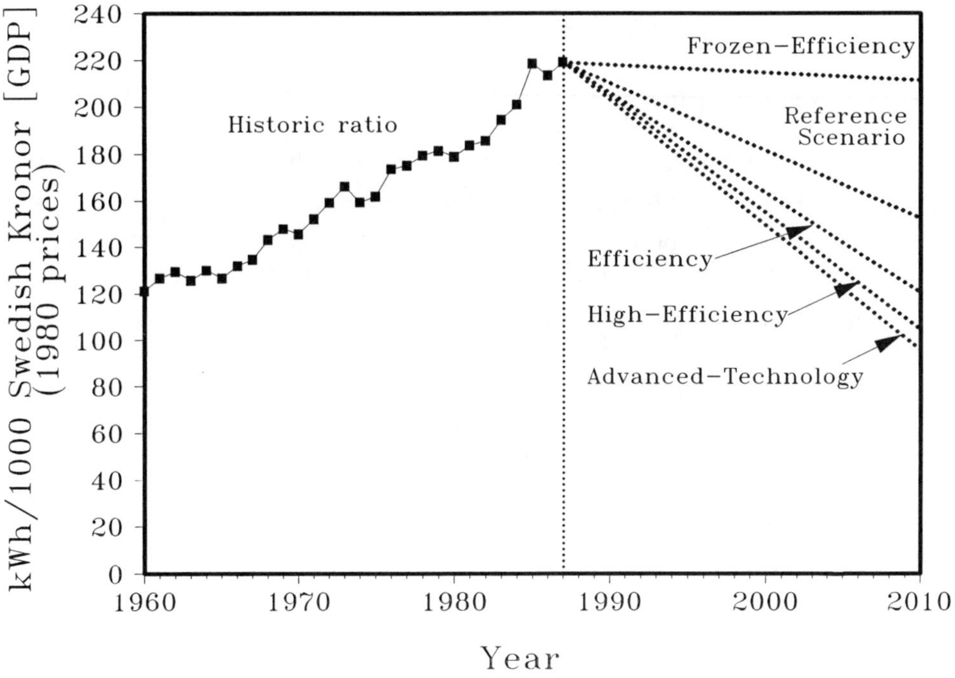

Figure 9. Translation of the scenario results into overall national efficiency trends. Here, efficiency is defined as electricity use per unit of gross domestic product. The historic trend shows how electricity use increased much more rapidly than overall economic growth up to 1987, due in part to the steady growth of electric heating. All scenarios are based on annual GDP growth of 1.9%. The values shown exclude interruptible power, up to 10 kWh/1000 kronor in recent years.

Other factors relate to the end-use technologies. The Efficiency and High-Efficiency scenarios are conservative insofar as a less-than-exhaustive list of efficiency technologies and strategies are included. There are many unexplored efficiency opportunities. The factors that could reduce electricity use further than in these scenarios include:

- Technological change can bring entirely new efficiency opportunities, e.g., aerogel window materials or electrochromic smart windows discussed in the chapter by Granqvist. Only a few such improvements are included in the Advanced-Technology Scenario.
- One-half of the double-paned windows in electric-resistance homes are upgraded to triple-glazing, but only those constructed up to 1981. Greater savings could be attained by installing a third pane treated with a low-emissivity coating.

- No operations-and-maintenance measures are included for the industrial and service sectors.
- Increased automation and controls can save large amounts of electricity but are not included here. Power-integrated Circuits (PICs), described in the chapter by Baldwin, will lead to a new generation of "smart motors" with efficiencies above those we have assumed (54).
- Increased use of pollution controls can reduce waste and thus indirectly increase efficiency due to increased material productivity.
- Due to lack of data on the existing stock, no low-cost measures such as water-efficient showerheads and thermostat controls are included. Also not included are heat pump water heaters.
- The only building insulation retrofits considered are those in homes heated with electric-only systems. Not only might additional retrofits be cost-effective in other buildings, but would also reduce the risk of unexpected demand growth if multi-fuel homes re-convert to electric heating.
- Except for the Advanced-Technology Scenario, no electric space-heating savings potential is assumed for vacation homes.
- No fuel-switching is assumed for residential electric appliances although a potential exists (stoves use 1.4 TWh, clothes washing and drying use 1 TWh, and water heating uses about 5 TWh in the Reference Scenario). No natural gas substitution is assumed for commercial cooking equipment.
- Air conditioning is an emerging end-use in new and existing service-sector buildings. We assume no efficiency improvement, although many options exist (55).

Just as the factors described above could contribute to efficiency gains beyond those incorporated in our scenarios, a similar list may be constructed to enumerate factors that could work to push demand upwards:

- Technological change could introduce new electricity end-uses or processes, e.g., microwave drying of lumber in industry.
- Of the buildings that have converted from electricity to fuel, some could later convert back to electricity for heating following higher fuel prices and the absence of balancing policies. The size of this "virtual load" is 6.9 TWh, in the Efficiency Scenario.
- We have incorporated only to a small extent the "take-back" of electricity savings in the form of increased energy services, e.g., increased lighting intensity or lamp on-time, higher interior temperatures, larger refrigerators, etc.

- Persistence of savings. The deterioration of technology efficiency has been taken account of in the residential appliances model (refrigerator and freezer efficiency declines by 1% annually) but not elsewhere.
- Increased utilization of pollution control equipment or increased ventilation to make "buildings healthier" over what is already assumed could require additional electricity use.
- Additional waste-heat recovery in industry could conserve fuel but drive electricity use upwards more than assumed.

There are distinct differences in the potential magnitude of impacts from the uncertainties in the factors listed above. Uncertainties related to the factors pertaining to the development of society as a whole could lead to scenarios of significantly higher or lower electricity demand. The latter two groups of factors – concerning specific technologies and their performance – would affect national electricity use less.

3 Electricity/Heat Supply Scenarios

3.1 Methodology

In this section, we develop electricity supply and heat scenarios to couple with the demand scenarios described above. We take the same conceptual approach to supply-side technology as we did in our demand-side analysis. Here the goal is to illuminate some of Sweden's supply options by incorporating best-available technologies as well as some which are still in the research and development phases but could be brought to the market well within the time-frames of a 1987-2010 planning period. We adhere to the system boundary, described earlier, that includes all of the services today provided by electric heating and other electric end-uses, industrial cogeneration of electricity and process heat, and district heating.

An important objective in electricity planning is to obtain a production system that supplies power at the lowest possible cost, given the accepted reliability requirements.[22]

[22] The maximum acceptable loss of load probability (LOLP) criterion in Sweden is 0.1 percent (9 hours per year over the entire system) and the maximum risk of energy shortage criterion is 3.0 percent (rationing will occur 3 years out of 100). Historically, LOLP has been very low in Sweden, thanks to the large share of hydro power in the system. However, the thermal power share has increased due to the nuclear program and along with it the issues of reliability. The

The stepped curves shown in Figure 1a show Sweden's current reliability criterion and the generating capacity available in Sweden as the nuclear plants are phased out. The planning task is to bring new generating capacity on-line by the year that the demand curve drops below the stepped curve. Depending on the assumed demand scenario and which phase-out schedule is assumed, new capacity would be required sometime between 1995 and 2010.

For new capacity, we utilize technology relevant to the Swedish context. To account for existing generation capacity, we use costs reported by Vattenfall and from the literature.

The integrated nature of heat and power production necessitates the explicit inclusion of district heating in our analyses. We assume that municipal district heating demand in 2010 is the same as that in 1987 (38 TWh_{th} including losses). This is based on efficiency improvements in existing buildings being offset by the entry of new customers. The district heating today provided by refuse-derived fuels (RDF) incineration, waste heat, grey-water heat pumps, etc., (5 TWh, or 13% of the total) is included in the preceding estimate. In industry, we include 15 TWh of industrial process heating, to be provided by a mixture of cogeneration and boilers.[23] The electricity production cost for a year is credited for the heat *sales* from cogeneration during the year (i.e. 35 TWh in the scenarios).

3.2 Electricity Supply Technologies

This section briefly describes the existing generating mix and various generation options and their economics. The variable cost estimates indicated for today's capacity are those used by Vattenfall.

3.2.1 Condensing power

Today's non-nuclear condensing power in Sweden is based on oil. There is currently a generation capacity for 20 TWh/year, but less than 1 TWh is typically utilized. The variable cost of this oil-fired capacity is 3.9 cents/kWh.

generating capacity, upon which reliability calculations are based now permits a demand level of 145 TWh/year (including the Vattenfall convention of including 5 TWh of imports) based on these criteria. The reliability factors are commonly summarized using the rule-of-thumb that demand must not rise above the level, defined in footnote 14.

[23] For district heating provided with natural gas fuel, we assume the current cost in Sweden of 2.7 cents/kWh. To this we add 0.35 cents/kWh to account for the transmission system cost (Vattenfall estimates, 85% boiler efficiency).

There are currently no gas- or coal-fired condensing power plants in Sweden. The total busbar costs for constructing and operating coal- and natural gas-fired plants in Sweden are estimated at 4.2 cents/kWh.

3.2.2 Combined heat and power (CHP) and industrial cogeneration

Today there is an installed CHP capacity in district heating systems of 6.5 TWh_e/year, and about half of this is utilized based on a mixture of oil and solid fuels (peat and coal), with a variable costs of 1.7 cents/kWh_e. The busbar costs for new capacity are estimated at 3.2 cents/kWh_e in Sweden. Pressurized fluidized bed combustion (PFBC) systems may provide both electricity and useful heat while emitting less sulfur oxides than in conventional coal-fired power systems. Discussion of the state-of-the-art in PFBC is provided by Pillai in this volume (56).

Industrial backpressure cogeneration systems have been in the Swedish system since before 1950. Today's installed capacity can produce 5 TWh_e/year, much of which is in the paper and pulp industries. Production in 1987 was 3 TWh and the variable production costs are 1.6 cents/kWh. We estimate the total busbar cost at 3.1 cents/kWh.

Steam-injected gas turbines (STIG) fueled with gasified biomass could potentially supply large quantities of power and heat in Sweden. As described in the chapter by Larson *et al.* (57), such systems can be configured with absorption heat pumps or as combined cycles to attain relatively high overall efficiencies. The total busbar costs of electricity generated by biomass-gas driven turbines for Sweden are estimated at between 4.5 and 5.6 cents/kWh, assuming fuel prices of \$3/GJ including the costs of drying, chipping, and delivery to the power plant. This range is based on an avoided fuel cost of \$2.0-\$5.0/GJ in determining the heat credit; the natural gas prices in Box 1 correspond to about \$3/GJ. In our analysis, we assume a higher fuel cost of \$3.6/GJ (58, 59) and that the busbar cost of this option will be 5.6 cents/kWh.

Due to the small size of many district heating systems in Sweden, turbine systems smaller than the 53 MW_e ones assumed in these calculations would have to be used in many cases. A similar situation holds for the 25 MW_e circulating fluidized bed (CFB) coal systems that we specify for cogeneration. For the biomass CHP systems, this effect increases the capital cost by up to 60% (at 5 MW_e), or ~0.7 cents/kWh. However, a higher heat credit is also attained by smaller systems and this will partially offset the cost increase. This effect is much smaller than the 50% increase of total busbar costs that we calculate in a sensitivity analysis of biomass cost in Section 4.2.

3.2.3 Gas turbines

The installed capacity in the Swedish system is currently 1,800 MW. These oil-fired turbines are used as peaking capacity only. Total busbar costs are 9.7 cents/kWh.

New technologies make it possible to use natural-gas-fired turbines in baseload applications. The steam-injected gas turbines (STIG) and intercooled steam-injected gas turbines (ISTIG) described by Williams & Larson in this volume have busbar costs that are competitive with traditional generating technologies and can be used in electric-only or combined heat and power configurations (60). Advanced gas-fired combined cycles are also an important technology, and may be based on natural gas, gasified coal, or gasified biomass.

3.2.4 Wind power

Sweden has sufficiently strong winds to institute the increased use of wind power. Land use restrictions favor the use of offshore plants. Cost estimates vary widely, depending on siting and the quality of the wind resource at the site. Experimental wind stations exist and provide some indication of the investment requirements.

According to a recent governmental committee report, sites have been identified that would permit a generation of 1.5–7 TWh/year at land-based sites and 22 TWh/year offshore (61). However, this report does not include any economic analysis. Vattenfall's current belief is that the likely utilization of wind is 0-3 TWh/year at a cost of 7.0 cents/kWh.[24]

3.2.5 Fuel cells & photovoltaics

Electricity and heat could be provided with fuel cells during the scenario time frame. Also, photovoltaic solar cells could perhaps provide some electricity during the summers late in the scenario time frame. The chapters by Blomen and Carlson discuss the state of the art and prospects for the future (62, 63).

[24] The estimated cost interval is 6.3–7.9 cents/kWh (personal communication, Göran Svensson, Vattenfall, February 10, 1989).

3.3 Resource Constraints and Potentials

Many new types of power generation systems could be constructed in Sweden. The introduction of systems based on natural gas is constrained by gas availability, prices, and the current lack of a gas transport network.

The potential for biomass-fired generation is also constrained by fuel availability. Approximately 50 TWh/year of existing presently unused biomass fuel could be recovered and utilized for power generation (64, 65). Gasified and then converted to electricity and heat in gas turbines, this resource could generate ~15 TWh of electricity and ~15 TWh of district heating. An additional 12-40 TWh of fuel could be cultivated in the form of coniferous mixed or deciduous forests on agricultural land that has either already been abandoned or is no longer needed (0.5-1.0 million hectares of fields and pasture land) (66). Field trials in Sweden indicate a higher sustainable biomass production rate (40-80 TWh/year) from this same land area (67, 68).[25]

The industrial cogeneration potential is a function of the need for process heat in industry. The heat need is in turn based on the introduction of heat-efficient processes. An estimate of the potential is provided in (69). We assume that half of this potential would be utilized by 2010, and that the systems used would have power-to-heat ratios of 0.7, based on steam-injected gas turbines. This will yield a potential electricity production of 3.5 TWh/year beyond today's level of 5 TWh.

3.4 The Supply Scenarios

To formulate various supply scenarios, it is necessary to adopt some decision rules governing the selection and ranking of electricity generation technologies and fuels. The key objectives are to minimize direct costs as well as non-economic factors such as emissions and energy-import dependency. To illustrate a range of possible outcomes, we construct three types of generation scenarios and apply them to the four demand scenarios developed above. In each scenario, we begin with an existing mix of 65 TWh hydro; 1 TWh each oil condensing and gas turbines; 5 TWh of industrial backpressure.[26] In the

[25] This corresponds to a productivity of 14.4 tons of dry matter per hectare per year.

[26] The hydro estimate includes approximately 2 TWh of additional output beyond the 1987 level. The remaining capacity in oil-condensing and gas turbine units (~25 TWh/year) is allocated to providing reliability reserves and is thus not included in these "typical-year" calculations. Because we dedicate these plants to this purpose, the construction of additional excess capacity is unnecessary to fulfill the reliability criterion. We construct our supply scenarios assuming no net power imports in a typical year.

Table 3. Generation scenarios: power and heat production, fuel use.

	Alpha (Power/Heat)	Busbar Cost (¢/kWh)	Economic-Dispatch Scenario Reference Scenario	Efficiency Scenario	Hi-Eff. Scenario	Adv-Tech. Scenario
Existing Capacity (TWh_e)						
Hydro		2.01	65.0	65.0	65.0	65.0
Combined heat & power (CHP)	0.5	4.94	6.5	6.5	6.5	6.5
Oil condensing power stations		5.85	1.0	1.0	1.0	1.0
Gas turbines		9.71	1.0	1.0	1.0	1.0
Industrial cogeneration (gas)	0.5	3.10	5.0	5.0	5.0	5.0
New Capacity (TWh_e)						
Industrial cogeneration (biomass)	0.7	5.30	0.0	0.0	0.0	0.0
Fluidized bed/CHP (coal)	0.5	3.19	5.3	5.3	5.3	4.8
Steam-injected cycle CHP (gas)	1.1	3.19	5.3	5.3	5.3	4.8
Steam-injected cycle CHP (biomass)	1.1	5.60	0.0	0.0	0.0	0.0
Central station (coal)		4.17	25.2	10.9	3.4	0.0
Combined cycle (gas)		4.17	25.2	10.9	3.4	0.0
Wind generators		7.00	0.0	0.0	0.0	0.0
Total Electric Generation (TWh_e)			139.5	110.9	95.9	88.0
Combined Heat and Power, Industrial Process Heat						
Industrial cogeneration (TWh_{th})			10.0	10.0	10.0	10.0
Municipal CHP (TWh_{th}) [includes 5 TWh RDF, etc.]			33.4	33.4	33.4	31.8
Supplemental district heating or process heat (TWh_{th})			9.6	9.6	9.6	11.2
Second-Order Fuel Use						
Fuel switching, to oil from electricity (TWh_e)			-3.0	6.9	6.9	7.9
Fuel switching from electricity to natural gas in industry (TWh_e)			0.0	2.1	3.6	3.6
Total Fuel Use (all energy services) (TWh)	Coal		96	60	41	31
	Oil		17	29	29	32
	Gas		86	55	42	33
	Biomass		0	0	0	0
	Total		199	144	112	96

Economic-Dispatch Scenario 6.5 TWh existing CHP is incorporated. In the other scenario existing CHP systems would be rebuilt/replaced using new technology with improved power-heat production ratios. This supply mix includes an excess production capacity of 20 TWh that is needed to insure the conventional planning criterion of 3% risk of energy shortage.

To the already existing generating capacity, we then add, in the specified order, additional capacity to satisfy the energy and power performance criteria. For cogeneration we assume that up to three-quarters of the municipal district heating needs are supplied through steam-injected gas turbines operated

Natural Gas/Biomass Scenario				Environmental-Dispatch Scenario			
Reference Scenario	Efficiency Scenario	Hi-Eff. Scenario	Adv-Tech. Scenario	Reference Scenario	Efficiency Scenario	Hi-Eff. Scenario	Adv-Tech. Scenario
65.0	65.0	65.0	65.0	65.0	65.0	65.0	65.0
0.0	0.0	0.0	0.0	0.0	0.0	0.0	0.0
1.0	1.0	1.0	1.0	1.0	1.0	1.0	1.0
1.0	1.0	1.0	1.0	1.0	1.0	1.0	1.0
5.0	5.0	5.0	5.0	5.0	5.0	5.0	5.0
3.5	3.5	3.5	3.5	3.5	3.5	3.5	3.5
0.0	0.0	0.0	0.0	0.0	0.0	0.0	0.0
18.3	18.3	4.4	0.0	5.0	5.0	0.0	0.0
13.0	13.0	13.0	9.5	26.3	26.3	17.4	9.5
0.0	0.0	0.0	0.0	0.0	0.0	0.0	0.0
29.7	1.1	0.0	0.0	29.7	1.1	0.0	0.0
3.0	3.0	3.0	3.0	3.0	3.0	3.0	3.0
139.5	110.9	95.9	88.0	139.5	110.9	95.9	88.0
15.0	15.0	15.0	15.0	15.0	15.0	15.0	15.0
33.4	33.4	20.8	13.6	33.4	33.4	20.8	13.6
4.6	4.6	17.2	24.4	4.6	4.6	17.2	24.4
–3.0	6.9	6.9	7.9	–3.0	6.9	6.9	7.9
0.0	2.1	3.6	3.6	0.0	2.1	3.6	3.6
0	0	0	0	0	0	0	0
4	17	31	31	4	17	11	12
124	60	32	23	96	32	23	23
50	50	50	50	90	90	84	68
178	127	113	104	190	139	118	103

in CHP mode. The amount of electricity cogenerated is determined by the system's annual average alpha value. When full heat output is not needed for district heating, the excess heat is used to raise steam which is injected into the gas turbine to enhance power production (70). The alpha value just cited refers to the ratio of total annual electricity produced to the total heat produced for district heating.

Detailed descriptions of the scenarios, economics, and emissions are shown in (71) and the generation mix for each scenario is shown in Table 3.

3.4.1 Economic-Dispatch Scenario

In order of increasing busbar cost, new power plants are constructed and used to meet the demand in excess of that provided by the existing plants. Here, the usage of the term Economic-Dispatch refers to long-term planning and is analogous to the short-term decision rule about which plants to operate. We assume a 50% coal, 50% natural gas generation mix. This is based on the assumption that market forces will cause natural gas prices to adjust until there is parity with the costs of gas-fired and between coal-fired electric power generation.

3.4.2 Natural Gas/Biomass Scenario

In this scenario, Sweden's existing *non-utilized* biomass resources (50 TWh fuel) are used to produce heat and electricity. The biomass is gasified and used first in steam-injected gas turbines to produce useful heat and electricity and then in district heating boilers if the electricity demand has been satisfied, and there are remaining biomass fuel resources. For industry, this includes 3.5 TWh of new biomass-fired cogeneration systems with alpha values of 0.7 as described by Larson *et al.* for steam-injected systems operating almost year-round at full heat output. We also include 3 TWh of wind-generated electricity to represent Vattenfall's estimate of the resource that might be exploited. Natural gas generation is then introduced according to economic dispatch, i.e., combined heat and power until the district heating demand is met and then conventional power stations.

3.4.3 Environmental-Dispatch Scenario

In this scenario, new power plants are constructed and used to meet the electricity and district heating demand in order of increasing CO_2 emissions. Biomass residues plus up to 40 TWh (fuel) production from "energy plantations" are utilized in combined heat and power stations, industrial cogeneration systems, and for district heating. Wind power is utilized as in the previous scenario (3 TWh). If needed, electricity and heat are then generated with natural gas according to economic dispatch.

4 Integrated Resource Planning

This section integrates the final supply and demand scenarios. We determine the total cost of operating the system in the various supply/demand scenarios just described, i.e., the combined cost of supplied and conserved electricity.

4.1 The Integrated Resource Supply Curve

The concept of supply curves is commonly used in energy planning, and is especially useful when a diversity of resources are to be considered. In a supply curve, the available resources are ranked in order of increasing cost. We use the concept of the traditional supply curve in a new way by combining *conserved* and *supplied* electricity – the sum of these two components is the total electric energy service of 194 TWh_{eq} defined earlier by the Frozen-Efficiency baseline.

A supply curve can be presented as a series of steps, with cumulative supply on the horizontal axis and the incremental cost of each supply addition on the vertical axis. For supply-side resources, the y-axis value is the levelized busbar cost (cents/kWh) and for demand-side resources it is the cost of conserved electricity (also cents/kWh). We developed supply curves that combine these two kinds of resources, based on a ranking of their costs.

Twelve different supply curves could be drawn for various combinations of the four supply scenarios and three generation scenarios presented in this chapter. Each would supply 194 TWh_{eq} of electricity services, but would do so with varying mixes of efficiency measures and types of power plants. Two such "integrated supply curves" are shown in Figures 10a and 10b. The first figure represents the Reference Scenario for demand and the Economic-Dispatch Scenario for supply, and the second figure shows the Efficiency Scenario and the Environmental-Dispatch Scenario. (Note that the supply curve is used here to rank and compare changes in electricity use resulting from some sort of capital investment. Therefore, the costs of fuels that replace electricity are included below instead of in the supply curves.) Sweden's district heating and industrial backpressure heat needs are also met in the scenarios but are not shown explicitly in the curves.

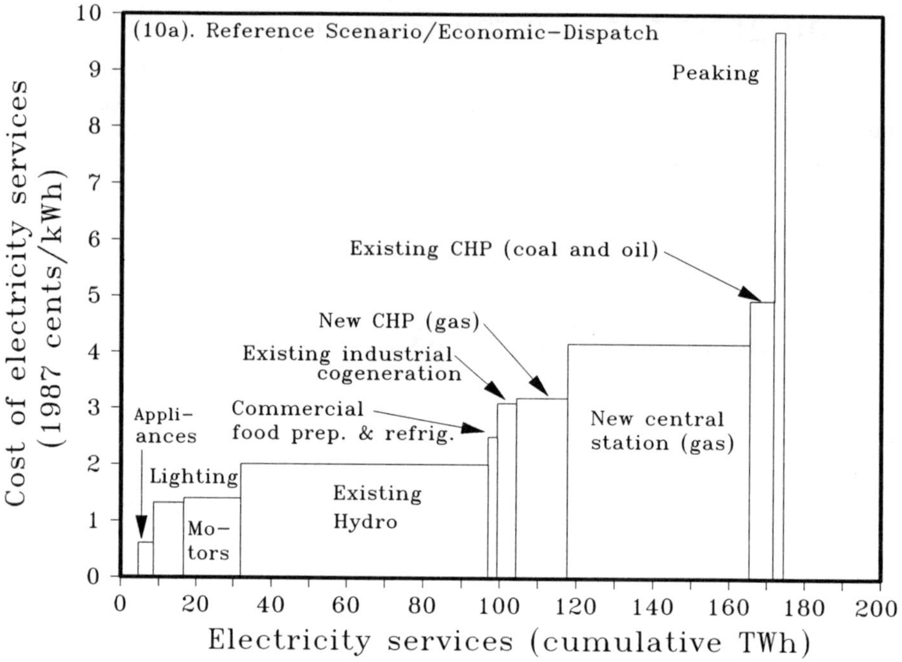

Figures 10a and 10b. "Integrated" electricity resource supply curves for Sweden. The curves rank scenario end-use technologies and new supply technologies (described in the chapter) according to their levelized per-kWh costs (6% real discount rate). The x-axis measures cumulative electricity conserved or supplied by 2010 with respect to the Frozen-Efficiency baseline and the y-axis measures the levelized cost for each technology. Fuel-switching is shown at zero cost because this is electricity use in buildings with existing multi-fuel systems that have an operating cost reduction at essentially no capital cost. The operating costs for fuel are counted in the section on total societal economics. The supply curves shown exclude efficiency savings for which we have not determined the investment cost (see the last footnote in Table 2).

To present another aspect of the results, in Figures 11a and 11b we show the monthly electricity demand for the same supply/demand scenario combinations shown in the previous two figures.[27] In Figure 11b, electricity is supplied mostly with domestic resources. The decreased seasonality of demand is also

[27] We have apportioned the weather-independent consumption for our scenarios evenly across the months, except for industrial use in July which we reduce to 67% of the annual average, due to the month-long national holiday (72).

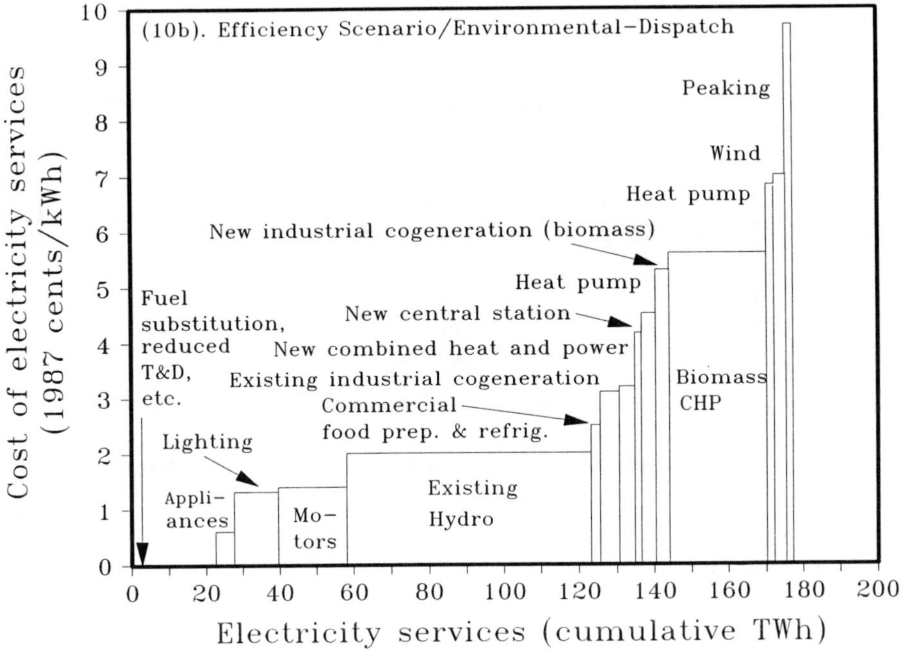

notable, mainly due to the assumed switching from electricity to fuel, primarily oil, for space heating. Hydro capacity is utilized differently in the two scenarios, in order to operate the CHP systems efficiently.

4.2 Scenario economics

Many types of economic costs are included within the system boundary that we have defined for the scenarios. We have tabulated the annualized costs of existing electricity capacity, new electricity capacity, district heating, natural gas transmission, fuels used in switching away from electric space heating and electric-based process heating in industry, and the investments in improved efficiency. Focusing first on electric energy services, we can define the total cost of these services as the annual investment in generating capacity, efficient

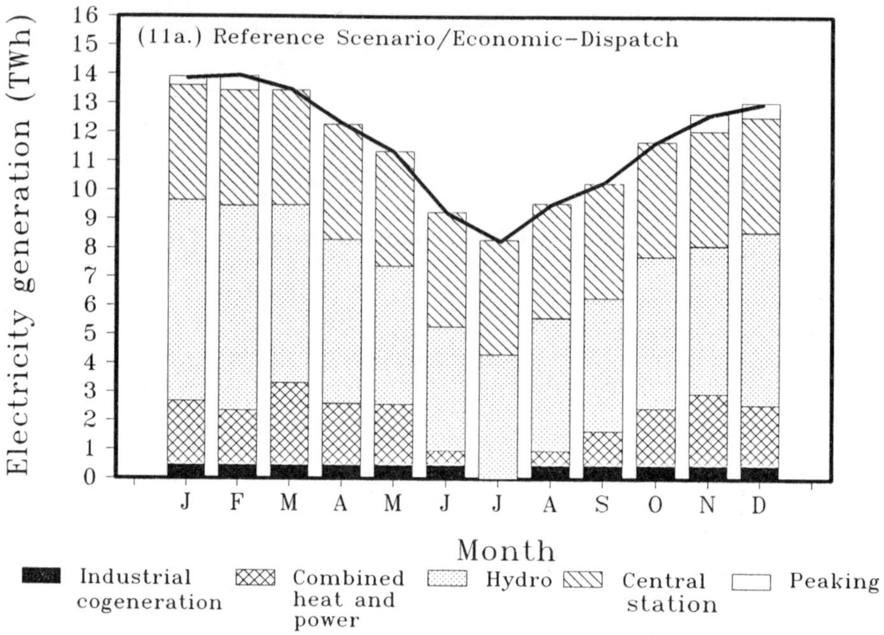

Figures 11a and 11b. Monthly electricity generation and supply mix corresponding to two demand scenarios. Figure 11a shows the Reference Scenario demand level and the Economic-Dispatch supply mix. Figure 11b shows the Efficiency Scenario and the Environmental-Dispatch supply mix with 90 TWh of biomass fuel availibility. The demand for lighting and space heating (including that in industry) is allocated to the months according to average-year heating degree-days in Stockholm. Water heating in one-fuel systems is spread evenly over the months. Because of the presence of time-of-use rates, electric water-heating in multi-fuel systems is assumed to occur in the six-month summer period of March-August. Based on historic statistics, industrial demand during the national holiday month (July) declines by 33%. Monthly hydro production in 2010 is based on Vattenfall's estimates for an average rainfall year. Power generation by cogeneration systems is curtailed during summer because of minimal need for district heating.

end-use technology, and fuel switching, divided by 194 TWh_{eq} (Table 4). In every case, total costs tend to decline as efficiency increases. This is because the average cost of conserved electricity is lower than the average supply cost.[28]

[28] It has not been possible to identify the costs for all of the efficiency measures in the scenarios. In the Efficiency Scenario, the supply curve includes costs for 66 out of the 83 TWh efficiency improvement beyond 1987. There is no reason to believe that the remaining TWh are more or less costly but we assume that they will cost 50% more per kWh. In the High-Efficiency Scenario, we assume that the unmonetized electricity services cost 100% more per kWh. No estimate is made for the Advanced-Technology Scenario.

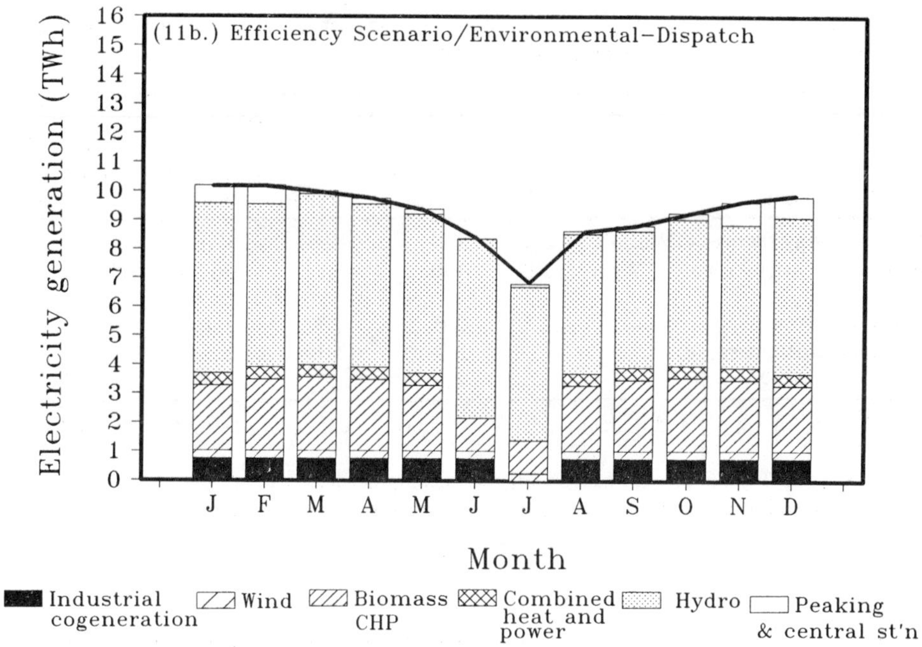

All of the energy services in our system (obtained by combining the costs of district heating and process heat with the results just discussed) amount to 247 TWh_{eq} (194 + 53). The costs per kilowatt-hour equivalent of these energy services are comparable to today's busbar costs of electricity and district heating in Sweden, both about 2.8 cents kWh. This busbar cost is, by definition, equal to the costs of electricity services in 1987. The annualized societal costs are summarized in Figure 12. The costs lie between \$5.5 and \$7.0 billion per year (2.2-2.8 cents/kWh_{eq}) over the full range of supply and demand scenarios.

The average cost of the new electricity supply would range from 3.2 cents/kWh to 5.8 cents/kWh_e (including wind) (Table 5). The marginal cost would be up to 5.6 cents/kWh, when biomass is the marginal resource.

We tested the sensitivity of the results shown in Table 4 to alternate assumptions about the cost of conservation:

(i) By increasing the cost of conserved electricity by 50%, the cost of electric energy services for the High-Efficiency/Environmental-Dispatch Scenario rises from 2.3 to 2.7 cents/kWh_{eq}, still less than the 1987 value of 2.8 cents/kWh_{eq}.

(ii) The Reference Scenario/Economic-Dispatch result rises from 2.6 cents/kWh_{eq} to 2.8 cents/kWh_{eq} given the same assumption.

Table 4. Total annualized costs for all 194 TWh_{eq} electric energy services (generating capacity, efficient end-use technology, and fuel switching).[a]

(cents/kWh_{eq})		<— Demand Scenario			—>
Supply Scenario	1987	Reference Scenario	Efficiency Scenario	High-Efficiency Scenario	Advanced-Technology Scenario
Electric energy services (TWh_{eq})[b]		194	194	194	194
of which electricity supply (TWh_e)	129	140	111	96	88
Economic-Dispatch	2.8[c]	2.6	2.3	2.1	---
Natural Gas/Biomass		2.7	2.4	2.3	---
Environmental-Dispatch		2.9	2.6	2.3	---

a All costs for supply and demand-side investments are calculated using the same 6% real discount rate.

b Excludes 53 TWh of process heat and district heating in each case.

c Joint Council of the Swedish Electric Power Producers (Kraftsam) estimate.

Table 5. Average cost of new capacity.

(cents/kWh_e)	<— Demand Scenario			—>
Supply Scenario	Reference Scenario	Efficiency Scenario	High-Efficiency Scenario	Advanced-Technology Scenario
Electricity Generation (TWh_e)[a]	68	39	24	15
Economic-Dispatch	4.0	3.8	3.6	3.2
Natural Gas/Biomass	4.4	4.5	5.3	5.8
Environmental-Dispatch	4.8	5.3	5.7	5.8

[a] In the Economic-Dispatch Scenario, new capacity construction is 6.5 TWh lower because existing CHP plants are utilized rather than reconstructing new, more-efficient plants.

We also tested the impact of 50% higher biomass fuel costs. This would result in biomass-STIG busbar costs of 7.6 cents/kWh, up from the base case value of 5.6 cents/kWh:

(i) For the Environmental-Dispatch Scenario, the total electricity service costs for the Reference Scenario and High-Efficiency Scenario demand levels were 3.2 cents/kWh and 2.6 cents/kWh, respectively, up from the base case values of 2.9 cents/kWh and 2.3 cents/kWh.

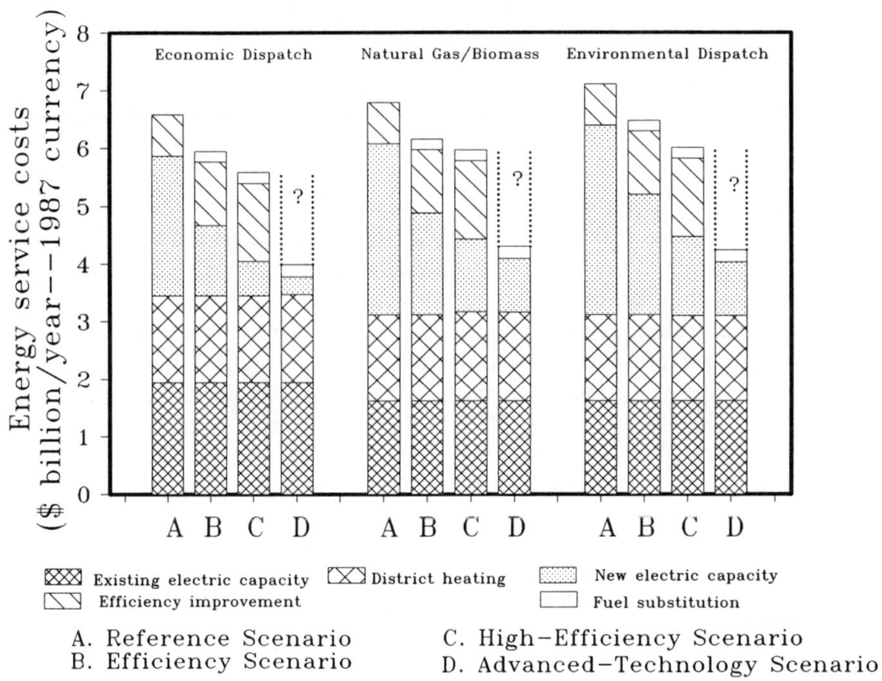

Figure 12. Energy service costs in 2010. The total annualized costs for 247 TWh_{eq} energy services today provided by direct electricity production, combined heat and power, industrial cogeneration, efficiency investments, and the costs of substitute fuels are included in the diagram. No conservation investment has been estimated for the Advanced-Technology Scenario. The values for existing capacity vary in some cases because some existing CHP capacity and industrial backpressure are replaced with biomass-fired generators in those scenarios. Costs are based on constant real fuel prices shown in Box 1.

(ii) For the Reference/Environmental-Dispach Scenario, the average cost for new electricity supply will rise from 4.9 to 5.8 cents/kWh_e. For High-Efficiency/Environmental-Dispach, the cost will rise from 5.7 to 7.5 cents/kWh_e.

Even given these assumptions, the fact remains that costs for electricity services in the Efficiency scenarios are lower than in the Reference scenario, as does, of course, the higher costs of biomass-using scenarios in relation to fossil-fuel-only scenarios.

5 Carbon Dioxide and Other Emissions

The quantitative assessment of the emissions associated with power and heat generation is playing an increasingly large role in energy systems analysis and energy policy. The most ominous manifestation of energy-related emissions is global warming, commonly known as the "greenhouse effect". For each demand and supply scenario, we have assessed the output of carbon dioxide, the main greenhouse gas.[29] In addition, we have tabulated the airborne emissions of certain heavy metals (cadmium and mercury), hydrocarbons, sulfur, and NO_x, plus solid waste materials (e.g., ash) left behind.

For each generation mix in the supply scenarios, we apply emissions factors for state-of-the art combustion technologies (73, 74). All new generating capacity in the scenarios complies with Sweden's 1995 guidelines for sulfur, NO_x, and particulates emissions (75).[30] In the Economic-Dispatch Scenario, we assume today's fuel mix in existing central heat and power production: approximately 70% coal, 30% heavy fuel oil. For the purpose of estimating emissions from the district heating sector and from industrial process heat, we assume a fuel conversion efficiency of 85% and exclusive use of natural gas, where biomass is not used.

Because each efficiency scenario incorporates some fuel substitution, we include the associated emissions from this increase in fuel use. Most of the effect is due to conversions from electricity to heating oil (6.9 TWh_e) and the replacement (upon retirement) of large-scale heat pumps with fuel-based district heating systems (2 TWh_e), but there is also increased natural gas use in the steel industry that displaces electricity now used in electric arc furnaces and other process heating tasks (~4 TWh_e).[31] In order not to underestimate the CO_2 effects, we assume that all conversions from electric space heating are to oil (while in reality some will likely be to wood fuel, which makes no net CO_2 additions to the atmosphere, or to natural gas). The Reference Scenario

[29] Other important greenhouse gases are methane, CFC-11, CFC-12, nitrous oxide, and trophospheric ozone. Carbon dioxide is thought to be responsible for about 50% of the warming effect.

[30] The 1995 emissions guidelines for power production facilities are: sulfur (0.05 g S/MJ fuel burned), NO_x (0.05–0.10 g NO_x/MJ, units over 300 MW_e) and 0.10–0.20 (g NO_x/MJ, units below 300 MW_e), particulates (20 mg/MJ, annual average for coal-based systems).

[31] We select a Plasmascrap process based on natural gas. Plasmascrap can also use 45 kg coal per ton of steel produced (76). To put this prospective emissions increase in perspective, if SSAB's use of coal (650 kg/ton) remains the same in 2010 and their production grows from 3 Mtons/year to 3.6 Mtons/year, then coal use will be 2,340 kilotons annually. Given Vattenfall's assumed production trends, steel output from electric arc furnaces would grow from 1.6 Mton/year to 1.9 Mtons/year. The corresponding incremental coal requirement in 2010 is 86 kilotons (at 45 kg/ton).

includes some switching from fuel to electricity, and we credit this to lower emissions from oil-fired heating systems.

As mentioned earlier, a present national goal is to freeze current CO_2 emissions. The current levels from the power and heat sector are low due to an electricity production mix that is based almost entirely on nuclear and hydroelectric power. Were there interest in constructing fossil-fuel-based power plants, the option always exists to seek offsets outside the power and heat sector, but this strategy may be unacceptable in the long run. Natural gas might be allowed as a transitional fuel if carbon dioxide emissions can be correspondingly reduced in other sectors. The same holds for other fossil fuels, although the reduction of CO_2 emissions in other sectors would then have to be larger because of the higher amounts of CO_2 released when using these other fuels.

In Figures 13a-13d the carbon emissions from the heat and power sector (247 TWh_{eq}) are shown for each generation and demand scenario. The curves show the total annual costs of providing the energy services. The results for carbon reveal an extremely broad range of emissions and yet very similar costs, regardless of the generation mix. The set of bars accompanying the Economic-Dispatch Scenario show that efficiency will indeed reduce carbon, but not nearly enough to meet the target. On the other hand, the Environmental-Dispatch Scenario may not suffice either. A combination of substantial efficiency improvement *and* low-carbon generation technologies is the only solution among the twelve combinations shown. And, as shown in Figure 13, the total societal economic costs of the high-CO_2 scenarios may be *higher* than the costs of the low-CO_2 scenarios.

As shown in Table 6, only three of the twelve scenario combinations result in per-capita carbon emissions – including those emissions outside of the power and heat sector – equal to or lower than those in 1986.[32] However, even the best case is nearly two times the world average of ~1 ton/capita annually (79).

In the highest emissions case, total carbon emissions increase to over 150% of their 1986 levels and those from the heat and power sector alone increase to ~500%. In contrast, the High-Efficiency/Environmental-Dispatch case results in a 10% reduction in CO_2 from today's levels. Emissions from the power and heat sector alone decline by 35% from today's levels (Table 7). In 1986, the power and heat sectors accounted for 15% of Sweden's CO_2 emissions whereas – assuming constant emissions from other sectors – in 2010 with coal/gas economic dispatch and efficiency improvements corresponding to the Reference Scenario, it would be the source of 50% of national emissions.

32 In calculating Sweden's total CO_2 emissions, we include the energy-related uses of coal, oil, and gas from (78).

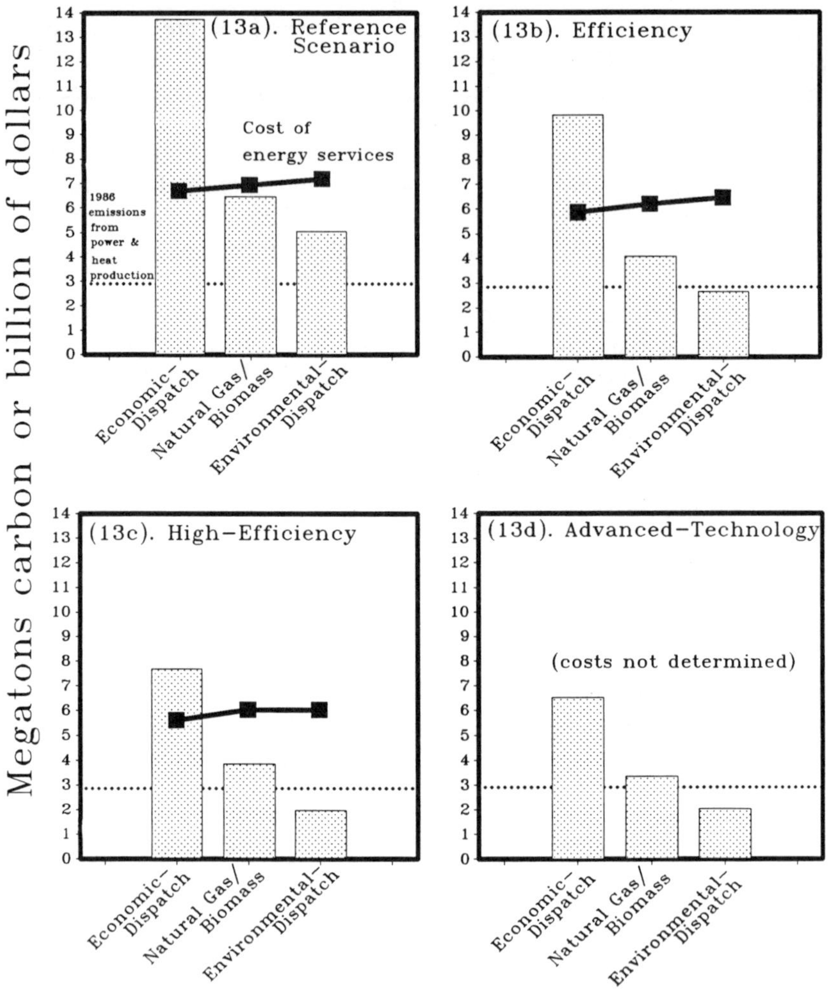

Figure 13a-13d. Annual carbon emissions from the production of heat and power within the system boundary, plus the use of substitute fuels in 2010. Results are shown for each generation scenario and demand scenario. The total costs of energy services from Figure 12 are given by the curves.

No CO_2 emissions from the transportation of fuels are included. For domestic biomass they will be very small, but this is not necessarily the case for the import of coal.

We calculated the offsets required for each scenario if carbon emissions from the power and heat sector are to remain at their present levels. In all but

Table 6. Per-capita carbon emissions (all sectors)[a].

(tons/capita-year) Supply Scenario	1986	Reference Scenario	Efficiency Scenario	High-Efficiency Scenario	Advanced-Technology Scenario
		<——— Demand Scenario ———>			
Economic-Dispatch	1.91	3.10	2.67	2.40	2.27
Natural Gas/Biomass		2.26	1.99	1.96	1.89
Environmental-Dispatch		2.10	1.83	1.73	1.74[b]

a Assumes constant 1986 levels from outside the power and heat sector. The global average emissions were 1.03 tons/capita-year in 1986. For derivation, see Bodlund *et al* (77).
b Results are higher than in the High-Efficiency Scenario due to greater amounts of fuel switching.

Table 7. Carbon emissions in 2010 (power and heat sector).

(Mtons/year) Supply Scenario	1986	Reference Scenario	Efficiency Scenario	High-Efficiency Scenario	Advanced-Technology Scenario
		<——— Demand Scenario ———>			
Economic-Dispatch	2.9	13.6	9.9	7.6	6.5
Natural Gas/Biomass		6.4	4.1	3.8	3.3
Environmental-Dispatch		5.0	2.7	1.9	2.0

three of the twelve cases, substantial offsets would be necessary. For reference, total 1986 carbon emissions from industry were approximately 4.2 Mtons/year; from residential space heating 3.8 Mtons/year, and from road transport 4.7 Mtons/year. Based on the emissions shown in Figure 13, achieving the highest offset of ~10.5 Mtons/year does not seem conceivably possible. Attaining full offsets in any of the Economic Dispatch cases would be quite difficult, as it would for the Reference Scenario demand for any of the three generation strategies. Substantial improvements in automobile efficiency and additional efforts to conserve heating oil in the buildings sector could in principle achieve sufficient offsets for some cases where the natural gas/biomass systems are used. However, it must be kept in mind that these offsets may be needed to meet future agreements, such as those outlined in the recent Toronto and Hamburg recommendations that call for 20%-50% reductions in CO_2 emissions by 2005 and 2015, respectively (80, 81).

Another alternative – the control of CO_2 emissions from fossil-fueled power plants – has been proposed, but in coal plants it is estimated that 16% of the power plant's capacity would be required to operate the control system and overall production costs would increase by 56% to 100%. The capital investment would range from 70% to 150% of the existing capital investment (82).

Some additional carbon reductions will occur over the short run if biomass plantations are used. These plant materials will absorb carbon from the atmosphere for as many years as the size or intensity of plantation area is being increased. We have not included this "biomass buffering" effect in our analysis.

We also evaluated other emissions. Figures 14a and 14b present a comparison of carbon dioxide and other emissions by showing an index for each by-product for each demand level, where 100 equals the Reference Scenario/Economic-Dispatch combination. In the fossil-based Economic-Dispatch generation scenario, nitrogen oxides, polyaromatic hydrocarbons, and heavy metals (cadmium and mercury) are approximately 40-70% lower in the Efficiency scenarios than in the Reference case. Cadmium levels are only slightly lower in the Environmental-Dispatch Scenario because those in coal are replaced by those in biomass. There is an important difference, however, in that coal mining mobilizes heavy metals otherwise bound in mineral form (and the greenhouse gas methane), thereby increasing the amount of heavy metals circulating in the bio-geological system.

For all pollutants, emissions will vary between years, primarily depending on the annual hydro-power production. Seen as an average over many years, the emissions presented here would be largely representative, but we have not made a detailed analysis of this.

Before concluding, we note that it is important to guard against adverse environmental effects that might result based on the assumptions of the scenarios. For example, it would be necessary to make use of environmentally acceptable ways of growing the biomass beyond that already available as waste products, improving buildings, designing gas turbine exhaust cleaning systems, avoiding CFCs and other greenhouse gases in all applications. The strategy described in this chapter is not in conflict with these guidelines. Concerning CFCs from heat pumps, we assume successful development of benign alternatives (see the chapter by Berntsson (83) in this volume).

6 Policy Implications

6.1 The Challenge of Choices

The preceding scenarios illustrate some of the choices available to Sweden in determining her electricity future. It is possible from a technical standpoint to

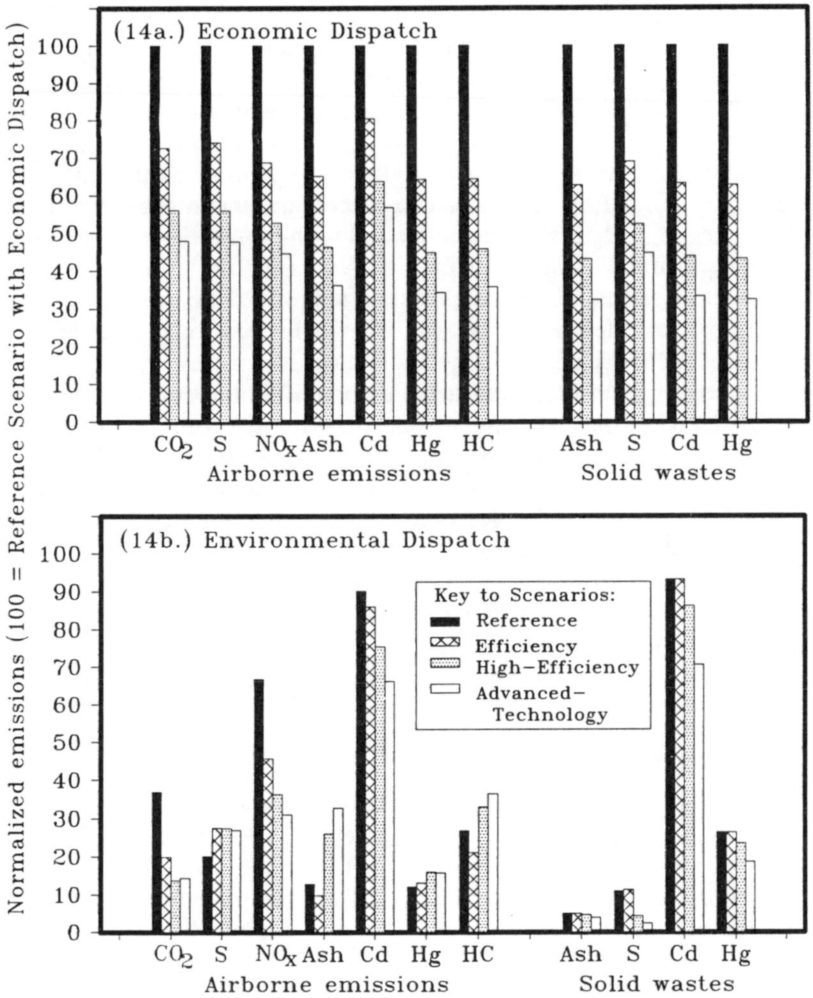

Figures 14a and 14b. Emissions indices. The height of the bars is shown as an index. In this way, the relative changes for many different materials can be easily compared. The following waste product data are for the Reference Scenario/Economic-Dispatch Scenario case, i.e., the one indexed to equal 100 in Figure 14a. They can thus be used to directly estimate the corresponding values for the other bars, according to their respective indices. *Annual Airborne Emissions:* net carbon emissions 13,644 ktons [C]; sulfur emissions 19 ktons [S]; NO_x emissions 89 ktons NO_x; airborne ash 6 ktons; cadmium emissions 52 kg [Cd]; mercury emissions 278 kg [Hg]; Polyaromatic hydrocarbon emissions 390 kg; *Annual Solid Waste Production:* ash waste material 1,720 ktons; sulfur waste material 105 ktons [S]; cadmium waste material 3,459 kg [Cd]; mercury waste material 1,106 kg [Hg]. (note; NO_x emissions are computed here as their NO_2 equivalents.)

find paths that satisfy the demand for electricity services associated with continued economic growth, uphold environmental quality standards, and expedite the phase out of nuclear power as planned. The scenarios have very different CO_2 emissions, while other environmental aspects meet present environmental standards. These low-CO_2 scenarios would require the increased use of efficient end-use technology in all sectors as well as the installation of new low-CO_2 generating facilities. Use of coal and natural gas would lead to large increases in CO_2 emissions compared with present levels. The economic differences are much smaller. In fact, the costs of the Efficiency and High-Efficiency/Environmental Dispatch scenarios would be similar to or lower than those for the Reference/Economic Dispatch scenario, as calculated in this chapter, based on today's knowledge about costs and prices.

Experience suggests that such a development would not happen by itself. Individual decision-makers tend to optimize for themselves rather than for society at-large. By definition, the Reference Scenario illustrates the development that one could expect under present operation of the marketplace. The difference between the Reference Scenario and the High-Efficiency Scenario – 44 TWh/year – can be thought of as one measure of the opportunities for alternative market dynamics. To capture some or all of these savings, policies would have to be shaped to make markets work better and stimulate actors in all sectors to make use of the technical opportunities we have described in the preceding pages. Information, education and training, rules, and regulations would have to reflect this by making the national good also be the actor's good.

The chapters by Geller (84), Williams (85), and de la Moriniere (86) describe how such policies might be designed, markets organized, and the necessary financing arranged, and the chapter by Geller *et al* (87) describes the potential for research and development to help accelerate the introduction of new technologies into the marketplace. These four chapters discuss many policy options, and the experiences that are available from testing some of them in the U.S. and in France.

It is evident that electricity prices are only one of several significant factors in the process of achieving a cost-effective total development of the energy system, or the electricity system, and that there is great uncertainty in price/demand relationships (88).[33] Electricity costs are often not very significant and consumers lack influence over how the equipment they purchase is designed. It is therefore not surprising to find that the effective

[33] For a discussion of electricity price responsiveness in Sweden, see Mills, E. 1988. *The role of energy prices in shaping electricity end-use structure and demand in Sweden.* Manuscript in preparation. Lund University. Department of Environmental and Energy Systems Studies. Report to Vattenfall's Uppdrag 2000 project.

rates-of-return demanded by consumers are often 50–100% or higher (89, 90), much above the interest rates commonly used in evaluating supply investment options. This suggests that the total system will not develop in a least-cost direction. Reflecting long-run marginal costs in rate structures (e.g., in inverted-block tariffs that convey the marginal cost incentive without increasing the total cost of electricity) is probably necessary but not sufficient to bring the options discussed in this book into reality.

Policy measures would also have to be formulated on the supply side, because costs of new generating capacity are higher than average generating costs. This is due to the large, low-cost hydropower production in Sweden. The average costs of new supplies would range up to 5.8 cents/kWh_e, well above the new system-wide average cost of ~3 cents/kWh_e. The problem is most pronounced in the High-Efficiency/Environmental Dispatch scenario. The extent of policies required to stimulate new generation under these circumstances depend on electricity pricing principles. With marginal cost pricing, costs would be covered, but otherwise one could not expect non-hydro-owning utilities or industries to invest in, e.g., cogeneration. An arrangement where central utilities purchase power from independent power producers on an avoided-cost basis would be one way to approach this dilemma.

As discussed in other chapters in this book, research and development work remain to be done on many complex issues related to some technologies included in the scenarios. This work should be done now, and not postponed, if these technologies are to contribute significantly by 2010.

6.2 Lead Times

A perceived dilemma in selecting the end-use efficiency improvements as a major component of an overall electricity strategy is the lack of data on the effectiveness of various policy mechanisms intended to secure these resources. Can power supply planning that fulfills the usual reliability criteria be accomplished without investing so much in supply capacity that the economic imperative to utilize this excess capacity becomes an obstacle for continued efficiency improvement?

The answer to this question lies partly in the fact that the new supply technologies have short lead times, as compared with traditional central-station generating facilities. Both combined cycles and the advanced gas turbines (described by Williams & Larson in this volume) would require about 3 years for construction and 2 years for licensing, assuming fuel supplies such as natural gas and biomass are available on the market in the relatively small quantities that would be needed for each additional unit. Coal and oil are readily available and can by and large be transported the point of use with the existing infrastructure. Natural gas would require a transmission system, and biomass is presently available only at the levels specified in the Natural Gas/Biomass Scenario. A large natural gas transmission system would require 5–10 years to build. To implement the Environmental-Dispatch scenario would involve a 5- to 7-year lead time for energy plantations. For both natural gas and biomass, time is needed to establish a market.

To the extent that electricity plans utilize short-leadtime strategies, the traditional problem facing utilities having to forecast electricity demand 8 or 10 years into the future – and increase the generating capacity sufficiently to mitigate the forecasting uncertainties – can be reduced. It thus becomes possible to follow load growth and the effectiveness of efficiency-improvement policies, and to respond to the realities as they develop, while maintaining the traditionally high reliability standards, but with far lower uncertainty (and thus expense) than has been the case historically.

Although all scenarios have abundant excess capacity, the use of steam-injected gas turbines offers an additional route to winter-time extra reserve power production capacity. If heat normally supplied in cogeneration using these units is shifted to the stand-by oil-fired boilers needed for backup in every district heating system for reliability reasons, this frees up steam that can augment power production. In this way, reserve boilers contribute to an improved reserve margin in the power system.

Demand-side opportunities can be utilized at the rate of capital turnover and expansion. This often means 10-20 years for full penetration. Fuel switching, on the other hand, can be much faster (a property that is disadvantageous if the switching is reversed at a later date).

When utilizing demand-side resources, it becomes necessary to account for the lead time involved with programs or other mechanisms intended to accelerate the introduction of end-use technologies. For scenarios that rely on commercially available technologies, a key component is the speed with which a given program can achieve a targeted participation rate. For scenarios that rely on technologies still under development, the key factor is the rate at which research and development proceed and thus bring new technologies to marketplace.

In any case, developing demand-side resources begins with identifying them. This chapter has provided an assessment that could be expanded upon. Vattenfall has initiated a market-research project called "Uppdrag 2000" to identify the cost-effective conservation opportunities in Sweden (91). Over the next few years, the project will spend $60 million and have a staff of roughly 70 people. The work will focus mostly on demonstration projects intended to obtain actual data on technical and behavioral issues in each sector. There remains a three- to four-year period of continued investigations before implementation begins.

6.3 Balance of Payments

Some of the preceding scenarios would improve the balance of payment situation in Sweden. In recent years, Sweden's balance of payments has oscillated from year to year between net debts and net credits with an annual balance of plus-or-minus $1 billion/year between 1983 and 1986 (92). The size of the impact for the scenarios would be primarily a function of the difference between the cost of no-longer imported uranium and newly-imported coal, gas, and/or oil. We find that for a demand level corresponding to the Reference Scenario, and a fossil-based supply mix, that the impact on balance of payments would be adverse whereas the two Efficiency plus Environmental-Dispatch scenarios indicate strong benefits.[34] At one extreme (Reference/Economic-Dispatch) the costs of imported coal for power generation and of oil for district heating would be $1.4 billion/year. At the other extreme (High-Efficiency/Environmental-Dispatch) the cost of imports would be $0.25 billion/year.

The cost of electricity services, especially to industry, is an important component of competitiveness and the balance of the payments equation. The closure of some industrial firms in the Reference Scenario is intended to reflect their sensitivity to the cost of electricity. In Table 4 we saw that for several of the scenarios, the costs of producing electricity services were comparable to or lower than today's.

[34] Imported uranium today costs $425 million/year, based on Vattenfall's estimate of 0.652 cents/kWh for uranium fuel in 1987 and 65 TWh of nuclear-based generation. Let us consider two simplified cases. At one extreme, the fossil fuel imports required to generate 74 TWh_e/year (the difference between the Reference Scenario demand and projected hydroelectric production) would be $1,600 million at 0.86 cents/kWh (fuel) coal import price and 40% conversion efficiency in the power plant. Thus, an adverse impact on the balance of payments would result. At the other extreme, in a low-demand scenario with biomass fuel (but accounting for 15 TWh fuel switching to imported oil at $20/bbl), the cost of fossil fuels would be only $250 million. In addition, were district heating to be provided from biomass fuel, oil imports could be further reduced by approximately $950 million. We have not estimated the effects of new power plant construction and conservation-related capital investment, but most of these things can be produced in Sweden. It is clear that efficiency and a domestic-based fuel mix could have a significant positive impact on Sweden's balance of payments in the future.

7 Summary and Conclusions

We have presented four electricity demand scenarios for Sweden in the year 2010, based on different levels of electricity end-use efficiency. We combined the results with three generation scenarios, based on the utilization of various fuels and generation technologies. All of the scenarios assume 1.9% annual growth in gross domestic product and the corresponding amount of new buildings construction, industrial output, appliances purchased by households, etc. Each scenario combination provides all the energy services associated with this economic growth. Based on the final demand levels and supply mixes, we evaluated the total costs of energy services and the associated emissions of carbon dioxide and other by-products of heat and power generation. We did not evaluate the impact of higher or lower GDP trends nor we did we test the implications of different lifestyle developments.

There is a conceptual difference between some of the demand scenarios. The Reference Scenario is based on expectations of market behavior in the absence of new policy measures but with a 50% increase in real electricity prices. This scenario also assumes some price-induced industrial plant closures in the steel and electrolysis subsectors. Two other scenarios illustrate the results of more intensive use of electricity-efficient end-use technologies. These technologies are cost-effective in comparison with new electricity generation, based on a 6% real discount rate, and no direct taxes. In the last scenario, "Advanced-Technology", we assume the use of technologies still under development.

In the scenarios, direct use of electricity contributes between 88 and 140 TWh of Sweden's 194 TWh equivalent demand for electric energy services in 2010. The reductions in electricity use are brought about largely by improvements in lighting, motors, miscellaneous office equipment, and (except in the Reference Scenario) substitution of electric space heating by fuel. The additional oil use from fuel conversions is equivalent to 4% of Swedens's total oil imports in 1987 (based on replacing 6.9 TWh of electricity).

All the scenarios satisfy present environmental guidelines, but, obviously, some of these scenarios do not meet the environmental ambition of keeping CO_2 emissions constant. Given presently known technologies, we find that to achieve the economic growth mentiond above, fulfill national ambitions of keeping CO_2 emissions at or below their present level, and provide energy services requires a *combination* of energy-efficiency and low-carbon electricity supply strategies. Depending only on one or the other will not suffice.

In order to keep carbon emissions at or below the present level, bio-fuels have to be used to avoid the five-fold increase in carbon emissions from the power and heat sector that would result from using natural gas and coal to meet a demand level of 140 TWh. Furthermore, with two of the low-electricity scenarios, Sweden could provide a large fraction of its heat and power with domestic resources. A prerequisite for using bio-fuels in this way is completion of biomass gasification technology development.

These Efficiency scenarios could be attained by 2010 at an overall cost of electricity services of 2.6 cents/kWh_{eq} or less, versus a busbar cost of 2.8 cents/kWh in 1987, and a marginal cost of 4.2 to 5.6 cents/kWh_e (depending on whether natural gas or biomass is the marginal resource), including all investment in end-use technologies. On the supply side, the use of biomass is more expensive than that of natural gas or coal, and this extra cost can be seen as a price for reducing CO_2 emissions. On the other hand, we have chosen measures on the demand side that reduce all emissions (including CO_2) and deliver the electricity services at a cost *less than* that of constructing and operating fossil fuel power plants.

In evaluating the economic costs and considering the uncertainties, we are reluctant to say that either scenario is more expensive than the others in supplying all the energy services within the system we study. However, we submit that the societal costs per unit of energy services in the Efficiency scenarios do not increase from their current levels and CO_2 emissions are 35% lower in two cases.

Because the development of the preceding scenarios is based on a national perspective, the associated benefits would not necessarily be attractive to individual actors. These actors may or may not find it in their interest to pursue the technical opportunities that comprise the scenarios, e.g. industrial cogeneration and biomass recovery. To deal with this, policies will have to be shaped to bring individual actors' interests in line with the national interest. Some policy mechanisms could be implemented by the Swedish utilities as they become full energy service companies in the sense that Edison intended, but many other entities also have key roles to play.

From the point of view of a utility, a new type of uncertainty enters the supply planning process because future demand-side policies and their impacts are hard to predict. Whereas U.S. utilities and local governments have gained considerable experience with demand-side planning – thereby reducing the associated planning uncertainties – there is a great lack of experience in the Swedish context. Also, many of these policies would have to be formulated and implemented by non-utility actors, while the utilities would retain responsibility for providing the required electricity.

* * * *

The problems and opportunities we have described here are not unique to Sweden. Bringing down primary energy use by increasing end-use efficiency is the key strategy to deliver society's desired energy services and at the same time contribute to the amelioration of major global problems (93). The efficient use of energy illustrated in the scenarios would enable Sweden to develop her energy system in ways compatible with solving global problems.

As the twenty-first century approaches, the Swedish society is planning for an expanding use of services from electricity and energy. The great business opportunity for utilities and entrepreneurs is to make the electric energy services their business activity. This will be an exciting era, with a key role for high-technology in both the supply and demand arenas. The new challenges are in policy-making and in the marketplace.

Acknowledgments

We thank Vattenfall's Market Analysis Group, Gunnar Larsson (Head), Matti Malinen, Drasko Markovic, and Anders Sjögren for their assistance with demand modeling and for review comments. Useful discussions and review comments were also provided by Enno Abel, Harry Albinsson, Thore Berntsson, Lars Brinck, Birgitta Dangården, Sven Eketorp, Arne Elmroth, Lars Gertmar, Claes Göran Granqvist, Anders Hedenstedt, Lars Jacobsson, Jon Koomey, Eric Larson, Gunnar Leman, Måns Lönnroth, Jørgen Nørgård, John Robinson, Peter Steen, Robert Williams, and Bo Wiman.

One of us (TBJ) acknowledges the gracious support of the Swedish Energy Research Commission (Efn).

Notes to Box 2

[a] The percentages exclude changes in saturation, i.e., they are calculated based on the change in the electricity consumption per appliance. Appliance savings shown here are weighted by the use in single- versus multi-family homes. The appliance intensities are from Vattenfall and Nørgård (8). Nørgård's per-appliance estimates for freezers and combination refrigerator/freezers are adjusted upwards by 26% (single-family) and 11% (multi-family) to account for the anticipated increase in use of automatic defrost.

[b] For the residential sector, we consider the replacement of existing incandescent lamps with compact fluorescent lamps using McGowan's technology estimates (9). Penetration is based on Vattenfall's assumptions about the number of lamps that can be replaced (only 5 out of 23 lamps per house) due to limiting factors such as frequent cycling. The Vattenfall model assumes substantial "takeback" of savings as customers increase lighting on-time by about 30%. [This is assumed to occur because information supplied with the lamps instructs the users to operate them for 4 hours at a time.] In the last two Scenarios, we assume that most of the incandescent lamps not converted to compact fluorescent lamps are instead replaced with new, efficient incandescents (20% savings). For the services sector, we assume that 30% of current lighting energy use (i.e., a much lower percentage of the lamps) is by incandescent lamps and that

compact fluorescent replacements are applicable; 5% are flood and area lighting suitable for IRF and metal-halide lamps; and 65% are fluorescent fixtures suitable for some mixture of electronic ballast retrofits, occupancy and daylighting sensors, and better lamps. Thus, the savings shown in the Box are weighted averages for the three types of lighting systems (76%, 35%, and 58%, respectively). In industry, the end-use share assumptions are 10%, 20%, and 70% for the three lighting system types and the respective per-fixture savings assumptions are the same as those used for services. Many documented case studies of lighting retrofits now exist in the literature. For example, one recent installation of 1,000 dimming controls resulted in 25% savings at a cost of conserved energy of 2.3 cents/kWh (10).

[c] These savings primarily include those for more efficient motors, compressors, etc., as well as for adjustable-speed drives applied to systems that have energy losses due to part-load operation, and savings from improved system design, e.g., piping or ventilation duct sizing to minimize losses. Baseline service sector electricity use is given within the Vattenfall model. To define the relevant energy use in Swedish industry, we use the end-use data for motors, etc., in (11) and other sources. We have partitioned electricity used in industrial motors, pumps, etc., into two categories – one that is likely to be eligible for adjustable-speed drives and the other that has either implemented them already or has no opportunities. Efficiency options still exist within the second group (e.g., for system optimization and idling controls), but savings are less certain. We have treated mechanical pulping separately (about 5.6 TWh/year in 1987) and assume 10%, 20%, and 30% savings in the last three scenarios, based on (12) and (13). Savings from controls that reduce idling time may be as great as 40% in the Swedish paper and pulp industry, (personal communication, Bengt Lindahl, Scanpump, October 12, 1988). Measured savings of 22% from such controls have been documented for industrial applications in Japan (14). According to Baldwin's chapter, savings for ASDs alone can range from 25%-50% depending on the application (15).

[d] Detailed data for seven different technology improvements (retrofits and new appliances) in food preparation are given in (16); savings vary from 10-50%, depending on the measure and its application. Some of the savings would come from adjustable-speed drives, more-efficient compressors, refrigerator doors, and thicker insulation. Convective and microwave ovens contribute to savings in food preparation and ohmic heating reduces the electricity requirements for preserving certain food products (17). We assume that technology change is made at the time of equipment replacement as opposed to by retrofit.

[e] Office electronics electricity uses include computers, copiers, and printers. The savings estimates are based on the chapter by Norford *et al* in this volume (18). The continuing trend towards miniaturization and towards low-power transistors in computing equipment and input/output devices will lead to great efficiency gains. Other trends, such as the use of color monitors and laser printers are more electricity intensive than the processes they replace. Note Norford *et al's* warning that nameplate ratings tend to overstate actual power requirements by a factor of 2-4 for personal computers and by a factor of 4-5 for printers. The authors also estimate that in the U.S. approximately 40% of office equipment is left running at night.

[f] The savings estimate in the efficiency scenarios is from the chapter by Abel in this volume. The stock average non-electric-heated commercial building today uses 100 kWh/m^2 annually. (For comparison, the California Title-24 standard for new buildings requires a maximum average electricity intensity (including air conditioning) of ~100 kWh/m^2 for new buildings for

large office buildings.) The difference between 140 kWh/m^2 and 50 kWh/m^2 corresponds to 1.2 TWh/year by the year 2010.

[g] In Sweden's industrial sector, by far the most important segment in terms of overall electricity demand is paper and pulp, which used 38% of the 46 TWh consumed by industry in 1987. Most of this was for motors and grinding equipment. Improvements to the grinding process are critical as are the opportunities for introducing new thermo- or chemo-mechanical pulping processes that are less electricity intensive. Improved water-removal techniques are also important.

[h] Examples of simple, fast-payback measures are given in (19). Reference Scenario estimates include the effect of the closure of some electric arc furnaces (EAFs) which affects the weighted averages given here. EAF savings for the other scenarios are counted separately in footnote "j" and do not assume any closures. In the Efficiency Scenario, before computing efficiency-related improvements, we incorporate 2.1 TWh_e fuel switching to natural gas for process heating (and 2.8 TWh_e in the last two Scenarios), based on (20) and on a review of this source by the Swedish Gas Association (Memorandum from Ulf Norhammar and Göran Engström, January 8, 1988. Stockholm). In the last two scenarios for EAFs, 0.6 TWh_e is substituted for gas.

[i] We adopt the chemical industry estimate of savings potential (15%) (personal communication, Hans Wernborg, Swedish Chemical Industry Association (*Kemikontoret*), January 11, 1989). A recent case study shows even higher savings, based on actual savings from a retrofit at Gränges Aluminum in Sweden, which produces 24% of Sweden's aluminum. From an investment of $43 million, product yield was increased from 13,500 tons/year to 22,500 tons/year and electricity intensity was reduced from 19 MWh/ton to 14 MWh/ton. The investment costs included non-energy improvements such as increased floor area and emissions controls. (Personal communication, Per-Olof Aronson, Gränges Aluminum. April 14, 1988.) The cost of conserved electricity for this case study is 2.6 cents/kWh. The overall potential savings for the sub-sector are uncertain because the remaining aluminum is produced in a plant today consuming approximately 17 MWh/ton, but the Gränges case is included in all but the Reference Scenario. The Reference Scenario includes some plant closures.

[j] In this volume, Eketorp describes many opportunities for improvements. In the Efficiency Scenario, savings result from electricity-efficiency measures (scrap pre-heating, improved transformers) and from increased quantitites of oxygen and natural gas introduced into the furnace. The electricity requirement for manufacturing the oxygen (0.7–1.0 kWh/m^3) is accounted for in the final savings estimate of 20% beyond today's furnaces, which averaged 590 kWh_e/ton in 1987 in Sweden. In both the High-Efficiency and Advanced-Technology Scenarios, we assume the use of the Plasmascrap process for recycled scrap steel currently melted in electric-arc furnaces, EAFs. Plasmascrap requires one-quarter as much electricity as do Sweden's current EAFs. Pilot plants are in operation in Sweden, producing 1 ton molten steel per hour. Commercialization is possible within five years (personal communication, Sven Santen, Ovako Steel, February 2, 1989). Commercial units are expected to melt scrap steel with 150 kWh/ton.

[k] A large fuel-switching potential exists due to the flexibility of the Swedish stock of space-heating equipment. In single-family homes with multi-fuel heating systems, roughly 6 TWh of electricity are used for space heating. We assume that 90% of the homes respond in some way: of the respondents, four-fifths switch fuels and one-fifth (the highest-use households) install

earth-source heat pumps. We thus assume 90% x 80% = 72% penetration of fuel switching. The heat pump COPs are 2.7 for the first efficiency scenario and 3.3 for the second and third, based on Berntsson's chapter. In multi-family buildings, 2.6 TWh are consumed in multi-fuel systems. We assume 90% of these systems convert away from electricity. We also assume that 90% of the electricity used in multi-fuel systems in service sector buildings (0.9 TWh) is substituted for oil.

[l] Those homes that have no fuel flexibility consume about 13 TWh for electric space heating. One-third of these homes – by floor area – are served by hydronic systems, and consume 4.3 TWh. In our scenario, we convert 90% of these hydronic systems to earth-source heat pumps, with COPs same as in note "k". Great efficiency improvements have been made in recent years and are likely to continue. Only 3.5% of the existing single-family homes in Sweden today have electric heat pumps (21).

[m] Most large heat pumps for district heating are assumed to be replaced with fuel-based systems at the end of their useful lifetimes, except in the Advanced-Technology scenario where all are replaced.

[n] We consider wall and ceiling insulation for those homes heated with direct electric resistance systems. In 1987, the demand for space heating electricity in these homes (single-family only) was roughly 8.7 TWh. Based on the estimates of existing insulation levels of: ceiling 0.27 [W/m^2-C] and walls 0.38 [W/m^2-C], from the chapter by Elmroth, our scenarios upgrade U-values to 0.10 (ceilings) and 0.17 (walls) and assume 20% of the homes implement wall retrofits and 50% install roof insulation. Our assumption is based on limited technical opportunities (e.g., number of homes with room for insulation in attic spaces and the number of homeowners re-siding or re-roofing their houses before 2010). We estimate savings assuming a simple "U-value times surface area" calculation, based on Stockholm weather conditions. We adopt Elmroth's estimate that 0.7 TWh could be saved by adding a third pane to existing double-pane windows in homes with electric-resistance heating systems, and assume 45% of the homes find it possible to make the retrofit. Based on the large number of electric-resistance homes using over 30,000 kWh/year, we include the savings from earth-source heat pumps for those homes in this group that do not implement either wall or roof insulation. These homes and ~25,000 fuel-heated homes have separate electric water heaters, for which we assume 25% savings as the best-available models (more insulation in the tank wall) replace existing ones. We reduce these savings by two-thirds to compensate for increased space-heating needs and for limited penetration. In the Efficiency Scenario, 10% and 20% savings are assumed to result from insulating walls and roofs of multi-family buildings with direct electric heating systems or electric-only boilers (these homes consumed 1.4 TWh in 1987). For commercial buildings, we assume 20% and 15% savings, respectively, based on total use of 2.3 TWh in 1987. In direct-electric buildings, an additional 5% savings are assumed in the last two scenarios.

[o] This performance level is based on the chapter by Elmroth in this volume. For comparison, a home constructed to comply with the current Swedish Building Norm (SBN) standard would use about 5,000 kWh/y for space heating and 3,000 kWh/year for water heating. The primary physical differences between these two house types are in low-emissivity coatings or gas fill in the triple-glazed windows, reduced air infiltration, and higher quality control during construction.

References

1. Swedish State Power Board. 1988. *The Vattenfall three-year plan: 1989-1991*. Stockholm. (In Swedish)
2. Government of Canada. 1988. The changing atmosphere: implications for global security. Conference statement. Toronto, June 27–30.
3. Swedish Ministry of Finance. 1987. *Long-term economic planning report*. Swedish Government Official Report (SOU). Stockholm. (In Swedish)
4. Swedish State Power Board. 1988. *Scenarios for long-term planning*. Stockholm. (In Swedish)
5. Swedish State Power Board. See ref. 1.
6. Swedish Ministry of Finance. 1987. See ref 3.
7. Bodlund, B., Mills, E., Karlsson, T. and Johansson, T.B. 1989. *Technical support document for the challenge of choices: technology options for the Swedish electricity sector*. Lund University.
8. Nørgård, J. 1989. Low electricity appliances – options for the future. In this volume.
9. McGowan, T. 1989. Energy-efficient lighting. In this volume.
10. MacInnis, R. 1988. DEC aims dimmable ballasts at $15,525 savings, pp. 6. *Energy User News*.
11. Swedish Ministry of Energy and Environment. 1987. *Electricity: Efficiency in the 1990s*. Swedish Government Official Report (SOU)/Report of the Expert Group. Stockholm. (In Swedish)
12. Swedish Ministry of Energy and Environment. 1987. See ref 11.
13. National Swedish Board for Technical Development (STU). 1988. *Effective electricity use in industry – research and development needs*. pp 14–16. Report No. 677. Stockholm. (In Swedish)
14. Tsuchiya, H. 1987. Electricity conservation in Japan. *Proceedings of the NATO Advanced Study Institute Meeting on Energy Efficiency in Buildings*, Portugal.
15. Baldwin, S.F. 1989. Energy-efficient electric motor drive systems. In this volume.
16. Usibelli, A., Greenberg, S., Meal, M., Mitchell, A., Johnson, R., Sweitzer, G., Rubenstein, F., and Arasteh, D. 1985. *Commercial-sector conservation technologies*. Lawrence Berkeley Laboratory Technical Report No. 18543.
17. Schaefer, H. and Rudolf, M. 1989. New electrothermal processes. In this volume.
18. Norford, L.K., Rabl, A., Harris, J.P., Roturier, J. 1989. Electronic office equipment: the impact of market trends and technology on end-use demand for electricity. In this volume.
19. Tsuchiya, H. 1987. See ref. 14.
20. Swedish Ministry of Energy and Environment. 1987. See ref 11.
21. Swedish Central Statistics Bureau. 1988. *Energy statistics for one- and two-dwelling houses in 1987*. Report E 16 SM 8801. See tables 11-24 for statistics on electric heating equipment. (In Swedish)
22. Swedish State Power Board. 1988. See ref 4.
23. Gardiner, B.L., Piette, M.A., and Harris, J.P. Measured results of energy conservation retrofits in non-residential buildings: an update of the BECA-CR data base. 1984. *Proceedings of the 1984 ACEEE Summer Study on Energy Efficiency in Buildings*. Also as Lawrence Berkeley Laboratory Technical Report No. 17881.
24. Abel, E. 1989. Use of electricity in commercial buildings. In this volume.
25. Piette, M.A., Wall, L.W., and Gardiner B.L. 1986. Measured performance. *Ashrae Journal* 28(1).

26. Mills, E., Harris, J.P., and Rosenfeld, A.H.. Developing demand-side energy resources in the United States: trends and policies, pp.169-193. *Energie Internationale*, 1988-1989 L'institut D'economie et de Politique de L'energie. Grenoble, France. (in French) Available in English as Lawrence Berkeley Laboratory Technical Report No. 24920.
27. Elmroth, A. 1989. Building design and electricity use in single-family houses. In this volume.
28. Granqvist, C.G. 1989. Energy-efficient windows: options with present and forthcoming technology. In this volume.
29. Berntsson, T. 1989. Heat pumps. In this volume.
30. Eketorp, S. 1989. Electrotechnologies and steelmaking. In this volume.
31. Santen, S.O. and Feinman, J. 1988. Pilot plant studies of the PLASMASCRAP process. *Iron & Steelmaker*. pp. 24-28. August.
32. Schaefer, H. and Rudolf. 1989. See ref. 17.
33. Eketorp, S. 1989. See ref 30.
34. Local Government's Research Institute on Public Finance and Administration. (Amtskommunernes og Kommunernes Forskningsinstitut, AKF). 1988. Danish research project: potentials for electricity conservation.
35. Petäjä, J. and Hausen, H. 1988. *The possibilities for electricity saving and decentralized power generation during* 1986-2020. Helsinki University of Technology. Report 24. (In Finnish)
36. The SERI Solar/Conservation Study Group. 1981. *A new prosperity: building a sustainable energy future*. Andover, Mass. Brickhouse.
37. Williams, R.H. 1987. A low-energy future for the US. *Energy the International Journal*. 12:929-944.
38. Krause, F.L., Rosenfeld, A.H., Levine, M.D., Brown, J., Connell, D., du Pont, P., Greely, K.M., Meal, M., Meier, A.K., Mills, E., and Nordman, B. 1987. *Analysis of Michigan's demand-side electricity resources in the residential sector*. Lawrence Berkeley Laboratory Technical Report No. 23025.
39. Hunn, B.D., Baughman, M.L., Silver, S.C., Rosenfeld, A.H., and Akbari, H. 1986. *Technical potential for electrical energy conservation and peak demand reduction in Texas buildings*. Report to the Public Utility Commission of Texas. Center for Energy Studies, The University of Texas at Austin, Conservation and Solar Energy Program.
40. Usibelli, A., Gardiner, B., Luhrsen, W., and Meier, A.K. 1983. *A residential conservation data base for the Pacific Northwest*. Report to the Bonneville Power Administration. Lawrence Berkeley Laboratory Technical Report No. 17055.
41. Meier, A.K., Wright, J., and Rosenfeld, A.H. 1983. *Supplying energy through greater efficiency: the potential for conservation in California's residential sector*. Berkeley. University of California Press.
42. Geller, H.S., de Almeida, A., Barkovich, B., Blumstein, C., Goldstein, D., Meier, A.K., Miller, P., de la Moriniere, O., Rosenfeld, A.H. and Schuck, L. 1986. *Residential conservation power plant study: phase 1 – technical potential*. Prepared for the Residential Conservation Services Department, Pacific Gas and Electric Company. American Council for an Energy-Efficient Economy.
43. Marbek Resources Consultants. 1987. *Electricity conservation supply curves for Ontario*. Report prepared for the Ontario Ministry of Energy. Toronto.
44. Robinson, J.B. 1987. Insurmountable opportunities? Canada's energy efficiency resources. *Energy the International Journal*. 12(5):403-417.
45. *Home Energy*, 6(1):24. 1989
46. MacInnis, R. 1988. See ref. 10.

47. McRae, M.R., Rufo, M.W., Guddat, R.O., and Baylon, D.A. 1988. Service life of energy conservation measures. *Ashrae Journal*, 12:25-28.
48. Usibelli *et al.* See ref. 16.
49. Isaksson, T. 1988. *Electric heating in single-family houses*. Stockholm. Swedish State Power Board, Task 2000 Project. Stockholm.
50. Nørgård. 1989. See ref 8.
51. Organization for Economic Cooperation and Development. 1987. OECD National Accounts, Main Aggregates Volume 1, pp. 62–63.
52. Swedish Central Statistics Bureau. 1988. *Statistical abstract of Sweden*, pp. 230. Stockholm. (In Swedish)
53. Rosenfeld, A.H. and Koomey, J.G. 1988. Promoting efficiency investments in new buildings. Testimony before the Wisconsin Public Service Commission. Docket No. 05-EP-5: Hearings on Advance Plan 5.
54. Baldwin, S. 1989. See ref. 15.
55. Usibelli et al. 1985. See ref. 16.
56. Pillai, K. 1989. Pressurized fluidized bed combustion. In this volume.
57. Larson, E.D., Svenningsson, P., and Bjerle, I. 1989. Biomass gasification for gas turbine power generation. In this volume.
58. Hildeman, A., Rutegard, G., Tibblin, G. 1986. *The tree-section method: integrated production of forest energy and pulp feedstock*. The Swedish University of Agricultural Science. Department of Forest-Industry-Market Studies. Report no. 19. Uppsala. (In Swedish)
59. Parikka, M. 1988. *Economic analysis of energy forest plantations – a "future-oriented" calculation*. The Swedish University of Agricultural Science. Department of Forest-Industry-Market Studies. Uppsala. (In Swedish)
60. Williams, R.H. and Larson, E.D. 1989. Expanding roles for gas turbines in power generation. In this volume.
61. Official Report of the Swedish Government (SOU). 1988. *Sites for wind power*. Stockholm. Allmänna Förlaget. (In Swedish)
62. Blomen, L. 1989. Fuel cells. In this volume.
63. Carlson, D. 1989. Low-cost power from thin-film photovoltaics. In this volume.
64. Swedish State Power Board. 1988. Biofuels. Report Number UM-1988/40. Project number 93078. Stockholm. (In Swedish)
65. Larson et al. 1989. See Ref. 57.
66. Swedish State Power Board. 1988. See ref. 64.
67. Olsson, T. 1986. *Production results in Swedish energy forest plantations*. Swedish University of Agricultural Science. Department of Ecology and Environment. Energy Plantations Group. Uppsala. (In Swedish)
68. Christersson, L. 1985. Intensively grown woody biomass production on sandy soil. *Biomassa och Energi*. (In Swedish)
69. Johansson, T.B., Steen, P., Bogren, E., Eriksson, L.R., Karlsson, S., and Svenningsson, P. 1985. *Perspectives on energy*. 1985. Stockholm. Liber Press. (In Swedish)
70. Larson *et al.* 1989. See ref 57
71. Bodlund et al. 1989. See ref. 7.
72. Swedish State Power Board. 1988. *Market information from ME.*, pp. 18. Stockholm. (In Swedish)
73. Swedish State Power Board. 1984. *Natural gas, health, environment*, pp. 104. Stockholm. (In Swedish)

74. Stjernquist, I., Brinck, L., Schlyter, P., Svenningsson, P., Johansson, T.B. 1986. *Environment and energy*. Prepared for the National Environmental Protection Board. Report 3238. (In Swedish)
75. Swedish National Energy Administration and National Environmental Protection Board. 1987. *New sulfur restrictions from combustion sources*. Report 3412.
76. Santen and Feinman. 1988. See ref 31.
77. Bodlund et al. 1989. See Ref. 7.
78. International Energy Agency. 1988. *Energy balances of OECD countries: 1985/1986*. Paris.
79. International Institute for Environment and Development and the World Resources Institute. 1988. *World Resources 1987*, pp. 248 and 335. Basic Books, Inc., New York.
80. World Congress on Climate and Development. 1988. Draft "Hamburg Manifesto."
81. Government of Canada, 1988. See ref 2.
82. Cheng, H.C., and Steinberg, M. 1986. A study on the systematic control of CO_2 emissions from fossil-fueled power plants in the U.S.. *Environmental Progress*. 5(4):245-255.
83. Berntsson, T. 1989. See ref. 29.
84. Geller, H. 1989. Implementing electricity conservation programs: progress towards least-cost energy services among utilities in the U.S. In this volume.
85. Williams, R.H. 1989. Innovative approaches to marketing electric efficiency. In this volume.
86. de la Moriniere, O. 1989. Energy service companies: the French experience. In this volume.
87. Geller, H., Harris, J.P., Ledbetter, M., Levine, M.D., Mills, E., Mowris, R., and Rosenfeld, A.H. 1989. The importance of government-supported research and development in advancing energy efficiency in the United States buildings sector. In this volume.
88. Bohi, D.R. and Zimmerman, M.B. 1984. An update on econometric studies of energy demand behavior. *Annual Review of Energy*. 9:105-154.
89. Ruderman, H., Levine, M.D., and J.E. McMahon, J.E. 1987. The behavior of the market for energy efficiency in residential appliances including heating and cooling equipment. *The Energy Journal*. 8:1101-124.
90. Williams. 1989. See ref 85.
91. Swedish State Power Board, Uppdrag 2000 – Electricity Conservation Project at Vattenfall. 1988. *Electricity conservation: a resource for both Vattenfall and its customers*. Stockholm.
92. Swedish Central Statistics Bureau. 1988. *Statistical abstract of Sweden*, 1987, pp. 149. (In Swedish)
93. Goldemberg, J., Johansson, T.B., Reddy, A.K.N., and Williams, R.H. 1988. *Energy for a sustainable world*. New Delhi. Wiley-Eastern Limited.

Index

A

absorption heat pump cycle 181
absorption heat pumps 187
  advanced 211
absorption heat transformer 181
absorptivity 325
acid deposition 505
adjustable-speed drives (ASDs) 34, 36
  advantages of 46
  applications 46
    scenarios for Sweden 894
  case studies 46
  classification of 37
  constant flow applications 44
  intelligent controls 45
  part-load or variable flow operation 44
  system load 45
advanced gas turbines
  environmental aspects 535, 538, 539
  natural gas resources 540
  Texaco gasifier/advanced combined cycle, (Texaco/ACC) 523, 542
aeroderivative turbines 504, 523
air-blown Lurgi fixed-bed gasifier 533
air-tight buildings 246
alnicos 41
aluminum (see industry, aluminum)
ambient radiation, solar constant and spectrum 91
amorphous alloys of FeNiBSi 41
amorphous silicon (see also photovoltaics)
  diffusion length 607
  drift length 607
  light-induced effects 608
  optical absorption 605
amorphous silicon power modules, performance of 612
amorphous silicon solar cells 595
appliances and lights, electricity requirements for 240
appliances
  low electricity 125
  scenarios for Sweden 164, 894, 901 – 902
arc furnaces 270, 305, 493, 499, 894
arc heating 298
atmospheric absorption 91
attic insulation 243
avoided cost 846

B

balance temperature 230
ballasts
  electromagnetic 67, 73
  electronic high-frequency 67, 73, 74
best available technology 130
bidding for conservation 749, 845, 851
biomass applications 714
biomass costs 723, 732
biomass gasification 697
  alkalai compounds 712
  back-pressure steam turbines 721
  biomass supply 723
  bubbling fluidized-bed gasifier 706
  char conversion 700
  circulating fluidized-bed (CFB) gasifier 706, 707
  emissions 698, 713, 714
  feedstocks 702, 708
  fixed-bed gasifiers 704, 705
  fluidized-bed gasifiers 705
  fundamentals 700
  gas clean up 712
  gasifier designs 704, 710
  moving-bed gasifier 706
  potential for in Sweden 728
  pyrolysis 700

biomass residue production 733
biomass-based electricity production 717, 725
  central station power 718, 725
  possible applications for Sweden 915
  cogeneration 719, 726
  lifecycle costs 719, 720, 729
biomass-gasifier gas turbine systems 698
bipolar junction transistors 35
Bonneville Power Administration (BPA) 838
British Coal Utilisation Research Association (BCURA) 562
buildings
  all-air air-conditioning systems 228
  design and electricity use in single-family homes 235
  electricity use in commercial 217
  space-conditioning requirements 218, 240
  space-heating equipment in Sweden 888
built-in heating elements 152

C

California Energy Commission 433
carbon dioxide emissions in Sweden 928 – 32
  parliamentary guidelines 884
carbon scrubbing in central station power plants 931
carbon taxes 852
cavity resonator 336
cement (see industry, cement)
central heating system 155
CFC gas 132
CFC problem 211
channel furnaces 311
chemical recuperation 532
chemically recuperated steam-injected gas turbine 504
chemicals (see industry, chemicals)
Cheng cycle 529
chromogenic materials 95
Cicchetti-Curkendall proposal 844
Cicchetti-Hogan proposal 848
circulation pump 156
Clean Coal Technology Demonstration Program 534
clothes drying
  spinning 143
  substitution of electricity for 149
clothes drying technologies 143, 148, 149
clothes washing 140
cloud gels 115
CO_2, cost of low emission futures 930
CO_2-lasers 345
coal gasification
  allothermal operating mode 670, 675
  autothermal operating mode 670
  Bergbau-Forschung test plant 675
  coal properties 677
  combined cycle power stations 666
  efficiencies 688
  entrained flow 665, 673, 685
  fixed bed 665, 682
  fluidized-bed combustion 665, 683
  gas composition 678
  gas purification 679
  gas turbine power stations 666
  NO_x and alkalai problem 681
  plant costs 692
  plant size 690
  pressure and temperature setting 677
  pyrolysis 668
  reactor types 672
  technologies for combined cycle power generation 665
coal-fired STIG/ISTIG 533
coefficient of performance (COP) 134
cogeneration 379, 504, 513, 797, 846
  scenarios for Sweden 915
cold washing 142
combined refrigerator/freezers 135, 138
commercial building energy use
  electricity 217
  influences 370
Compagnie General de Chauffe 816
Compact Integral Compact Flourescent Lamp 78, 79
Component Test Facility (CTF) 562
compression heat pump cycles 175
compression/absorption heat pump 209
compressors 133
  reciprocating and rotary 134
computer monitors 445

computer-integrated-manufacturing (CIM) 612
controls, industry 845
conversion of solid fuels into gas, theory of 667
cooking technologies 150 – 155
Cool Water technology 523
Cost-of-Light 69
cupola 311
cupola furnace 313
current-source inverter (CSI) 36
cycloconverters 38

D

Danish demonstration projects 787
daylighting 60
deep-welding effect 344
demand-side competetive bidding 749, 845, 851
detergents
  dish washing 146
  enzymes 143
developing countries, gas turbine power for 547
dielectric heating 298, 301, 331
direct heating 297
  high implicit 853
  social 833
diseconomy of scale 510
dish washing technologies 144
Divisia decomposition 466
Divisia index 461
dynamic conservatism 776

E

economies of scale 510
efficacy 62
efficiency
  advanced technology 130
  innovative approaches to marketing 831
electric arc furnaces (EAF) (see arc furnaces)
electric kettles 152
electric motor speed control, classification of 34
electric motors
  asynchronous (AM) 409
  classification and characteristics of 24
  convertors 411
  fraction of United States electricity generated devoted to 22
  induction 409
  inverters 411
  stock of American 419
  S.C. 410, 416
  variable-speed drives 411
Electric Power Research Institute (EPRI) 428, 866
electric utility industry
  appliance efficiency 775
  combustion technology 772
  electric motor efficiency 774
  environmental concerns 769
  financial incentives for 756
  gas turbines 772
  grid stability 778
  incentive programs 744
  lighting efficiency 774
  loan programs 750
  photovoltaics and fuel cells 773
  restructuring of 765, 779
  technological exploration 770
electrical generators 393
  current density 394
  S.C. 394, 416, 417
electrical steels
  grain-oriented 40
  non-oriented 40
electricity conservation
  and CO_2 emissions 928 – 932
  Denmark 787, 793
  programs 353, 741
electricity consumption
  by sector for Sweden 1987 901
  per unit of GDP in Sweden 911
  by sector for Sweden, OECD, and North America 418
  end-use categories 792
  historical development in Sweden 898
  potentials for Sweden in 2010 883, 897 – 903

role of sectoral shifts in
manufacturing 461
electricity efficiency 128
electricity generation
by source for Sweden, OECD, and North
America 418
coal
bituminous 573
efficiency 556
emissions 556
lignite 573
waste products 556
historical development in Sweden 898
renewable energy 800
roles for gas turbines in 503
routes for 556
scenarios for Sweden in 2010
917 – 920
costs 923
emissions 928
wind potentials in Sweden 916
electricity in industry (see industry)
electricity intensity 128
electricity planning
end-use 794
lead times 935
supply 796
electricity savings (see also saving potentials)
electrotechnologies and steelmaking 261
implementation of 804
electricity sector, fraction of United States
primary energy devoted to 21
electricity service scenarios 795, 892
electricity services 127, 787
electricity, substituting 128
electricity transmission and distribution
power cables 398
power plant operations 407
power transformers 401
spinning reserve 407
superconducting technologies 398, 401,
416
electricity versus GDP ratio in Sweden,
historic and scenario 911
electrochromism 105
electromelting 311
electron gun 343
electronic office equipment 427
commercial buildings 427
complementary metal-oxide semiconductor
(CMOS) technology 444
electricity use of 427
estimates of United States stock and
sales 430
fax machines 430
laser printers 432
microelectronics 444
minicomputers 429
office automation technology 454
office copiers 430
personal computers 428
automatic switching 447
disk drives 438
laptop 435
market saturation 452
operating costs 448
potential efficiency
improvements 450
printers and plotters 430
rated and measured loads 435
scenario for Sweden in 2010 894
usage profiles 440
workstations 429
electronics and efficiency 163
electron-beam
drilling 344
heating 298, 305, 342
perforation 345
welding 344
electrotechnologies and steelmaking
a steel plant with 100% electricity 291
decreasing electrical energy 266
direct casting 287
energy efficiency 267
energy structures 264
radically new methods 287
replacing electrical energy 274
electrothermal processes, new 297
Empresa Nacional de Electricidad S A
(ENDESA) 579
end-use efficiency 741
end-use technology 130
energy budgets 866
energy planning 788
energy service companies 811, 843
centralization of 812
contracts 817
energy-management contracting 812

industry and consumer points of view 827
performance-based contracts 811
technology-oriented 844
third-party financing 812
energy services
annualized cost of in Sweden, 2010 scenarios 892
least-cost 741, 755
lifecycle cost of 832
energy-efficient houses, characteristics of 241
energy-guzzler taxes 852
enzymes 142
EPRI, Electric Power Research Institute 428, 866
escalation in nuclear power costs 509
evaporative regeneration 520
excimer-laser 345

F

fan and boiler feedpumps, typical energy savings 46
fans
affinity laws for 44
characteristic pressure flow and curves for 31
inlet vanes 33
outlet dampers 33
fans, pumps, and control systems, electricity requirements for 240
fan-coil systems 228
Federal Energy Regulatory Commission 846
ferrites 41
float glass 96
fluidized-bed combustion 555
ash 573, 576, 585
atmospheric (AFBC) 557, 561
atmospheric (AFBC) bed temperatures 562
bad coals 555
bed material 561
catalytic cracking units 585
Ca/S molar ratio 573
clean coal technology 555
coal feeding techniques 571, 585
compact plants 556
efficiency 561, 565, 589
hot gas cleaning 566, 590
turbine entry temperature 566
emissions 559, 573, 575
gas/solids motion 560
hydrodynamic and thermodynamic properties of 558
load changing 572
low temperature combustion 559
market for 587
once-through boiler concept 572
PFBC combined cycle, schematic diagram 564
PFBC cycles 563
PFBC design features
bubbling bed 566
recycling of fines 566
refractories 568
tube bundle 571
PFBC plant sizes 573
PFBC process
advantages of 561 – 563
corrosion 562
erosion of blades 562
PFBC relative economics 582
PFBC risk assessment 582
pressurized (PFBC) 522, 555, 556
repowering with 580
risks 586, 587
sulphur capture 559
tube erosion 571
fluidized-bed furnace 330, 331
fluorescent lights
dimming controls for 873
phosphors 866
solid-state ballasts for 868, 871
forecasting (see industry, forecasting)
form electrodes 331
fossil fuel use, legislation 556
free heat gains 128
freezers 131, 137
thermal insulation of cabinets 132
fuel cell catalyst availability 658
fuel cell electricity production, comparison of costs 656
fuel cell system design

design efficiency 641
flowsheets 641, 642
fuel processor 641
internal reforming 640
process integration 641, 642
fuel cell systems
advantages of 630, 631
capital costs for different 600 MW power plants 654
cogeneration 631
commercialisation of 659
cost of electricity (COE)
calculating 650
cogeneration 631
gas engine 632
TAG methodology 650
cost reduction factors 649
economics of 646, 661
economy of scale 661
efficiency of 641, 644, 645
environmental considerations 633, 634
environmental regulations 658
gas turbine 633
market penetration 661
modular construction 636
reliability 661
serial production 636
stack costs 655
steam reformer 631
subsystems of 628
fuel cells
alkaline (AFC) 634, 636, 638
decentralized electricity generation 627
direct methanol (DMFC) 639
hydrogen 627
hydrogen rich gas 636
main characteristics of 637
molten carbonate (MCFC) 633, 636, 638
phosphoric acid (PAFC) 633, 636, 638
solid oxide (SOFC) 639
solid polymer (SPFC) 639
stacks 627, 640
system efficiency 628
fuel-bound NO_x 713

G

Gas Research Institute 866
gas turbine power generation 201, 666, 697
gasifier-gas turbine systems 697
gas/steam turbine combined cycle 504
Gate Turn-Off thyristors 35
glass melting 322
greenhouse effect 505, 928

H

halogen lamp cooker 154
halophosphate phosphors 75
hayboxes 154
heat distribution systems 155
heat exchangers 134
air-to-air 248
condenser 134
evaporator 134
heat pumps 173
ASD driven 50
COP 175
diesel/gas-driven 186
heat sources for 185
heat storage systems connected to 186
industrial 197
optimal annual operating time for 191
two-stage electric compression 199
vapour recompression 198
heat recovery 159
heating systems, electric-resistance 235, 252
heating ventilation and air-conditioning systems (HVAC) 217
heat-recovery systems, electrically-driven 237
heavy-duty industrial turbines 525
hidden heating 891
high-intensity smelting (HI smelt) 277
hot forming 313, 314
hot gas cleanup 697

household appliances and lighting 251
hydrogen for reduction 292

I

illuminance 63
Illuminating Engineering Society of North America (IESNA) 59, 69
illumination levels 60
independent power producers 846
indoor air quality 159
indoor comfort 159
indoor pollutants 159
induction coils 314, 315, 319
induction motor losses 23
induction motors, efficiency 23
induction-channel furnace 313
induction-crucible furnace 311
induction-surface heating 319
inductive hardening 305, 317, 320, 321
inductive heating 298, 301, 303, 315
inductor 317
industry
  aluminum 498
  automation in motor vehicles manufactu ring 500
  capacity utilization 476
  case studies 491
  cement 497
    grinding processes 499
    modernization 498
    wet process clinker kilns 498
  chemicals 469, 498
  conservation 498, 499
  consumption per unit of output 489
  dry process kilns 498, 499
  electrical intensity 461, 489
  electricity costs 462, 463, 499
  electricity-demand scenarios for Sweden in 2010 901 – 903
  electricity-intensive sectors 475
  electrification 500
  environmental regulations 490
  forecasting 461, 465
    electricity demand models for Sweden 887
  induction heating 499
  iron and steel (see also electrotechnologies and steelmaking)
  iron and steel 469, 481, 492, 493
    blast furnaces 495
    electric arc furnaces (EAF) 493
    variable-speed drives 495
    waste heat and gas recovery 495
  manufacturing 461, 463
  material throughput 499
  materials consumption 478, 481
  mechanical paper finishing 499
  motor vehicle manufacturing 498
  natural gas 462
  as electricity substitute, scenarios for Sweden in 2010 894
  paper and pulp 469, 481, 496
    heat recovery 496
    pulping processes 496
    thermochemical pulping scenarios for Sweden in 2010 894
    variable-speed motors 497
  petroleum 462, 469
  plastics 481
  pollution control 494, 497
  product mix 464, 489, 493, 496
  production data 471, 472, 854
  productivity 490
  rationalization of production 490, 495, 499
  refining (see industry, petroleum)
  scenario design 465
  sectoral disaggregation 468, 469
    level of 470
  steel (see also industry, iron and steel)
    continuous casting 499
    direct rolling 499
  structural change 475
    recession 476
  technological change 489, 498
  technological versus process change 499
  truly new technologies 500
  waste paper 499
  water throughput 499
infrared reflecting film (IRF) 68
infrared (IR)
  drying 325
  furnace 327

heating 301, 325, 327
oven 325
radiators 325, 328
insulated gate bipolar transistor 35
integrated resource planning 921
intensity index 128, 164
intercooled steam-injected gas turbine (ISTIG) 504, 531, 533, 699
intercooling 520
internal combustion engines 179
International Commission in Illumination (CIE) 61
International Gas Trade Project 544
inverters 205
iron and steel (see industry, iron and steel)

J

jet engine technology 520
Joule's heat 327, 333

K

Kaling cycle 513
lamp burnout factor (LBO) 85

L

lamp lumen depreciation (LLD) 85
lamps
biaxial 76
compact fluorescent 76
electrodeless 79
fluorescent 66, 72, 73, 857
fluorescent color rendering index 75
GLS 69, 77
halogen 68
high pressure sodium (HPS) 66, 80
high-intensity discharge (HID) 66, 80, 82
incandescent 66, 68
integral compact fluorescent 77
IRF incandescent 69
low voltage incandescent 71
mercury 67
metal-halide (MH) 66, 67, 80
MR16 72
SL-18 fluorescent 79
laser welding 346
laser-beam heating 298, 305, 345, 347
least-cost approach to electricity planning 833, 836
least-cost utility planning 741, 755
light loss factor (LLF) 85
light pipes 84
light source 65
lighting 59
chromaticity 75
contrast 63, 65
controls 74
efficacy 65
energy consumption per capita 61
energy efficient 59
fluorescent systems 74
lumen hours 85
maintenance 84
scenario for Sweden in 2010 894
system efficacy 67
system elements 64
line harmonics 48, 49
liquid-crystal-based materials 115
load duration curve 381
load-commutated inverter (LCI) 36
loans 741, 750
LOCUS-system 799
low-temperature heat 128
lumens 63
lumen-hours 61, 63
luminaire dirt depreciation (LDD) 85
luminous efficacy 86
Lurgi dry-ash gasifier 707
lux 63

M

magnet applications 412, 414
magnet energy storage, S.C. (SMES) 405
magnetic cores 40
magnetic power-system stabilization 408
magneto-hydrodynamic generators 412

magnetron 335
magnetron sputtering 875
maintenance costs of gas turbines 527
material growth 168
mediate heating 297, 303
metal oxide reduction in converters 280
metal tube radiator 328
metallurgical tools, converters operating with injection 261
Metglas 41
microwave heating 298, 331, 335
microwave oven 339
MOS-controlled thyristor 250
motor drive systems
  conventional design 29
  energy-efficient electric 21
  power factor 28
  safety margin for sizing 31
  sizing 30
  system load 29
motor driven systems
  part- and variable-load operation 32
  system control strategies 32
motor stock, characteristics of United States 32
motors
  direct current (DC) 23
  energy-efficient 23
    side-effects of 27
    transient starting currents 27
  induction 23
  oversizing 28
  synchronous 23

N

naptha feed 650
natural ventilation 161
Nd-YAG-lasers 345
nonazeotropic mixtures 207
Normalized Annual Consumption (NAC) index 363
Northwest Power Planning Council 842
no-losers test 842
nuclear power
  phase-out plans in Sweden 883

O

ohmic heater 341
ohmic heating 298, 303, 317
oil taxes 852
opportunity cost for capital 833
Organization for Economic Co-operation and Development (OECD) 866
oxygen-blown Texaco gasifier 533

P

pans
  electric 152
  insulated 154
paper and pulp (see industry, paper and pulp)
Peak Activated Rate and Real Time Pricing experiments 382
peak load 155
peak load management strategies 381
penetration measures 318
permanent magnet motors, electronically commutated or brushless (PM) 41
permanent magnets 41
  cobalt in 43
  demagnetization 43
  energy products for 42
  harmonics 43
  $Nd_2Fe_{14}B$ 41
  overloads 43
  permiability 43
  rare-earth (REPMs) 41
  world market for 43
phosphors
  high efficiency 75
  two photon 79
photochromic additives 115
photochromic glass 98
photovoltaic effect 596
photovoltaics (see also amorphous silicon)
photovoltaics (PV) 595
  cadmium telluride 604
  conversion efficiency 597
  copper-indium diselenide 603
  grid-connection systems 622
  manufacturing costs 616 – 621

stand-alone applications 622
thin-film 595, 601
Pinch Technology 200
Plank spectrum 91
plasma, the fourth state of matter 282
plasma
cutting 341
injections 341
welding 342
Plasmascrap process 284
scenario for Swedish steel industry 894
plasma-beam heating 298, 341
plate electrodes 331
power factor controllers 38
power hybrid 38
power integrated circuits 38
pressure cookers 154
pressurized biomass gasification 697
pressurized coal iron gasification (P-CIG) process 285
pressurized fluidized-bed combustor (see fluidized-bed combustion, pressurized)
primary energy requirement 307
Princeton Scorekeeping Method (PRISM) 357, 364
printers, ion-deposition 446
PRISM applications
commercial and office buildings 369
multifamily buildings 368
single-family houses 365
PRISM-CO (cooling only) 373
PRISM-HC (heating and cooling) 372
PRISM-HO (heating only) 373
Public Utility Regulatory Policies Act of 1978 (PURPA) 513
pulse-width modulated (PWM) drives 36
pumping efficiency 44
pumps
affinity laws for (centrifugal) 44
throttling valves 33
pyrolytic coatings 877

Q

quality workmanship 250
quartz tube radiator 328

R

radiant power 65
radiation spectrum 325
radio-frequency (RF)
drying 331, 335
heating 298, 331
applications 333
rebates 741, 747, 852
red-hot tubular rings 152
refining (see industry, petroleum)
refrigerator compressors, high efficiency 868
refrigerators 131, 135
thermal insulation of cabinets 132
regeneration 520
regulation of energy performance 852
reheat 520
research and development
appliances 878
buildings 863, 865
government support of 852
heat pumps 880
high-efficiency refrigerator 879
space heating and cooling 878
United States industry 865
water heating 878, 879
resistance heating 298
Swedish buildings sector 1987 888
rinsing 140
rotor resistance 27

S

saving potentials
appliances 164
cost of conserved electricity 904, 921
electrotechnologies and steelmaking 261
fan and boiler feedpumps 46
in buildings 217, 235
lighting 59
personal computers 450
scanning-type hardening 321
scorekeeping (see also PRISM)
scorekeeping 353, 849
annual-cycle storage 379
cogeneration 379

cross-sectional models 377
direct utility control 383
diversity factor 381
electricity use 376
evaluation 356
fuel switching 378
household electricity use 354
load factor 380
non-congruent metering 377
normalization 355
peak demand 379
photovoltaic and wind peaking power 384
physical models 359
power factor 377
socio-economic models 359
thermodynamic quality 377
weather adjustment 360
scrap melting without direct use of electrical energy 278
scrubbing processes, coal gasification 679, 680
seasonal performance factor (SPF) 184
shaft furnace 322, 324
shift-reactors 640
silica aerogel 113
simple-cycle gas turbine 515
single-shot hardening 321
single-stage coal-oxygen process 277
slip energy recovery systems 38
small household equipment 163
small-scale heat pumps 184
smart windows 95, 118
smelting reduction 275
soft magnetic materials 39
solid transparent insulation materials 110
space and water heating
electricity used for 235
scenarios for Sweden in 2010 901, 921
Sweden 1987 901
space-conditioning requirements, buildings 218, 240
specific electrical energy use 219
stacked-junction solar cells 599
static Kramer drives 38
static Scherbius drives 38
steam injection 520
steam-injected gas turbine (STIG) 504, 528, 531, 534, 653, 699, 718
steam-reforming 520
steel (see industry, steel)
stray-field electrodes 331, 332
superconducting machines 394
superconducting materials 390
S.C. ceramic materials 393
superconductivity 389
time-aspects of 421
superinsulated houses 158
supply curves of conserved electricity 921
surface hardening 321
surplus heat 218
sustainable energy development 806
Sweden in 2010, electricity supply and demand scenarios 883
Swedish single-family homes, electricity requirements for 238
switched reluctance motor 39

T

tank furnaces 322
technology, options for the Swedish electricity sector 883
Texaco coal gasifier 523, 542
thermal NO_x 713
thermochemical heat treatment 330
thyristor 35
thyristor controlled shunt capacitors 409
time-of-day electricity cost controls 48
tri-phosphors 75

U

ultrasonic washing 143
uninterruptable power supply systems (UPS) 638
utility programs 741
utility-sponsored auctions 846

V

vacuum insulation 133
vapour recompression cycle 176

ventilation systems 158
  constant-volume 231
  electrically-driven 237
  mechanical 255
  mechanical exhaust and supply fan 248, 250
  variable-air-volume (VAV) 231
visibility 64
visual display unit (VDU) 65
voltage-source inverter (VSI) drive 36

W

walls, thermal conductivity 242
washing machines 140
water heaters, immersion 152
water heating 251
  electricity requirements for 240
wind power, potentials in Sweden 916
window transmittance
  standard window glass 97
  anti-reflection treatments 108
  doped oxide semiconductor coatings 102
  electrochromic coatings 105
  energy-efficient 89
  evacuated 113
  gases for thermal insulation of 109
  low-emissivity coatings for 868, 874
  low-emittance 93
  maintenance 254
  noble-metal based coatings 100
  quantitative performance parameters for energy-efficient 95
  smart 95
  solar control 93
  thermal conductivity 93, 245
  thermochromic coatings 107
  vacuum 114
World Energy Conference (WEC) 507